4th International Conference on Innovative Materials, Structures and Technologies (IMST 2019)

IOP Conference Series: Materials Science and Engineering
Volume 660

Riga, Latvia
25 - 27 September 2019

Editors:

Sandris Rucevskis
Diana Bajare
Daira Toma
Laura Vitola

ISBN: 978-1-7138-0915-9
ISSN: 1757-8981

Printed from e-media with permission by:

Curran Associates, Inc.
57 Morehouse Lane
Red Hook, NY 12571

Institute of Physics
Dirac House, Temple Back
Bristol BS1 6BE UK

Phone: 44 1 17 929 7481
Fax: 44 1 17 920 0979

techtracking@iop.org

Additional copies of this publication are available from:

Curran Associates, Inc.
57 Morehouse Lane
Red Hook, NY 12571 USA
Phone: 845-758-0400
Fax: 845-758-2633
Email: curran@proceedings.com
Web: www.proceedings.com

TABLE OF CONTENTS

CONSTRUCTION TECHNOLOGY AND PROJECT MANAGEMENT

ENERGY EFFICIENT BUILDING AND INDOOR ENVIRONMENTAL SYSTEMS

FLOW STRUCTURE INTERACTION

INNOVATIVE CEMENT COMPOSITES

INSPECTIONS, ASSESSMENT, REPAIR

ROAD AND BRIDGE ENGINEERING

SMART, MULTIFUNCTIONAL, COMPOSITE MATERIALS AND STRUCTURES

SUSTAINABLE MATERIALS AND WASTE MANAGEMENT

THERMAL ENERGY STORAGE MATERIALS AND SYSTEMS

IOP Conf. Series: Materials Science and Engineering **660** (2019) 011001 doi:10.1088/1757-899X/660/1/011001

Preface

The 4th International Conference on Innovative Materials, Structures and Technologies (IMST 2019) took place on 25-27 September 2019 in Riga, Latvia. The conference was organised by the Faculty of Civil Engineering of the Riga Technical University in cooperation with Brno University of Technology in Czech Republic. Following the established tradition, the aim of the conference was to promote and discuss the latest results of industrial and academic research carried out in the following engineering fields: analysis and design of advanced structures and buildings; innovative, ecological and energy efficient building materials; inspection, assessment and monitoring methods; construction technology and project management; sustainable and safe transport infrastructure; waste management; civil engineering education and BIM.

The conference provided an excellent opportunity for researchers, representatives of the industrial community, engineers, managers and students to share the latest achievements, discuss recent advances and highlight the current challenges. IMST 2019 attracted over 120 scientists from 22 countries. After rigorous reviewing, 78 technical papers were accepted for publication in the conference proceedings. On behalf of the organizing committee I would like to thank all the speakers, authors, session chairs and reviewers for their efficient and timely effort.

Finally, I would like to thank all those who helped to make this event happen. Special thanks go to prof. Diana Bajare, Daira Toma and Laura Vitola for their major contribution to organizing the conference.

Dr. Sandris Ručevskis

Editor-in-Chief
Faculty of Civil Engineering
Riga Technical University

IOP Conf. Series: Materials Science and Engineering **660** (2019) 011001 doi:10.1088/1757-899X/660/1/011001

Editors

Prof. Diana Bajare
Faculty of Civil Engineering, Riga Technical University; e-mail: diana.bajare@rtu.lv

Daira Toma
Faculty of Civil Engineering, Riga Technical University; e-mail: daira.toma@rtu.lv

Laura Vitola
Faculty of Civil Engineering, Riga Technical University; e-mail: laura.vitola_1@rtu.lv

Organised by:

1862
RIGA TECHNICAL UNIVERSITY

Supported by:

eit
RawMaterials Hub
Regional Center Baltic

BALT GRAF

BRNO UNIVERSITY OF TECHNOLOGY

PART OF EUROPEAN BIOTECH WEEK
INNOVATION IS IN OUR GENES

Peer review statement

All papers published in this volume of *IOP Conference Series: Materials Science and Engineering* have been peer reviewed through processes administered by the proceedings Editors. Reviews were conducted by expert referees to the professional and scientific standards expected of a proceedings journal published by IOP Publishing.

IOP Conf. Series: Materials Science and Engineering **660** (2019) 012001 doi:10.1088/1757-899X/660/1/012001

Shrinkage of the alkali-activated slag mortars containing alternative activator

B Kucharczyková[1], V Bílek[2] and H Šimonová[1]

[1]Brno University of Technology, Faculty of Civil Engineering, Veveří 331/95, 602 00 Brno, Czech Republic

[2]Brno University of Technology, Faculty of Chemistry, Purkyňova 46/118, 612 00, Brno, Czech Republic

E-mail: barbara.kucharczykova@vutbr.cz

Abstract. The paper deals with the experimental determination of the shrinkage process of the alkali-activated slag mortars (AASM). Two mortars which differed especially in the type of used aggregate were mixed for the experiment. The slag-aggregate ratio was 1:3 for both mortars. The waste sludge, obtained from the water glass production, was used as an activator. The desired workability of the fresh mixtures was reached by addition of water. Because of the high water demand of waste sludge, the water-slag coefficient (*w/s*) was relatively high – 0.98 for the mortar with the natural sand and 0.73 for the mortar with the standard sand. All test specimens were removed from the moulds at the age of 24 hours and left to dry freely in the laboratory conditions with a temperature of 21 ± 2°C and relative humidity of 50 ± 10% during the whole time of measurement. The results show that AASMs which contain waste sludge exhibit rapid shrinkage with a high magnitude independent on the type of aggregate. During the first 40 hours of ageing, both mortars reached approx. 50% of the final value of shrinkage. The high shrinkage was strengthened by a high *w/s* coefficient needed to maintain the workability of the fresh mixtures.

1. Introduction

The research of utilization of industrial byproducts in building industry is a very actual topic around the world due to severe ecological and economical demands of the production of Portland clinker (PC). One traditional and widely used possibility is blending of PC with secondary raw materials such as slag and fly ash to produce blended cements. Another option is the production of wide range of non-cement alternative binders [1].

Alkali-activated materials (AAMs) have potential to replace PC-based materials in some applications because of their mechanical properties or durability [2, 3]. However, they suffer from shrinkage and related high cracking tendency, particularly those based on alkali-activated slag (AAS). It was reported that the drying shrinkage of AAS is several times higher compared to PC-based materials [4–7]. Moreover, the cost of alkaline activators can also be an issue as well as their usually non-ecological production.

The solution can be using of some alternative activators. In this study, the waste sludge from water glass production was applied. Because of significant amounts of dissolved alkalis and silicates, it is dangerous waste and its disposal is very expensive. On the other hand, it has noticeable potential for alkaline activation. Moreover, it contains significant amounts of solid particles and rest of siliceous sand used for water glass production, which would save natural fine aggregate. Therefore, this study is focused on physical-mechanical characterization of mortars where only this alternative activator was used for slag activation and simultaneously as a partially replacing material for natural and standard siliceous sand.

2. Experiment details

The pilot study presented herein is a part of the comprehensive pilot study which main objective is to obtain a start-up information about physical, mechanical and fracture characteristics of slag mortars activated by a waste sludge produced in the Czech Republic. This article focuses especially on the shrinkage process which is a general and still actual problem in all cases of the alkali-activated slag materials.

2.1. Materials

Ground granulated blast furnace slag with predominant content of amorphous phase and Blaine fineness of 400 m^2/kg was used as a precursor for AAS-based mortars. Chemical composition of slag is given in Table 1. It was activated by the waste sludge from water glass production which contained 4.1% Na_2O, 7.3% soluble SiO_2, 16.5% water and 72.1% of non-reacted siliceous sand. The dose of activator was adjusted to 8% Na_2O with respect to the slag weight. Two kinds of mortars with a sand to binder ratio of 3:1 were prepared. The mortars differed in the type of the used sand. For one mortar, natural sand with grain size up to 4 mm was used, and the second one was formulated using standardized CEN siliceous sand with maximum grains of 2 mm consisting of three fractions according to EN 196-1 [8]. The type of sand was used to distinguish between the two mortar series: *N* (natural) and *S* (standard). The water to slag (*w/s*) ratio including water from activator was 0.98 for *N* mixture and 0.73 for *S* mixture. Such high values of the water to slag ratio are caused by high water demand of the solid particles in an activator since they are relatively fine and porous. These solid particles introduced into the mixtures within the activator were also calculated to the dose of the above mentioned sand to binder ratio, i.e. they partially replaced the used sands. For *S* series, they replaced the whole finest fraction and a part of the middle one of the CEN sand, while for *N* series its total amount was reduced accordingly. Replacement of fine fraction of CEN sand was preferred because its particle size distribution is the closest to relatively fine particles from activator (see Figure 1). It was possible because, contrary to natural sand, individual fractions of CEN sand were weighed separately.

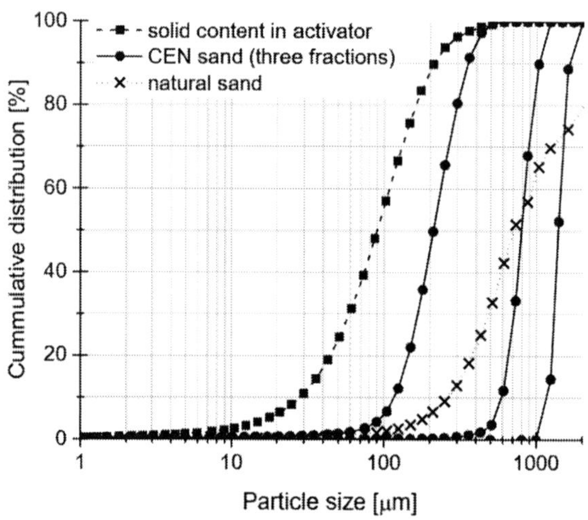

Figure 1. Particle size distribution of the used sands as determined using laser diffraction (combined with sieve analysis for particles higher than 1 000 mm).

Table 1. Chemical composition of slag as determined using X-ray fluorescence.

CaO	SiO_2	Al_2O_3	Fe_2O_3	MgO	Na_2O	K_2O	SO_3	Mn_2O_3	TiO_2	P_2O_5
39.6	38.9	8.9	0.27	8.8	0.39	0.40	1.4	0.83	0.32	0.01

2.1.1. Specimens preparation and curing. Mixing was performed using Hobart mixer and took approximately 8 minutes. First, the activator and water were mixed together to prepare homogenous activating solution. Then, the slag and finally the sand were added. The fresh mortar was then tested in terms of consistency using a flow table test with a base diameter of 100 mm, density and air content using the device according to EN 1015-7 [9]. The mortar was cast into the polyethylene moulds (Hakorite) for preparation of prisms with dimensions of 40 × 40 × 160 mm and sealed for 24 hours at a temperature of 21 ± 2 °C. Then the specimens were demoulded and stored in an air conditioned room where the temperature of 21 ± 2 °C and relative humidity of 50 ± 10% were maintained. The informative compressive strength was determined for both mortars at the age of 28 days (see Figure 2b). The test procedure follows the standard EN 196-1 [8]. The basic characteristics of fresh and hardened mortars represented by the average value and sample standard deviation (SSD) in brackets are introduced in Table 2.

Figure 2. Moulded test specimens (a); Test of informative compressive strength (b); Arrangement of the shrinkage measurement (c).

Table 2. Basic properties of fresh and hardened mortars: Mean value, (SSD).

mixture	fresh state				hardened	
	spread (mm) after		bulk density (kg/m^3)	air content (%)	bulk density (kg/m^3)	$f_{c,28days}$ (MPa)
	lifting the cone	15 jolts				
N	120	170	2070	4.2*	1900 (10.7)	14.9 (0.53)
S	160	235	2130	3.2*	1990 (16.5)	17.9 (0.30)

* The air content was measured for both mortars at the start and at the end of their pouring into the moulds. Both mortars exhibited a decrease in the air content from 4.8 to 3.6 % for mixture *N* and from 4.0 to 2.4 % for mixture S.

2.2. Measurement of shrinkage and mass losses

The shrinkage process of the AAS mortars was determined based on the measurement of the length changes of the prismatic specimens with dimensions of 40 × 40 × 160 mm equipped with the special markers embedded into the ends of the test specimens (in the longitudinal axis of the specimens). The measurement was started right after demoulding of the test specimens at the age of 24 hours.

The changes in length were measured in regular intervals using the measurement gauge equipped with the digital probe with resolution of 0.001 mm (see Figure 2c).

Simultaneously, the actual mass of each test specimen was registered using the laboratory balance with resolution of 0.5 g. The changes in the mass were used for determination of the process of mass losses which is one of the indicator used for prediction of the drying shrinkage process.

All measurements were performed in an air-conditioned laboratory with a temperature (T) of 21 ± 2 °C and relative humidity (RH) of 50 ± 10%. Three test specimens made from mortar *N* and *S* were taken for determination of the shrinkage and the mass losses process.

3. Results and discussion

The results of measurement are shown in following figures. The steady-state values are related to the conditions maintained in the laboratory during measurement (T = 21 ± 2 °C; RH = 50 ± 10%).

Figure 3 shows the process of shrinkage determined for mortar *N* and *S*. Both figures display results obtained from three independent measurements. As shown, the absolute value of shrinkage is very high for both mortars with the maximum of about 16.100 mm/m and 16.700 mm/m for mortar *N* and *S*, respectively. The high value of shrinkage is related to a high *w/s* ratio which is needed to maintain the workability of the mortars. In spite of the fact that the *w/s* for mortar *S* was about 25% lower in comparison to the mortar *N*, the final steady-state values of shrinkage are almost identical.

The differences in the shrinkage process of both mortars can be seen during the first 7 days of the measurement, when the mortar *S* started to shrink rapidly approx. 1 day after demoulding of the test specimens (see Figure 4a).

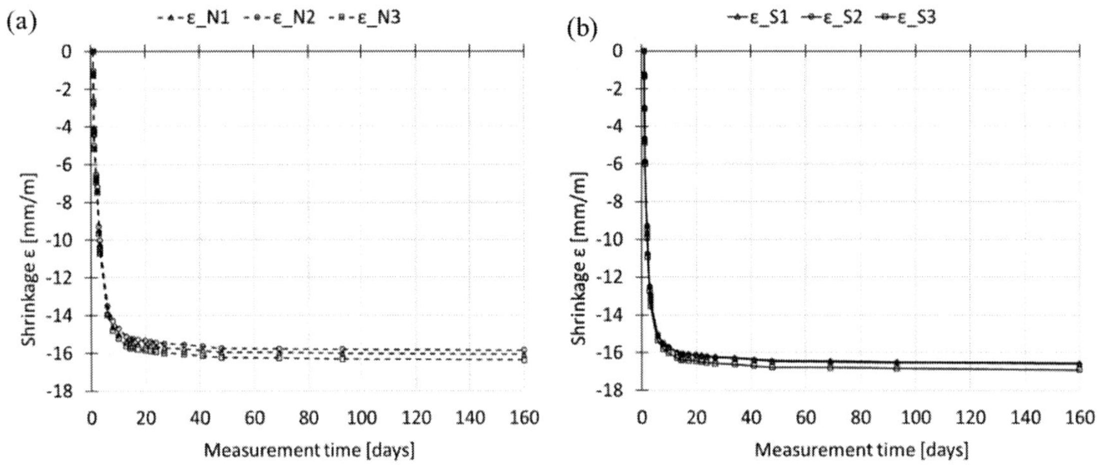

Figure 3. Shrinkage process for mortar *N* (a) and *S* (b).

The reason for the differences in short-term shrinkage behavior would be different porosity and pore size distribution of the prepared mortars, which is supported by the different *w/s* ratios, air content (Table 2) as well as the mass loss (Figure 4b). The combination of these parameters suggests that mortar *N* had higher content of larger pores and more permeable structure which resulted in more intense drying during the first several days and simultaneously in lower shrinkage extent. After this stage, drying of finer pores inside the mortar *N* accompanied with higher shrinkage took place and consequently the differences in the shrinkage extent of both tested mortars partially diminished after seven days.

The cracks formation is another factor which influences the shrinkage process. The formation of the cracks is reflected in the shape of shrinkage curve as retardation of the shrinkage. The wider the cracks are, the more visible effect on the curvature of shrinkage is recorded. To verify the existence of the cracks in the internal structure of both mortars, a visual inspection of the test specimens subjected to the quasi-static fracture test of the specimens with an initial notch was made. The visual inspection confirmed the existence of the macro-cracks in specimens made from mortar *N* (see Figure 5a). No visible cracks were observed in the case of specimens made from mortar *S* (see Figure 5b).

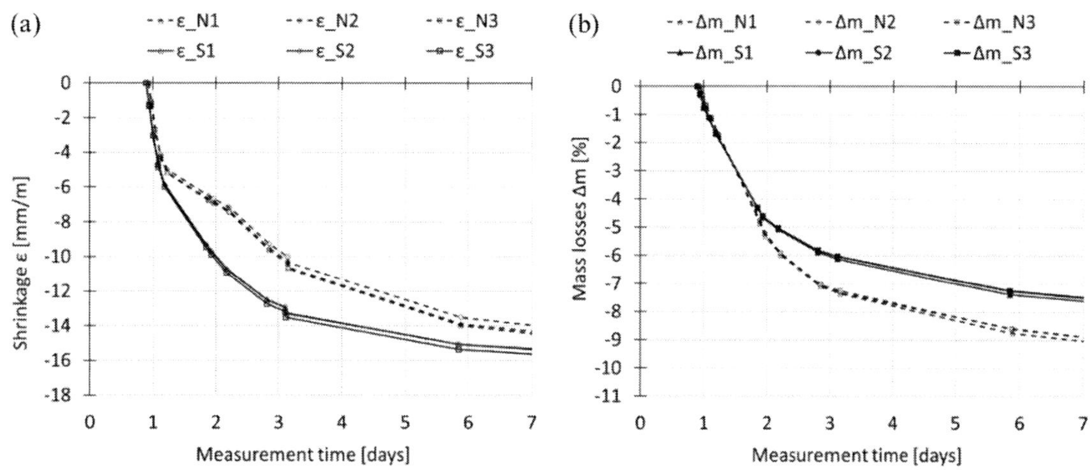

Figure 4. Initial process of shrinkage (a) and mass losses (b) for mortar *N* and *S*.

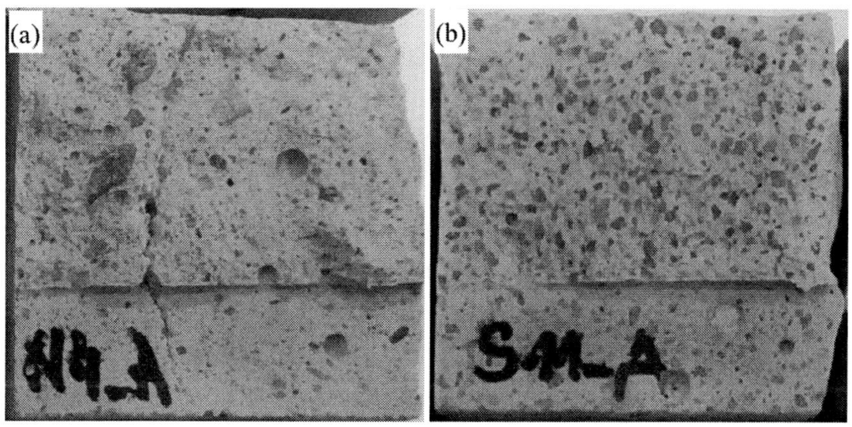

Figure 5. Cracks in the internal structure of the specimens made from mortar *N* (a).

The process of mass losses displayed in Figure 6 shows that the steady-state value is about 10.5% and 8.9% for mortar *N* (with higher *w/s*) and *S*, respectively. The difference in mass losses of the mortars is about 16%.

Differences in the process of shrinkage (ε) and mass losses (Δm) is also visible from the (Δm/ε)-curve displayed for both mortars in Figure 7. There is an observable phenomenon, in the time interval from 25 to 53 hours after specimens demoulding, when the (Δm/ε)-curve changes the shape due to the retardation of shrinkage in the case of mortar *N*.

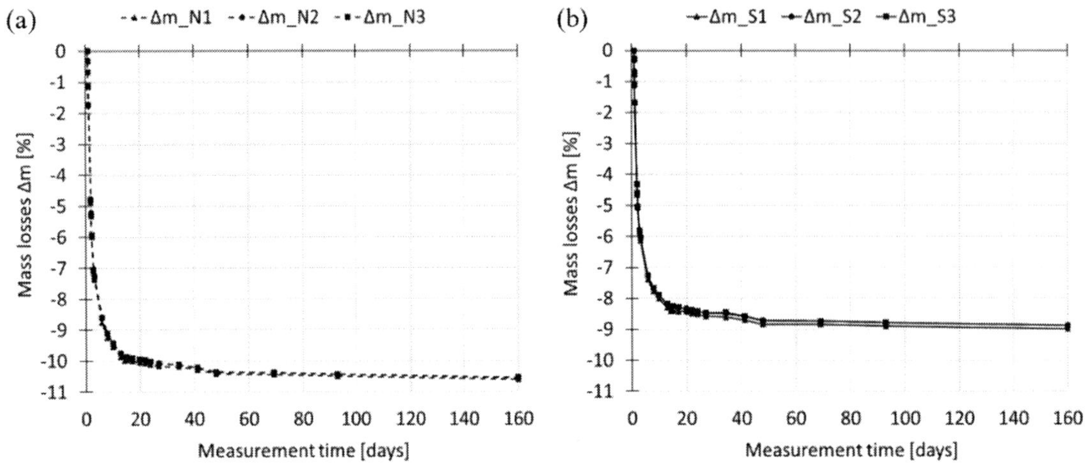

Figure 6. Mass losses process for mortar *N* and *S*.

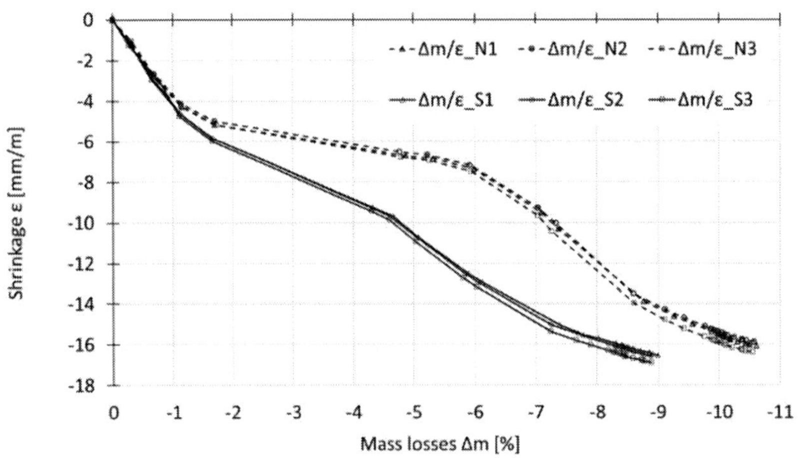

Figure 7. Shrinkage – Mass losses relation (ε/Δm) for mortar *N* and *S*.

4. Conclusions

The results of the pilot study show particular problems which can occur during the production and maturation of the mortars made from slag activated by the waste sludge from water-glass production.

The results show that the high water demand of the waste sludge due to the relatively high content of fine and porous particles is one of the main disadvantages of this materials. The high *w/s* ratio affect negatively the pore size distribution, process of desiccation and shrinkage. The process of desiccation is accompanied also by the cracks' formation, which influences the strength parameters. The shrinkage is for at least 10 times higher than in the case of ordinary cementitious mortars. The results of this study serve as the basis of the ongoing research which focuses especially on the ways for decreasing the *w/s* and shrinkage value at the preservation of required workability during the manufacturing process.

Acknowledgments

This work was supported by the Czech Science Foundation [project No. GA17-14302S and No. 18-12289Y].

IOP Conf. Series: Materials Science and Engineering **660** (2019) 012001 doi:10.1088/1757-899X/660/1/012001

References

[1] Juenger M C G etal. 2011 *Cement and Concrete Research* **41**(12) pp 1232–1243
[2] Provis J L and Van Denvenetr J S J eds 2014 *Alkali activated materials: stateof-the-art report, RILEM TC 224-AAM* (Dordrecht: Springer)
[3] Shi C, Krivenko, P and Roy D 2006 *Alkali-activated cements and concretes* (London: Taylor & Francis)
[4] Krizan D and Zivanovic B 2002 *Cement and Concrete Research* **32**(8) pp 1181–1188.
[5] Melo Neto A A, Cincotto M A and Repette W 2008 *Cement and Concrete Research* **38**(4) pp 565–574
[6] Collins F and Sanjayan J G 2000 *Cement and Concrete Research* **30**(9) pp 1401–1406
[7] Palacios M and Puertas F 2007 *Cement and Concrete Research* **37**(5) pp 691–702
[8] ČSN EN 196-1 2016 *Methods of testing cement – Part 1: Determination of strength* (Prague: ÚNMZ)
[9] ČSN EN 1015-7 1999 *Methods of test for mortar for masonry – Part 7: Determination of air content of fresh mortar* (Prague: ÚNMZ)

Alkali-aggregate reaction in alkali-activated cement concretes

P Krivenko[1], O Gelevera[1], O Kovalchuk[1], G Bumanis[2] and A Korjakins[2]

[1]Kyiv National University of Construction and Architecture, Scientific research institute for binders and materials, V.D. Glukhovskii, Povitroflotskii ave. 31, Kyiv 03680, Ukraine
[2]Department of Building Materials and Products, Institute of Materials and Constructions, Faculty of Civil Engineering, Riga Technical University, Kalku Str. 1, Riga, LV 1658, Latvia

E-mail: kovalchuk.oyu@gmail.com

Abstract. The paper discloses the results of a study on structure formation processes in the interfacial transition zone "cement stone – aggregate" of the alkali-activated cement concretes. The results indicate the importance of the content of Al_2O_3in the cement and aggregate. Together with components that are able to interact actively with alkalis in the presence of reactive SiO_2, the processes taking place during the alkali-aggregate reaction can have weather destructive or constructive effect. The constructive effect is caused by incorporating the corrosion products into alkaline aluminosilicate compounds, thus retarding the internal corrosion of the concrete and providing the higher durability even in case of alkali-activated cements and alkali-susceptible aggregates with relatively high alkali content.

1. Introduction

Internal corrosion of concrete resulted from alkali-aggregate reaction (AAR) is one of the less studied and "hidden" reasons of reduction of durability. The present study was dedicated to this issue with a final goal to develop methods of regulation of the reaction "alkali–silicic acid" and technological recommendations on how to prevent the concrete from this corrosion in order to meet the norm-specified requirements as to durability of buildings and structures. Corrosion of concrete as a result of reaction between alkalis of the cement and silica of the aggregate, is, as a rule, a long-term process. Sometimes, in the process of inspection of the structures this type of corrosion is identified in a wrong way, since its external signs (features) are similar to those from the processes of frost destruction and corrosion induced by aggressive environment.

The most widely known explanation of the reasons for AAR-induced corrosion is an interaction of the alkali metal compounds (of sodium and potassium) with active silica of the aggregate [1]–[4].
An alkali can be brought into a concrete with the cements containing increased quantities of sodium and potassium oxides (up to 1…2% by mass calculated as Na_2O-equivalent). A quantity of the alkalis brought into the concrete can be increased as a result of the use of chemical, mineral and organo-mineral admixtures and additives containing alkalis [5]–[7]. The alkalis also can come from outside, for example, with sea water or deicing reagents [8], [9]. Alkali-activated cements (AAC) like slag alkali-activated cements, alkali-activated portland-, slag portland-, pozzolana cements, geocement-based bonds, etc., become more common in construction industry while the content of alkalis (expressed as Na_2O) could reach between 1.5 and 5% by mass [10]–[14]. As a result of interaction of alkalis with active silica of the aggregates the hydration products causing deformations, strength

decline and deterioration of concrete are formed on the surface of the interfacial transition zone (ITZ) "aggregate – cement stone" [15]. In these cases, a network of cracks appears on the surface of the concrete followed by full destruction of the concrete [16], [17].

The compressive strength of the concrete does not guarantee durability of concrete by itself [18]–[20]. High durability is guaranteed, first of all, by high compaction of the concrete as well as by composition of the hydration products, both the cement stone itself, as well as hydration products in the ITZ "aggregate – cement stone" [21]–[23].

In Figure 1 is a classical schematic representation of the AAR in case of alkali- susceptible aggregates (ASA) with damage of the ITZ "aggregate – cement stone". A gel-like shell, due to accumulation in it of the reaction products, tends to increase in volume, initiating the occurrence in the concrete of the expansion deformations and in certain conditions can result in damage of the concrete structures. Moreover, the gel-like CSH- shell acting as a semi-permeable membrane can allow for the water and alkali to penetrate and reach the aggregate grain. As a result of this, the osmotic pressure is produced in the ITZ of the aggregate with the cement stone which can reach rather high values. Since, for example, the osmotic pressure of the sodium silicate gel may reach 20 MPa and tensile strength of the concrete is only 3-5 MPa, this may result in the occurrence of micro cracks and gradual deterioration of the concrete [24]–[26].

As a result of the wider use in the practice of construction of the AAC, theoretical background of which has been developed by V.D.Glukhovsky and which are being developed by his followers [27], durability of concrete in case of the ASA is of first priority and concern.

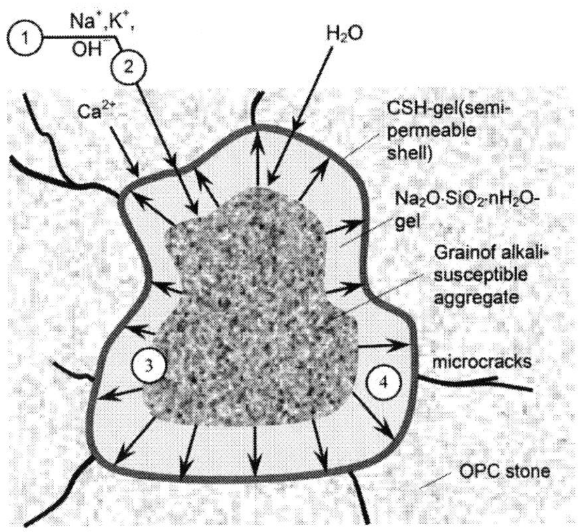

Figure 1. Schematic representation of the alkali - silicic acid reactions in the OPC stone [6]:1 – formation of alkali- metal hydroxides; 2 – diffusion of ions to place of reaction; 3 –ion-exchange and ion reaction; 4 – water absorption and swelling.

As a result of the mechanism of hydration and hardening of the AAC which are completely different from those of the OPC, the processes of interaction of the environment of the hardening AAC stone with reactive aggregates differ in principal from those taking place in case of OPC [26], [28]. In particular, an effect of "encapsulation" of the grains of the ASA with a film composed from hydration products, which retard, though not always completely exclude, the corrosion process.

The present research study resistance of the aggregates from alkali-susceptible rocks in the environment of the cement stone using different types of cements, including AAC, and to assess the degree of influence of the cement matrix and different additives on the rate of reaction between the ASA and the alkali of the cement matrix as well as to reveal how it may be reflected on service properties and durability of the concrete structures. The following factors which affect the "internal corrosion" of the concrete have been taken into account: aggregate type, aggregate particle size, quantity of alkali and type of alkaline component, cement type, type of modifying active additive and its quantity. The influence of each factor taken separately and of combination of factors was studied.

2. Materials and Methods

The following cements were used in the research on behavior of ASA:

- OPC CEM II 42.5N A-S (Na_2O-equivalent = 0.22% by mass);
- OPC CEM II 42.5N A-S with the increased content of Na_2O = 1.3% by mass;
- slag alkaline (alkali activated) cement (Type LCEM I, strength Grade 42.5 [11]) with Na_2O-content = 2.5% by mass;
- alkaline (alkali activated) slag portland cement "OPC clinker:slag"= 40:60 (Type LCEM IV, strength Grade 42.5 [11]) with the Na_2O-content = 2.5% by mass.

Anhydrous soda ash was used as Na_2O equivalent. The soda ash was introduced in a dry state in the process of joint grinding of cement components. In order to compare influence of alkaline component on cement characteristics, in paragraph 3.3 was used sodium metasilicate replacing soda ash, keeping constant alkalinity of the system as Na_2O equivalent. The potentially susceptible aggregates examined were aggregates from natural rocks with characteristics given in Table 1. Quartz crystallites content was set according to complex (quantitative and qualitative) XRD analysis.

Table 1. Characterization of the aggregates under study.

Origin	Rock type	Oxide composition, % by mass										Acidity modulus	Glassy phase, wt.%
		SiO_2	Al_2O_3	TiO_2	FeO	Fe_2O_3	CaO	MgO	Na_2O	K_2O	Other		
Austria	Olivine rock	44.99	15.56	2.25	–	10.53	10.42	8.98	2.42	2.4	0.10	3.06	5-7
Ukraine	Basalt rock	50.42	12.47	2.48	9.45	8.5	9.34	5.74	2.22	0.92	–	4.17	17-19
Ukraine	Andesite rock	54.16	18.03	0.43	9.39		8.57	4.65	1.06	2.06	–	5.46	39-41

The ASA were used as fine gravel of polyfractional composition with the particle size varying between 0.5 and 8 mm. A river quartz sand with a grading factor M=1.2 was taken as aggregate in the preparation of the reference specimens. The cement to aggregate ratio was 1:3. W/C ratio was in range from 0.39…0.4. In order to suppress destructive processes in the ITZ the metakaolin was added in a quantity of 15% by mass of the cement.

Prismatic specimens 4×4×16 cm and 2.5×2.5×28.5 cm with copper benches (for control of deformations) were prepared for flexural and compressive strength test. After 2 days the specimens were taken from the molds and placed for long-term steam treatment for 28 days (T=70°C, RH=100%) – accelerated procedure [25], [28] with test ages at 14 and 28 days. In compliance with the accelerated procedure used in the present study the aggregates can be classified after 28 days of continuous steam treatment with regard to suitability (conformance) for the use in concretes as the following:

a) expansion less than 1.5 mm/m – an aggregate is not alkali- susceptible one (no need for further testing);
b) expansion between 1.5 and 2.0 mm/m – an aggregate is alkali-susceptible one (preventive measures should be taken to reduce effect of reactive silica);
c) expansion higher than 2.0 mm/m – an aggregate is alkali-susceptible one and is to be not allowed for the use in concrete (alternative aggregates should be recommended).

A measurement was taken preliminary on the specimens intended for control of deformations. The measurements of deformations were done using a digital device with a value of the scale division of 0.01 mm. The ITZ was studied using a digital microscope on thin sections of the specimens.

3. Results and Discussion

Initially, the size of the quartz crystallites present in the aggregates was measured to determine aggregate reactivity. Determination was done using three characteristic peaks of quartz: d=0.426(8); 0.334(10); 0.182(6) nm. The results are given in Figure 2. An evident fact is the less size of crystallites, the higher solubility, hence- the reactivity, may be expected [5], [12]. Thus, coming from the size of quartz crystals in the rocks under study, in reactivity the rocks can be arranged as the following: andesite rock (65.7 nm)> basalt rock (114.1 nm) > olivine rock (124.8 nm).

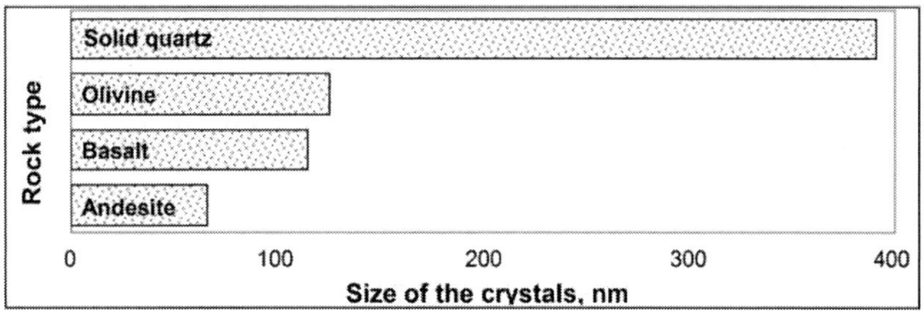

Figure 2. The size of the quartz crystals in different rocks.

3.1. OPC-based concretes

The presence of alkali ($Na_2O \leq 0.6\%$ by mass) in the cement composition is not dangerous as no expansion was found in all test specimens (Figure 3). Thus, there is no need to apply technological measures or use additives or admixtures. As it follows from Figures 3b and 4a, the increase in the Na_2O content in the OPC up to 1.3% by mass accelerates the destructive processes in the ITZ and in case of the andesite rock results in critical expansion exceeding 2.0 mm/m (+2.43 mm/m). The signs of corrosion are clearly seen in the microphotos of the ITZ (Figure 4a). The expansion of other compositions is found to fall within the admissible limits.

The introduction of the metakaolin additive is found to allow the use of the andesite rock as well (Figure 3b), the ITZ being clear and free from the products of corrosion (Figure 4b).

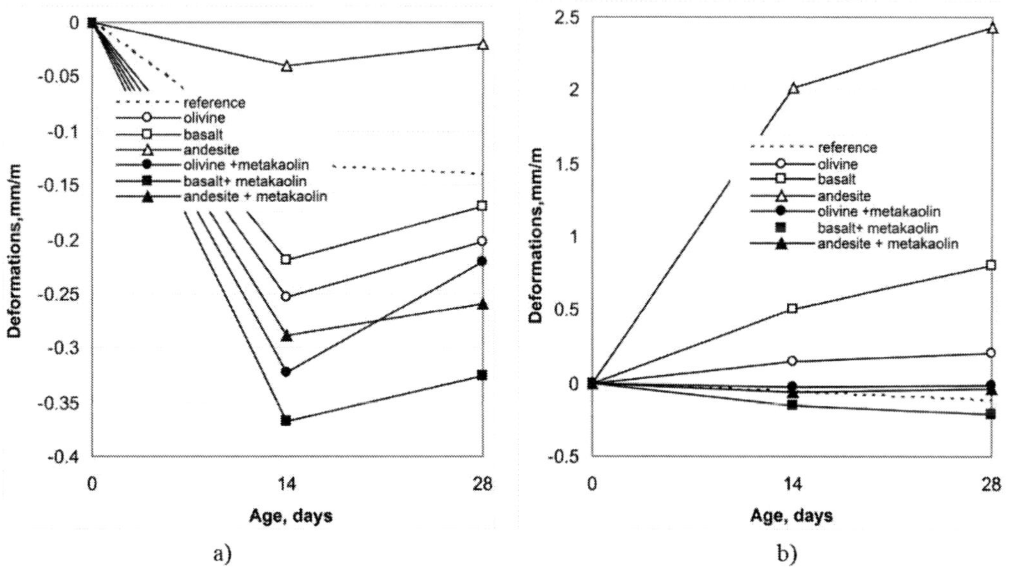

Figure 3. a) Deformations of the concretes made with ASA: a) OPC CEM II 42.5N (Na_2O = 0.22% by mass). b) OPC with the increased content of alkali (Na_2O = 1.3% by mass).

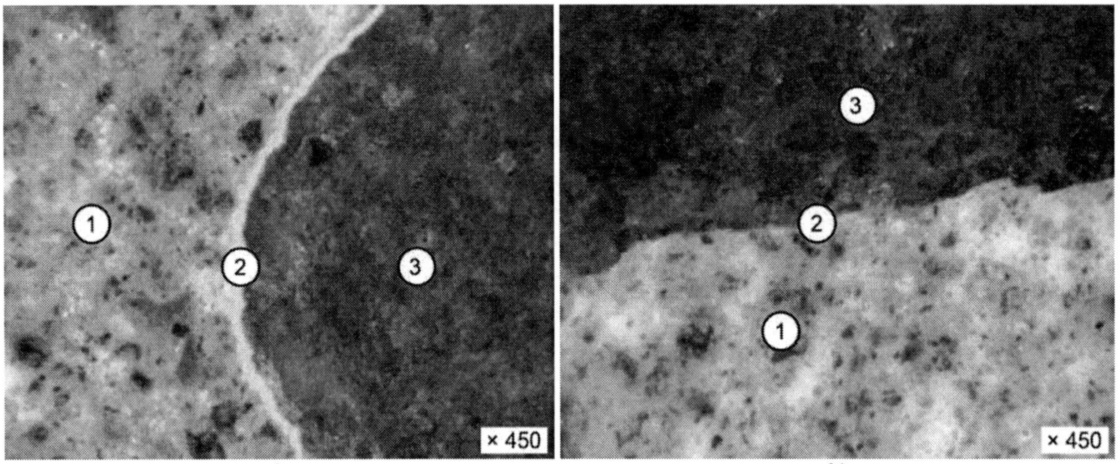

a) b)

Figure 4. The ITZ "andesite aggregate-cement stone" in case of OPC (Na_2O=1.3% by mass); a) without the metakaolin; b) with the metakaolin;1 – cement stone; 2 – ITZ; 3 – alkali-susceptible aggregates.

3.2. Slag alkali-activated cement-based concrete and alkali-activated slag portland cement-based concrete

In case of the slag alkali-activated cement all specimens without the additive (with the exception of the reference one) showed the expansion from +0.12 to+0.316 mm/m, what is not a critical one (Figures 5 and 6). The metakaolin additive allowed to reduce further the expansion (–0.14...+0.01) mm/m (Figure7). No any products of corrosion induced by AAR were detected in the ITZ (Figure7, a, b).

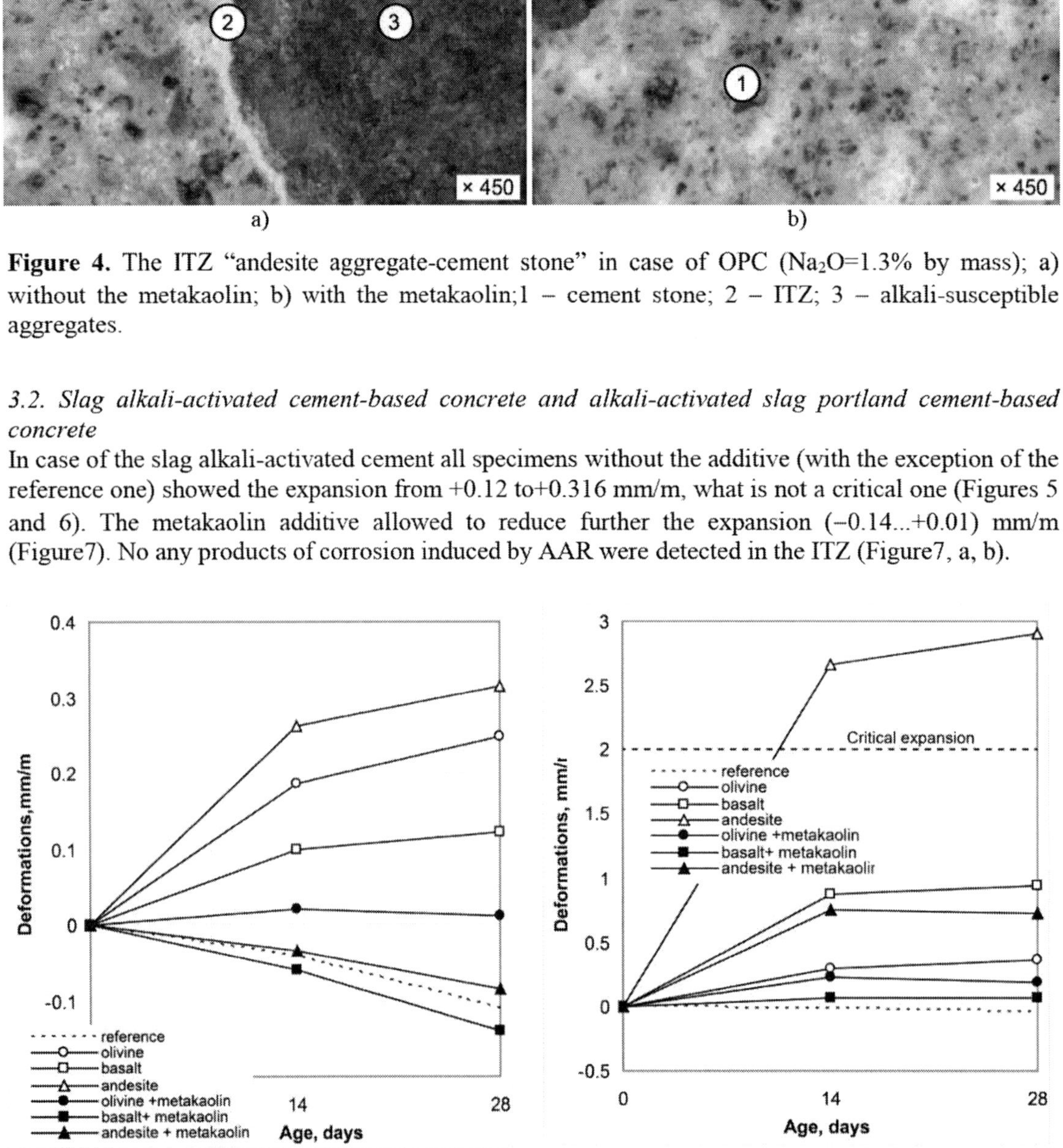

a) b)

Figure 5. Deformations of the compositions made with ASA: a) slag alkali-activated cement (Na_2O = 2.5% by mass). b) alkali-activated slag portland cement (Na_2O=2.5% by mass) Type LCEM IV "OPC clinker:slag"= 40:60.

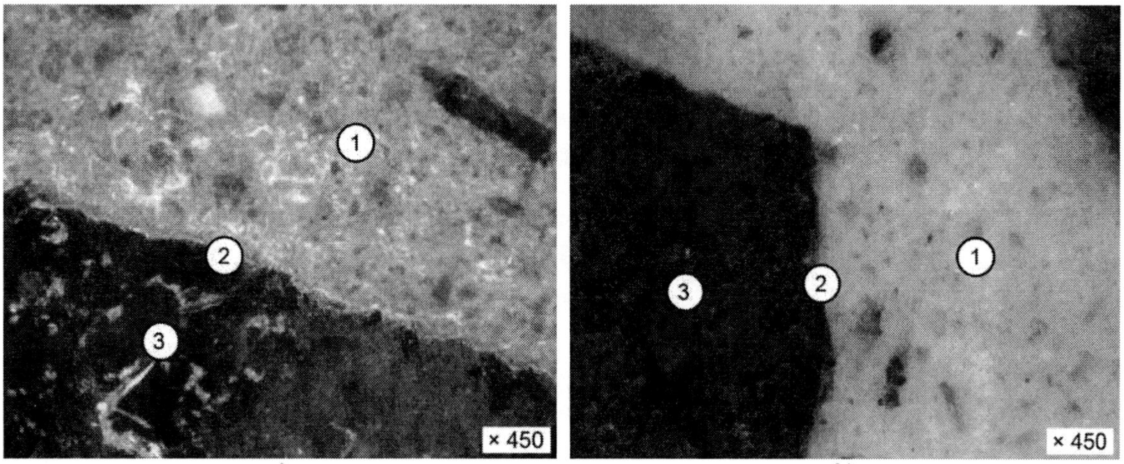

a) b)

Figure 6. Interfacial transition zone "andesite – cement stone" (slag alkali-activated cement (Na_2O= 2.5% by mass):a) without the metakaolin; b) with the metakaolin;1 – cement stone; 2 – ITZ; 3 – alkali-susceptible aggregate.

A behavior of the ASA in the concretes based on alkali-activated slag portland cement (Figures 6b and 7) is somewhat similar to that of these aggregates in the presence of OPC with the increased content of Na_2O (Figures 3 and 4), but with the higher values of expansion. The largest expansion was reported of the specimen with the andesite aggregate without the additive ±2.91 mm/m, thus much exceeding the maximal admissible limit. The specimens prepared using other aggregates under study showed the expansion within the range from +0.36 to 0.947 mm/m.

Such sensitivity of the andesite aggregate can be attributed to the presence in both cases of Ca^{2+} ions in rather large quantities except for the Na^+ ions and to the higher content of active silica in the andesite aggregate – of around 40% by mass (Table 1).

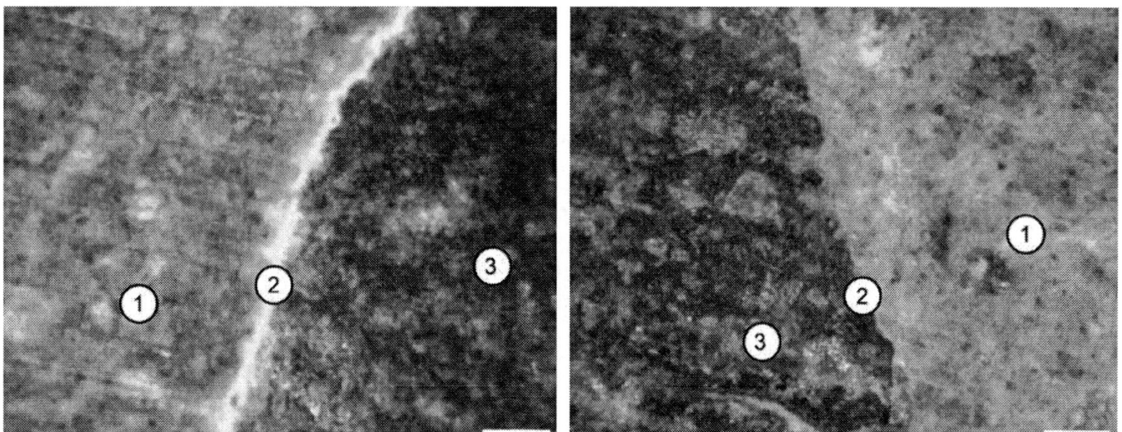

a) b)

Figure 7. The interfacial transition zone "andesite aggregate – cement stone" in case of alkali-activated slag portland cement (Na_2O = 2.5% by mass). a) without the metakaolin; b) with the metakaolin; 1 – cement stone; 2 – ITZ; 3 – alkali-susceptible aggregate.

A reason of such "favorable (positive)" behavior of the ASA in case of slag alkali- activated cement may be attributed to the presence in the finely ground slag side by side with amorphous SiO_2, active Al_2O_3 in sufficient quantities, thus creating conditions for synthesis of the alkaline zeolite-like hydro aluminosilicates represented by a formula – $(Na_2,Ca)O \cdot Al_2O_3 \cdot mSiO_2 \cdot nH_2O$.

This, first of all, enables to bind/attract excessive quantities of alkali into insoluble or poorly insoluble hydration products, and secondly, promotes the formation around grains of the alkali-susceptible aggregates of almost impermeable and strong shell, which prevents the alkaline ions from diffusion to the aggregate and following development of destructive processes in the ITZ "cement stone – ASA". This also enables to enhance adhesion between the cement stone and aggregate. The introduction of active alumina in a form of metakaolin gives further increase of this effect (Figure 3, 5).

3.3. Effect of size of grains of the aggregates and type of alkaline component

A sensitivity/ susceptibility of the aggregates in question to the AAR depending upon size of their grains were studied. The results are given in Figure 8a. A conclusion was made that the grains of size from 1.25 to 5 mm had the largest sensitivity to alkali.

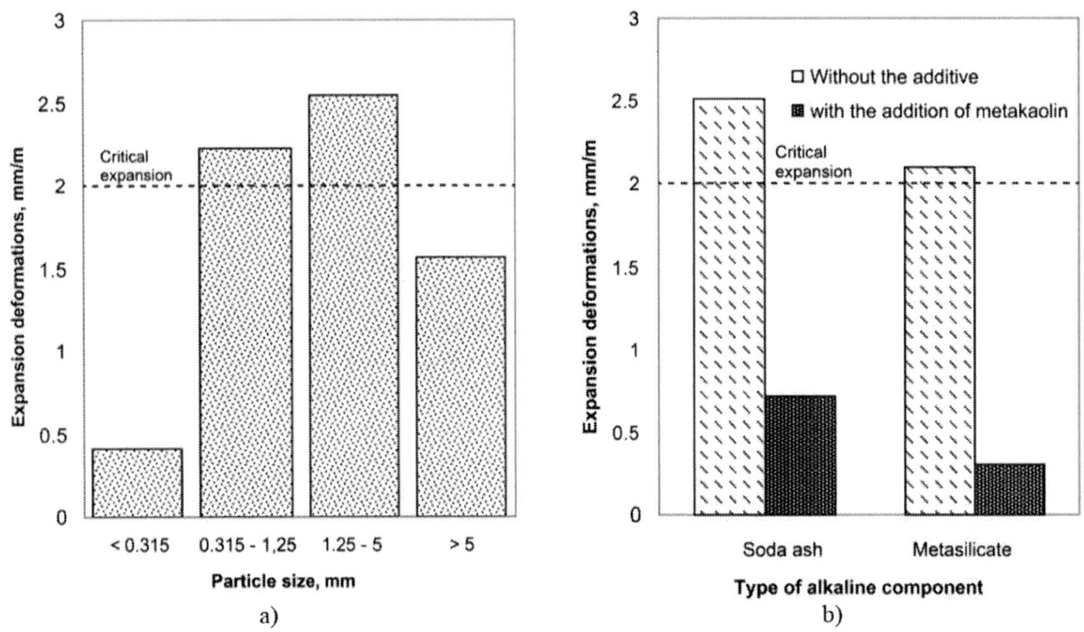

Figure 8. Expansion deformations of the concrete made with alkaline slag portland cement; a) The effect of particle size of the andesite aggregate; b) the effect of type of alkaline component of expansion deformations of the concrete.

An influence of type of alkaline component on deformation of expansion is given in Figure 8b. A conclusion was made that the concretes in which sodium metasilicate was used showed somewhat low shrinkage values compared to those of the concretes made with the soda ash. Evidently, this may be attributed to the presence in the metasilicate of active SiO_2, which is capable even at early stages of hydration to bind partially free calcium of the cement, thus resulting in the formation of not enough dense semi-permeable membrane composed of the CSH-gel and reduction of the quantity of the sodium silicate gel in the ITZ "cement stone – ASA".

3.4. Effect of quantity of metakaolin additive

A full-factor experiment was used to assess effect of quantity of the metakaolin additive and alkali on the rate of reaction "alkali – silicic acid". An adequate equation of regression was produced as a numeral model. The results are given in Figure 9. A conclusion can be drawn that increase in quantity of the alkali in the concrete and decrease in quantity of the metakaolin additive are proportional to the growth of expansion deformations, that is, development of degradation of the ITZ. If, coming from recommendations given in [26], an expansion value below 1.5 mm/m to accept as a safety value, then, with the typical content of alkalis in the alkali-activated cements of 2.5...4% by mass (as Na_2O), introduction of the metakaolin additive in a quantity from 7 to 10% by mass seems to be an adequate measure.

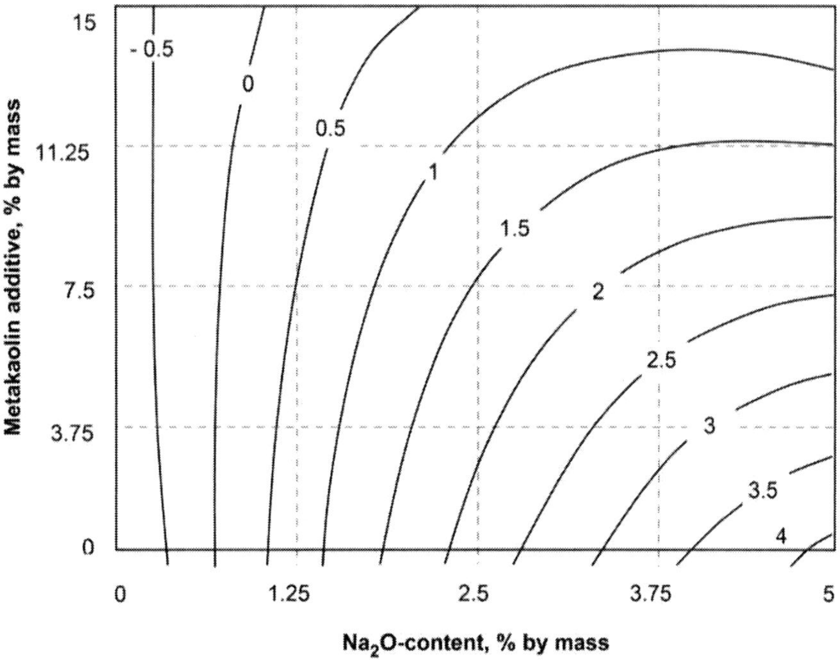

Figure 9. Isolines of expansion deformations, mm/m, vs quantities of the metakaolin additive and alkali content of the cement. Cement – alkali-activated slag portland cement (OPC clinker:slag=40:60); aggregate – andesite rock.

3.5. Mechanism of AAR-related corrosion

In case of the soda ash as an alkaline component, the reactions in the ITZ can evolve as the following:

$$Na_2CO_3 + Ca(OH)_2 \rightarrow CaCO_3 + 2NaOH\ (Na^+, OH^-) \quad (1)$$

$$2NaOH + SiO_2 + H_2O \rightarrow Na_2O{\cdot}SiO_2{\cdot}nH_2O\ (gel\ of\ alkali\ metal\ silicate) \quad (2)$$

Under condition of large quantities of calcium side by side with sodium (that is, in case of OPC and slag portland cement), the reaction can evolve further:

$$Na_2O{\cdot}SiO_2{\cdot}nH_2O + Ca(OH)_2 \rightarrow 2NaOH + CaO{\cdot}SiO_2{\cdot}nH_2O\ (CSH\text{-}gel) \quad (3)$$

The gel of alkali metal silicate itself is not able to produce an expansion pressure. It only breaks the contact "cement stone – aggregate", what is not good, but yet not critical. The presence of calcium enables to synthesize a dense gel-like hydro silicate of calcium (CSH), which is present in the outer part of the reaction zone and acts as a semi-permeable membrane to permit access of new portions of water, ions of Na+ and OH– to the place of reaction, but not allowing removal of the reaction products and thus leading to the increase of pressure in the ITZ "cement stone – ASA" (see Figure 1). Besides, an alkali released in the end of reaction, takes part in the deterioration of the deeper layers of a grain of the ASA. That's why, finally, the available quantities of $Ca(OH)_2$ can be, to some extent, a restricting (limiting) factor for damping the reaction. This may explain the reasons of more breaking destructive effect in the presence of excessive quantities of not only the Na+-ions, but of the Ca^{2+} -ions compared to action of solely Na^+-ions.

Thus, a conclusion can be drawn that formation in the ITZ of the silicate gel of alkali metal or alkali-earth metal taken each separately is not able to produce a breaking pressure and result in degradation of the ITZ. Of danger is a combination of these reaction products and their combined action.

The metakaolin additive allowed to reduce an expansion of the specimens made with the andesite aggregate from a dangerous (+2.91 mm/m) to admissible level (+0/719 mm/m) (Figure5). The expansion of other specimens with the metakaolin additive was found to be not appreciable one (under the limits from +0.063 to +0.193 mm/m).

A scheme of reaction "alkali – silicic acid" with so-called "positive effect of corrosion" in AAC stone is given in Figure 10. An active alumina can get into the AAC composition together with active mineral substances such as granulated blast-furnace slag, fly ash, and metakaolin. A gel of the alkali metal ($Na_2O{\cdot}SiO_2{\cdot}nH_2O$-gel), formed as a result of reaction between the alkali and ASA, is an intermediate product and a supplier for synthesis of the zeolite-like ($Na_2O{\cdot}Al_2O_3{\cdot}mSiO_2{\cdot}nH_2O$) dense, strong (solid) and impermeable shell around the aggregate which stops further development of the destructive reaction.

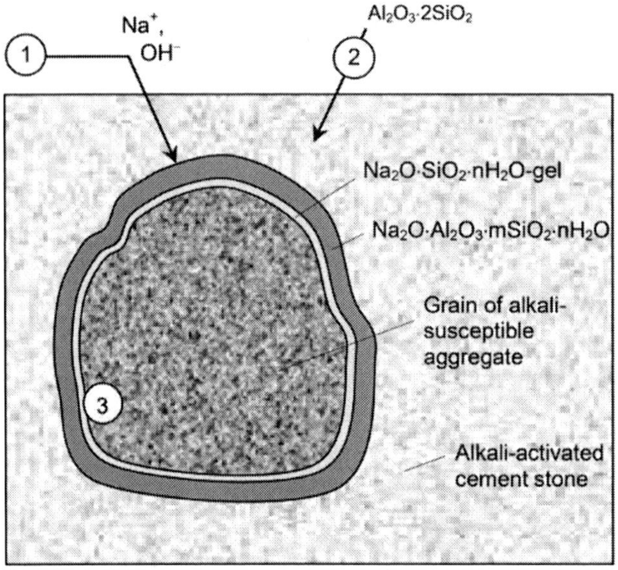

Figure 10. Schematic representation of reaction "alkali-silicic acid" with "positive" effect of corrosion in the alkali-activated cement stone: 1 – diffusion of ions from the alkaline activator solution to the place of reaction; 2 – metakaolin; 3 – ion-exchange and ionic reaction.

4. Conclusions

The results of study allowed to demonstrate a possibility to use the alkali-susceptible aggregates (ASA) in the concretes made from the alkali-activated cements (AAC). The introduction of the active alumina into the compositions of the AAC had a positive effect allowing to keep the structure formation processes in the interfacial transition zone "cement stone – ASA" under control by reducing the expansion deformations down (below 1.5 mm/m) to admissible levels or even completely suppress them. To reach a safety value of expansion limit for the alkali-activated cements (with 2.5...4% Na_2O_{eq}), it was found that introduction of the metakaolin additive in a quantity from 7 to 10% by mass seems to be an adequate measure. It was shown that use of soda ash is more dangerous for alkali-aggregate reactions comparing to the compositions with sodium metasilicate. Andesite rock seams to be the most active type of ASA and the most active fraction of aggregate is 0.315/5.0 mm. The following studies will be directed on developments and investigation of properties of concrete compositions using ASA.

Acknowledgement
The research leading to these results has received the funding from the State budget of Ukraine (project 4DB-2019).

References
[1] Lapovska S, Petropavlovskyi O, Kovalchuk O, Krivenko P and Pasko A 2018 Design of the composition of alkali activated portland cement using mineral additives of technogenic origin *Eastern-European J. Enterp. Technol.* vol 4 138176
[2] Pavel K, Oleg P, Hryhorii V and Serhii L 2017 The Development of Alkali-activated Cement Mixtures for Fast Rehabilitation and Strengthening of Concrete Structures *Procedia Engineering* vol 195 pp 142-146
[3] Garcia-Lodeiro I, Taboada V C, Fernández-Jiménez A and Palomo Á 2017 Recycling Industrial By-Products in Hybrid Cements: Mechanical and Microstructure Characterization *Waste and Biomass Valorization* vol 8 (5), pp 1433-1440
[4] Bumanis G, Bajare D, Locs J and Korjakins A 2013 Alkali-silica reactivity of foam glass granules in structure of lightweight concrete *Constr. Build. Mater.* vol. 47 pp 274–281
[5] Stark J 2013 *Alkali-Kieselsaeure-Reaktion* Weimar: Bauhaus-Universitaet 2008
[6] Stark J and Wicht B 2013 Dauerhaftigkeit von Beton - Der Baustoff als Werkstoff *Beton- und Stahlbetonbau*
[7] Bhuvaneshwari B, Sasmal S, Baskaran T and Iyer N R 2012 Role of Nano Oxides for Improving Cementitious Building Materials *J. Civ. Eng. Sci.*
[8] Runova R, Rudenko I, Gots V, Lastivka O and Konstantynovskyi O 2018 The efficiency of plasticizing surfactants in alkali-activated cement mortars and concretes *MATEC Web Conf.* vol 230, 03016
[9] Obremsky D V, Kovalchuk A V, Konstantynovskyi O P, Rudenko I I and Nikolainko M V 2018 Efficiency of Redispersible Polymer Powders in Mortars for Anchoring Application Based on Alkali Activated Portland Cements *Key Eng. Mater.*
[10] Kochetov G, Prikhna T, Kovalchuk O and Samchenko D 2018 Research of the treatment of depleted nickel-plating electrolytes by the ferritization method *Eastern-European J. Enterp. Technol.*
[11] National standard of Ukraine 2009 *Alkaline cements Specifications*
[12] Horpynko O, Kruts T, Geviuk I, Kropyvnytska T and Sanytsky M 2018 Design of Rapid Hardening Quaternary Zeolite-Containing Portland-Composite Cements *Key Eng. Mater.*
[13] Krivenko P 2017 Why alkaline activation - 60 years of the theory and practice of alkali-activated materials *Journal of Ceramic Science and Technology*
[14] Kazanskaya L F and Smirnova O M 2018 Supersulphated Cements with Technogenic Raw Materials *Int. J. Civ. Eng. Technol.* vol. 9 (11) pp 3006–3012
[15] Bernal S A and Provis J L 2014 Durability of alkali-activated materials: Progress and perspectives *J. Am. Ceram. Soc.* vol. 97 pp 997–1008
[16] Rudenko I, Runova R, Kryvenko P, Omelchuk V and Skorik V 2017 Analysis of plasticizer effectiveness during alkaline cement structure formation *Eastern-European J. Enterp. Technol.*

[17] Kovalchuk O, Grabovchak V and Govdun Y 2018 Alkali activated cements mix design for concretes application in high corrosive conditions *MATEC Web Conf.*
[18] Bumanis G, Bajare D and Goljandin D 2017 Performance Evaluation of Cement Mortar and Concrete With Incorporated Micro Fillers Obtained By Collision Milling in Disintegrator *Ceram. - Silikaty*, vol. 61 (3) pp 1–10
[19] Krivenko P, Petropavlovskyi O and Kovalchuk O 2018 A comparative study on the influence of metakaolin and kaolin additives on properties and structure of the alkali-activated slag cement and concrete *Eastern-European J. Enterp. Technol.* vol. 1 no. 6 (91) pp 33–39
[20] Shakhmenko G, Korjakins A and Bumanis G 2010 Concrete with microfiller obtained from recycled lamp glass *10th International Conference Modern Building Materials, Structures and Techniques*
[21] Plugin D, Plugin A, Pluhin O, Borziak O and Dudin O 2017 The effect of structural characteristics on electrical and physical properties of electrically conductive compositions based on mineral binders *MATEC Web Conf.*
[22] Velandia D F, Lynsdale C J, Provis J L and Ramirez F 2018 Effect of mix design inputs, curing and compressive strength on the durability of Na2SO4-activated high volume fly ash concretes *Cem. Concr. Compos.*
[23] Omelchuk V, Ye G, Runova R and Rudenko I I 2018 Shrinkage Behavior of Alkali-Activated Slag Cement Pastes *Key Eng. Mater*, vol 761 pp 45–48
[24] Fernández-Jiménez A, Garcia-Lodeiro I, Maltseva O and Palomo A 2019 Hydration mechanisms of hybrid cements as a function of the way of addition of chemicals *J. Am. Ceram. Soc.*
[25] Krivenko P, Drochytka R, Gelevera A and Kavalerova E 2014 Mechanism of preventing the alkali-aggregate reaction in alkali activated cement concretes *Cem. Concr. Compos.*
[26] Ke X, Criado M, Provis J L and Bernal S A 2018 Slag-Based Cements That Resist Damage Induced by Carbon Dioxide *ACS Sustain. Chem. Eng.*
[27] Runova I I, Kochevyh RF, Rudenko M O 2005 On the slump loss problem of superplasticized concrete mixes *Proc. Int. Conf. Admixtures - Enhancing Concr. Perform.* pp 149–156
[28] Krivenko P, Kovalchuk O and Pasko A 2018 Utilization of Industrial Waste Water Treatment Residues in Alkali Activated Cement and Concretes *Key Eng. Mater.* vol 761 pp 35–38

The effect of alkali activation solutions with different water glass/NaOH solution ratios on geopolymer composite materials

N Doğan-Sağlamtimur[1], H Öznur Öz[2], A Bilgil[2], T Vural[1] and E Süzgeç[1]

[1]Department of Environmental Eng., Niğde Ömer Halisdemir University, Niğde, Turkey
[2]Department of Civil Engineering, Niğde Ömer Halisdemir University, Niğde, Turkey

E-mail: oznurozell@gmail.com

Abstract. In this study, geopolymer materials were produced from fly ash (FA) supplied from İsken Sugözü Thermal Power Plant located in Adana, Turkey. FA and Rilem Cembureau Standard Sand were used together with the ratio of 0.50. At first, two different alkaline solution/material ratio (FA+standard sand) (L/M) were selected as 0.20 and 0.40 for the design parameters. In the production of geopolymer composite material, sodium silicate (Na2SiO3) and sodium hydroxide (12 M NaOH) were used together within the ratio of 1, 1.5, 2, 2.5 and 3 by weight, respectively. A totally of 20 mixes were cured at 70 and 100 oC for 24 hrs, respectively and thereafter kept in room temperature until testing age. Physical characteristics of hardened mortar were determined via the bulk density, water absorption and porosity values at 28 days while the strength of geopolymers was obtained on the results of compressive strength and flexural strength tests conducted at 7, 28 and 90 days. Considering the testing parameters, geopolymer material with the highest compressive strength was found as 76.0 MPa (28-days) on the mixture produced with L/M ratio of 20% by weight; the alkaline solution consisted of a mixture of Na2SiO3 and 12 M NaOH in weight ratio of 2 by curing at 70 oC for 24 hrs. However, test results showed that there was an optimum limit for the alkaline solution ratio, such that exceeding this limit gave the reverse effect for the strength characteristics of the geopolymer material.

1. Introduction

The awareness of sustainable development in the world has led to the evaluation of the environmental effects as well as the technical properties of the construction materials in the design. Conventional concrete produced with ordinary Portland cement (OPC) is the most widely used material in the construction sector. After mining and grinding raw materials that include limestone and clay, the Portland cement can be produced via a sintering temperature as high as 1500 °C in a cement kiln for desired strength characteristics. High energy consumption is required for the grinding of clinker as well as the production stage. Additionally, the burning of fossil fuel used in the production results high levels of carbon dioxide (CO_2) emissions associated warming of the atmosphere as well as the small amount of other harmful gases. It is reported that cement production for 1 ton causes approximately 0.9 ton of CO_2 emission [1-2]. Due to the fact that the annual cement production in the world is over 3.5 billion tones and the cement sector is responsible for 5-8% of the man-made CO_2 emissions, the researchers have been forced to search the new construction materials as an alternative to conventional concrete in the terms of the environmental effects [2-3].

Fly ash (FA) is driven out of the boiler with flue gases in coal-fired power plants and then generally captured from flue gases by some special equipment (e.g. electrostatic precipitators or other particle filtration equipment) before the flue gases reach the chimneys. The high quantity of FA is currently disposed to landfills by creating a threat to the environment due to the effects to ground water as well as biodiversity [3-4]. The compositions of FA vary noticeably depending on the source of the coal being burned, but the main components are SiO_2 (40-60%), Al_2O_3 (20-30%), Fe_2O_3 (4-10%) and CaO (5-30%) [5]. In order to comply with sustainable development concept, it is very important to minimize the negative environmental impacts of waste materials. On the other hand, FA that is a by-product from thermal power plants can be used as a supplementary cementitious material through the pozzolanic reaction instead of OPC. Another alternative which has been discussed as a substitute for OPC concrete is the geopolymer concrete that can be easily produced with FA by synthesizing pozzolanic compounds or aluminosilicate source materials with highly alkaline solutions [6-7]. Geopolymers consume significantly less energy and reduce the emission of greenhouse gases. The use of FA as the cementitious source in the manufacture of geopolymer concrete has been intensively investigated for not only mechanism of geopolymerization but also mechanical properties of the resulting concrete. It has been shown that in general Class-F FA that is the one of the most suitable widely used aluminum silicate sources for the production of geopolymers as the final product exhibits superior mechanical and durability properties when exposed to aggressive media and under thermal loading [8-10].

The objective of this study is to examine the potential of FAs that are supplied from İsken Sugözü Thermal Power Plant located in Adana, Turkey in the manufacture of geopolymer and develop a FA-based geopolymer composite material with its enhanced physical-mechanical properties. In this study, FA and Rilem Cembureau standard sand (CEN reference, as fine aggregate) were used with the ratio of 0.50. Two different alkaline solution/material ratio (FA+standard sand) (L/M) were selected as 0.20 and 0.40 for the design parameters. In the production of geopolymer composite material, sodium silicate (Na_2SiO_3) and sodium hydroxide (12 M NaOH) were used together within the ratio of 1, 1.5, 2, 2.5 and 3 by weight, respectively. This paper presents the findings of various laboratory tests on the behavior of geopolymers to be used as composite materials.

2. Materials and Test Methods

2.1 Materials

An ASTM C618[11] Class-F FA supplied from İsken Sugözü Thermal Power Plant in Adana (Turkey) was used as a geopolymerizing material in this study. Chemical compositions of FA obtained from X-ray fluorescence (XRF) test were presented in Table 1. The specific gravity of FA was obtained as 2.25. Rilem Cembureau standard sand complied with TSE EN 196-1 [12] specification was used in the production of mortar mixtures. Sieve analysis of this special sand given in Table 2 is found to be in agreement with the limits of standard. The activator was a combination of Na_2SiO_3 known as water glass or liquid glass and 12 M NaOH. The Na_2SiO_3 solution was consisted of Na_2O=14.7%, SiO_2=29.4% and H_2O=55.9%. Additionally, 12 M NaOH solution was prepared by dissolving of NaOH existing in the form of pellets.

Table 1. Chemical compositions of FA (%).

Oxides (%)	SiO_2	Al_2O_3	Fe_2O_3	CaO	Na_2O	MgO	K_2O	SO_3	BET (m^2/g)	LOI (%)
FA	62.28	21.46	7.01	1.53	0.26	2.37	3.81	0.07	2.26	1.78

Table 2. Properties of Rilem Cembureau standard sand.

Diameter of sand grain (mm)	0.08	0.16	0.5	1.0	1.6	2.0
Remaining (%)	99	87	72	34	6	0
Limits of specification (%)	99±1	99±5	67±5	33±5	7±5	0

2.2 Mix Design

A total of 10 mixes were designed in which different L/M and Na_2SiO_3/NaOH ratios were utilized, respectively. The FA/sand ratio in the geopolymer mortar was kept constant at 0.50 (400/800). At first, two different L/M ratios were selected as 0.20 and 0.40. In addition, Na_2SiO_3 and 12 M NaOH solution were used together to support the dissolution of solid FA particles within the ratios of 1, 1.5, 2, 2.5 and 3 by the weight, respectively. Therefore, the ratio of SiO_2/Na_2O was changed between 0.57-1.09 in each alkali activation solution. The mixing ratios of geopolymer mortar are presented in Table 3.

Table 3. Mix proportion of geopolymer mortars.

FA (g)	**Standard Sand (g)**	**Alkaline Solution/Materials Ratio (L/M)**	**Na_2SiO_3/NaOH Ratio**	**NaOH (g)**	**Na_2SiO_3 (g)**	**Total Water/Materials Ratio**
400	800	0.20	1	120	120	0.18
			1.5	96	144	
			2	80	160	
			2.5	68	172	
			3	60	180	
		0.40	1	240	240	0.20
			1.5	192	288	
			2	160	320	
			2.5	136	344	
			3	120	360	

The mixing procedure used for the FA-based mortar is the same as the cement-based mortar. In the first stage of the mortar formation, FA and standard sand were mixed with a mixer for 30 seconds to obtain homogeneity. Pellets of NaOH were dissolved in the water to obtain the required concentration before mixing. Then, Na_2SiO_3 and 12 M NaOH were mixed for the desired content of the alkaline solution and added to the mixture. The workability of fresh geopolymer mortar was tested with respect to ASTM C1437 [13]. The target slump flow diameter determined as 14±2 cm was obtained by producing the different test batches for each type of mortar. At the end of the production, the prepared mortar immediately fed into molds with the dimensions of 40x40x160 cm and vibrated until getting full compaction. Finally, geopolymer mortars were cured at 70 and 100 °C by using the protecting cover to prevent evaporation of water, for 24 hrs, respectively. Thus, a total of 20 different hardened geopolymers were obtained. As seen in Figure 1, the hardened samples were kept under room temperature until the testing age for the physical and mechanical tests.

Figure 1. Geopolymer samples at laboratory condition.

The hardened geopolymer mortars were coded based on the design parameters such as L/M ratio, curing temperature and Na_2SiO_3/NaOH ratio. For example, when the geopolymer was manufactured with the ratios of L/M=0.40, Na_2SiO_3/NaOH=1 and cured at 70 °C, this mixture was denoted as 40FA-70-1.

2.3 Experimental Study

The bulk density, water absorption (by mass) of the hardened mortars were measured according to ASTM C642 [14] while TS EN 772-4 [15] was used to determine the porosity of samples at 28 days. Compressive and flexural strengths of geopolymer mortars were tested with in the line of ASTM C 349-14 [16] and ASTM C 348 [17] at 7, 28 and 90 days, respectively. The average of six measurements for each test is reported at each testing age.

3. Test Results and Discussion

The variations of test results for physical characteristics of geopolymer are given in Figure 2. As seen in Figure 2, the bulk densities at 28 days were changed between 1820-2079 and 1798-2012 kg/m^3 for hardened mortars cured at 70 and 100 °C for 24 hrs, respectively. Increasing the ratio of Na_2SiO_3/NaOH resulted an increment for the bulk density of the specimens cured at 70 °C while the bulk density of geopolymers cured at 100 °C were decreased by the increased ratio of Na_2SiO_3/NaOH. This may be attributed to fact that the presence of aggregate induced thermal expansion between the aggregate and paste, thus increasing curing temperature as well as Na_2SiO_3/NaOH ratio caused the disintegration in the polymer network as well as the expansion of specimens [18].

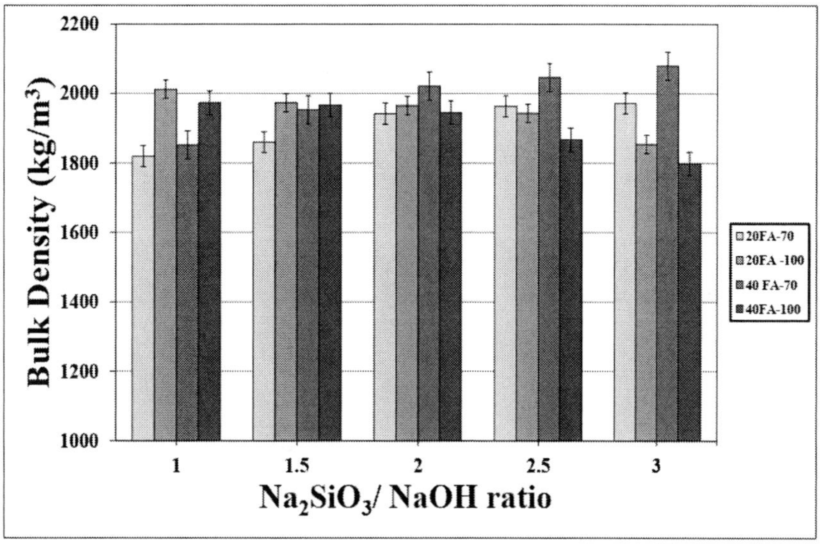

Figure 2. Bulk density of hardened geopolymers at 28 days.

Figure 3 and 4 shows the results of water absorption and porosity in units of percent (%) for geopolymer mortars at 28 days, respectively. Decreasing porosity reduced the water absorption of geopolymers as a result of decreasing pore size. It was observed that geopolymer mortar specimens produced with the ratio of $Na_2SiO_3/NaOH$ up to 2 and cured at 100 °C resulted lesser values of water absorption and porosity when compared to one's cured at 70 °C. 40FA-70-1 specimen had 24.55% water absorption and 38.43% porosity whereas 20FA-100-2 showed comparatively lower corresponding values of 8.31% and 17.0%, respectively. This conclusion was supported by the researchers showed that high dissolution of the FA particles in the mixture obtained by higher alkali content resulted in denser microstructure in geopolymers [19-20].

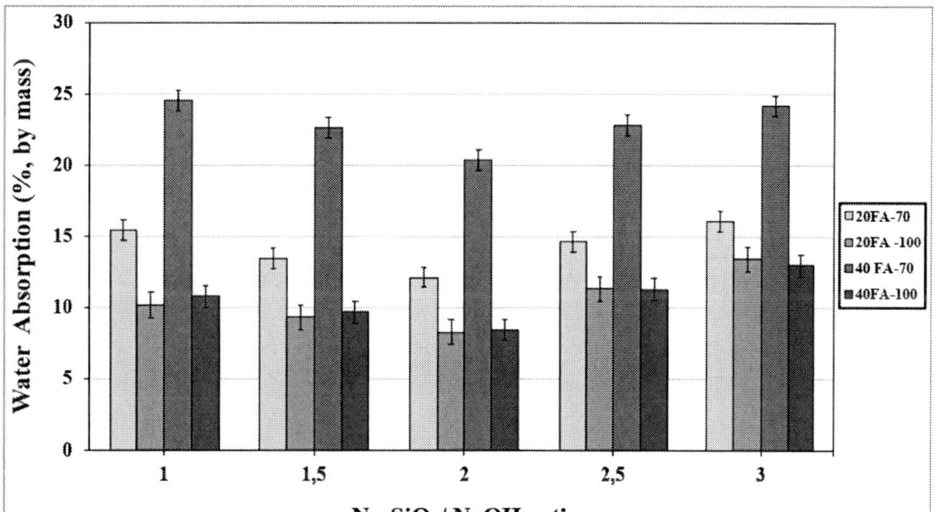

Figure 3. Water absorption of hardened geopolymers at 28 days.

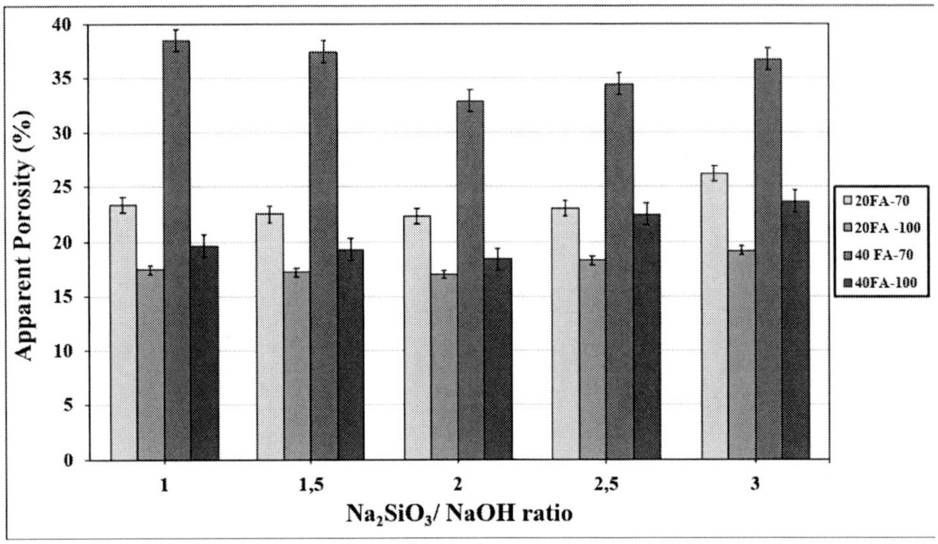

Figure 4. Porosity of hardened geopolymers at 28 days.

Compressive and flexural test results with respect to $Na_2SiO_3/NaOH$ ratio are indicated in Figures 5 and 6 at 7, 28 and 90 days, respectively. As observed from Figures, the testing age does not have an important effect on strength characteristics of geopolymer after the curing process was completed [18]. Increasing the ratio of $Na_2SiO_3/NaOH$ gave an increment of strength characteristics up to an optimum point ($Na_2SiO_3/NaOH=2$). Generally, it was accepted that higher concentration gave the better strength. But, there was an optimum limit for the alkaline solution ratio, such that exceeding this limit gave the reverse effect for the strength characteristics [3-21]. Thus, the highest compressive and flexural strength values were found 76.0 MPa and 10.5 MPa for 20FA-70-2 at 28 days, respectively. However, the compressive and flexural strength were reduced by the increasing ratio of L/M. This effect can be explained by Yang et al. [22] stated that similar to hydraulic Portland cement paste, higher water-to-cementitious-materials ratio can cause a reduction on strength values. Moreover, the hardened mortars designed with the ratio of L/M=0.40 and cured at 70 °C had the lowest compressive and flexural strength. This result agrees with the finding of Bhowmick and Ghosh [23] showed that when higher sand content was used in the geopolymerization process, the quantity of the geopolymer gel may not be sufficient to bind all loose aggregates; thus, the reduction in strength was observed. Additionally, for the same ratio of $Na_2SiO_3/NaOH$, increasing curing temperature caused deterioration in compressive and flexural strength. Similarly, Atiş et al. [24] indicated that the available water in the mix instantly vaporized by heat curing process before dissolution of FA in the alkaline environment.

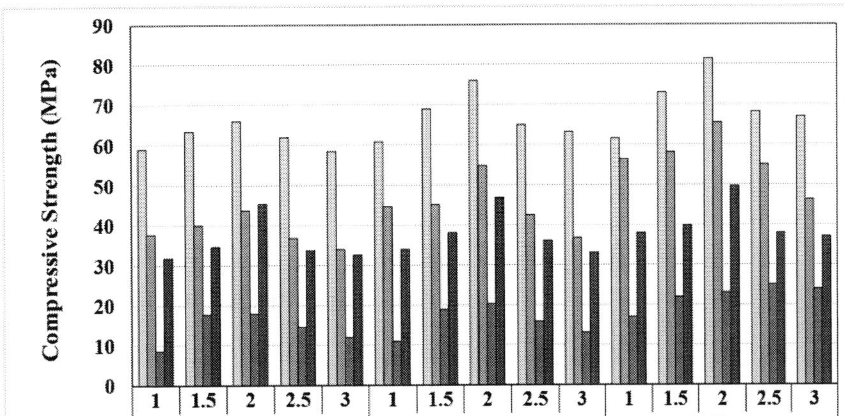

	7 Days					28 Days					90 Days				
	1	1.5	2	2.5	3	1	1.5	2	2.5	3	1	1.5	2	2.5	3
20FA-70	59.0	63.4	66.0	62.0	58.5	60.9	69.0	76.0	65.0	63.2	61.6	73.0	81.4	68.2	67.0
20FA-100	37.7	40.0	43.7	36.8	34.0	44.7	45.1	54.8	42.5	36.8	56.4	58.2	65.5	55.1	46.3
40FA-70	8.6	17.8	18.0	14.6	12.0	11.0	19.0	20.4	15.9	13.1	17.1	22.0	23.1	25.1	24.0
40FA-100	31.8	34.6	45.4	33.7	32.6	34.0	38.1	46.8	36.1	33.1	38.0	39.8	49.6	38.0	37.1

Figure 5. The variations of compressive strengths with respect to $Na_2SiO_3/NaOH$ ratio for hardened geopolymers.

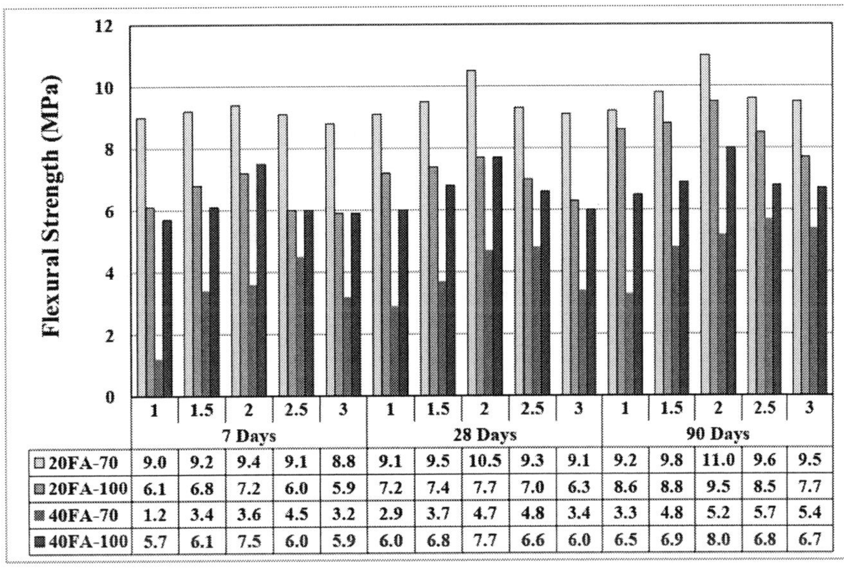

	7 Days					28 Days					90 Days				
	1	1.5	2	2.5	3	1	1.5	2	2.5	3	1	1.5	2	2.5	3
20FA-70	9.0	9.2	9.4	9.1	8.8	9.1	9.5	10.5	9.3	9.1	9.2	9.8	11.0	9.6	9.5
20FA-100	6.1	6.8	7.2	6.0	5.9	7.2	7.4	7.7	7.0	6.3	8.6	8.8	9.5	8.5	7.7
40FA-70	1.2	3.4	3.6	4.5	3.2	2.9	3.7	4.7	4.8	3.4	3.3	4.8	5.2	5.7	5.4
40FA-100	5.7	6.1	7.5	6.0	5.9	6.0	6.8	7.7	6.6	6.0	6.5	6.9	8.0	6.8	6.7

Figure 6. The variations of flexural strengths with respect to $Na_2SiO_3/NaOH$ ratio for hardened geopolymers.

4. Conclusions

The following conclusions can be drawn from this experimental study:

- Workable geopolymer mortar with an acceptable strength can be manufactured by activating locally available Class-F FA with different alkaline solution content comprised of Na_2SiO_3 and 12M NaOH solution.
- The bulk density of the specimens cured at 70 °C was increased by the increasing the ratio of Na_2SiO_3/NaOH. However, the increment of curing temperature resulted a decrease for the bulk density due to the thermal expansion of the aggregate in geopolymer matrix.
- Water absorption and porosity values were decreased by the ratio of Na_2SiO_3/NaOH up to 2 as well as the increasing curing temperature.
- Geopolymer mortars had a compressive strength ranged between 11.0-76.0 MPa at 28 days were produced. Increasing the ratio of Na_2SiO_3/NaOH gave an increment of strength characteristics up to an optimum point obtained as 2. Thus, the highest compressive and flexural strength values were found as 76.0 MPa and 10.5 MPa for the mixture with L/M ratio of 0.20 cured at 70 °C at 28 days, respectively. Moreover, the compressive and flexural strength were reduced by the increasing ratio of L/M due to the geopolymer gel can be sufficient to bind all sand in the mixture.
- It can be concluded that geopolymers produced in this study with an acceptable strength could be used in construction sector such as bricks blocks, paving stone or kerbstone as well as prefabricated materials. Geopolymer production not only provides an opportunity to use FA as a binder by alkaline activation but also contributes the environmentally sustainable development by avoiding the disposal problems of FA.

Acknowledgments

This study is a part of the project "Development of eco-friendly composite materials based on geopolymer matrix and reinforced with waste fibers" funded by the European Commission, within the 7th Framework Programme for Research and Technology Development (FP7), Topic #02: Waste management, recycling and urban mining (Project No. ELAC2015/T02- 0721) under the ERANet-LAC: Latin America, Caribbean and European Union and supported by a grant of the Turkish Scientific and Technological Research Council of Turkey (TÜBİTAK). The Turkish project number is 116Y549. We thank Turan Şevki Köker, Arife Gözde Akan, Pervin Nas and Betül Dikkaya for valuable contribution in experimental parts of this study.

References

[1] Erdoğan, ST 2014 Jeopolimerler çevre dostu mudur? *Çimento ve Beton Dünyası* **107** 50-55, in Turkish.

[2] Gümüş A 2016 Geopolimer beton özelliklerine termal kür prosesinin etkisi (MSc Thesis), *Afyon Kocatepe University*, Afyonkarahisar, Turkey, in Turkish.

[3] Ling Y 2018 Proportion and performance evaluation of fly ash based geopolymer and its application in engineered composites (PhD Thesis) *Iowa State University, Iowa, USA* https://lib.dr.iastate.edu/cgi/viewcontent.cgi?article=7405&context=etd. Accessed 3 June 2019

[4] ACI Committee 232 2004 *232.2R-18:* Use of fly ash in concrete *Amer. Concr. Inst.* https://www.concrete.org/store/productdetail.aspx?ItemID=232218&Language=English&Units=US_AND_METRIC. Accessed 3 June 2019

[5] Zhuang X Y, Chen L, Komarneni S, Zhou C H, Tong D S, Yang H M, Yu W H and Wang H2016 Fly ash-based geopolymer: clean production, properties and applications *J. Clean Prod.* **125** 253-267

[6] Sujeong L, Seo M D, Kim Y J, Park H H, Kim T N, Hwang Y and Cho S B 2010 Unburned carbon removal effect on compressive strength development in a honey comb briquette ash-based geopolymer *Int J. Miner Process.* **97(1-4)** 20-25

[7] Singh P R and Shah N D 2014 Impact of coal combustion fly ash used as a binder in pavement *J. Civ. Eng. Environ. Tech.* **1(2)** 57–60
[8] Albitar M, Mohamed Ali M S, Visintin P and Drechsler M 2015 Effect of granulated lead smelter slag on strength of fly ash based geopolymer concrete *Constr. Build. Mater.* **83** 128-135
[9] Hardjito D, Rangan B V2005Development and properties of low-calcium fly ash-based geopolymer concrete. In: Research Report GC 1, Curtin University of Technology, Perth, Australia
[10] Van Deventer J S J, Provis J L, Duxson P 2012 Technical and commercial progress in the adoption of geopolymer cement *Miner. Eng.* **29** 89-10
[11] ASTM C618-00 2000 Standard specification for coal fly ash and raw or calcined natural pozzolan for use as a mineral admixture in concrete. *ASTM International.* https://www.astm.org/DATABASE.CART/HISTORICAL/C618-00.htm. Accessed 3 June 2019
[12] DS/EN 196-1 2016 Methods of testing cement-Part 1: Determination of strength. https://webstore.ansi.org/standards/ds/dsen1962016. Accessed 3 June 2019
[13] ASTM C1437-15 Standard test method for flow of hydraulic cement mortar. *ASTM International.* https://www.astm.org/Standards/C1437. Accessed 3 June 2019
[14] ASTM C642-13 Standard test method for density, absorption and voids in hardened conrete. *ASTM International.* https://www.astm.org/Standards/C642. Accessed 3 June 2019
[15] TS EN 772-4 Methods of test for masonry units-Part 4: Determination of real and bulk density and of total and open porosity for natural stone masonry units. *TSE.* https://intweb.tse.org.tr/standard/standard/Standard.aspx?081118051115108051104119110104055047105102120088111043113104073099081049053120103119090080100047. Accessed 3 June 2019.
[16] ASTM C349-18 Standard test method for compressive strength of hydraulic-cement mortars (using portions of prisms broken in flexure). *ASTM International.* https://www.astm.org/Standards/C349. Accessed 3 June 2019
[17] ASTM C348-18 Standard test method for flexural strength of hydraulic-cement mortars. *ASTM International.* https://www.astm.org/Standards/C348.htm. Accessed 3 June 2019
[18] Kong D and Sanjayan J 2010 Effect of elevated temperatures on geopolymer paste, mortar and concrete *Cem. Concr. Res.* **40(2)** 334-339
[19] Thokchom S, Ghosh P and Ghosh S 2009 Effect of water absorption, porosity and sorptivity on durability of geopolymer mortars *ARPN J Eng. Appl. Sci.* **4(7)** 28-32
[20] Görhan G and Kürklü G 2014 The influence of the NaOH solution on the properties of the fly ash based geopolymer mortar cured at different temperatures *Compos. Part B-Eng.* **58** 371-377
[21] Khale D and Chaudhary R 2007 Mechanism of geopolymerization and factors influencing it development: a review *J. Mater. Sci.* **42(3)** 729-746
[22] Yang T-R, Chang T-P, Chen B-T, Shih J-Y and Lin W-L 2012 Effect of alkaline solutions on engineering properties of alkali-activated GGBFS paste *J. Mar. Sci. Tech-Japan.* **20(3)** 311-318
[23] Bhowmick A and Ghosh S 2012 Effect of synthesizing parameters on workability and compressive strength of fly ash based geopolymer mortar *Int. J. Civ. Struct. Eng.* **3(1)** 168-177
[24] Atis C D, Görür E B, Karahan O, Bilim C, Ilkentapar S and Luga E 2015 Very high strength(120 MPa) class f fly ash geopolymer mortar activated at different NaOH amount, heat curing temperature and heat curing duration *Constr. Build. Mater.* **96** 673-678

IOP Conf. Series: Materials Science and Engineering **660** (2019) 012004 doi:10.1088/1757-899X/660/1/012004

Influence of curing conditions on mechanical and fracture properties of alkali activated slag concrete

H Šimonová[1], B Kucharczyková[1], V Bílek Jr.[1, 2], L Topolář[1] and D Kocáb[1]

[1] Brno University of Technology, Faculty of Civil Engineering, Veveří 331/95, 602 00 Brno, Czech Republic

[2] Brno University of Technology, Faculty of Chemistry, Purkyňova 464/118, 602 00 Brno, Czech Republic

E-mail: simonova.h@vutbr.cz

Abstract. The aim of this work is to determine the physical, mechanical and fracture characteristics of concrete specimens with alkali activated binder cured at different conditions. The concrete mixture with alkali activated ground granulated blast furnace slag was selected for the purpose of the experiment. The binder was activated by the liquid sodium silicate with the silicate modulus of 2.0. Natural sand and gravel were used as a fine and coarse aggregate, respectively. The maximum aggregate size was 16 mm. The binder : aggregate : water ratio was 1 : 3 : 0.46. During the specimens ageing, the dynamic modulus of elasticity was determined using a resonance method. The fracture characteristics were determined based on the results of the three-point bending test of specimens provided with an initial central edge notch. The fracture experiments were conducted at the age of 3 and 38 days. The loading force (F), the displacement measured in the middle of the span length (d), and the crack mouth opening displacement ($CMOD$) were continuously recorded during the test. The records of fracture tests were subsequently evaluated using the effective crack model, work-of-fracture method and double-K fracture model. The fracture tests were also accompanied by the monitoring of acoustic responses using the acoustic emission method.

1. Introduction

Alkali-activated materials are formed via dissolution of a suitable aluminosilicate precursor in presence of an alkaline activator, typically sodium hydroxide or sodium water glass. The dissolved species are polycondensated and they are thus rearranged to a binding phase. Their composition depends on the nature of activator and on the composition of the aluminosilicate precursor. Activation of low-calcium precursors like fly ashes and metakaolin result in the formation of the three-dimensional structure of sodium-aluminate-silicate-hydrates (NASH) while activation of high-calcium precursors, typically slag, leads to the formation of calcium-aluminium-silicate-hydrates (CASH) which are more similar to CSH phases formed during the Portland cement hydration. Differences in the structure and composition result also in the differences in their properties [1].

Alkali-activated slag (AAS) is popular because of its high early age strengths easily obtained at room temperatures but can suffer from the fast setting time and particularly from shrinkage and cracking. Despite many approaches to reduce shrinkage have been done more or less successfully, among others e.g. partial slag substitution by mineral admixtures [2], internal curing [3] or curing at elevated temperatures [4] as well as the use of shrinkage-reducing admixtures [5], high cracking tendency of AAS is still an issue.

Ideally, cracking does not take place but if it does, there is an option to reduce its extent and consequences thanks to self-healing effect. For Portland cement-based materials improvement of self-healing ability various approaches including chemical admixtures, bacteria or microencapsulation has been used [6]. On the other hand, the self-healing phenomenon in AAS-based materials has been poorly investigated, despite it has certain potential. Moreover, AAS is very sensitive to curing conditions and therefore in the present study fracture parameters of AAS with different curing conditions and cracking development were determined.

2. Material and specimens

Ground granulated blast furnace slag with Blaine fineness of 400 m^2/kg was used as a precursor for alkaline activation. According to X-ray diffraction, about 90% was amorphous phase and the rest were akermanite and calcite. For slag activation water glass having silicate modulus of 2.0 and solid content of 46.8 was used at the dose corresponding to 8% of Na_2O with respect to the slag weight. Three fractions, i.e. 0–4, 4–8 and 8–16 mm, of natural aggregate at the proportions given by Füller granulometric curve were used. Slag : aggregate : water ratio was 1 : 3 : 0.46 including water contained in water glass. Before the addition of solids, water glass with water were mixed together. Then the slag was added followed by an addition of aggregate. Total mixing time since the addition of slag to water was 6 minutes.

The AAS concrete mixture was cast into the polyethylene moulds with dimensions of $100 \times 100 \times 400$ mm to prepare prismatic specimens intended for the fracture tests. The moulds were covered with a thin PE foil and stored in the laboratory with ambient temperature of 21 ± 2 °C and relative humidity of 50 ± 10 % for 24 hours. After that, all specimens were demoulded and left to dry freely at the same laboratory conditions. After another 24 hours when specimens were drying freely, the cracks caused by shrinkage appeared at specimen's surfaces, see figure 1 (left). At this age (48 hours), the specimens were divided into two groups. One group of the test specimens was left to dry freely for the entire time of aging. The second group was immersed in the water-bath for the predetermined curing time, after which the specimens were again removed from the bath and left to dry freely until the testing time. This procedure was chosen to prevent the crack development and to start the self-healing process in AAS concrete specimens.

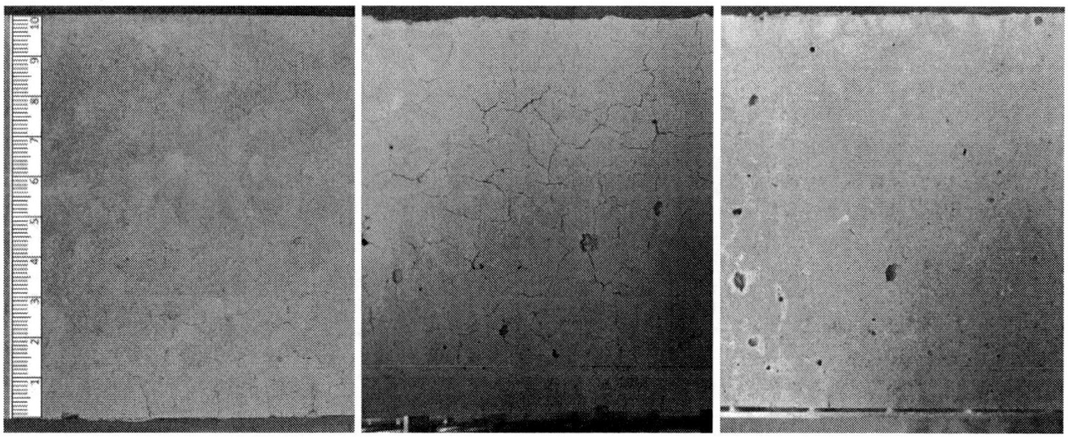

Figure 1. Specimen's surface at the age of 2 days (left), 27 days stored at air (in the middle) and 27 days stored under water conditions.

The results of the fracture tests of the specimens stored at the air conditions performed at the age of 3 days were taken as the initial values of investigated parameters. To examine the influence of the curing conditions, the fracture tests were performed for both sets of specimens at the age of 38 days. The specimens cured in water were removed from the water bath at the age of 27 days and were left to dry freely until the testing time. Based on the curing conditions and the age of specimen when fracture tests were performed, the sets of specimens are designated as follows: 3_days_air, 38_days_air and 38_days_water. Each testing set consists of three specimens.

3. Methods

3.1. Resonance method

During the specimens ageing, the dynamic modulus of elasticity was monitored and determined using the resonance method according to the standard ČSN 73 1372 [7]. The natural frequency of longitudinal vibration was determined for each specimen. The vibration was generated by a mechanical impulse using the impact hammer and the natural frequency was measured by an oscilloscope Handyscope HS4 with a piezoelectric sensor. The dynamic modulus of elasticity E_{dyn} was then calculated from this natural frequency according to the standard [7].

3.2. Fracture test

The fracture characteristics were determined based on the results of the three-point bending test of specimens provided with an initial central edge notch. The notch was made with a diamond blade saw and its depth was approximately equal to 1/3 of the depth of the specimen. The span length was equal to 350 mm. The fracture experiments were performed at the age of 3 and 38 days. For this purpose, a very stiff multi-purpose mechanical testing machine LabTest 6.250 with the load range of 0–250 kN was used, see figure 2. The loading force (*F*), the displacement measured in the middle of the span length (*d*), and the crack mouth opening displacement (*CMOD*) were continuously recorded during the test with a frequency of 5 Hz. All performed fracture test were accompanied by the measurements using the acoustic emission technique.

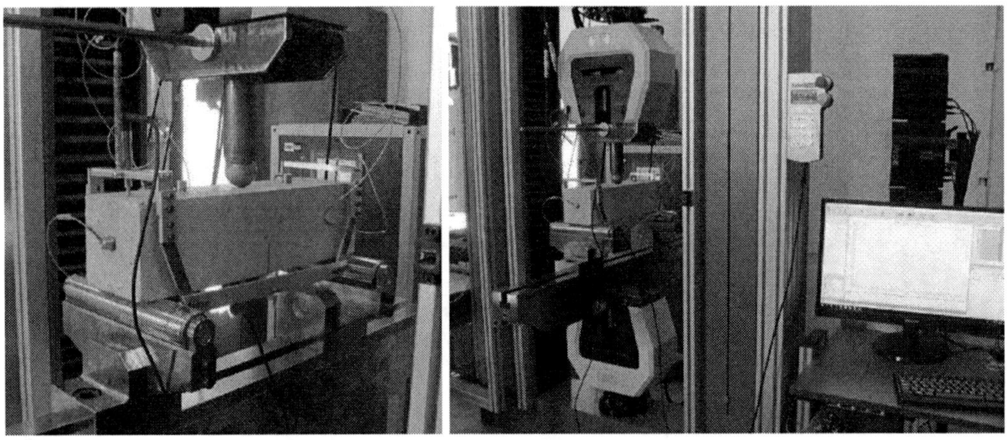

Figure 2. The three-point bending fracture test configuration (left); overall view of the testing machine.

3.2.1. Processing of measured data. At the beginning of loading, small-sized fluctuations in the measured values of monitored parameters are often recorded. This effect is caused by the crushing of small protrusions on the specimen's surface due to the pressure at the support and loading points. Therefore, it is advisable to correct the beginning part of the measured diagram in order to obtain the appropriate estimation of the modulus of elasticity value.

In the case of specimen made of composite with a brittle matrix, subjected to the displacement-controlled loading, stability loss can occur due to the insufficient stiffness of the test equipment in comparison to the specimen's stiffness. This stability loss leads to an interruption in record of the *F–d* and *F–CMOD* curves. Therefore, a procedure has been developed to identify and correct the data discontinuity in the case of fracture tests of specimens based on the brittle matrix [8]. The work-of-fracture and fracture energy values determined from measured diagrams without this correction could be overestimated.

The processing of measured diagrams including the above-mentioned phenomena was performed in GTDiPS software [9] which uses advanced transformation methods to process extensive point sequences. The correction of the measured diagrams in this case included primarily the shifting of the origin of the coordinate system, the smoothing of the diagram and the reduction of the number of points. For illustration, the figure 3 introduces the *F–d* and *F–CMOD* diagrams after the advanced correction.

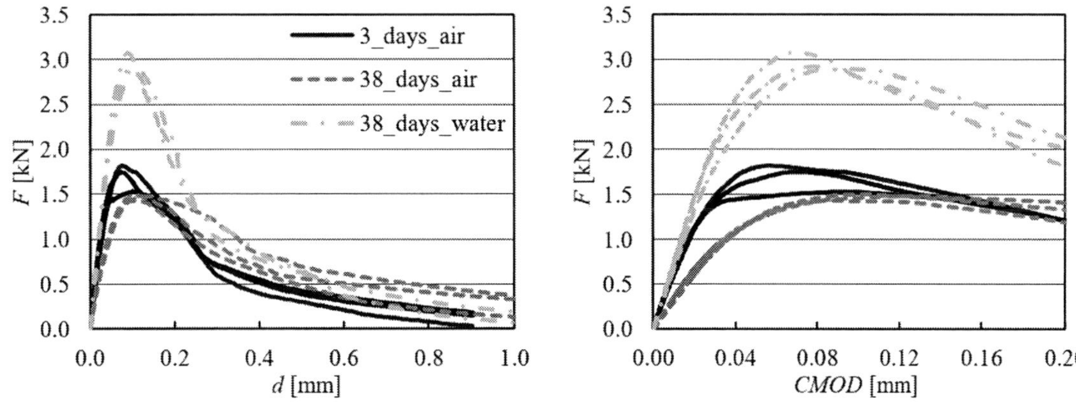

Figure 3. The corrected *F–d* and *F–CMOD* diagrams.

3.2.2. Evaluation of F−d diagrams. After the above described processing of recorded diagrams, at first the initial parts of *F–d* diagrams were used to estimate the modulus of elasticity E_c value according to [10]. The fracture toughness value K_{Ice} was determined using the effective crack model [10] which combines the linear elastic fracture mechanics and crack length approaches. To determine K_{Ice}, the effective crack length a_e, corresponding to the maximum load F_{max} and mid-span displacement *d*, had to be calculated first. The work-of-fracture W_F value was obtained from the whole *F–d* diagrams according to the RILEM method [11] and corresponds to area under the *F–d* diagram. Subsequently, the specific fracture energy G_F value was determined, that the W_F values were divided by area of ligament (cross-section of specimens through which the crack grows).

3.2.3. Evaluation of F−CMOD diagrams. The *F–CMOD* diagrams (a part until the peak load) were subsequently evaluated using the double-*K* fracture model [12]. The advantage of this model is that the double-*K* fracture criterion can predict the crack initiation, stable crack propagation, and unstable fracture process during the crack propagation in quasi-brittle material. According to this criterion, the two size-independent parameters termed as initial cracking toughness K_{Ic}^{ini} and unstable fracture toughness K_{Ic}^{un} can predict different stages of the fracture process. This model is based on combination of the concept of cohesive forces acting on the faces of the fictitious (effective) crack increment with a criterion based on the stress intensity factor (details can be found in numerous publications – e.g. in Kumar and Barrai [12]). Finally, the value of the load F_{ini} is determined according to [12]. This value can be defined as the load level at the beginning of stable crack propagation from the initial crack/notch.

3.3. Acoustic emission method

The acoustic emission (AE) method can be used for monitoring the fracture process of specimen during the three-point bending test. AE belongs to the non-destructive testing techniques [13]. Unlike other techniques, AE is a passive control method that can examine the entire volume structure of construction, structural member or – as in this case – of the specimen. The AE method detects only active damages, which begin to develop during the test. The AE activity arises from the rapid release of energy in a material because of stimulation by internal or external stress [14, 15]. To describe acoustic emission signals formed during the fracture test, the number of events N_{AE}, duration D_{AE}, amplitude A_{AE} and energy E_{AE} of AE signals were monitored and recorded.

3.4. Basic mechanical parameters determination

The informative compressive strength f_c value was determined on the fragments remaining after the fracture experiments of AAS composites had been performed. The flexural strength f_f value was estimated based on the maximum force recorded during the three-point bending fracture test. The tensile strength f_t value was identified from *F–d* diagrams using the artificial neural network (ANN) based on the inverse analysis method [16, 17].

4. Results and discussion

The average value (obtained usually from 3 independent measurements) and the coefficient of variations CoV of selected mechanical, fracture and acoustic emission parameters of AAS concretes cured at the different conditions are summarized in following tables.

Table 1 introduces the value of dynamic modulus of elasticity E_{dyn} during the specimens ageing cured at the different conditions. All sets of test specimens exhibit the same E_{dyn} value at the age of 2 days before they are placed into different curing condition (see Table 1). In the case of air-stored specimens, the E_{dyn} value slightly increases with specimens age (until the age of 7 days). Further desiccation and development of shrinkage cracks cause a gradual decrease in the value of this parameter (see Table 1). At the age of 34 days the value of E_{dyn} was almost the same as at the age of 3 days. In the case of specimens cured in the water bath, the E_{dyn} value gradually increases up to more than two times in comparison to specimens stored at air conditions. This increase is caused by presence of water and mainly by the self-healing process of AAS concrete. After the specimens are removed from the water bath at the age of 27 days and left to dry freely until the age 38 days, when the fracture tests are performed, the E_{dyn} value decreases slightly. However, it is still about 75% higher than the value of the air-stored specimens.

Table 1. The average value of dynamic modulus elasticity (CoV in %) of AAS concrete.

E_{dyn} [GPa]	Specimen's age					
	2 days	7 days	16 days	23 days	28 days	34 days
3_days_air	11.6 (2.0)	–	–	–	–	–
38_days_air	11.6 (3.1)	13.8 (1.2)	12.6 (1.3)	12.2 (1.5)	12.2 (2.5)	11.9 (1.4)
38_days_water	11.8 (1.9)	22.1 (1.5)	26.3 (1.0)	27.9 (0.7)	24.6 (0.6)	21.1 (0.4)

Table 2. The average value of basic material properties (CoV in %) of AAS concrete.

	Set of specimens		
Parameter/Unit	3_days_air	38_days_air	38_days_water
γ [kg/m^3]	2280 (0.9)	2270 (0.3)	2270 (0.3)
f_c [MPa]	30.3 (0.3)	40.2 (3.5)	72.8 (1.4)
f_f [MPa]	1.98 (9.4)	1.80 (7.6)	3.72 (5.3)
f_t [MPa]	1.75 (9.6)	1.69 (1.6)	2.55 (2.2)

IOP Conf. Series: Materials Science and Engineering **660** (2019) 012004 doi:10.1088/1757-899X/660/1/012004

Table 2 introduces the basic physical and mechanical parameters, namely the bulk density γ, compressive f_c, flexural f_f and tensile f_t strengths values. The bulk density was determined before the fracture tests were performed. The value of bulk density is not influenced by curing conditions and age of specimens. In the case of air-stored specimens, the compressive strength value increases with the specimen's age by about 10 MPa. On the contrary, the tensile and flexural strength values slightly decrease up to 10% because of further development of cracks caused by shrinkage. In the case of specimens cured in the water bath, the value of compressive, flexural and tensile strength increases by about 80, 110 and 50%, respectively in comparison with air-stored specimens.

Table 3 introduces the selected mechanical fracture parameters determined based on the above introduced non-linear fracture models. In the case of air-stored specimens, gradual development of cracks caused by shrinkage has a negative effect on the modulus of elasticity value E_c determined from the fracture test. The E_c value decreases by about 40% at the age of 38 days. On the other hand, the water curing has a positive effect and the E_c value is more than two times higher in comparison with the air-stored specimens. The fracture toughness value was determined using two different fracture models. Both models give the same results. In the case of air-stored specimens, the fracture toughness value decreases by about 15% with specimens age. Comparing the results at the age of 38 days, the fracture toughness and specific fracture energy values are by about 80% and 65% higher for specimens stored in the water-bath. The resistance to stable crack propagation, in this case expressed by K_{Ic}^{ini} / K_{Ic}^{un} ratio, does not change with specimen's age in the case of air-cured specimens. In the case of the water-cured specimens, the resistance to stable crack propagation is more than two times higher in comparison with the air-cured specimens. A similar trend is observed in the case of the load ratio F_{ini} / F_{max}, i.e. the ratio between the load level at the beginning of stable crack propagation and maximum load obtained during the test. The presence of cracks caused by shrinkage in air-stored AAS concrete leads to beginning of the stable crack propagation at earlier level. From obtained results, it is evident that water curing has a positive effect on both stable and unstable crack propagation and also on the post peak behaviour of AAS concrete. It is related to the self-healing process in the AAS concrete, when the cracks caused by early-age shrinkage are gradually healed up due to the presence of curing water.

It is well-known that alkali-activated materials need long time to achieve their final properties. However, they also need sufficient amount of water because water is a necessary medium for the dissolution and condensation reactions responsible for their hardening. Therefore, if they are left to dry, the mentioned reactions leading to the formation of the main binding phase (CASH) are suppressed and the properties of such material do not significantly change in time or even rather decrease as a consequence of ongoing shrinkage and other degradative processes, e.g. carbonation. On the other hand, if the specimens are returned to water, their hydration can continue which results in the further formation of CASH. It can grow in the pores but also inside the previously developed cracks and thus we expect that the origin of the observed self-healing effect lies in the formation of CASH. Hence, ongoing hydration itself but also the related self-healing are the main reasons for the increase in all fracture parameters of the investigated concretes immersed in water after the cracking.

Table 3. The average value of mechanical fracture parameters (CoV in %) of AAS concrete.

	Set of specimens		
Parameter/Unit	3_days_air	38_days_air	38_days_water
E_c [GPa]	9.9 (11.6)	6.1 (9.7)	12.5 (7.8)
K_{Ice} [MPa·m$^{1/2}$]	0.550 (15.5)	0.471 (3.3)	0.842 (3.3)
G_F [J/m^2]	82.9 (6.2)	90.6 (10.5)	149.2 (9.9)
K_{Ic}^{un} [MPa·m$^{1/2}$]	0.558 (11.1)	0.462 (11.9)	0.833 (4.1)
K_{Ic}^{ini}/K_{Ic}^{un} [–]	0.182 (30.8)	0.191 (48.6)	0.410 (8.2)
F_{ini}/F_{max} [–]	0.322 (24.8)	0.309 (40.3)	0.595 (2.0)

Table 4 introduces the selected parameters of AE signals obtained during the fracture tests. In the case of air-stored specimens, the number of AE events N_{AE} decreases with specimens ageing. It is related to the nature of AE method, which detects only the active damages. The higher amount of cracks (damage of material) caused by shrinkage during specimens ageing leads to easier crack propagation through the specimens and therefore the detected number of AE event is lower. On the other hand, in the case of specimens stored in the water-bath, the number of AE events is higher because of the more compact structure of material and therefore a higher number of new cracks is initiated and propagated before the fracture. The maximum amplitude of AE signal A_{AE} is connected with the intensity of AE event source. In general, the AE amplitude corresponds to the extent of fracture, whereas the energy emitted from the AE source depends on the crack propagation; small displacement emits waves of low energy, while large crack propagation generates higher amounts of energy [18]. This corresponds to obtained results, the lowest value of amplitude and energy of AE signals is detected in the case of air-stored specimens at the age of 38 days because of the material damage caused by shrinkage.

Table 4. The average value of acoustic emission parameters (CoV in %) of AAS concrete.

	Set of specimens		
Parameter/Unit	3_days_air	38_days_air	38_days_water
$\boldsymbol{N}_{AE}$ [–]	38 (10.5)	8 (50.0)	64 (21.7)
$\boldsymbol{D}_{AE}$ [μs]	1355 (1.4)	1648 (3.7)	1376 (4.8)
$\boldsymbol{A}_{AE}$ [mV]	1191 (0.8)	861 (26.5)	1561 (15.9)
$\boldsymbol{E}_{AE}$ [10^{-3}V·s]	17.0 (22.9)	4.7 (34.3)	41.4 (17.7)

5. Conclusions

The aim of this study was to quantify the effect of different curing conditions on the physical, mechanical and fracture characteristics of concrete specimens with alkali activated binder. One group of the test specimens was left to dry freely for the whole time of ageing. The second group was immersed in the water-bath for a predetermined curing time, after which the specimens were again removed from the bath and left to dry freely until the testing time. This procedure was chosen to prevent the crack development caused by shrinkage and to start the self-healing process in AAS concrete specimens.

From the presented experimental results, the positive effect of water curing on all monitored parameters is evident. Additional fracture tests will be performed after the dynamic modulus of elasticity value of the water-cured specimens will be stabilized.

Acknowledgments

This outcome has been achieved with the financial support of the Czech Science Foundation under project No 18-12289Y.

References

[1] Provis J L and van Deventer J S J (eds.) 2014 *Alkali Activated Materials: State-of-the-Art Report, RILEM TC 224-AAM* (Dordrecht: Springer Netherlands)
[2] Aydin S 2013 *Construction and Building Materials* **43** pp 131–138
[3] Sakulich A and Bentz D 2013 *Materials and Structures* **46** pp 1355–1367
[4] Aydin S and Baradan B 2012 *Materials & design* **35** pp 374–383
[5] Palacios M and Puertas F 2007 *Cement and Concrete Research* **37** pp 691–702
[6] Nguyen H H etal. 2018 *Construction and Building Materials* **165** pp 801–811
[7] ČSN 73 1372 2012 *Non-destructive testing of concrete – Testing of concrete by resonance method* (Prague: ÚNMZ)
[8] Frantík P, Průša J, Keršner Z and Macur J 2007 About stability loss during displacement-controlled loading *Fibre Concrete* (Prague) pp 99–102

[9] Frantík P and Mašek J 2015 *GTDiPS software* http://gtdips.kitnarf.cz/
[10] Karihaloo B L 1995 *Fracture Mechanics and Structural Concrete* (New York: Longman Scientific & Technical)
[11] RILEM TC-50 FMC Recommendation 1985 *Materials & Structures* **18** pp 285–290
[12] Kumar S and Barai S V 2011 *Concrete Fracture Models and Applications* (Berlin: Springer)
[13] Grosse C U and Ohtsu M 2008 *Acoustic Emission Testing* (Heidelberg: Springer)
[14] Ohtsu M 2015 *Acoustic emission (AE) and related non-destructive evaluation (NDE) techniques in the fracture mechanics of concrete* (Cambridge: Woodhead Publishing)
[15] Sagar R V 2009 *Proc. of the National Seminar & Exhibition on Non-Destructive Evaluation* (Tiruchirappalli, India) pp 225–228
[16] Novák D and Lehký D 2006 *Engineering Application of Artificial Intelligence* **19** pp 731–740
[17] Lehký D, Keršner Z and Novák D 2014 *Advances in Engineering Software* **72** pp 147–54
[18] Aggelis D E etal. 2011 *Construction and Building Materials* **25** pp 4126–4131

IOP Conf. Series: Materials Science and Engineering **660** (2019) 012005 doi:10.1088/1757-899X/660/1/012005

A Comparative Study of Mechanical Properties of Fly Ash-Based Geopolymer Made by Casted and 3D Printing Methods

K Korniejenko[1], M Łach[1], SY Chou[2], WT Lin[2], J Mikuła[1], D Mierzwiński[1], A Cheng[2] and M Hebda[1]

[1] Institute of Materials Engineering, Faculty of Mechanical Engineering, Cracow University of Technology, Jana Pawła II 37, 31-864 Cracow, Poland.
[2] Department of Civil Engineering, National Ilan University, No. 1, Section 1, Shennong Road, Yilan City, Yilan County, 260 Taiwan.

E-mail: kinga.korniejenko@mech.pk.edu.pl

Abstract. Currently additive manufacturing (AM) is a rapidly developing industrial sector and a disruptive technology. It is an answer for new challenges such as resources saving and energy effectiveness as well as response for circular economy needs. Unfortunately, the full exploitation of 3D printing technology for ceramic is currently limited due to the in-process and in-service performance of the available materials' sets, especially in application in construction industry. The main aim of the article is comparison mechanical properties such as compressive and flexural strength between casted and injected samples (simulation for 3D printing process). The same geopolymer mix, plain as well as with flax fibres, based of fly ash class F was casted and cured in the same temperature prior to its mechanical test with 'printed' samples.

1. Introduction

3D printing is a rapidly developing industrial sector and, potentially, a disruptive technology that have a lot of advantages such as resources saving and energy efficiency compared to subtractive technologies [1]. Unfortunately, the full exploitation of 3D printing technology for ceramic is currently limited due to the in-process and in-service performance of the available materials' sets, especially in application in construction industry: 'While 3D printing techniques have been successfully applied in a wide range of industries such as aerospace and automotive, its application in concrete construction industry is still in its infancy' [2]. These achievements attract the attention of different industries and science fields about 3D printing technology. 3D printing technology used in the concrete is currently being studied in the construction industry applications and academia, such as military constructions or rapid rehabilitations [1].

With respect to the construction materials, additive technologies for concrete have been found to be the most viable option of widespread use in automated construction processes in the near future. Cementitious materials, such as concrete or mortar, are well understood and have unique fresh and hardened properties as well as an extensive variety of readily available admixtures to customize its performances [3, 4]. In conclusion, the requirements and properties of materials which are being used with a layer based automated construction approach. However, in most cases only a limited amount of powder has been used to fabricate structures. Shakor was developed for Z-Corporation's three-dimensional printing process with a water-based binder using a unique mix of cements [5]. Khalil was indicated that a 3D printable mix made out of Calcium Sulfo-aluminate cement and ordinary Portland cement was developed and the extrudability and buildability was done and tested with a compressive strength of 79 MPa for printed specimens [6]. Ma was applied to accommodate the rapid development of 3D printing using geopolymer. It was proved the well bonding and coordination of the micro-cable and geopolymer [7]. In addition, use of 3D printing could reduce 35–60% of the overall costs of concrete construction due to the removal of formwork [8].

The further development of this technology required improvements to design new materials, to assess materials performance and to improve processing strategies. The research in the area in case of concreates have been done, but the investigation connected with other ceramic materials such as geopolymers are still new area [9, 10]. Contemporary only some prototype solution in the area of geopolymers materials were designed in Australia, China and Russia [11]. The current state-of-art in this area is presented in table 1.

Table 1. The state-of-art for AM technologies [12].

Country	Existing technology	Main limitation	Source of information
Russia	AM technology - 3 D printing for geopolymers is developed by RENCA company. The company has working prototype. It was tested in relevant environment.	The technology is presented as working solutions, but according to the information presented by the RENCA company on the conference in France (2018). The technology is not stable and required mixing manually. Recipes are individually created for each customer.	https://youtu.be/pKekH5beIZU https://youtu.be/hXdE8ozDfhg www.geobeton.ru, www.apis-cor.com, http://renca.org/ Information presented by RENCA on GEOCAMP 2018
China	- Printing mainly from powder. - Printing based on long steel fiber. - Addition short fiber glass.	Some information about theoretical concept and information about laboratory prototype.	[13, 14]
Italy	The project WASP declared development 3D technology form different materials.	There is lack of detailed information about geopolymers.	www.wasproject.it
Australia	Swinburne University in Melbourne work on the technology. Laboratory prototype have been done.	Lack of information about possible applications.	https://www.youtube.com/watch?v =6v8sYuJdRqw
Singapore	Cooperation in consortium. Laboratory prototype have been done .	Lack of information about possible applications.	https://www.youtube.com/watch?v =1ZyMd8CaONQ
Belgium	KULeuven created the group called SREMat. The prototype has not been created yet.	The group starts his work in 2018.	https://www.mtm.kuleuven.be/Ond erzoek/sremat
Finland	Project financed by UIA.	Project started in November	https://www.uia-

The prototype has not been created yet.	2017.	initiative.eu/en/city/lappeenranta

Some researchers have conducted to include short fibers in cementitious materials to achieve desired toughness and ductility enhancement for 3D printing [15, 16]. It also indicated the freeform additive manufacturing processes was useful to acquire sufficient mechanical capacities as the rebar reinforced structures.

Previous studies [17] have demonstrated that the performance of cement-based mixtures in construction-scale 3D printing mixtures is characterized in terms of print quality, shape stability, and printability. The main aim of the article is comparison mechanical properties between mechanical properties such as compressive and flexural strength for casted and injected samples (simulation for 3D printing process).

2. Materials and methods

2.1. Materials

Geopolymers were made from fly ash from the CHP plant in Skawina (Poland) and sand in ratio 1:1. Additionally half of the samples with 1% by mass of green tow flax fibers.

The fly ash is rich in oxides such as SiO_2 and Al_2O_3. The detailed contains of oxides is presented in table 2. The morphology of the particles of fly ash was typical of such by-products of coal combustion and suitable for the process of alkali-activation [18].

Table 2. Chemical composition of the fly ash [18].

	SiO_2	Al_2O_3	Na_2O	CaO	Fe_2O_3	MgO	K_2O	SO_3	TiO_2	P_2O_5	BaO
[% mass]	55.89	23.49	0.59	2.72	5.92	2.61	3.55	0.16	1.09	0.82	0.2

The fibres have been purchased from the Institute of Natural Fibres and Medicinal Plants in Poland. Tow is a coarse, broken fiber, removed during flax processing. They are usually shorter than 30 cm and are semi-products for textile fibers production [18]. Green tow flax fibres (figure 1 and 2) – the fibre is similar to tow flax fibre after dew retting process, but it is more stiff. The fibre for samples were 3-5 cm long.

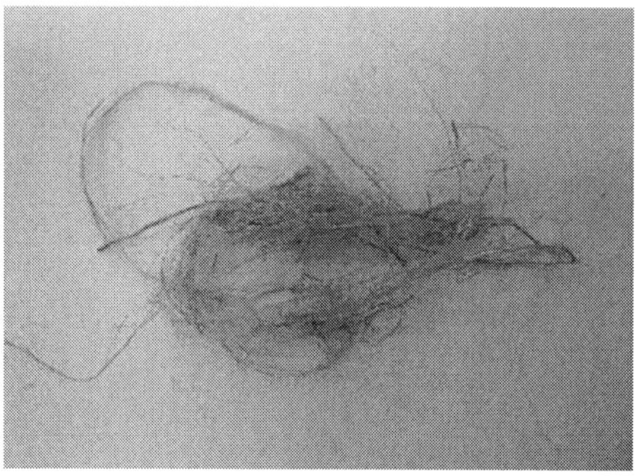

Figure 1. Green tow flax fibre [19].

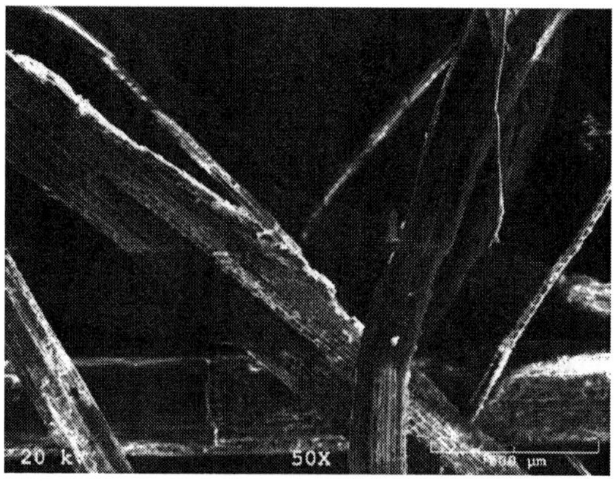

Figure 2. Green tow flax fibre – SEM picture.

The process of activation has been made by 10M sodium hydroxide solution combined with the sodium silicate solution. The ratio of liquid parts to dry parts was 0.4 by mass.

2.2. Samples preparation

The all samples were prepared using sodium promoter, fly ash, sand and half of them with using flax fibers (1% by mass of the composite). The main goal of fibers additive was reinforcement, especially increasing the flexural strength [20]. The process of activation has been made by 10M sodium hydroxide solution combined with the sodium silicate solution (the ratio of liquid glass - 1:2.5). To produce geopolymers, flakes of technical sodium hydroxide were used and an aqueous solution of sodium silicate (R-145) which molar module was 2.5 and density was about 1.45 g/cm^3 and the tap water. The alkaline solution was prepared by means of pouring the aqueous solution of sodium silicate over the solid sodium hydroxide. The solution was mixed and left until its temperature became stable and the concentrations equalized, which took, about 2 hours. The fly ash, sand, alkaline solution and fibers were mixed about 10 minutes by using low speed mixing machine (to receive the homogeneous paste).

Next, the two methods of production were implemented: injection molding made manually and traditional method (figure 3 and 4), were the material was put into two sets of plastic moulds (for compressive strength and flexural strength tests). The injection molding was made by using a syringe.

Figure 3. Plain sample (without fibres) made by injection method.

Figure 4. Sample with 1 % of green tow flax fibres, made by traditional casted method.

The following samples has been performed:

- Plain samples made by injected process (simulation for 3D printing process) PS3D (figure 5),
- Plain samples made by traditional casted technology – PSC (figure 6),
- Samples with 1% of green tow flax short fibres made by traditional casted technology - FSC,
- Samples with 1% of green tow flax short fibres made by injected process (simulation for 3D printing process) – FS3D.

Figure 5. Plain sample made by injected process (simulation for 3D printing process).

Figure 6. Sample made by traditional casted technology.

The samples were NOT subjected to vibratory. Tightly closed molds, by using foil, were heated in the laboratory dryer for 24h at 75 °C. Then, the samples were unmolded. They were investigated after 7 days.

2.3. Methods

Compressive strength tests were carried out according to the methodology described in the standard EN 12390-3 ('Testing hardened concrete. Compressive strength of test specimens'), because of the lack of separate standards for geopolymer materials. The tests involved 6 cubic samples: 50 x 50 x 50 mm. Tests were performed on an concreate press - MATEST 3000kN with speed 0,5 [MPa/s].

Flexural strength tests were carried out according to the methodology described in the standard EN 12390-5 ('Testing hardened concrete. Flexural strength of test specimens'), because of the lack of separate standards for geopolymer materials. The tests involved 3 prismatic samples: 50 x 50 x 200 mm (space between supporting points 150mm). Tests were performed on an universal testing machine - MATEST 3000kN with speed 0,05 MPa/s. The calculations were based on following equation:

$$f_{cf} = \frac{Fl}{d_1 d_2^3} \quad (1)$$

where:
f_{cf} – compressive strength, MPa
F – maximal load, N
l – space between supporting points, mm (for conducted tests:140 mm)
d_1, d_2 – sample dimensions, mm

3. Results and discussion

3.1. Compressive strength test

The results of the compressive strength test are shown in table 3. The results of both methods and samples with fiber reinforced and plain (without fibers) specimens are presented.

Table 3. Compressive strength test after 7 days.

Sample	MPa	standard deviation
PSC	44.9	5.1
FSC	34.7	5.3
PS3D	47.1	5.2
FS3D	39.6	5.6

Better results were attained for plain samples than ones with fibers, regardless of the technology of samples' manufacturing. The technology of injection gives slightly better results than traditional casting method. However, the difference between mechanical properties of the samples produced by different methods could be caused by lack of vibrating during the traditional method and removing the air bubbles in injection process. The best results have been achieved for plain samples made by injected process (simulation for 3D printing process).

3.2. Flexural strength test

The results from the flexural strength tests are shown in table 4. The results of both methods and samples with fiber reinforced and plain (without fibers) specimens are presented.

Table 4. Flexural strength test after 7 days.

Sample	MPa	standard deviation
PSC	6.1	0.9
FSC	6.0	0.5
PS3D	7.4	0.3
FS3D	7.1	0.7

Similarly to the compressive strength tests, better results were attained for plain samples than ones with fibers, regardless of the technology of samples' manufacturing. However, in case of flexural strength this differences are statistically not important. The best results give the injection process (simulation for 3D printing process). It could be caused by lack of vibrating during the traditional casting method.

4. Conclusions

The development of the AM technologies requires: improvements to design new materials, to assess materials performance and to improve processing strategies. The achieved results show that mechanical properties of the new composites made by injection methods (simulation of 3D technology) are comparable with the results achieved by traditional casting process. It could by caused by lack of vibrating during the traditional casting method and because of that some air-bubbles may remain, however macro observations do not shows any voids in material.

The results for compressive and flexural strength are better for plain samples than ones with fibers, regardless of the technology of samples' manufacturing. It shows that short fibers addition could influence similarly on both of the technologies. However, this topic required future research, including fiber orientation and distribution change depending on 3D printing parameters (flow seed, nozzle diameter, fiber length, trajectory etc.) and how it influence the flexural, tensile, compressive strength and fracture energy.

Acknowledgements

1. This work has been financed by the by Polish National Agency for Academic Exchange w under the International Academic Partnership Programme within the framework of the grant: *E-mobility and sustainable materials and technologies EMMAT*.
2. This work has been supported by *The PROM Programme - International scholarship exchange of PhD candidates and academic staff* (POWR.03.03.00-00-PN13/18) and co-financed by the European Social Fund under the Knowledge Education Development Operational Programme for Poland.
3. This work has been supported by the ERANet-LAC 2nd Joint Call (http://www.eranet-lac.eu) funded by the National Centre for Research and Development in Poland, within the framework of the grant: *Development of eco-friendly composite materials based on geopolymer matrix and reinforced with waste fibers*.

References

[1] Tay YWD, Panda B, Paul SC, Mohamed NAN, Tan MJ and Leong KF 2017 *Virtual and Physical Prototyping* **12** **(3)**, 1
[2] Nematollahi B, Xia M and Sanjayan J 2017 Current Progress of 3D Concrete Printing Technologies (34th International Symposium on Automation and Robotics in Construction - ISARC 2017), p 1- 8
[3] Kazemian A, Yuan X, Meier R, Cochran E and Khoshnevis B 2017 Construction-Scale 3D Printing: shape stability of fresh printing concrete (12th International Manufacturing Science and Engineering Conference - (MSEC 2017), Los Angeles, CA
[4] Perrot A, Rangeard D and Pierre A 2016 Materials and Structures 49 (4), 1213
[5] Shakor P, Sanjayan J, Nazari A and Nejadi S 2017 Construction and Building Materials 138, 398
[6] Khalil N, Aouad G, Cheikh KE and Rémond S 2017 Construction and Building Materials 157, 382
[7] Ma G, Li Z, Wanga L and Bai G 2019 Materials Letters 235, 144
[8] Lloret E, Shahab AR, Linus M, Flatt RJ, Gramazio F, Kohler M and Langenberg S 2015 Comput. Aided Des. 60, 40
[9] Panda B, Paul SC, Mohamed NAN, Tay YWD and Tan MJ 2017 Measurement 113, 108
[10] Panda B, Paul SC, Hui LJ, Tay YWD and Tan MJ 2017 Journal of Cleaner Production 167, 281
[11] Xia M and Sanjayan J 2016 Materials and Design 110, 382
[12] Korniejenko K 2019 The Urban infra revolution project Journal n 02 Project led by the city of Lappeenranta
[13] Lim JH, Panda B and Pham QC 2018 Construction and Building Materials 178, 32
[14] Panda B and Tan MJ 2018 Ceramics International 44, 10258
[15] Christ S, Schnabel M, Vorndran E, Groll J and Gbureck U 2015 Materials Letters 139, 165
[16] Panda B, Paul SC and Tan MJ 2017 Materials Letters 209, 146
[17] Kazemian A, Yuan X, Meier R and Khoshnevis B 2019 Performance-Based Testing of Portland Cement Concrete for Construction-Scale 3D Printing 3D Concrete Printing Technology, ed JG Sanjayan, A Nazari and B Nematollahi (Butterworth-Heinemann: Elsevier) chapter 2, pp 13-35
[18] Mierzwiński D, Korniejenko K, Łach M, Mikuła J and Krzywda J 2018 IOP Conference Series: Materials Science and Engineering 416, 012035
[19] Korniejenko K, Łach M, Hebdowska-Krupa M and Mikuła J 2018 IOP Conference Series: Materials Science and Engineering 379, 012023
[20] Korniejenko K, Frączek E, Pytlak E and Adamski M 2016 Procedia Engineering 151, 388

A Study on the Physicochemical Properties of Different Post-Process Wastes from Thermal Processes

M Łach[1], K Korniejenko[1], N Komar[2] and J Mikuła[1]

[1] Institute of Materials Engineering, Faculty of Mechanical Engineering, Cracow University of Technology, Jana Pawła II 37, 31-864 Cracow, Poland.
[2] Ekologia Przedsiębiorczość Innowacje Spółka z o.o., Kuhna 17, 42-256 Olsztyn k/Częstochowy, Poland.

E-mail: michal.lach@mech.pk.edu.pl

Abstract. The main aim of the article is determinate the safety for use the anthropogenic raw materials such as ashes and slags produced in waste incineration plants and possibilities of they using for manufacturing building materials. The article presents the results of research on physicochemical properties of ashes and slags produced in waste incineration plants. The research methods used were: water leaching, radioactivity, dioxins and furans content, oxide and phase composition. The major treatment for investigated waste is a high level of leaching of harmful substances and the content of dioxins and furans. It is necessary to take appropriate measures to develop procedures for safe processing and storage of this type of materials. As shown in the research, ashes from the two investigated waste incinerators do not meet the requirements for leaching levels for hazardous waste landfills. These wastes must undergo processing to reduce to an appropriate level of leaching. Referring to the sum of dioxins in soils in urban areas, which is from 5 to 10 ng / kg, it can be stated that the waste tested contains very high dioxin contents and is hazardous waste. Working with this kind of waste requires special precautions. The ashes from waste incineration plants exceed this limit several dozen times (eg 221 ng I-TEQ / kg). Due to very high values of leaching and dioxin content in the raw materials, even the rinsing process does not guarantee the possibility of obtaining waste that could be stored in hazardous waste landfills. The article suggests the further way of safety processing this kind of materials.

1. Introduction

Every year, the expenditure for the planned and implemented investments related to the construction of municipal waste incineration plants keeps on increases throughout the world. This procedure enables a significant reduction in the volume of the waste generated in the extent from 80 to 95%. However, also as a result of the combustion process, various post-process materials are created – ashes and slags that contain heavy metals and other dangerous substances; these materials should be properly solidified or processed in such a way that enables their safe economic use [1].

As a result of municipal waste incineration, per each ton of burnt waste approximately 0.25 ton of slags are generated and approx. 0.075 ton of fly ash, as well as dust from dust removal, filter cakes and gypsum from exhaust gas cleaning processes [2]. At present, intensive research is being carried out all over the world, focusing on improving the existing technologies and developing new ones, which will enable the safe use or storage of waste generated as a result of the incineration of municipal and other waste.

Among the most popular immobilisation processes, we can distinguish the following ones [2]:

- cementation
- bituminisation
- vitrification
- other technologies, i.e. Synrock or Geodur – based on mixes dedicated to a given type of waste
- geopolymerisation

According to literature data regarding the currently used methods of stabilisation, it is not possible to carry out an effective stabilisation process of fly ash generated from municipal waste incineration without proper pre-treatment – e.g. rinsing [1,3]. It is widely known that the stabilisation of this type of waste in cement matrices without its pre-treatment does not make it possible to reduce leaching of chlorides and sulphates to the levels that are permitted by law. In addition, chlorides and sulphates have a negative impact on the durability of cement matrices. Serving as an alternative solution may be a matrix based on geopolymers. The use of individual solutions, including geopolymers, requires comprehensive testing of processed waste. As already mentioned, there are many technologies for processing hazardous materials generated in thermal processes, while post-process residues generated in the energy industry are valuable secondary raw materials that have many different applications. It is important, however, to accurately identify the properties of such by-products, as they can significantly differ in chemical composition and properties. Research by means of SEM, XRD, DTG, pycnometry, etc. allows for a comprehensive definition of the properties of post-process waste which may be the basis for making decisions regarding their practical implementation [4].

In identifying the composition and properties of by-products of thermal waste treatment, among the methods that serve to be particularly useful there is, for example, the conjugated heat transfer analysis [5,6].

Understanding the properties of waste is of particular importance when it comes to eluviation (leaching) of certain elements. In the case of bottom ashes, the stability of such leaching is established, but the impurities exceed the legal limits for e.g. copper, antimony, chloride and sulphate. Therefore, research should focus on the reduction of these specific pollutants in the examined bottom ash [7]. Also with regards to leaching, bottom ashes should be used as aggregates on roads, since metal leaching is less significant than in the case of aggregates used in concretes [8].

Knowledge of the characteristics of waste originating from thermal processes, including the recognition of its leachability, or the content of metallic elements, facilitates not only the selection of the right technologies for their processing, but it also allows for an optimisation of their composition in order to reduce their harmfulness to the environment. As shown by studies [9], the addition of the metallic Fe decreases Pb (among others) elution.

The conducted research presented in this paper was aimed at determining the basic physicochemical properties of waste in view of choosing their processing technology in the future. A further goal of the conducted research was to obtain information on the possibilities of safe application and processing of waste in technological processes.

2. Methodology

This article only presents selected studies of waste from several selected thermal installations. Dioxin measurements were performed in the Track Tests Laboratory at the Cracow University of Technology. The morphology of the samples shown below was examined using a JEOL JSM-820 scanning microscope. The samples were properly prepared in advance. A small amount of ash was dried into a solid mass and then placed on a carbon substrate ensuring the discharge of the load from the sample. The materials were sprayed with a thin layer of gold using the JEOL JEE-4X sprayer. Natural radioactivity tests were carried out using a PI-MAZAR 01 device. The sample weight in each case amounted to 2400 g. Measurement time: 30 cycles of 2000 seconds each.

Leaching tests were performed in accordance with PN-EN 12457-4: 2006. The pH value of the obtained water extract was determined as per PN-EN ISO 10523: 2012 using the accredited method.

The designations and description of the test samples have been shown in Table 1 below.

Table 1. Symbols and descriptions of the materials tested.

Symbol	Description of the waste
WASTE1	The fly ash proceeding from a waste incineration plant in the city of Klaipeda in Lithuania,
WASTE2	The fly ash from a combined heat and power plant in Jedlicze (steam boilers fired with fuel oil, natural gas and culm),
WASTE3	The fly ash proceeding from a waste incineration plant in Katowice (Poland) MPGK Sp. z o.o.; 19 01 07 * Solid wastes from treatment
WASTE4	The slag proceeding from a waste incineration plant in Katowice (Poland) MPGK Sp. z o.o.; 19 01 11 * Slags and bottom ash containing hazardous substances

Four different types of samples were tested whereby the waste materials came from 3 different installations, and, whereas in the case of the incineration plant in Katowice, it was the slag and ash that were tested (WASTE 3 and 4).

3. Results

Table 1 presents the results of the investigation of the oxide composition of the examined waste. Samples were characterised by the highest content of calcium oxide CaO ranging from 5% for WASTE2 waste to more than 40% for WASTE4 waste. The content of Na_2O, K_2O, MgO, TiO_2 in the examined waste groups ranged from 0.09-5.88%. These contents (due to low values) should not affect the course of processing this waste). Roasting losses ranged from 13% to 58%.

Table 2. The oxide composition of the investigated waste.

Symbol	SiO_2 %	Al_2O_3 %	Fe_2O_3 %	Na_2O %	K_2O %	CaO %	MgO %	TiO_2 %	SO_3 %	LOI %
WASTE1	1.39	0.57	0.64	0.70	3.34	35.23	0.16	0.34	9.65	30.92
WASTE2	13.16	6.08	10.14	0.21	1.71	5.07	0.66	0.95	1.90	58.55
WASTE3	0.71	0.23	0.05	1.47	0.53	34.72	0.09	0.27	1.53	35.07
WASTE4	17.58	4.34	3.83	1.01	0.38	40.30	0.84	5.88	7.58	13.03

The figures 1-4 below show the morphology of the particles of the examined waste materials. The samples differed significantly in shape, degree of fragmentation and surface development. A significant porosity of the WASTE2 material was noted. These are probably particles of unburnt carbon, which is also confirmed in the LOI amounting to 58.55%. The WASTE1 material occurs as a very fine particle agglomerate, similarly to WASTE3. The degree of fragmentation of the WASTE3 material is larger (it has smaller particles).

Figure 1. Morphology of the sample WASTE1.

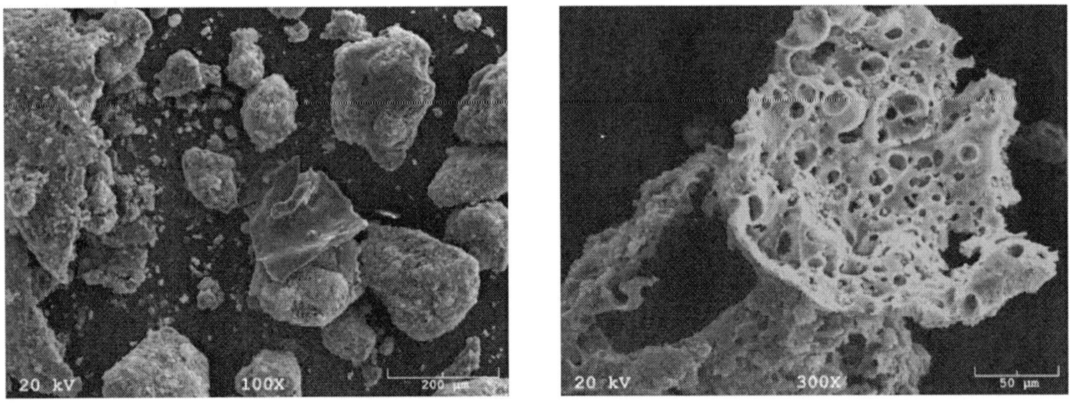

Figure 2. Morphology of the sample WASTE2.

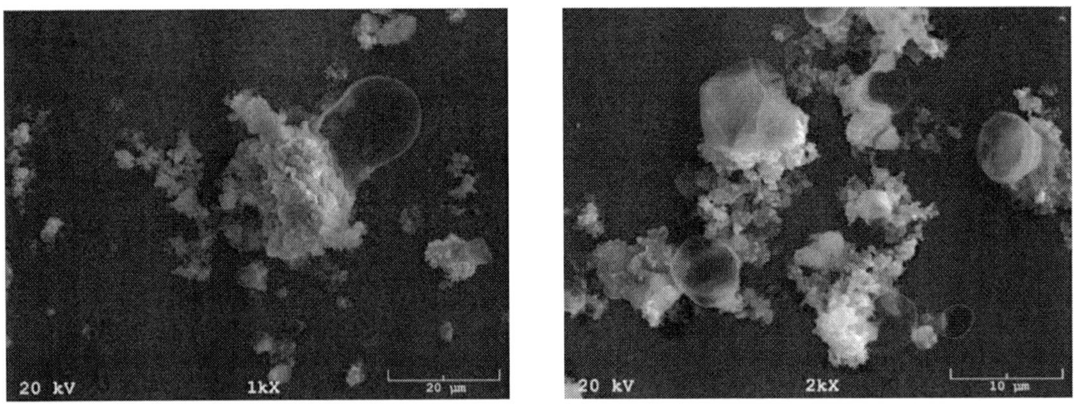

Figure 3. Morphology of the sample WASTE3.

Figure 4. Morphology of the sample WASTE4.

In order to obtain information on the possibilities of safe application and treatment of waste obtained from incinerators in technological processes, the content of dioxins and furans present in the materials in question was examined. The results have been shown in Table 3. Upon referring these obtained contents to the level of the sum of dioxins measured in soils in urbanised areas, which amounts to 5 to 10 ng/kg, it can be stated that the examined waste has a very high dioxin content and must therefore be considered hazardous waste. Working with this kind of waste requires special precautions.

Table 4 presents the results of the natural radioactivity test. These tests were conducted only for two of the four waste type examined. The results showed that the materials meet the requirements regarding the content of natural radioactive isotopes in relation to raw materials and building materials used in buildings intended as shelter for humans and livestock (f1 <1.2, f2<240 Bq/kg).

$f1 = \frac{C_K}{3000\,Bq/kg} + \frac{C_{Ra}}{300\,Bq/kg} + \frac{C_{Th}}{200\,Bq/kg}; f2 = C_{Ra};\ C_{K,}\ C_{Ra,}\ C_{Th}$ - isotope concentrations

Table 3. Total content of dioxins in the tested wastes.

	WASTE1	WASTE2	WASTE3	WASTE4
The sum of dioxin (I-PCDD/F-TEQ) [ng/kg]	221±57	124 ± 32	5518±1435	14±3.7

Table 4. Results of the study of natural radioactivity of waste using the Gamma Spectrometry method

Parameter		WASTE1	WASTE4
Potassium Sk		926.92 ± 56.66	100.71 ± 13.39
Radium SRa		12.93 ± 3.91	10.34± 2.86
Thorium STh		3.94 ± 2.15	8.72 ± 1.78
Indicators	f_1	0.37 ± 0.03	0.11 ± 0.02
	f_2	12.93 ± 3.91	10.34 ± 2.86

Table 5 presents the results of the leachability test. The waste from the WASTE1 and WASTE3 groups does not meet the requirements for leaching levels for hazardous waste landfills. This waste must undergo processing to adapt it to an appropriate level of leaching. The waste from the WASTE2 group is allowed to be stored in a hazardous waste landfill in view of its exceeded selenium leaching (Se) (exceeded for landfills other than hazardous and inert waste). Similarly, the WASTE4 slag is permitted to be stored in a hazardous waste landfill in view of its exceeded Bar (Ba) leaching for landfills other than hazardous and inert waste.

Table 5. Comparison of the results of leaching waste water studied

	WASTE1 mg/kg	**WASTE2 mg/kg**	**WASTE3 mg/kg**	**WASTE4 mg/kg**
Arsenic (As)	<0.50	0.2	<1.0	<0.10
Bar (Ba)	76.5	1.7	113	120
Cadmium (Cd)	0.25	<0.010	<0.050	<0.010
Total chrome (Cr)	0.26	0.17	0.27	<0.050
Copper (Cu)	225	<0.050	<0.25	0.1
Mercury (Hg)	<0.010	<0.010	<0.010	<0.010
Molybdenum (Mo)	1.5	0.25	0.33	0.19
Nickel (Ni)	<0.050	<0.050	<0.10	<0.050
Lead (Pb)	365	<0.050	33.4	<0.050
Antimony (Sb)	<0.50	0.13	<1.0	<0.10
Selenium (Se)	<0.10	0.58	<0.50	<0.10
Zinc (Zn)	44.5	<0.50	160	<0.50
Chlorides (Cl-)	162 000	<50	304 000	2 000
Fluorides (F-)	44	37	28	1.4
Sulphates (SO42-)	27 800	441	11 700	41
Dissolved organic carbon (DOC)	24	19	740	70
Solid dissolved compounds (TDS)	360 000	1 460	544 000	20 300
Chrome 6+	<1.0	<0.20	<1.0	<0.10
pH	>12.0	8.8	>12.0	>12.0

The waste examined is characterised by significant leaching values of TDS and chlorides. In view of the very high leaching values of these substances in the starting material, probably even the rinsing process does not guarantee the possibility of obtaining waste that can be stored in inert waste landfills. This process can only support further processing.

Conclusions

Thermal processes of waste treatment allow for a reduction of significant amounts of waste, but they generate considerable volumes of post-process waste containing hazardous substances. Such materials should be properly disposed of using the best possible technologies and simultaneously maintaining economic efficiency. To make this possible, post-process waste must be thoroughly examined with regards to its physicochemical properties and its harmfulness to humans and the environment. It is only on the basis of comprehensive research that it is possible to apply the appropriate method of waste treatment and to select subsequent stages of the procedure. Post-process waste generated in thermal waste treatment processes differs significantly depending on the type of waste incinerated, installations and many other variables.

As shown in the studies described in this paper, waste originating from thermal processes contains very high dioxin volumes and is therefore considered hazardous waste. Working with this kind of waste requires special precautions. Due to substance leaching, the majority of waste can be stored only on landfills for hazardous waste, or due to exceeding the permissible values, some of them should be processed to obtain an appropriate leaching level.

Acknowledgments
1. This work has been supported by Intelligent Development Operational Program 2014-2020, 1.1.1: Industrial research and development works carried out by enterprises, funded by the National Centre for Research and Development in Poland, within the framework of the grant: 'Development and implementation of eco-innovative post-process waste stabilization technology' ('Opracowanie i wdrożenie ekoinnowacyjnej technologii stabilizacji odpadów poprocesowych'), grant no. POIR 01.01.01-00-0740 / 16-00.
2. This work has been supported by the ERANet-LAC 2nd Joint Call (http://www.eranet-lac.eu) funded by the National Centre for Research and Development in Poland, within the framework of the grant: 'Development of eco-friendly composite materials based on geopolymer matrix and reinforced with waste fibers'.

References

[1] Ferone C, Colangelo F, Messina F, Santoro L and Cioffi R 2013 *Materials* **6**, 3420
[2] Mikuła J, Łach M and Mierzwiński D 2017 *Ecological Engineering* **18 (3)**, 37
[3] Łach M, Mikuła J and Hebda M 2016 *J Therm Anal Calorim* **125**, 1035
[4] Shi HS and Kan LL 2009 *J Hazard Mater* **164**, 750
[5] Klupa A 2016 *Journal of Sustainable Mining* **15** 143
[6] Feng-Yim Ch and Ming-Yen W 2006 *J Hazard Mater* **B138**, 594
[7] Tang P, Florea MVA, Spiesz P and Brouwers HJH 2015 *Constr Build Mater* **83**, 77
[8] Allegrinia E, Vadenbo C, Boldrin A and Astrup TF 2015 *J Environ Manage* **151**, 132
[9] Oehmig WN, Roessler JG, Zhang J and Townsend TG 2015 *J Hazard Mater* **283**, 500

Mechanical Properties of Geopolymer Concretes Reinforced with Waste Steel Fibers

R Gailitis[2], K Korniejenko[1], M Łach[1], J Sliseris[2], J Morán[3], E Rodriguez[3] and J Mikuła[1]

[1] Institute of Materials Engineering, Faculty of Mechanical Engineering, Cracow University of Technology, Jana Pawła II 37, 31-864 Cracow, Poland.
[2] Faculty of Civil Engineering, Riga Technical University, 1 Kalku Street, LV-1658, Riga, Latvia.
[3]Area de Materiales Compuestos Estructurales (CET), Instituto de Investigaciones en Ciencia y Tecnología de Materiales (INTEMA), Facultad de Ingeniería, UNMDdP-Conicet, Diagonal J. B. Alberdi 2695, 7600 Mar del Plata, Argentina.

E-mail: kinga.korniejenko@mech.pk.edu.pl

Abstract. The article presents the research that try to determinate the possibilities of utilization the waste came from used tires to create the composites based on geopolymer matrix. The tire is multicomponent construction. It mainly consists of elastomer (rubber), metal and textile fibres such called textile cord. A lot of components causes difficulties in the tire recycling process. The main aim of the research was determinate the possibilities of recycling the waste steel from used tires in geopolymer composites and develop the eco-friendly material for construction industry. The matrix based on fly ash from power station located in city named Skawina (Poland) and fine sand at a ratio of 1:1. The process of activation was made by 10M sodium hydroxide solution combined with the sodium silicate solution. In order to manufacture these composites the addition of 2% and 3.5% of waste steel fibres by mass was applied. Also specimen without steel fiber reinforcement were made to get reference specimens. The waste steel fibres came from recycling company from Argentina – 'Regomax'. The specimens were prepared according to the methodology described in the standard EN 12390-1. The research methods used were: microstructure research, tensile strength and compressive strength tests as well as analysis of breakthroughs.

1. Introduction

Cement is categorized as indispensable material in the construction industry all over world, special in developing countries. Due to this increased cement consumption there are intense negative effects, such as release of carbon dioxide into the atmosphere [1]. In comparison to the traditional materials, such as Portland concrete, geopolymers have a number of advantages, especially connected with reduction of footprint and eco-friendly character. Manufacturing of this class of materials compared to the traditional concretes is economically more beneficial including the low energy consumption.

Additional environmental benefit is connected with using to production process waste materials: for example, fly ashes and mine tailings. Coal power stations contributes to 25-30% of world's energy production. Consequences to this is 800 million tons of fly ash generated worldwide every year by power stations. Only half of this amount is recycled. This recycled amount can be increased by manufacturing environmentally friendly binders such as geopolymer [2]. Also it has to be acknowledged that production of Portland cement causes significant amount of CO_2 emissions. Portland cement production every year causes around 5-7% of the total CO_2 anthropogenic emissions. Cement production reached a distressing value of about 4200 million tons in 2016 that contributes approximately to 3570 million tons of CO_2 [3].

Geopolymer is comparable in performance to ordinary Portland cement [4]. Geopolymer concrete main advantage is its contribution to the environment. It is estimated that carbon footprint made by geopolymer concrete manufacturing is 26 to 46% less than Portland cement concrete if in concrete mix Portland cement is replaced completely [4, 5]. It is reckoned that production of 1 tonne of caolin geopolymer contributes to 0.180 tonnes of CO_2, that is 6 times less than manufacturing of Portland cement concrete [6].

Geopolymer belongs to a group of novel three-dimensional inorganic materials. This novel material got multiple beneficial properties such as low density, low cost, environmentally friendly nature and high mechanical performance. However like traditional brittle materials, geopolymer shows poor tensile and flexural properties and appalling fracture behavior [7]. As a composite material geopolymer concrete is two or more constituent material arrangement. A continuous called matrix and the dispersed phase or phases, either fibers or particulates, in order to develop another material with desired combination of properties [5, 8]. A significant increase of tensile strength fracture energy can be achieved by adding fibers to geopolymer matrix [9, 10].

In terms of sustainable raw material management, it is crucial to recycle industrial waste as much as possible and also to develop new technologies that not only reduces industrial waste landfills but also produce materials with new added value [2, 11]. Contemporary, every year approximately 17 million tons of waste tires, which have no further use [12, 13]. This waste is categorized as serious contaminant to environment, therefore, recycling of tires is extremely important [14, 15].

This study shows how two secondly used components interacts with each other and what kind of properties has got developed material. The aim of this study is to show how different amount of reused steel fiber reinforcement can improve or disapprove material properties in compression and tensile loads.

2. Materials and methods

Cubic (70x70x70 mm) and prismatic (50x50x200 mm) specimens were prepared with 2% and 3,5% by mass steel fibers from recycled car tire cords and without steel fiber reinforcement. The matrix was based on fly ash from power plant located in Skawina city (Poland). This kind of fly ash is suitable for manufacturing geopolymers because of proper physical properties and chemical composition. The fly ash contains of spherical aluminosilicate particles in different sizes: > 0.0039 in. [>100 μm] – ca.3%, 0.0028 – 0.0039 in. [71-100 μm] – ca. 12%, 0.0025 – 0.0028 in. [63-71 μm] – ca. 10%, 0.0022 – 0.0025 in. [56-63 μm] – ca. 15% and <0.0022 in. [<56 μm] – ca. 60%. This fly ash is rich in oxides such as SiO_2 (47.81%), Al_2O_3 (22.80%). High value of SiO_2 and Al_2O_3 is advantageous for geopolymerization.

Steel fibers were obtained from Argentinian company "Regomax", that recycles old tires to get milled rubber for synthetic grass and other rubber produce production. Steel cords from tires are recycling process byproduct that has no particular market as reusable material so they have scrap value - http://www.regomax.com/.

Specimens were prepared using sodium promoter, fly ash, sand (ratio sand and fly ash – 1:1) and steel fibers (2% and 3.5%). The process of activation has been made by 10M sodium hydroxide solution combined with the sodium silicate solution (liquid glass at a ratio of 1:2.5). In order to manufacture the composites the technical sodium hydroxide in flakes were used and water solution of sodium silicate R–145 (modulus 2.5, density 0.052 $lb/in.^3$ - 1.45 g/cm^3). Tap water was used instead of the distilled one. The alkaline solution was prepared by means of pouring the aqueous solution of sodium silicate and water over solid sodium hydroxide. The solution was mixed and left until its temperature became stable and the concentrations equalized about 2 hours. The fly ash, sand, alkaline solution and steel fibers were mixed about 15 minutes by using low speed mixing machine (to receive the homogenous paste). Next, it was poured into two sets of plastic molds. The specimens were hand-formed and then subjected to

IOP Conf. Series: Materials Science and Engineering **660** (2019) 012007 doi:10.1088/1757-899X/660/1/012007

vibratory removal of air bubbles. Tightly closed molds were heated in the laboratory dryer for 24h at 75 °C. Then, the specimens were unmolded. The prepared specimens had following dimensions: per each testing batch 3 cubes 70x70x70 mm and prisms 50x50x200 mm.

3. Results and discussion

3.1. Microstructure research

The SEM observations were made for plain samples (figure 1) as well as for composition reinforced by fibres (figure 2 and figure 3). The images were made at various magnifications - between 20 – 220x. The different magnification allows to observe of microstructure of composites, including fibres distribution as well as it gives a preliminary information about the coherency of fibres (reinforcement) with the geopolymer matrix.

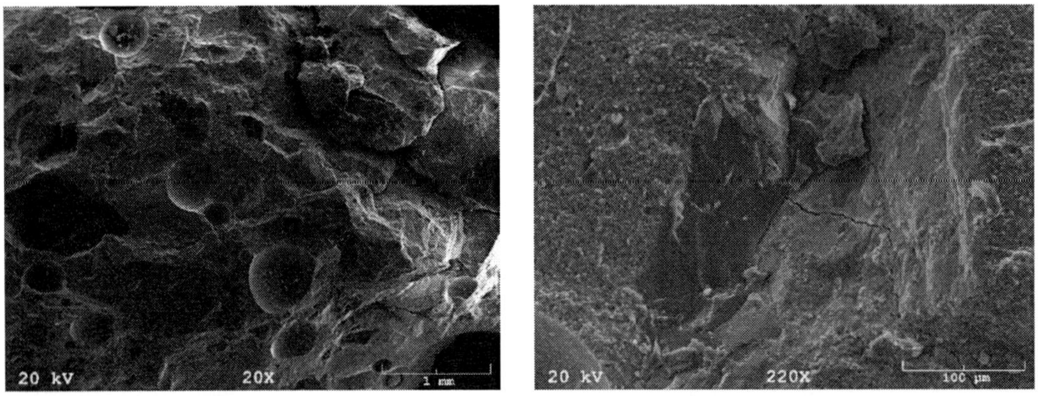

Figure 1. SEM scan of non-reinforced geopolymer sample.

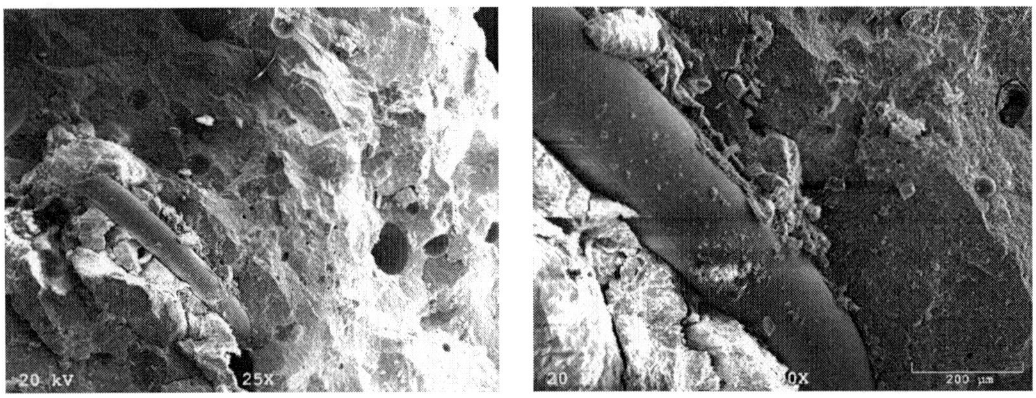

Figure 2. SEM scan of geopolymer sample reinforced with 2% steel fibers.

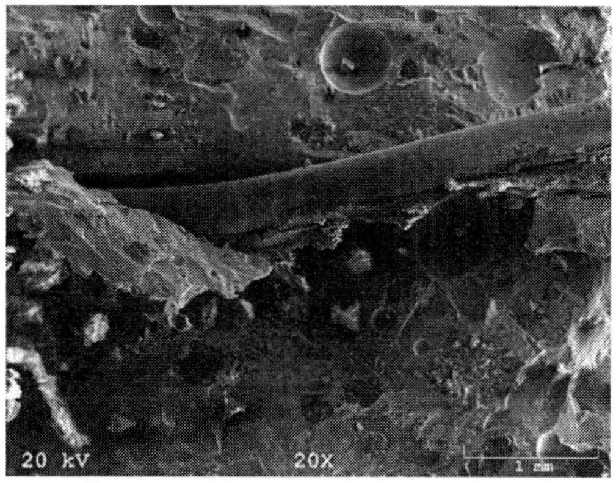

Figure 3. SEM scan of geopolymer sample reinforced with 3.5% steel fibers.

The microstructural observation allow to notice that the structure is coherent - good adhesion the steel fibres to the matrix. The contact zone are visible on figure 3.

3.2. Compressive strength
Table 1 and figure 4 show the compressive strength for geopolymer concrete cubes 28 days after they were made.

Table 1. Compressive strength of geopolymer concrete.

Specimen type	Specimen number	Specimen size			Compressive strength, kN	Compressive strength, MPa
		Width, mm	Height, mm	Length, mm		
Geopolymer with 3.5% steel fibers	1	71.55	71.20	70.86	581.7	114.7
	2	72.17	71.42	71.38	605.0	117.4
	3	71.19	71.26	71.98	562.7	109.8
Geopolymer with 2% steel fibers	1	71.54	71.53	71.15	398.0	78.2
	2	71.48	71.18	71.30	407.6	80.0
	3	71.44	71.65	71.16	431.9	85.0
Geopolymer without steel fibers	1	70.22	71.58	71.12	362.0	72.5
	2	72.38	71.08	71.34	475.2	92.0
	3	71.62	71.20	71.70	467.7	91.1

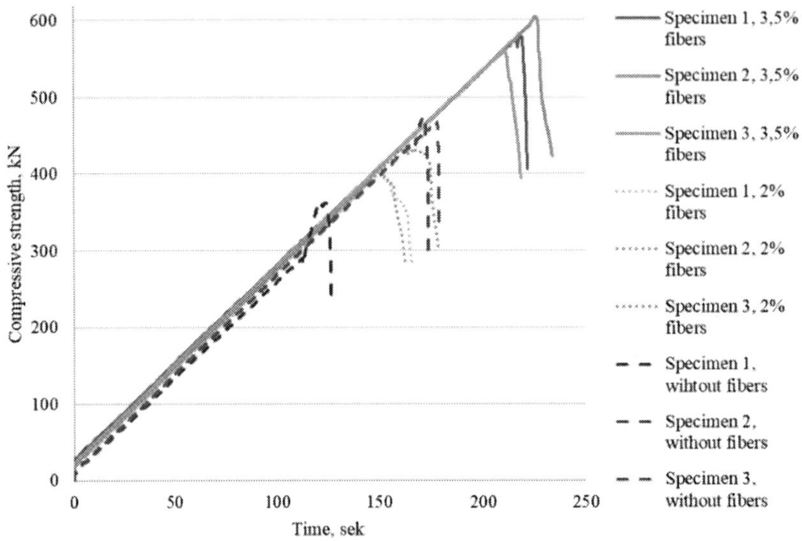

Figure 4. Geopolymer cubic specimen compressive strength loading graph.

For specimens with largest amount of steel fiber reinforcement compressive strength is considerably higher than all other specimens. Furthermore it is interesting that compressive strength of specimens without steel fiber reinforcement is not the lowest.

3.3. Tensile strength

Table 2 shows the tensile strength for geopolymer concrete spherical specimens 28 days after they were made. For specimens with largest amount of steel fiber reinforcement compressive strength is considerably higher than all other specimens. As well as in for cubes the tensile strength of specimens without steel fiber reinforcement is not the lowest.

Also in figure 5 it is shown that specimens with 3.5% steel fiber reinforcement after first crack development in bended part still holds and increases load capacity after it breaks.

Table 2. Tensile strength of geopolymer concrete.

Specimen type	Specimen number	Specimen size			Compressive strength, kN	Compressive strength, MPa
		Width, mm	Height, mm	Length, mm		
Geopolymer with 3,5% steel fibers	1	50.46	50.03	206.67	5.8	10.2
	2	50.56	50.37	203.33	5.9	10.4
	3	50.21	51.56	206.67	6.5	11.3
Geopolymer with 2% steel fibers	1	50.84	50.14	210.00	4.7	8.4
	2	51.45	50.16	205.00	4.3	7.5
	3	50.23	50.21	210.00	3.6	6.4
Geopolymer without steel fibers	1	50.08	50.22	210.00	4.2	7.5
	2	49.69	50.25	208.30	4.8	8.6
	3	49.54	50.22	208.30	5.6	10.1

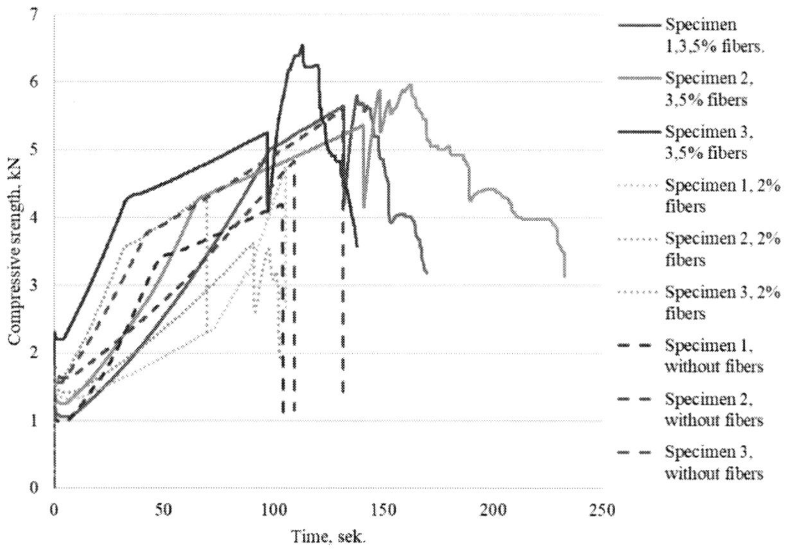

Figure 5. Geopolymer spherical specimens tensile loading graph.

As it is visible in figure 6, figure 7 and figure 8 only specimens with 3.5% steel fiber reinforcement keeps carrying load after crack appearance. All other specimens (with 2% fiber reinforcement and without fibers) fail after crack appearance.

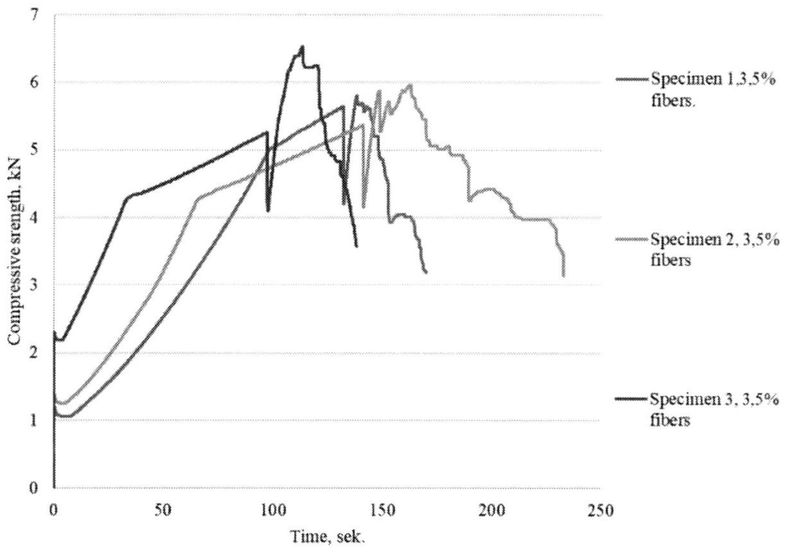

Figure 6. Geopolymer spherical specimens with 3.5% steel fiber reinforcement tensile loading graph.

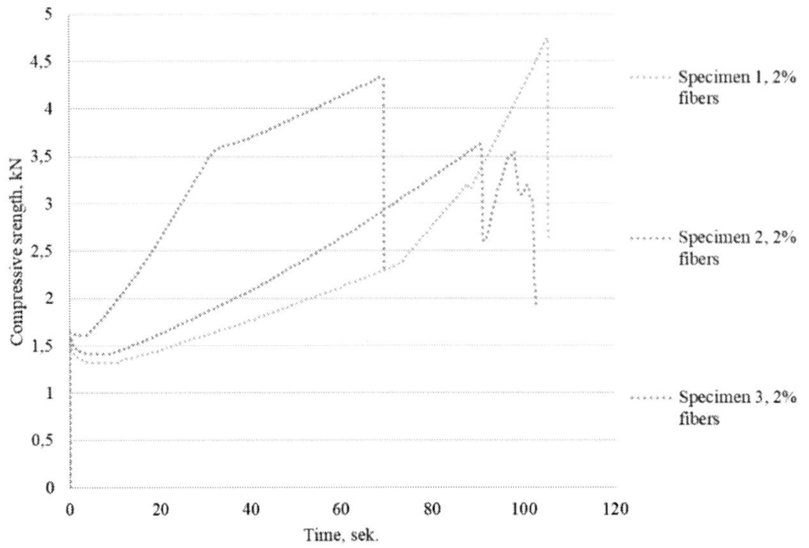

Figure 7. Geopolymer spherical specimens with 2% steel fiber reinforcement tensile loading graph.

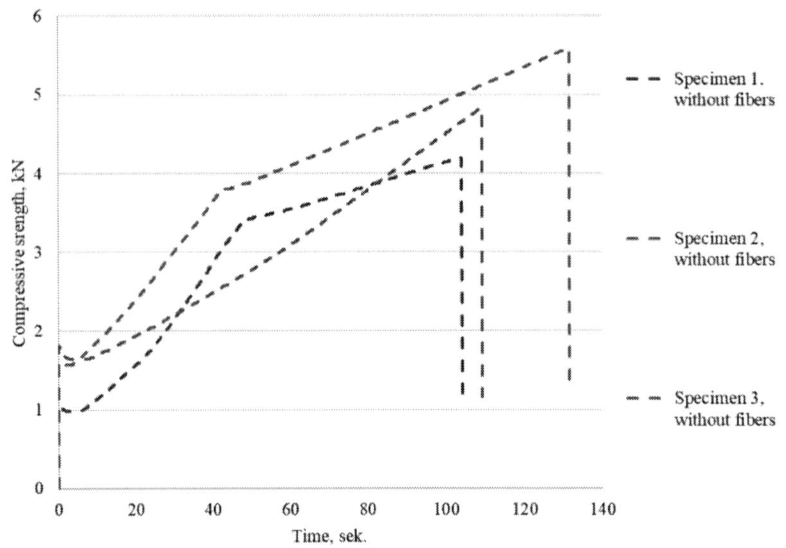

Figure 8. Geopolymer spherical specimens without steel fiber reinforcement tensile loading graph.

4. Conclusions

Regarding tensile strength specimens with fiber reinforcement of 3.5% has further load capacity even if some of geopolymer matrix has failed in stretched part of specimen. For specimens with 2% fiber reinforcement there are not enough fibers in stretched part so they could not carry all the load and specimen fails. In compressive strength cubic specimens with steel fiber reinforcement fails without significant late increase in load capacity as the specimens without fiber reinforcement do.

The load-bearing capacity of geopolymer cubes with 3.5% steel fiber reinforcement is 29% higher than specimens with 2% steel fiber reinforcement and 26% higher than specimens without reinforcement. Furthermore it can be aknowladge that for geopolymer cubes steel reinforcement gives increase in compressive strength when steel reinforcement is 3.5%. For less reinforcement amount there is decrease in specimen compressive strength.

For tensile strength there is similar conclusion. Specimens with 3.5% steel fiber reinforcement has got 30% higher tensile strength than specimens with 2% steel fiber reinforcement and 18% higher tensile strength than specimens without fibers. Furthermore specimens without fibers has got 15% higher tensile strength than specimens with 2% steel fiber reinforcement. It could only mean, that for geopolymer matrix reinforced with waste tire steel cord fibers has to be at least 3.5% from mass of the mix to contribute to specimen strength increase.

The decrease in tensile and compression strength for geopolymer concrete with 2% steel fiber reinforcement can be because the fibers could be surfaced to specimens top not bottom side where tensile loads are. Due to this the reinforcement has not worked as it should have and most on tensile load was carried by geopolymer matrix not steel fiber reinforcement with geopolymer matrix together. Furthermore, the collapse of specimens with 2% steel fiber reinforcement and specimens without reinforcement (Fig.5, Fig.7 and Fig.8) is similar and could indicate, that for specimen with 2% steel fiber reinforcement the steel fibers have not been arranged evenly through the cross section of specimen.

Acknowledgments

1. This work has been supported by the Latvian Council of Science within the scope of the project 'Long-term properties of innovative cement composites in various stress-strain conditions' No. lzp-2018/2-0249.
2. This work has been supported by the ERANet-LAC 2nd Joint Call (http://www.eranet-lac.eu) and funded by Ministry of Science, Technology and Productive Innovation (MINCYT) in Argentina and the National Centre for Research and Development in Poland, within the framework of the grant: 'Development of eco-friendly composite materials based on geopolymer matrix and reinforced with waste fibers'.

References

[1] Celik A, Yilmaz K, Canpolat O, Al-Mashhadani MM, Aygörmez Y and Uysal M 2018 *Constr Build Mater* **187**, 1190
[2] Mucsi G, Szenczi A and Nagy S 2018 *Journal of Cleaner Production* **178**, 429
[3] Novais RM, Carvalheiras J, Seabra MP, Pullar RC and Labrincha JA 2017 *Journal of Cleaner Production* **166**, 343
[4] Luna-Galiano Y, Leiva C, Villegas R, Arroyo F, Vilches L and Fernández-Pereira C 2018 *Materials Letters* **233**, 1
[5] Lokuge W, Wilson A, Gunasekara C, Law DW and Setunge S 2018 *Constr Build Mater* **166**, 472
[6] Nguyen KT, Le TA and Lee K 2018 *Constr Build Mater* **169**, 462
[7] Yan S, He P, Jia D, Wang J, Duan X, Yang Z, Wang S and Zhou Y 2017 *Composites Part B* **114,** 289
[8] Sprince A, Pakrastinsh L and Vatin N 2016 *IOP Conf. Ser.: Mater. Sci. Eng.* **123 (1)**, 012050
[9] Sliseris J 2018 *Mechanics of Composite Materials* **5 (5)**, 1
[10] Nikolenko SD, Sushko EA, Sazonova SA, Odnolko AA and Manokhin VY 2017 *Magazine of Civil Engineering* **7**, 3
[11] Vishwakarma V and Ramachandran D 2018 *Constr Build Mater* **162**, 96
[12] Łach M, Kiszka A, Korniejenko K and Mikuła J 2018 *IOP Conf. Ser.: Mater. Sci. Eng.* **416,** 012089
[13] Smithers Rapra 2017 *The Future of Tire Manufacturing to 2022,* https://www.smithersrapra.com/market-reports/tire-industry-market-reports/the-future-oftire-manufacturing-to-2022 (access: 23.03.2018)
[14] Molanorouzi M, Mohaved SO 2016 *Polymer Degradation and Stability* **128**, 115
[15] ETRMA, *Statistics - Edition 2017 2017 The European Tyre and Rubber Industry in figures*, http://www.etrma.org/library-2 (access: 23.03.2018)

IOP Conf. Series: Materials Science and Engineering **660** (2019) 012008 doi:10.1088/1757-899X/660/1/012008

Effect of Curing conditions on the Compressive Strength of Spent Fluid Catalytic Cracking Catalyst (SFCCC)-based Geopolymers

G Ruiz[1,2], R Aguilar[2,*], J Nakamatsu[3] and S Kim[2]

[1]Universidad de Piura, Department of Civil Engineering, Av. Ramón Mugica 131, Urb. San Eduardo, Piura, Peru
[2]Pontificia Universidad Católica del Perú - PUCP, Department of Engineering, Av. Universitaria 1801, San Miguel, Lima 32, Peru
[3]Pontificia Universidad Católica del Perú - PUCP, Department of Sciences, Av. Universitaria 1801, San Miguel, Lima 32, Peru

[*]E-mail: raguilar@pucp.pe

Abstract. This research evaluates the effect of the curing conditions on the mechanical properties of geopolymers containing spent fluid catalytic cracking catalyst (SFCCC) generated as a waste by the Peruvian oil-refinery industry. For this purpose, two curing conditions were evaluated: i) sample sealing during oven curing to avoid water loss, and ii) the time before curing (pre-curing time or delay time, before heating). The curing temperature was varied from 25°C up to 100°C. The performance of samples was assessed measuring the compressive strengths and the bulk densities. The results reveal that there are optimal curing conditions associated with the higher compressive strength of the SFCCC-based geopolymers. This optimal condition occurred when water loss was reduced by sealing, and samples were allowed to rest before curing (pre-curing). These two conditions assured not only a higher compressive strength but also a denser and more homogeneous product. The results indicate that the higher compressive strengths (up to 34.5 MPa) and bulk densities (up to 1.6 g/cm^3) were obtained when the specimens were sealed with plastic bags during the curing process to avoid moisture loss and when they were subjected to a pre-curing period of 24 hours before heating in an oven at 50°C for 1 day followed by curing at room temperature for 6 days.

1. Introduction

Geopolymers and alkali-activated binders are developed as a friendly environmental alternative of Ordinary Portland Cement (OPC) [1] due to its reduced CO_2 emissions compared to the cement industry [2] and other properties [1,3]. Geopolymers are inorganic polymers synthesized by alkaline activation from materials rich in silica and alumina at slightly elevated temperatures [1,3,4]. The aluminosilicate material is used as a solid precursor and mixed with an activating solution to produce a paste-like OPC's.

Besides an optimum composition of the activating solution required for an adequate reaction of the precursor [5,6], generally, geopolymers require heat to achieve better mechanical properties [7]. In this concern, curing conditions can be defined by temperature and time to promote the geopolymerization reactions and the protection system used to avoid water loss.

However, the effect of curing on compressive strength is still unclear [8,9]. Heating is necessary for geopolymerization for the proper dissolution and to favor the polymerization process [8,10,11], but an excess of temperature or an excessive time at these conditions can affect adversely the mechanical properties of the product [7,8,12–14]. In general, it is recognized as an optimum temperature the one that produces the highest compressive strength of the geopolymer [8,12,14]. Temperature is also related to the appearance of cracks and micro-cracks in the samples. Al Bakri Abdullah et al. found that a curing temperature of 60 °C produced denser samples without micro-cracks when compared to samples cured at room temperature for fly ash foamed geopolymers [15]. Similarly, it has been observed that prolonged curing at high temperatures can result in the breakdown of the granular structure of the geopolymer mixture. This can result in fast dehydration and excessive shrinkage due to the contraction of the gel, which does not transform into a more semi-crystalline form [12]. In addition, it has been mentioned that temperature and curing time can result in an increase of the compressive strength but only if enough water is present in the environment during the reaction [10]. Another important parameter in the pre-curing time at room temperature; it has been found to be beneficial for strength development [8]. It is thought that pre-curing, this is, the time before application of heat, induces dissolution of silica and alumina from the solid precursor and the formation of a continuous matrix phase, increasing, therefore, the homogeneity of the geopolymer materials [8,12,16]. All these conditions need to be determined by experimentation.

Spent fluid catalytic cracking catalyst (SFCCC) is a by-product from the petroleum industry used to improve the efficiency of the process [17]. The total amount of SFCCC generated worldwide is estimated to be about 170 000 t/year [17] causing waste disposal problems [18]. An oil refinery in Piura, northern Perú, discards up to 4 ton of SFCCC daily as solid wastes. Moreover, the SFCCC is considered a hazardous waste material due to its content of heavy metals (Mo, Co, Ni, V, S) and other contaminants as coke and residual oil [17]. However, this catalyst is a highly reactive aluminosilicate with excellent pozzolanic properties [19–22] and thus, the investigation of its use as a raw material to produce alkaline-activated composites is very attractive. In this line, stable geopolymer binders and mortars and alkali-activated mortars have been developed with this SFCCC by [6,18,22]. Tashima et al. have synthesized stable geopolymer binders and mortars with compressive strengths in the range of 8.52–68.34 MPa using SFCCC [6]. Alkali-activated mortars with up to 80 MPa of compressive strength have been also developed using SFCCC wastes [22]. In both studies, sodium silicate and sodium hydroxide were used to synthesize the SFCCC-based binders. These results were obtained applying 65 °C curing temperature during 3 days in samples protected with plastic bags.

The aim of this paper is to investigate the effects of several curing regimes on the compressive strength and bulk density of the SFCCC-based geopolymers. The use of this waste material could bring a reduction in cement used as a construction material and also help reduce the need for hazardous waste disposal, being both aspects of great environmental impact in the construction industry.

2. Experimental program

2.1. Materials

The SFCCC was obtained as a solid powder residue from a local oil refinery in Piura, northern Perú. The chemical composition analysis of SFCCC by X-ray fluorescence (Table 1) revealed that it has slightly more SiO_2 and less Al_2O_3 than similar materials previously reported [6,18,22] and a SiO_2/Al_2O_3 ratio greater than one. Lanthanum and vanadium oxides were identified as contaminants present in the zeolite structure during the catalyst synthesis [18]. The mineralogical information obtained from powder X-ray diffraction (XRD) analysis showed the presence of several crystalline phases together with an important quantity of amorphous components. Area analysis performed with Bruker TOPAS 5 indicated that the total amorphous components are approximately 64 - 66% (w/w). The crystalline phases were identified as a partially dealuminated Y-zeolite (PDF file 79-1479), mullite (79-1457), a silicate with the ZSM-5 zeolite structure (PDF file 44-0002) and quartz (PDF 86-2237). Rietveld fit with Bruker TOPAS 5 indicated that Y-zeolite dominated the crystalline phases (roughly 85% w/w) with minor amounts of other minerals (mullite 8-9%; ZSM-5 zeolite and quarts being below 3% each) The chemical composition is in agreement with the presence of these minerals. Its specific gravity was 2.6 g/cm^3.

IOP Conf. Series: Materials Science and Engineering **660** (2019) 012008 doi:10.1088/1757-899X/660/1/012008

The SFCCC was subjected to a milling process to reduce its particle size and increase its reactivity [9,12,14,21] using a laboratory ring pulverizing mill. Particle size analysis was determined by sedimentation process. The particle size range was 1.2-52 μm with a mean particle diameter of about 22 μm, which is slightly larger compared to what has been reported in previous studies [6,18,22]. The particle shape of SFCCC before and after milling were examined by SEM microscopy observation. Before milling, particles are spherical and have a roughness texture and after milling, particles are angular and laminated. The distribution of the particle size of the SFCCC and shape of the particles after milling are presented in Figure 1.

Table 1. Chemical composition on an oxide basis of SFCCC from X-ray fluorescence analysis (wt%).

Oxide	SiO_2	Al_2O_3	Fe_2O_3	MgO	CaO	Na_2O	K_2O	TiO_2	La_2O_3	V_2O_5	P_2O_5	LOI	Others
SFCCC	49.61	46.23	0.70	0.19	0.09	0.70	0.06	0.72	2.93	0.91	0.20	1.01	0.47
[18]	46.94	48.40	0.59	0.17	0.11	0.31	0.02	1.20	1.36	0.01	0.01	0.50	0.38
[6,22]	46.04	47.47	0.58	0.17	0.11	0.30	0.02	--	--	--	-	--	5.31

An activating solution (AS) based on a mixture of sodium silicate (SS) and sodium hydroxide (NaOH) solutions were used as suggested by Tashima et al. [22]. Sodium hydroxide pellets (99% purity) were dissolved in deionized water and a commercial sodium silicate solution (water glass, 28% SiO_2; 8% Na_2O: 64% H_2O) was used as provided.

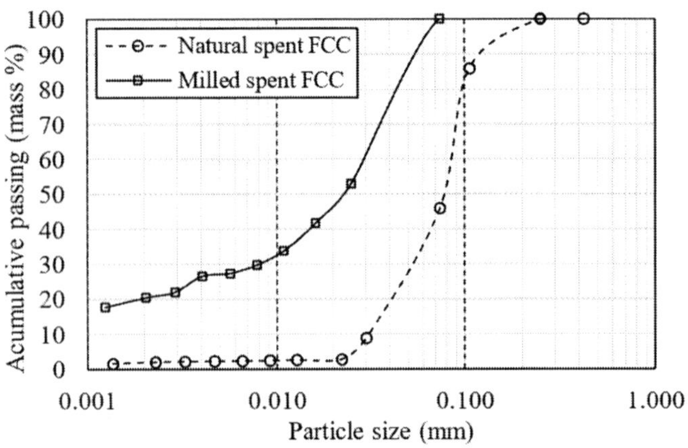

(a)

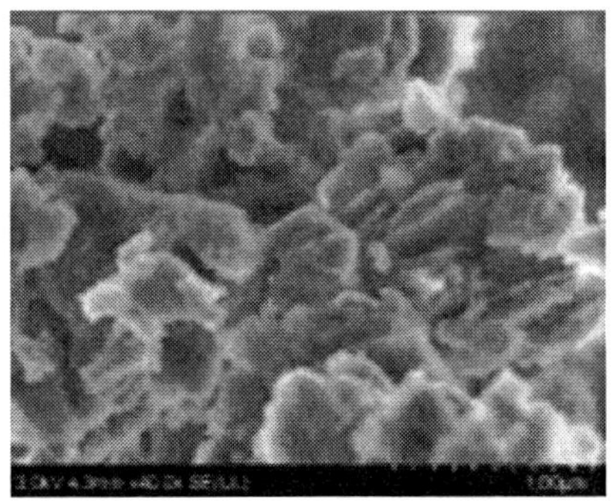

(b)

Figure 1. Particle size distribution and shape of the spent FCC. (a) Particle size distribution comparing in natural and after milled using the laboratory ring mill. (b) Laminar particles of SFCCC after milled.

2.2. Specimens preparation

Based on previous reports [6,18,22], the mixture was prepared using a 15 M NaOH solution and SS in a proportion of 1.25 in weight (Ms=0.85) and AS to SFCCC weight ratio of 1.20 (Na_2O 18.8%, SiO_2 15.6%, H_2O/SFCCC mass ratio of 0.74, H_2O/total solids mass ratio of 0.52). Geopolymer samples were prepared by mixing dry SFCCC powder with AS at approximately 25 °C for 10 minutes. The fresh paste was cast immediately in forged bronze cubic moulds (5 cm) that meet recommendations for testing the compressive strength of hydraulic cement mortars of ASTM C109 standard [23] and were subjected to vibration for 1 minute to remove air voids. Bulk density and compressive strengths of cubic samples were determined after 7 days. The reported values were the average of three samples.

2.3. Experimental program for curing conditions

To evaluate the curing conditions, two factors were considered as the curing regime: use of covering protection for the samples to avoid loss of water and a pre-curing period at room temperature for 24 hours. In this way, two conditions for protection against water loss were evaluated, covered (C) and uncovered (U) samples. Covered samples were sealed with plastic bags. Uncovered samples were let in the moulds without covering during curing inside the oven or at ambient conditions. The pre-curing consisted of letting the samples sit at room temperature for 1 day after been poured into the mould before heating. A curing temperature in a range of 25 °C to 100 °C for 24 hours was then applied for all cases. Ambient conditions were in the range of 25±2°C and 55% of humidity. Samples cured at high temperatures were stored in a ventilated oven with no humidity control.

The samples were subjected to three different regimes of curing conditions as a combination of the two factors mentioned before, as follows:

(1) Uncovered condition without pre-curing regime (UT). The demoulded samples were put in the oven without any protection. For temperatures higher than 25 °C, samples were placed in the oven immediately after being poured in the moulds for 1 day. After this time, the samples were taken out from the oven and let cool down to room temperature (25±2°C) until the time of testing.

(2) Covered condition without pre-curing regime (CT). The samples were covered with plastic bags immediately after being poured in the moulds and heat was applied immediately after for 1 day. After this time, samples were taken out from the oven and let cool down to room temperature (25±2°C) until testing.

(3) Covered condition with pre-curing regime (CPT). The samples were let to sit at room temperature (25±2°C) in the moulds without covering for 1 day. After that, samples were demoulded, sealed with plastic bags and cured. In this case, covered protection was applied for 6 days. The curing temperature was applied for 1 day after which samples were taken out from the oven and allowed to cool down to room temperature (25±2°C) until testing.

3. Results and discussion

3.1. Effects of curing regime on the compressive strength

The results of the compressive strengths of geopolymer binders are shown in Figure 2. For UT regimen, the temperature in an oven application seems to have an adverse effect on the compressive strength when covering protection is not applied. All the samples presented cracking after curing, even though at room temperature. At room temperature, cracks had a depth of about 0.1 mm and oriented the failure pattern during the compressive test. On the exterior of this depth, the material was dense and had a shiny appearance; inside, the matrix had a white and powdering texture. This result can be attributed to an accelerated process of polymerization of the sample from the outside to the inside. This also produces a high strength layer on the exterior but reduces the homogeneity of the material. Despite the cracks, the compressive strength of the samples cured at room temperature was considerable compared to the samples cured at higher temperatures. The humidity in the environment during polymerization might have been responsible for the increase in the strength. For higher temperatures, the material presented a lower compressive strength but a more ductile fracture than the sample exposed to room temperature. This indicates that a better evolution of the polymerization process occurs. This behavior is consistent with the drying process and can be attributed to the lack of water when the temperature of the oven is increased. In this case, the samples had a more homogeneous texture, inside and outside, than that obtained at room temperature. However, samples cured at 100 °C presented a reduced compressive strength unless the homogeneous texture. This indicates a poor geopolymerization process of the samples by an excess of temperature and explains the low compressive strength measured. So, considering the integrity and homogeneity of the material, an optimal temperature near 50 °C can be considered for this regimen, at which the compressive strength reaches an acceptable level.

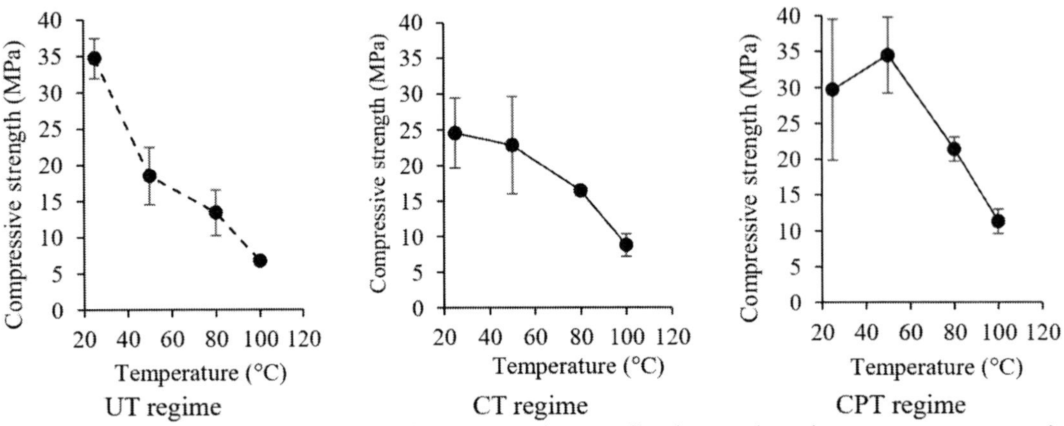

Figure 2. Effect of covering protection, pre-curing application and curing temperature on the compressive strength. (UT) Uncovered condition without pre-curing regime. (CT) Covered condition without pre-curing regime. (CPT) Covered condition with pre-curing regime.

In the CT regime, the samples were covered with plastic bags the whole time, right after moulding and until testing. In this case, a significant increase in the compressive strength was observed for temperatures above 25 °C. At room temperature, a reduction of the compressive strength is observed, when compared to the previous regime (UT). For this condition, the use of plastic bags reduces the compressive strength. This may occur because, at low temperatures, the excess of water inside the bag reduces the velocity of the polymerization, reducing the compressive strength compared to the uncovered condition (UT), since the reaction is an equilibrium process which needs to be displaced by water removal. But at higher temperatures, heating compensates this excess of water, accelerating the reactions and increasing the strength. This confirms that heating can be unfavorable to the final properties of geopolymers even at high humidity conditions [24] and the reduction of the loss of water results in a denser and more homogeneous material. For this regime, room temperature can be considered the optimal temperature.

Pre-curing, which is the time before the application of heat, was also evaluated (CPT). An optimal temperature was identified in this regime for the compressive strength. It is clear that pre-curing has no influence in the compressive strength at elevated temperatures, but it is of great consideration at the optimal temperature. These results confirmed the importance of pre-curing before application of heat as Nazari, Bagheri, and Riahi concluded [12]. This previous time is required to allow the dissolution of silica and alumina in the solid precursor without an excessive acceleration of polymerization. Compared to the previous two curing regimes, an increase of the compressive strength was observed when pre-curing was applied in addition to water loss protection. This increase was observed in all the samples with respect to CT regimen, although this increment is not significative for higher temperatures.

3.2. Influence of curing conditions on bulk density

For technological applications, it is often important to get the value of the apparent density of the finished product. The bulk densities of the samples versus the temperature of curing for the curing regimen are presented in Figure 3. It can be observed a severe reduction of the bulk density when no plastic bags were used (UT) compared to the other two regimens (CT and CPT). This can be attributed to the loss of water at the beginning of the geopolymerization, avoiding its incorporation at the framework structure. The water lost also includes the water required to form the OH^- ions needed for the synthesis of the geopolymer compounds [10]. This explains the reduction of the compressive strength observed in this condition. The same effect is observed when high temperatures are applied even if the sealed bags are used (CT and CPT).

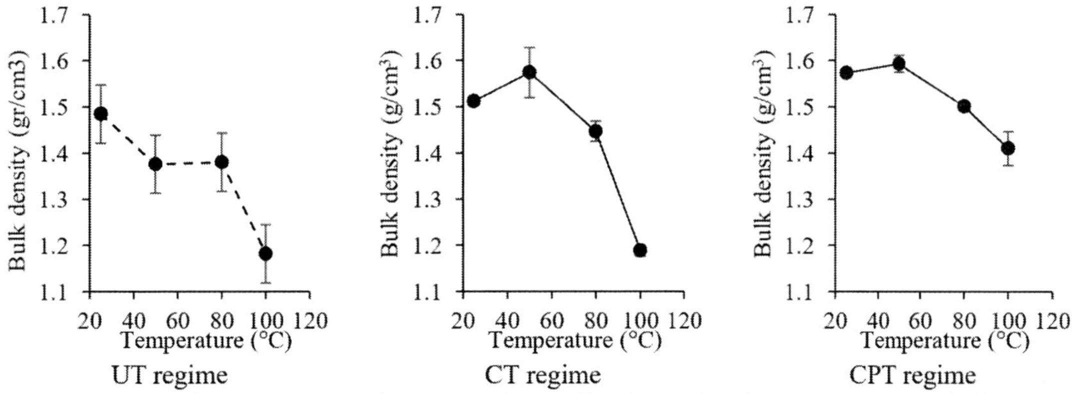

Figure 3. Effect of covering protection, pre-curing application and curing temperature on bulk density. (UT) Uncovered condition without pre-curing regime. (CT) Covered condition without pre-curing regime. (CPT) Covered condition with pre-curing regime.

Using sealed bags (CT and CPT), an optimal temperature for a maximum bulk density was observed and match to the temperature for the maximum compressive strength. At low temperatures, the excess of water cannot be eliminated quickly and remains in the material during solidification, creating microporous in the material which results in a slight reduction of its density but does not affect its compressive strength. At higher temperatures, the velocity of the reaction is higher than the velocity needed to eliminate the excess of water, increasing the microporosity of the matrix and reducing its density. Cracking development on the material needs to be taken into consideration as a factor for the reduction of the compressive strength. While porous and homogeneity of the material influence the density, the presence of cracks determines the compressive strength, reducing it. However, this process is different if a pre-curing period is applied. This can be explained because the amount of water remaining after pre-curing is lower, making the velocity of the reaction similar to the velocity needed to eliminate the water, improving the homogeneity of the material and its bulk density.

Individual values of the compressive strength and their corresponding bulk densities are plotted in Figure 4. A tendency to increase the compressive strength with the bulk density is observed. Even though the mechanical properties of geopolymers do not depend on the bulk density of the materials [24], curing conditions and pre-curing regimes influence the geopolymerization process and the framework structure of the matrix, promoting a denser and more homogeneous material. The homogeneity of the material can contribute to a higher compressive strength.

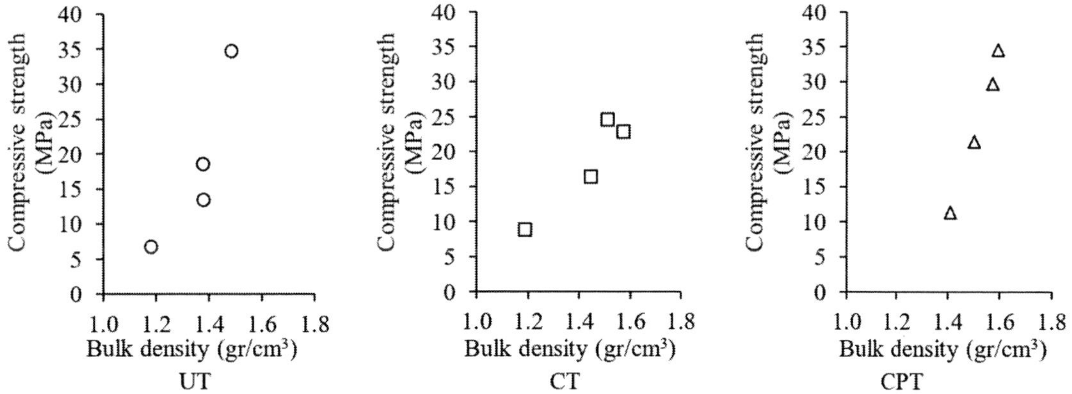

Figure 4. Relationship between bulk density and the compressive strength in SFCCC based geopolymer binders evaluated in this research: (UT) Uncovered condition without pre-curing regime. (CT) Covered condition without pre-curing regime. (CPT) Covered condition with pre-curing regime.

4. Conclusions

The temperature of curing, the pre-curing regime and the covering protection are key factors that determine the final compressive strength of the SFCCC based geopolymers. No covering protection of the samples can contribute to a higher compressive strength upon certain curing temperatures but cannot assure the homogeneity of the material nor the cracking of the sample.

The use of sealed plastic bags helps to reduce the loss of water and helps to control the cracking of the samples, producing a denser and more homogeneous material. In consequence, it is an efficient system to improve the homogeneity when heat is applied and to increase the compressive strength at room temperature.

Pre-curing before heating is of great importance to assure the correct geo-polymerization process. Nevertheless, it has no influence in the compressive strength at elevated temperatures.

Optimal temperature for curing in order to get a maximum compressive strength is only possible when pre-curing is applied together with a convenient water loss protection. The optimization is not possible if no covering or no pre-curing is used.

The curing regime has a significant effect on bulk density. The higher values for the bulk density are associated with a homogeneous material and can be increased if sealed plastic bags and a pre-curing regime are applied. This denser and more homogeneous matrix results in a higher compressive strength material.

For SFCCC-based geopolymers, the highest compressive strength (about 34 MPa) can be obtained with a 24 hours pre-curing period at room temperature followed by curing at 50 °C for one day in sealed bags. However, at 25 °C, the same pre-curing period and water protection system, an acceptable compressive strength of 29.7 MPa can be reached for practical purposes.

Finally, it can be concluded that SFCCC can be used effectively to produce geopolymer binders and, in this way, contribute to reduce the environmental impact of its disposal.

Acknowledgments

This work was supported by CONCYTEC and SENCICO PERU under the project: "GeoBloque: Desarrollo de bloques de construcción ultraligeros con geopolímeros" (Contract No. 105-207-FONDECYT). Authors acknowledge the support of Compañía Minera de Agregados Calcáreos S.A. (COMACSA) for the use of its facilities and equipment. Also, the special contribution of Dr. Luis Ortega, Associate Professor of the Department of Sciences at Pontificia Universidad Católica del Perú. The first author also acknowledges Universidad de Piura for the scholarship support for her PhD studies, also, the technical personnel of the same institution.

References

[1] Duan P, Yan C, Zhou W and Luo W 2016 *Construction and Building Materials* **111** 600–10
[2] Khankhaje E, Hussin M, Mirza J, Rafieizonooz M, Salim M, Siong HC and Warid M 2016 *Materials & Design* **89** 385–98
[3] Moraes J, Tashima M, Akasaki J, Melges JP, Monzó J, Borrachero M, Soriano L and Payá J 2016 *Construction and Building Materials* **124** 148–54
[4] Novais R, Ascensão G, Buruberri L, Senff L and Labrincha J 2016 *Materials & Design* **108** 551–9
[5] Ibrahim M, Megat Johari M, Maslehuddin M, Rahman M, Salami B and Mohamed H 2019 *Construction and Building Materials* **201** 186–95
[6] Tashima M, Akasaki J, Melges J, Soriano L, Monzó J, Payá J and Borrachero M 2013 *Fuel* **108** 833–9
[7] Part W, Ramli M and Cheah C 2015 *Construction and Building Materials* **77** 370–95
[8] Riahi S, Nazari A, Zaarei D, Khalaj G, Bohlooli H and Kaykha M 2012 *Materials & Design* **37** 443–9
[9] Riahi S and Nazari A 2012 *Ceramics International* **38** 4467–76
[10] Zuhua Z, Xiao Y, Huanjun Z and Yue C 2009 *Applied Clay Science* **43** 218–23

[11] Zhuang X, Chen L, Komarneni S, Zhou C, Tong D, Yang H, Yu W and Wang H 2016 *Journal of Cleaner Production* **125** 253–67
[12] Nazari A, Bagheri A and Riahi S 2011 *Materials Science and Engineering* **528** 7395–401
[13] Ahmari S and Zhang L 2012 *Construction and Building Materials* **29** 323–31
[14] Nazari A, Riahi S and Bagheri A 2012 *Materials & Design* **35** 296–302
[15] Al Bakri Abdullah M, Hussin K, Bnhussain M, Ismail K, Yahya Z and Razak R 2012 *Int J Mol Sci* **13** 7186–98
[16] Detphan S and Chindaprasirt P 2009 *International Journal of Minerals, Metallurgy and Materials* **16** 720–6
[17] Marafi M and Stanislaus A 2008 *Resources, Conservation and Recycling* **52** 859–73
[18] Rodríguez E, Bernal S, Provis J, Gehman J, Monzó J, Payá J and Borrachero M 2013 *Fuel* **109** 493–502
[19] Mas M, Tashima M, Payá J, Borrachero M, Soriano L and Monzó J 2015 *Key Eng Materials* **668** 411–8
[20] Payá J, Monzó J and Borrachero M 1999 *Cement and Concrete Research* **29** 1773–9
[21] Payá J, Monzó J, Borrachero M and Velázquez S 2003 *Cement and Concrete Research* **33** 603–9
[22] Tashima M, Akasaki J, Castaldelli V, Soriano L, Monzó J, Payá J and Borrachero M 2012 *Mat. Letters* **80** 50–2
[23] ASTM C109/C109M-16a. *Test Method for Compressive Strength of Hydraulic Cement Mortars (Using 2-in. or [50-mm] Cube Specimens).* (West Conshohocken, PA: ASTM International)
[24] Davidovits J 2008 *Geopolymer Chemistry & Applications* 2nd ed. (France: Geopolymer Institute) p 623

IOP Conf. Series: Materials Science and Engineering **660** (2019) 012009 doi:10.1088/1757-899X/660/1/012009

Synthesis of a Geopolymer Binders Using Spent Fluid Catalytic Cracking (FCC) Catalyst

G Ruiz[1,2], R Aguilar[2,*], J Nakamatsu[3] and S Kim[4]

[1]Universidad de Piura, Department of Civil Engineering, Av. Ramón Mugica 131, Urb. San Eduardo, Piura, Perú
[2]Pontificia Universidad Católica del Perú - PUCP, Department of Engineering, Av. Universitaria 1801, San Miguel, Lima 32, Perú
[3]Pontificia Universidad Católica del Perú – PUCP, Department of Science, Chemistry Division, Av. Universitaria 1801, San Miguel, Lima 32, Perú
[4]Pontificia Universidad Católica del Perú – PUCP, Department of Engineering, Av. Universitaria 1801, San Miguel, Lima 32, Perú

*E-mail: raguilar@pucp.pe

Abstract. A synthesis of a geopolymer binder using spent fluid catalytic cracking (FCC) catalyst is evaluated in this research. Spent FCC catalyst is a solid waste product from the oil refining industry worldwide. It is composed mainly by aluminium and silicon oxides. The silica modulus (Ms), NaOH molar concentration, and liquid/solid (AS/FCC) weight ratio were studied as key factors of the mixture composition for the evaluation of the compressive strength and bulk density of geopolymers. Additional total ratios of SiO_2/Al_2O_3, Na_2O/SiO_2, and H_2O/total solids, which consider chemical composition of the final mixture, were studied. Results indicate that there is an optimum value for each Ms, NaOH molar concentration and AS/FCC that produces the maximum compressive strength. These values also match with the optimized total ratios of SiO_2/Al_2O_3, N_2O/SiO_2 and H_2O/total solids in the mixture, which can be considered as alternative control parameters for the formulation process. The bulk density is more related to AS/FCC value. A geopolymer binder with a compressive strength of 24.7 MPa after curing for one day at 65°C and six days at room temperature was produced using this spent FCC waste.

1. Introduction

Spent fluid catalytic cracking (FCC) catalyst is a by-product from the petroleum industry used for improving the process efficiency [1]. The activity of this catalyst declines with time and when this happens, the "old inactive" catalyst is replaced by a "new active" catalyst [1]. This generates huge amounts of solid wastes; as an example, an oil refinery in Piura, northern Perú, discards about 1 400 ton of spent FCC every year as solid wastes. The total quantity of spent FCC generated worldwide is estimated to be about 170 000 t/year [1,2] and it will continue increasing as the environmental requirements of cleaner fuels increase [1]. Unfortunately, unless reuse or regeneration treatments are applied to reduce the quantity of spent catalyst discarded [1], after some cycles of regeneration, further regeneration may not be economically feasible, and the spent catalyst must be discarded in landfills [1]. The storage, transportation, treatment and final disposal of the spent catalyst require compliance with stringent environmental regulations because of their hazardous nature [1]. FCC contains mainly SiO_2 and Al_2O_3 together with some minor impurities. The spent catalyst discarded additionally contains 7-

20% V+Ni, 15-25% coke, 7-15% Sulphur and 5-10% residual oil together with active metals (Mo and Co or Ni) [1].

The use of FCC as raw materials in the production of by-products is an attractive option from the environmental and economic points of view [1]. Besides of the preparation of abrasives, synthetic aggregates, glass-ceramics, and refractory bricks and cement [1], spent FCC has been used in the Portland clinker production [4]. Evaluation of this spent FCC as pozzolanic material in cement binder production [3,5] and cement mortars [6,7] have been carried out. Other researches have focused the fine aggregate replacement in concrete [8], and the improvement of its durability [9]. In transportation, FCC catalysts have been studied as filler and fine aggregates in asphalt mixtures [10,11]. More recently, preliminary evaluations of FCC catalysts for the development of geopolymers and alkali activated binders has been developed [2,12,13]. Geopolymers and alkali activated binders are binders developed as a friendly environmental alternative of Ordinary Portland Cement (OPC) [14] due to its reduced CO_2 emissions [15] and other properties [14,16]. Tashima et al. [12,13] synthetized stable geopolymer binders and mortars with compressive strengths in the range of 8–68 MPa using spent FCC catalyst. Alkali-activated mortars with up to 80 MPa of compressive strength were developed using this FCC waste [13]. In both studies, sodium silicate and sodium hydroxide were used to synthetize the FCC-based binders. Additionally, spent FCC has been evaluated to produce geopolymeric roof tiles [17] and concrete blocks [18].

An optimum concentration of the alkaline solution is required for an adequate activation of the solid precursor or raw materials [13,19,20], which is controlled by the silica modulus, sodium concentration and liquid alkaline solution/solid precursor weight ratio. In previous research, a strong dependence of compressive strengths values with the silica modulus has been observed [13] and other research suggest the existence of an optimum value [19,20,21]. It has been also determined that this FCC residue requires a solution of NaOH and waterglass for activation [12] and that there is an optimum Na_2O concentration (in this case, expressed as a percentage in weight of the FCC in solid state) associated to an optimum silica modulus [13]. Other researchers suggest other parameters for the optimization of the formula for alkali-activated materials. For example, Varela et al. [22] proposed the used of total SiO_2/Al_2O_3, Na_2O/Al_2O_3 and Na_2O/SiO_2 molar ratios, which include the chemical compounds of both the alkaline solution and the solid precursor instead of the precursor alone. On the other hand, the concentration of the alkaline solution used for activation influences the bulk density of hardened geopolymer. Bumanis et al. [23] found that density of metakaolin-based geopolymers increase by increasing the activator´s concentration because of the porous microstructure produced by evaporation of the free water in the material. The aim of this paper is to evaluate the influence of chemical composition parameters on the compressive strength and bulk density of the geopolymer binder using the spent FCC catalyst produced by a local oil-refinery plant in Piura, Perú. For this purpose, different range values of sodium concentration, silica modulus and the liquid/solid weight ratio were analyzed, using the compressive strength and bulk density as parameters of control. The use of this waste material would have two positive aspects for the environment, it could replace OPC in some construction applications and a avoid disposing of the material in scarce landfills.

2. Experimental program

2.1. Materials

The spent FCC was supplied by Incablock SAC, a local concrete precast company. The chemical composition of the spent FCC catalyst was obtained by X-ray fluorescence spectroscopy (see results in table 1). The spent FCC catalyst has slightly more SiO_2 and less Al_2O_3 than similar materials [e.g. 3,12,13], and its SiO_2/Al_2O_3 molar ratio is more than one. Lanthanum and vanadium oxides were identified as contaminants present in the zeolite structure during the catalyst synthesis [3]. The mineralogical composition was gathered with an X-ray diffraction (XRD) analysis. The total amorphous components represent between 64% to 66% of the total. Crystalline zeolite phases as Y-zeolite, and ZSM-5 zeolite are present in the catalyst residue. Less than 3% of other aluminosilicates such as mullite, and quartz were also observed. The spent FCC has a specific gravity of 2.6 g/cm^3, measured using the standard water pycnometer method [24].

The spent FCC was subjected to a milling process to reduce the size of the particles and increase its reactivity [2,4,25-27]. The particle size range of the spent FCC was 1.3-150 µm with a mean particle diameter of about 75 µm. After milling, the particle size range was reduced to about 1.2-52 µm with a mean particle diameter of about 22 µm, which is slightly larger compared to what is reported in previous studies [3,12,13]. The water needed for a normal consistency of the FCC [28] was about 77% (or 0.77 mass ratio).

Table 1. Chemical composition of the spent FCC catalyst used in this study (wt%).

SiO_2	Al_2O_3	Fe_2O_3	MgO	CaO	Na_2O	K_2O	TiO_2	La_2O_3	V_2O_5	P_2O_5	LOI	Others
49.61	46.23	0.70	0.19	0.09	0.70	0.06	0.70	2.93	0.91	0.20	1.01	2.52

The alkaline solutions (AS) were prepared by mixing sodium silicate (SS) and sodium hydroxide solution [12]. Sodium hydroxide pellets (99% purity) was dissolved in deionized water to prepare the sodium hydroxide solution and a commercial sodium silicate solution (water glass, 28% SiO_2; 8% Na_2O: 64% H_2O) was used as provided. The AS were prepared at least 2 hours prior to their use to allow the solution to cool to room temperature (20 ± 2°C). This step is necessary to avoid the premature setting of the geopolymer mixture and a reduction on the final properties [25,27].

2.2. *Sample preparation and testing procedure*

Geopolymer samples were prepared by mixing the dry spent FCC catalyst powder with the AS at approximately 20 °C. The fresh paste was cast immediately in cubic molds (5 cm x 5 cm x 5 cm) and subjected to vibration for 1 minute to remove air voids, as recommended by ASTM C109 standard [29]. Samples were cured at 65°C for 1 day after which they were demolded and left at room temperature (20±1°C) for the rest of the time until testing. The compressive strength tests were performed by using a hydraulic testing machine MTS model Exceed 45.105 controlled by displacement, according to ASTM C109 standard [29]. The bulk density was estimated as a weight in air per volume unit; size of the samples was measured using a Vernier caliper. The bulk densities and compressive strengths of cubic samples were determined after 7 days. The reported values were the average of three samples.

2.3. *Mixtures proportion*

Three parameters were considered in the preparation of the alkaline activating solution: the silica modulus (Ms), NaOH concentration (NaOH) and liquid alkaline solution/solid precursor mass ratio (AS/FCC). Ms is related to the molar ratio Na_2O/SiO_2 in the activating solution which has a significant effect on the strength of the geopolymers. This parameter determines the amount of soluble silicates present and it is crucial in controlling the dissolution rate and the gelation process during geopolymerization [30]. NaOH is the molar concentration of Na^+ of sodium hydroxide in the solution in mol/kg H_2O and has relevance to assure the dissolution of the solid precursor. AS/FCC controls the consistency and helps in the workability of the fresh mixture and the uniformity of the final hardened product.

As shown in Table 2, the study was divided into three steps parts, described as follows:

(1) Analysis of the silica modulus (Ms). 15 M NaOH solution was combined with different quantities of the SS solution. Starting from an Ms value around 0.85, as suggested by [13], five mixtures were evaluated (SERIES A).

(2) Analysis of the NaOH concentration. For a fixed Ms molar ratio of 0.85, four samples with different NaOH molar concentrations of 7.5 M, 10 M, 12.5 M and 15 M were evaluated (SERIES B).

(3) Analysis of liquid alkaline solution/solid precursor ratio (AS/FCC). With both, the Ms and NaOH molar concentration fixed, the AS/FCC was varied from 0.8 to 2.0 to change the consistency of the mixture. Five mixtures for two series of 15M (SERIES C) and 10M (SERIES D) respectively were used to verify the influence of NaOH molarity on the AS/FCC value.

Table 2. Mix proportions of FCC geopolymeric samples used for evaluation.

SERIE	NaOH solution (mol/L)	Ms (molar ratio)	AS/FCC (mass ratio)	SiO_2/Al_2O_3 (total) (molar ratio)	Na_2O/SiO_2 (total) (molar ratio)	H_2O/total solids (total) (wt.)
A1	15	0.71	1.10	2.39	0.33	0.49
A2	15	0.85	1.18	2.50	0.32	0.51
A3	15	1.08	1.32	2.68	0.30	0.55
A4	15	1.18	1.37	2.76	0.29	0.57
A5	15	1.27	1.42	2.84	0.28	0.59
B1	15	0.85	1.18	2.50	0.32	0.51
B2	12.5	0.85	1.26	2.50	0.32	0.57
B3	10	0.85	1.39	2.50	0.32	0.66
B4	7.5	0.85	1.58	2.50	0.32	0.79
C1	15	0.85	0.80	2.28	0.24	0.39
C2	15	0.85	1.00	2.39	0.28	0.45
C3	15	0.85	1.10	2.45	0.30	0.49
C4	15	0.85	1.18	2.49	0.32	0.52
C5	15	0.85	1.30	2.56	0.34	0.55
D1	10	0.85	1.10	2.36	0.27	0.56
D2	10	0.85	1.35	2.48	0.31	0.65
D3	10	0.85	1.39	2.50	0.32	0.66
D4	10	0.85	1.73	2.67	0.37	0.76
D5	10	0.85	2.08	2.84	0.42	0.86

Four additional parameters are included for the evaluation in table 2. Valera et al. [22] suggested that molar ratios of the mixtures need to include the SiO_2 and Na_2O species provided by both the alkaline solution and the solid precursor. In this study, SiO_2/Al_2O_3, Na_2O/SiO_2, Na_2O/Al_2O_3 and H_2O/total solids ratios of all samples were calculated using the chemical concentration of the species in the final mixture. A possible correlation between the compressive strength and these parameters are evaluated.

3. Results and discussion

3.1. Effect of silica modulus (Ms)

The results show that the compressive strength of the FCC based geopolymer binder increased with the Ms until 0.85. This optimum value can be identified in the figure 1a. Higher Ms values limits the dissolution of Si and Al species due to the lack of Na^+ [31]. For lower values of Ms, an excessively high concentration of NaOH resulted in premature precipitation of the geopolymer gel due to the excessive OH^- ions [22,31]. The 0.85 Ms value is very close to what was obtained by [13], for a similar spent FCC residue.

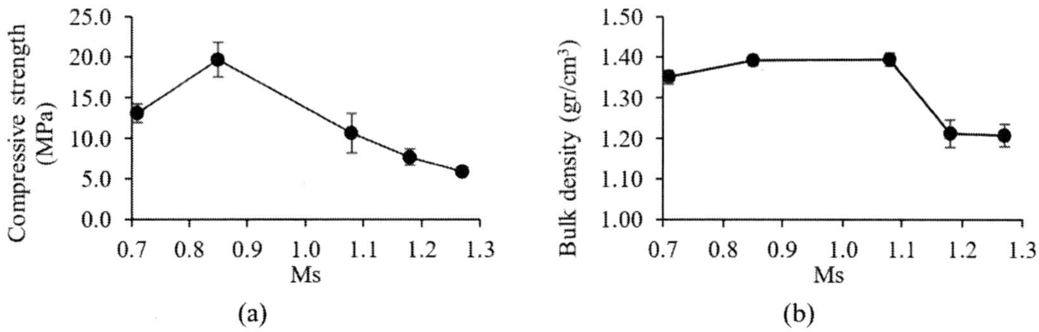

Figure 1. Effect of the Ms modulus on (a) the compressive strength and (b) the density of FCC based geopolymer binder.

In respect to the bulk density, a decreasing tendency related to the optimum Ms can be observed (figure 1b), unless this variation is not clear. It can be attributed to the large amount of water (H_2O/Na_2O ratio) present for high Ms ratios. This may be explained since free water occupies a space in the volume of the sample and as the water is not part of the crosslinked aluminosilicate network, it evaporates, leaving empty spaces which can be recognized as porosity with a low specific weight [30]. This behavior confirms that the bulk density is not affected by the Ms [32].

3.2. Effect of NaOH concentration

The compressive strengths of the FCC based geopolymer binders increased with increasing the NaOH molar concentration from 8 to 12.5 M in the mixture; when 15 M NaOH solution was used, the compressive strength remained relatively constant (only 5.3% increase) (figure 2a). A higher NaOH concentration may produce a higher compressive strength due to the better dissolution of the Si and Al species during the polymerization reaction, as found by other researches [30,31]. However, an excessive NaOH concentration can also dissolve the stable species formed initially, limiting the formation of the geopolymerization products and reducing the compressive strength [13].

As shown in figure 2b, the relationship between the bulk density and NaOH concentration is not clear. Again, water played a role only during the first stages of the geopolymerization until its elimination during the polymerization process, generating the porosity that explains the density.

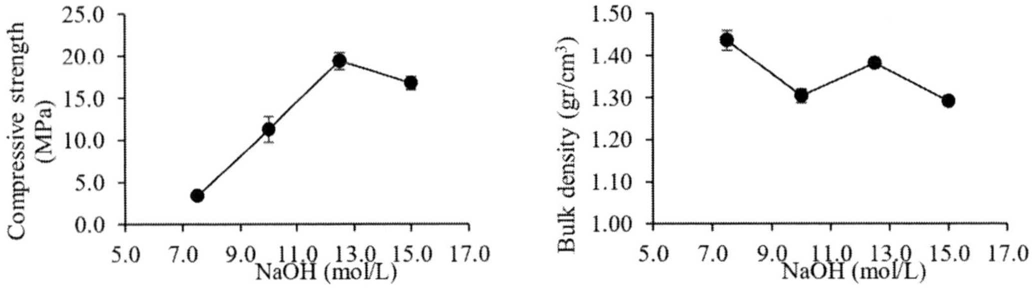

(a) (b)

Figure 2. Effect of the [NaOH] on (a) the compressive strength and (b) the bulk density of the FCC based geopolymer binder.

3.3. Effect of liquid alkaline solution/solid precursor ratio (AS/FCC)

An optimum value of AS/FCC ratio of 1.10 produced the highest compressive strengths with both 10 M and 15 M NaOH concentration (figure 3a). For lower values of AS/FCC ratios, the mixtures were too dry and difficult to introduce in the molds. The low compressive strengths of the dry mixtures can be also attributed to the lack of enough NaOH or liquid medium in the mixture. For AS/FCC values higher than 1.20, the mixtures had a liquid consistency. However, a reduction of the compressive strength was observed for these mixtures which could be attributed to the excess of activator and the inhibition of the polymerization process [13].

In general, these results confirm that the workability of the mixture is important because it influences the compressive strength [18] and the AS/FCC value does not depend on the NaOH concentration. AS/FCC assures the adequate accommodation in molds and the homogeneity of the samples. This same behavior was observed in previous studies and confirms the attack of the precursor and the chemical combination of Na^+, OH^- and silicate ions in the geopolymeric matrix [12]. In conclusion, an AS/FCC ratio between 1.0 to 1.37 produced mixtures with adequate consistency and workability that made them easy to accommodate in the molds without much effort.

The bulk density is more directly related to the AS/FCC ratio (figure 3b). For the 15 M NaOH alkaline solution, the optimum AS/FCC ratio produced the highest bulk density. Similar behavior can be observed for the 10 M NaOH solution, where the optimum AS/FCC ratio produced the lowest density. This confirms that AS/FCC value has an effect in the bulk density [33].

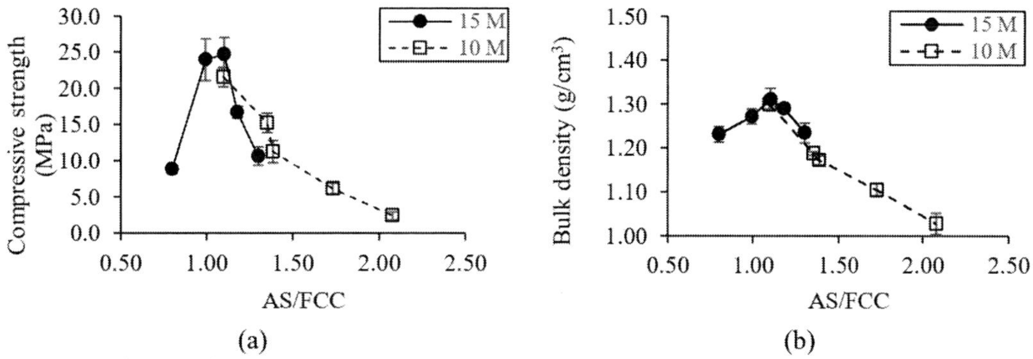

Figure 3. Effect of the AS/FCC weight ratio on (a) the compressive strength and (b) bulk density of the FCC based geopolymer binder.

It can be concluded that the optimum formulation for geopolymerization of spent FCC catalyst requires an alkaline activator solution prepared with 15 M NaOH solution and a sodium silicate of Ms of 0.85, in a proportion of AS/FCC close to 1.1, approximately. The optimum AS/FCC has 17.25 % of Na_2O and 18.72% of SiO_2, that is close to the range suggested by [13]. The difference can be attributed to the type and quantity of the amorphous components and finesse of the precursor material, although complete data was not reported.

3.4. Analysis of the total ratios SiO_2/Al_2O_3, Na_2O/SiO_2, Na_2O/Al_2O_3 and H_2O/total solids

A very strong correlation between the compressive strength and the chemical composition on the final mixture can be observed in the figure 4. These parameters were calculated as the chemical species on the total mixture (alkaline solution and FCC residue). For the three key factors evaluated previously (NaOH concentration, Ms and AS/FCC) there is a correspondence to the optimum values of the total ratios. A SiO_2/Al_2O_3 final molar ratio between 2.4 and 2.5 produces the highest 7 days compressive strength (figure 4a) which corresponds to both the optimum Ms and AS/FCC values. In the same way, an optimum Na_2O/SiO_2 final molar ratio close to 0.30 (figure 4b) corresponds to both the optimum Ms and AS/FCC values. These results are consistent with previous researches. For example, [6] found optimum SiO_2/Al_2O_3 and Na_2O/SiO_2 total molar ratios for fly ash and rice husk bark ash based geopolymers that produced the maximum compressive strengths. For the Na_2O/Al_2O_3 molar ratio the optimum values are in a range between 0.6 and 0.7. Additionally, H_2O/total solids ratios between 0.4 and 0.7 were related to the highest compressive strength (figure 4d), which is consistent with the range of 0.4 and 0.6 values for H_2O/FCC obtained by [12] for a similar FCC. Similar finding was reported by [34] in a class C fly ash geopolymer mortar, reaching a compressive strength of 10.29 MPa at a H_2O/total solids ratio of 0.40.

3.5. Compressive strength and bulk density

The bulk densities and compressive strengths of the mixtures at the age of 7 days are plotted in figure 5. Bulk densities of geopolymer binders are between 1.08 and 1.47 g/cm^3 which are relatively low compared to other geopolymer binders [35,23]. There is a region where the maximum compressive strength is related to a range of density between 1.25 g/cm^3 and 1.35 g/cm^3. For lower values, the compressive strength is lower too. This region is related to high values of AS/FCC that represents an excess of solution for a certain weight of FCC residue. This excess of solution is also associated to a high amount of water that produces a porous microstructure, reducing both the density and the compressive strength. But, when this high AS/FCC value is associated to high NaOH concentrations, the bulk density increases due the presence of the total solid species in the mixture, but the excess of Na^+ cations inhibits the formation of the stable species during the polymerization process reducing the compressive strength. These values correspond to the region of densities higher than 1.35 g/cm^3.

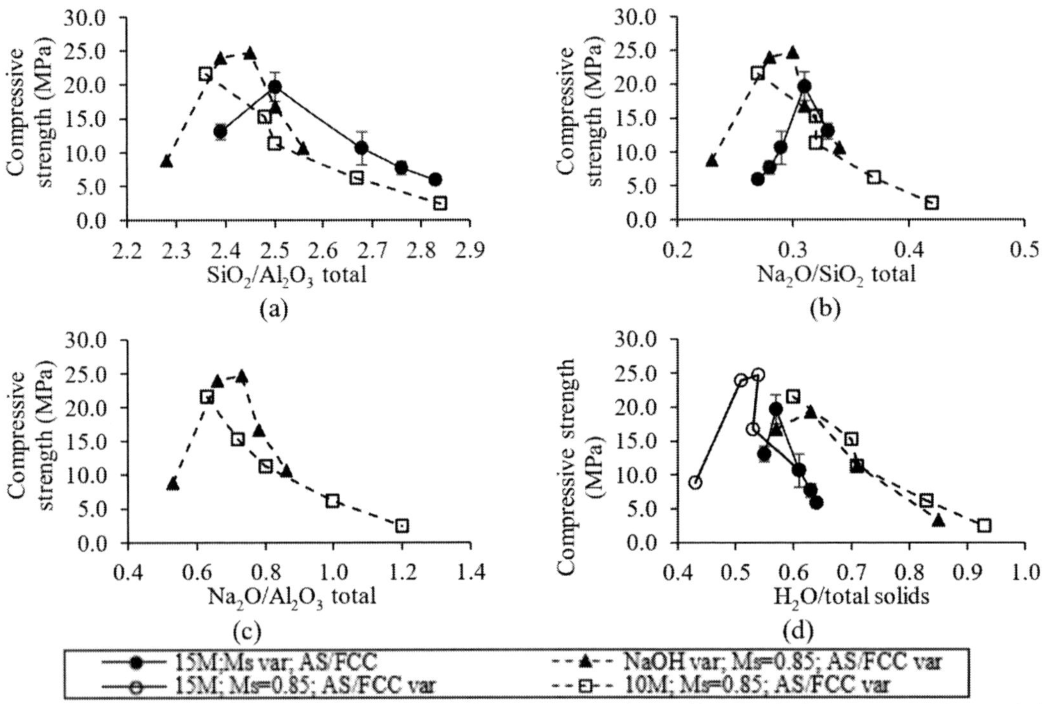

Figure 4. Influence of the total ratios in the final mixture (including both the FCC residue and the alkaline solution) on the compressive strength (a) SiO_2/Al_2O_3, (b) Na_2O/SiO_2, (c) Na_2O/Al_2O_3 and (d) H_2O/total solids.

For low AS/FCC values, there is low amount of water in the mixture that contributes to increase the density but also limits the compressive strength because of the reducing in the workability of the mixture. This lack of workability does not allow the polymerization process. In this region, the key factors evaluated in previous sections play an important role for the development of the compressive strength. This finding confirms that, considering an adequate range of AS/FCC value, the compressive strength is not related to the density but to the quality of bonds formed in the dimensional network of the geopolymer [33] as a consequence of the Ms and NaOH concentration [35,23].

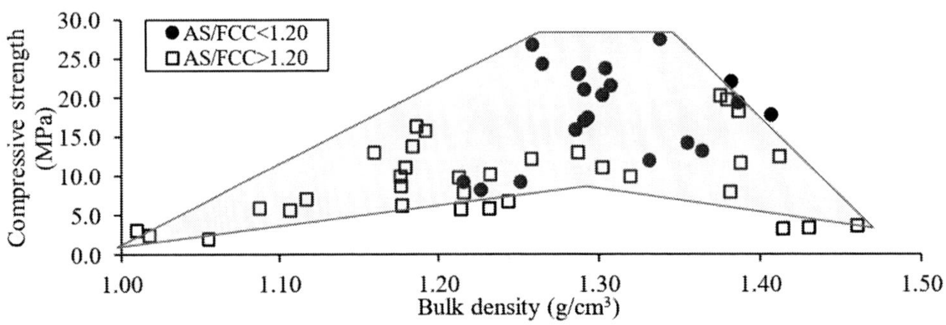

Figure 5. Comparison of the bulk density and the compressive strength in FCC based geopolymer binders for all the mixtures evaluated in this research.

4. Conclusions

The compressive strength depends on some key parameters of formulation such as Ms, NaOH concentration and AS/FCC value. This study allowed reaching a maximum 7 day-compressive strength of 24.7 MPa when a Ms modulus of 0.85, 15 M NaOH solution and a AS/FCC value of 1.10 was used. The results also indicate that total ratios in the mixture (SiO_2/Al_2O_3, Na_2O/SiO_2, Na_2O/Al_2O_3 and H_2O/total solids) that includes the species of both the solid precursor and the alkaline solution, can be used to control the optimum formulation for geopolymers production. The bulk density of the geopolymer binders were between 1.08 and 1.47 g/cm^3, which is lower than other geopolymer binders. The bulk density has a strong correlation to the AS/FCC value used in the formulation. Finally, it can be concluded that spent FCC catalyst from the oil refinery industry is suitable to produce geopolymer binders. This new application for the residue may contribute to reduce the environmental impact of its disposal.

Acknowledgments
This work was supported by CONCYTEC and SENCICO PERU under the project: "GeoBloque: Desarrollo de bloques de construcción ultraligeros con geopolímeros" (Contract No. 105-207-FONDECYT). Authors acknowledge the support of Compañía Minera de Agregados Calcáreos S.A. (COMACSA) for the use of its facilities and equipment. The first author also acknowledges Universidad de Piura for the scholarship support for her PhD studies, also, the technical personal of the same institution.

References
[1] Marafi M, Stanislaus A 2008 *Resources, Conservation and Recycling* **52** 859–73
[2] Rodríguez E, Bernal S, Provis J, Gehman J, Monzó J, Payá J and Borrachero M 2013 *Fuel* **109** 493–502
[3] Payá J, Monzó J, Borrachero M 1999 *Cement and Concrete Research* **29** 1773–9
[4] Al-Dhamri H, Melghit K 2010 *J Hazard Mater* **179** 852–9
[5] Payá J, Monzó J, Borrachero M and Velázquez S 2003 *Cement and Concrete Research* **33** 603–9.
[6] Chen H-L, Tseng Y-S and Hsu K-C 2004 *Cement and Concrete Composites* **26** 657–64
[7] Su N, Fang H-Y, Chen Z-H and Liu F-S 2000 *Cement and Concrete Research* **30** 1773–83
[8] Su N, Chen Z-H and Fang H-Y 2001 *Cement and Concrete Composites* **23** 111–8
[9] Costa C, Ribeiro MS and Brito N 2014 *Mater. Science and Applications* **05** 905–14
[10] Alshamsi K, Baawain M, Aljabri K, Taha R and Al-kamyani Z 2012 *Procedia - Social and Behavioral Sciences* **53** 326–34
[11] Dulaimi A, Nageim HA, Ruddock F and Seton L 2017 *J. Mater. Civ. Eng.* **29** 4017048
[12] Tashima M, Akasaki J, Castaldelli V, Soriano L, Monzó J, Payá J and Borrachero M 2012 *Mat. Letters* **80** 50–2
[13] Tashima M, Akasaki J, Melges J, Soriano L, Monzó J, Payá J and Borrachero M 2013 *Fuel* **108** 833–9.
[14] Duan P, Yan C, Zhou W and Luo W 2016 *Construction and Building Materials* **111** 600–10
[15] Khankhaje E, Hussin M, Mirza J, Rafieizonooz M, Salim M, Siong HC and Warid M 2016 *Materials & Design* **89** 385–98
[16] Moraes J, Tashima M, Akasaki J, Melges J, Monzó J, Borrachero M, Soriano L and Payá J 2016 *Construction and Building Materials* **124** 148–54
[17] Mas M, Tashima M, Payá J, Borrachero M, Soriano L and Monzó J 2015 *Key Eng Materials* **668** 411–8
[18] Lizán J, Bouzón N, Payá J, Borrachero M, Bonilla M and Monzó J 2014 *Proc. II Congreso Internacional de Estudios del Desarrollo* Huelva Spain
[19] Lampris C, Lupo R and Cheeseman C 2009 *Waste Manag* **29** 368–73
[20] Wu H-C and Sun P 2007 *Construction and Building Materials* **21** 211–7
[21] Silva G, Castañeda D, Kim S, Castañeda A, Bertolotti B, Ortega-San-Martín L, Nakamatsu J and Aguilar R 2019 *Construction and Building Materials* **215** 633-43

[22] Varela B, Tavares P, Teixeira A and Castro-Gómes J 2013 *Jour. of Mat. Science and Engineering with Advanced Technology* **7** 1-17
[23] Bumanis G, Bajare D, Locs J 2014 *Key Eng Materials* **604** 169-172
[24] ASTM D854-14. *Test Methods for Specific Gravity of Soil Solids by Water Pycnometer.* (West Conshohocken, PA: ASTM International)
[25] Nazari A, Bagheri A and Riahi S 2011 *Materials Science and Engineering* **528** 7395–401
[26] Nazari A, Riahi S and Bagheri A 2012 *Materials & Design* **35** 296–302
[27] Riahi S and Nazari A 2012 *Ceramics International* **38** 4467–76
[28] ASTM C305-14. *Practice for Mechanical Mixing of Hydraulic Cement Pastes and Mortars of Plastic Consistency.* (West Conshohocken, PA: ASTM International)
[29] ASTM C109/C109M-16a. *Test Method for Compressive Strength of Hydraulic Cement Mortars (Using 2-in. or [50-mm] Cube Specimens).* (West Conshohocken, PA: ASTM International)
[30] Part W, Ramli M and Cheah C 2015 *Construction and Building Materials* **77** 370–95
[31] Jeeva Chithambaram S, Kumar S and Prasad M 2018 In: Das BB, Neithalath N, editors. *Sustainable construction and building materials* (New York NY, Springer Berlin Heidelberg) p 651–658.
[32] Davidovits J 2008 *Geopolymer chemistry and applications* 2nd ed. (France: Geopolymer Institute) p 623
[33] Mahboob S 2014 Development of Low Carbon and Low Energy Geopolymer-based Cement free Construction Materials [Thesis]. London: Brunel University
[34] Wardhono A 2019 *MATEC Web Conf.* 258 (4 Part E) 5032
[35] Çelikten S, Sarıdemir M 2018 *Proc. IV International Ceramic, Glass, Porcelain Enamel, Glaze and Pigment Congress* (Eskişehir, Turkey) 295-300

IOP Conf. Series: Materials Science and Engineering **660** (2019) 012010 doi:10.1088/1757-899X/660/1/012010

Interactive learning medium in contemporary engineering education accented for practical training of highly skilled personnel

Z Veide[1] and V Strozheva[2]

[1] Assistant Professor, Riga Technical University, Department of Computer Aided Engineering Graphics, Riga, Latvia,
[2] Lecturer, Riga Technical University, Department of Computer Aided Engineering Graphics, Riga, Latvia

E-mail: zoja.veide@rtu.lv

Abstract. The contemporary generation of students reluctantly uses traditional teaching tools such as textbooks, which makes it necessary to create interactive training materials using modern technologies. This article presents the experience of creation of interactive educational materials for graphical engineering disciplines, which is aimed for students of colleges and vocational schools. In the material developing, the didactic experience of technical universities teachers from five countries has been summarized. Elaborators of this course have got joint goals of improvement of quality of education and solving similar problems of fast and efficient presentation of educational material acceptable to the modern generation of students.

1. Introduction

In the modern world, the development of information technologies happens every day. Technological developments, which occupy an important place in people's lives, lead to changes in all spheres of our life, including education [1, 2, 3]. The rapid development of technology not only makes life easier for people but also increases the interaction with computers and the Internet. The era in which we live is a period of time, which is called the digital era when computers and the Internet occupy a central place.

Computer-aided technologies such as CAD, CAM, CAE, etc. as well as Building Information Modelling are widely used in engineering. The full use of the potential of these systems is possible only when the "language of engineers" is understood - engineering graphics. Currently, users have a huge lack of knowledge in the field of engineering graphics. There is a lot of different information on the Internet, but most of the teaching materials use the third-angle projection method to create drawings, it lacks an integrated approach and it does not consider various styles of student learning.

Data analysis showed that most intending students of vocational schools, colleges and universities do not even have basic knowledge of technical graphics. Teachers of vocational schools, teachers of colleges and universities have a difficult task in a very short period to help students master the graphical subject, develop spatial skills, without which it is impossible productive creative activity. From the trend of reducing the number of contact hours in the class, students spend more time to study the subject independently. Because of a deficiency of basic theoretical knowledge in geometry the first-year students experience difficulties in independently studying theoretical material of engineering graphics [4]. Analysis of Internet resources, articles, and conference reports suggests that there are not enough educational materials prepared according to ISO standards that can help students learn graphical subjects, motivate and promote students interest in the study and support their learning [5, 6].

The article describes the pre-results of the project "Development of Interactive and Animated Drawing Teaching Tools" DIAD-tools. The aim of the project is to create teaching tools which would facilitate students of vocational schools, colleges and universities to master engineering graphical subjects teaching materials better and increase self-motivation to study [7,8]. Working with this interactive teaching material will help students to memorize necessary information and apply it practically.

2. Project description

In the project DIAD-tools participating members of eight organizations from five countries - Latvia, Lithuania, Poland, Slovakia and Estonia. The duration of this project is thirty months, the beginning of the project is October 2017. Target groups of the project are students of vocational schools, vocational teachers, students of higher schools and colleges, lecturers of higher schools and colleges and scientists and specialists in the field of Technical Graphics, Descriptive Geometry, Engineering Computer Graphics, Civil Engineering, Mechanical Engineering.

The DIAD-tools project partners were selected according to their work experience, competencies, skills that are necessary for the activities of the project. The following partners are participating in the project: Public Institution Vilnius Builders Training Centre - the applicant of the project, coordinator; the Slovak University of Technology in Bratislava - responsible for creation of training materials; Ida-Virumaa Vocational Education Centre is responsible for dissemination and valorisation activities; Aleksandras Stulginskis University - responsible for performance of methodological materials analysis and creation of training materials, the Lithuanian Society of Engineering Graphics and Geometry - monitoring the progress, quality and achievement of project activities; Riga Technical University - responsible for creation of training materials; Panevėžys University of Applied Sciences is responsible for preparation of testing methodology of project training materials, processing of testing results and creation of recommendations for improvement of the training materials; Silesian University of Technology - responsible for creation of training materials.

The objective of the project - creation of interactive and animated drawing teaching tools to help students the understanding the topics of engineering graphical subjects and motivates them to learn. The project gives the possibility for the partners to share their experiences and structure the conventional drawing symbols and drawing elements according to ISO international standard. Specialists of universities from four European Union countries (Lithuania, Latvia, Poland and Slovakia) are working on the creation of training materials [9]. Considering the need of project partners to improve technical and didactical competences of drawing there will be prepared the four chapters of interactive teaching tools:

1. Execution of drawings. Geometric drawing.
2. Basics of projection drawing. Images. Sections. Cuts.
3. Joints of parts. Working drawings of parts.
4. Construction drawing.

The creation of these materials takes place in several stages: 1. analyzing the content, the content description (English version); 2. the collection, structuring according to the content and creation of the materials (English version); 3. placement of prepared materials on the project website (English version); 4. translation of textual content of the training materials from English into the languages of project partners (Lithuanian, Estonian, Latvian, Polish, Slovak); 5. placing of translated teaching materials on the website of the project and preparation for the testing; 6. the testing of the prepared training material in the partner's countries; 7. processing the test results and making recommendations for improving the training materials; 8. enhancing the materials, the creation of the final version and publishing the project training materials on the project website.

3. Needs Analysis Study

On the stage of preparing the content of the training materials, the Needs Analysis Study was conducted. The survey was organized in five countries: Estonia, Latvia, Lithuania, Poland, Slovakia. The goal of this survey was to investigate the methodology of teaching the graphical subjects in countries of the project partners, the relevance of the training materials chapters, the standards used, labour market requirements to preparation level of vocational school graduates. Three questionnaires for different target groups were prepared: 1. teachers, lecturers, scientists; 2. students and 3. employers, labour market representatives. The 56 filled questionnaires from employers were received, 149 – from teachers and 349 - from students. The great part of questionnaires from employers and teachers were gotten from Lithuania, most of the student questionnaires were received from Latvia.

Comparing the first two target groups answers to questions about the difficulties arising from study certain themes of technical graphics, it can be concluded that the opinions of students and teachers are the same. Both teachers and students called the dimensioning in the shaping section the most difficult task for understanding students. Students also see difficulties in the theme of scales. Smooth connections and curves, according to teachers and students, are the topics of geometric construction that are most difficult for students. According to students and lecturers, the complicated themes in the mastering of the projection drawing are the creation of cuts and the identification of objects from an orthogonal projection drawing. For students, the creation of a section is also most often mentioned as the more difficult task of this section.

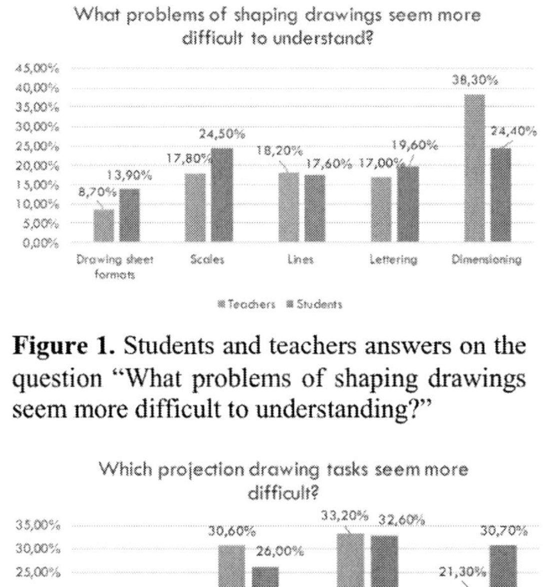

Figure 1. Students and teachers answers on the question “What problems of shaping drawings seem more difficult to understanding?”

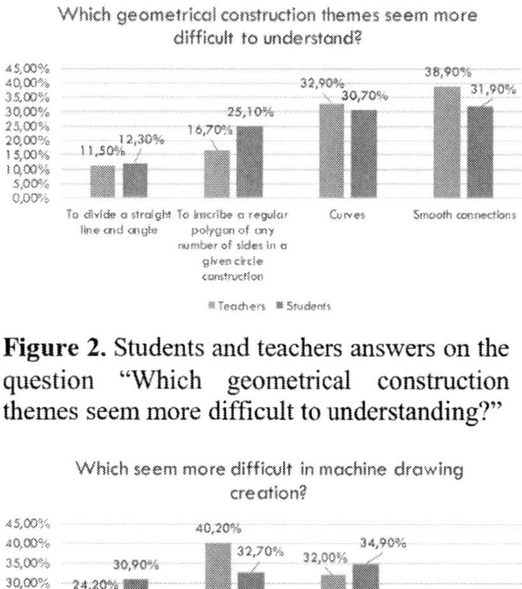

Figure 2. Students and teachers answers on the question “Which geometrical construction themes seem more difficult to understanding?”

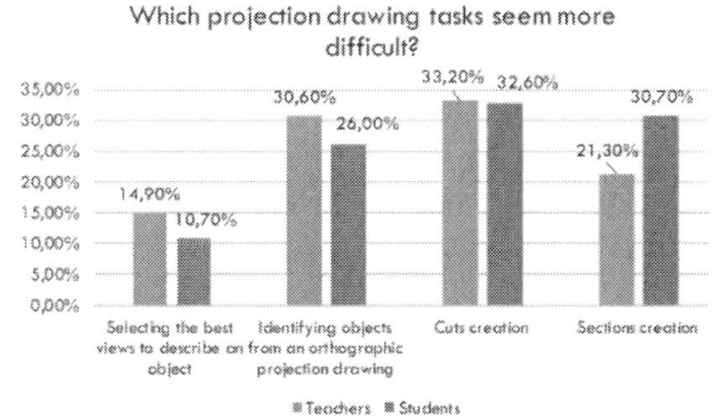

Figure 3. Students and teachers answers on the question “Which projection drawing tasks seem more difficult?”

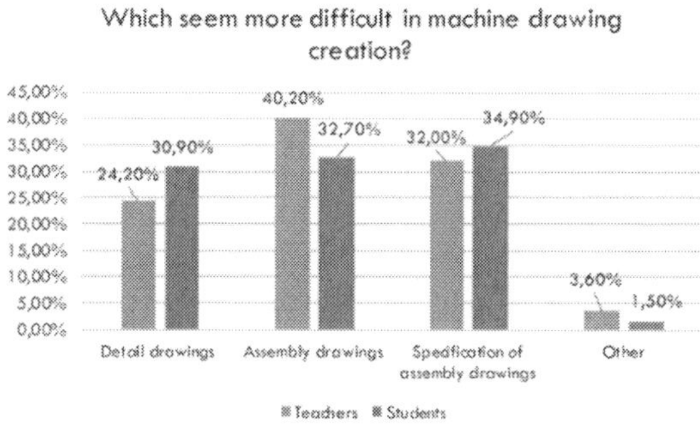

Figure 4. Students and teachers answers on the question “What seem more difficult in machine drawing creation?”

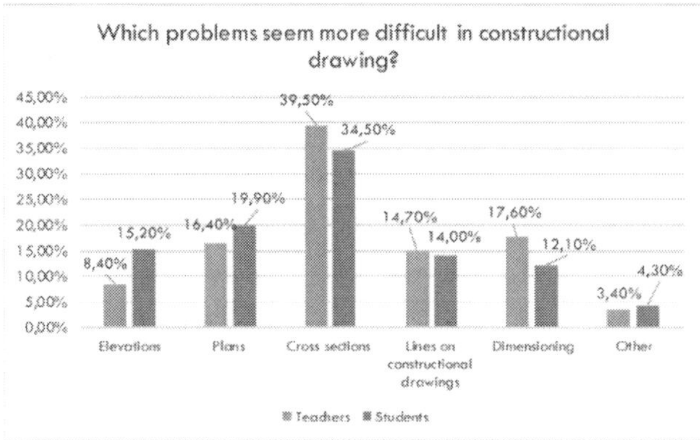

Figure 5. Students and teachers answers on the question "Which problems seem more difficult in constructional drawing?"

Educators and students in the field of mechanical engineering drawing identified the topics of assembly drawings and specifications of assembly drawings as causing problems. In the construction drawing section, teachers and students identify drawing in cross section as the most difficult task for students.

After processing the survey results, it turned out that most of the teachers and students (78% of teachers and 63% of students) were in favour of the compulsory study of technical drawing in high schools. In accordance with the opinion of teachers, the lack of students' knowledge and reluctance to learn are two of the main difficulties they encounter.

According to the opinion of employers, vocational and high schools' graduates should have more practical abilities, the skill to personalize (customize) the CAD systems, understanding BIM programs and be able to use its. It is necessary for graduates also to have basic knowledge in the field of production technology, tolerance and fit, roughness, machine construction, which is necessary for correct design and reading of machine parts list. Freehand sketching is a desirable skill in the job market.

After analyzing the above-mentioned results, we can conclude that preparing the training material, more attention should be paid to scales, and dimensioning; inscribing polygons in a circle, curves and smooth connections; identifying objects from an orthographic projection drawing, cuts and sections creation; detail and assembly drawing creation, the specification of assembly drawings; drawings of elevations, plans and cross-sections.

4. Training materials

Taking into account the findings of the study, the content of each part of the interactive and animated drawing teaching tools was approved at the 2nd Partnership project meeting. The project partners made an agreement on the following content:

1. **Execution of drawings. Geometric drawing. Standards, their importance in plotting drawings:** 1.1. Scales; 1.2. Lines on engineering drawings and their application in practice; 1.3. Dimensioning; 1.4. Curves and smooth connections and their application in practice; 1.5. Dividing the circles into equal parts and drawing polygons.

2. **Basics of projection drawing. Views. Sectional Views. Sections:** 2.1. Projection methods. Orthographic representation. 2.2. Identifying objects from an orthographic projection drawing; 2.3. Cuts creation. 2.4. Sections creation. 2.5. Geometric body involutes and their application in practice.

3. **Joints of parts. Working drawings of parts:** 3.1. Threads and Threads Representation in Drawings; 3.2. Treaded fastenings; 3.3. Separable and permanent joints; 3.4. Making training assembly drawings; 3.5. Reading and Detailing of Assembly Drawing.

4. **Construction drawings:** 4.1. Architectural and construction drawings - general principles, creation of building images (elevations, plans, sections, scale marking in drawings); 4.2. Conventional

signs in building drawings, conventional markings of materials (practical assignments); 4.3. Principle of dimension on architectural drawings and application of the dimensional common modular system in building drawings (practical assignments); 4.4. Components of a construction project (practical tasks); 4.5. Examination of a construction project, reading of the drawings (practical assignments).

Researchers and teaching staff of the Computer Aided Engineering Graphics Department of RTU are responsible for the creation of the third part of training materials "Joints of parts. Working drawings of parts". In the course of implementation of the project, five interactive materials (MP4 format) were developed. The duration of one teaching material is from 6 to 9 minutes. Within the framework of the approved content of this training material, the following issues were considered for selected topics. Threads and Threads Representation in Drawings: terms of screw threads, making threads, external thread representation, internal thread representation, threads designations, vanish threads, necks and chamfers. Threaded Fastenings: bolted joints, studed joints, screw joints, threaded fastenings drawing (simplified and conventional representation of threaded fastenings). Separable and Permanent Joints: keyed joints, splined joints, keyed joint and splined joint drawing, welded joints (basic terms of a welded joint, designation of the weld on the drawings, example of working drawing of welded joints). Assembly Drawing: steps to make an assembly drawing from component drawings, parts list, part numbers on the drawing, simplification and conventions for assembly drawings, training assembly drawing. Detailing Assembly Drawing: steps of making the detail drawing, limits, tolerances and fits, surface characteristics, parts drawings.

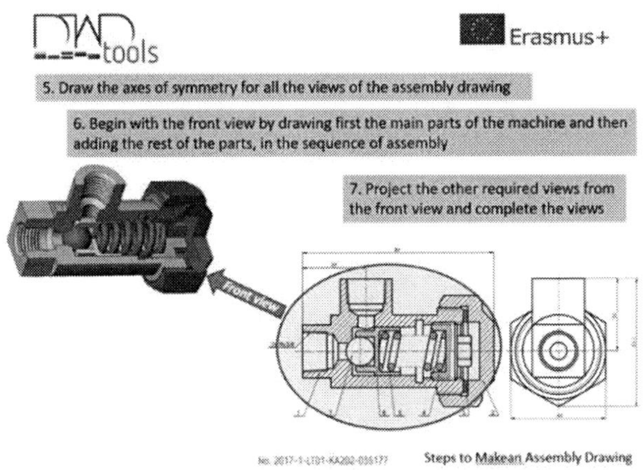

Figure 6. Example of the training material "Assembly Drawing".

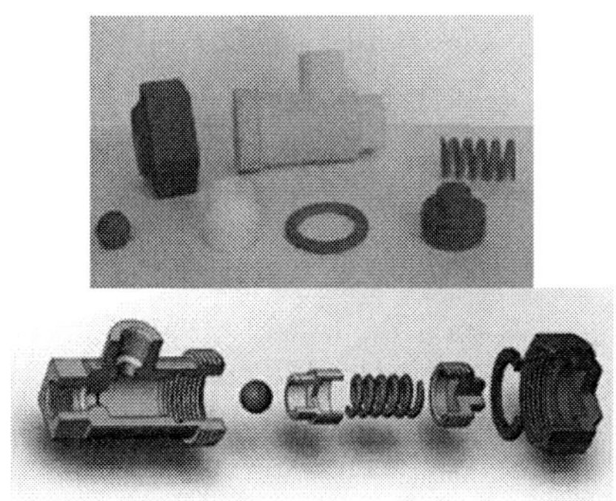

Figure 7. Example of the training material "Assembly Drawing".

These materials can be used as additions to classroom lectures. Educators will be able to demonstrate these training materials, complementing the topics of their curricula. They can give extra study time to students who cannot fully understand the course material through classroom activities and supporting materials such as a textbook. Students can view and learn interactive material as often as they want until they understand the material [10]. This learning resource is especially important in teaching many students when the teacher does not have the opportunity to devote time to each student. Video materials can provide the student audience with more complex study material that may be more interesting for many students [11]. Students of vocational schools will have the opportunity to independently study the material, as well as to test their knowledge performing the exercises and tasks. This material will be especially useful for students of vocational schools because not only theoretical material of the relevant topic is presented here, but also various processes, such as threading, welding, parts assembly into an industrial product.

5. Conclusions

The use of video materials has a positive effect on increase of student motivation and gives positive feedback from students.

It provides more fast transfer of primary necessary information to the students in comparison to the lecture form or independent work with educational literature.

The development of interactive training materials requires the determination of its optimal content, volume and duration of the video to achieve maximum efficiency of perception of the learning material.

Training materials are an auxiliary tool that allows lecturers to optimize their contact hours working with large groups of students and to focus on practical work that develops the skills and abilities required to train qualified professionals.

The video course allows students to master the study material in a convenient time for themselves and in a comfortable environment, creating a positive attitude to towards the subject being studied.

The possibility of repeated repetition, rewinding or forward solves the problem of individual pace of learning. These features make learning more personalized.

It should be noted that the developed educational video is not intended to replace completely the traditional forms of verbal and written presentation of the training topics of the course of graphical disciplines, but is an additional tool to improve the efficiency of the educational process.

References

[1] Borrego M, Foster M J, Froyd J, Systematic literature review in engineering education and other developing interdisciplinary fields. *Journal of Engineering Education*, 103(1), pp. 45–76, 2014

[2] Chien J, How digital media and Internet transforming education. *Proceedings of E-Learn 2012--World Conference on E-Learning in Corporate, Government, Healthcare, and Higher Education 1 Montréal, Quebec, Canada: Association for the Advancement of Computing in Education (AACE)*. pp. 69-77, 2012

[3] Berk R A, Teaching strategies for the net generation. *Transformative Dialogues: Teaching & Learning Journal*, Volume 3 Issue 2 November, pp. 1-24, 2009

[4] Veide Z, Strozeva V, Dobelis M, Problems of Motivation of Students to Study Compulsory Subject "Engineering Graphics". *Scientific Proceedings of the 12th International Conference on Engineering Graphics BALTGRAF 2013* June 5-7, Riga, Latvia, pp. 237-242, 2013

[5] Boelens R, De Wever B, Voet M, Four key challenges to the design of blended learning: A systematic literature review, *J. Educational Research Review*, 22, pp.1–18, 2017 http://doi.org/10.1016/j.edurev.2017.06.001

[6] Edited by Rahman A, Ilic V, Blended Learning in Engineering Education: Recent Developments in Curriculum, Assessment and Practice. CRC Press, 322p. ISBN 9781315165486, 2018

[7] Safiyeh Rajaee Harandi, Effects of e-learning on students' motivation, *The 3rd International Conference on Leadership, Technology and Innovation Management. Procedia - Social and Behavioural Sciences* 181, pp. 423 – 430, 2015

[8] Siklandar P, Kangas M, Ruhalahti S, Korva S, Exploring triggers for arousing interest in the online learning. *Proceedings of INTED 2017 Conference* 6th–8th March, Valencia, Spain, INTED 2017 Conference, pp. 9081–9089, 2017

[9] García-Peñalvo F J, Colomo-Palacios R, Innovative teaching methods in Engineering. *International Journal of Engineering Education (IJEE)*, 31(3), pp.689-693, 2015

[10] Stošić L, The importance of educational technology in teaching. *(IJCRSEE) International Journal of Cognitive Research in Science, Engineering and Education*, Vol. 3, No.1, pp.111-114, 2015.

[11] Bravo E, Amante B, Simo P, Enache M, Fernandez V, Video as a new teaching tool to increase student motivation. *2011 IEEE Global Engineering Education Conference (EDUCON) – "Learning Environments and Ecosystems in Engineering Education"*4-6 April 2011, Amman, Jordan Publisher: IEEE, 638-643 pp, 2011

Composition of Engineering Drawing through Communication Skills and Social/Material Interactions from the Semiotic Aspect

H Lille[1] and A Ruus[2]

[1]Estonian University of Life Sciences, Fr. R. Kreutzwaldi Street, 51014 Tartu, Estonia
[2]Tallinn University of Technology, Puiestee 78, 51008 Tartu, Estonia

E-mail: harri.lille@emu.ee

Abstract. The report analyses how to better proceed from general discussion on graphic literacy in engineering education, through communication between the triad components: concept (problem, engineering idea, existing object), teacher and student, from the semiotic point of view, using Peirce's triadic model whose the basic claim is that signs consist of three inter-related parts: object, interpretant and represent. A pedagogical model is presented in the form of a triad, where problem solving is focused on form: object-problem identification or existing product; interpretant-teacher and interpretant-student who form teacher concept (TC) and teacher representation (TR), and student concept (SC) and student representation, i.e. engineering drawing (SD). At all stages of the learning process the teacher must rely on interpreting students' representations as the evidence of their understanding. Successful application of representational images (engineering drawings) is not a simple task; it is something that must be trained and used in professional practice, as manufacturing of real products is based on engineering drawings.

1. Introduction

In the long history of visualization in engineering graphics, one has gained (presented) new ideas for solving problems [1]. Because in ancient times drawing was not considered an expressive category but rather a fact of "technical" nature, or a form of language imposed by available tools, it is one of the oldest engineering disciplines. The development of Engineering Graphics (EG) is influenced (besides other theories) by semiotics - the theory of signs and symbols and, to some agree, also the theory of meaning-making. Peirce's triadic model can help encode and decode signs – in our case (subject) the engineering drawing, for which there is international agreement that excludes different interpretations and includes the study of how meaning is constructed and understood [2,3]. Engineering and technology students learn basic knowledge, in order to compose engineering drawings as reflections of (physical) existing objects or concepts of future objects (without existing structure), within the EG course. The activity of composing engineering drawings involves also the semiotic activity of classroom participants, who can have different cultural and language backgrounds. A scheme of a triadic pedagogical model is presented in which, according to Peirce's triadic model – object (product identification), interpretant (teacher, student) and represent (teacher, student) – the activity (information processing) is continually transformed through the interpretative collaboration and elaboration by teacher and students (communication), through listening to their feedback and sharing their views (according to teacher's verbal instructions regardless of linguistic limitations, i.e. so-called visual/verbal learning) and explanations. A model teacher and student in information processing and communication as one of the possibilities for engineering students to solve specific technical problems are presented. One of principal objectives of engineering education is that the acquired knowledge of engineering drawings, based on formal semiotics and graphic conventions, allows to encode (and decode) a technical engineering idea into graphic representations or a 3D model as a medium through which visual images (without existing structure) in the mind of the designer are converted into the real object (product).

2. Visualization of engineering ideas in engineering drawing

The technical engineering idea (product identification), i.e. the internal representation, is usually generated in the head of the engineer, serendipitously (Figure 1).

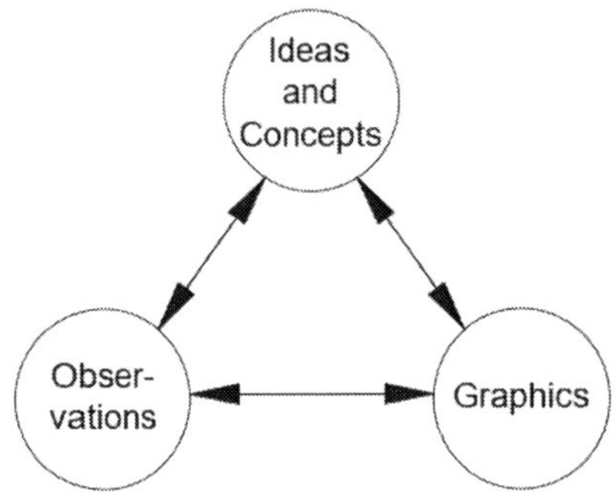

Figure 1. Triangular model of observations, ideas and graphics [4].

It means that through spatial thinking in the process of visualization (sketching) and implementation (conceptual design) of engineering ideas identification of a product is understood and its function and form (solid and surface modelling, design analysis) are planned, which ends up with manufacturing (manufacturing simulation, rapid prototypes, technical drawings and documentation) of real technical products and marketing [5].

In order that the design idea is unambiguously transmitted (communicated), usually by means of the engineering drawing, the external representation or the so-called design (intercultural) language is acquired within Descriptive Geometry – a theory of geometric shaping and aims to study and understand objects in three-dimensional space (3D) from their two-dimensional representations (2D) in one or more planes; and within EG - creation of a technical documentation course which serves as a basis for developing the principles and methods of expression of engineering drawing. Today, both of these subjects are mainly realized by means of computer aided drawing.

Within the EG course engineering students grasp the fundamentals of graphics: sketching, graphics projection, sectioning, dimensioning, engineering drawing. EG involves correct expression of the engineering design, learning the standards related to working and assembly drawings. After completing the course, the students are able to read and design professional drawings using traditional or modern technology. Concisely: forming of ideas, testing and modifying them, and communication of the result of implementation (conceptual design). It means that the designer works within a spectrum of signs. Parts of these signs are depictions, pictorial – representational images; others are scriptorial, involving words – verbal images. The development of EG is influenced (besides other theories) by semiotics - the theory of signs and symbols and, to some agree, the theory of meaning - making. Thus, whenever meaning-making is involved, semiotics can be applied and its models and methods can be use when one enters the subject field of EG. Peirce's triadic model can help encode and decode signs – in our case (subject) the engineering drawing for which there is international agreement that excludes different interpretations and includes the study of how meaning is constructed and understood [2].

The acquired knowledge of engineering drawings, based on formal semiotics and graphic conventions, allows to encode (and decode) technical ideas into graphic representations or the graphic model as a medium through which visual images (without existing structure) in the mind of the designer are converted into the real object (product).

3. From Peirce's triadic model to the semiotic activity of the design process

Peirce's triadic model, the sign–vehicle, mediates between the object and the interpretant and can be applied for transferring the data of the depicted object. Peirce's basic claim is that signs consist of three inter-related parts: the object, the interpretant and the represent [3, 6].

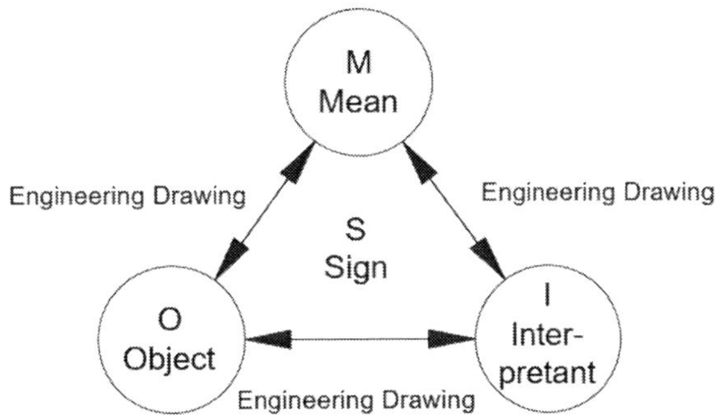

Figure 2. Peirce's triadic model (edited by Bense 1983[6]) and Lille et al 2015 [7]) for transferring the data of the design object.

When editing this model for EG, in our earlier study, we regarded the representamen in its narrow sense as the sign itself. In the present context it comprises icons – three-dimensional (3D) drawings or printing, rapid prototyping; indexes – working and assembly drawings; and symbols – graphic symbols of engineering (e.g. technical drawings) [5].

Peirce's notion of the sign offers a good methodological model to the EG student for describing how engineering drawings and depicted objects (details) are inter-related. It is evident that for the existence of signs, there must be somebody who creates and interprets them. Considering the structure of semiosis, the sign is the first, the object that it evokes is the second, and the interpretant is the third, all of which comprise Peirce's ontological categories, see also [8]. The model of a sign (object, representamen, interpretant) defines meaning-making as an infinite process proceeding through the "interpretant", which is based on personal and social experience. Thus it can be also stated that meanings are formed during use. Hence, the interpretant plays a fundamental role in the process of studying EG, establishing a dynamic relationship between the object and the representation and acting through the triadic model [9]; for our adapted version, see Figure 3. Thus interpretation can be influenced both by depicting what is being interpreted and by guiding the interpetant student.

As noted above, engineering drawings as a visual language (for us, characteristically, graphically standardized language) can be interpreted as signs and can be examined also from the semiotic point of view. According to Eco [10], "semiotics resembles medicine in so far as it is a confederation of specialities". Therefore it is not a pure theory, including also representational modes other than language, and as such, only becomes useful when applied to specific contexts, particularly visual design – art, film but also engineering drawing as a specialized branch of it. Yet drawings should not be interpreted differently by different people but should be focused on conveying information in a simple manner. According to the semiotician Y. Lotman, the basic issues in semiotics are the following. What did the author want to convey? Did you understand it [11]? The same applies to the graphic language: "one can draw" and "one can understand a technical drawing" [12], which means that "I" language and "you" language coincide.

4. Semiotic activity in the classroom: teacher and student in information processing and communication

However, in the case of different social groups (for whom the language of instruction in the classroom is a foreign or a second language), they create the meaning of the idea and its representations differently. Culture determines both the meaning of concepts and utterances, which can coordinate an existing internationally accepted pattern of meanings.

According to a Chinese proverb, one picture is worth a thousand words. Still, visual information is often difficult if not impossible to communicate in verbal images (form), e. g. the mathematical notational form (1+1=2), or one plus one equals two. The cornerstone of the designer's work is legality of drawings, which means that any mistake may render the drawing useless. The unambiguity of drawings is of crucial importance as they are usually not addressed to anybody, with the exclusion of an increasing number of cases where the external design is directed to catching customers' attention. Thus the drawing depicts succinctly and laconically what is to be conveyed for manufacturing the product. It is not a piece of art created for enjoyment. At the same time, it improves the readability of the drawing when, in addition to functionality, it has an aesthetic value. This trend has been characteristic of architectural drawings (construction drawings) throughout history.

The activity of composing engineering drawings constitutes also the semiotic activity of classroom participants – the activity which is continually transformed through the interpretative collaboration and elaboration by the teacher and students (communication), through listening to their feedback and sharing their views (according to teacher's verbal instructions regardless of linguistic limitations, i.e. so-called visual/verbal learning) and explanations (interaction between students and between students and the teacher) Figure 3. The presented triadic pedagogical model serves as one option for engineering students to efficiently read and compose unambiguous engineering drawings.

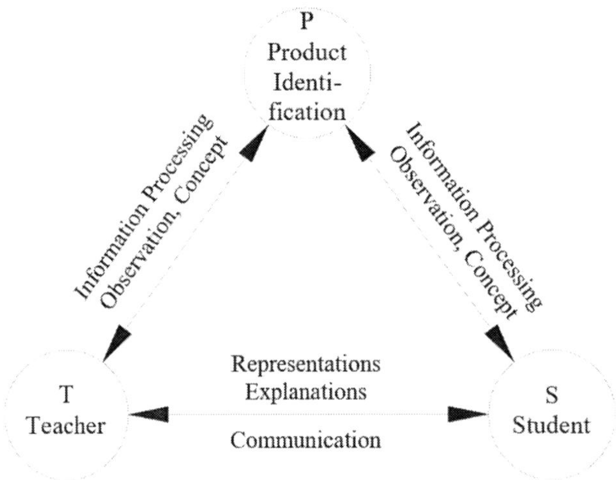

Figure 3. Teacher and student in information processing and communication Waldrip et al 2010 [9], (edited by Lille).

Successful application of representational images is not a simple task; it is something that must be trained and used in professional practice. Here we suggest a pedagogical model presented in Figure 3, developed as a triadic model in Figure 4 where problem solving is focused on form. Let us have a problem (P) (product identification) or an existing (physical) object. We have an interpretant-teacher and an interpretant-student who form teacher concept (TC) - teacher representation (TR), and student concept (SC) - student representation (SR), respectively.

First, we consider the behaviour of the model at interpretation of the assembly drawing and the reverse engineering (RE) concept, for analysing the design of an existing (physical) object or so-called reading and writing the drawing [13]. Let the teacher choose some detail. Composing of a drawing on the basis of a detail, an accepted practice in EG, poses a problem for a freshman engineering student. He or she has a problem, teacher representation and student conception (P, TR, SC). The teacher focuses on the essential planes of the detail during which the student gains an understanding of the detail's meaning (where it used) and of the principles of its function. Further, the student develops representational competence which allows him or her to create one's own representation and to learn how to compose adequate drawings of the presented detail.

Second, we consider the design process as the augmented reality (AR) complex of spatial objects, their perceiving and visualization, and consequent writing of the engineering drawing [14]. Advanced engineering students face the problem how to understand a mental representation of an engineering idea (concept, problem), which would create meaning (individuals need mental power within social and institutional contexts in which meaning-making occurs) and which in turn is expressed by a sign. This suggests use of a triad consisting of the problem, student conception and consequent student drawing (P, SC, SD), as well a triad consisting of the problem, teacher conception and teacher drawing, e.g. rough draft (P, TC, TD). Note, that the representation of an object (future product) can take different forms: icons - three-dimensional (3D) drawings; indexes – working (2D) drawings; and symbols – highly schematized pictures [15].

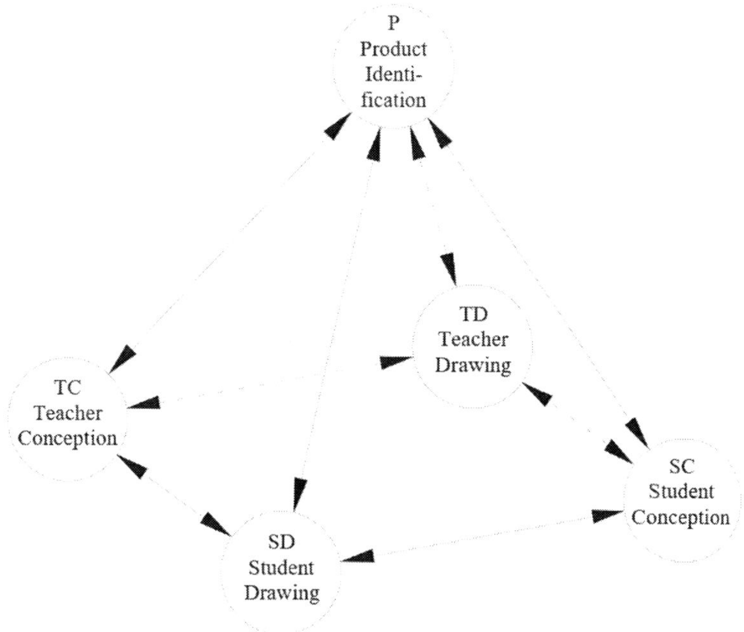

Figure 4. Triadic pedagogical model of Waldrip et al 2010 [9] (edited by Lille).

Here always arises the question how the teacher and student appreciate the adequacy of a representation, since a single representation cannot cover all possible aspect of an object. Consequently, the student needs to learn how to select the representations of the pertinent parts of an object required for manufacturing the real product. Student representations and their revision can function variously as exploratory tools for initial thinking, i.e. scaffolding for building understanding, and records of new thinking and reasoning, depending on the purposes of representations. At all stages of the learning process, the teacher must rely on interpreting students' representations as the evidence of their understanding [13]. It should be noted that what makes a sign the sign is meaning; thus signs and meanings are created as the functionality of objects that are depicted (existing) or that are to be designed (planned).

Peirce's triadic model can help encode and decode signs for which there is international agreement that excludes different interpretations. It is evident that semiotics has the capacity to give meaning to the design process during which something new can be created. The engineering drawing is a clear and coded message relating the design concept to the meaning and communication of the design product. It means that engineering working drawings need to be composed according to an international standard, which will minimize the risk of mistakes. Consequently, they are representations which act through codes and other regularities which do not derive from general laws of physics but are not discordant to them, being historically established. These codes and regularities represent a strictly organized system, originating from the collective memory of a number of one-time engineers (memory always depends on the needs of society or community), which can be divided into different substructures, but their unambiguity has nevertheless preserved [16].

When teaching engineering drawing, this kind of approach may help proceed from the semiotic point of view, as a process of reading and composing of graphic signs among a group of engineering students with different cultural and language backgrounds. Such an approach will definitely regulate all activity and communication, which could impoverish the modes of expression in one's native language; free exchange of opinion and unexpected debate would be suppressed. Yet, as happens in everyday life, some things remain unutterable, whatever the language, and understanding may be acquired without words, in a visual form, as conveyed by the above Chinese proverbs. Moreover, as far as various conflicts are associated with representation and interpretation in communication, semiotic methods and models could be of assistance in settling or even preventing conflicts.

Conclusions

A communication model involving teacher and student in information processing as one of the possibilities for engineering students (also with different cultural and language backgrounds) to efficiently read and compose unambiguous engineering drawings is presented.

The drawing is treated as a semiotic sign by applying Peirce's well knowing triadic model involving the object, the interpretant, and the representant. This serves as the basis for a triadic pedagogical model: object – product identification; inrerpretant – teacher, student; and representation – teacher drawing, e.g. rough draft, student drawing.

The activity of composing engineering drawings involves also the semiotic activity of classroom participants using the triadic pedagogical model – the activity (information processing) which is continually transformed through interpretative collaboration, representations and explanations by teacher and students (communication).

Semiosis as a process of semiotic activity allows realizing the dual function of the semiotic sign by the following designation: the drawing as a sign that denotes an object and the drawing as a sign that evokes making of the object (manufacturing product).

Acknowledgement

We would like to express our gratitude to Ms Ester Jaigma for linguistic help with this paper.

References

[1] Plans and architectural drawings in Ancient Egypt http://www.ucl.ac.uk/museeums-static/digitalegypt/architecture/plans.html
[2] http://www.newworldencyclopedia.org/entery/Semiotics
[3] Peirce's Theory of Signs. https://plato.stanford.edu/entries/peirce-semiotics/
[4] Bedward J, Wiebe E N, Madden L, Carter M, Minogue J 2009 *Graphic literacy in elementary science education: Enhancing inquiry, engineering problem solving and reasoning skills* Presented at the ASEE Annual conference, Austin, TX. http://www4.ncsu.edu/~wiebe/articles/GEES-ASEE-Annual09F-ew0404.pdf
[5] Barr R and Freshman E A 2018 *Book of Abstracts of the 18th Int. Conf. on Geometry and Graphics(Milano)* ed L Cocchiarella p 345
[6] Bense M 1983 Das Universum der Zeichen: Essays Über die Expansionen der Semiotik (Baden-Baden: Agris Verlag Germany) (in German)
[7] Bense M 1983 Das Universum der Zeichen: Essays Über die Expansionen der Semiotik (Baden-Baden: Agris Verlag Germany) (in German)
[8] Liu K 2000 *Semiotics in information systems engineering,* Cambridge University Press pp13-17
[9] Waldrip B, Prain V and Carolan J 2010 *Res. Sci. Educ.* **40** 65 https:/doi.org 10.1007/s11165-009-9157-6
[10] *Semiotics* 2018 ed S Salupere and K Kull (Tartu: University of Tartu Press) 564p (in Estonian) https://www.tyk.ee/semiootika/00000012540
[11] Lotman Y 2006 *Culture Semiotics: Text – Literature – Culture* (Tallinn: Olion) (in Estonian)
[12] Iwata R and Hirano S 2010 14th International Conference on Geometry and Graphics (Kyoto) ed H. Suzuki paper No312 *icgg2010.graphicscience.jp/*
[13] Barr R 2012 *Eng Design Graphics J* **76** (3) pp 8-12 http://www.edgj.org
[14] http://www.engineering.com/DesignerEdge/water/ArticleID/14409/3D-Models-Augmented-Reality-Take-Boredom-Out-of-Engineering-Subjects.aspx
[15] Lille H and Ruus A . 2019 *Proc. of the 18th Inter. Conf. on Geometry and Graphics (Milano)* ed L Cocchiarella vol 809 (Springer, Cham) p 2201 https://doi.org/10.1007/978-3-319-95588-9_202
[16] *ISO Standards Handbook* 2002 Technical drawings vol 1 Technical drawings in general vol 2 Mechanical engineering drawings Construction drawing Drawing equipment

3D computer drawing competition CADrina

A Nõmm[1], R Mägi[2] and K Meister[3]

[1] Academy of Architecture and Urban Studies, Tallinn University of Technology, Tallinn, Estonia,
[2] Estonian Maritime Academy, Tallinn University of Technology, Tallinn, Estonia
[3] Academy of Architecture and Urban Studies, Tallinn University of Technology, Tallinn, Estonia

E-mail: aurika.nomm@taltech.ee

Abstract. CADrina is IT-festival, where takes place 3D computer drawing competition, which has been organized for 9 years by Kadrina Secondary School. In this competition AutoCAD program is used. Participants will create 3D objects in AutoCAD during a given period according to the specifications that will be provided. They will be judged on accuracy and speed. The competition took place in several rounds - the preliminary round and the final round. The event provides opportunities to showcase students' knowledge and skills as well as to be a role model and inspiration for the future choice of young people. 3D computer drawing competition is a good chance to develop team-working skills under pressure among students.

1. Introduction

No one is born as an engineer [1]. Future engineer must learn. But for study, the topic of interest is also necessary. How to raise interest in engineering? Already in childhood games are the foundation of future engineers.

What language engineers interact – English, Russian, Chinese...? Drawing is a language of engineering - it is understandable for all different nationalities. Although engineering graphics are initially hard to understand for students, the subject can be animated with a variety of games and humor [2].

Technical drawing should already start in schools. A good example is the Kadrina Secondary School [3], which was created 1902 in Estonia. There are long traditions in this school to teach technical drawing. Since 2010, there are computer graphics competitions named CADrina organized in Kadrina Secondary School. The CADrina is a technology, IT and smart technologies festival, aimed to promote engineering studies among secondary school students. It ends up with a final show, where under the eyes of the spectators, a thematic engineering project - skateboard, bridge, song arch, etc. - is designed using AutoCAD software. Nearly a whole school will take part of the final show, and the evening will always end with a concert.

IOP Conf. Series: Materials Science and Engineering **660** (2019) 012012 doi:10.1088/1757-899X/660/1/012012

2. 3D computer drawing competition CADrina

This article focuses on the competition that took place on 23 November 2018 [4]. Every competition has a specific theme. This time the theme was "On the Way to the Future" and the task was to draw a space shuttle (Figure 1).

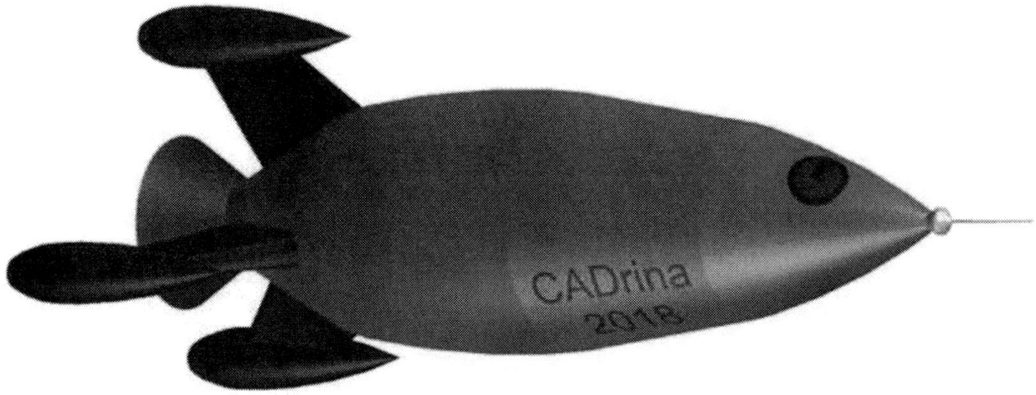

Figure 1. Space shuttle – CADrina 2018 competition task.

2.1. The prelude

The prelude, which starts at 10 a.m., consists of three tasks. There is a maximum of 40 minutes to solve one task. In the preliminary stage, the team consisted of three members will solve tasks together. Each team member is once behind the computer, while the other two can guide and advise while sitting next to each other (Figure 2).

Figure 2. Team is working.

Competitors will receive a tutorial (Figure 3) and a source file (Figure 4) to store the solution.

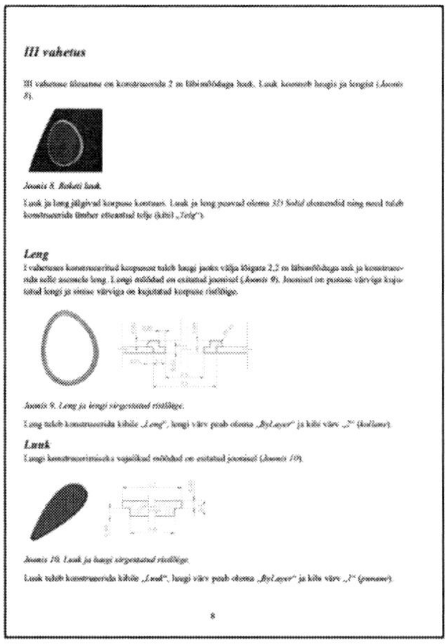

Figure 3. One page of instructional material.

Figure 4. The view of source fail in preliminary stage.

Good knowledge of the following commands was helpful in solving tasks: *align, arc, arraypolar, circle, copy, ellipse, explode, extrude, extend, help, insert, layout, line, loft, move, osnap, pagesetup, pedit, pellipse, planesurf, pline, purge, revolve, rotate, rotate3d, save, saveas, scale, subtract, surftrim, trim, ucs, union.*

2.2. The final

The final starts at 6 p.m. and there turns out who are the four teams to participate. In the final each member will solve the task independently. Solving takes place in front of the audience, where the solution can be monitored on the screens (Figure 5). At the same time, other teams and members are interviewed by the host (Figure 6).

Figure 5. Team member is solving the task in the final.

Figure 6. Other members are interviewed.

All contestants will receive the task on paper (Figure 7) one hour before, during which they will be able to discuss with each other and consult with the supervisor. There is only 10 minutes to solve one task. The source file already contains some basic data that would make task solving faster and more enjoyable for the audience (Figure 8). Before solving, participants will see an accelerated video of the task solution. It provides useful hints for quick and correct solution of the task [5].

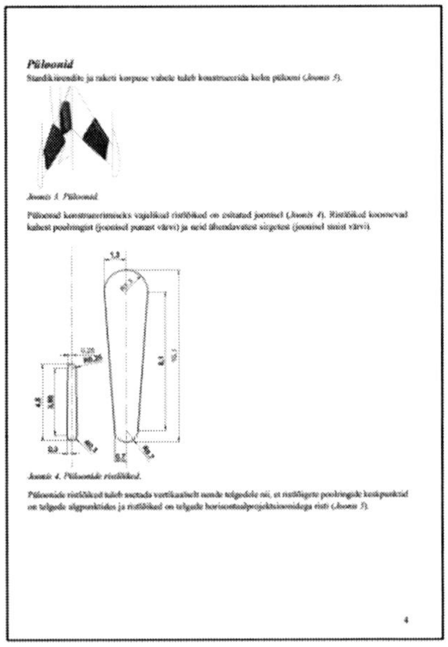

Figure 7. One page of instructional material in the final.

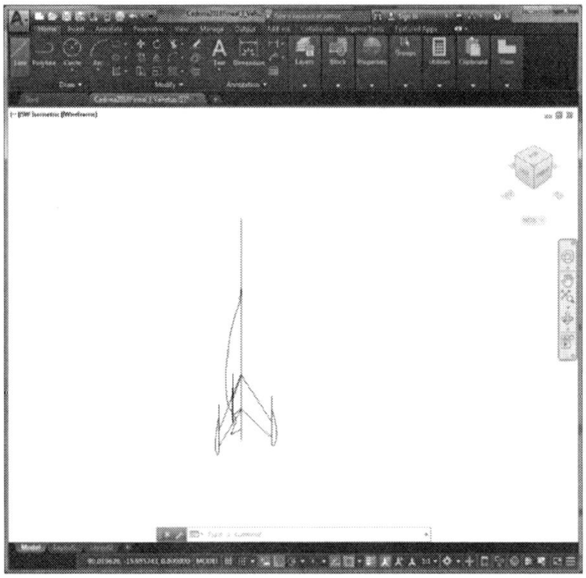

Figure 8. The view of source fail in final stage.

2.3. *Evaluation*

The competition takes place over time and quality. Quality defects are recalculated into fines and added to the time required to solve the problem. If the problem is resolved correctly and the time is less than 40 minutes, then the time taken is calculated in minutes, the seconds are rounded up to the full minute.

The problem is solved in high quality when all the required elements are drawn or modelled exactly, are located on the specified layers, have the colour specified in the task and are placed exactly in the right place in the room. The process of solving and the methods chosen are not evaluated. In order to solve tasks, the competitor may choose a different approach, different commands, etc.

2.4. *Participants and awards*

This time, 8 teams were participating, of which 3 from secondary schools, 3 teams from universities and 2 from vocational schools [6]. There were gift cards as awards given to the best teams. The winner will also receive a traveling trophy (Figure 9).

Figure 9. CADrina's traveling trophy.

2.5. Supporters and sponsors

The event was supported by the European Regional Development Fund as well as local businesses and public authorities. The number of supporters is impressive, with nearly 20 who believe in the sustainability of the event.

3. Conclusion

3D computer drawing competition CADrina introduces exciting and youthful way technical education, educates young people with technical intelligence and entrepreneurial spirit [7]. The competition will develop and test students' engineering graphics skills as well as 3D computer skills. The competition tasks have been complex enough from year to year to provide challenges to competitors.

This time, 8 teams were participating, of which 3 from secondary schools, 3 teams from universities and 2 from vocational schools from Estonia. In the future, the aim is to give the event an international dimension and participants from outside Estonia are welcome. Hopefully there will be enough enthusiastic teachers who teach and guide students to orient themselves in the 3D world.

References

[1] Mägi R 2006 Inseneriks ei sünnita *Õpetajate Leht* **9** 13

[2] Mägi R 2013 Engineering Graphics and Humor *Proc. Int. Conf. on Engineering Graphics BALTGRAF 2013* vol 12 ed M Dobelis (Riga: Riga Technical University) pp 141-148

[3] Kadrina Secondary School Online: https://www.kadrina-kool.edu.ee/eng/ [Accessed 26 February 2019]

[4] Results of CADrina 2018 Online: https://www.kadrina-kool.edu.ee/est/uudised/cadrina/1092-cadrina 2018-tulemused [Accessed 26 February 2019]

[5] Cadrina 2018 Finaal Online: https://youtu.be/e_n3eyEjn14 [Accessed 26 February 2019]

[6] CADrina Online: https://www.cadrina.ee/autocad [Accessed 26 February 2019]

[7] Parijõgi M 2016 CADrina kasvatab tulevasi insenere *Õpetajate Leht* **39** 10

IOP Conf. Series: Materials Science and Engineering **660** (2019) 012013 doi:10.1088/1757-899X/660/1/012013

How to boost the students' interest to engineering graphics?

M Dobelis[1], M Sroka-Bizon[2], and T Branoff[3]

[1] Faculty of Civil Engineering, Riga Technical University, Kipsalas St. 6A, Riga, LV-1048, Latvia
[2] Faculty of Civil Engineering, Silesian University of Technology, Krzywoustego St 7, 44-100, Gliwice, Poland
[3] Department of Technology, Illinois State University, 215E Turner Hall, Campus Box 5100, Normal, IL, 61790-5100, USA

E-mail: Modris.Dobelis@rtu.lv

Abstract. Evident boost in Architecture, Engineering and Construction (AEC) industry now is possible using BIM concept – Building Information Modelling. A similar workflow under a term Product Lifecycle Management (PLM) is known for quite a time. Successful introduction of these technologies require extensive application of 3D geometric models. Development of spatial imagination and 3D models by means of CAD systems should be tackled in the engineering courses. Universities have to participate more actively in the process of faster implementation of these technologies into life. A common background knowledge required to handle this problem is a development of spatial visualization skills since in the backbone of the both these concepts there is a three-dimensional (3D) geometric model or database. The former classic university courses like descriptive geometry and engineering graphics used in the last century have undergone considerable reduction in credit points, sometimes caused by administrative orders or experienced decent content changes dictated by IT achievements. However, development of the spatial comprehension and visualization skills should not be under estimated in the future syllabuses, because they are crucial for 3D modelling in a wide range of engineering applications. The paper will share an experience about recent research about teaching spatial comprehension, 3D visualization and CAD communication skills in technical universities. Results of the experience gained from a 2011 Fulbright Scholarship teaching and research project about the measuring the engineering graphics literacy skills and further application in practice are represented as well. The paper will focus on a review of adopted and tested teaching methods and approaches rather than detailed analysis of the situation in general.

1. Introduction

A fundamental engineering graphics education in a curricula should be acknowledged based on its role in a modern project evolution. Nowadays any project development involves several stages, such as conceptual design, detailing, analysis, iterative re-design, manufacturing, assembling, testing, management, etc. Both in a typical engineering product development and building industry scenarios most stages are performed with Computer Aided Design (CAD) systems. Vast amount of new Information Technology (IT) tools and media appear every year with increasing speed. However, the links between individual stages of CAD design processes very often are scheduled manually. Evident boost in Architecture, Engineering and Construction (AEC) industry now is possible using BIM concept – Building Information Modelling [1]. For the design of industrial and consumer products, a similar workflow under term of Product Lifecycle Management (PLM) is known for quite a time [2]. Recent studies show that fundamental PLM functionalities can create a synergy with the BIM-enabled AEC industry [3]. Successful introduction of these technologies starting from the early stages of the project development require extensive application of 3D geometric models. Questions about the development of spatial imagination and creation of 3D models by means of CAD systems should be included in the courses of engineering graphics.

There are significant challenges for these technologies and software tools and how to bring them into service as fast as possible to replace the traditional use of CAD applications in a wide range of companies. Universities also have to participate more actively in the process of faster implementation of these technologies into life. A common background knowledge required to tackle this problem is a development of spatial visualization skills since in the backbone of the both these concepts there is a three-dimensional (3D) geometric model or database. The former classic university courses like descriptive geometry and engineering graphics used in the last century have undergone considerable reduction in credit points, sometimes caused by administrative orders or experienced decent content changes dictated by IT achievements. However, development of the spatial comprehension and visualization skills should not be under estimated in the future syllabuses, because they are crucial for 3D modelling in a wide range of engineering applications.

2. Blended Learning

New hardware and digital tools introduced by technological developments have significantly changed the education methods, teaching process, its access and use, and availability of present information. Rapid development of information technologies have stimulated researches in the field of education and educational technologies since the methods of the 21st century are technology-oriented methods.

Over the last decade the students' attitudes and behaviors have changed drastically with changes in ways of studying and communicating. A completely new generation of young people, which grew up in an era of high-speed internet and numerous advanced technological gadgets, are enrolling the universities already for several years. Many university students come with a setup "Why should I study this because I do not need this now?" forgetting about gaining practical marketable skills that will help them now and later. It is extremely difficult to hold their interest on subject matter in engineering classes, which require much attention.

A current staff from the Baby Boomer, X, and Millennial generations face problems of controlling the study process in new environment and convey the experience to a new Generation Z. It does not work in a way it did just a decade ago, when verbal face-to-face or written communication was a standard. Messaging apps that "don't leave a paper trail," where communications are sent and quickly gone, are not the best methods to gain practical skills in what industry expects from graduates. Recent research shows that much effort is spent in many teaching areas on how to trigger the Generation Z's interest in studies and how to enhance students' readiness, expectancy, acceptance, self-regulation, collaboration, problem solving, and joy of learning [4]. It is beyond a doubt that the use of IT in education can support student-centered and collaborative learning and develop thinking skills and creativity, but they have to be used in a reasonable balance. Contemporary research shows that triggering interest can enhance students' spatial imagination skills, creativity, joy of learning, improved everyday performance and the outcome [4].

AR applications enable faster comprehension of complex spatial problems and relationships, which makes them useful during the learning process in engineering graphic courses. An AR didactic toolkit AR-DEHAES [5] was successfully adopted to promote a student-centered learning in engineering graphics courses [6]. This entertaining IT tool is based on the application of AR and requires a computer, webcam, and specially designed set of materials for learning. The main purpose of fostering an interest in acquiring spatial comprehension with modern IT tools was achieved. About 81% of the respondents admitted that working with AR-DEHAES allows perceiving and visualizing spatial objects much easier.

Blended learning might be one of the solutions, which refers to a course of study in which the teacher in the traditional classroom delivers parts of some topics, while the learner independently of the teacher, preferably on-line, undertakes other parts. Blended learning will work out only if a creative atmosphere may be set up for practice in an exciting and stimulating way. Five areas of possible change are identified that deserve attention and can be integrated into the classroom without a complete revamp of the curriculum [7]. They are as follows: integrate active and problem based learning (1); help students extract answers from an ocean of information (2); assess often and provide feedback (3); engage creativity (4), and help students make connections (5).

Diverse knowledge background of the students enrolling into engineering programs leads to assorted experiences and expectations in terms of problems of teaching and learning with and without engineering graphics basics from the high school. The diversity of the enrolling students will remain in the future in terms of prior practical experiences and it demands innovative methods that would help to decrease the high dropout rate and promote effective learning.

3. Pencil Aided Drafting or Computer Aided Design?

Spatial reasoning skills strongly correlate to the mathematical abilities and to the success in many technical fields, which require the application of CAD software in industry. This suggests that parents and schools should pay more attention to the pupils' creative skills development while playing with such everyday items as Legos, construction toys, or even video games that require visualization and mental rotation skills. Educational computer games can be developed for training spatial imagination in specific aspects of engineering graphics, however it appears to be studied just occasionally [8]. In general, spatial reasoning skills are frequently associated with higher levels of academic achievement and creativity where the content creation requires it. These skills are measured with the ability to visualize and manipulate 2D and 3D objects, and are considered as being related directly to the achievements in technology, engineering, math, and science [9]. However, recently reported results prove that spatial reasoning skills are often not identified as areas of aptitude in children and as a result, they may not be taught, valued, or assessed in an academic setting.

A review [10] about the changes in engineering graphics education, dictated by IT advancements, proposes a set of student learning outcomes in teaching communication graphics for the next generation of globally competitive engineers. Results of the survey conducted in 2004 and 2012 among 24 faculty indicate that new computer graphics tools and techniques are now preferred and well accepted mode of graphical communication in engineering education in USA. The summarized in the table 1 results of the research show the identified before outcomes in engineering graphics and the changes in their impact from the point of view of the faculty. These results indicate that long-standing techniques of descriptive geometry and manual geometric construction were no longer well recognized outcomes with respect to the computer aided methods. However, some of the traditional geometry problem solving skills and ability to use hand drawing tools in 2012 have been re-appreciated in comparison to 2004.

Results of the students' self-assessment survey [10] indicated that significant gain in all outcomes can be achieved in a graphics course, albeit in a self-reported manner. The conclusions drawn from this research state that modern engineering graphics education should be considered an important form of engineering communication, and should focus on three main areas of instruction. This instruction should include engineering graphics fundamentals (1), computer graphics modelling (2), and computer model applications to digital analysis, manufacturing, and design projects (3). The expected outcomes are in well agreement with the ABET (Accreditation Board for Engineering and Technology) criterion calling for "effective communication" in contemporary engineering education. Specifically, 3D computer modelling, assembly modelling, and digital model application to design and manufacturing all received significant notices in the survey results [10].

The experience of several schools worldwide proves that the most effective way is if this field uses a contemporary approach of teaching the traditional and essential engineering graphics skills by integrating them with 3D CAD modelling content. A fundamental textbook [11] is adopted in many leading universities not only in the USA, but also worldwide, and perfectly covers the basic visualization and sketching techniques that enable students to create and communicate graphic ideas effectively.

Table 1. The Summary Results from a Graphics Faculty Outcomes Survey (N=24) [10]. Importance of the outcomes are listed from highest to lowest ranking (Scale 5) in 2012.

No. of outcome	The identified from the literature review the outcomes engineering graphics	Rank in 2004	Rank in 2012	Changes
1.	Ability to Create 3D Solid Computer Models	4.75	4.75	-
2.	Ability to Sketch Engineering Objects in the Freehand Mode	4.67	4.54	↓
3.	Ability to Visualize 3D Solid Computer Models	4.46	4.54	↑
4.	Ability to Create 3D Assemblies of Computer Models	4.29	4.54	↑
5.	Ability to Create Dimensions	4.38	4.38	-
6.	Ability to Create Section Views	4.13	4.33	↑
7.	Ability to Generate Engineering Drawings from Computer Models	4.33	4.29	↓
8.	Ability to Analyse 3D Computer Models	3.71	4.13	↑
9.	Ability to Create 2D Computer Geometry	4.21	4.08	↓
10.	Ability to Perform Design Projects	3.96	4.08	↑
11.	Knowledge of Manufacturing and Rapid Prototyping Methods	3.42	3.63	↑
12.	Ability to Create Presentation Graphics	3.42	3.46	↑
13.	Ability to Solve Traditional Descriptive Geometry Problems	2.29	2.75	↑
14.	Ability to Create Geometric Construction with Hand Tools	2.13	2.71	↑

4. Contemporary Communication Graphics

A 3 ECTS course BTG701 "Basics of Communication Graphics" was designed particularly for the program “Intellectual Robotic Systems” at RTU in a response to the program holders' wish to include a conventional engineering graphics course to acquaint the students with the basics of engineering design. The course is based on an approach used at the North Carolina State University (NCSU) in a similar course GC350 "Applied CAD and Geometric Controls", which was observed in details during the exchange of experience in spring 2011 Fulbright Scholarship teaching and research project.

One half of the course deals with the basic concepts of descriptive geometry and engineering graphics using conventional pencil technique. The second part covers the creation of 3D parts, virtual assembly of components and documentation using parametric feature based modelling software SolidWorks. However, this part is merged with classic pencil technique from the third week and the same home assignments are required to be completed in both techniques to have a computer solution for assistance. Midterm exam is concluded in a pencil technique without CAD assistance and includes the sectioning of simple geometric primitives with plane and representing them in multi-view drawing. The second part is focused on 3D model creation from different type of information: from a pictorial view with given dimensions, from a multi-view drawing when a form and size of the geometric primitives has to be analysed and determined, and from assembly drawing with bill of materials and only installation and overall dimensions given. An assembly drawing with 6-8 parts in a semester project masters the skills of "reading", and proves the understanding by "writing" in the form of 3D models.

As the privilege of this method is a possibility for the students to see the result in 3D and check it much faster, especially if the teacher has prepared an automated assessment. If the same task is assigned to all students, concerns of plagiarism prevention issue still remain. A preparation of numerous variants of individual tasks in a session is a common practice for conventional engineering graphics courses.

A drawback of the use of the classic pencil technique while mastering engineering drawing "reading and writing" literacy skills is poor hand drafting practices, which cannot be mastered in short courses.

A lack of students' patience in combining an imaginary spatial image of an object while navigating through the "forest of lines" in the corresponding multi-view drawing is another hard to beat factor. It often causes when the students try to dimension parts in the drawings but do it in a chaotic way between neighbouring lines without any spatial reasoning. A requirement to create a feature based model forces to analyse the 3D geometry with caution, determine the proper size and location of geometric primitives, and reproduce them in a valid virtual model. Requirement to use fully defined sketches allows the students to self-assess each modelling step promptly.

The developed course is in use since fall 2013 and later was successfully acknowledged by another study program "Adaptronics". It is interesting to mention, that lately a program "Automobile Transport" requested to design a new course "Mechanical Engineering Drafting" which includes practicing the drawing of advanced machine elements using pencil aided technique. Just a couple of informative overall AutoCAD demonstration classes were requested with the suggestion that the students might want to practice it by themselves from a numerous available tutorials including video tutorials in a native language, if it is needed. Similar manual engineering graphics and drafting courses were is use in a majority of technical universities worldwide just two or three decades ago, but them they were restructure at a different speed depending on the on the countries' IT welfare and universities' hardware capacity. Classic manual drafting courses were gradually abolished in favour to emerging courses dealing with general computer skills, like word processing, spreadsheet calculation and in rare cases some programming in languages, like Pascal etc.

5. Measuring Engineering Graphics Literacy

Engineering graphics educators have been studying students' modelling strategies and methods of evaluating models for approximately twenty years. Recently conducted studies tried to identify how the understanding of engineering graphics literacy correlates with interpreting an assembly drawing, visualizing the various parts, and then modelling these parts within a constraint-based modelling software. These studies revealed that the modelling test used was related to other measures of student success in courses – spatial visualization ability, score on the final project, and score on the final exam.

An attempt to measure an engineering graphics literacy was made in a Fulbright Scholarship teaching and research project in spring 2011 at NCSU [12]. It was supposed that by means of advanced "technologic toy" – a contemporary 3D parametric modelling CAD software – is possible to engage the students into entertaining experiment. The students were required to prove their spatial comprehension and the skills of the interpretation of the assembly drawing by modelling in a limited period as many individual parts as possible.

Several devices consisting from 6-10 parts with different levels of complexity were selected and modeled to check the model validity. Typical elements of machine parts like threads, chamfers, fillets, notches, grooves, springs, etc. were present in these parts. Multi-view assembly drawings with parts list were printed and used for practical training and pilot test studies. This approach insured that the assembly drawings prepared this way will be most suitable for "reading and writing" test and they could not be misinterpreted easily for being not clear enough.

Educators from several universities worldwide were addressed and suggested to perform this "reading and writing" test using parametric feature based modeling software like SolidWorks, SolidEdge, Inventor, etc. [12]. Alternatively they could organize the test in a modified way even with 2D drafting software and submitting their rules and final results for comparison. A positive response was received from two universities in Latvia, one in Croatia, and one in Russian Federation and they joined the research to test the suggested methodology conducted from NCSU. The staff from these universities received the set of drawings for training and actual test. The suggested quantitative evaluation method of "reading and writing" test combined with 3D parametric modeling could be used for accurate measuring of graphic literacy level of the engineering students. One of the main drawbacks is a concern about the ability to scale-up the project to handle more students because the time required assessing the student files is very high.

6. Reverse Engineering Projects

Many engineering graphics educational programs use some form of project activities in comnbination with teaching the graphics theory, to inspire the students with hands-on applications, e.g. Reverse Engineering (RE) project. RE project is a systematic methodology for analysing the design of an existing device or system. The project consists of student teams selecting and reverse engineering a mechanical device through a dissection and documentation process [10]. This concept has been used in a 4.5 ECTS free choice course BTG450 at RTU. The course starts with an overview of PLM concept in mechanical engineering. An analysis of the geometry of typical machine parts is covered and a practical training is performed in a context with parametric feature based design using SolidWorks. Parametric sketches, constraints and their use are discussed in hands-on training examples covering the most widely used basic modelling features like Extrude, Revolve, Loft, and Sweep. Advanced topics like design library, Toolbox, hole wizard, threads, etc. are included in much faster pace. Assembly modelling and document production in a Drawing mode with the creation of views, sections, aligned and crop views, dimensioning, and creation of Bill of Materials concludes the practical training. Home works are assigned as a rule to master practical skills. The checking of individual home works requires extensive faculty work on opening each file and is very time-consuming. Although the students are warned to submit their own work, rare cases of plagiarism warn about aspects, which have to be kept in mind.

In an attempt to search for automation of the checking process, an existing SolidWorks Add-In Graderworks v.2.37 [13] was tested on the compulsory home works. Using this tool it is easy to catch the plagiarism, but rather difficult to assess the obtained result against the expected or the correct answer using the provided options for automatic checking of the mass properties of model and metadata from files. Current version of GraderWorks allows to design a custom rubric of items to be analyzed in students' files and even check if each sketch is fully defined. However, the students have so many different and diverse cognitive approaches that it is not possible to identify the only one supposedly correct track of the solution, which can be easily checked with the tool. No other mention of the use of this tool for similar purposes was find in the literature.

The BTG450 course is concluded with a RE project. Generally, the projects are organized as a teamwork [10], but in the particular course at RTU the students choose individual projects. A mechanical device consisting from 7-10 parts, has to be disassembled, analyzed and the parts measured the parts, and afterwards 3D models created and assembled. Some students get so excited to create projects with even more than 100 components, while some others hardly meet the minimum requirements. Final grade takes into account both the individual semester performance and project presentation results. It is quite difficult to compare and assess the diverse and complex projects quantitatively; therefore, when in spring 2018 a considerably increased number of students selected the course it was decided to choose an alternative method of assessment and motivation – a certification.

7. Certification as a Motivator

Over the last decade, leading CAD software developers have designed and launched certification exams for many products. A Dassault Systèmes company developed Certified SolidWorks Associate – Academic (CSWA – Academic) certification program, which is intended for a student with a minimum of six to nine months of SolidWorks experience and basic knowledge of engineering and fundamentals and practices. Worldwide interest in the certification is constantly increasing over ten years [14].

To boost the students' interest, a "motivating carrot" in the form of CSWA – Academic exam was suggested as an alternative to the RE project and a course exam. A challenge to receive an internationally recognized certificate with personal listing on the CSWA Academic directory triggered a positive attitude from all students, and the students chose this suggested option instead of RE project. In order to cover all the topics required for certification a slightly modified course included wider range of topics, which was not possible in the previous course because of limited absorbing capacity resulting from insufficient interest. In addition to this much faster training pace was expected – all training had to be completed in one study semester or four months instead of six month claimed by certification experts.

Some training exercises used before in the course for practice and home works were modified and put on Moodle for automated checking to mimick the actual certification tests as close as possible.

This "carrot motivating" approach was tested on 65 students from "Medical Engineering and Physics" program at RTU in the course BTG450 by a lecturer in a class with 30 workstations in two sessions. There were 33 first year students, 24 second year and 8 fourth year students, but no gender differences will be analysed in this paper. They all had taken a compulsory course BTG131 "Descriptive Geometry and Engineering Graphics" in the very first semester on the studies. Inspired by the students' interest and a wish to facilitate reaching their goal, the teacher dubbed his on-site class hands-on tutorials with separately recorded and edited video tutorials for part and assembly modelling exercises. After several weeks, the number of students in the class slightly decreased. An answer to the question "Why this is happening?" was "We can do it at home when we want!" This proved to be a wrong assumption because at home there were always too many distractions. This was an unexpected feedback from the experiment, which raises doubts about the usefulness of such an approach in the future.

To earn free CSWA exam vouchers a free sample exam had to be completed with at least 90 points. CSWA – Academic exam version in two parts with 90 min each was preferred to a single three-hour version based on the previous observations that some students are having difficulties to stay focused for long time. To get a certificate both parts of the exam had to be completed with at least 80 points each from a maximum 120 possible. The CSWA certification outcome summarized in the table 2 shows that 29 students or 44.6% succeeded. To achieve this result only three of the 2nd year students had to retake only one part of the real exam. For about one quarter of the first and second year students the pace seemed to be a little too high, so they did not make any attempt to qualify for the certification and switched to a slow track "stick mode" with the same compulsory home assignments. To get the maximum grade in the course, which is equal to those with the CSWA certificate, they had to prepare and defend the RE project. The results were administered in a spreadsheet with single formula where the number of completed home works and all points earned in the three parts of the certification were also taken into account to ensure a quantitative and fair assessment. Much lower but still passing grade was possible to receive without RE project provided some CSWA credits were earned as well. Measurable elements of the performance used in the assessment and strict rules announced in the beginning of the course allowed to avoid a possible complains about the grading.

Table 2. The Summary of the CSWA Exam Results in spring 2018. (N=65).

Year of Study	No. of Students	Sample Exam Attempts	% of Attempts	Passed Sample Exam	Repeated Sample Exam	Average Points in Sample Exam	Passed CSWA Exam	Retake one part of CSWA	Average Points in CSWA Exam	Made RE Project	% from all	Quit the course
1st	33	26	78.8	20	4	105	12	0	212	7	36.4	1
2nd	24	18	75.0	16	0	107	12	3	219	9	50.0	1
4th	8	8	100.0	8	0	129	5	0	214	0	62.5	0
Total	65	52	84.6	44	4	114	29	3	215	16	49.6	2
	100.0%	80.0%		67.7%	6.2%	87.7%	44.6%	4.6%	89.6%	24.6%		3.1%
					max	130		max	240			

The distribution of the number of CSWA certificate holders with increasing a study year quite well characterizes a constantly ascending positive attitude to the self-sufficient learning. About one third (36.4%) of the 1st year students, one half of 2nd year and almost two thirds (62.5%) of 4th year students passed the certification exam with a score close to the average 215 points (89.6%) out of 240 possible.

8. The Role of Competitions

To contribute an interest in comprehension of engineering drawings beyond academic courses in more challenging activities, competitions were organized for well-motivated students who want to prove their knowledge. The annual Olympic CAD & PAD contests were organized in two tracks – CAD and PAD – Pencil Aided Design. The type of tasks varied from year to year and was always different for each track, but never exceeded two astronomic hours.

A typical task for the CAD track was to check the students ability to "read and write" or interpret an assembly drawing of some mechanical device consisting from 6-11 components. SolidWorks was used as a "writing" tool to prove their understanding to prepare exact models of the components, the form and size of which had to be determined from the assembly drawing. No pictorial representation was given but after the contest the students were acquainted with the real part they modelled just a while ago and they could instantly assess what they missed interpreting the multi-view drawing. Figure 1 renders an idea about the complexity of the part to be modelled – how many and what kind of views and section views were given in the contest in fall 2015.

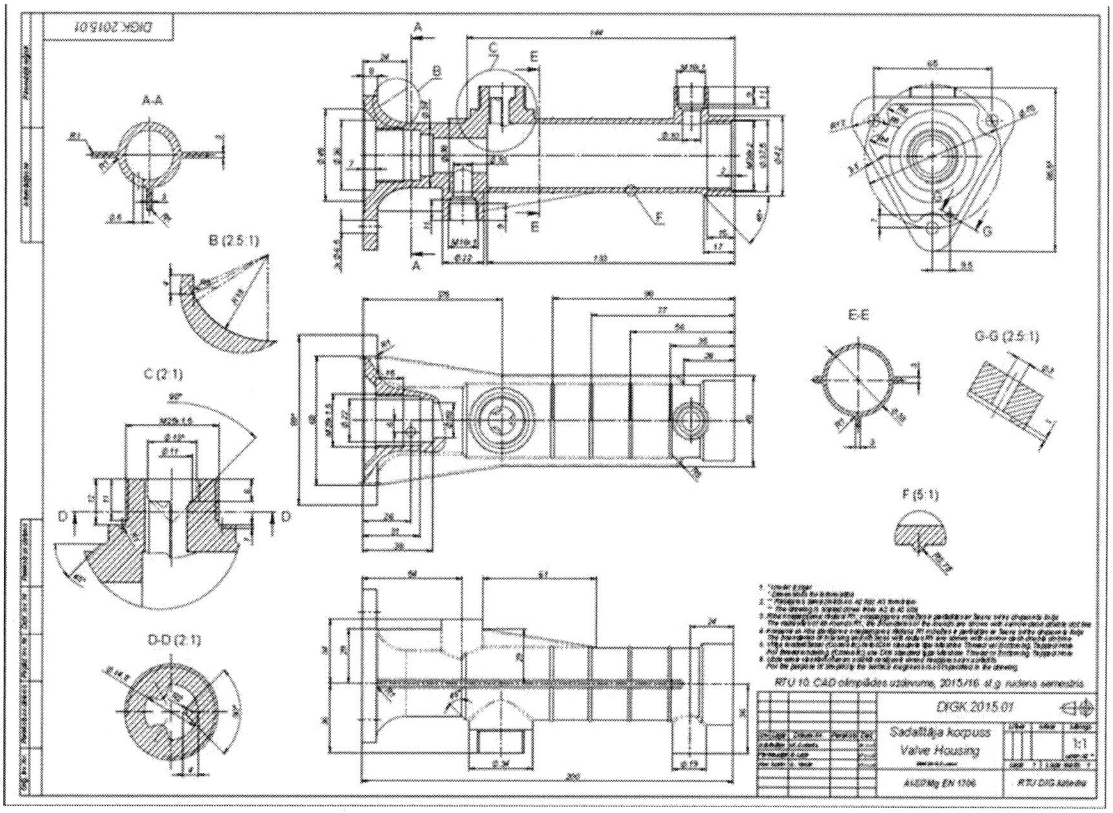

Figure 1. A complexity of a part used in an Olympic CAD contest in fall 2015.

Certificates acknowledging participation in the contest and small prizes for the best was only a minor motivator. Several students reported the participation as self-motivated challenge to test their skills in what they thought might be closely related with their future career. Graphic competitions were terminated after ten years of activities in 2015 the as the organizers from student side switched to other type of competitions, like the design of wafer towers, the construction of paper or pasta bridges, etc.

9. Conclusions

The mode of instruction in the areas of engineering and technology has changed. The traditional lecture setting where the teacher is the primary deliverer of information is no longer the preferred method of instruction.The students have to take more responsibility for their own learning using IT to gather relevant information at unprecedented pace and quality, which was not available before.

With wide access to the information and educational opportunities that IT has enabled, the teacher's role can be shifted to be a "Guide on the Side" if one can find motivating tools. Inclusion of professional certification of the skills and organizing rewarding contests in the courses are just a few of these tools.

Many schools and universities across the world are constantly redesigning learning spaces to enable new models of education, foster more interaction and small group work, and the use of IT as cognitive facilitator. Technologies like video tutorials, Augmented Reality (AR) applications, instant automated checking of home works and gaming setup are used for training spatial comprehension and developing practical skills in engineering education.

10. References

[1] Sacks R Eastman C Lee G and Teicholz T 2018 *BIM Handbook: A Guide to Building Information Modeling for Owners, Designers, Engineers, Contractors, and Facility Managers* (Wiley 3rd ed) p 688

[2] Immonen A and Saaksvuori A 2013 *Product Lifecycle Management* (Springer Science & Business Media) p 223

[3] Aram S and Eastman C 2013 Integration of PLM solutions and BIM systems for fhe AEC industry *30th Int. Symp. on Automation and Robotics in Construction and Mining (ISARC 2013): Building the Future in Automation and Robotics* 11–15 Aug 2013 Montréal, Canada pp 1582–91

[4] Sun L P Siklander P and Ruokamo H 2018 How to trigger students' interest in digital learning environments: A systematic literature review *Seminar.Net (Electronic Materials vol 14) pp 62–84*

[5] Martín-Gutiérrez J Saorín J L Contero M and Alcaniz M AR_Dehaes: An educational toolkit based on augmented reality technology for learning engineering graphics *2010 10th IEEE Int. Conf. on Advanced Learning Technologies* 5–7 Jul 2010 Sousse, Tunisia pp 133–37

[6] Veide Z and Strozheva V 2019 Triggering the students' positive attitude for the studies of engineering graphics courses through the augmented reality content *Proc. of the 12th Int. Scientific and Practical Conf. "Environment. Technology. Resources"* 20–22 June 2019 Rezekne, Latvia pp 242-46

[7] Moor K Jones K and Frazier S R 2017 Engineering education for generation Z *American J. of Engineering Education* **8** pp 111–26

[8] Jurane I 2013 Engineering education improvement opportunities using computer games *Proc. of the Int. Scientific Conf. "Society, Integration, Education" Book Series: Society Integration Education* Vol I 24–25 May 2013 Rezekne, Latvia pp 100–9

[9] Branoff T J and Dobelis M 2013 The relationship between students' ability to model objects from assembly drawing information and spatial visualization ability as measured by the PSVT:R and MCT *Proc. of the 120th ASEE Annual Conference and Exposition* 23–26 Jun 2013 Atlanta, Georgia, USA pp 1–9

[10] Barr R 2012 Engineering graphics outcomes for the global engineer *Proc. of the 15th Int. Conf. on Geometry and Graphics* 1–5 Aug 2012 Montreal, Canada pp 54–63

[11] Hartman N W Bertoline G R Wiebe E N and Ross W A 2018 *Loose Leaf for Fundamentals of Graphics Communication* (McGraw-Hill Education) p 792

[12] Dobelis M Branoff T J and Nulle I 2012 Quantitative assessment of the students' engineering graphics literacy via modeling objects from assembly drawing information *Proc. of the 15th Int. Conf. on Geometry and Graphics* 1–5 Aug 2012 Montreal, Canada pp 136–147

[13] Garland A P 2018 Graderworks. Your SolidWorks grading assistant (*Electronic Materials*)

[14] Webster R and Ottway R 2018 Computer-Aided Design (CAD) certifications: Are they valuable to undergraduate engineering and engineering technology students? *J. of Engineering Technology* **35** (2) pp 22–32

IOP Conf. Series: Materials Science and Engineering **660** (2019) 012014 doi:10.1088/1757-899X/660/1/012014

Occurrences in Czech Railway Infrastructure and their Relation to the Economic Efficiency of Investments

V Hromádka[1], J Korytárová[1] and E Vítková[1]

[1]Brno University of Technology, Faculty of Civil Engineering, Veveří 331/95, 602 00 Brno, Czech Republic

E-mail: hromadka.v@fce.vutbr.cz

Abstract. The paper focuses on the evaluation of impacts of occurrences in the Czech railway transport system on social and economic benefits and financial efficiency of investment projects in railway infrastructure. The evaluation of financial and economic efficiency of investment projects financed from the funds of the European Union is based on the Guide to Cost-Benefit Analysis of Investment Projects by European Commission from 2014. In the Czech Republic, this Guide has been complemented, for the purposes of economic and financial evaluation of projects in transport infrastructure, by the Departmental Guideline for the Evaluation of Economic Effectiveness of Transport Construction Projects by the Ministry of Transport of the Czech Republic which specifies economic evaluation of projects of roads and highways, railways and waterways important for traffic. The Departmental Guideline defines benefits which can be included into the economic evaluation, e.g. savings in the operation costs, the decrease in costs connected with accidents, the decrease in costs connected with the travel time or negative externalities usually related to impacts on the environment. However, in the case of the railway infrastructure, benefits connected to the increase in safety and fluency of the transport due to introducing security and information systems are not taken into account, even if these investments form a standard part of most railway infrastructure investment projects. The objective of the paper is to identify both the number and the structure of railways occurrences, which can influence effective functioning of railway transport for the purpose of their subsequent research in order to take into account decrease in their use in economic analysis. The methodology of the research is based on detailed analysis of individual occurrences which happened on Czech railways during past three years. The output of this paper provides a detailed statistic of occurrences including damage caused and propose recommendations for the next steps which should lead to the quantification of relevant benefits.

1. Introduction

The implementation of public transport infrastructure projects is always linked to the need of large amount of financial resources. In the Czech Republic, these funds are usually provided by the State Transport Infrastructure Fund. In order to provide these funds, the investor (e.g. the Road and Motorway Directorate, the Railway Infrastructure Administration or the Directorate of Waterways) must, among other duties, prove the effectiveness of the project which funding is being applied for. Methodologically, the processing of economic evaluation of projects in transport infrastructure is covered by the Departmental Guideline of the Ministry of Transport of the Czech Republic. This Guideline is detailed and addresses many aspects of economic analysis of transport infrastructure projects in all transport modes, i.e. road, rail and water. However, some aspects are not dealt with in more detail. This concern, among other things, the economic evaluation of railway infrastructure projects, in particular evaluation of the benefits related to the increase in safety and reliability of the railway infrastructure. The implementation of measures leading to the increase in safety and reliability of the railway infrastructure forms an integral part of most railway transport constructions, however, the present methodological documents do not allow to adequately evaluate the benefits related to the implementation of these measures. This article can be considered as an introductory text to a broader project task which aim will be the evaluation of the benefits mentioned and their reflection in the already existing methodological materials. This paper deals with the analysis of extraordinary events (occurrences) which happened in the 2015–2017 period in the relation to the railway infrastructure in the territory of the Czech Republic which were registered and investigated by the Railway Inspectorate of the Czech Republic. The purpose of the research is to identify key types of occurrences, determine their socio-economic impacts and define the amount of damage that could be prevented by the timely implementation of measures aimed at increasing safety and reliability of the railway infrastructure. Part of the results of the research is designing the subsequent procedure within the topic which the research deal with.

2. Literature review

While the issue of assessing socio-economic benefits in the form of savings in costs related to the occurrences related to the increase in the level of security and reliability of the railway infrastructure is not currently being dealt with in detail, resent publication have activity paid significant attention to the issue of railway occurrences and their causes. "There have been numerous international studies that have identified error types using an alternative approach to the analysis of occurrence reports. These studies describe and analyse railway workers' tasks and consequently identify and classify worker errors and factors associated with those errors" [1] There exists „the motivation to better understand why these accidents occur, and how they can be prevented" [2]. The paper [3] focused on Application of Human Factors Analysis and Classification System (HFACS) to the rail safety of the line incidents and concludes that „Minor safety incidents on the railway cause disruption, and may be indicators of more serious safety risks" [3]. The relations between latent and active errors are the subject of the paper [4]. The authors are of the opinion that "the effectiveness and the safety of railway operations depend on many factors including rail traffic rules, equipment reliability, general and safety management, and other human factors, and latent errors pose the greatest risk to the system safety" [4]. The paper [5] concentrates on the conditions affecting safety on the railway from the point of view of train drivers' experiences and perceptions and concludes that the most beneficial for the railway safety are rail traffic management systems which "were seen as the primary safeguards within the railway system and were considered the predominant reason for the decrease in crashes" [5] Klockner [6] looks on the railway occurrences from the socio-economic point of view and states that "the new assumption and challenge calls for the understanding that accidents are complex processes involving the entire socio-technical system" [6].

It is clear from the above-mentioned quotations that railway occurrences are analysed in detail in order to find the crucial causes and consequently the means to prevent them. It can be assumed that the reduction in the number of occurrences will lead to more efficient operation of the railway transport and that measures aiming at reducing the consequences of occurrences can be quite effective from an economic point of view. This article focuses specifically on identifying the possibilities how the effectiveness of measures aiming at reducing the number of occurrences can be determined.

In general, the Guide to Cost-Benefit Analysis of Investment Projects [7] issued by the European Commission deals with the public sector project efficiency assessment. This document provides basic CBA analysis implementation procedures for public sector projects, especially projects funded by the European Union. The specifics of CBA analysis implementation for individual public sector areas are demonstrated by several case studies. The key document for the projects implemented in the Czech Republic in the field of transport constructions is the Departmental Guideline for the Evaluation of the Economic Effectiveness of Transport Construction Projects [8] issued by the Ministry of Transport of the Czech Republic. This document builds directly on the Guide [7] and develops CBA principles, particularly for the implementation of transport infrastructure projects including road and highway projects, railways and waterway transport projects.

3. Methodology

The paper focuses on the issue of evaluating the benefits related to the increase in safety and reliability of the railway infrastructure. The subject of the article is an initial analysis focused on assessing the socio-economic impacts of railway occurrences. The outcome of the text is the interpretation of the results obtained and formulation of recommendations for further procedure for evaluating the benefits associated with the implementation of the measures with the purpose to increase reliability and safety of the railway infrastructure. These measures can be in the form of the station security equipment, the track-side security equipment and the train security equipment. All these measures become common part of the project in the railway infrastructure, however benefits connected with these measures are not methodologically included in the economic analysis.

The basic data source was the database of final reports from the investigation of occurrences registered and investigated by the Railway Inspectorate of the Czech Republic [9]. For the purpose of the article, final reports from the 2015–2017 period were used. The authors created a database based on these final reports including the occurrence registration number, the description of the occurrence, the number of people injured or killed, the material damage caused within the occurrence and expert estimate whether the occurrence would not happen if equipment increasing safety and reliability was installed in the corresponding railway section. This expert estimate were carried out by members of the project team participating on the project cited in the Acknowledgement according to the experiences with occurrences in the Czech railway system and the description of individual occurrences taken from final reports from the investigation of occurrences. An expert estimate was made by assigning it by one of the following descriptions:

- Certainly not,
- Probably not,
- Maybe yes,
- Probably yes,
- Definitely yes.

Depending on the description of the occurrence, the investigated events were divided into five groups according to the prevailing cause of their emergence. The following groups of occurrences have been identified:

- Unauthorised train journey,
- Trains collision,
- Collision with a person,
- Technical problem,
- Derailment of the railway vehicle.

Such structured occurrences were further analysed. Firstly, a comprehensive analysis of the occurrence frequency and their impacts (injury, death, material damage) divided according to the occurrence predominant cause was made.

Secondly, only those occurrences from the database, for which the above-mentioned expert estimate was identified as "probably yes" and "certainly yes" were taken into account. The data analysis identified the frequency of these occurrences and the amount of damage to health and property that could have been prevented by installing equipment for improving safety and reliability of the railway infrastructure.

Thirdly, the evaluation of the damage to the health was carried out using the data determined in the subchapter [8] and the total amount of socio-economic damage related to the occurrence emergence, which could have been prevented by introducing appropriate measures for increasing safety and reliability of the railway infrastructure, was identified.

4. Results and discussion

Occurrences on the Czech railways recorded and investigated by the Rail Safety Inspection which is responsible for investigating the causes and circumstances of the occurrence emergence were analysed within the case study. The analysis carried out within the case study included all the investigated occurrences from the 2015–2017 period, except for occurrences in the form of a collision of the railway vehicle with the other road users at the railway crossings. In total, 56 occurrences were included in the analysis (27 from 2017, 11 from 2016 and 18 from 2015). As a result of these occurrences, a total property damage amounted at CZK 380,846,492 was created during the 2015–2017 period, six people died and 89 people were injured. Table 1 lists the damage caused by occurrences divided by the predominant cause of the occurrence emergence.

Table 1. Damage caused by occurrences in the 2015–2017 period.

Predominant cause of the occurrence emergence	Number of occurrences	Number of deaths	Number of injured	Material damage (CZK)
Unauthorised train journey	17	2	38	236,637,007
Trains collision	15	1	42	74,099,216
Collision with a person	5	3	2	5,400
Technical problem	4	0	7	26,662,065
Derailment of the railway vehicle	14	0	0	43,461,274

Source: authors' own work according to [9]

The paper focuses on the evaluation of the benefits related to the implementation of equipment increasing safety and reliability of the railway infrastructure. For this reason, in the case of the occurrences mentioned above, it was researched whether these events would not happen if the measures to improve safety and reliability of the railway infrastructure were implemented in the corresponding area. The above-mentioned occurrences were assessed by the authors as to whether or not the above-mentioned measures for increasing safety and reliability could have prevented the occurrence from happening on the scale defined in the section 3.

For further analysis, only those occurrences with results "probably yes" and "definitely yes" in the evaluation were taken into account. A summary of the information on these occurrences is given in Table 2.

Table 2. Occurrences influenced by the implementation of measures for increasing safety and reliability.

Predominant cause of the occurrence emergence	Number of occurrences	Number of deaths	Number of injured	Material damage (CZK)
Unauthorised train journey	6	1	34	203,494,987
Trains collision	4	0	22	35,752,315
Collision with a person	0	0	0	0
Technical problem	0	0	0	0
Derailment of the railway vehicle	0	0	0	0

Source: authors' own work according to [9]

Building on the materials and methodologies for cost-benefit analysis (CBA) processing [7], [8], a transformation of socio-economic impacts in the form of health damage or loss of life on cash flows could be made. As already mentioned in the methodological section, the economic loss shown in Table 3 can be assumed in a simplified way in relation to the damage to health or life.

Table 3. Occurrences influenceable by implementation of the measures for increasing safety and reliability.

Damage as a result of the occurrence	Socio-economic impact (CZK)
Death	20,790,000
Damage to health	942,053

Source: authors' own work according to [7, 8]

In relation to the previous Tables 1 to 3, it is possible to determine the overall socio-economic impacts caused by occurrences in the 2015–2017 period for all investigated occurrences as well as for all occurrences which could have been prevented by the timely implementation of measures leading to increase in safety and reliability of the railway infrastructure. These impacts are shown in Table 4.

Table 4. Overall socio-economic impacts caused by occurrences in the 2015-2017 period.

Predominant cause of the occurrence emergence	Total socio-economic impacts (CZK)	Socio-economic impacts which could be prevented (CZK)
Unauthorised train journey	314,015,021	256,314,789
Trains collision	134,455,442	56,477,481
Collision with a person	64,259,506	0
Technical problem	33,256,436	0
Derailment of the railway vehicle	43,461,274	0
Total occurrences	589,447,679	312,792,270

Source: authors' own work according to [7], [8], [9]

From the information listed in Tables 1 and 2 it can be deduced how many occurrences emerged during the researched period, what damage to health and life occurred and what damage could have been prevented if measures to increase safety and reliability had been implemented in the past. From the point of view of the number of occurrences, there was no significant reduction, previously implemented measures for increasing safety and reliability would have prevented about 18% of occurrences. In terms of the total material damage, however, the impact was more significant, in the case of implemented measures for increasing safety and reliability of the railway infrastructure, 62% of the material damage, one loss of life and 63% of injuries would not happen.

After the conversion to the socio-economic impacts given in Table 4, it can be concluded that as a result of the previous implementation of the measures for increasing safety and reliability of the railway infrastructure, a 53% reduction in the socio-economic costs related to the occurrence emergence on the railway could be expected.

The results of the analysis carried out can be taken as informative only, as the analysis itself did not cover all the occurrences that happened on the railway. The analysis included only those occurrences that were investigated by the Railways Inspectorate of the Czech Republic and thus were considered more important. However, it is clear from these results that the implementation of measures for increasing safety and reliability of the railway infrastructure can bring a large number of socio-economic benefits, which consist in particular in the savings related to damage resulting from occurrences in relation with the health and property. Evaluation of these benefits can therefore bring a very interesting input into the economic analysis to assess the effectiveness of railway projects. The evaluation of these benefits and their reflexion into economic analysis will be the aim of further research in this field. However, the research will not focus only on the information obtained from the Railway Inspectorate of the Czech Republic, but attention will be paid also to a much more detailed database of occurrences managed by the Railway Infrastructure Administration.

5. Conclusions

The paper focuses on the socio-economic evaluation of measures dealing with the railways infrastructure which aim at increasing safety and reliability of the railway infrastructure. The article represents an initial article within a larger research task and brings partial conclusions in relation to the analysis of occurrences in relation to the railway infrastructure and identification of possible socio-economic benefits of implementing measures for increasing its safety and reliability. The data from the final reports on the investigation of the occurrences, which was carried out by the Railway Inspectorate of the Czech Republic in the 2015–2017 period, were used for the analysis. The analysis shows not only the frequency of the investigated occurrences during the period under research and the amount of damage caused by these occurrences, but also the number of occurrences and the level of socio-economic impact that could have been prevented by the possible implementation of measures for increasing safety and reliability of the railway infrastructure. This paper brings partial conclusions in the researched area and proposes further procedure for dealing with the benefits evaluation related to the implementation of measures for increasing the safety and reliability of the railway infrastructure proposed and implemented within the framework of the railway infrastructure projects. The final objective of the research is to complete the methodology for the assessment of the effectiveness of projects in the railway infrastructure.

Acknowledgement

This paper has been worked out under the project of the Technology Agency of the Czech Republic "TL02000278 Evaluation of increased safety and reliability of railway infrastructure after its modernization or reconstruction".

References

[1] Baysari M, Mcintosh A and Wilson J Understanding the human factors contribution to railway accidents and incidents in Australia. *Accident Analysis and Prevention* [online], Elsevier, 2008, **40** (5), 1750-1757 [cit. 2019-02-18]. DOI: 10.1016/j.aap.2008.06.013. ISSN 0001-4575

[2] Read G J M, Lenné M G and Moss S A Associations between task, training and social environmental factors and error types involved in rail incidents and accidents, *Accident Analysis & Prevention,* Volume 48, 2012, Pages 416-422, ISSN 00014575, DOI: 10.1016/j.aap.2012.02.014

[3] Madigan R, Golightly D and Madders R Application of Human Factors Analysis and Classification System (HFACS) to UK rail safety of the line incidents, *Accident Analysis & Prevention,* Volume 97, 2016, Pages 122-131, ISSN 0001-4575, DOI: 10.1016/j.aap.2016.08.023

[4] Zhou J-L and Lei Y Paths between latent and active errors: Analysis of 407 railway accidents/incidents' causes in China, *Safety Science*, Volume 110, Part B, 2018, Pages 47-58, ISSN 0925-7535, DOI: 10.1016/j.ssci.2017.12.027

[5] Forsberg R Conditions affecting safety on the Swedish railway – Train drivers' experiences and perceptions, *Safety Science*, Volume 85, 2016, Pages 53-59, ISSN 0925-7535, https://doi.org/10.1016/j.ssci.2015.12.015

[6] Klockner K and Toft Y Railway accidents and incidents: Complex socio-technical system accident modelling comes of age. *Safety Science* [online]. Elsevier, 2018, 110, 59-66 [cit. 2019-02-18]. DOI: 10.1016/j.ssci.2017.11.022. ISSN 0925-7535

[7] Sartori D *Guide to Cost-benefit Analysis of Investment Projects, Economic appraisal tool for Cohesion Policy 2014-2020.* European Commission, Directorate-General for Regional and Urban policy. 2014. ISBN 978-92-79-34796-2.

[8] Ministry of Transport of the Czech Republic *Departmental Guideline for the Evaluation of*

Economic Effectiveness of Transport Construction Projects. 2017. Retrieved from http://www.sfdi.cz/pravidla-metodiky-a-ceniky/metodiky/
[9] Rail Safety Inspection Office. *Database of investigated accidents and incidents and final reports*. Retrieved from http://www.dicr.cz/zaverecne-zpravy-z-mimoradnych-udalosti.

IOP Conf. Series: Materials Science and Engineering **660** (2019) 012015 doi:10.1088/1757-899X/660/1/012015

Application of multi-criteria method to assess the usefulness of a hydrotechnical object for floating housing

E Miszewska[1] and M Niedostatkiewicz[2]

[1] Gdańsk University of Technology, Faculty of Civil and Environmental Engineering, Department of Construction Management and Earthquake Engineering, 11/12 Narutowicza Street, Gdańsk, Poland
[2] Gdańsk University of Technology, Faculty of Civil and Environmental Engineering, Department of Concrete Structures, 11/12 Narutowicza Street, Gdańsk, Poland

E-mail: emiurban@pg.edu.pl

Abstract. This publication presents the analysis of three hydrotechnical objects located in the Municipality of Gdańsk with a view of mooring Floating Houses. The assessment of the adaptation of a hydrotechnical object has been carried out by a multi- criteria method AHP and using the main criteria such as: mooring system, communication with the mainland, availability of the utility networks, waste disposal and location of the parking spaces. The analysis allowed to conclude that there are no hydrotechnical objects in Gdańsk, which would be fully adapted to mooring of Floating Houses without additional infrastructure investments. The most adapted site out of the three ones considered in the Municipality of Gdańsk is location no. 1 – the Rybackie Pobrzeże.

1. Introduction

Both in Western Europe and in the United States of America, the phenomenon of living in Floating Houses has gained popularity at the beginning of the 1980s. It was the reaction to the growth in the wealth of societies whilst rising in the prices of land. The only exception is the Netherlands, where the popularity of Floating Houses has been increasing in the manner forced by nature. About 60% of the land territory of the Netherlands is below sea level [1, 2]. The consequence of the greenhouse effect is the climate change which causes sea level rise and more violent storms. That is precisely why in the Netherlands it has been decided that they will not fight the sea, but use its strength and adjust to it. The Dutch Government has implemented provisions for supporting floating housing.

In Poland floating housing is gradually growing more popular. The first Floating House was located in Wrocław in 2010. Since then, there has been many changes and public awareness has become great enough not only to accept Floating Houses, but to awaken an interest of increasing number of people.

The objective of this work is the answer to the question if the existing hydrotechnical infrastructure in Poland is prepared for the new type of construction, namely Floating Houses. The selection of the location of Floating Houses was considered in terms of the most important criteria for users: mooring system, communication with the mainland, availability of the utility networks (water installation, sewage, electricity), waste disposal and location of the parking spaces. The publication analysis three selected hydrotechnical objects (waterfronts) located in the centre of Gdańsk, applying multi-criteria method AHP (Analytic Hierarchy Process). AHP is one of the multi-criteria decision making technique and is used to solve complex problems in various areas for example in political science, sociology and management process to evaluate multiple types of projects or in complex technical-economic issues. Due to combining the concepts of mathematics and psychology it is also one of the fastest developing as well as the most famous methods in the world. One of the advantages of AHP is pairwise comparisons of selected elements and attributing them to a scal [3].

2. The technical parameters and construction of Floating House

The object concerned is the floating dwelling, whereas the construction features described are most commonly found amongst the objects of this type in Poland.

A Floating House is settled on a special floating system, which is based on pantoons with dimensions of 2,5- 3.0 meters in width and 10-12 meters in length, pooled in longitudinal or lateral setup. The frame of "the surface" part is made out of hot tinned steel mouldings, protected by a corrosion- resistant coating as well as impregnated and multilayer glued wood. The hull of the fame building kit can be made out of "aqua panels", DLH laminboard, panels of powder coated aluminium sheets, treated wood or its conglomerate of PCV. The roof was designed as flat, thermally insulated with panels of Styrodur, covered with water insulation. Rainwater from the roof is conveyed to drainpipes hidden in the facade. The door and window carpentry is made of wood or powder coated aluminium profies. Insulating is made of Styrodur, according to the thermal regulations for the residential houses. The board of the floating object is made of tropical wood or wood conglomerate and PCV. The rails protecting against falling overboard are made of stainless steel with wooden or steel handrails. The object is equipped with:

- electrical installation,
- heating,
- water and sewage system [2, 4].

3. The description of the muli-criteria decision method AHP

The method of the hierarchical analysis of a problem (AHP- Analytic Hierarchy Process), developed by Saaty [5-7], is mostly used to support the analysis connected with selecting a decision option. The options could be the physical objects, such as machines, products, as well as some states represented by design or implementation variants leading to achievement of a certain state (quality, safety, risk). By AHP method it is also possible to carry out a diagnostic or comparative assessment of the objects in question, because the assessment of variants constitutes a pre-decision-making stage.

AHP method recognizes a multi - criteria approach based on compensatory strategy of preferences modelling and on the assumption of variants comparability. Taking into account the preferences of an assessor, which decides about the subjectivity of the assessments, underpins the multi- criteria approach that treats these preferences as a natural phenomenon for the assessments made by a human, in contrast to the measurement which is objective.

AHP method considers the specificity of psychological evaluation processes which have a relational and hierarchical nature. Numerous applications of the method in supporting economical, technical and social decisions confirm its usefulness, especially in the applications where most criteria are qualitative and the experience of an assessor constitutes the main source of the subjective assessments [8].

AHP methods popularity and strong position in the largest international databases of high-scoring peer-reviewed journals, such as, for example, ScienceDirect, Wiley, Scopus. When comparing the frequency of articles on AHP with other decision support methods, it can be concluded that references to the AHP method are most common [9].

4. Modelling using hierarchical method

Analytic Hierarchy Process (AHP) consists of 4 stages:

Step 1. Prioritizing a problem - the aim of this stage is the detailed description of a problem, identification of the participants, defining the main objective and expectations. Thereafter, a decomposition of a problem is undertaken in the form of the primary objective, the main and partial factors and variants considered, which generate some fulfillment of aims function on particular levels of the hierarchical model.

Step 2. The assessment of the criteria by comparison with pairs - it is conducted by a decision- maker, who compares the pairs with each other using the criteria and the criteria in relation to the primary objective on the basis of subjective decision as to which of the criterion and to what extent is more important than the other. The relations between the particular elements are established on the basis of 9- point scale presented in Table 1 [7, 10].

Table 1. Fundamental scale of absolute numbers [11].

Identysity of Importance	Definition	Explanation
1	Equal Importance	Two activities contribute equally to the objective
2	Weak or slight	
3	Moderate importance	Experience and judgment slightly favor one activity over another
4	Moderate plus	
5	Strong importance	Experience and judgment strongly favor one activity over another
6	Strong plus	
7	Very strong or demonstrated importance	An activity is favoured very strongly over another; its dominance order of affirmation
8	Very, very strong	
9	Extreme importance	The evidence favouring one activity over another is of the highest possible order of affirmation
1.1-1.9	When activities are very close a decimal is added to 1 to show their difference as appropriate	A better alternative way to assigning the small decimals is to compare two close activities with other widely contrasting ones, favouring the larger one as needed a little value over the smaller one using the 1-9 values.
Reciprocals of above	If activity *i* has one of the above nonzero numbers assigned to it when compared with activity *j*, then *j* has the reciprocal value when compared with *i*	A logical assumption
Measurements from ratio scales		When it is desired to use such numbers in physical applications. Alternatively, often one estimates the ratios of such magnitudes by using judgment

To make pair comparison of the selected criterions it is essential to put them in the diagonal matrix type (n x n). The comparison is made by identifying the impact of the element on the left side of the matrix to the elements at the top of the matrix. Below the main diagonal there are the inverse of the pairwise comparisons, the formula of matrix A [3]:

$$A = \begin{bmatrix} 1 & a_{12} & \dots & a_{1n} \\ \frac{1}{a_{12}} & 1 & \dots & a_{2n} \\ \vdots & \vdots & \vdots & \vdots \\ \frac{1}{a_{1n}} & \frac{1}{a_{2n}} & \dots & 1 \end{bmatrix}, \tag{1}$$

Due to the inverse of the pairwise comparisons, the i-th row is the inverse of the i-th column, so there is a relation:

$$Aw = nw \tag{2}$$

where:

w – vector of weights $w_1, w_2, \ldots, w_n$

Elements of the vector of weights w are priorities vector of the various criteria because of the overall goal:

$$w = \sum_{j=1}^{n} w_j a_{ij} \tag{3}$$

Step 3. Defining mutual preferences (scales) with respect to criteria and decisional variants- having built the matrix, the calculations of criteria scales are made. The standardized rows of matrix are aggregated and the proprietary vector of matrix is calculated:

$$\lambda_{max} = \frac{1}{w_i} \sum_{j=1}^{n} a_{ij}\, w_j \tag{4}$$

A is consistent if and only if [3, 12]. The second factor necessary to obtain the AHP method is the CI (Consistency Index). It is the negative average of the other roots of the characteristic polynomial of A:

$$CI = \frac{\lambda_{max} - n}{n-1} < 0{,}10 \tag{5}$$

The last factor is CR (Consistency Ratio). If the ratio of CI is significantly small, the estimate of w can be accepted. CR is determined by the formula:

$$CR = \frac{CI}{RI} \tag{6}$$

where:
RI- random index [3, 8, 13].

In the case where λ max = n and CI = 0, CI index is calculated with regard to random index RI, which is the average value of CI for a large number of random generated matrix of comparisons. The value of RI is provided in the tabulated form [7, 10].

Table 2. Value of index RI [8, 10].

n	2	3	4	5	6	7	8
RI	0	0,52	0,89	1,12	1,25	1,40	1,45

Step 4. The analysis of the selected results – choosing the best variant, which would address the main objective.

5. Selection criteria for the hydrotechnical objects
The hydrotechnical objects appointed for further analysis using AHP method were selected based on three key criteria, which are as follows:

- the mooring zone- the object should be situated in the living area,
- the owner of the hydrotechnical object- it is recommended that the owner is a municipal or state entity,
- the access to a floating object from land- there should be easy and unrestricted access provided to an object.

In the light of the criteria stated above, three locations were selected:
- location no. 1 - The Rybackie Pobrzeże,
- location no. 2 - Stara Motława,
- location no. 3 - Nowa Motława.

Figure 1. Map of the location options [14].

The locations presented in Figure 1 are situated in the city center, in the living area. In all three cases, a municipal or state entity is the owner of the quay.

The Rybackie Pobrzeże is located by Targ Rybny in Gdańsk, on the left bank of the river Motława Gdańsk. It has a plate construction set up on the sheet piling type Larssen II and a mooring trestle made of prefabricated piles. It is equipped with moorage piles, electrical and water installations, however, there are no installation of sewage collection from a floating object. The quay is sited by a frequented fairway and is used as a boulevard by the inhabitants. It is a densely urbanized area under protection of the conservator of monuments [2].

Stara Motława is positioned on the waterfront, between Most Krowi and Most Zielony on the left bank of the river Stara Motława, which has been made navigable again since 2012. The girt structure quay of 99 meters in length. The upper edge of the quay is covered with a steel angle. The quay is equipped with a steel railing, a discharge pipe probably used as rainwater drainage system, continuous wooden buffer lin and two falls in the quay, one of which was used for mooring of a water tram stop. The site is not developed with any utility networks, it only holds some moorage piles [2].

Nowa Motława is a quay located between Podwale Przedmiejskie Street and Strągiewna Street, along Szopy Street. It is the girt structure quay, which is not equipped with any installation. In the quay there is a fall equipped with the moorage piles and protective railings, which hinder free communication with the mainland [2].

6. Preliminary assumptions and description of the criteria

In order to compare the assessment criteria of the examined hydrotechnical objects, the main criteria had been selected, then they were divided into partial criteria [5]. The main criteria were demonstrated based on the technical solutions and the expectations of a Floating House user. These are as follows:
- the mooring system
- communication with the mainland
- availability of utility networks
- the place for municipal waste disposal
- a parking space [2].

The first main criterion is safe mooring, that is mounting a vessel or other floating object to the mooring devices of hydrotechnical objects by means of a mooring line. There are various mooring systems available, these are:
- mooring to steel piles, which are driven into the seabed,
- mooring using mooring booms, that is special beams, which hold the house in the safe distance from the shore,
- system of anchors Seaflex allows the mooring of a floating object by placing four concrete blocks around a Floating House on the bottom of the reservoir. They are connected with a floating platform using a system of resilient ties, which limit the impact on wave motion and hold the object in a stable position in relation to the shore or a mooring pier [2].

Another important criterion is communication between the users of a vessel and the mainland. Communication with the land is secured through:
- additional devices, for example a gangway used if the distance from a vessel to the edge of the quay is significant or if the difference between the level of the quay and the deck of a Floating House prevents independent communication with the mainland,
- falls in the quay, applied in a case of the mooring to "high quay", where the difference between the level of the deck and the edge of the quay is significant, in addition the construction of a hydrotechnical object has a special decrease in the construction in order to allow the mooring of the smaller vessels,
- supplementary devices or the decrease in the quay- used when a decrease in a quay does not ensure free and safe communication with the mainland,
- without the decrease in the construction of a quay and additional devices.

Providing users with the access to utility networks, satisfying their basic needs such as, access to electricity, heating, water and sewage disposal is vital and should be guaranteed in the place of permanent residence.

Therefore, a hydrotechnical object ought to be equipped with:
- electrical installation- a possibility of connecting to the network
- water and sewage system- on floating objects water and sewage system is treated separately as water intake and sewage discharge. Whilst during the summer season water intake does not raise any difficulties, in wintertime most water installations are emptied for technical reasons, so that freezing water does not destroy them. In such cases a user should have an opportunity of water supply delivered by the water trucks. Sewage can be disposed of via a special installation at quays or by storing waste in containers on a vessel and collecting it by a slurry spreader when the containers are filled.

Each household regardless its location, size or number of people, generates municipal waste. It is required by law to sign an agreement concerning waste disposal. A user of a floating house should have an opportunity to select a waste collection point to service the collection of municipal waste.

Currently, in every flat/house there is a minimum of one car and the inhabitants of floating objects have the need to use and own such means of transport too. It is not essential for functioning to ensure a parking space nearby a vessel, but it significantly improves the comfort of life [2].

7. Hierarchical model

The analysis using AHP method was conducted in Excel spreadsheet.

7.1. Prioritisation of a problem
A decision- making problem in AHP method is presented in a hierarchical way, which illustrates grouping all the criteria into the main and the partial ones. The following table presents 5 main and 15 partial criteria for the analysed issue.

Table 3. A summary of main and partial criteria.

1.	The mooring system
1.1.	Mooring Piles
1.2.	Mooring Booms
1.3.	Seaflex anchors
2.	Communication with mainland
2.1.	Using auxiliary devices
2.2.	Decrease in the construction of a quay, configuration with or without a gangway
2.3.	Decrease in the construction of a quay, configuration with or without a gangway, with an obstacle in a form of a railing
3.	Utility networks
3.1.	Water installation
3.2.	Sewage
3.3.	Electricity
4.	A waste collection point
4.1.	An agreement combined with a nearby property
4.2.	Necessity to provide underground rubbish bins
4.3.	Opening an individual waste collection point
5.	Parking spaces
5.1.	Free public car park
5.2.	Paid public car park
5.3.	No car park in close vicinity to the quay

7.2. The assessment of criteria by comparison using pairs
The matrices of pairwise comparisons of the criteria given by the specialists of floating objects construction, sailors and users of floating houses is shown in Table 4 [5], and for the matrix for level 1 is shown in Table 5 [5].

7.3. The analysis of the selected results
As demonstrated by the results of the AHP analysis, it has been proven that The Rybackie Pobrzeże-location no. 1 is the most usefulness of a hydrotechnical object for floating housing. The judges pointed out that the most important criterion in the order are: utility networks, the mooring system and communication with mainland. Criteries like parking spaces and a waste collection point has the least effect.

Table 4. Comparision matrices and local priorities – level 2 [5].

The mooring system	1.	2.	3.	Priority vector	Communication with mainland	1.	2.	3.	Priority vector
1.	1	7	4	0.69552	1.	1	$\frac{1}{6}$	$\frac{1}{5}$	0.07796
2.	$\frac{1}{7}$	1	$\frac{1}{4}$	0.07543	2.	6	1	3	0.63484.
3.	$\frac{1}{4}$	4	1	0.22905	3.	5	$\frac{1}{3}$	1	0.28720
λ_{max}=3.07642, CI=0.0382, CR=0.07348					λ_{max}=3.09402, CI=0.0470, CR=0.09040				
Utility networks	1.	2.	3.	Priority vector	A waste collection point	1.	2.	3.	Priority vector
1.	1	8	6	0.76116	1.	1	$\frac{1}{4}$	3	0.21092
2.	$\frac{1}{8}$	1	$\frac{1}{3}$	0.07261	2.	4	1	7	0.70494
3.	$\frac{1}{6}$	3	1	0.16623	3.	$\frac{1}{3}$	$\frac{1}{7}$	1	0.08414
λ_{max}=3.0735, CI=0.0368, CR=0.07069					λ_{max}=3.0324, CI=0.0162, CR=0.03112,				
Parking spaces	1.	2.	3.	Priority vector					
1.	1	$\frac{1}{3}$	$\frac{1}{7}$	0.08096					
2.	3	1	$\frac{1}{5}$	0.18839					
3.	7	5	1	0.73064					
λ_{max}=3.0626, CI=0.0313,CR=0.06024									

Table 5. Pairwise comparison for level 1 [5].

	1. The mooring system	2. Communication with mainland	3. Utility networks	4. A waste collection point	5. Parking spaces	Priority vector
1.	1	3	$\frac{1}{2}$	5	8	0.3034
2.	$\frac{1}{3}$	1	$\frac{1}{4}$	3	5	0.1413
3.	2	4	1	6	9	0.4534
4.	$\frac{1}{5}$	$\frac{1}{3}$	$\frac{1}{6}$	1	3	0.0683
5.	$\frac{1}{8}$	$\frac{1}{5}$	$\frac{1}{9}$	$\frac{1}{3}$	1	0.0335
λ_{max}=5.155098, CI=0.0388,CR=0.03462						

8. Summary

The AHP method although time - consuming is the most popular tool for the analysis of muli-criteria issues [3]. Its advantages such as flexibility, easiness of use, objectivity of variant selection, comparison of both - qualitative and quantitative factors [10], outweigh the disadvantages which are reflection of reality through the hierarchical model, relativity of experts' opinions and problems with application of 9 - point scale of Satty.

This publication analysed three hydrotechnical objects located in the Municipality of Gdańsk. The main criteria adopted for the analysis by a multi- criteria method AHP were: mooring system, communication with the mainland, availability of the utility networks, waste disposal and location of the parking spaces. The locations were selected based on a location in the urban zone, the owner of a public or municipal institution and easy access to the quay. The assessment was carried out in relation to the technical solutions and the expectations of the user. The above analysis demonstrated that in the area of the Municipality of Gdańsk concerned, there are no hydrotechnical objects which would be fully adapted to mooring of Floating Houses without additional infrastructure investments. Out of the main criteria, the most relevant was utility networks, and the least important was parking spaces. As a result, it has shown that the Rybackie Pobrzeże is the most adapted site for mooring Floating Houses out of the three ones considered in the Municipality of Gdańsk.

References

[1] Zaremba K Dostęp (access 2013) Co to jest Dom Na Wodzie? (What is a house on the water?) www.domynawodzie.pl
[2] Miszewska-Urbańska E 2013 Młodzi naukowcy dla Polskiej nauki (Young scientists for Polish *science)* vol. 10 (Kraków: CREATIVETIME) pp 75-83
[3] Stoltmann A 2016 Application of AHP Method for Comparing the Criteria Used in Locating Wind Farm *Acta Energetica* **3** pp 144-149
[4] WATER HOME Floating Real Estate (access 2013) *www.waterhome.pl*
[5] Saaty T L 1990 How to Make a Decision: The Analytic Hierarchy Process Eur. J. Oper. Res. 48 pp 9-26
[6] Saaty T L 1994 Fundamentals of Decision Making and Priority Theory with the Analytic Hierarchy Process (Pittsburgh PA: RWS Publications)
[7] Saaty T L 1996 The Analytic Hierarchy Process: Planning, Priority Setting, Resource Allocation (Pittsburgh PA: RWS Publications)
[8] Downarowicz O, Krause J, Sikorski M and Stachowski W 2000 Wybrane metody ergonomii i nauki o eksploatacji (*Selected methods of ergonomics and operating science*) ed Downarowicz O (Gdańsk: Gdańsk University of Technology) pp 7-42
[9] Tułecki A and Król S 2007 Modele decyzyjne z wykorzystaniem metody Analytic Hierarchy Process (AHP) w obszarze transportu (Decision models with the application of Analityc Hierarchy Process (AHP) method in the transportation area) *Problemy Eksploatacji (Eksploatacion problem)* **2** pp 171-179
[10] Prusak A and Stefanów P 2014 AHP – analityczny process hierarchiczny. Budowa i analiza modeli decycyjnych karok po kroku (AHP - analytic hierarchical process. Construction and analysis of decision-making models step by step) *Warszawa Wydawnictwo C.H. Beck (Warsaw Publisher. C.H. Beck)*
[11] Saaty T L 2013 On the Measurement of Intangibles. A Principal Eigenvector Approach to Relative Measurement Derived from Paired Comparisons *Notices of the American Mathematical Society* **60** pp 192-208
[12] Babic Z and Plazibat N 1998 Ranking of enterprises based on multicriterial analysis *International Journal of Production Economics* **97** pp 29-35
[13] Adamus W and Gręda A 2005 Wspomaganie decyzji wielokryterialnych w rozwiązywaniu wybranych problemów organizacyjnych i menedżerskich (Multiple criteria decizion suport in orgazniational and mamnagement chosen problems solving) *Badania Operacyjne i Decyzje* (Operations Research and Decisions) **2** pp 5–36
[14] Google earth map (access 2019) *https://earth.google.com/web/@54.34937825,18.66029902,0.0059695a,1817.22706919d,35y,0h,0t,0r*

IOP Conf. Series: Materials Science and Engineering **660** (2019) 012016 doi:10.1088/1757-899X/660/1/012016

A review study on specific requirements for refurbishment of military buildings in cold climates

A Prozuemnts[1], A Borodinecs[1] and M Krizmane[1]

[1]Institute of Heat, Gas and Water Technology, Riga Technical University, Latvia

E-mail: anatolijs.borodinecs@rtu.lv

Abstract. In most cases there are no specific and clear design requirements defined in local building codes, the responsibility for appropriate design of military buildings therefore lies primarily on third party design service contractors that are hired upon a short term agreement, whereas government is primarily involved in financing stages and military personnel is solely involved in long term exploitation of these structures. Authors of this paper focus on analyzing and isolating specific requirements for refurbishment of military buildings with regards to their structural reliability, safety, endurance, fire and explosion-resistance, heating, ventilation and air-condition systems performance and efficiency, as well as adding energy efficiency, building performance optimization recommendations and measures in moderate and cold climate conditions.

1. Introduction

This study aims to define main parameters of military buildings to developed guidance for refurbishment of military buildings in cold climates. US army experience shows that energy efficient measures can be simply implemented also in military buildings and ensure energy reduction of almost 65% depending on climate zone [1,2] [3]. In addition, research [4] have shown that such building as military building have a strong potential to significantly reduce energy consumption. Average energy consumption of unclassified buildings in Latvia is 240 kWh/m^2. Some of the typical refurbishment and new construction solutions can be easily used also in some of military buildings while such specific types of buildings as shooting ranges, ammunition storage require different solutions. In addition to final energy consumption, the more advanced energy sources should be developed for such buildings [5] [6].

By its definition, a military building is any structure designed to house functions performed by a military unit. General types include:

- Administrative Facilities;
- Ammunition Storage Facilities;
- Fortifications, Defense structures;
- Hospitals;
- Housing, Barracks;
- Training Facilities;
- Utility Structures;
- Vehicle Repair, Maintenance, & Storage Facilities;
- Weapons and Ammunition Storage and Production Facilities and etc.

Military buildings and barracks, with composite deck floors lightly built on joists and beams, rigidly fixed to the columns resist the effect of blast (and seismic and hurricane load). The sheeting can be frangible (that is, will bend, deflect, wrap around its supporting steel) but not fall to bits and fly around.

Where blast resisting glass is used it must be held in place with steel mullions and transoms fixed strongly to defense structures frames and designed to bend plastically without fracturing.

Fortifications and Defense Structures are defense contractors for blast resistant military buildings and other critical infrastructure, specializing in purpose made force protection against a variety of threats including bomb blast protection and explosion mitigation solutions.

This category also includes defensive barracks, headquarters buildings, blast walls, cover from view screens, gates, hardened hangars, mortar resistant roofs, observation posts, hangars, rocket screens, road barriers, observation posts, dangerous goods stores, vehicle checkpoints and workshops.

Force protection is an important factor in peacekeeping and peacemaking operations. Defensive structures are therefore necessary. The military buildings should provide an outstanding safety for Headquarters, Barracks, Hospitals and Weapon and Ammunition Storage Facilities.

Blast protection of these facilities and the personnel using and protecting them can be enhanced with defensive structure techniques. A Blast Wall consists of steel supporting structures and steel reinforced bolt-on concrete panels. The concrete is placed lying flat in the prefabricated panels which are then lifted into position and bolted on.

Nothing can give total safety against forced access, but strong steel blast gates can improve security. A blast gate should preferably open outwards, and close to form a V-type barrier which will jam if forced from outside. They should be fully clad to give cover from view; or clad in solid steel plate to give protection against small arms and artillery shrapnel.
The blast gate (Figure 1) should be covered from an observation post on one side or both sides. It would be better if the gate can be surrounded by blast deflecting walls. It would be better still if the blast walls had a labyrinth or dual lock gate, so that in coming trucks enter, and stop surrounded by blast wall and gate. Such dual lock blast gates would be able to save many lives if methods of attack used in the past were repeated.

Defense Structures chain hedgehogs are a lightweight, transportable and effective standby gate.

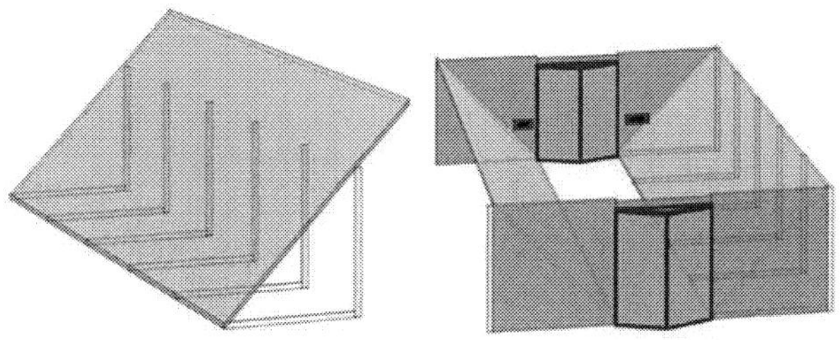

A blast wall **Lock gates with blast walls.**

Figure 1. Blast Wall solutions.

According to Latvia rules on Cabinet Regulation No 1001 Regulations Regarding the Acquisition, Registration, Recording, Possession, Transportation, Conveyance, Carrying, Sale of Weapons and Ammunition and Possession of Collections of Weapons, the external walls of the weapons depository shall be equivalent in durability to 200 mm thick reinforced concrete walls or brick external walls the thick-ness of which is not less than 510 mm. This issue significantly reduced possibilities to use wood frame constructions. However Cross-laminated timber CLT wood construction can meet before mentioned requirements. Structural insulated panel also can be used in some type of military constructions. Mainly structural insulated panel can be with steel or OSB skins.

Study [7] analyzed SIP panels with steel skin. This study has shown that application of steel skin thickness SIP significantly improves construction blast resistant.

OSB Structural insulated panel has a significantly lower mechanical properties in comparison to steel skin structural insulated panel. However their properties can be significantly improved by application of extra steel or fiberglass mesh. Study [8] has shown that strengthening the OSB skin SIP panel with 3 mm-thick glassfibre can increase the projectile impact resistance capacity to wind velocity 32 m/s.

2. Review of structural solutions

Explosions and blast can produce, in a very short time, an overload much greater than the design load of a building. Explosives or projectiles can cut or deform structures with chemical energy or kinetic energy. In spite of this buildings can and do survive such effects without collapse, if correctly designed to do so. On the other hand, structures which are not so designed can suffer rapid cumulative collapse. Cladding and glass can be detached and fly around, forming lethal weapons. Such debris is often the biggest cause of injury and death. Steps should be taken to maximize the distance from any attack using gates, barriers, chicanes and such like. Nothing can be guaranteed to eliminate all risks; but if the following design features were to be incorporated, many lives could be saved, and many structures and businesses would survive. In addition to blast risk, neighborhood fires can affect such buildings. In order to minimize risks of forest fires, buildings should be kept 42–43 meters from the forest fires [9].

Floors must be prevented from 'falling off' their supports. If pre-cast concrete planks are used they should have sufficient bearing; but they should not depend on bearing and gravity to stay in place: they should be made continuous with rebars between adjacent planks and preferably be made continuous with the supporting beams, using shear connectors. However, a more robust detail is to pour continuous concrete slabs on to composite style decking which is itself continuous over 3 or so joists; such slabs should be poured so that they encapsulate the main beam to which the joists are fixed, and around the columns. For fast erection of such type of buildings, prefabricate panel can be used. For example metal sandwich panel can insure necessary strengthens and energy efficiency [7] [10]. Also SIP panels can be adopted for these purposes [8].

Joists should be made continuous themselves, through every main beam and wherever they coincide with outer columns. The joints should exceed the plastic capacity of the joists so that, if they fail, it is by plastic hinge and not by joint failure. Where joists are attached into the webs of outer beams no moment resistance is possible but there should be sufficient bolts to make shear failure unlikely before plastic hinges form in the outer main beam.

Main beams should be continuous across the structure and should have connections to the outer columns which exceed the plastic capacity of the main beam. This means that in the case of overload the beams deform, forming hinges, absorbing energy and taking time. Blast or shock loads will diminish in a very short time.

The main outer columns should remain elastic and strong enough to carry likely loads even when main beams attached to them form plastic hinges. Care should be taken that the shear capacity of the column should not be exceeded within the moment connection zone by the moment in the beam: this almost always requires hunched beam-to-column connections. In addition load bearing metal beam should be covered by fire resistant layer [11].

The ground to first floor columns carry the heaviest loads. They are always more vulnerable to attack. These columns are almost always longer than columns on other floors. They often have less stability because of gaps between them. And they often have no continuity below, as they sit on 'pinned' feet. So special care has to be taken: they need to be stronger; to have barriers to protect them; to have continuity at footings level with ground beams or slabs.

If all this continuity is achieved, even if a column or two are cut or deformed, the grillage of beams and joists and slabs at each floor throughout the building will continue to carry the loads. They may well deform substantially, joists and beams may well bend and form plastic hinges or act as a catenary net to share loads; but it would be exceptionally difficult to demolish such a building.

3. Analysis of HVAC requirements in indoor shooting ranges

Primary purpose of ventilation is to remove contaminants created during the firing of a weapon from the respiratory zones of those occupying the range. Exploding primers containing lead styphnate and friction from the lead slug against the gun barrel create airborne lead. Carbon monoxide and other contaminants are also created during the firing of a weapon. Protecting users from the inhalation of combustion-related contaminates is the key concern when designing an indoor shooting range [12]. The American Conference of Government Industrial Hygienists recommends a minimum airflow velocity of 0.25 m/s across the firing line and away from the shooter's breathing zone. Traditional indoor ranges use dedicated outdoor air systems utilizing 100% outside air ventilation system to provide this flow rate.

ANSI/ASHRAE Standard 62-2001, allows a dilution ventilation strategy to maintain clean and breathable air for users in an energy-efficient manner. Therefore, to provide sufficient ventilation and to comply with energy efficiency strategy it is feasible to partially recirculate range air. However, this method imposes significant indoor air quality (IAQ) challenges due to high emission of lead substance from bullets. Lead may cause serious long-term health problems. Mechanical filtration effectively removes airborne lead. Study [12] suggest in their study to use three stage mechanical filtration as an optimum solution to combine good IAQ and energy efficiency in the indoor shooting ranges.

At the first stage, a pre filter is used to filter return and exhaust air at 30% efficiency. Next stage is the secondary filtration, that removes 90-95% of contaminants and particles. All recirculated air is additionally cleaned through 99.99% HEPA filters for maximum lead removal.

Carbon monoxide is the most prevalent gaseous component of combustion in all cases, and is an ideal surrogate for firing range activity. CO is a toxic gas, however, it cannot be removed by filtration, therefore it must be diluted with outside air to obtain acceptable concentration levels.

The range exhaust and outdoor air intake rates are controlled by the building automation system to ensure that carbon monoxide level is below the setpoint limit of 25 ppm (8-hours exposure limit according to OSHA/ACGIH Guidelines for CO). The carbon dioxide level is also limited in accordance with ASHRAE recommendations. If the human activity within the range increases, the variable speed exhaust fan increase the flow to extract additional gaseous concentrations and eliminate the human exposure to contaminant gases.

It is also important to note that in indoor shooting ranges a slightly negative pressure is to be maintained, or so-called underpressure of at least 5 Pa, so that the indoor air is being extracted continuously. Contaminants need to be contained in the range space. This will prevent the ingestion of these harmful particles, and also keep the non-range spaces and surfaces of the building free of contamination. This in turn will ensure the health and safety of all customers and employees. This factor is especially essential when there is a high activity in the range, as lead and carbon monoxide levels may increase significantly.

Providing negative pressure in the range is accomplished by designing more exhaust than supply air. Creating laminar air flow at the firing line will protect a shooter's respiratory zone, but it will not keep contaminants created in the range from entering other areas of the building. The industry standard for this design is 10% greater exhaust than supply. The tighter the range area of the building is constructed, the less differential is necessary to maintain the proper pressure differential. Some ranges have been designed with the 10% differential, but the exhaust is filtered and there was no method of modulation for the exhaust based on filter loading.

If the shooting range is not a stand-alone structure the ventilation system serves more areas, therefore more precautions are applied with regards to the ventilation system. All adjacent areas that are not exposed to lead and CO emissions should be served by a conventional constant or, if necessary, variable volume ventilation system. The ventilation system serving indoor shooting range and the ventilation system serving other adjacent spaces are isolated from one another to prevent air leakage and, thus, cross contamination.

Temperatures and airflow in the shooting ranges are controlled with respect to user activity and clothing. Shooters move in various positions and are lightly active with a metabolic rate at or below 2.0 met. With respect to the temperature inside the shooting range, the temperatures should be kept in between 18-20 °C, that is somewhat cooler than lobby or classroom areas.

The temperature and the air flowrate that should be maintained in shooting ranges are not defined by local building codes in Latvia, therefore a common practice is to refer to The American Society of Heating, Refrigerating and Air-Conditioning Engineers (ASHRAE), Occupational Safety and Health Administration (OSHA) or National Air Filtration Association (NAFA) guidelines.

With regards to the air volume in the shooting ranges, if contamination emissions such as lead and CO cannot be accurately specified, the methodology is to use a recommended and critical airflow rates referring to NAFA recommended air velocity of 0,25 m/s during normal operation, and maximum air velocity of 0,5 m/s during critical operation.

The best method for supplying airflow to an indoor shooting range is from a full-length perforated plenum wall located behind and parallel to the firing line to ensure a steady laminar flow that both mixes with the indoor air and displaces the contaminated air. Supply air should be discharged from the ceiling, as close to the back wall as possible.

The exhaust grilles are to be located on the opposite end in the target zone, however there are more aspects possible with regards to the placement of the exhaust grilles. In all cases air flow pattern should be maintained in the direction of the target zone. In most of the existing shooting ranges in Latvia exhaust grilles are placed either on the ceiling or side wall in the target zone. However, during high activity shooting practices and trainings high smoke contamination may occur even if the fans operate at full or critical speed (which in shooting range is defined as 30%-50% higher air velocity and flowrate above normal operation mode). Therefore, it is recommended to locate exhaust line approximately 5 m away from shooting positions. In shooting ranges that are above 50 m in length (distance between the end walls) there might be several shooting positions, at 25 m and 50 m distance from the target, therefore additional exhaust grilles are recommended to be located along those two positions.

The principle of combined air exchange model proposed by National Air Filtration Association (NAFA) guidelines.

4. Energy efficiency solutions

In order to set guidelines for comfort levels and optimal energy consumption passive house criteria can be used as a threshold. Due to the required strength and other above mentioned criteria related to building envelopes, good and very good energy consumption criteria can be achieved. Table 1 sets out passive house certification criteria and a basic evaluation of the importance of such criteria for unclassified military buildings. Such a basic evaluation is applicable to most unclassified building types. When looking for a more detailed evaluation on the criteria, specific building types must be assessed separately.

Table 1. Passive house criteria.

Criteria	Result to be achieved	Importance level
Space heating energy demand	max 15 kWh/m2	medium
Primary energy demand	max 60 kWh/m2	high to medium
Airtightness	max 0,6 1/h (50 Pa)	high
Thermal comfort	max 10 hours of indoor air temperature above 25°C	high to low

Space heating energy demand is rated as medium due to the possible struggles within cold climates to reach such low energy efficiency and considering the specific other requirements for building structures and protection elements that can directly relate to thermal bridges and therefore harden the potential for achieving such a low energy efficiency. Aiming for 15 kWh energy demand is much more important than achieving it.

Primary energy efficiency can be highly important for specific cases where no conventional heating methods can be used, and no network electricity supply is available. In such areas, firstly, it is crucial to minimize the energy demand and secondly renewable energy technologies can be used to meet the demand. Study [13] has shown importance of different building systems integration in order to achieve better level of energy efficiency. However it is not visible that in nearest future military buildings can be self-sufficient with application of existing technologies [14].

Airtightness is important criteria not only in terms of energy efficiency, but also other measures, therefore importance of airtightness is rated as high. For example, airtightness relates to resilience of structures as well as acoustic requirements and comfort.

5. Conclusions

Shooting ranges and army training facilities require a separate ventilation system, if they are not situated in standalone structures, that complies with strict health and safety requirements. Air exchange is ensured by mechanical ventilation system only with three stage filtering, if part of exhaust is recirculated. Underpressure should be maintained in shooting ranges in order to maintain adjacent spaces contaminant-free. Typically supply air should account for 90% of exhaust air.

It is important to maintain laminar air flow and avoid turbulent air flow pattern to ensure that contaminated air plume is not flowing backwards. The optimum air velocity range for shooting ranges is between 0.25 m/s in normal operation mode and 0.5 m/s in critical operation mode, when intense shooting trainings take place. With regards to the temperature, no specific requirements are defined, as the shooting range is a space where a person typically spends less than 2 hours. However, authors suggest that the temperature is kept at a level between 18 C and 20 C. During intense shooting training sessions the indoor temperature might be lower.

As military structure stock in Latvia is relatively outdated and a lot of existing structures need renovation and refurbishment, it is essential to define solid roadmap along with the guidelines when it comes to structural issues, energy efficiency measures and healthy indoor climate within those buildings. Another driving factor to define such guidelines is the fact that Latvian military buildings stock will be expanding in the coming years. In many cases basic requirements can be used in tenders for such buildings, outlining the quality criteria for design & build tenders, this way ensuring a single responsible party for achieving quality results.

Acknowledgements

This study was supported and financed by European Regional Development Fund projectNr.1.1.1.1/16/A/048 "NEARLY ZERO ENERGY SOLUTIONS FOR UNCLASSIFIED BUILDINGS".

References

[1] Liesen R, Ellis P, Zhivov A and Herron D 2012 Extremely low energy design for army buildings: Barracks *ASHRAE Transactions* vol 118 pp 767–89

[2] Deru M, Langner R, Zhivov A, Liesen R, Herron D and Smith V 2012 Extremely low-energy design for army buildings: Dining facility *ASHRAE Transactions* vol 118 pp 790–815
[3] Langner R, Deru M, Zhivov A, Liesen R and Herron D 2012 Extremely low-energy design for army buildings: Tactical equipment maintenance facility *ASHRAE Transactions*
[4] Booth S, Barnett J, Burman K, Hambrick J and Westby R 2010 Net Zero Energy Military Installations : A Guide to Assessment and Planning *NREL Tech. Rep.*
[5] Barsi D, Costa C, Satta F, Zunino P and Sergeev V 2018 Feasibility of mini combined cycles for naval applications *MATEC Web Conf.*
[6] Barsi D, Perrone A, Qu Y, Ratto L, Ricci G, Sergeev V and Zunino P 2018 Compressor and Turbine Multidisciplinary Design for Highly Efficient Micro-gas Turbine *J. Therm. Sci.*
[7] Chen W, Hao H, Chen S and Hernandez F 2015 Performance of composite structural insulated panel with metal skin subjected to blast loading *Mater. Des.*
[8] Meng Q, Hao H and Chen W 2016 Laboratory test and numerical study of structural insulated panel strengthened with glass fibre laminate against windborne debris impact *Constr. Build. Mater.*
[9] Fryanova K O and Perminov V A 2017 Impact of forest fires on buildings and structures *Mag. Civ. Eng.*
[10] Vedishcheva I S, Ananin M Y, Al Ali M and Vatin N I 2018 Influence of heat conducting inclusions on reliability of the system "sandwich panel -metal frame" *Mag. Civ. Eng.*
[11] Gravit M V., Golub E V. and Antonov S P 2018 Fire protective dry plaster composition for structures in hydrocarbon fire *Mag. Civ. Eng.*
[12] Morgenthaler A B and Shumway D F 2002 Indoor shooting range *ASHRAE J.*
[13] Petrichenko M R, Nemova D V., Kotov E V., Tarasova D S and Sergeev V V. 2018 Ventilated facade integrated with the HVAC system for cold climate *Mag. Civ. Eng.*
[14] Titkov V V., Bekbayev A B, Munsyzbai T M and Shakenov K B 2018 Construction of autonomous buildings with wind power plants *Mag. Civ. Eng.*

Energy performance of temporary shelters

A Borodinecs[1], A Geikins[1] and S Smirnov[2]

[1] Institute of Heat, Gas and Water Technology, Riga Technical University, Latvia
[2] Institute of Mathematics and Computer Science, University of Latvia, Riga, Latvia

E-mail: anatolijs.borodinecs@rtu.lv

Abstract. This paper analyzes methods for calculating energy consumption of temporary shelters for the purpose of reducing energy consumption for heating and improving the level of comfort. Mobile shelters are widely used by rescue emergency teams. Often such type of buildings has a limited access to reliable source of energy supply. Thus it is vitally important to make correct heat losses analysis to select proper capacity of energy source. The paper presents analysis of heat losses of rubber/fabric shelters without extra thermal insulation.

1. Introduction

Mobile temporary shelters are widely used for camps and rescue operations. Usually such shelters are placed in remote areas without connection to electrical grid or district thermal energy network. Mobile energy sources, mainly diesel engine with an electric generator are used for energy supply. Such solution has limited amount of energy which can be produced onsite with out extra supply. Thus, is a vitally important to make precis calculation of heating and cooling loads in order to ensure long term energy supply for mobile temporary shelters. Especially this is very important for rescue operation where optimal thermal comfort should be insured for victim care. In should be noted that not only thermal comfort is important, the human is health affected by CO and CO_2 concentration [1] in shelters. The main reason is insufficient ventilation due to limited energy source. Nowadays existing mobile energy solution can be supplied with extra storage [2]. The capacity of such units is limited in comparison to traditional energy sources [3]. Thus, thermodynamic properties of shelters' envelope constantly improves [4] [5] which give possibility to improve overall energy performance and insure better thermal comfort and indoor air quality. Currently there is a lack of publicly available energy consumption calculation tools for extremely light temporary structures. Study [6] use ANSYS Fluent for natural convection inside the air gap for less energy consumptions. However it should be mentioned that not all well know approaches on passive cooling [7] can be efficiently used in temporary shelter. ANSYS Fluent is pretty complicated software which is not commonly used by engineers. Tre IDA-ICE tool can be used for more simplified energy consumption calculations in tents [8]. In scope of this study the analysis of methods for calculating heat losses from temporary shelters was performed. Our purpose is to reduce energy consumption for heating and improve the level of comfort. Key ideas for calculating heat losses from shelters were comes from the Technical Report Natick/TR-79/017 [9]. In essence, these methods do not differ from the methods used for calculating heat losses for buildings and based on laws of thermodynamics [10] [11]. These methods slightly modified and extended in order to use for new materials and adopt results for dynamic energy simulations. The basis of all calculation methods is an irreversible process: heat transfer which may consist of three separate types of disseminating thermal energy (conduction, convection and radiation) in various combinations. The structure of the paper is the following: first, we are going to look at methods for determining heat losses from shelters and provide a concrete example of calculation. Finally, we are going to compare the data with the values that were obtained using measuring instruments [12] and we will adjust some coefficients.

2. Materials and methods

There are two key types of heat loss from shelters: heat transmission through external structures (walls, ceiling, floor), and infiltration or air drainage towards the outside and influx of cold air inside. Strictly speaking, there is also the third type of heat loss - heat reflection from shelter surfaces but considering the relatively low surface temperature, its intensity is low (in accordance with the Stefan–Boltzmann law). Certainly, heat reflection should not be completely disregarded, however we are going to assume that heat reflection intensity is included in heat flux density from the shelter surface to the surrounding environment characterized by the heat transfer coefficient. Total heat loss can be calculated as sum of loss through heat transmission (including the radiation component) and heat loss through infiltration. Since our goal is to determine heat loss characteristics that do not depend of weather conditions at a certain moment, we are going to divide all expressions by ΔT and thus acquire heat losses per one degree.

$$L = \frac{\sum_n (Q_C)_n + Q_A}{\Delta T} = \sum_n (U_t\, A)_n + \rho\, c\, V_a$$

where n is the number of shelter surfaces that should be addressed separately.

Heat losses from shelters are influenced by temperature as well as other weather conditions such as wind and clouds. Therefore, we are going to analyze the weather conditions that result in maximum heat losses, and weather conditions that result in minimum heat losses. Subsequently, we are going to compare the calculated data with the values that were obtained using measuring instruments. It could be assumed that maximum heat losses take place during nighttime with clear sky and moderate wind weather conditions. It can be assumed that shelter surface temperature in such conditions in equal to external air temperature. Infiltration takes place through doors that have not been closed tightly as well as through shelter material.

We are going to assume that minimum heat losses take place at cloudy sky with no wind weather conditions. Windless conditions mean that infiltration will be insignificant and heat transfer coefficients between the surface and external and internal air will be identical, i.e. Uo=Ui. Shelter floor should be analyzed separately. In our example, shelter floor is ground, therefore it is necessary to evaluate ground temperature and apply the necessary heat transfer coefficient. Ground temperature in the winter can be assumed as 0°C It is important to understand that, additionally to the floor, a shelter has quite a high vertical temperature gradient and therefore temperature in the bottom part of the shelter must be used for calculating heat losses through ground. Additionally, we are going to assume that heat losses through ground practically coincide at both maximum and minimum heat losses.

3. Thermodynamic properties of shelters' external envelope

Shelters are collapsible temporary structures that can be assembled quickly. Shelters serve as a refuge from adverse weather conditions, temporarily accommodating and housing human occupants, performing works, household needs as well as storing various materials in field conditions. In terms of structure shelters can be categorized as follows framed; Non-framed; Inflatable. Inflatable shelters require a very short assembly time. Framed shelters structures easily bear snow loads of up to 15-20 kg/m2 and wind loads of up to 30 m/sec.

Shelters can also be systematized according to their form, volume, material and number of layers. See Figure 1 for various forms of shelters. Wind resistance of shelters depends not only on the shelter material and the number of layers, but also on their shape.

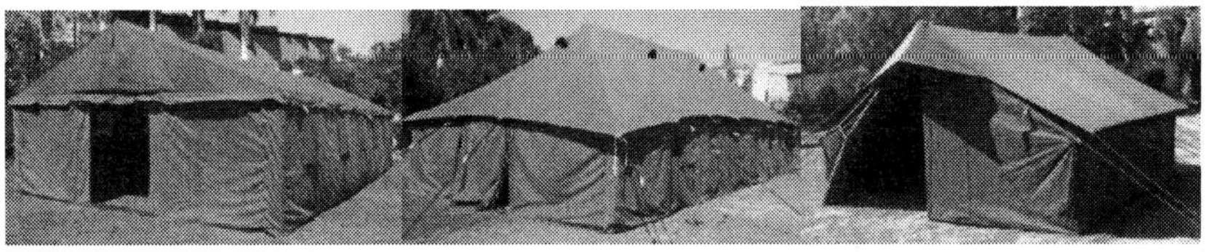

Hospital Shelter Frame Shelter Gable Shelters

Figure 1. Types of Shelter.

Thermal conductivity and water resistance of shelters depend on the shelters' material and number of layers. One-layer and two-layer enclosing fabric is commonly used. Multi-layer fabric is also used in shelters intended for use in winter. However, one must take into account that the assembly time of such shelters is longer, and they are typically used for long-term stationing. Thermal resistance of a modern non-insulated one-layer shelter is 0.33 m²K/W [13]. Correspondingly, the heat transfer coefficient is - 2.98 W/m²K. Ac-cording to source data, thermal resistance of such shelters is 0.124 m²K/W. Thermal resistance of two-layer shelters is 0.60 m²K/W. Respectively, the heat transfer coefficient is – 1.68 W/m²K [11]. For example, thermal resistance of standard TEMPER insulated fabric is 0.564 m²K/W [12]. According to the data of another source [13], thermal resistance of one-layer shelter fabric is 0.21 m²K/W. It is possible to increase thermal resistance to 0,39 m²K/W by means of using an additional layer of fabric and creating a fixed air layer. In order to improve a shelter's thermo-mechanical properties, it is possible to use multi-layer fabric with vacuum heat insulation.

The use of aerogel enables to substantially decrease heat losses during winter, and limit heat influx during summer. In such a case, thermal resistance is 1.814 m²K/W. An additional option is to use spray polyurethane foam insulation. Polyurethane foam is sprayed on the shelter's external surfaces, and it increases the existing structure's thermal resistance to 1.25 m²K/W as it solidifies. Using foam is significantly cheaper than using aerogel fabric. However, it is no longer possible to relocate a shelter after applying polyurethane foam insulation, and thermal insulation works must be carried out by means of engaging an outsourced service provider. Also extra thermal insulation material should be carefully choses, taking into account exploitation conditions. Extra water leakage can significantly reduce thermal conductivity [14]. Also application of light weight metal structure can be considered as reliable solution [15] for fast camp assembly.

4. Evaluation of heat losses

Let us analyze an average general-purpose shelter (Figure 2).

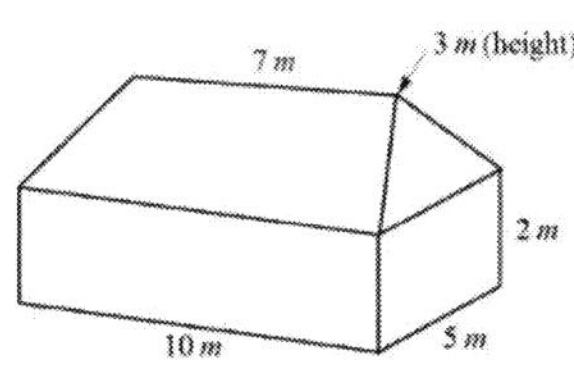

Figure 2. Average general-purpose shelter with dimensions.

Shelter floor is the ground, ground area is $A_g = 50\ m^2$, wall area is $A_w = 60\ m^2$, roof area is $A_r = 2 \times 4.5 + 2 \times 22.1 = 53.2\ m^2$. The proportion of heat transfer coefficient of the shelter material to shelter material thickness [10] is $\frac{\lambda_c}{l} = 38\ \frac{W}{m^2K}$.

Let us assume that the difference between average indoor temperature and average outdoor temperature is ΔT=22.8°C In turn, the difference between average indoor temperature and the temperature in the shelter's bottom part is ΔTg=5.5°C. Heat transfer coefficient between indoor air and ground surface in this case [12] is Ug =11.4 W/m2K. Thus, heat losses through the ground

at both maximum and minimum heat losses are $L_g = \frac{U_g\, A_g\, \Delta T_g}{\Delta T} = \frac{11.4\times 50\times 5.5}{22.8} = 137.5\ \frac{W}{K}$. Heat transfer coefficients between the shelter surface and internal and external air in windless weather

conditions equal $U_o=U_i=8.5W/m^2K$. Thus $U_t = \frac{1}{\frac{1}{U_i}+\frac{l}{\lambda_c}+\frac{1}{U_0}} = \frac{1}{\frac{1}{8.5}+\frac{1}{38}+\frac{1}{8.5}} = 3.8\ \frac{W}{m^2K}$ and heat losses

from walls and roof are $L_{w,r} = U_t(A_w + A_r) = 3.8 \times (60 + 53.2) = 430.2\ \frac{W}{K}$. Since infiltration at minimum heat losses may be disregarded, the result is: L_{min}=135.5+430.2=567.7 W/K.

In the maximum heat losses case, we must consider infiltration. $L_A = \frac{Q_A}{\Delta T} = \rho\, c\, V_a$, where $\rho\, c = 1.2\ \frac{kg}{m^3} \times 1004.2\ \frac{J}{kg\, K} = 1205\ \frac{J}{m^3\, K}$. Let us assume that the shelter's maximum air permeability is $0.0005\ \frac{m}{s}$, shelter surface area is $113.2\ m^2$ but in this case we are going to consider only one half of the surface area. Further, we are going to assume that air discharge through gaps between shelter walls and ground is 0.0045 m^2/s shelter perimeter is 30 m. Therefore V_a = (0.0005x113.3)/2+0.0045x30=0.163 m^3/s. Thus, $L_A = \rho\, c\, V_a = 1205 \times 0.163 = 196.4\ \frac{W}{K}$. At maximum heat losses $\frac{1}{U_0} = 0$ and $U_t = \frac{1}{\frac{1}{8.5}+\frac{1}{38}} = 6.9\ \frac{W}{m^2K}$. Thus, heat losses from walls and roof are $L_{w,r} = U_t(A_w + A_r) = 6.9 \times (60 + 53.2) = 781.1\ \frac{W}{K}$. Thus, we acquire maximum heat losses $L_{max} = L_g + L_A + L_{w,r} = 137.5 + 196.4 + 781.1 = 1115\ \frac{W}{K}$.

As we can see, the difference between the minimum heat losses L_{min}=567.7 W/K and the maximum heat losses L_{max}=1115 W/K is significant. So, the weather impact on heat losses is very important.

5. Brief analysis and adjustment of obtained results

Let us analyze the functional connection of heat losses through ground L_g from the difference between the average indoor temperature and temperature in the bottom part of the shelter, i.e. let us analyze the function $L_g(\Delta T_g)$, where ΔT_g will be considered as the variable. Evidently, the functional connection is linear (Figure 4 right).

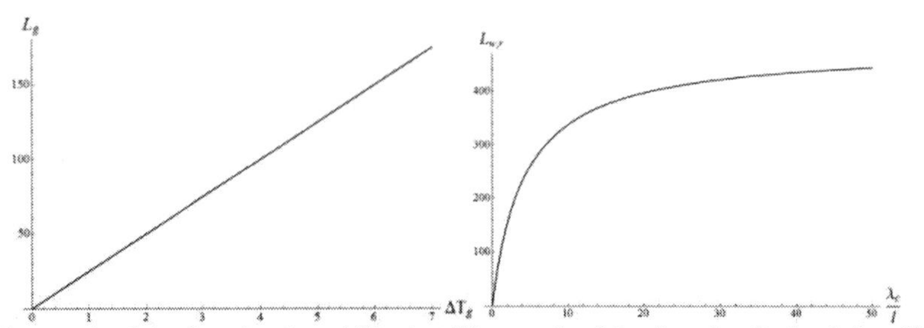

The graph of the function L_g (ΔT_g) The graph of the function L_(w,r) (λ_c/l)

Figure 3. Analysis of heat loss specifics.

As we can see in the diagram, the larger the difference ΔT_g (or, the larger the vertical temperature gradient), the larger are the heat losses. Thus, vertical gradient must be decreased in order to decrease

IOP Conf. Series: Materials Science and Engineering **660** (2019) 012017 doi:10.1088/1757-899X/660/1/012017

heat losses through the ground. It can be achieved by, for example, applying forced indoor air motion in the shelter.

Let us discuss the functional connection of heat losses through external structures $L_{w,r}$ on the proportion λ_c/l (heat transfer coefficient of the material vs. shelter material thickness), i.e., let us discuss a function $L_{w,r}(\lambda_c/l)$ (at minimum heat losses). In this case, the functional connection is not linear (Figure 3 left).

As we can see from the diagram, for values of up to ≈ 10 of the proportion $\frac{\lambda_c}{l}$ heat losses increase rapidly, but if $\lambda_c/l > 20$, then the heat loss increase is low. Interval $10 < \lambda_c/l < 20$ can be considered as the transition interval. Thus, if the aim is to reduce heat losses, the proportion λ_c/l must be decreased. It is especially important in the interval $10 < \lambda_c/l < 10$ Evidently, in order to decrease the proportion λ_c/l, shelter material thickness must be increased and/or materials with lower heat transfer coefficients λ_c must be used.

Let us analyze the functional connection of heat losses through infiltration L_a on the coefficient that characterizes air discharge through the gaps between the shelter's walls and ground. If this coefficient is expressed as k_s, then the function $L_a(k_s)$ should be analyzed. The functional connection is linear, but the function's diagram does not move through the starting points of coordinates due to the air permeability of the material. Evidently, if the aim is to reduce heat losses, then the coefficient k_s must be reduced. This can be achieved by minimizing gaps, for example. Heat losses can also be reduced by choosing materials with lower air permeability.

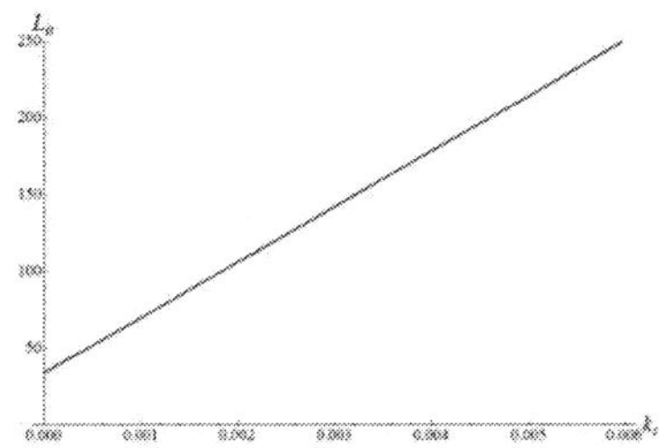

Figure 4. The graph of the function L_a (k_s).

Now let us compare the calculated data L_{min}=567.7 W/K and L_{max}=1115 W/K with the values that were obtained using measuring instruments [8, 9], respectively 454 W/K and 766 W/K. Heat transfer coefficients that were taken from the tables have an impact on the final result. Therefore, using the data obtained from measuring instruments, we are going to adjust heat transfer coefficients in the case of minimum heat losses. The same heat transfer coefficient $8.5\ \frac{W}{m^2K}$ was used for shelter walls and roof, both inside and outside. However, warm air collects under the roof surface. Therefore, heat transfer intensity changes because poor air circulation reduces convection. From the expression $L_{min} = 454 = L_g + L_{w,r} = 137.5 + L_{w,r}$ it follows $L_{w,r}$ = 454-137.5=316.5 W/K. It can be separately calculated $L_w = 3.8 \times 60 = 228\ \frac{W}{K}$ and thus L_r=316.5-228=88.5 W/K. From this, it follows 1.66 W/m^{2}K. From expression $U_t = \frac{1}{\frac{1}{U_i}+\frac{l}{\lambda_c}+\frac{1}{U_0}}$ let us express U_i. Thus $U_{i\,r}$ = 2.18 W/(m^{2}K). The heat transfer coefficient for roof U_{ir} = 8.5 W/m^{2}K was acquired by means of calculation, but using the measured data, this coefficient can be adjusted, and the value 2.18 W/m^{2}K can be obtained. All other coefficients can be adjusted in a similar manner.

6. Conclusions

There are no any publicly available calculation tools for heat losses estimation in such temporary shelters as light weight tents. The commercially available software such a IDA-ICE can be used for estimation of thermal comfort and annual energy consumption. However, it doesn't consider such parameter as liner air permeability. The heat losses through the non-insulated floor also should be addressed in more detailed way. The method of calculations of heat losses from shelters was discussed. Concrete example of calculations was considered. Obtained results were analyzed in order to reduce energy consumption for heating and improve the level of comfort. The calculated data was compared with the values obtained using measuring instruments and adjustment of some coefficients was made.

Acknowledgements
This study was supported and financed by by European Regional Development Fund projectNr.1.1.1.1/16/A/048 "NEARLY ZERO ENERGY SOLUTIONS FOR UNCLASSIFIED BUILDINGS".

References
[1] Leigh-Smith S 2004 Carbon monoxide poisoning in tents - A review *Wilderness Environ. Med.*
[2] McCulley J K and Professor A 2015 Portable solar power: Mobile elemental power plant II *2015 IEEE Conference on Technologies for Sustainability, SusTech 2015*
[3] Barsi D, Perrone A, Qu Y, Ratto L, Ricci G, Sergeev V and Zunino P 2018 Compressor and Turbine Multidisciplinary Design for Highly Efficient Micro-gas Turbine *J. Therm. Sci.*
[4] Thrall A P and Quaglia C P 2014 Accordion shelters: A historical review of origami-like deployable shelters developed by the US military *Eng. Struct.*
[5] Vilder I De, Buyle G and Roel Gijsbers 2016 *MEMBRANES FOR SHELTERS a comparative study to assist humanitarian workers in selecting suitable textile materials*
[6] Prabhu V V and Languri E M 2018 Shelter energy use reduction through buoyancy-driven air flow manipulation: A numerical and experimental study *Sustain. Energy Technol. Assessments*
[7] Hlaichi Mohammed A, Al-zuwaini H, Sergeev V, Socolova E, Skulkin S and Mukhametova L 2019 Article ID: IJMET_10_02_143 Mukhametova, Passive Cooling by Integrate Solar Chimney with Earth to Air Heat Exchanger *Int. J. Mech. Eng. Technol.*
[8] Cornaro C, Sapori D, Bucci F, Pierro M and Giammanco C 2015 Thermal performance analysis of an emergency shelter using dynamic building simulation *Energy Build.*
[9] Malcolm N. Pilsworth J 1978 *The Calculation of Heat Loss from Tents*
[10] Agami Reddy T, Jan F Kreider, Peter S, Curtiss A R 2016 *Heating and Cooling of Buildings: Principles and Practice of Energy Efficient Design* (CRC Press)
[11] Morris Grenfell D 2004 *Building Heat Transfer* ed L John Wiley & Sons
[12] Chaloux P N 1964 *Heat Retention Test on Tent Liners.AMXRE-MED-896*
[13] Anon 2017 *PVC FABRIC CLADDING.The EFASS covering membrane forms the protective barrier between the environment and the valuable equipment and personnel inside*
[14] Vatin N I, Pestryakov I I, Sultanov S T, Ogidan T, Yarunicheva Y A and Kiryushina A P 2018 Water vapour by diffusion and mineral wool thermal insulation materials *Mag. Civ. Eng.*
[15] Rybakov V A, Kozinetc K G, Vatin N I, Velichkin V Z and Korsun V I 2018 Lightweight steel concrete structures technology with foam fiber-cement sheets *Mag. Civ. Eng.*

IOP Conf. Series: Materials Science and Engineering **660** (2019) 012018 doi:10.1088/1757-899X/660/1/012018

The procedure for determination of the dependence of the cost of insulation materials on their thermophysical characteristics

O S Gamayunova[1], A E Radaev[2] and M R Petrichenko[3]

[1-3] Peter the Great St. Petersburg Polytechnic University, Polytechnicheskaya, 29, St.Petersburg, 195251, Russia

E-mail:gamayunova@inbox.ru

Abstract. Optimal insulation for the walls of the house is selected based on the material of construction and must meet several requirements. This article describes the method of justifying the dependence of the cost of heaters on their technical characteristics, based on a statistical analysis of price and thermophysical characteristics, used later in the development of optimization models for the selection and use of thermal insulation materials (heaters). To assess the quality of the resulting economic and mathematical model, the coefficient of determination was used. In order for the number of regressors (factors) not to affect the statistics of this coefficient, an adjusted coefficient of determination was also calculated, which gives a "penalty" for the factors included in the model.

Introduction

A good home is a warm home. That is why today apply a variety of insulation in the construction and renovation of buildings and structures.

The issues of energy saving and, in particular, the insulation of the external walls of buildings and structures have always attracted the attention of Russian and foreign researchers [1-20].

Livchak V.I. and Gorshkov A.S. made a description of the evolution and development of regulatory requirements for enclosing structures over an almost 100-year period. They showed how the calculated formulas, terms, dimensions of physical quantities changed and transformed [1]. Livchak V.I. gives specific suggestions for improving the methodology for calculating the heat balance of residential and public buildings, makes recommendations on how to certify buildings when conducting energy audits in accordance with international standards [2].

Gorshkov A.S. and Nemova D.V. paid much attention to the issue of improving the energy efficiency of buildings, as well as the economic efficiency of energy-saving measures [3, 4].

Yelchishcheva T.F. and Slivin R.V. presented a feasibility study of wall insulation options on the example of a 9-storey residential building [5]

The thermophysical parameters of enclosing structures and the justification of measures to improve the energy efficiency of external walls were carried out by such Russian and foreign researchers as: Petrichenko M.R., Kotov E.V., Nemova D.V., Vatin N.I., Korniyenko S.V., Tarasova D.S., Sergeev V.V., Zaborova D.D., Kukolev M.I., Mussorina M.T., Borodinecs A. and etc. [6-17]. Some of these researchers considered the energy efficiency of individual series of residential buildings [18-20].

The requirements of standards for thermal insulation are formed on the basis of economic feasibility, while the required values of heat transfer resistance of enclosing structures should be achieved without a special increase in the cost of construction. But it turns out that in conditions of a temperate climate, and even more so cold, single-layer walls with thickness calculated from the bearing capacity are simply cold. This will require that the thickness of the wall (layer) must be increased in order to achieve the requirements of the standard for heat saving.

An alternative is the use of insulation. On the modern construction market there are many heat-insulating materials, differing in cost, thickness and other technical characteristics. In this case, many questions arise: what insulation to choose? What are the characteristics? But how will the characteristics and cost of insulation affect the payback period of energy saving measures?

1. Materials and Methods

Within this section, a description will be given of the procedure for determination of the dependence of the cost of insulation materials on their technical characteristics.

The main provisions of the procedure:

- The object of the study is the dependence of the specific cost of the heater (per 1 m^2) on its technical characteristics (thickness, thermal conductivity, density, etc.).
- To substantiate this dependence, various samples of insulation materials from mineral (basalt) wool, polystyrene foam and extruded polystyrene foam are considered.
- It is assumed that the specific cost of a heater depends on its technical characteristics, which determines the expediency of using a linear multifactor regression model to form a dependence. This model has the form:

$$\hat{y} = a_0 + \sum_j a_j \cdot x_j, \qquad (1)$$

  where $\hat{y}$ – the dependent characteristic, a_0 – free member of the regression model; a_j – regression coefficients; x_j – factor j which influences the dependent characteristic.
- Justification of the dependence of the cost of insulation on its technical characteristics assumes the justification of the model parameters - coefficients $\{a_j\}$ - according to the criterion of minimizing the sum of squares of deviations of the actual values of the cost of insulation samples from the predicted values determined with known technical characteristics $\{x_j\}$ of samples.
- To assess the adequacy of the regression model, it is supposed to use the coefficient of determination R^2, the corresponding interval of values $(-\infty;1]$ and its derived indicators.

Baseline data of the developed procedure are presented in Table 1.

Table 1. Name, units and designations of source data and indexes.

No	The name of the characteristics	Designation	Unit rev.
1	The number of samples of insulation	n	units
2	Number of insulation parameters	m	units
3	Insulation sample index	$i=1,\dots,n$	-
4	Insulation parameter index	$j=1,\dots,m$	-
5	The value of parameter j for sample insulation i	x_{ji}	-
6	The actual values of the cost of insulation i	y_i	rub.

The structure of the developed procedure is described by a sequence of stages, each of which involves the calculation of a specific quantitative characteristic. The list of characteristics being presented in the chronological order of their calculation is given in Table 2.

Table 2. Name, units and formulas for calculating the calculated characteristics.

No	The name of the characteristics	Unit rev.	Formula
1	The element of the main matrix of the system of equations of the regression model corresponding to the row l (l = 1, 2, ..., m) and column k (k = 1, 2, ..., m)	-	$A_{lk} = \begin{cases} n, & if\ l=1, k=1; \\ \sum_{i=1}^{n} x_{il}, & \text{if } l>1, k=1; \\ \sum_{i=1}^{n} x_{ik}, & \text{if } l=1, k>1; \\ \sum_{i=1}^{n} x_{il} \cdot x_{ik}, & \text{in other case.} \end{cases}$
2	The matrix element of the system of equations for the parameter j' of the predictive model (j' = 0, 1, ..., m), corresponding to the row l (l = 1, 2, ..., m) and column k (k = 1, 2, ..., m)	-	$A_{lk}^{j'} = \begin{cases} \sum_{i=1}^{m} y_i, & \text{if } l=1, k=j'; \\ \sum_{i=1}^{m} y_i \cdot x_{il}, & \text{if } l>1, k=j'; \\ A_{lk}, & \text{in other case.} \end{cases}$
3	The value of the parameter j' (j' = 0, 1, ..., m)[a]	-	$a_{j'} = \frac{\left\lvert A_{lk}^{j'} \right\rvert}{\left\lvert A_{lk} \right\rvert}$
4	Forecast (calculated) values of the cost of insulation i	rub.	$\hat{y}_i = a_0 + \sum_{j=1}^{m} a_j \cdot x_{ij}$
5	The average value of the actual cost of insulation in the amount of n	rub.	$\bar{y} = \frac{1}{n} \cdot \sum_{i=1}^{n} y_i$
6	Coefficient of determination	-	$R^2 = 1 - \frac{\sum_{i=1}^{n} (y_i - \hat{y}_i)^2}{\sum_{i=1}^{n} (y_i - \bar{y})^2}$
7	Number of parameters, including free member	units	$k = j + 1$
8	Adjusted coefficient of determination	-	$R_{adj}^2 = 1 - (1 - R^2) \cdot \frac{n-1}{n-k}$

[a] The value j'=0 corresponds to the value constant; otherwise, the value determines the proportionality coefficient of the specific cost of the insulation to the value of the parameter j=j'.

It is important to note that the main characteristic that determines the adequacy of the developed predictive model for the dependence of the cost of insulation on its technical characteristics (determined by the expression of item 4 of Table 2) is the coefficient of determination R^2 reaching its highest value

– 1 – in the case when the predicted values of the cost are equal to actual ones for all considered samples of insulation. In general case, the coefficient R^2 has the value that is less than 1, but the condition of acceptability of the developed predictive model for further calculations is determined by the expression

$$R^2 \geq 0.6 \tag{2}$$

At the same time, factors such as the number of thermal insulation characteristics taken into account, as well as the number of thermal insulation samples considered, have a great importance on the adequacy of the predictive model.

For taking into account of the first of these factors, on the basis if the statement that the value of the coefficient of determination R^2 does not decrease when additional factors are added to the model, it was also proposed to calculate the corrected coefficient of determination R^2_{adj} (item 8 in Table 2), which gives a "penalty" for the additional factors included and, accordingly, more reliably assesses the model.

The second factor is connected to the law of large numbers, which determines the direct relationship between the number of observations and the representativeness of the sample set, i.e. the degree of compliance of the characteristics of the sample composition with the phenomenon or conformity under investigation). According to the overwhelming number of literary sources in the field of statistics, the sample set is representative if the number of elements included in its composition is at least 30 units.

2. Results and Discussion

The developed tools were implemented on a practical example (determination of insulation parameters for external walls). All calculations were made in the program Microsoft Excel. The initial data for testing the analytical method – model names, values of technical characteristics and cost for 30 insulation samples sold in the North-West Federal District of the Russian Federation – are presented in Table 3. As these characteristics were considered the following:

- Thickness.
- Coefficient of thermal conductivity.
- Average density.
- Water absorption in 24 hours,% by volume.
- Flammability group.

It is important to note that the "Flammability group" parameter is discrete and for its quantitative description it was decided to use a set of positive integer values.

The elements of the main matrix of the system of equations of the regression model (formula 3) were calculated in accordance with item 1 of Table 2.

In a similar way, the particular determinants of the system of equations for the parameters a_0 , ..., a_5 were calculated. The remaining parameters of the computational model (Table 4) are calculated by the formulas items 3 – 8 of Table 2.

As a result, the model will take the form:

$$\hat{y} = -887.86 + 4273.02 \cdot x_1 - 30137.84 \cdot x_2 + 2.66 \cdot x_3 + 1295.8 \cdot x_4 + 392.76 \cdot x_5 \tag{3}$$

As can be seen from expression (3), the cost constant, as well as the specific cost for the parameter of the thermal conductivity coefficient, turned out negative. The negative value of the value constant is determined by the decrease in the actual value of the insulation cost in comparison with objective value, determined only by the values of technical characteristics, which, in turn, is possible during the process of goods' selling in conditions of market with significant competition either within the country or at the interregional level.

The inverse dependence of the cost of insulation on the coefficient of thermal conductivity is obvious: the worse the insulation conducts heat, the greater its operational efficiency.

It is important to note that the standard and adjusted coefficients of determination for the developed predictive model are equal to 0.938 and 0.918, respectively, which indicates the high adequacy of the specified model, and, as a consequence, its high practical significance. As for the absolute deviations of the predicted values of the cost for the different models (samples) of insulation being considered from the actual ones, the largest deviation was 93.39 rubles / m^2, which is about 28,6% of the average cost of 1 m^2 of insulation within the considered sample set. The specified deviation is acceptable according to the standard scope of work on the warming of structural elements of buildings and structures and the amount of funding allocated for these purposes.

Table 3. Initial data and characteristics.

No	Heater	Thickness	Coefficient of thermal conductivity	Average density	Water absorption in 24 hours,% by volume	Flammability group	Actual cost
n		$\delta_i = x_{1i}$	$\lambda_i = x_{2i}$	$\rho_i = x_{3i}$	$\mu_i = x_{4i}$	$\varphi_i = x_{5i}$	$C = y_i$
-		m	W/(m·°C)	kg/m³	%	-	rub./m²
1	PENOPLEX Wall, 1185x585x50 mm	0.05	0.032	20	0.5	3	280
2	PENOPLEX Comfort, 1200x600x20 mm	0.02	0.03	30	0.4	3	110
3	PENOPLEX Comfort, 1200x600x30 mm	0.03	0.03	30	0.4	3	167.5
4	PENOPLEX Comfort, 1200x600x40 mm	0.04	0.03	30	0.4	3	217
5	PENOPLEX Comfort, 1200x600x50 mm	0.05	0.03	30	0.4	3	263
6	PENOPLEX Comfort, 1200x600x60 mm	0.06	0.03	30	0.4	3	277
7	PENOPLEX Comfort, 1200x600x80 mm	0.08	0.03	30	0.4	3	368
8	PENOPLEX Basis, 1200x600x20 mm	0.02	0.03	30	0.4	3	92
9	PENOPLEX Basis, 1200x600x30 mm	0.03	0.03	30	0.4	3	138
10	PENOPLEX Basis, 1200x600x40 mm	0.04	0.03	30	0.4	3	184
11	PENOPLEX Basis, 1200x600x50 mm	0.05	0.03	30	0.4	3	232
12	PENOPLEX Basis, 1200x600x60 mm	0.06	0.03	30	0.4	3	276
13	PENOPLEX Basis, 1200x600x80 mm	0.08	0.03	30	0.4	3	368
14	PENOPLEX Basis, 1200x600x100 mm	0.1	0.03	30	0.4	3	440
15	BASWOOL Standard 50, 1200x600x100 mm	0.1	0.038	50	1	1	207
16	BASWOOL Standard 60, 1200x600x50 mm	0.05	0.038	60	1	1	116
17	BASWOOL Standard 60, 1200x600x100 mm	0.1	0.038	60	1	1	231
18	BASWOOL Standard 70, 1200x600x50 mm	0.05	0.038	70	1	1	129
19	BASWOOL Standard 70, 1200x600x100 mm	0.1	0.038	70	1	1	257
20	BASWOOL Facade 110, 1200x600x50 mm	0.05	0.038	110	1	1	218
21	BASWOOL Façade 110, 1200x600x100 mm	0.1	0.038	110	1	1	436
22	BASWOOL Façade 110, 1200x600x150 mm	0.15	0.038	110	1	1	653
23	BASWOOL Facade 120, 1200x600x50 mm	0.05	0.038	120	1	1	236
24	BASWOOL Façade 120, 1200x600x100 mm	0.1	0.038	120	1	1	473
25	BASWOOL Facade 120, 1200x600x150 mm	0.15	0.038	120	1	1	709
26	BASWOOL Facade 140, 1200x600x50 mm	0.05	0.038	140	1	1	246
27	BASWOOL Facade 140, 1200x600x100 mm	0.1	0.038	140	1	1	491
28	BASWOOL Façade 140, 1200x600x150 mm	0.15	0.038	140	1	1	737
29	BASWOOL Facade 160, 1200x600x100 mm	0.1	0.038	160	1	1	490
30	BASWOOL Façade 160, 1200x600x150 mm	0.15	0.038	160	1	1	735
	Sum	2.26	1.03	2150.00	21.70	58.00	9776.50
	Average	0.08	0.03	71.67	0.72	1.93	325.88
	Minimum	0.02	0.03	20.00	0.40	1.00	92.00
	Maximum	0.15	0.04	160.00	1.00	3.00	737.00

The main determinant of the system of equations:

$$\Delta = \begin{vmatrix} n & \sum_{i=1}^{n} x_{1i} & \sum_{i=1}^{n} x_{2i} & \sum_{i=1}^{n} x_{3i} & \sum_{i=1}^{n} x_{4i} & \sum_{i=1}^{n} x_{5i} \\ \sum_{i=1}^{n} x_{1i} & \sum_{i=1}^{n} x_{1i}^2 & \sum_{i=1}^{n} x_{1i}\cdot x_{2i} & \sum_{i=1}^{n} x_{1i}\cdot x_{3i} & \sum_{i=1}^{n} x_{1i}\cdot x_{4i} & \sum_{i=1}^{n} x_{1i}\cdot x_{5i} \\ \sum_{i=1}^{n} x_{2i} & \sum_{i=1}^{n} x_{2i}\cdot x_{1i} & \sum_{i=1}^{n} x_{2i}^2 & \sum_{i=1}^{n} x_{2i}\cdot x_{3i} & \sum_{i=1}^{n} x_{2i}\cdot x_{4i} & \sum_{i=1}^{n} x_{2i}\cdot x_{5i} \\ \sum_{i=1}^{n} x_{3i} & \sum_{i=1}^{n} x_{3i}\cdot x_{1i} & \sum_{i=1}^{n} x_{3i}\cdot x_{2i} & \sum_{i=1}^{n} x_{3i}^2 & \sum_{i=1}^{n} x_{3i}\cdot x_{4i} & \sum_{i=1}^{n} x_{3i}\cdot x_{5i} \\ \sum_{i=1}^{n} x_{4i} & \sum_{i=1}^{n} x_{4i}\cdot x_{1i} & \sum_{i=1}^{n} x_{4i}\cdot x_{2i} & \sum_{i=1}^{n} x_{4i}\cdot x_{3i} & \sum_{i=1}^{n} x_{4i}^2 & \sum_{i=1}^{n} x_{4i}\cdot x_{5i} \\ \sum_{i=1}^{n} x_{5i} & \sum_{i=1}^{n} x_{5i}\cdot x_{1i} & \sum_{i=1}^{n} x_{5i}\cdot x_{2i} & \sum_{i=1}^{n} x_{5i}\cdot x_{3i} & \sum_{i=1}^{n} x_{5i}\cdot x_{4i} & \sum_{i=1}^{n} x_{5i}^2 \end{vmatrix} = \tag{4}$$

$$= \begin{vmatrix} 30 & 46054 & 43525 & 2150 & 43667 & 58 \\ 46054 & 0.2158 & 0.0803 & 196.3 & 1.839 & 24898 \\ 43525 & 0.0803 & 0.0358 & 78.46 & 0.78 & 1.874 \\ 2150 & 196.3 & 78.46 & 221100 & 1906 & 2970 \\ 43667 & 1.839 & 0.78 & 1906 & 18.33 & 33.1 \\ 58 & 24898 & 1.874 & 2970 & 33.1 & 142 \end{vmatrix} = -8E-14$$

Table 4. Model parameters.

No	Parameter name	Unit rev.	Identified / Formula	Value
1	Cost constant	rub./m^2	$a_0 = \Delta a_0 / \Delta$	-887.861
2	The coefficient of proportionality of the unit cost of insulation value of the parameter δ	rub./m^3	$a_1 = \Delta a_1 / \Delta$	4273.02
3	The coefficient of proportionality of the unit cost of insulation value of the parameter λ	$\frac{\text{rub.}\cdot{}^\circ\text{C}}{\text{m}\cdot\text{W}}$	$a_2 = \Delta a_2 / \Delta$	-30137.83
4	The coefficient of proportionality of the unit cost of insulation value of the parameter ρ	$\frac{\text{rub.}\cdot\text{n}}{\text{kg}}$	$a_3 = \Delta a_3 / \Delta$	2.66
5	The coefficient of proportionality of the unit cost of insulation value of the parameter μ	$\frac{\text{rub.}}{\text{m}^2\cdot\%}$	$a_4 = \Delta a_4 / \Delta$	1295.8
6	The coefficient of proportionality of the unit cost of insulation value of the parameter φ	rub./m^2	$a_5 = \Delta a_5 / \Delta$	392.758
7	Coefficient of determination	-	R^2	0.932
8	Number of parameters, including free member	units	k	6
9	Adjusted coefficient of determination	-	R^2_{adj}	0.918

3. Conclusions

Thus, the implementation of the developed analytical methodology on a practical example confirmed the adequacy of the obtained predictive model, which led to the conclusion of the high practical importance of the developed tool. At the same time, the structure of the formed predictive model, which determines the analytical dependence of the cost of insulation on its technical characteristics, can be effectively used to solve problems connected to determination of the optimal characteristics for insulation for a given surface area of the insulated element according to the criterion for minimization of the insulation cost with taking into account of the constraints for calculated values of thermal resistance. Relevant procedures are planned for the next stages of the study.

References

[1] Gorshkov A S and Livchak V I 2015 History, evolution and development of regulatory requirements for enclosing structures *Construction of Unique Buildings and Structures* **3 (30)** pp 7-37

[2] Livchak V I Proposals to develop methods for conducting power engineering investigation 2015 *Chief power engineer* **5-6** pp 35-43

[3] Gorshkov A S, Vatin N I, Rymkevich P P and Kydrevich O O 2018 Payback period of investments in energy saving *Magazine of Civil Engineering* **2 (78)** pp 65-75

[4] Nemova D V, Gorshkov A S, Vatin N I, Kashabin A V, Tseytin D N and Rymkevich P P 2014 Technical and economic assessment on actions for heat insulation of external envelops external walls of apartment building with the double-skin façade *Construction of Unique Buildings and Structures* **11 (26)** pp 70-84

[5] Yelchishcheva T F and Slivin R V 2012 Technical and economic assessment of exterior wall insulation (using the example of a 9-storey residential building) *Globalnaya nauchnaya integratsiya* **6** pp 62-64

[6] Vatin N I, Ivanov A Y, Rutman Y L and Chernogorskiy S A 2017 Earthquake engineering optimization of structures by economic criterion *Magazine of Civil Engineering* **8 (76)** pp 67-83

[7] Petrichenko M R, Kotov E V, Nemova D V, Tarasova D S and Sergeev V V 2018 Numerical simulation of ventilated facades under extreme climate conditions *Magazine of Civil Engineering* **1 (77)** pp 130-140

[8] Korniyenko S V, Vatin N I, Gorshkov A S, Olshevskiy V Y and Pestryakov I I 2018 Designed thermophysical parameters of walls made of autoclaved aerated concrete blocks *Construction of Unique Buildings and Structures* **6 (69)** pp 35-58

[9] Statsenko E A, Ostrovaia A F, Musorina T A, Kukolev M I and Petritchenko M R 2016 The elementary mathematical model of sustainable enclosing structure *Magazine of Civil Engineering* **8 (68)** pp 86-91

[10] Korniyenko S V, Vatin N I and Gorshkov A S 2016 Heat and power characteristics analysis carried out for the residential building made of autoclaved aerated concrete blocks *Construction of Unique Buildings and Structures* **12 (51)** pp 45-60

[11] Korniyenko S V, Vatin N I and Gorshkov A S 2016 Assessment of moisture conditions of walls with facade's thermoinsulation composite systems with external mortar layers *Construction of Unique Buildings and Structures* **6 (45)** pp 34-54

[12] Ivanova E S and Gorshkov A S 2016 Calculation of energy consumption of the building with the minimum requirements for thermal protection *Construction of Unique Buildings and Structures* **4 (43)** pp 58-72

[13] Zaborova D D, Kukolev M I, Mussorina M T and Petritchenko M R 2016 The simplest mathematical model of the energy efficiency of layered building envelopes *St. Petersburg polytechnic university journal of engineering science and technology* **4 (254)** pp 28-33

[14] Musorina T A, Gamayunova O S and Petrichenko M R 2018 Thermal regime of enclosing structures in high-rise buildings *Vestnik MGSU* **8 (119)** pp. 935-943

[15] Musorina T A, Gamayunova O S and Petrichenko M R 2017 Substantiation of design measures to increase energy efficiency of exterior walls *Vestnik MGSU* **11 (110)** pp 1269–1277
[16] Sergeev V V, Petrichenko M R, Nemova D V, Kotov E V, Tarasova D S and Nefedova AV 2018 The building extension with energy efficiency light-weight building walls *Magazine of Civil Engineering* **8 (84)** pp 67-74
[17] Sergeev V V, Petrichenko M R, Nemova D V, Kotov E V, Tarasova D S, Nefedova A V and Borodinecs A 2018 The building extension with energy efficiency light-weight building walls. *Magazine of Civil Engineering* **84 (8)** pp 67-74
[18] Tseytin D N, Vatin N I, Nemova D V, Rymkevich P P and Gorshkov A S 2016 Feasibility study for renovation of facades heat insulation of residential buildings of the first mass series *Construction of Unique Buildings and Structures* **1 (40)** pp 20-31
[19] Martynenko E A, Staritcyna A A and Rybakov V A 2016 Reconstruction of the Residential District of St. Petersburg Historic Center *Construction of Unique Buildings and Structures* **1 (40)** pp 32-42
[20] Gumerova E, Gamayunova O and Meshcheryakova T 2018 Energy efficiency upgrading of enclosing structures of mass housing of the soviet union *Advances in Intelligent Systems and Computing* **692** pp 432-439
[21] Malyuk V I, Radaev A E and Silkina G Yu 2018 Procedure for determining the characteristics for development of industrial enterprises using optimization modeling tools *St. Petersburg State Polytechnical University Journal. Economics* **11 (6)** pp 195-211
[22] Radaev A E and Kobzev V V 2016 Optimizing the configuration for the supply network of industrial enterprises based on the probabilistic process model *St. Petersburg State Polytechnical University Journal. Economics* **6 (256)** pp 166-178

Energy saving potential of ventilation systems with exhaust air heat recovery

J Zemitis[1] and A Borodinecs[1]

[1] Riga Technical University; Heat, Gas and Water technology institute; Riga, Latvia, Kipsalas street 6A

E-mail: Jurgis.Zemitis@rtu.lv

Abstract. The paper presents the results of data gathering regarding the heat recovery efficiencies of different heat recovery types at various air flows supplied by some of the most common manufacturers. The study showed that the average heat recovery efficiency of modern, commonly used rotary heat exchangers is around 83% while for counter-flow exchangers around 86%. Also, in the paper, the simulation results of heat energy consumption and recovered heat potential for the single-family house are presented. The simulations are performed in IDA-ICE software and show that for the specific building it is possible to reduce the heating energy up to 84% if ventilation with heat recovery is used. The results also showed that in case the internal heat loads decrease, then it slightly affects the heat recovery potential. In the case of no internal gains, this value decreases by 4%.

1. Introduction

Modern, newly built buildings, as well as renovated ones, are quite often equipped with some sort of mechanical ventilation system, whether it would be hybrid type, full supply/exhaust system or local, personalized ventilation system [1],[2]. This is done to increase indoor air quality (IAQ) and, at the same time, to maximally reduce the heat losses that occur through the natural ventilation system. In existing researches [3] it is stated that ventilation heat recovery systems can give substantial final energy reduction, but the primary energy benefit depends strongly on the type of heat supply system. However, for renovated buildings or some lower priced newly built buildings, it could either not be possible or feasible to install such and simpler solutions could be chosen. This leads to the necessity to carefully choose the appropriate ventilation system type to maximize energy savings while providing good indoor air quality.

The adapted European norms require that all new buildings starting from the year 2021 must be meet the requirements to be considered nearly zero energy buildings (nZEB). The definition of nZEB as stated in Latvian Regulations for Energy Certification of Buildings Nr. 383 (Ministru kabineta noteikumi Nr.383 "Noteikumi par ēku energosertifikāciju") [4], regulates that the building can be classified as an almost zero energy building if it meets all the following requirements: the energy performance of the building for heating corresponds to Class A (residential buildings < 40 $kWh/m^2/year$; non-residential < 45 $kWh/m^2/year$), at the same time, providing indoor microclimate in accordance with the construction, hygiene, and labor protection requirements; total primary energy consumption for heating, hot water supply, mechanical ventilation, cooling, and lighting is not more than 95 $kWh/m^2/year$; high-efficiency systems are used in the building. These systems must include heat recovery from ventilation with efficiency at least 75%%; some usage of renewable energy sources; no installation of low-efficiency fossil fuel heating equipment in the building.

One of the most important engineering systems to achieve before mentioned parameters is to recover the heat from ventilation systems. As of now the combination of infiltration and ventilation is responsible for approximately 50% of the total heat losses in well-insulated buildings for a moderate Europe climate [5]. To reduce this, mechanical ventilation with heat recovery is used. Such systems can also help to significantly save primary energy, at the same time ensuring good indoor air quality while the airtightness of new and renovated buildings is increasing [3],[6],[7]. The ventilation systems have always been present in the buildings, but the exact solutions have changed during the years. The ventilation systems have evolved starting from simple natural ventilation with supply through construction cracks or windows to fully mechanical ducted supply/exhaust ventilation system with various in-between solutions. For example, integration of ventilation system in ventilated facades show promising results [8]. This makes the choice of the optimal ventilation solution difficult in each specific situation as each type ensures different comfort and control level but also varies in installation and running costs [9]. To select the optimal solution projected payback period for investments in energy saving must be calculated, taking into account the size of the investment, the estimated or actual value of the achieved energy saving effect, the dynamics of energy carriers tariff growth, the discounting of future cash flows, and also a value and a period of loan repayment [10].

Several different heat recovery types can be applied to ventilation systems [11], but especially in cases of nearly zero-energy buildings it is vitally important to take into consideration that the exhaust temperature after heat exchanger for fully mechanical supply/exhaust ventilation systems must be limited to 0° to +5°C, depending on the heat recovery type, as for such buildings more often the aim is to have as high heat recovery efficiency as possible [7]. In general, there are 3 types of ventilation heat recovery systems – rotary, plate heat exchangers (can be divided in cross-flow or counterflow) and run-around coils [12]. The heat recovery units can also be divided whether they are recuperative or regenerative. Recuperative heat exchanger means that the heat exchanger is intended to transfer thermal energy from one air stream to another without moving parts. Examples of these include a plate or tubular heat exchanger with the parallel flow, cross flow or counter flow, or a combination of these, or a plate or tubular heat exchanger with vapor diffusion. Example of the regenerative heat exchanger is a rotary heat exchanger which consist of parts like rotating wheel, a drive mechanism, a casing or frame, and seals to reduce bypassing and leakage of air from one stream or another. Such heat exchangers, if made from a special material which allows latent heat transfer, have varying degrees of moisture recovery capability depending on the material used [13]. For example, some manufacturers have the heat exchanger made with aluminum core with a silica gel-based coating. If such technology is applied then it can increase the IAQ as during the winter it is vitally important to recover as much moisture as possible, especially in the cold climatic conditions of Latvia. The existing studies have proven the benefit of such solutions [14].

Nowadays one of the most used ventilation heat recovery types is rotary. It can be applied both in domestic and commercial air handling units as it can be in various sizes. For large air volumes, a rotary heat exchanger can consist of multiple smaller units, therefore, saving space. Other types of heat recovery who have the risk of freezing consume extra energy for preheating, defrosting or reduce the supply air volume, therefore, compromising the IAQ. At the same time, one of the disadvantages of rotary heat exchangers are the potential of air cross-contamination from exhaust and supply, however, this can be reduced if special purging section is applied. On average the air overflow is in range of 0.04% to 5%. Estimated efficiency can be up to 85%.

Some years ago, the cross-flow plate-type heat exchanger was the most popular type of energy recuperation because of its relatively low cost and that it can be installed in a simple single-component ventilation system. It can be also be used in both domestic and commercial ventilation systems. Cross-flow heat exchangers do not have any modulation possibility, therefore, a risk of overheating or introducing extra heat in cooling period exists if no bypass is installed. In general, this type of recuperator has very high efficiency for extract and supply air separation - up to 99.9%, therefore it can serve bathrooms and office rooms in one system. As the heat recovery efficiency of such units is only up to 70%, they have lost their market share as they do not comply with the eco-design regulations.

Counter-flow plate type heat exchangers have the highest maximum energy efficiency. They can be installed in both domestic and commercial ventilation systems. The efficiency of such type of heat recovery units can reach up to 95-98% at outside air temperatures around -5° to 0°C. However, a drawback of such devices is their need for preheating and defrosting. At lower outside temperatures the heat exchanger starts to freeze up and to defrost it the fresh air supply is closing and blowing warm air through the heat exchanger in the recirculation mode, at this point the units become louder and no or reduced air exchange is ensured. To avoid this a preheating of the outdoor air to -5° / 0°C must be provided.

In addition to these, it is possible and, in many cases, more appropriate to use smaller decentralized, room-based ventilation systems as they give the possibility to install ventilation system with heat recovery in confined spaces. However, some additional parameters and climatic data must be accounted for. A study analyzing indoor relative humidity in apartments with room-based ventilation system with rotary heat exchanger indicated that it is suitable for single-room ventilation of dry rooms, such as living rooms and bedrooms, while excessive moisture from kitchens and bathrooms provided a mold risk [15]. Other publications indicate that a specific type of decentralized ventilation system with two heat exchangers, which are hydronically connected in parallel, is an appropriate solution for use in hot and humid climates because more cooling and dehumidification capacity is available [16].

A study [17] suggests that in renovated apartment buildings with natural passive stack ventilation, the indoor air quality is quite bad and has high CO2 concentration and relative humidity level. Similar data on IAQ problems in multi-apartment buildings shows research [18] done in Estonia. Some authors [19] have already performed analysis on how ventilation rates vary depending on the chosen system. Others [20] have analyzed the system influence on IAQ. Another aspect to take into consideration is the necessity to ensure the necessary ventilation or VOC control. The existing standards regarding ventilation volume usually do not account for this and that can lead to bad and even harmful IAQ.

The ventilation system heat recovery yearly efficiency is dependent on the specific project and the internal heat gains as well as designed air flows. To study how these parameters, affect the energy amount, which is recovered, dynamic simulations must be performed, and the obtained results analyzed.

2. Methods

The study in this paper is divided into two subtasks – data gathering of actual various ventilation system type heat recovery efficiency and dynamic energy consumption simulations of the specific building to analyze the energy saving potential and influence on comfort level.

At first the heat recovery efficiency of existing, in the market available, AHUs with various heat recovery types and air volumes was compared. This was done to gather data about the average heat recovery efficiency to correctly assume what level of heat recovery can be taken into dynamic simulations. In the last few years, more attention has been paid to energy efficiency including the minimal allowed heat recovery efficiency by AHUs. The Commission Regulation (EU) No 1049/2001 Requirement 1253/2014 implements eco-design requirements for ventilation equipment. The eco-design is a requirement that does not permit the design and installation of equipment with lower parameters, which is signed in this regulation. According to these requirements, the minimal heat recovery efficiency starting from the year 2018 is 73% for heat recovery systems without indirect medium while for run-around type systems the lower limit is 68%. This means that it is necessary to check the actual heat recovery efficiency to ensure that the building can be considered an nZEB.

To study how different technical solutions can influence the annual energy consumption for the whole building, various ventilation strategies in accordance with other parameters were tested with dynamic simulation software. The used software for the simulations was IDE-ICE which is widely used by professionals and researchers as the software enables is based on dynamic multi-zone simulations of thermal indoor climate as well as the energy consumption of the entire building. The main advantage of the software package is the possibility to obtain a detailed report for each of the building zones as well as for the whole building. The results include heat and airflows, maintained temperatures, sources of heat losses and energy costs to maintain a comfortable temperature.

The analyzed building is a two-story-high private residential building with a total heated area of 466.5 m^2. Both stories have similar planning and area. The ceiling height for premises located on the 1st floor is 2.85 m, while for the 2nd floor the height differs from 2.4 to 3.0 m, as the roof is slightly sloped. For the purpose of analysis, it is assumed that the building is located in Latvia and weather data of the city of Kolka is used. The U-values of the external envelope elements are as follows: External wall - 0.17 W/(m^2·K); Windows – 0.82 W/(m^2·K) and g-value of 0.5; Floor on slab - 0.12 W/(m^2·K); Roof - 0.11 W/(m^2·K). All of these correspond to the required values of the Latvian building code. The temperature setpoint for the heating season was set to be +21°C, while for cooling season to be at +25°C.

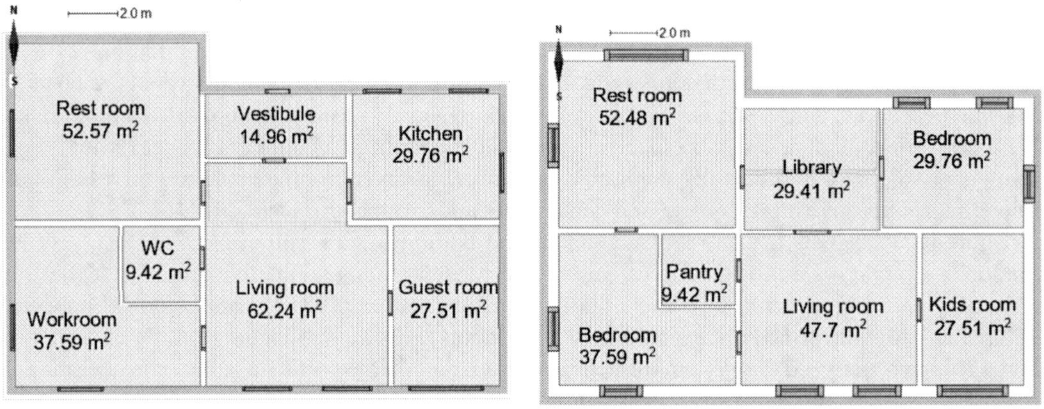

Figure 1. Floor plans of the analyzed residential building.

For the first simulation scenario, it was assumed that the building is equipped with a constant air volume ventilation system which ensures close to 0,5 air exchange rate for the whole building or in total 640 m^3/h. The assumed heat recovery efficiency of AHU is set at 85%, which represents close to the average value according to the previously performed study (Table 1). Also, the internal heat gains from people, lighting, and equipment were simulated. For each of the rooms, a specific occupancy profile was created. For all the living rooms it was assumed that the persons will be present at home from 17:00 in the evening until 8:00 in the next day at weekdays and all the time on holidays. For the bedrooms, the occupancy schedule was set to be occupied from 22:00 in the evening to 6:00. As the building is relatively large the maximal number of occupants reach 16 persons (Figure 2). Although this would most likely not be realistic, such value is left as 30 m^2 per person is still quite representative.

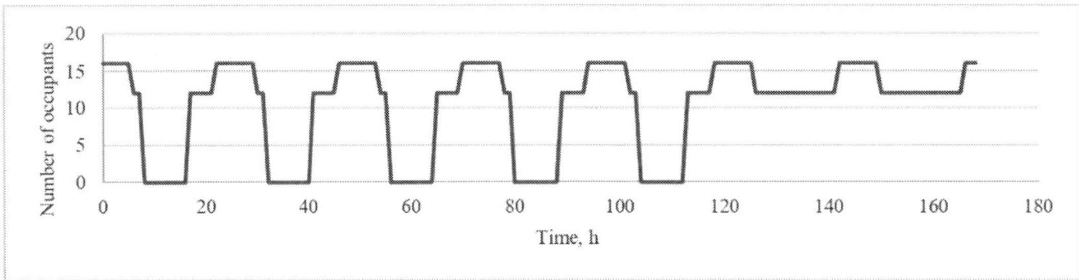

Figure 2. Weekly overall occupancy schedule for simulated building.

The schedule of equipment used is also adjusted according to occupancy. It is foreseen that equipment is used from 6:00 to 8:00 and from 17:00 to 23:00 on weekdays, while on Saturdays and Sundays it is used all the time. The lighting schedule is a bit more complex and is dependent on the time of the year. For the period from 1st September to 30th April the light is on from 6:00 to 8:00 and from 17:00 to 23:00, while for the rest of the year the light is switched on later – at 20:00.

3. Results and discussion

3.1. Heat recovery efficiency of different heat recovery types

To check how the eco-design regulation has affected the improvement of heat recovery units, data of typically used ventilation units with specified air volumes were gathered. The analyzed manufacturers include Komfovent, Systemair, Flaktwood and IV Produkt. For each of them, a unit ensuring 400, 700, 1000 and 5000 m^3/h has been looked at and the specified heat recovery value noted. The parameters to perform the energy efficiency calculations at winter regime for all of the devices were: outdoor temperature -21°C and RH 80%, while indoor temperature +20°C and RH 40%. This represents the data for a cold region like Latvia. The results are shown in Table 1 and Table 2.

Table 1. Comparison of heat recovery temperature efficiency for different AHU models with a rotary heat exchanger at various air flow volumes.

Manufacturer	Rotary Heat Exchanger			
	Model name	**400 m³/h**	**Model name**	**700 m³/h**
Komfovent	DOMEKT-R-400-H-R1-M5/M5-C6-L/A	81	DOMEKT-R-700-V-R1-M5/M5-C6-L/A	81
Systemair	SAVE VTR 500 L	84	SAVE VTR 700 L	83
Flaktwoods	eCO Top 03	88.8	eQ Prime 005	91.5
IV Produkt	ACER-04-AA-N1-01	85.3	ACER-06-AA-N1-01	85
Average value:		**84.8**		**85.1**
Manufacturer	**Model name**	**1000 m³/h**	**Model name**	**5000 m³/h**
Komfovent	VERSO-R-1300-UH-CW-R1-F7/M5-C5.1-L/A	81.5	VERSO-R-5000-H-W-R1-F7/M5-C5.1-L/A	75.2
Systemair	TOPVEX_TR03_HWH-L-CAV_M0	82.6	TOPVEX_TR12_HWH-L-CAV_M0	78
Flaktwoods	eQ Prime 005	88.5	eQ Prime 018	85
IV Produkt	ACER-06-AA-N1-01	83.4	EMMT-05-240-50-2-2200	79.5
Average value:		**84.0**		**79.4**

Table 2. Comparison of heat recovery temperature efficiency for different AHU models with a counter-flow heat exchanger at various air flow volumes.

Manufacturer	A counter-flow heat exchanger			
	Model name	**400 m³/h**	Model name	**700 m³/h**
Komfovent	DOMEKT-CF-500-F-R2-M5/M5-C6-X	90	DOMEKT-CF-700-F-R1-M5/M5-C6-X	87
Systemair	SAVE VTC 300 L	90	SAVE VTC 700 R	92
Flaktwoods	Eco Premium 1	84.1	eQ Prime 005	87.9
IV Produkt	-	-	ATEM-04-AA-EN-CI-H-00	80.5
Average value:		**88.0**		**86.9**
Manufacturer	Model name	**1000 m³/h**	Model name	**5000 m³/h**
Komfovent	VERSO-P-10-C-H-PM/IE5/1.4/1.4-F7-M5-X-X-R1-C5.1-X	73.4	VERSO-P-50-2-H-PM_IE5_2_9_2_9-F7-M5-X-X-R1-C5_1-X	73.5
Systemair	Topvex SX/C03 EL-R-CAV	90	Topvex SC11 HW-L-CAV	90
Flaktwoods	eQ Prime 005	85.3	eQ Prime 018	83
IV Produkt	ATEM-06-AA-EN-CI-H-00	80.6	EMMT-05-240-50-3-3800	78.8
Average value:		**82.3**		**81.3**

The results of the comparison of heat recovery efficiency show that for both the rotary type and counter flow plate type exchangers the values are very similar (Figure 3). The total average of all models and all air flow volumes showed that in the case of a rotary heat exchanger the temperature efficiency is around 83% while for counter-flow exchangers around 86%. This means that both solutions are a viable option and can be used in cases of new, low energy consuming buildings. The analysis of the

IOP Conf. Series: Materials Science and Engineering **660** (2019) 012019 doi:10.1088/1757-899X/660/1/012019

results showed that the models supplied by Komfovent have a bit lower heat recovery efficiency in comparison to others, but this can vary case by case and is strongly depended on the actual airflow volume and how close it is to the next type size for the specific manufacturer.

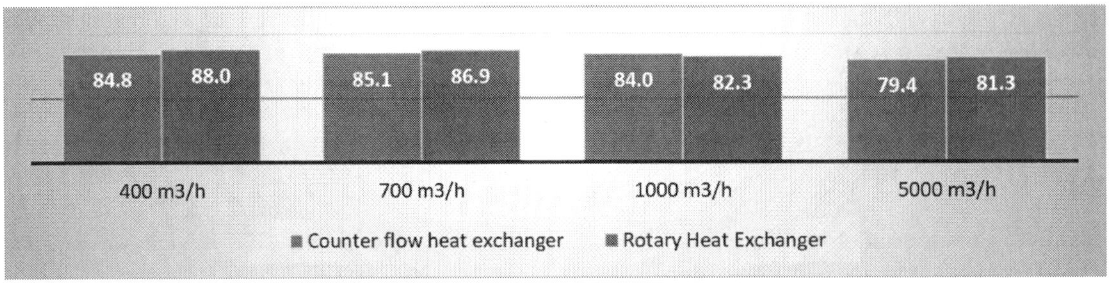

Figure 3. Comparison of average heat recovery temperature efficiencies at various air volumes for counter-flow and rotary heat exchangers.

3.2. Simulation results of heat recovery energy potential for single family house

The following simulations can be divided into two separate types. One of them presents the results for the maximal needed heating/cooling power. These are calculated for the most critical day of the year when the temperature is lowest for more days in a row and no heat gains are accounted for or when the temperature is high, and the sun is shining the most in case of a cooling load. The second type of calculations show the results for the whole year simulation. They represent the total amount of consumed and supplied energy as well as energy balance.

The results show that the simulated building in the first scenario would need design heating power of 7.47 kW. This consists of 6.48 kW that are necessary to compensate for transmission heat losses and 0.94 kW that are needed for ventilation air heating. The temperatures in all the premises are close to set point of +21°C, except two rooms that do not have a heating element in them and are heated by the enclosing rooms. As for cooling load, the simulations show that the maximal necessary cooling energy is on 15th September. At this date, the cooling load reaches 11.4 kW.

For the whole year simulation, the results show the energy balance for the situation, which most closely resembles the real-life operation of the building. The total energy for heating, in this case, amounts to 4178 kWh/year or 8.92 kW/h/m^2/year. The energy consumption for cooling in the given case is 8465 kW or 18.1 kW/h/m^2/year.

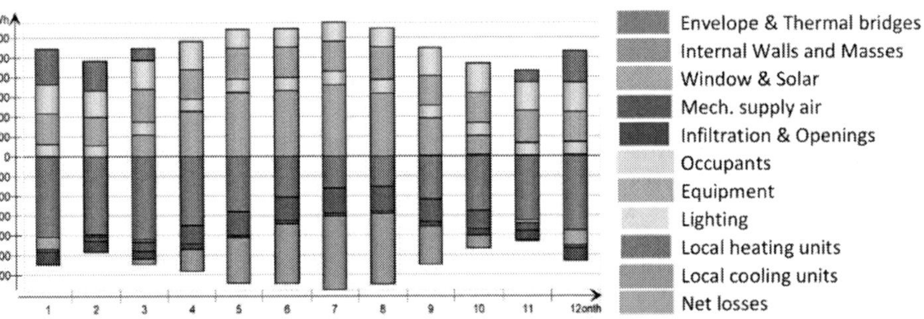

Figure 4. Division of energy balance for a year at various months.

The energy balance shows (Figure 4) the division of different internal heat gains and losses. The results show that for given building the internal heat gains from equipment is the major source of heating. Also, the solar gains and energy from lighting are responsible for a large part of passive heating energy. In total the annual heating energy for the building is 32.74 MWh. Of this only 3 MWh are supplied through the heating system while the rest 29.67 MWh or 93.8 % are supplied by solar and internal heat gains. Also, the recovered energy by the AHU was simulated and the results showed that for the whole year period in total 23328.8 kWh was recovered. The division by the months is shown in Figure 5.

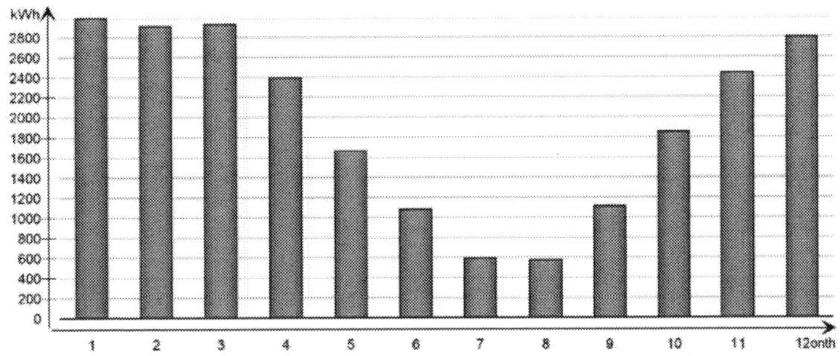

Figure 5. Recovered energy by the AHU heat recovery for each month.

To see what the results for the building would be if no internal heat sources would be present a comparative simulation was performed. Only in this case the lighting, occupancy, and equipment are take out. The results show that in such a case the total heating energy would amount for 14152.7 kWh/year or 30.25 kWh/m^2/year. The cooling energy would be decreased to 1542 kW or 3.3 kWh/m^2/year. Also, the annual recovered heating energy by the AHU decreased by 800.4 kWh to 22528.4 kWh.

If a cooling system is added that the AHU can also recover cooling energy. The simulation results showed that in total 0.00368 kWh of cooling energy can be recovered. Therefore, the potential for this is small. If the AHU is equipped with indirect evaporative cooling section than the recovered heating energy is only 14674.8 kWh while recovered cooling energy reaches 44.17 kWh. In case the heat recovery type is changed to rotary type with potential to recover also moisture than the yearly recovered energy amounts to 16471.9 kWh but the necessary energy for heating is 1479.1 kWh while for cooling 1017.3 kWh.

4. Conclusions

The study showed that the average heat recovery efficiency of modern, commonly used rotary heat exchangers is around 83% while for counter-flow exchangers around 86%. Therefore, it can be concluded that they achieve the requirements of eco-design. Also, the results suggest that it could be viable to increase the required minimal heat recovery efficiency to achieve nearly zero energy building status according to Latvian Regulations for Energy Certification of Buildings Nr. 383.

The simulation results for a single-family residential building with a floor area of 467 m^2 and ventilation air volume of 640 m^3/h show the potential of heat recovery from ventilation exhaust air. The total heating energy for a year of the simulated building is 4178.1 kWh, if the AHU is equipped with a heat recovery unit with an efficiency of 85%. For the whole year period in such a case, the 23328.8 kWh was recovered. Therefore, it can be concluded that if no heat recovery would be present than the energy consumption would rise to 26964.8 kWh or 57.8 kWh/m^2. The results also showed that in case the internal heat loads decrease, then it slightly affects the heat recovery potential. In case no internal gains are present, it decreases by 4%.

In future, it is necessary to perform further simulations regarding heat recovery potential of ventilation systems under various circumstances to find the highest potential work case. Also, it would be necessary to simulate how the variable air volume systems would affect the energy consumption for heating of supply air.

Acknowledgments

This work has been supported by the European Regional Development Fund within the Activity 1.1.1.2 "Post-doctoral Research Aid" of the Specific Aid Objective 1.1.1 "To increase the research and innovative capacity of scientific institutions of Latvia and the ability to attract external financing, investing in human resources and infrastructure" of the Operational Programme "Growth and Employment" (No. 1.1.1.2/VIAA/1/16/033).

References

[1] Kim M K and Baldini L 2016 Energy analysis of a decentralized ventilation system compared with centralized ventilation systems in European climates: Based on review of analyses *Energy Build.* **111** 424–33
[2] Naumov A L, Tabunshchikov I A, Kapko D V. and Brodach M M 2015 Research of the microclimate formed by the local DCV *Energy Build.* **90** 1–5
[3] Dodoo A, Gustavsson L and Sathre R 2011 Primary energy implications of ventilation heat recovery in residential buildings *Energy Build.* **43** 1566–72
[4] Anon 2013 Ministru kabineta noteikumi Nr.383 Noteikumi par ēku energosertifikāciju
[5] Laverge J and Janssens A 2012 Heat recovery ventilation operation traded off against natural and simple exhaust ventilation in Europe by primary energy factor, carbon dioxide emission, household consumer price and exergy *Energy Build.* **50** 315–23
[6] Silva M F, Maas S, Souza H A de and Gomes A P 2017 Post-occupancy evaluation of residential buildings in Luxembourg with centralized and decentralized ventilation systems, focusing on indoor air quality (IAQ). Assessment by questionnaires and physical measurements *Energy Build.* **148** 119–27
[7] Vinha J, Manelius E, Korpi M, Salminen K, Kurnitski J, Kiviste M and Laukkarinen A 2015 Airtightness of residential buildings in Finland *Build. Environ.* **93** 128–40
[8] Petrichenko M, Nemova D V, Kotov E V, Tarasova D and Sergeev V V 2018 Ventilated facade integrated with the HVAC system for cold climate *Mag. Civ. Eng.* **77** 47–58
[9] Zemitis J, Borodinecs A and Kalamees T 2018 Analysis of Various Ventilation Solutions for Residential and Non-residential Buildings in Latvia and Estonia *Springer Proceedings in Energy* pp 51–61
[10] Gorshkov A, Vatin N, Rymkevich P P and Kydrevich O O 2018 Payback period of investments in energy saving *Mag. Civ. Eng.* **78** 65–75
[11] Alonso M J, Liu P, Mathisen H M, Ge G and Simonson C 2015 Review of heat/energy recovery exchangers for use in ZEBs in cold climate countries *Build. Environ.* **84** 228–37
[12] CIBSE 2004 *CIBSE Guide F: Energy efficiency in buildings*
[13] Anon Commission Regulation (EU) No 1049/2001 Requirement 1253/2014
[14] Zhang M, Qin M, Rode C and Chen Z 2017 Moisture buffering phenomenon and its impact on building energy consumption *Appl. Therm. Eng.*
[15] Smith K M and Svendsen S 2016 The effect of a rotary heat exchanger in room-based ventilation on indoor humidity in existing apartments in temperate climates *Energy Build.* **116** 349–61
[16] Kim M K, Baldini L and Leibundgut H 2013 Advanced Decentralized Ventilation Systems for Cooling and Dehumidification with Compact Heat Exchanger *11th REHVA World Congress CLIMA 2013* (Prague, Czech Republic)
[17] Dimdina I, Krumins E and Lesinskis A 2014 Indoor Air Quality in Multi-Apartment Buildings before and after Renovation *Constr. Sci.* 4–10
[18] Voll H, Kõiv T-A, Kuusk K, Maivel M and Mikola A 2010 Indoor Climate and Energy Consumption in Residential Buildings in Estonian Climatic Condition *WSEAS Trans. Environ. Dev.* **6** 247–56
[19] Santos H R R and Leal V M S 2012 Energy vs. ventilation rate in buildings: A comprehensive scenario-based assessment in the European context *Energy Build.* **54** 111–21
[20] Ben-David T and Waring M S 2016 Impact of natural versus mechanical ventilation on simulated indoor air quality and energy consumption in offices in fourteen U.S. cities *Build. Environ.* **104** 320–36

IOP Conf. Series: Materials Science and Engineering **660** (2019) 012020 doi:10.1088/1757-899X/660/1/012020

Numerical investigation of influence of the temperature and relative humidity of air on the drying process of porous building

M Wasik[1], P Łapka[1], Ł Cieślikiewicz[1], M Seredyński[1], P Furmański[1], K Pietrak[1], M Kubiś[1], T Wiśniewski[1] and M Jaworski[1]

[1]Warsaw University of Technology, Faculty of Power and Aeronautical Engineering, Institute of Heat Engineering, 21/25 Nowowiejska St, 00-665 Warsaw, Poland

E-mail: michal.wasik@itc.pw.edu.pl

Abstract. The impact of the relative humidity and temperature of air on drying rates of the porous building material was investigated by the non-equilibrium heat and moisture transfer model implemented in the ANSYS Fluent software. Simulations of 24 hours drying process of a brick initially saturated in 97% by water were conducted for two drying air temperature, i.e., 23.8°C and 40°C and relative humidity in the range from 10% to 50%. The relative moisture content at the end of simulations and times needed to reach the relative moisture content bellow 0.3 were determined. Not only the temperature but also the relative humidity of drying air had impact on the drying rate during the first drying period when water evaporated from the sample surface. The temperature of drying medium was more important than the relative humidity during the second drying period when water evaporated inside the brick. The drying process was faster in the higher temperature than in the lower one, even for the high relative humidity.

1. Introduction

Masonry walls and building foundations are under threat of presence of excessive moisture. Natural disasters like floods and storms, the high level of ground water and the lack or worn out of the waterproof insulations are typical reasons of accumulation of excessive moisture in the building structure. Consequences of increased damp in walls are very serious, i.e., it deteriorates the physical and chemical conditions of the wall [1, 2] and increases heat gains and losses to the surroundings which results in the generation of higher cooling and heating costs [3, 4]. Moreover, excessive moisture in the wall encourages growth of mildew and microorganisms, which are dangerous for occupants [5]. Hence, renovation of damp walls, i.e., drying and preventing against water re-penetration, is very important.

There are several methods of moist solids drying, e.g., convective [6, 7], microwave [8, 9] and vacuum [10]. For damp walls the thermo-injection method [11-13] which relies on the convective drying is very effective. The method was developed in Poland in the nineties of the previous century. It is based on drilling an array of boreholes in the damp wall. In the first step, these boreholes are used for heating and drying of the wall by special heating probes. Then in the second step, boreholes are used again for injection of a special fluid mixture which creates waterproof membrane. The membrane is an effective and durable way of preventing against water re-penetration through the wall. The dryness and temperature of the wall during injection of the hydrophobic fluid have an impact on the effectiveness of the membrane creation, i.e., the lower moisture content and the higher temperature of the wall, the better membrane quality. However, temperature during the membrane formation should be lower than approx. 60°C because the cross-linking of the hydrophobic mixture occurs for temperature above 60°C.

The drying process of masonry walls is characterized by long duration and high energy consumption. Hence, the renovation and protection of walls against water re-penetration is very expensive. Preconditioning of air used as drying medium allows to speed up the process but increases energy consumption due to air heating or dehumidification. On the other hand, long operation of drying devices generates the spending and decreases the income. All these aspects should be considered and balanced to get optimal economic conditions for the drying of building structure. The numerical modeling and optimization of drying process may help to achieve this objective.

Combined heat and moisture transfer in porous medium is common not only in building materials but also in insulations [14, 15], protecting clothing [16-19], wood drying [20] and food processing [21]. In the field of simulation of transport phenomena in building materials two main groups of models are distinguished, i.e., building energy simulation (BES) models as well as heat and air and moisture (HAM) transfer models. The commercial software like TRNSYS, ESP-r and EnergyPlus are based on BES models in which heat and moisture transfer is treated in a simplified way. On the other hand, the WUFI and Delphin software utilize simplified HAM models. To achieve better results HAM models are combined with computational fluid dynamics (CFD) models. For example, equilibrium CFD-HAM models with two equations, i.e., energy and moisture balance equations, were proposed in [21-24]. These models predict the total moisture content and then by applying additional relations allow for computation of vapor and water fractions in the porous material.

In this paper the novel non-equilibrium heat and moisture transfer model is presented and was applied to find the impact of air parameters, i.e., the temperature and relative humidity on drying rates of porous building material. The new model was configured to simulate one-dimensional (1D) transport process in the brick. The analysis carried out showed utility of the model for investigation and optimization of the drying process of building materials and structures.

2. Non-equilibrium mathematical model of transport phenomena in the porous building material

Moisture which resided in the porous building material was assumed to consist of two phases, i.e., water vapor and free liquid water (in funicular and pendular state). Bound water at the surface of the solid component was neglected. The sum of volume fractions of all phases satisfied the following condition:

$$\varepsilon_s + \varepsilon_l + \varepsilon_g = 1 \tag{1}$$

where: subscripts g, l and s denote moist air, water and solid component, respectively and ε is the volume fraction. The volume fraction of solid (ε_s) was constant during analysis, while the sum of volume fraction of water and moist air was equal to the volume fraction of pores ($\varepsilon_p = \varepsilon_l + \varepsilon_g$).

For considered problem three balance equations were formulated, i.e., water vapor, liquid water and energy balance equations instead of four balance equations as in [25, 26]. The pressure in pores was assumed constant and equal to $p = 101325$ Pa, which allowed for simplification of mathematical description and elimination of the dry air balance equation.

2.1. Balance of moisture

The amount of moisture in pores changed due to the diffusion and capillary pressure gradient. Moreover, the amount of vapor and water varied due to mass transfer between phases, i.e., evaporation and condensation. Hence, balance of vapor and liquid water were given by the following equations:

$$\frac{\partial}{\partial t}\left(\varepsilon_g \rho_v\right) = \nabla\left(D_{v,ef} \nabla \rho_v\right) + \dot{m}_{lv} \tag{2}$$

$$\frac{\partial}{\partial t}\left(\varepsilon_l \rho_l\right) = -\nabla\left(K_l \nabla p_c\right) - \dot{m}_{lv} \tag{3}$$

where: subscript v denotes vapor, $D_{v,ef}$ is the vapor effective diffusivity in pores, K_l – liquid water permeability, $\dot{m}_{lv}$ – intensity (mass flow rate per unit volume) of evaporation/condensation, p_c – capillary pressure and ρ – density. Application of two separated balance equations for gaseous and liquid moisture allowed for formulation of the model with non-equilibrium between moisture phases. Moreover, evaporation and condensation were assumed to have finite rate and their intensities were described by following equations:

$$\dot{m}_{lv} = \begin{cases} h_{vl} a_s \dfrac{\varepsilon_l}{\varepsilon_p} \left(\rho_{v,sat} - \rho_v\right) & \text{– for evaporation} \\ h_{vl} a_s \left(\rho_{v,sat} - \rho_v\right) & \text{– for condensation} \end{cases} \tag{4}$$

where: a_s denotes the pores area per unit volume of the porous medium, h_{vl} – mass transfer coefficient between vapor and water in the porous medium and $\rho_{v,sat}$ – vapor density for the saturation conditions. The $\varepsilon_l/\varepsilon_p$ ratio denotes the part of the pores area (volume) occupied by liquid moisture. The mass flow rate per unit volume had positive value for evaporation and negative one for condensation.

2.2. Balance of energy

In the model the thermal equilibrium between different components and phases in the porous medium was assumed, i.e., all constituents had the same temperature. Moreover, heat transfer due to moisture movement (i.e., the diffusion and capillary transport) was neglected. The energy balance equation in the porous building material was following:

$$\frac{\partial}{\partial t}\left[(\rho c)_{ef} T\right] = \nabla \cdot \left(k_{ef} \nabla T\right) - \dot{m}_{lv} \Delta h_{lv} \tag{5}$$

where: k_{ef} is the effective thermal conductivity, T – temperature and Δh_{lv} – latent heat of evaporation. The effective heat capacity was described by the formula:

$$(\rho c)_{ef} = \varepsilon_s \rho_s c_s + \varepsilon_l \rho_l c_l + \varepsilon_g \rho_a c_{p,a} + \varepsilon_g \rho_v c_{p,v} \tag{6}$$

where: subscript a stands for dry air and c and c_p denote the specific heat and specific heat at constant pressure, respectively.

Table 1. Thermophysical properties assumed in simulations [27].

Property	Symbol	Value
Universal gas constant	B (J/mol/K)	8.314
Dry air specific heat	c_a (J/kg/K)	1005.0
Water specific heat	c_l (J/kg/K)	4192.1
Brick specific heat	c_s (J/kg/K)	840.0
Vapor specific heat	c_v (J/kg/K)	1875.2
Water vapor resistance diffusion factor	C_{dry}	24.79
Average pore diameter	d_{av} (m)	$1.6 \cdot 10^{-5}$
Mass transfer coefficient	h_{vl} (m/s)	10^{-4}
Dry air molecular mass	M_a (kg/kmol)	28.86
Vapor molecular mass	M_v (kg/kmol)	18.0
Capillary moisture content	W_{cap} (kg/m^3)	130.0
Brick porosity	ε_p	0.13
Latent heat of evaporation	Δh_{lv} (J/kg)	$2.6 \cdot 10^6$
Water density	ρ_l (kg/m^3)	1000.0
Brick density	ρ_s (kg/m^3)	2087.0

3. Numerical simulations

Preparation of the numerical model consisted of the following steps: creation of the computational geometry, generation of the mesh and implementation of transport equations in the commercial software

ANSYS Fluent which based on the Finite Volume Method. Advanced customisation interfaces, i.e., the User Define Function (UDF), User Define Scalar (UDS) and User Define Memory (UDM) were used during development of the numerical model. The computational model was constrained to two-dimensional (2D) space.

The developed 2D numerical model was configured to simulate 1D heat and moisture transport in the brick of 3 cm thickness by applying symmetry boundary conditions at lateral walls. The mesh had 90 elements along the thickness of the brick and was refined to the top wall. The time step size was equal to 0.5 s. This allowed to carry out simulations in reliable time.

3.1. Material properties and closing relationships

Computations were conducted for the ceramic brick whose thermophysical properties used in simulations are given in table 1. Dry and moist air as well as vapor parameters were calculated using the ideal gas relationship. Closing relationships for the system of governing equations, eq. (2), (3) and (5), were following [27]:

- Pores area per unit volume:

$$a_s = \frac{6}{d_{av}(1-\varepsilon_s)} \tag{7}$$

- Vapor diffusivity in pores:

$$D_{v,ef} = \frac{2.61\cdot 10^{-5} M_v \left[1-\left(\frac{W}{W_{cap}}\right)\right]}{C_{dry} BT \left[0.503\left(1-\frac{W}{W_{cap}}\right)^2 + 0.497\right]} \tag{8}$$

- Effective thermal conductivity of the moist brick:

$$k_{ef} = k_s + 0.0047W \tag{9}$$

- Water permeability in the brick:

$$K_l = \frac{1.1437\cdot 10^{-9}}{\left[1+\left(1.76\cdot 10^{-5} p_c\right)^{4.3}\right]^{1.6}} \tag{10}$$

- Vapor saturation pressure:

$$p_{v,sat} = 614.3\exp\left(17.06\frac{T-273.15}{T-40.25}\right) \tag{11}$$

- Modified saturation pressure:

$$p^{*}_{v,sat} = p_{v,sat}\exp\left(-\frac{p_c}{\rho_l BT}\right) \tag{12}$$

- Volumetric moisture content:

$$W = \varepsilon_l \rho_l + \varepsilon_v \rho_g \tag{13}$$

- Retention curve which is applied to find the capillary pressure:

$$W(p_c) = W_{cap}\left\{0.846\left[1+\left(1.394\cdot 10^{-5} p_c\right)^4\right]^{-0.75} + 0.154\left[1+\left(0.9011\cdot 10^{-5} p_c\right)^{1.69}\right]^{-0.408}\right\} \tag{14}$$

- Water vapor density for saturation conditions:

$$\rho_{v,sat} = \frac{p^*_{v,sat} M_v}{BT} \tag{15}$$

where: B is the universal gas constant, C_{dry} – water vapor resistance diffusion factor, d_{av} – average pore diameter, M_v – molecular mass of vapor and W_{cap} – capillary moisture content.

3.2. Boundary and initial conditions

For simulated 1D case only the top wall was in the contact with flowing (drying) air. The bottom wall was adiabatic and impermeable for moisture. Side walls were symmetry planes. On the top wall the following convective boundary condition with liquid moisture evaporation accounted for was applied:

$$q = h_T \left(T_{amb} - T_s\right) + \Delta h_{lv} \mathbf{j}_{l,w} \tag{16}$$

where: subscript w denotes the wall, h_T is the convective heat transfer coefficient (h_T = 22.5 W/m^2/K), T_{amb} – temperature of drying air and $\mathbf{j}_l$ – liquid moisture flux which evaporated/condensed at the surface.

For the first period of drying (evaporation from the surface) when liquid moisture in the pendular state is present at the boundary (the liquid saturation s is higher than the minimal saturation for water in the pendular form $s > s_{min} = 0.25$), mass fluxes of vapor and liquid moisture were following:

$$\mathbf{j}_{l,w} = \left(-K_l \nabla p_c\right)_w = h_m \left(Y_e - \frac{\rho_{v,w}}{\rho_{g,w}}\right) \quad \text{and} \quad \mathbf{j}_{v,w} = \left(D_{v,ef} \nabla \rho_v\right)_w = 0 \tag{17}$$

where: h_m is the mass transfer coefficient (h_m = 0.0258 kg/m^2/s) and Y_e – vapor mass fraction in flowing air. When liquid saturation dropped below minimum saturation (i.e., $s < s_{min} = 0.25$) boundary conditions were following:

$$\mathbf{j}_{l,w} = \left(-K_l \nabla p_c\right)_w = 0 \quad \text{and} \quad \mathbf{j}_{v,w} = \left(D_{v,ef} \nabla \rho_v\right)_w = h_m \left(Y_e - \frac{\rho_{v,w}}{\rho_{g,w}}\right) \tag{18}$$

Initial conditions were following: the uniform brick temperature of T_{init} = 23.8°C and water saturation of s_{init} = 97% which corresponded to the volumetric moisture content of W_{init} = 126.1 kg/m^3.

Table 2. Boundary conditions assumed in simulations.

Case no.	Temperature (°C)	Relative humidity (%)	Humidity ratio (g H_2O/kg dry air)
1	23.8	10	1.848
2	23.8	20	3.706
3	23.8	30	5.576
4	23.8	40	7.457
5	23.8	50	9.349
6	40	10	4.649
7	40	20	9.367
8	40	30	14.158
9	40	40	19.021
10	40	50	23.960

All together 10 cases were analysed with various drying air temperature (i.e., 23.8 and 40°C) and relative humidity (i.e., from 10 to 50%). Duration of the drying process was 24 h. First four cases corresponded to the low drying medium temperature equal to the initial temperature of the brick. In the next five cases the higher temperature than the initial one was assumed which speeded up the drying process. The assumed air parameters for each case were presented in table 2.

Table 3. Relative moisture content after 24 h and drying time to the relative moisture content of 0.3.

Case no.	Relative moisture content after 24 h	Drying time to the relative moisture content of 0.3 (h)
1	0.237	13.35
2	0.245	14.04
3	0.269	14.94
4	0.277	16.17
5	0.284	17.92
6	0.166	10.89
7	0.180	11.75
8	0.197	12.81
9	0.218	14.17
10	0.241	15.95

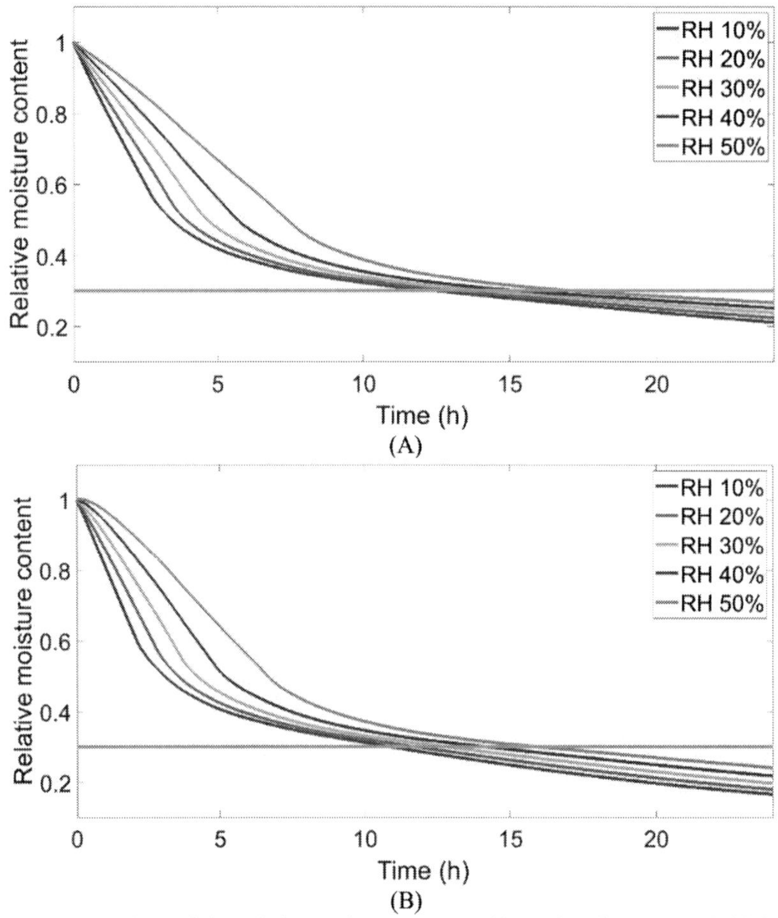

Figure 1. Temporal variations of the relative moisture content (the ratio of current to initial moisture content) for different temperatures and relative humidities of drying air: A) 23.8°C and B) 40°C.

4. Results and discussion

Two parameters were monitored during simulations: the relative moisture content (the ratio of current to initial moisture content) after 24 h of drying process and the time instant for which the relative moisture content dropped below 0.3, which corresponded to 2% of the mass fraction of moisture in the brick – see table 3. It was found that the temperature of drying medium had significant impact on the drying rate. This influence was more important than the impact of humidity ratio (relative humidity).

For example, cases no. 5 and 7 had almost the same humidity ratio (see table 2) but large difference in the drying rates between these cases was found. For case no. 7 the relative moisture content of 3% was achieved 6 h earlier than for case no. 5. On figure 1 temporal variations of the relative moisture content are presented against variable temperature and relative humidity of drying air. It is noticed that in the first drying period almost 50% of water was removed from the sample. Moreover, drying rates for the first drying period, when water evaporated from the surface and specimen temperature decreased, were larger for the higher temperature of drying air due to more intensive evaporation from the sample surface. It is also interesting that for case no. 9 and 10 initial temperature of the sample was lower than the wet bulb temperature for drying air. Hence, at the beginning of the drying process vapor from air was condensed at the surface of the brick – see short plateau on figure 1 B for the relative humidity of 40 and 50%. When the brick surface temperature raised above the wet bulb temperature the relative moisture content in the specimen started decreasing. During the second drying period, when the temperature of the specimen started rising and the drying process was controlled by the diffusion of vapor in the specimen, differences between drying rates for analysed cases were not so large but the higher drying rates were found for the higher temperature of drying air. The relative humidity had minor influence on drying rates in this period.

5. Conclusions

In this paper the new non-equilibrium HAM model for porous building materials was presented. The model was then applied for analysis of drying process of the brick. Influence of the temperature and relative humidity of drying air was investigated. Both the temperature and relative humidity of drying air were found to have impact on drying rates. However, the temperature more significantly influenced on the drying rates than the relative humidity especially in the first drying period. Drying rates in this period were controlled by the water evaporation from the surface of the sample. The intensity of evaporation was larger in higher temperatures. The relative humidity of drying air also affected this intensity as it influenced on the wet bulb temperature. During the second drying period, when the drying process was controlled by the diffusion of vapor in the brick, slight influence of the drying air temperature and very weak influence of the relative humidity on drying rates were found. Concluding, to obtain high drying rates, warm and dry air should be used in the first drying period, while warm with the normal relative humidity in the second one. The ambient air, not preheated, may be also used for the drying but the drying rate will be lower than for warm air.

In the presented work calculations were conducted applying the simplified model and geometry. Therefore, the results obtained should be used for the initial optimization of the drying process. In the future the model will be applied for the more complex 2D and 3D cases with the air flow accounted for.

Acknowledgments

This work was supported by the National Centre for Research and Development (Poland) under grant no. POIR.04.01.02-00-0099/16 "Development of innovative technology of drying and moisture sealing of masonry walls, DryWall".

References

[1] Lourenço P B, Luso E and Almeida M G 2006 Defects and moisture problems in buildings from historical city centres: A case study in Portugal *Build. Environ.* **41** 223–34
[2] Karagiannis N, Karoglou M, Bakolas A, Krokida M and Moropoulou A 2017 Drying kinetics of building materials capillary moisture *Constr. Build. Mater.* **137** 441–9
[3] Mendes N, Winkelmann F C, Lamberts R and Philippi P C 2003 Moisture effects on conduction loads *Energy Build.* **35** 631–44
[4] Osanyintola O F and Simonson C J 2006 Moisture buffering capacity of hygroscopic building materials: Experimental facilities and energy impact *Energy Build.* **38** 1270–82
[5] Shoemaker R C and House D E 2006 Sick building syndrome (SBS) and exposure to water-damaged buildings: Time series study, clinical trial and mechanisms *Neurotoxicol. Teratol.* **28** 573–88

[6] Castro A M, Mayorga E Y and Moreno F L 2018 Mathematical modelling of convective drying of fruits: A review *J. Food Eng.* **223** 152–67
[7] Boutelba I, Zid S, Glouannec P, Magueresse A and Youcef-ali S 2018 Experimental data on convective drying of potato samples with different thickness *Data Br.* **18** 1567–75
[8] Das I and Arora A 2018 Alternate microwave and convective hot air application for rapid mushroom drying *J. Food Eng.* **223** 208–19
[9] Szadzińska J, Kowalski S J and Stasiak M 2016 Microwave and ultrasound enhancement of convective drying of strawberries: Experimental and modeling efficiency *Int. J. Heat Mass Transf.* **103** 1065–74
[10] Zecchi B, Clavijo L, Martínez Garreiro J and Gerla P 2011 Modeling and minimizing process time of combined convective and vacuum drying of mushrooms and parsley *J. Food Eng.* **104** 49–55
[11] Kaliszuk-Wietecka A and Matuszewski T 2015 The new approach to drying and waterproofing process in the thermoinjection method *Mater. Bud.* **1** 24–5
[12] Matuszewski T 2009 Sposób osuszania i zabezpieczania murów przed ponownym zawilgoceniem *Polish Pat.* **PL 219284**
[13] Matuszewski T and Olędzki M 2012 Sposób osuszania murów i zabezpieczenia ich przed ponownym zawilgoceniem *Polish Pat.* **PL 223771**
[14] Fan J and Wen X 2002 Modeling heat and moisture transfer through fibrous insulation with phase change and mobile condensates *Int. J. Heat Mass Transf.* **45** 4045–55
[15] Cheng X and Fan J 2004 Simulation of heat and moisture transfer with phase change and mobile condensates in fibrous insulation *Int. J. Therm. Sci.* **43** 665–76
[16] Łapka P and Furmański P 2018 Verification and validation of an advanced model of heat and mass transfer in the protective clothing *Heat Mass Transf.* **54** 2461–74
[17] Łapka P, Furmański P and Wiśniewski T 2017 Assessment of thermal performance of protective garments. The advanced numerical model, *Int. J. Numer. Method. H.* **27** 1078–97
[18] Furmański P and Łapka P 2017 Evaluation of a human skin surface temperature for the protective clothing-skin system based on the protective clothing-skin imitating material results *Int. J. Heat Mass Transf.* **114** 1331–40
[19] Łapka P and Furmański P 2017 Application of a biequational bioheat transfer model in the skin for assessment of thermal characteristics of protective garments *Przem. Chem.* **96** 343–47
[20] dos Santos G H and Mendes N 2009 Heat, air and moisture transfer through hollow porous block *Int. J. Heat Mass Transf.* **52** 2390-8
[21] Defraeye T 2014 Advanced computational modelling for drying processes - A review *Appl. Energy* **131** 323–44
[22] Defraeye T 2011 Convective heat transfer coefficients for exterior building surfaces: Existing correlations and CFD modelling *Energy Conv. Manag.* **52** 512–22
[23] Defraeye T, Blocken B and Carmeliet J 2012 Analysis of convective heat and mass transfer coefficients for convective drying of a porous flat plate by conjugate modelling *Int. J. Heat Mass Transf.* **55** 112–24
[24] Belleghem M Van, Steeman M, Janssen H, Janssens A and Paepe M De 2014 Validation of a coupled heat, vapour and liquid moisture transport model for porous materials implemented in CFD *Build. Environ.* **81** 340–53
[25] Łapka P, Wasik M, Furmański P, Seredyński M, Cieślikiewicz Ł, Pietrak K, Kubiś M, Wiśniewski T S and Jaworski M 2018 Preliminary mathematical and numerical transient models of convective heating and drying of a brick *MATEC Web Conf.* **240** 01022
[26] Wasik M, Cieślikiewicz Ł, Łapka P, Furmański P, Kubiś M, Seredyński M, Pietrak K, Wiśniewski T S and Jaworski M 2019 Initial credibility analysis of a numerical model of heat and moisture transfer in porous building materials *AIP Conf. Proc.* **2078** 020106
[27] Van Belleghem M, Ameel B, Janssens A and De Paepe M 2011 Modeling heat and moisture transport in porous materials with cfd for building applications *HEFAT2011, 8th Int. Conf. on Heat Transfer, Fluid Mechanics and Thermodynamics* (Pointe Aux Piments) pp 455–63

IOP Conf. Series: Materials Science and Engineering **660** (2019) 012021 doi:10.1088/1757-899X/660/1/012021

Development of the experimental stand for investigation of the drying process in moist walls

Ł Cieślikiewicz[1], P Łapka[1], R Mirowski[1], M Wasik[1], M Kubiś[1], K Pietrak[1], P Furmański[1], M Seredyński[1] and T Wiśniewski[1]

[1]Warsaw University of Technology, Faculty of Power and Aeronautical Engineering, Institute of Heat Engineering, 21/25 Nowowiejska St, 00-665 Warsaw, Poland

E-mail: lukasz.cieslikiewicz@itc.pw.edu.pl

Abstract. The paper presents development of the experimental stand for investigation of drying process in a part of a wall which consisted of about twenty bricks joined by mortar. The thermo-injection method was applied for drying. In the first stage of this method several holes are drilled in the moist wall and are used to mount heating probes. These probes supply air to holes and at the same time heat air and the wall up. Then hot air leaves the hole flowing along its boundaries and heating and drying the wall. In the second stage holes are used for impregnation of the wall with hydrophobic fluid which creates waterproof membrane. The stand was based on a precise platform balance applied to measure the variation of global moisture content in the specimen. Moreover, local variations of moisture content were measured applying TDR method. The temperature measurements were carried out using RTDs which were placed in different locations in the wall. Additionally, the IR thermography was applied to register temperature on the specimen surface. The stand was closed in an insulation chamber to decrease the impact of the surroundings. The paper contains the preliminary results of measurements which test operation of the stand.

1. Introduction

Many sources of water in building materials might be found. The rising of groundwater level, condensation of water vapour from humid air in partitions as well as atmospheric precipitations are typical examples. The existence of moisture in building materials is highly unfavourable for the building structure. Moisture changes mechanical properties of building materials, induces biological corrosion and salts decay and deteriorates properties of building thermal insulation [1-3]. Moreover, it influences on the indoor microclimate conditions as well as affects the comfort of the occupants and their health [4]. Moisture in building structures might be accumulated for a long time, especially in historical buildings [5, 6]. Due to mentioned problems, drying, waterproofing and renovation of walls are required in many buildings.

Not only in building materials drying process is of importance. Many industrial processes which use different types of porous media are struggling with this process, e.g., paper production, food processing, drug manufacturing and crops drying. Moisture transfer is also important in the clothing [7-9]. Wide range of different types of dried porous materials encountered in the industry constitutes large field for complex research of the drying phenomena. Generally, drying process might be classified in the following main groups:

- Convective drying [10-14].

- Vacuum drying [14].
- Microwave drying [13, 15-17].

The convective drying process is widely used particularly in the food processing for which the influence of the physical and chemical pre-treatment of products on the drying kinetics was found important [10]. Castro et al. [11] noticed that, the problem in modelling of convective drying, especially in fruits, is related mainly to availability of material properties, which may depend on the pre-treatment. These parameters should be determined experimentally. Moreover, the convective drying might be combined with the microwave or vacuum drying to obtain the higher drying ratio. It was found that the combined convective and vacuum drying attains good drying kinetic at relatively low drying temperature which depends on the product [14]. Depending on drying material, combination of these technologies allows for minimization of total drying time and for avoiding negative effects on final quality of temperature-sensitive products. The convective drying with microwave assistance was examined in [13, 16] and showed good result in increasing the drying kinetics. However, the main disadvantage of the microwave drying process is significant increase of the medium temperature, which might be harmful for sensitive products, e.g., for food, and negatively affects the final product quality [16]. The automatic temperature control during the microwave drying was proposed in [17] in order to avoid quality damage or undesirable changes of food colour and texture. The proposed microwave system kept drying rates constant in the middle stage-drying period by using the computer-aided infrared thermography system to adopt temperature levels during the process. The combined convective and ultrasound drying was also examined [16]. Although, better convection parameters and final product quality were obtained, the energy efficiency of the ultrasound was low. The ultrasound action caused the vibration effect which contributed to improve mass transfer rates. Concluding, when combining the convective drying with additional energy sources, e.g., microwave or ultrasound, an extremely important is their dosing (operation) time [16].

Convective drying is most valuable method in the case of drying of building materials and existing buildings. In this case heat and moisture transfer between flowing air and building material or structure depends on local climate conditions, e.g., on temperature, velocity and relative humidity of air [1]. In laboratory conditions the drying process of building materials was studied in free convection conditions or with low velocity forced convection in a climatic chamber [18, 19]. For high speed forced convection, special wind tunnels were built. These tunnels were of either open-flow [13, 20] or closed-loop [1, 21, 22] type. In the previous papers convective drying process was usually investigated for relatively small specimens (e.g., samples of dimensions of 3×3×3 cm [1] or 1×3×9 cm [20]) with one surface of the sample in contact with drying air [1, 20]. The whole brick placed in the wind tunnel was also investigated experimentally [21, 22]. Additionally, in [21] possibility of analysis of drying process in the part of the wall located in the wind tunnel was presented. Experiments described in the literature were conducted with air flow velocities up to 10 m/s, relative humidity in the range 10-90% and temperature up to 60°C [21, 22].

One of the methods of convective drying and waterproofing of building materials and structures is the thermo-injection method [23-27]. This method was developed in Poland in the nineties of the previous century and is based on drilling an array of holes in the wet wall with a pitch of several centimetres. In the first stage of this method special heating probes are mounted in holes. These probes supply air to holes and at the same time heat the wall and flowing air up. The hot air leaves the hole flowing along its boundaries and heats and dries the wall. In the second stage, when the moisture content of the wall is lower than 4% weight special hydrophobic fluid is injected to holes. This process creates the membrane inside the wall which is impermeable for water and prevents from rewetting of the wall [23-27].

Investigations of drying process applying the thermo-injection method were never carried out before. Therefore, this method is for the first time studied in this work. The paper presents the development of the experimental stand for the investigation of heating and drying process of the part of the wall made of several red bricks by applying the thermo-injection method. The test measurement has been carried out and preliminary results are shown and discussed.

2. Description of the experimental stand

2.1. General assumptions
The developed stand should fulfil several requirements. Main are following:
- Separation from the surroundings which will prevent the specimen from external interactions, i.e., external forced convection as well as heat and water vapour exchange.
- Measurement of temporal variation of the mass of the whole specimen.
- Measurement of temporal variation of the local temperature in several points in the sample.
- Measurement of temporal variation of the local moisture content in several points in the sample.
- Measurement of temporal variation of air parameters in the chamber and at the inlet to drying device.
- Measurement of temporal variation of energy consumption by the drying device.

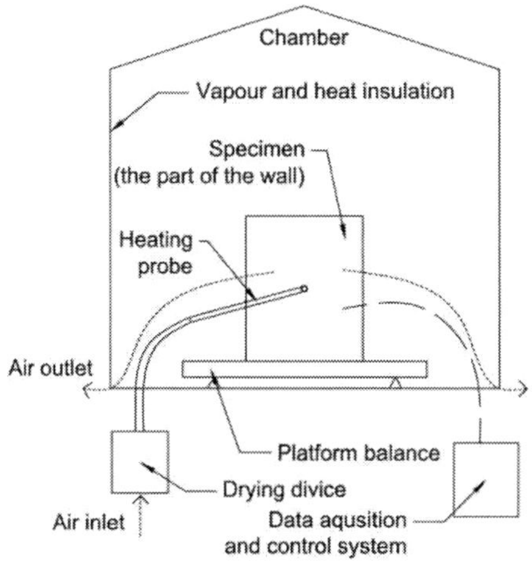

Figure 1. Simplified schematic of the experimental stand.

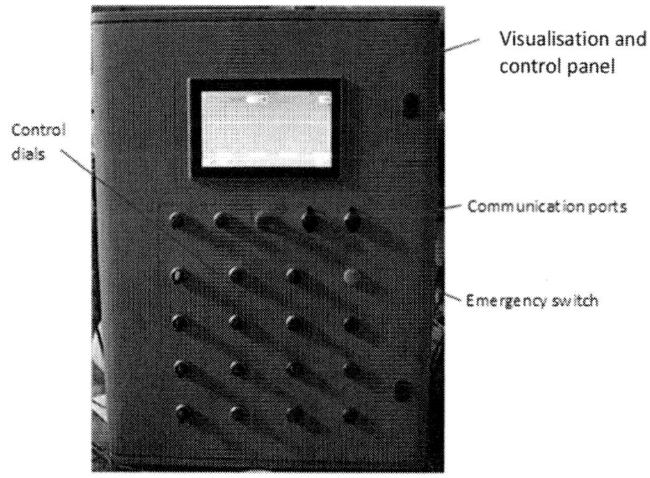

Figure 2. Control and data acquisition device.

2.2. Concept of the stand
The concept of the stand is presented in figure 1. The insulation from the surroundings consisted of a supporting structure made of steel tubes, a water vapour impermeable foil made of polyethylene and a heat insulation made of mineral wool. A little higher pressure than atmospheric one was maintained inside the chamber to avoid the infiltration of ambient air. Moreover, the excess of air in the chamber was flowing out to the surroundings by openings located close to the bottom of the chamber. To ensure measurement of the variation of the total mass of the sample the platform balance (RADWAG HY10 HRP150) with high precision and resolution on the level of 1 g was used. The balance was connected to the acquisition system (see figure 2) via RS232 cable. This configuration allowed for mass variation measurement within small time intervals. Temperature variations in the specimen were measured in two ways. Firstly, the surface distribution of temperature was registered with the IR thermography. Secondly, temperature variation inside the specimen was recorded using several 4.5 mm diameter sheathed resistance temperature detectors (RTDs, PT1000) made from platinum wire with resistance equal to 1000 Ω in 0°C and with 3-wire connection – see figure 3 for locations of RTDs in the specimen. The variation of the local moisture content was measured by applying time domain reflectometry (TDR) method which is based on high frequency electric signals (E-TEST TDR/MUX/MPTS with FP/mts probe). Temperature and relative humidity of air inside the chamber as well as temperature and relative humidity of ambient air provided to the drying device were measured with use of thermohydrometers (APAR AR252). Additionally, the power consumed by the drying device was recorded by using the

power meter. All measurement devices were connected to the programmable logic controller which recorded, stored and visualized measured data as well as controlled the drying process (see figure 2).

3. Test measurement

In order to verify assumed parameters of the stand the one 40 W heating probe was placed in the specimen and used during test measurement. Air for drying process was taken by the drying device from the room where the stand was located without any conditioning before providing it to the probe.

3.1. Specimen preparation

The specimen was built as the part of the wall which consisted of 21 ceramic red bricks joined by mortar with approx. 81.5 kg of dry mass. The schematic structure of the specimen is presented in figure 3. In the middle of the specimen 20 cm depth and 20 mm diameter hole was drilled and then the heating probe was placed inside it. Additional four drillings, around drying hole, of 10 cm depth and 4 mm diameter each were prepared for RTDs. Moreover, for measurement of the local moisture content four drillings of 5 cm depth and 4 mm diameter each for the TDR detector were made close to the highest and lowest location of RTD holes. The considered building structure was conditioned in the surrounding for one month to obtain the proper mortar binding. After this period the specimen was fully submerged in water for another month. Shortly before measurement the considered building structure was removed from water and covered with water and water vapour proof foil made of polypropylene. Mass of absorbed water was equal to approx. 11.1 kg which correspond to moisture content of 13.5% weight. The foil covered all sides except the side with drillings. Free spaces between surface of holes and measurement probes were filled up with a mixture of brick powder and water. Subsequently, the building structure was positioned on the platform balance and measurement began. The photography of the part of the wall with heating and measurement probes is shown on figure 4.

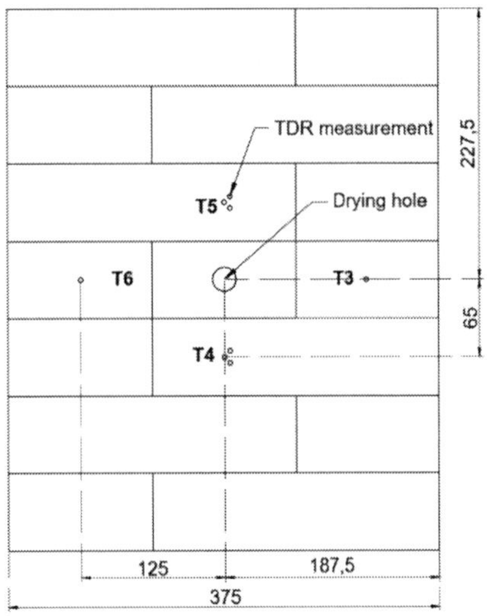

Figure 3. Schematic of the specimen with heating and measurements probes.

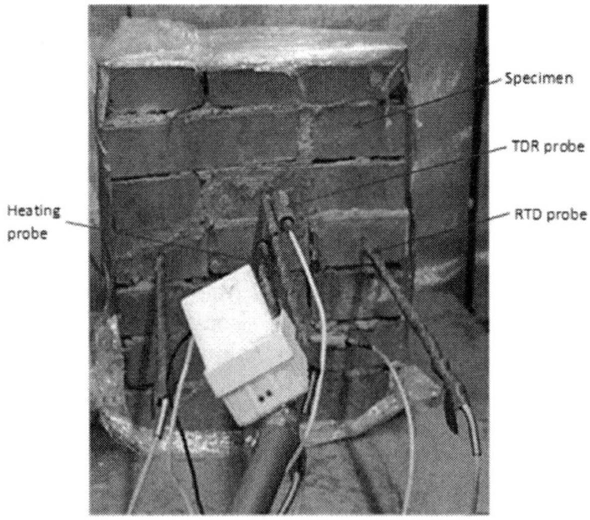

Figure 4. Photography of the specimen with heating and measurement probes.

3.2. Results of test measurement

Temperature variations in four points in the wall (i.e., T3, T4, T5 and T6 – see figure 3) registered by RTDs during test measurement are presented in figure 5. Slightly lower temperatures registered by T3 and T6 detectors than by T4 and T5 ones were caused by larger distances from drying hole of T5 and T6 probes than T4 and T5 ones. In the first stage of the drying process rapid drop of temperature registered by all RTDs was observed. This effect was caused by the intensive evaporation of water from the surface of the specimen to the air in the chamber. The specimen before the beginning of the measurement was saturated with water and then shortly after removing from the water bathing was mounted on the stand. Next, the specimen temperature increased slowly and after approx. 40 h of the drying process the temperature again started to decrease to saturation parameters in the drying material – the second stage of drying began. This second decrease of temperature corresponded to the evaporation of water contained deeper inside the specimen. After approx. 60 h of the drying process the temperature started rising slowly. Temperature fluctuations observed in the second stage of the drying were effects of variation of parameters of the ambient air used as drying medium. The change of the total mass of the specimen and the local relative moisture content (defined as a ratio of current to initial local moisture content) are presented in figure 6 and 7, respectively. The moisture lost rate is monotonic through the whole measurement. This is a consequence of using of only one heating probe for which the continuous evaporation of liquid moisture in different parts of the sample occurred through the whole measurement. The change in the local moisture content corresponded to the change in the local temperature what was especially visible during the first several hours of the process when rapid evaporation from the surface took place and the TRD probe registered sudden drop of the local moisture content. Because of only one heating probe was applied the specimen was not fully dried and process was stopped after 150 h. In practise, during drying of real masonry walls by applying thermo-injection method the heating probes are mounted densely with horizontal pitch of approx. 10 cm.

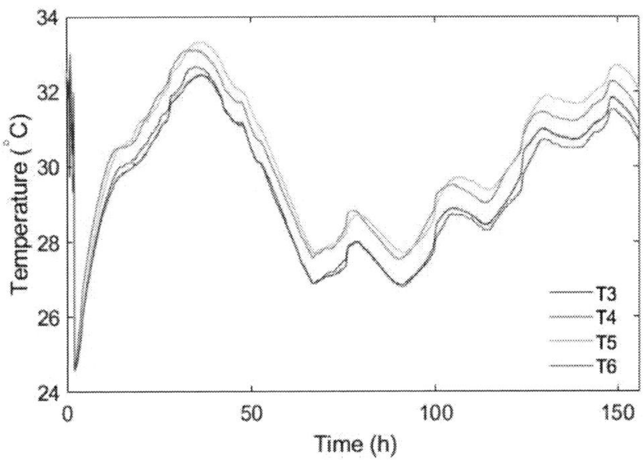

Figure 5. Temporal temperature variations in the specimen measured by RTDs.

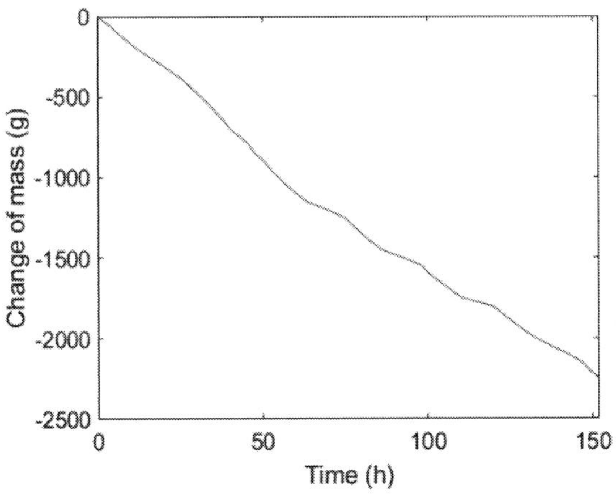

Figure 6. Lost of moisture mass in the specimen in time.

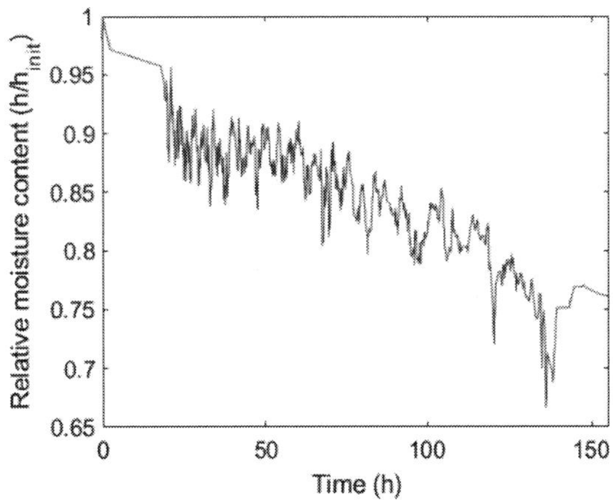

Figure 7. Variation of local relative moisture content measured by TDR close to T5 hole.

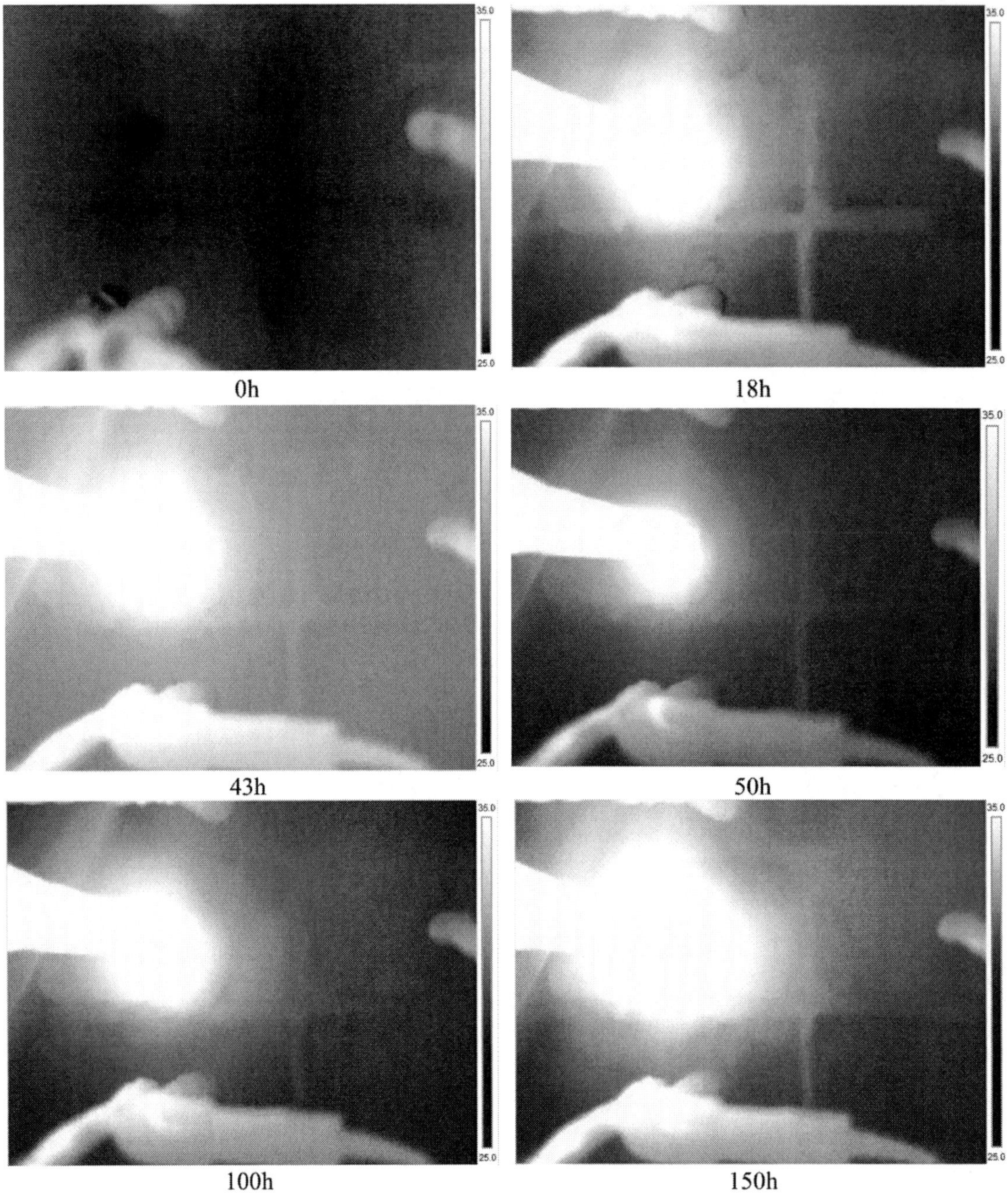

Figure 8. Thermograms showing temperature distribution on the surface of the specimen for different time instants during drying process (minimum and maximum on the scale are 25°C and 35°C, respectively).

Figure 8 presents contours of temperature on the surface of the sample during measurement. Infrared camera was positioned 1 m from the specimen surface. The specimen surface emissivity was assumed equal to 0.95. The first stage of drying in which surface evaporation was dominating is presented on thermograms which corresponds to time instants of 0 and 18 and 43 h of the drying process. In the second stage drop of surface temperature caused by evaporation of water in deeper parts of the specimen is presented on thermograms which corresponds to time instants of 50 and 100 and 150 h of the drying process. The brightest point on thermograms corresponded to the heating probe surface which was at temperature equal approx. 80°C.

4. Conclusion

The paper presents development of the experimental stand for measurements of drying process in the part of the wall which was made of several bricks. The stand in the form of the chamber with the specimen inside with one hole for placing the heating probe was build and equipped with platform balance, several RTDs and TDR probe. The test measurement was carried out and the results obtained led to the following conclusions. Better controlling of the drying process is required to eliminate fluctuations of the specimen temperature. This might be achieved by the stabilisation of air parameters (i.e., temperature and relative humidity) inside the chamber and at the inlet to the drying device. Moreover, number of holes and heating probes in the specimen should be increased to intensify drying process and reduce time of measurements. This action should be accompanied with increase of the number of RTDs and TRD probes. In addition, the effect of surface evaporation should be eliminated or controlled by conditioning the specimen, after removing it from the water bathing, in controlled conditions before the start of measurement. After proposed modifications of the stand and measurement procedure, the stand will be applied in the process of validation of mathematical and numerical models of heat and moisture transfer in porous building materials which are currently under development [28, 29].

Acknowledgements
This work was supported by the National Centre for Research and Development (Poland) under grant no. POIR.04.01.02-00-0099/16 *Development of innovative technology of drying and moisture sealing of masonry walls*, DryWall.

References

[1] Karagiannis N, Karoglou M, Bakolas A, Krokida M and Moropoulou A 2017 Drying kinetics of building materials capillary moisture *Constr. Build. Mater.* **137** 441–9
[2] Lourenço P B, Luso E and Almeida M G 2006 Defects and moisture problems in buildings from historical city centres: A case study in Portugal *Build. Environ.* **41** 223–34
[3] Trochonowicz M 2010 Moisture in buildings objects. Humidity testing problems *Bud. i Archit.* **7** 131–44
[4] Shoemaker R C and House D E 2006 Sick building syndrome (SBS) and exposure to water-damaged buildings: Time series study, clinical trial and mechanisms *Neurotoxicol. Teratol.* **28** 573–88
[5] Falchi L, Slanzi D, Balliana E, Driussi G and Zendri E 2018 Rising damp in historical buildings: A Venetian perspective *Build. Environ.* **131** 117–27
[6] Gueli A M, Garro V, Liuzzo M, Margani G, Pasquale S, Politi G and Stella G 2018 Effects of moisture on historical buildings TL ages *Meas. J. Int. Meas. Confed.* **118** 289–97
[7] Łapka P and Furmański P 2018 Verification and validation of an advanced model of heat and mass transfer in the protective clothing *Heat Mass Transf.* **54** 2461–74
[8] Łapka P, Furmański P and Wiśniewski T 2017 Assessment of thermal performance of protective garments. The advanced numerical model, *Int. J. Numer. Method. H.* **27** 1078–97
[9] Furmański P and Łapka P 2017 Evaluation of a human skin surface temperature for the protective clothing-skin system based on the protective clothing-skin imitating material results *Int. J. Heat Mass Transf.* **114** 1331–40
[10] Junqueira J R de J, Corrêa J L G, de Oliveira H M, Ivo Soares Avelar R and Salles Pio L A 2017 Convective drying of cape gooseberry fruits: Effect of pretreatments on kinetics and quality parameters *LWT - Food Sci. Technol.* **82** 404–10
[11] Castro A M, Mayorga E Y and Moreno F L 2018 Mathematical modelling of convective drying of fruits: A review *J. Food Eng.* **223** 152–67

[12] Boutelba I, Zid S, Glouannec P, Magueresse A and Youcef-ali S 2018 Experimental data on convective drying of potato samples with different thickness *Data Br.* **18** 1567–75
[13] Das I and Arora A 2018 Alternate microwave and convective hot air application for rapid mushroom drying *J. Food Eng.* **223** 208–19
[14] Zecchi B, Clavijo L, Martínez Garreiro J and Gerla P 2011 Modeling and minimizing process time of combined convective and vacuum drying of mushrooms and parsley *J. Food Eng.* **104** 49–55
[15] Kurik L, Kalamees T, Kallavus U and Sinivee V 2017 Influencing factors of moisture measurement when using microwave reflection method *Energy Procedia* **132** 159–64
[16] Szadzińska J, Kowalski S J and Stasiak M 2016 Microwave and ultrasound enhancement of convective drying of strawberries: Experimental and modeling efficiency *Int. J. Heat Mass Transf.* **103** 1065–74
[17] Cuccurullo G, Giordano L, Metallo A and Cinquanta L 2018 Drying rate control in microwave assisted processing of sliced apples *Biosyst. Eng.* **170** 24–30
[18] Bianchi Janetti M, Colombo L P M, Ochs F and Feist W 2017 Determination of the water retention curve from drying experiments using infrared thermography: A preliminary study *Int. J. Therm. Sci.* **114** 271–80
[19] Janssen H, Scheffler G A and Plagge R 2016 Experimental study of dynamic effects in moisture transfer in building materials *Int. J. Heat Mass Transf.* **98** 141–9
[20] Belleghem M Van, Steeman M, Janssen H, Janssens A and Paepe M De 2014 Validation of a coupled heat, vapour and liquid moisture transport model for porous materials implemented in CFD *Build. Environ.* **81** 340–53
[21] Cieślikiewicz Ł, Łapka P, Wasik M, Kubiś M, Pietrak K, Wiśniewski T S, Furmański P and Seredyński M 2018 Development of the experimental stand for investigation of heating and drying phenomena in the porous building materials with one surface of the sample exposed to the flowing air *E3S Web Conf.* **70** 03003
[22] Cieślikiewicz Ł, Wasik M, Kubiś M, Łapka P, Bugaj M, Pietrak K, Wiśniewski T S, Furmański P and Seredyński M 2019 Development of the experimental stand with centrally located specimen for the investigation of heat and moisture phenomena in porous building materials *Civ. Environ. Eng. Reports* **29** 53–65
[23] Kaliszuk-Wietecka A and Matuszewski T 2015 The new approach to drying and waterproofing process in the thermoinjection method *Mater. Bud.* **1** 24–5
[24] Olifierowicz J, Tomborowski T and Trzecianowsk A 1982 Sposób osuszania i zabezpieczania murów przed ponownym ich zawilgoceniem *polish pat.* **PL 139041**
[25] Matuszewska M and Matuszewski T 1997 Sposób równomiernego ogrzewania otworów w murze w trakcie jego osuszania i zabezpieczania przed ponownym zawilgoceniem oraz element grzejny do realizacji tego sposobu *polish pat.* **PL 184012**
[26] Matuszewski T 2009 Sposób osuszania i zabezpieczania murów przed ponownym zawilgoceniem *polish pat.* **PL 219284**
[27] Matuszewski T and Olędzki M 2012 Sposób osuszania murów i zabezpieczenia ich przed ponownym zawilgoceniem *polish pat.* **PL 223771**
[28] Łapka P, Wasik M, Furmański P, Seredyński M, Cieślikiewicz Ł, Pietrak K, Kubiś M, Wiśniewski T S and Jaworski M 2018 Preliminary mathematical and numerical transient models of convective heating and drying of a brick *MATEC Web Conf.* **240** 01022
[29] Wasik M, Cieślikiewicz Ł, Łapka P, Furmański P, Kubiś M, Seredyński M, Pietrak K, Wiśniewski T S and Jaworski M 2019 Initial credibility analysis of a numerical model of heat and moisture transfer in porous building materials *AIP Conf. Proc.* **2078** 020106

IOP Conf. Series: Materials Science and Engineering **660** (2019) 012022 doi:10.1088/1757-899X/660/1/012022

Measurement of thermal, hygric and physical properties of bricks and mortar common for the Polish market

K Pietrak[1], M Kubiś[1], Ł Cieślikiewicz[1], P Furmański[1], M Seredyński[1], M Wasik[1], T Wiśniewski[1] and P Łapka[1]

[1] Warsaw University of Technology, Faculty of Power and Aeronautical Engineering, Institute of Heat Engineering, 21/25 Nowowiejska St, 00-665 Warsaw, Poland

E-mail: kpietrak@itc.pw.edu.pl

Abstract. Hygric, thermal and physical properties were determined for building materials typical for the Polish market. One type of mortar and two types of masonry bricks were studied. Thermal diffusivity and specific heat of dry materials were found using the flash and differential scanning calorimetry methods, respectively. Water vapour permeability was tested using the cup method for a single temperature of 23°C and for two ranges of the air relative humidity (from 0 to 50% and from 50 to 94%). The permeability tests were performed according to ISO 12572:2016. Measurements of the sorption isotherms were carried out using the environmental chamber method, according to ISO 12571:2013. The results of measurements of the apparent and true density of the considered materials were used to determine their porosity. The presented measurements concerned materials involved in physical and numerical experiments aimed at quantitative and qualitative analyses of heat and moisture transfer during drying of building materials and walls.

1. Introduction

The knowledge of building material properties is especially important to face of current challenges related to economy, health care and environmental protection. Detailed models of heat and mass transfer help to design the user-friendly, environment-friendly, sustainable and energy-efficient buildings, but they need material properties of building materials as the input data. The number of measurement of these properties available in the literature is high and rapidly growing due to the extensive interest and funding. Additionally, the properties for many building materials are catalogued in the engineering databases and in computer programs such as WUFI® [1] and can be used to model heat and moisture transport in building components. Nevertheless, there is a continuous need for new measurements of the material properties which may be justified by three main reasons.

Firstly, each material sample is unique in terms of its transport properties, which depend on the details of manufacturing process. It has been shown that manufacturing technique affects microstructural parameters such as porosity, pore size distribution and dominant pore orientations [2] which have the profound impact on heat and mass transport properties of materials such as bricks. The second reason having influence on the properties is the type of base material and additives used in production, which are strongly location dependent. This is especially true for the recently promoted materials with sustainable, organic components that are different for different regions where the material is manufactured. Because of this, the literature results cannot be universally applied.

When equipped with reliable measurement techniques, laboratories should be able to obtain consistent measurement results. Nevertheless, discrepancies in the published data can be found, which are especially characteristic for measurements of the moisture transport properties, which is the third reason justifying the need to measure the properties of each building material involved in scientific research and construction of new buildings [3]. The main difficulty in measurement of hygro-thermal properties is associated with long time required to obtain the equilibrium or steady state conditions. The required stabilization times may be counted in weeks and months. Due to that, the measurement errors and sources of uncertainty are relatively slowly eliminated. Therefore, the results obtained by different laboratories for the same material may be significantly differ, especially if the water transport properties were measured by un inexperienced personnel [3].

The literature regarding the measurement of the hygric, thermal and physical properties of building materials is relatively rich [4-6]. Classical methods in that field are considered reliable and many of them are standardized by ISO, DIN and ASTM [5, 6]. The majority of standardized methods for the measurement of the air and water transport properties is based on the steady state assumption. In recent years, methods utilizing the unsteady state conditions have been intensively developed [7, 8]. In general, they allow for faster measurements, but mathematical description and modeling which accompanied them is much more complex than for steady state methods. Some authors emphasize that in many applications the building materials are subjected to dynamic environmental conditions for which the properties measured in the steady state are not accurate [7, 8]. On the other hand, classical methods are based on simple principles which make the measurements easier to design and the results easier to interpret. Due to these reasons, steady state methods are still globally used, as the basic means of measurement, and as the reference for results obtained using other methods.

This paper concerns the measurement of thermal, hygric and physical properties of bricks and mortar common at the Polish market. Apart from previously listed reasons to carry out measurements of the building material properties it is also worth to note that studies demonstrating large number of properties for the single material are seldom presented. The attempt to comprehensively characterize materials of interest was motivated by the need to verify mathematical and numerical models of heat and moisture transport in the building materials which were developed within the "DryWall" project (no. POIR.04.01.02-00-0099/16). The aim of the ongoing project is optimization of the thermo-injection drying technique [9]. In this project, experimental stands devised to investigate drying of walls and single bricks were built alongside with the numerical and analytical models of drying phenomena [2, 10-15]. This paper presents measurement results for materials used in the project, i.e., the fired red brick, silicate brick and one type of a mortar.

2. Materials

Objects of further investigation were three types of the masonry materials: red brick, silicate brick and mortar which are commonly available on the market in Poland.

3. Properties

Different material properties such as hygric, physical and thermal were investigated for three types of materials of samples in selected direction only. In the case of samples cut for vapor permeability measurements the standard ISO 12572:2016 limits possible ways of preparation of samples as it required relatively large samples of particular geometry. In the case of red and silicate bricks the direction through the height of the brick was the only one which fulfil ISO 12572:2016 standard requirements. Moreover, this direction is of interest because moisture is usually transported from the ground along the height of the wall. In the case of measurement of thermal properties of the red and silicate bricks, the investigation of anisotropy of their thermal conductivities was shown in the previous work [2]. However, in this paper thermal diffusivities and thermal conductivities of bricks were presented only in the direction through the thickness of the brick. This direction was selected because of the highest temperature gradients and heat flux occur between inner and outer side of the wall.

3.1. Hygric

3.1.1. Vapor permeability. The cup method was used to measure vapour permeability of samples made of the mortar and masonry bricks: red and silicate one. For bricks the measurements were carried out for direction along the height of the brick. The measurements were performed according to the standard ISO 12572:2016. The principle of this method is water vapor transmission through the investigated sample sealed to an open dish (figure 1). The dish was made of the high-temperature-resistant glass and the openings between the dish and the sample were sealed using the building silicone and technical stearin.

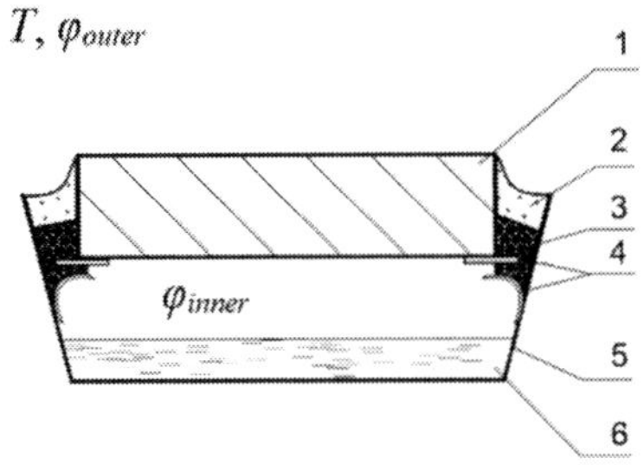

Figure 1. Assembly of the testing dish and sealed sample.

A difference in the partial vapour pressure p_v between the inner space of the dish and controlled atmosphere of the environmental chamber causes water vapour to penetrate the sample in through-thickness direction. Change of the mass of the whole assembly was measured. This procedure allows to determine water vapour flux G, water vapour flux density g, corrected water vapour flux density g_{me}, water vapor transmission w_{me}, as defined in the ISO 12572 standard. The correction of calculated water vapour flux density was performed according to the ISO 12572 standard, because the assembly contained occluded edges. Two different sources were separately applied to enforce water vapour transmission through the samples. Calcium chloride (desiccant) and aqueous solution of potassium nitrate assured the relative humidity of 0% and 94% inside the testing assembly, respectively. The environmental chamber was used to assure stable relative humidity of 50% and average temperature 23°C outside of the assemblies, so that the water vapour was transmitted in both directions: outside and inside of testing dishes. The results of those measurements (vapor permeability δ_{me}) are showed in table 1.

Table 1. Results of vapor permeability investigation.

Sample type	Red brick		Silicate brick		Mortar	
Relative humidity inside the dish	0%	94%	0%	94%	0%	94%
G (kg s^{-1})	$8.535\cdot10^{-9}$	$-1.270\cdot10^{-8}$	$-1.405\cdot10^{-6}$	$-6.999\cdot10^{-6}$	$2.304\cdot10^{-9}$	$-4.070\cdot10^{-9}$
g (kg m^{-2} s^{-1})	$6.732\cdot10^{-7}$	$-9.632\cdot10^{-7}$	$2.759\cdot10^{-7}$	$-7.076\cdot10^{-7}$	$1.756\cdot10^{-7}$	$-3.077\cdot10^{-7}$
g_{me} (kg m^{-2} s^{-1})	$7.142\cdot10^{-7}$	$-1.022\cdot10^{-6}$	$3.099\cdot10^{-7}$	$-7.945\cdot10^{-7}$	$1.877\cdot10^{-7}$	$-3.285\cdot10^{-7}$
w_{me} (kg m^{-2} s^{-1} Pa^{-1})	$5.087\cdot10^{-10}$	$-8.273\cdot10^{-10}$	$2.207\cdot10^{-10}$	$-6.431\cdot10^{-10}$	$1.337\cdot10^{-10}$	$-2.659\cdot10^{-10}$
δ_{me} (kg m^{-1} s^{-1} Pa^{-1})	$1.136\cdot10^{-11}$	$-1.941\cdot10^{-11}$	$5.104\cdot10^{-12}$	$-1.467\cdot10^{-11}$	$4.082\cdot10^{-12}$	$-7.991\cdot10^{-12}$

The properties presented in table 1 were calculated according to formulas mentioned in the standard ISO 12572:2016. Sets of samples consisted of four assemblies of each materials type and source used during the test. The permeabilities were investigated in pairs with a set of assemblies containing different source (i.e., φ = 0% and φ = 94%), so that opposite directions of water vapor transmission were obtained.

Despite the differences between results obtained for each pair of sets, their orders of magnitude are similar.

3.1.2. Sorption. Investigation was performed according to the standard ISO 12571:2013 with use of environmental chamber. Ten samples of each material type were cut from the inner part of bricks and mortar block and subsequently subjected to drying with the vacuum assisted furnace. The test was performed on samples with a mass of approximately 30 g. Precisely controlled parameters by the environmental chamber were set as follows: $T = 23^{\circ}C$ +/- $0.5^{\circ}C$; $\varphi_1 = 30\%$, $\varphi_2 = 53\%$, $\varphi_3 = 76\%$ and $\varphi_4 = 90\%$. Change in a mass of samples during conditioning were determined with the analytical balance. In order to calculate the amount of the absorbed water, dry samples were exposed to the inner condition of the environmental chamber. According to the standard an equilibrium state of the sample was assumed to be valid as soon as its mass change of three consecutive measurement did not exceed 0.1%. Then relative humidity was changed to the next value. The results of measurements were shown in figure 2.

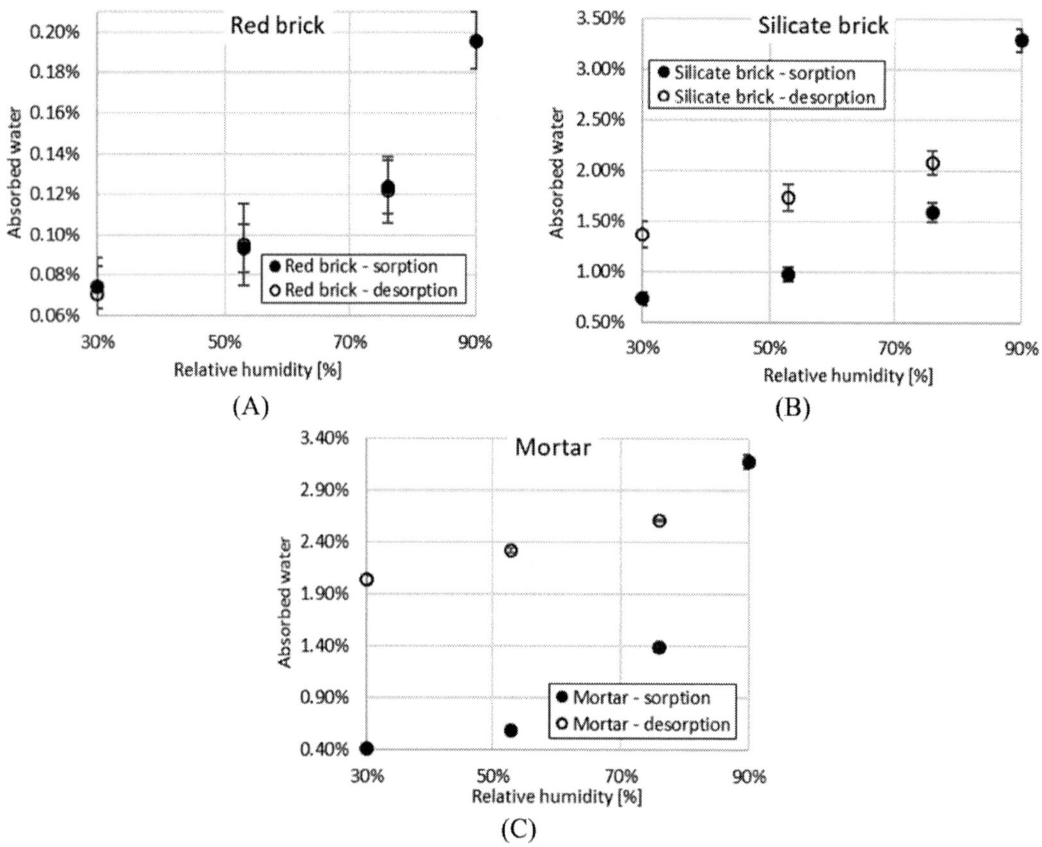

Figure 2. Sorption and desorption characteristics for: A) red brick, B) silicate brick and C) mortar.

Figure 2 shows characteristics of the absorbed water vs the relative humidity as an average determined from ten samples of each investigated material. The black symbols in the figures denote the datapoints obtained during the first stage – sorption process – where relative humidity was increased to the next humidity point after samples reached the equilibrium. The circular symbols denote the datapoint obtained during the second stage – desorption process. The relative humidity of $\varphi_4 = 90\%$ was the common point for two mentioned stages. In figure 2B and 2C the hysteresis of sorption-desorption process can be observed. Exception is the red brick (figure 2A), for which the datapoints of both stages overlap in the range of their standard deviation.

3.2. Physical

The pycnometer and geometric method were used to determine the apparent density and real density of the samples, respectively. Four samples of each type were investigated, and the averaged results of measurements were collected in table 2. Determination of apparent density ($\rho_{apparent}$) required the samples to be fragmented into a powder which was performed with a mechanical mixer. Such sample preparation leads to pores opening and causes the pycnometric liquid to infiltrate and immerse fragmented particles more easily. Before each measurement, the pycnometer was calibrated with the deionised water (< 0.10 S m^{-1}). Samples in the form of powder had been dried in a vacuum-assisted furnace with following parameters: $T = 110$°C, $t = 72$ hours and vacuum $p_{vac} = 95$ kPa. Then, the sample was put into the pycnometer, mixed with water and degassed under the same vacuum level for 30 min in order to assure good infiltration of each particle. Additionally, all assembly was left for 5-6 hours in room temperature.

The real density (ρ_{real}) of the samples was determined according to geometric method based on measurement of the mass and sample dimensions. In order to minimalize errors quite large (i.e., 100×100×150 mm), cuboidal shaped specimens were prepared. The slide calliper and analytical balance were utilized to determine the sample volume and mass.

Determination of both densities allowed to calculate the sample porosity P according to following formulae:

$$P = \left(1 - \frac{\rho_{real}}{\rho_{apparent}}\right) \cdot 100\% \tag{1}$$

Results of both measurements are shown in table 2.

Table 2. Physical properties of samples.

Sample type	Number of samples	ρ_{real} (g cm^{-3})	Std dev.	$\rho_{apparent}$ (g cm^{-3})	Std dev.	P (%)
Red brick	4	1.763	0.015	2.585	0.030	31.8
Silicate brick	4	1.740	0.026	2.588	0.029	32.8
Mortar	4	1.795	0.031	2.579	0.033	30.4

3.3. Thermal

3.3.1. Thermal diffusivity. Thermal diffusivity was determined with the use of the flash method (LFA 447, Netzsch) at three temperatures: $T_1 = 30$°C, $T_2 = 45$°C and $T_3 = 60$°C. Samples of three materials were cut from the inner part of bricks and dried before the measurement. For bricks the measurements were carried out for direction along the thickness of the brick. The results are shown in figure 3A.

3.3.2. Specific heat. Specific heat was determined with use of the differential scanning calorimeter DSC 7 (Perkin Elmer). Samples of three materials were cut from the larger parts of bricks and dried before investigation. Measurements were performed in the argon atmosphere (10 ml/min) with the heating rate $v_r = 10$ K/min. In figure 3B variation of the specific heat with temperature was shown for selected temperature values.

3.3.3. Thermal conductivity. Thermal conductivity of each type of building materials was calculated according to the indirect method based on the following formula:

$$k = a c_p \rho_{real} \tag{2}$$

where: a is the thermal diffusivity, c_p is the specific heat and ρ_{real} is the real density. The average value of each particular property was used in calculation. The results are shown in figure 3C.

The highest thermal diffusivity (figure 3A) is observed for samples made of the mortar $a = 0.969$ mm^2 s^{-1} at $T = 30$°C to $a = 0.918$ mm^2 s^{-1} at $T = 60$°C. Values lower by approximately 4% characterize samples cut from the silicate bricks. Thermal diffusivity of the red brick is lower, ranging from $a = 0.458$ mm^2 s^{-1} at $T = 30$°C to $a = 0.431$ mm^2 s^{-1} at $T = 60$°C.

In the case of specific heat, similar values can be observed for the red brick and silicate brick. The largest difference between them does not exceed 3%, whereas the specific heat of samples made of the mortar is higher than this of the silicate brick by approximately 12% (figure 3B). The specific heat increases with temperature for each sample type.

Thermal conductivity was calculated according to eq. (2). It required the direct measurement of thermal diffusivity, specific heat as well as the sample density. It must be noted that the last property was determined in the room temperature and was assumed to be constant over the investigated temperature range. The lowest thermal conductivity was observed for samples cut from the red brick which values are in the range from 0.596 W m^{-1} K^{-1} to 0.624 W m^{-1} K^{-1}. Thermal conductivity of the second sample (the silicate brick) was found higher than that of the red brick by approximately 87%. In the case of mortar, its thermal conductivity is approximately 130% of the thermal conductivity of the red brick.

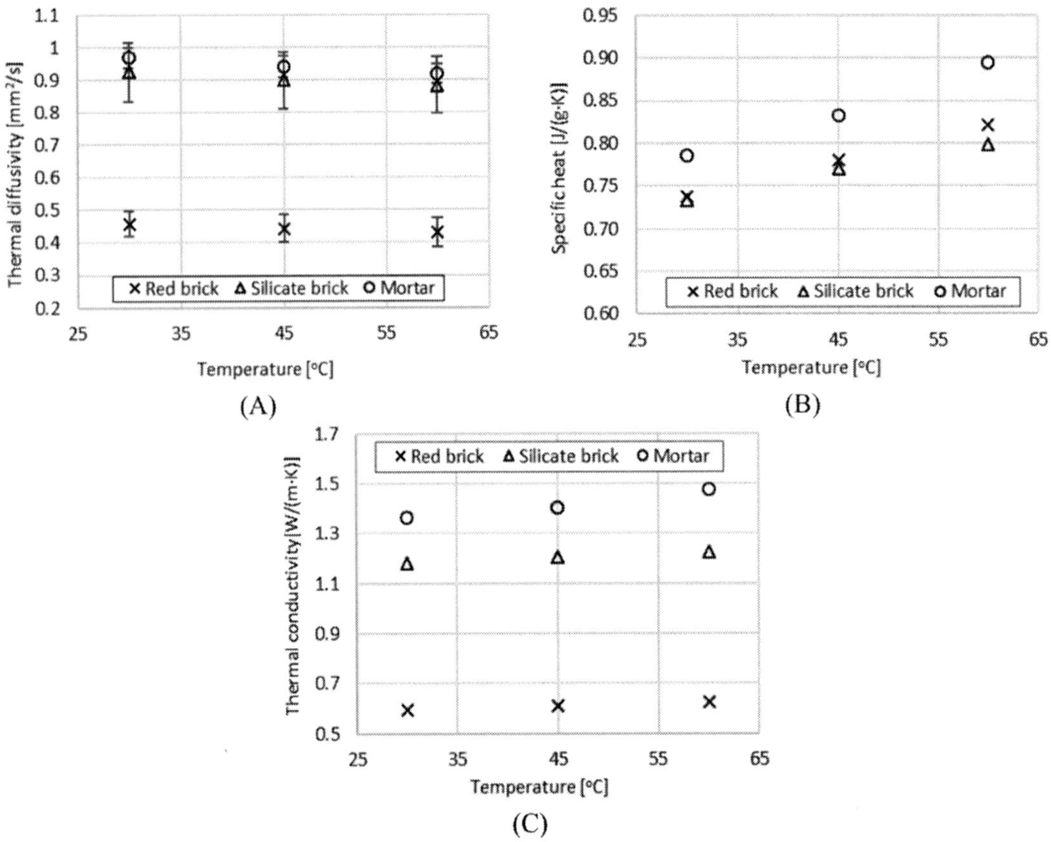

Figure 3. Thermal properties of samples made of red brick, silica brick and mortar: A) thermal diffusivity, B) specific heat and C) thermal conductivity.

4. Summary

In this paper, three types of the masonry materials: red brick, silicate brick and mortar were measured to find their hygric, physical and thermal properties. Determination of these properties is essential as many studies associated with prediction and modelling of heat and moisture transport during drying of building walls are conducted. Building materials common for the Polish market were comprehensively characterized using classical methods. The water vapour permeability was measured with the cup method according to the standard ISO 12572:2016. The next hygric parameter that was investigated in this paper was the sorption ability performed in accordance with the standard ISO 12571:2013 using the environmental chamber. Moisture absorption in the hygroscopic range is a long-term process as equilibrium state for each moisture point was obtained in approximately 10-12 days. Four samples of each material type were prepared and investigated. It was observed that the sorption-desorption process is characterized by the hysteresis in the case of samples cut from the silicate bricks and mortar. The data points obtained for the red brick in this two-stage test did not show this behavior.

Measurements of physical properties such as the apparent density and real density allowed to calculate the porosity of materials according to the eq. (1). The obtained values of the real densities were used in the subsequent calculation of thermal conductivity. Difference in the porosities of all investigated materials was small. Their values were in the range from 30.4% to 31.8%.

The thermal conductivity of each masonry material was determined according to the indirect method that is based on measurements of the thermal diffusivity, specific heat and real density. The investigation of thermal properties was carried out at three temperatures covering the range of working temperatures in the experimental drying stand. The stand was developed and subsequently modified to study the heat and moisture transport during drying process in a building wall [14, 15]. The results of measurements on thermal properties showed that samples made of silicate bricks and mortar possess thermal conductivity higher than red bricks, respectively by 87% and 130%.

Acknowledgments

This work was supported by the National Centre for Research and Development (Poland) under grant no. POIR.04.01.02-00-0099/16 "Development of innovative technology of drying and moisture sealing of masonry walls, DryWall".

References

[1] Zirkelbach D, Schmidt T, Kehrer M and Künzel H M 2019 *WUFI ® Pro 4 Manual* (Stuttgart: Fraunhofer IBP)

[2] Kubiś M, Pietrak K, Cieślikiewicz Ł, Łapka P, Furmański P, Wiśniewski T S, Seredyński M and Wasik M 2018 Investigations on thermal anisotropy of ceramic bricks *Proc. 5th Int. Conf. on Contemporary Problems of Thermal Engineering CPOTE 2018 (Gliwice)* pp 451–9

[3] Roels S, Carmeliet J, Hens H, Adan O, Brocken H, Cerny R, Pavlik Z, Hall C, Kumaran K, Pel L and Plagge R 2004 Interlaboratory comparison of hygric properties of porous building materials *J. Therm. Envel. Build. Sci.* **27** 307–25

[4] Nilsson L O (ed) 2018 *Methods of Measuring Moisture in Building Materials and Structures: State-of-the-Art Report of the RILEM Technical Committee 248-MMB* (Cham: Springer)

[5] Mukhopadhyaya P and Kumaran MK (ed) 2007 *Heat-Air-Moisture Transport: Measurements on Building Materials* (West Conshohocken: ASTM International)

[6] Künzel H M 1995 Simultaneous Heat and Moisture Transport in Building Components. One- and Two-dimensional Calculation Using Simple Parameters (Stuttgart: Fraunhofer IRB Verlag)

[7] Janssen H, Scheffler G A and Plagge R 2016 Experimental study of dynamic effects in moisture transfer in building materials *Int. J. Heat Mass Transf.* **98** 141–9

[8] Busser T, Berger J, Piot A, Pailha M and Woloszyn M 2018 Dynamic experimental method for identification of hygric parameters of a hygroscopic material *Build. Environ.* **131** 197–209

[9] Dreyer J and Hecht C 2001 Injection methods for retrofitting of moisture damaged constructions *WIT Trans. Built Environ.* **55** 517–26

[10] Cieślikiewicz Ł, Wasik M, Łapka P and Kubiś M 2019 Numerical investigation of flow stabilization in a channel with rapid change of flow direction *AIP Conf. Proc.* **2078** 200901

[11] Wasik M, Cieślikiewicz Ł, Łapka P, Furmański P, Kubiś M, Seredyński M, Pietrak K, Wiśniewski

T S and Jaworski M 2019 Initial credibility analysis of a numerical model of heat and moisture transfer in porous building materials *AIP Conf. Proc.* **2078** 020106

[12] Furmański P, Seredyński M, Łapka P, Wiśniewski T S, Pietrak K, Cieślikiewicz Ł, Kubiś M and Wasik M 2018 A model of heat and moisture transfer in a building material during drying based on phase equilibrium in the moist region *Proc. 5th Int. Conf. on Contemporary Problems of Thermal Engineering CPOTE 2018 (Gliwice)* pp 663–73

[13] Łapka P, Wasik M, Furmański P, Seredyński M, Cieślikiewicz Ł, Pietrak K, Kubiś M, Wiśniewski T S and Jaworski M 2018 Preliminary mathematical and numerical transient models of convective heating and drying of a brick *MATEC Web Conf.* **240** 01022

[14] Cieślikiewicz Ł, Łapka P, Wasik M, Kubiś M, Pietrak K, Wiśniewski T S, Furmański P and Seredyński M 2018 Development of the experimental stand for investigation of heating and drying phenomena in the porous building materials with one surface of the sample exposed to the flowing air *E3S Web Conf.* **70** 03003

[15] Cieślikiewicz Ł, Wasik M, Kubiś M, Łapka P, Bugaj M, Pietrak K, Wiśniewski T S, Furmański P and Seredyński M 2019 Development of the experimental stand with centrally located specimen for the investigation of heat and moisture phenomena in porous building materials *Civ. Environ. Eng. Reports* **29** 53–65

New numerical model of heat and moisture transfer in the wet brick

M Seredyński[1], P Furmański[1], P Łapka[1], M Wasik[1], Ł Cieślikiewicz[1], K Pietrak[1], M Kubiś[1], T Wiśniewski[1] and M Jaworski[1]

[1]Warsaw University of Technology, Faculty of Power and Aeronautical Engineering, Institute of Heat Engineering, 21/25 Nowowiejska St, 00-665 Warsaw, Poland

E-mail: msered@itc.pw.edu.pl

Abstract The subject of this study is numerical modelling of the fully coupled heat and moisture transfer in the building material, i.e., in a brick. The new mathematical model of equilibrium transport of heat and moisture in the form of continuous liquid medium (funicular region), discontinuous liquid medium (pendular region) and gaseous phase with dry air is proposed. As independent variables, the total mass moisture content and temperature are chosen. Both energy and moisture balance equations are mutually coupled. It is due to dependence of capillary and vapor pressures on the moisture content and temperature. Additionally, vaporization and condensation phenomena are also accounted for in the energy balance equation. Both transport equations are supplemented with boundary conditions expressing the coupled heat and moisture transfer at the dried wall or heat transfer at impermeable walls. Balance equations are discretized on the rectangular control volume grid in the framework of the in-house computational code. A grid-based method of generation of the control volume mesh is proposed, with 'half' control volumes related to boundary grid nodes created at domain walls. This approach enables deeper insight into implemented code as well as simplifies analysis of influence of selected terms in balance equations and implementation of complex hygro-thermal boundary conditions.

1. Introduction

There are several sources of moisture in building materials. One can mention capillary suction from foundations, rain infiltrating roofing, water leakages due to water installation failure, etc., so it is a common problem. Water can appear in such materials in three phases, namely as vapor, liquid and solid. Presence of water in building materials is generally detrimental. It increases heat losses, can lead to destruction of the building structure due to icing or its degradation and can intensify growth of fungi and molds.

In recent years the considerable focus on experimental and numerical investigation of moisture processes in building materials is observed. The first theoretical and experimental analysis of simultaneous energy and moisture transfer was carried out by Luikov [1]. The macroscale description of transport processes based on the volume averaging of microscale transport balance equations was derived by Whittaker [2]. Wide spectrum of problems related to heat and mass transport across porous materials was presented in the monograph by Kaviany [3]. Coupled heat and mass transfer processes accompanying condensation and transport of liquid in pendular state was investigated by Ogniewicz and Tien [4]. Teriku at al. [5] developed computational model of transient transport of coupled heat, air and moisture through a multilayered porous media. As the independent variables, temperature, relative humidity and total humid air pressure were chosen. The model was successfully benchmarked using analytical, numerical and experimental cases. The role of uncertainty in heat and moisture transport properties, related to measurement errors and heterogeneity of material, on results of numerical simulation was investigated numerically by Defraeye et al. [6]. The fully equilibrium model of heat, dry air and water transport was considered. The Monte-Carlo method was used to analyze several cases corresponding to many randomly disturbed physical properties of building material. The results indicated that uncertainties in the taken heat and moisture transport properties resulted in significant differences in drying behavior properties. The fully equilibrium model of heat, water and moist air transport was presented by Van Belleghem et al. [7] for fully funicular liquid water flow. Authors investigated the role of boundary conditions for the 2D drying case and compared results of simulations with experimental data.

Presented model is a continuation of earlier works performed by authors of this paper (e.g., [8-10]). In these papers computational models were developed twofold, as a fully equilibrium [8] or nonequilibrium one [9, 10]. In this submission the equilibrium model of heat and water transport is presented and further developed. The model is capable to capture two forms of liquid water flow, pendular, when liquid water is continuous and convective transport is present, and funicular, when liquid water is discontinuous, and diffusive transport of gases prevails. Transport of water is inseparably related to transport of heat. The temperature variation across the building material changes conditions of evaporation and condensation of water. Evaporation of water at the surface of building material decreases its temperature and this, in turn retards the moisture transport to surroundings. So, the heat and mass transfer equations should be taken into account in the reliable model of moisture transfer across the building material. Additionally, the total pressure of the humid air can be variable across the porous building material, due to changes in temperature and related evaporation/condensation processes, but this effect is assumed here to be negligible, so it is neglected here. In this paper the equilibrium numerical model of heat and moisture transport is presented, in the simple 1D problem statement. Performed simulations show the impact of the initial volumetric water (moisture) content and relative humidity of the ambient air on dynamics of the drying process, namely the temporal changes of temperature and water content at the surface of the wet brick. Additionally, the role of value of the limiting level of volumetric water content, W_{min}, on progress of drying process is investigated.

2. Model of moisture and energy transport

A brick has a very complex porous structure. Its pores are open and connected, so gases and fluids present in it can be freely transported across. Three phases are being identified: (a) solid structure of the brick, (b) liquid phase (water), and (c) vapor phase (gaseous mixture of dry air and vapor).

Their volumetric fractions are ε_s, ε_l and ε_g, respectively. They satisfy the following condition:

$$\varepsilon_s + \varepsilon_l + \varepsilon_g = \varepsilon_s + \varepsilon_p = 1 \tag{1}$$

where ε_p is the porosity of the porous material.

In the proposed model the volumetric water content, namely total mass of water related to volume of material, is considered. It is the sum of volumetric water content in liquid and vapor phases:

$$W = W_l + W_g = \rho_l \varepsilon_l + \bar{\rho}_v^g \varepsilon_g \tag{2}$$

where $\bar{\rho}_v^g$ and ρ_l are densities of water vapor and liquid, respectively.

The sum of volumes of liquid water and vapor are equal to volume of pores, so:

$$\varepsilon_p = \frac{W_l}{\rho_l} + \frac{W_g}{\bar{\rho}_v^g} \tag{3}$$

Due to condensation and evaporation processes mass of liquid and vapor phase changes. Perfect equilibrium between liquid and vapor phases is assumed, so volumetric moisture contents in liquid and vapor phases is simply related to total moisture volumetric content, W:

$$W_l = \frac{\varepsilon_p - W/\bar{\rho}_v^g}{1/\rho_l - 1/\bar{\rho}_v^g} \quad \text{and} \quad W_v = \frac{W/\rho_l - \varepsilon_p}{1/\rho_l - 1/\bar{\rho}_v^g} \tag{4}$$

Liquid water can be present in pores of the brick in continuous (pendular) or discontinuous (funicular) form. It means that if the content of liquid water is sufficiently high, it can form a continuous structure and transport of water is driven mainly by the gradient of the capillary pressure. Otherwise, the only mode of water transport is the diffusive transport of gases. The funicular form liquid water is attained if the total water content is lower than some predefined value, W_{min}, i.e., $W < W_{min}$. Otherwise, the pendular form is present.

2.1. Transport of moisture

Mass transport equations for vapor and liquid water are following:

$$\frac{\partial}{\partial t}(\rho_l \varepsilon_l) + \nabla \cdot \mathbf{j}_l = -\dot{m}_{lv} \quad \text{and} \quad \frac{\partial}{\partial t}(\bar{\rho}_v^g \varepsilon_g) + \nabla \cdot \mathbf{j}_v = \dot{m}_{lv} \tag{5}$$

where $\mathbf{j}_v$ and $\mathbf{j}_l$ are mass fluxes of vapor and liquid water, respectively. These fluxes, for the uniform total pressure of gas in pores, p_g, and neglecting the impact of gravity, takes the following forms:

$$\mathbf{j}_l = \zeta \rho_l \frac{K_l}{\mu_l} \nabla p_c \quad \text{and} \quad \mathbf{j}_v = -\bar{\rho}_g^g D_{v-a,ef} \nabla \left(\frac{\bar{\rho}_v^g}{\bar{\rho}_g^g} \right) \tag{6}$$

where: $D_{v-a,ef}$ is the effective diffusion coefficient of water vapor in dry air, K_l is the permeability of the porous medium for liquid water, μ_l is the liquid water viscosity and ζ accounts for mutual interactions of liquid and gaseous phases. Parameter ζ varies between 0 for the fully funicular and 1 for fully pendular liquid water transport. The relation between ζ and W in the range [W_{min} −1, W_{min}] is given with a sine-like function, which takes in these points values 0 and 1, respectively.

Both dry air and water vapor are treated as perfect gases, so their densities can be expressed from the following Clapeyron equation of state:

$$\bar{\rho}_a^g = \frac{(p_g - p_v) M_a}{BT}, \quad \bar{\rho}_v^g = \frac{p_v M_v}{BT}, \quad \bar{\rho}_g^g = \bar{\rho}_a^g + \bar{\rho}_v^g \tag{7}$$

where B is the universal gas constant and M_a and M_v are molar masses of dry air and water, respectively.

Partial pressure of vapor is related to actual relative humidity:

$$\varphi = \frac{p_v}{p_{sat}(T)} \tag{8}$$

and, according to the Kelvin's law, is related to the capillary pressure by:

$$p_c = -\rho_l \frac{BT}{M_v} \ln \varphi \tag{9}$$

Saturation pressure is a function of temperature only and is given with the following relation:

$$p_{sat}(T) = 611 \cdot \exp\left[\frac{17.08(T - 273.15)}{T - 38.97} \right] \tag{10}$$

After summation of eq. (5) the total moisture transport takes the following form:

$$\frac{\partial W}{\partial t} + \nabla \cdot (\mathbf{j}_l + \mathbf{j}_v) = 0 \tag{11}$$

All parameters present in the mass fluxes of moisture in eq. (6) are dependent on temperature, T, and total water content, W, so they can be presented as follows:

$$\mathbf{j}_l = D_l^T \nabla T + D_l^W \nabla W \quad \text{and} \quad \mathbf{j}_v = D_v^T \nabla T + D_v^W \nabla W \tag{12}$$

where the respective coefficients are defined as:

$$D_l^T = \zeta K_l \frac{\partial p_c}{\partial T},\; D_l^W = \zeta K_l \frac{\partial p_c}{\partial W},\; D_v^T = -\frac{M_a M_v}{M_g BT} D_{v-a,ef} \frac{\partial p_v}{\partial T},\; D_v^W = -\frac{M_a M_v}{M_g BT} D_{v-a,ef} \frac{\partial p_v}{\partial W} \tag{13}$$

The total water transport equation takes the following final form:

$$\frac{\partial W}{\partial t} + \nabla \cdot \left[\left(D_l^T + D_v^T \right) \nabla T + \left(D_l^W + D_v^W \right) \nabla W \right] = 0 \tag{14}$$

2.2. Energy transport in the moist material

The thermal equilibrium between all phases is assumed, so single energy balance equation can be expressed in the form:

$$\frac{\partial H}{\partial t} + \nabla \cdot \left(\mathbf{j}_l h_l + \mathbf{j}_v h_v \right) = \nabla \cdot \left(k_{ef} \nabla T \right) \tag{15}$$

where k_{ef} is the effective thermal conductivity of the porous medium.

H is the total enthalpy per unit volume of the multiphase medium, defined as:

$$H = \rho_s \varepsilon_s h_s + \rho_l \varepsilon_l h_l + \bar{\rho}_v^g \varepsilon_g h_v + \bar{\rho}_a^g \varepsilon_g h_a \tag{16}$$

Specific enthalpies are related to the triple point temperature as the reference one, and can be expressed as:

$$h_s = c_s \left(T[K] - T_{ref} \right),\; h_l = c_l \left(T[K] - T_{ref} \right),\; h_a = c_{p,a} \left(T[K] - T_{ref} \right),\; h_v = \Delta h_v + c_{p,v} \left(T[K] - T_{ref} \right) \tag{17}$$

where c_s, c_l, $c_{p,a}$ and $c_{p,v}$ are specific heats of components and Δh_v is the latent heat of evaporation. The reference temperature is equal to $T_{ref} = 273.15\text{K}$.

Substituting definitions of specific enthalpies, eq. (17), to the energy balance, eq. (15), it can be expressed using the temperature as independent variable. Then using liquid water and vapor mass fluxes expressed with eq. (12), the energy equation reduces to the following form

$$\begin{aligned} &(\rho c)_{ef} \frac{\partial T}{\partial t} + \left[\left(D_l^T c_l + D_v^T c_{p,v} + D_a^T c_{p,a} \right) \nabla T + \left(D_l^W c_l + D_v^W c_{p,v} + D_a^W c_{p,a} \right) \nabla W \right] \cdot \nabla T = \\ &= \nabla \cdot \left(k_{ef} \nabla T \right) - \dot{m}_{lv} \Delta h_v \end{aligned} \tag{18}$$

where $(\rho c)_{ef}$ is the effective volumetric specific heat given by:

$$\left(\rho c \right)_{ef} = \varepsilon_s \rho_s c_s + \varepsilon_l \rho_l c_l + \varepsilon_g \bar{\rho}_a^g c_{p,a} + \varepsilon_g \bar{\rho}_v^g c_{p,v} \tag{19}$$

and $\dot{m}_{lv}$ is the intensity of evaporation and condensation calculated with formula:

$$\dot{m}_{lv} = {dW_v}/{dt} \tag{20}$$

2.3. Boundary conditions and solution procedure

Two types of boundary conditions are considered. Across the surface of the brick, which is in contact with ambient air, both heat and moisture transport are possible. Fluxes of moisture and heat satisfy the following relations:

$$\left(\mathbf{j}_l+\mathbf{j}_v\right)\cdot\mathbf{n}\big|_b=\mathbf{j}_{v,amb}=h_m\left[\bar{\rho}_v^g\left(T_b\right)-\bar{\rho}_v^g\left(T_{amb}\right)\right]$$
$$\left(-k_{ef}\nabla T+\mathbf{j}_l h_l+\mathbf{j}_v h_v\right)\cdot\mathbf{n}\big|_b=h_T\left(T_b-T_{amb}\right) \tag{21}$$

where h_T, and h_m are heat and mass transfer coefficients, respectively, while T_b and T_{amb} are temperature of the wall and ambient air, respectively. If the wall is assumed insulated and additionally no mass transfer is possible, the right sides of eq. (21) are set zero.

The procedure of solution of coupled transport equations of moisture and energy is following. Primarily the moisture transport equation, eq. (14) is solved. Next, the moisture content in liquid and gaseous phases is calculated with eq. (4). Finally, the energy transport equation, eq. (18), is solved. This procedure is repeated until the convergence criteria are meet.

2.4. Material properties and closure relations

Solution of the presented above equations requires information related to material properties and closure relations, e.g., isotherm of sorption. Simulations were performed for the ceramic brick. Its material properties refer to those used in [7] and are listed in table 1.

Table 1. Thermophysical properties assumed in simulations [7].

Property	Symbol	Value
Density of water	ρ_l (kg/m^3)	1000.0
Density of brick	ρ_s (kg/m^3)	2087.0
Specific heat of dry air	c_a (J/kg/K)	1005.0
Specific heat of water	c_l (J/kg/K)	4192.1
Specific heat of brick	c_s (J/kg/K)	840.0
Specific heat of vapor	c_v (J/kg/K)	1875.2
Molecular mass of dry air	M_a (kg/kmol)	28.86
Molecular mass of vapor	M_v (kg/kmol)	18.0
Capillary moisture content	W_{cap} (kg/m^3)	130.0
Porosity of brick	ε_p	0.13
Latent heat of evaporation	Δh_{lv} (J/kg)	$2.6\cdot10^6$
Universal gas constant	B (J/kmol/K)	8314

Closure relationships for the system of equations are the same, as were used in [7]:

- Vapor diffusivity in pores:

$$D_{v-a,ef}=1.053\cdot10^{-6}\frac{1-W/W_{cap}}{0.503\left(1-W/W_{cap}\right)^2+0.497} \tag{22}$$

- Effective thermal conductivity of the moist brick:

$$k_{ef}=1+0.0047\,W \tag{23}$$

- Water permeability in the brick:

$$K_l=\frac{1.1437\cdot10^{-9}}{\left[1+\left(1.76\cdot10^{-5}p_c\right)^{4.3}\right]^{1.6}} \tag{24}$$

- Retention curve:

$$W = W_{cap}\left\{0.846\left[1+\left(1.394\cdot10^{-5}p_c\right)^4\right]^{-0.75}+0.154\left[1+\left(0.9011\cdot10^{-5}p_c\right)^{1.69}\right]^{-0.408}\right\} \quad (25)$$

3. Problem statement

Presented balance equations of the total moisture content, eq. (14), and energy, eq. (18), are discretized in the 1D domain (figure 1). The representation of a 1D control volume grid is based on the grid points, so outmost nodes are located on left and right walls and so two 'half' control volumes are generated at boundaries. The left boundary is adiabatic and impermeable, and the mass and energy transfer are possible across the right boundary, so boundary conditions given by eq. (21) are imposed there.

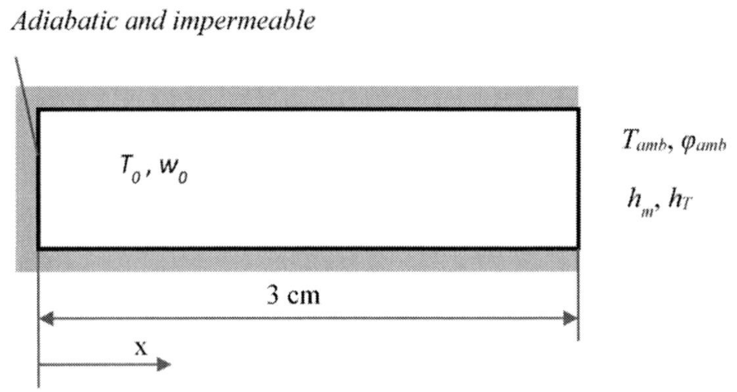

Figure 1. Schematic of the considered domain and boundary conditions.

Temperature of the ambient air, T_{amb}, and initial temperature, T_0, are equal to 23.8°C for all considered cases. The relative humidity, φ_{amb}, can be equal to 50 or 60%. The initial volumetric fraction of water, W_0, is equal to 80% or 90% of W_{cap}. The heat transfer coefficient, h_T, is equal to 25 W/(m^{2}K) for all cases. The mass transfer coefficient, h_m, is calculated according to the Chilton-Colburn analogy, for Lewis number, Le = 0.808, and is equal to 0.0287 m/s.

4. Results and discussion

The impact of initial water content in a brick and relative humidity of drying air on the drying process is investigated. Temporal variations of temperature and volumetric water content at the surface of the dried brick are presented in figure 2. The characteristic shape of the temperature curve is visible in figure 2A. Temperature drops in short time to its lowest value and is kept at constant value over some time period. Its length depends on both, W_0 and φ_{amb}. Obviously, the higher value of W_0 or φ_{amb} leads to longer the first period of drying, when temperature is at its lowest level. The value of the lowest temperature depends strongly on φ_{amb} and is practically W_0 independent (figure 2A). During the first period of drying, considerable changes in total water content are observed (figure 2B). After that time, temperature rises abruptly and variations in water content start to be much slower. It is due to change of the mechanism of water transport, from convective transport of liquid water to diffusion of water vapor. The value of the water content, at which water removal slows down is the same for all cases, equal to 15.4 kg/m^3, which corresponds to 11.8% of W_{cap}.

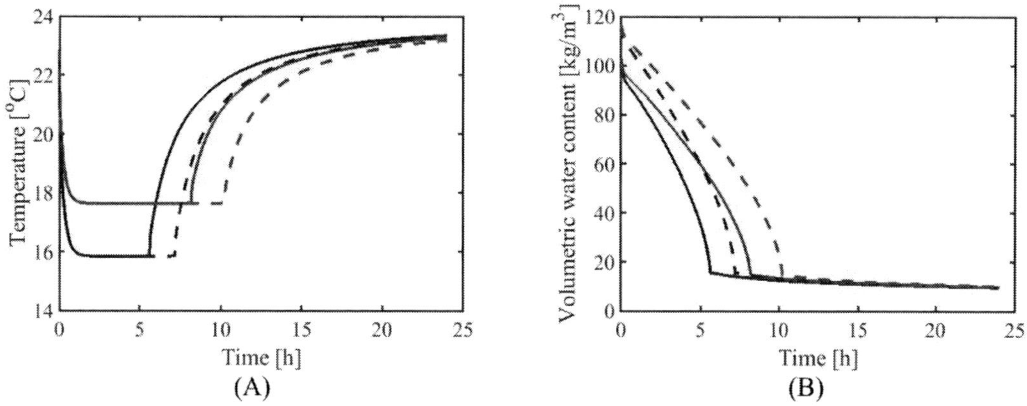

Figure 2. Temporal variation of: A) temperature and B) volumetric water content at the surface of the brick for initial water content of $W_0 = 80\% \cdot W_{cap}$ (continuous line) and $W_0 = 90\% \cdot W_{cap}$ (dashed line), for $\varphi_{amb} = 50\%$ (blue line) and $\varphi_{amb} = 60\%$ (red line).

The role of the existence of the funicular transport region, given with the limiting value of the volumetric water content, W_{min}, was performed for two values of W_{min}, namely 20kg/m^3 and 25 kg/m^3. Below that value the convective transport of liquid water is suppressed. Temporal variations of temperature and volumetric water content at the surface of the dried brick, predicted numerically for two values of W_{min}, and related to the case where liquid water transport is fully pendular is presented in figure 3. After the volumetric water content attains the W_{min}, it is kept approximately at the same level, so drying is considerably retarded. With increasing value of W_{min}, the rate of temperature increase is higher.

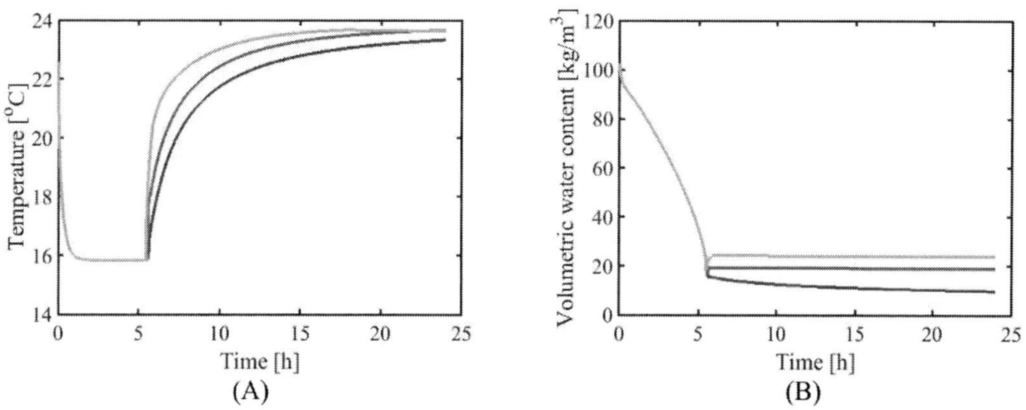

Figure 3. Temporal variation of: A) temperature and B) volumetric water content at the surface of the brick for initial water content of $W_0 = 80\% W_{cap}$ and φ_{amb}=50% for fully pendular transport (blue line) and for partially funicular transport; with W_{min} = 20kg/m^3 (red line) and W_{min} = 25kg/m^3 (yellow line).

5. Conclusions

In this paper the equilibrium model of heat and moisture transport was presented. It was applied to simulation of drying of uniform material, porous brick, in simple 1D geometry. The role if initial water content as well as relative humidity of drying air on variations of temperature and water content at the dried surface were investigated numerically. The temporal changes of temperature reveal characteristic shape (e.g., similar to these shown in [7]), which reflects two stages of drying. During the former, the convective transport of liquid water is dominant, temperature considerably drops, and is maintained at the constant level. Concurrently, the amount of moisture rapidly decreases. The second stage starts when the temperature starts to growth. The dominant mechanism of moisture transport changes, it is now driven mainly by diffusive transport of vapor. It results with rapid decrease of the rate of moisture changes at the surface. If the existence of funicular transport is taken into account, the higher values of water content are observed, corresponding to W_{min} values, and temperature growth rate is higher.

In this paper the simple model of heat and water (moisture) transport was presented. It will be generalized in future for cases involving changes in total pressure of the moist air and presence of nonequilibrium mass transfer between liquid and vapor phases. Numerical results will be also validated with experimental results, predicted with the facility presented in papers [11, 12].

Acknowledgments
This work was supported by the National Centre for Research and Development (Poland) under grant no. POIR.04.01.02-00-0099/16 *Development of innovative technology of drying and moisture sealing of masonry walls, DryWall.*

References

[1] Luikov A V 1966 *Heat and Mass Transfer in Capillary Porous Bodies* (Oxford: Pergamon)
[2] Whitaker S 1977 Simultaneous heat, mass and momentum transfer in porous media: Theory of drying *Advances in Heat Transfer* **13** 119–203
[3] Kaviany M 1995 *Principles of Heat Transfer in Porous Media* (New York: Springer)
[4] Ogniewicz V and Tien C L 1981 Analysis of condensation in porous insulations *Int. J. Heat Mass Transf.* **24** 421–29
[5] Tariku F, Kumaran K and Fazio P 2010 Transient model for coupled heat, air and moisture transfer through multilayered porous media *Int. J. Heat Mass Transf.* **53** 3035–44
[6] Defraeye T, Blocken B and Carmeliet J 2013 Influence of uncertainty in heat-moisture transport properties on convective drying of porous materials by numerical modelling *Chem. Eng. Res. Des.* **91** 36–42
[7] Van Belleghem M, Ameel B, Janssens A and De Paepe M 2011 Modeling heat and moisture transport in porous materials with CFD for building applications *HEFAT2011, 8th Int. Conf. on Heat Transfer, Fluid Mechanics and Thermodynamics* (Pointe Aux Piments) 455–63
[8] Furmański P, Seredyński M, Łapka P, Wiśniewski T S, Pietrak K, Cieślikiewicz Ł, Kubiś M and Wasik M 2018 A model of heat and moisture transfer in a building material during drying based on phase equilibrium in the moist region *Proc. 5th Int. Conf. on Contemporary Problems of Thermal Engineering CPOTE 2018 (Gliwice)* pp 663–73
[9] Łapka P, Wasik M, Furmański P, Seredyński M, Cieślikiewicz Ł, Pietrak K, Kubiś M, Wiśniewski T S and Jaworski M 2018 Preliminary mathematical and numerical transient models of convective heating and drying of a brick *MATEC Web Conf.* **240** 01022
[10] Wasik M, Cieślikiewicz Ł, Łapka P, Furmański P, Kubiś M, Seredyński M, Pietrak K, Wiśniewski T S and Jaworski M 2019 Initial credibility analysis of a numerical model of heat and moisture transfer in porous building materials *AIP Conf. Proc.* **2078** 020106
[11] Cieślikiewicz Ł, Łapka P, Wasik M, Kubiś M, Pietrak K, Wiśniewski T S, Furmański P and Seredyński M 2018 Development of the experimental stand for investigation of heating and drying phenomena in the porous building materials with one surface of the sample exposed to the flowing air *E3S Web Conf.* **70** 03003
[12] Cieślikiewicz Ł, Wasik M, Kubiś M, Łapka P, Bugaj M, Pietrak K, Wiśniewski T S, Furmański P and Seredyński M 2019 Development of the experimental stand with centrally located specimen for the investigation of heat and moisture phenomena in porous building materials *Civ. Environ. Eng. Reports* **29** 53–65

IOP Conf. Series: Materials Science and Engineering **660** (2019) 012024 doi:10.1088/1757-899X/660/1/012024

Numerical models for long-term performance assessment of lightweight insulating assemblies

M Birjukovs[1]*, I Apine[2] and A Jakovics[1]

[1]University of Latvia, Department of Physics, Riga, Latvia, Jelgavas iela 3, LV-1004

[2]University of Latvia, Botanical garden, Riga, Latvia, Kandavas iela 2, LV-1083

E-mail: mihails.birjukovs@lu.lv

Abstract. This paper presents the results of numerical modelling of hygrothermal processes in test buildings situated in the Botanical garden of the University of Latvia. Long-term performance of wall, floor and ceiling insulating assemblies for five different buildings was simulated using WUFI Pro 6.3. On the experimental side, measurements of temperature and relative humidity at key points of assembles have been accumulated over a period of roughly six years and material samples were taken from building envelope materials for biological analysis. We found that our models rather successfully reproduce the experientially observed temperature and relative humidity dynamics. In the process, we established material models for building envelope components, as well as different parameters that we previously unknown. We also compared mould growth risk predictions via the Sedlbauer critical curve model derived from both simulations and experiment against the results of lab tests conducted on materials samples - all three were found to be in good agreement, indicating that the Sedlbauer model is applicable to local climate conditions, as well as further validating numerical models.

1. Introduction

In light of increasingly stricter standards imposed on architects in terms of building performance such as indoor comfort and energy efficiency for new buildings, with energy reduction deadlines for existing ones due year 2020 [1, 2], as well as the ever present risks of mould growth [3, 4], condensation within wall, ceiling or floor insulation envelopes [3, 4] and degradation of materials [3, 5, 6], it is important to take care when designing with long-term exploitation and sustainability in mind. Water damage of building materials typically results in swelling, increased mould growth risk, altered material properties, such as thermal conductivity [3, 7]. Mould growth is detrimental in that it degrades building materials, altering their properties, inducing rot and decay in wood (but not exclusively), and rendering them non-functional and/or unreliable [3, 5, 6, 8, 9]. In addition, by-products of micro-organism population growth within building insulating and structural elements, as well as the micro-organisms themselves, are transported to building interior spaces (via liquid and vapour fluxes within materials, and air flow within ventilation systems and rooms) [3, 10], wherein they can exhibit harmful effects on inhabitants, such as allergic reaction, pronounced respiratory symptoms, intensified asthma, weakened immune system, etc. [11–13].

This and the great cost of potential mistakes [3] motivates numerical modelling, which allows one to assess the performance of designed solutions *in silico*, a risk-free environment. However, given that all of the above is dependent on generally complex temperature and relative humidity

(RH) dynamics within a building and that numerical models require many input parameters and dependencies, not all of which are known *a priori* or with sufficient precision, one cannot rely exclusively on simulations. Sometimes using even seemingly slightly different parameters for simulations can lead to poor design choices which can have many serious consequences, should erroneous designs be implemented [14]. This is why long-term building monitoring projects and test sites have been commissioned, for instance, in Finland [15], Spain [16] and Estonia [17].

In Latvia, roughly six years ago five experimental buildings with different insulating envelopes were constructed at a test site located in Riga, in the Botanical garden of the University of Latvia. Each building was equipped with an array of temperature, RH, heat flux and air velocity sensors connected to a data logging system [18,19]. One of the objectives was to establish which of the different assemblies present on site are better suited for Latvian climate, with its characteristic high RH levels, especially during winter with routinely occurring RH extrema of up to 0.9. While it is now clear which solutions are more optimal for local conditions, the large amount of data on temperature and RH dynamics within buildings and data from the meteorological station (located on site) regarding exterior climate conditions, all accumulated over a period of > 5 years, enables one to go further and use the experiment to verify numerical models. Moreover, samples of materials from different places within building envelopes were taken and analysed to study the degree of biological contamination by micro-organisms. Given that sensors are prone to occasional errors, the data from sensors, biological lab tests and simulations would complement one another to provide a clearer physical picture of how different materials and combinations thereof function under real operational conditions. This is important since, despite much effort from the local project team over the years, [20–24] and others, many aspects concerning actual initial conditions within buildings and their hygrothermal properties, and thus also long-term behaviour, are still unclear.

The objective of the present study is to first establish numerical models using *WUFI Pro 6.3* and compare their output to data from temperature and RH sensors. Then, mould growth risk predictions are derived from both experimental data and simulation results, and both are compared to lab test results. This way, we verify both our material models (either custom or taken from a certified database), interior climate models and the applicability of the standard Sedlbauer mould growth risk model to local climate [25,26].

2. The experiment

2.1. Test buildings

The test site houses five buildings, codenamed AER, CER, EXP, LOG and PLY, which stand for different materials used as insulators for wall envelopes. All five variations have identical roof and floor assemblies. Material layer order and thickness values are given in this section, below. A general overview of a test building (shapes are identical) is shown in Figure 1a. Detailed specifications for each building and sensor placement within are given in [27]. For our analysis, only certain temperature and RH sensors were of interest - these are indicated in Figure 2 for wall envelopes, and designated as "FLOOR" and "CEILING" in floor and ceiling assemblies, respectively (Figure 1a). These material layer and sensor arrangements are later translated to numerical models. Material parameters that were known due to our own studies are provided in [27], but some are also provided in section 4 with hygrothermal curves and coefficients that were available in databases (*Fraunhofer-IBP*, *MASEA*, *University of Technology Vienna*) and those deduced over the course of this study (those not showcased are available on-demand).

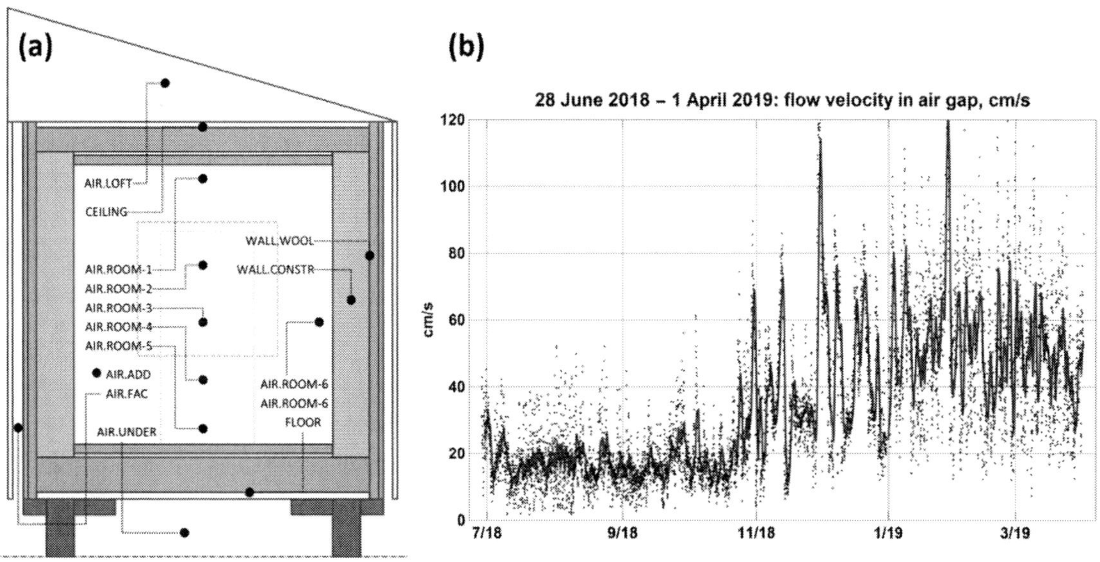

Figure 1. A layout of the sensor network within a test building (a), where black dots correspond to sensor locations, and typical air flow velocity readings from a sensor in the air gap in PLY (b), where black dots are 1-hour average measured air velocities and the red curve is a 1% length moving average of the dataset. Three types of sensors are installed: air velocity (interior, air gap, attic and outdoors), temperature and RH (at all indicated positions). In (a), "ROOM" represents sensors located within building interior, "LOFT" is the attic sensor, "AIR.FAC" is the flow velocity sensor within the air gap of the wall envelope, "WALL" and "FLOOR" are sensors embedded into walls and floor, respectively.

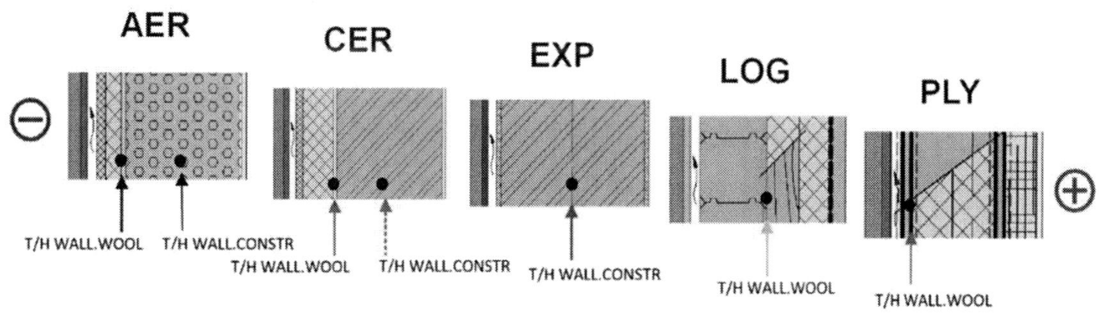

Figure 2. The five wall insulating assemblies with marked sensor locations (black dots). Here "-" indicates the outdoor side and "+" represents the indoor side of an assembly. "T" stands for temperature sensor, "H" is humidity sensor. "WOOL" means that the sensor is embedded into a soft material, and "CONSTR" are sensors place into hard structural materials.

All 5 wall envelopes have the following first three outer layers: Spruce 40 mm, $\rightarrow$ Plywood 6.5 mm, $\rightarrow$ Air gap, 30 mm. Inner layer structure, exterior to interior, is as follows:

- **AER**: Stone wool, 30 mm $\rightarrow$ Stone wool, 50 mm $\rightarrow$ Lime-cement plaster, 15 mm $\rightarrow$ Aerated concrete, 375 mm $\rightarrow$ Lime-cement plaster, 15 mm
- **CER**: Weather protection board, 3 mm $\rightarrow$ Stone wool, 120 mm $\rightarrow$ Lime-cement plaster, 15 mm $\rightarrow$ Perforated ceramic blocks, 440 mm $\rightarrow$ Lime-cement plaster, 15 mm

- **EXP**: Lime-cement plaster, 15 mm → Filled ceramic blocks, 510 mm → Lime-cement plaster, 15 mm
- **LOG**: Spruce, 200 mm → Stone wool, 200 mm → Vapour barrier, < 1 mm → Spruce, 40 mm
- **PLY**: Plywood, 20 mm → Stone wool, 200 mm → Plywood, 20 mm → Fibrolite, 75 mm → Cement plaster, 20 mm

Ceiling and floor assemblies have the following layouts:

- **Ceiling**: Plywood, 12 mm → Wood wool, 200 mm → Plywood, 4 mm → Stone wool, 50 mm → Vapour barrier, < 1 mm → Plywood, 6.5 mm
- **Floor**: Plywood, 21 mm → Vapour barrier, < 1 mm → Stone wool, 200 mm → Plywood, 21 mm → Stone wool, 50 mm → Vapour barrier, < 1 mm → Plywood, 21 mm

2.2. Processing of experimental datasets
Monitoring both buildings and weather conditions at the test site from *April 3, 2013* to *April 1 2019* yielded the following data:

- Outdoor temperature, RH, wind direction and speed, solar radiation intensity
- Temperature and RH values from sensors within test building (see Figures 1a and 2) and air flow velocity in wall air gap (Figure 1b)
- At the 5 year mark, building envelopes were opened and samples from key locations (including sensors) were taken to the lab for micro-organism content assessment

Data for outdoor climate was used both for interpretation of experimental results and as input for numerical simulations. In both cases outliers were removed and the remaining values were binned into 1-hour intervals, since characteristic time scales in building physics for these kinds of experiments are days/months/years. Also, in the case of model input, each hour was taken to be the average of the corresponding hours of the five separate yearly cycles, yielding a single year-long interval. This interval was then periodically extrapolated over the entire calculation period. The rationale is that no long term trends other than yearly cycles were present in gathered data, and such simplified representation of outdoor conditions allows one to observe the general behaviour of studied systems more clearly, which makes sense given the many unknown parameters.

Since *WUFI Pro 6.3* is only applicable to parts of assemblies with symmetries that admit 1D models, only data from sensors shown in Figure 2 and wall/floor sensors was used. After removing outliers, temperature and RH time series from these sensors were de-noised using a low-pass frequency filter with a threshold equal to the inverse of the 0.5 % of time series length, which is roughly 10 days.

3. Numerical simulations
The assemblies described in Figures 1 and 2 and in section 2.1 were simulated in *WUFI Pro 6.3*, a finite volume method (FVM) package that solves coupled heat and moisture transport equations under standard building physics assumptions [28]:

$$\frac{\partial H}{\partial T} \cdot \frac{\partial T}{\partial t} = \nabla \left(\lambda \nabla T + \frac{h_v \delta}{\mu} \nabla \left(\phi p_s \right) \right); \; T = T(\vec{r}, t); \tag{1}$$

$$\frac{\partial w}{\partial \phi} \cdot \frac{\partial \phi}{\partial t} = \nabla \left(D_\phi \nabla \phi + \frac{\delta}{\mu} \nabla \left(\phi p_s \right) \right); \; \phi = \phi(\vec{r}, t); \tag{2}$$

where T is temperature, ϕ is RH, $\lambda = \lambda(T,\phi)$, $\delta = \delta(T,\phi)$ and $p_s = p_s(T)$ are thermal conductivity, vapour diffusivity and vapour saturation pressure, respectively, μ is relative vapour diffusion resistance, $w(\phi)$ is water content density (given by the moisture storage function), $D_\phi = D_\phi(T,\phi)$ is water diffusivity and $H(T)$ is enthalpy. The left hand side of Equation (1) includes temperature dependent liquid-ice transition enthalpy and water/ice enthalpy flow, while vapour enthalpy flow is given by the second term on the right hand side of Equation (1). This and other constitutive relations for Equations (1) and (2) are given in [28].

Outdoor and indoor climate are prescribed by Robin boundary conditions for T and ϕ on the exterior and interior assembly surfaces:

$$\nabla T \cdot \vec{n} + h_t(T - T_0) = q(t) \tag{3}$$

$$\nabla(\phi p_s) \cdot \vec{n} + \beta_v \left(\phi p_s - (\phi p_s)_0\right) = 0 \tag{4}$$

where T_0 and $(\phi p_s)_0$ are exterior/interior temperature and water vapour pressure, respectively, $\vec{n}$ is the surface normal, h_t is heat transfer coefficient and β_v is moisture transfer coefficient (related to h_t via the Chilton–Colburn analogy). Exterior T_0 and ϕ_0 are given by the processed time series from the meteorological station, while interior conditions are given by *ISO 13788* or *EN 15026/DIN 4108/WTA 6-2* standards (herefrom referred to as simply ISO or EN) with variable mean temperature and RH class (moisture load) - the reason is that it is not yet entirely clear to what extent sensor data from within test building rooms are trustworthy and representative of actual inner surface conditions. No records of precipitation dynamics were available, so rain and snow were not accounted for. Solar radiation is accounted for by a heat source $q(t)$ added to Equation (3) at the outer boundary ($q(t) = 0$ at the inner boundary) and a set of transfer coefficients. Initial conditions for simulations were defined based on sensor measurements and construction/maintenance logs. Cloud index is unknown, so a standard coefficient suggested by *WUFI* for atmospheric counter-radiation is used. All of the above parameters and functions must be defined with care, since Equations (1) and (2) are highly non-linear. Establishing the unknowns among these is one of the primary concerns of this study. Note that in some cases initial conditions RH, to which Equations (1)-(4) are sensitive, were known only up to an interval.

Equations (1) and (2) were solved in a 1D approximation. Regarding the air gap in wall assemblies (Figure 1a and 2) and attic spaces (Figure 1a), data from a velocity sensor relocated into the PLY assembly air gap (Figure 1b) allowed to perform dimensional analysis and estimate the degree of ventilation. It was found that Grashof (Gr), Rayleigh (Ra), Reynolds (Re) and Richardson (Ri) numbers are $Gr \in [10^8; 10^{10}]$, $Ra \in [10^8; 10^{10}]$, $Re \in [10^3; 10^4]$, $Ri \in [0.1; 24]$ and on average $Ri < 1$, implying that highly turbulent forced convection dominates within the gap, homogenising air temperature and leading to temperature and RH that should be close to exterior conditions. Since the attic is ventilated and connected to the air gap in each building, we infer that the topmost layer of ceiling assemblies is also exposed to near-outdoor conditions. This allows one to eliminate the outer weather protection layer from numerical models of wall assemblies and ceiling envelope (and thus set $q(t) = 0$ in Equation (3)), provided that an appropriate effective heat transfer coefficient is specified. The floor insulation assembly is essentially an inverted roof (Figure 1a) and is directly exposed to outdoor air. The air gap below is sufficiently wide to disregard grass (periodically trimmed) and shading nullifies the effect of solar radiation. Thus, floor outer boundary is treated as an external wall with $h_t = 0.0588\ W/(m^2 \cdot K)$, while external boundaries of other assemblies are treated as interior walls and are assigned $h_t = 0.125\ W/(m^2 \cdot K)$.

Optimized numerical mesh consists of 750 cells over the assembly length with refinement at material and model boundaries (expansion factor 1.05). Time stepping is adaptive, with a maximum step of 1 h (exterior climate time step) and recursive refinement of up to 20x, each refinement level splitting the current time step level into 10 smaller steps, until step convergence is achieved.

4. Results and analysis

Sensor data and simulation results (temperature and RH time series from *April 3, 2013* to *April 1 2019*) were processed and compared. Virtual sensors were placed in *WUFI Pro 6.3* models in corresponding locations (1a and 1b) and output was processed in the same way as experimental values. Here, several representative results are be shown.

Starting with the ceiling envelope, one can see the results of *WUFI* modelling versus the experiment in Figure 3: Figures 3a and 3c, clearly show that temperature curves align nearly ideally for LOG, but there are considerable discrepancies in case of PLY starting with roughly the second half of 2018 (dashed red lines in Figures 3b and 3d).

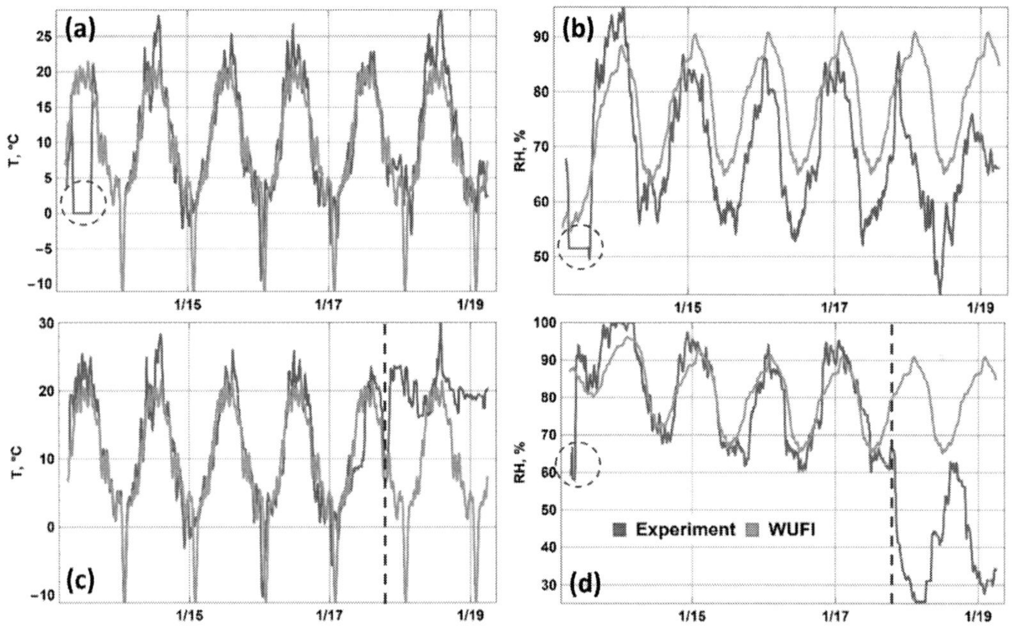

Figure 3. Experimental (light blue) and numerical (yellow) results for the ceiling envelope for temperature (a,c) and RH (b,d) dynamics over the entire monitoring period at the outer stone wool sensor location. Figures (a,b) correspond to LOG, data for PLY is shown in (c,d). Note the "flatlined" areas encircled with dashed red lines (a,b) - there, sensors have temporarily malfunctioned, resulting in data loss. Similarly, in (d) a RH sensor initially indicated unrealistic dynamics, so starting RH was chosen based on the further interval of the curve. For precise sensor locations, please refer to Figure 1.

Figures 3b and 3d indicate very good agreement for RH dynamics for PLY, but there is a considerable and systematic overshoot in case of LOG. Also note that radically different RH curves are predicted for PLY, again, from around the second half of 2018.

Interestingly, inspecting maintenance logs reveals that strong deviations in Figures 3c and 3d correspond to a prolonged series of conditioning system and phase change material (PCM) tests that lasted throughout the remainder of the time interval in question. The reason for RH deviations in LOG are not presently obvious. The established (or current best) initial and interior conditions for LOG and PLY are, respectively: $\phi_0 = 0.5$ globally and the EN interior climate with a high moisture load (HML); $\phi_0 = 0.9$ within the wood chip wool, $\phi_0 = 0.6$ elsewhere and EN-HML interior climate. In this and other cases, we are willing to share the established material and climate models and initial conditions on demand (corresponding author). For AER, CER and EXP, ceiling temperature sensors indicate excellent agreement with *WUFI*, but RH sensor data is, unfortunately, ripe with artefacts, which preclude analysis.

Temperature and RH dynamics for the PLY wall assembly are shown in Figure 4 - note the very good temperature curve fit. However, RH time series exhibit overshoots during winter and a deviation near the beginning. Although in Figure 4b initial conditions may look different, here $\phi_0 = 0.9$ was indeed set for stone wool and $\phi_0 = 0.6$ elsewhere, as per sensor data, but these initial values were removed during automatic data filtering. The EN-HML interior climate model was used.

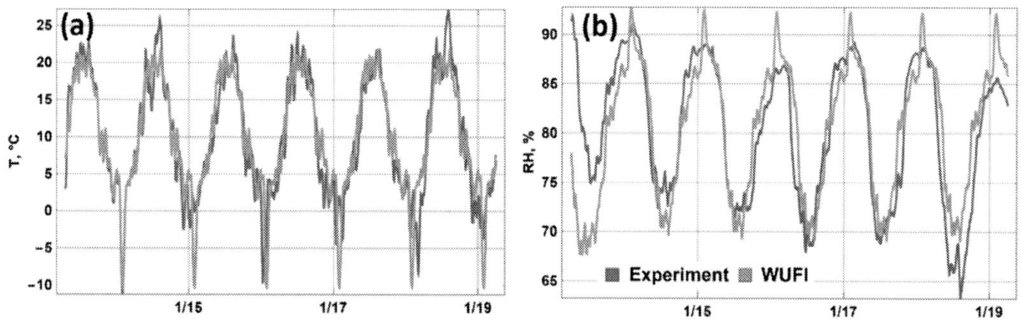

Figure 4. Temperature (a) and RH (b) dynamics for the PLY wall assembly at the stone wool sensor location.

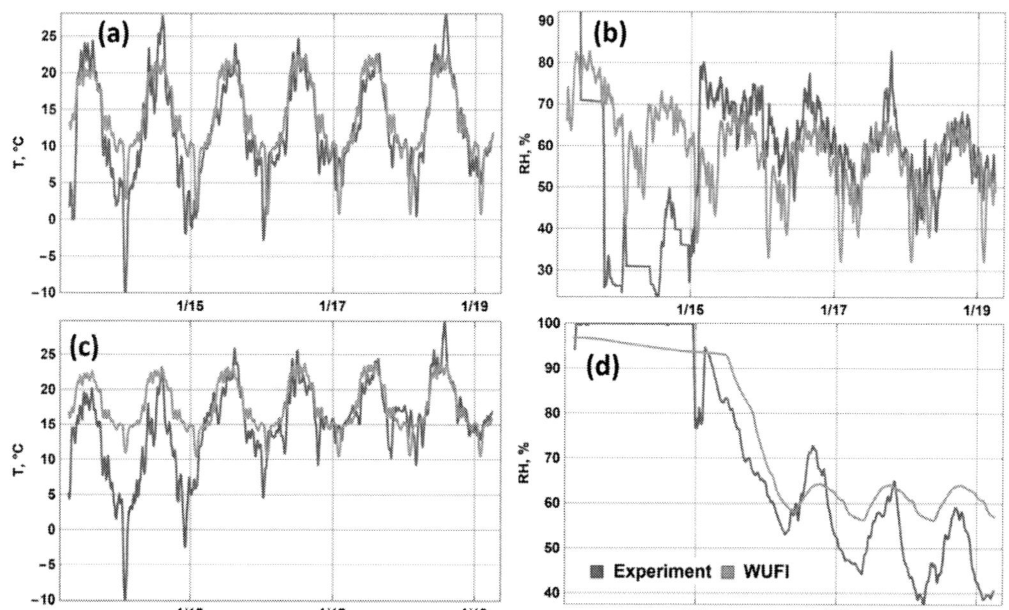

Figure 5. Temperature (a,c) and RH (b,d) dynamics for the AER wall assembly at the stone wool (a,b) and aerated concrete (c,d) sensor location.

Results obtained for the AER assembly wall are presented in Figure 5, wherein Figures 5a and 5b represent temperature and RH dynamics at the stone wool sensor, while 5c and 5d correspond to the sensor within the concrete slab. Concrete and wood are some of the more challenging materials to model in *WUFI*, often exhibiting hysteresis in sorption curves. Given that WUFI provides no means of accounting for this, it is pleasantly surprising that, as seen in Figures 5b and 5d, the general trends are captured successfully. While the undershoot seen in 5b could be explained by a higher true initial RH value, it is not yet clear why there is a delay in drying seen in 5d. The likely explanation is that the sorption curve is incorrect within the high RH interval. The interior climate also very strongly affects drying. The initial conditions established thus far are $\phi_0 = 0.95$ for aerated concrete and $\phi_0 = 0.5$ elsewhere. Again, the EN-HML interior climate seems to best fit the experimentally observed dynamics.

Figure 6 displays the comparison between mould growth risk assessment derived from sensor readings (Figure 6a) and predicted by *WUFI* (Figure 6b) for the PLY wall at the location of the sensor placed within the stone wool layer. Note the very similar pattern in both cases. Both experimental and simulation data suggest moderate risk for biodegradable materials such as wood wool and minor risk for more resilient materials, such as the stone wool in question, with very little organic content.

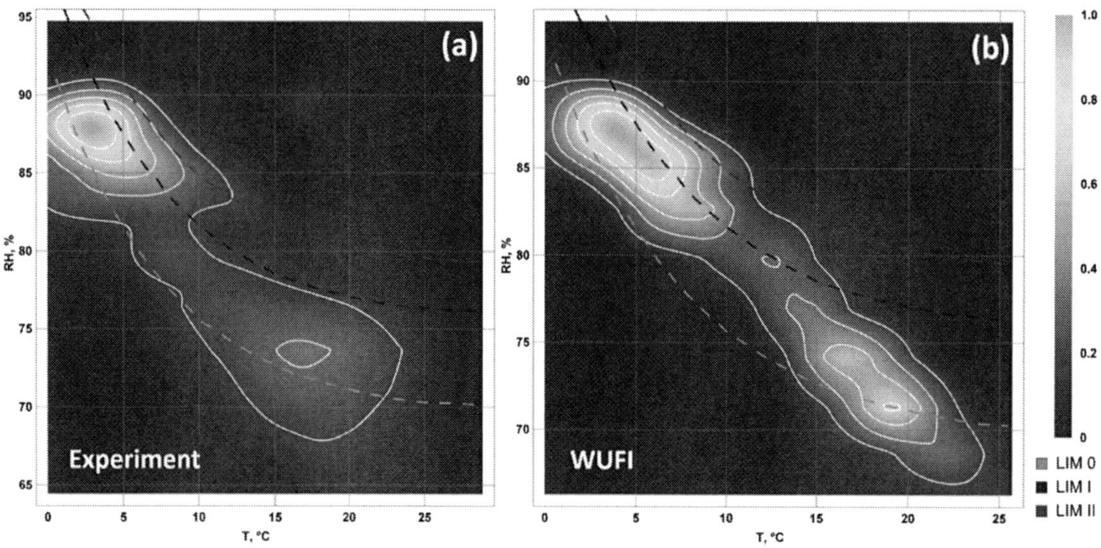

Figure 6. Mould growth risk assessment from *April 3, 2013* to *April 1, 2019* for the PLY wall at the location of the sensor placed within the stone wool layer. Temperature and RH pair (event) density is color coded and normalized to maximum (see the color bar to the right). White contours are event density isolines. LIM (lowest isopleth for mould) curves are due to the standard Sedlbauer mould growth risk model. Higher density above the LIM curves implies greater risk of mould growth.

A very important point is that this correlates with both visual inspection (no mould was detected) and lab tests conducted for samples taken from the stone wool surface. The latter revealed that micro-organisms concentrations of $13400 \pm 3454\ CFU \cdot g^{-1}$ (colony forming units per unit sample mass). Compared to $\sim 50\ CFU \cdot g^{-1}$ for visually clean surfaces and $\sim 55000\ CFU \cdot g^{-1}$ for mould covered surfaces, this suggests that mould growth risk was indeed only mild to moderate throughout the monitoring period. Interestingly, by far the greatest contributor to mould growth was the mesophilic it Cladosporium, which is typically found when RH $\in [0.8, 0.9]$ - this fits very well with the density maxima seen in Figures 6a and 6b.

Several material models were verified or validated. **Plywood board**: density $\rho = 500\ kg/m^3$, porosity $f = 0.5$, specific heat capacity $c = 1500\ J/(kg \cdot K)$, $\lambda = 0.1\ W/(m \cdot K)$, $\mu = 700$. **Wood wool (blowing birch)**: $\rho = 230\ kg/m^3$, $f = 0.999$; $c = 2100\ J/(kg \cdot K)$, $\lambda = 0.053\ W/(m \cdot K)$; $\mu = 4$. **PAROC eXtra (stone wool)**: $\rho = 28\ kg/m^3$, $f = 0,95$; $c = 850\ J/(kg \cdot K)$; $\lambda = 0,034\ W/(m \cdot K)$; $\mu = 1$. **Fibrolite board**: $\rho = 400\ kg/m^3$, $f = 0,74$, $c = 2100\ J/(kg \cdot K)$, $\lambda = 0,068\ W/(m \cdot K)$, $\mu = 44$. For hygrothermal curves, please see Figures 7, 8, 9 and 10.

Due to the article page limit, we are unable to provide the verified material model for cement lime plaster - however, the corresponding author will provide data on demand (values, data files). While it is tempting to showcase material models for the AER wall, as evident from Figure 5d, the hygrothermal model for aerated concrete must be further improved before the data is ready to be made public. This is also true for CER, EXP and Floor models. Once these are improved, we plan to publish the rest of established material models.

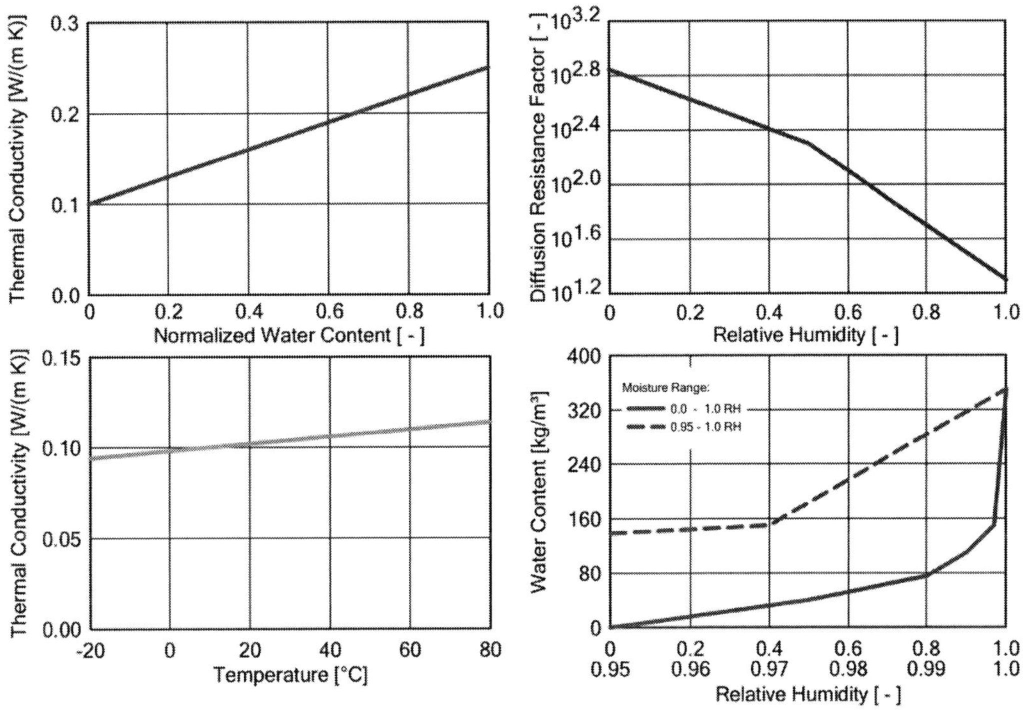

Figure 7. Hygrothermal curves for plywood board. Thermal conductivity plots (a) and (c) correspond to $\lambda(T, w) = \lambda_{dry}(T) \cdot (1 + c_1 w(\phi))$, where the dependency in (a) gives c_1 and (b) gives $\lambda_{dry}(T)$. Normalized water content is the ratio of water content to saturation value.

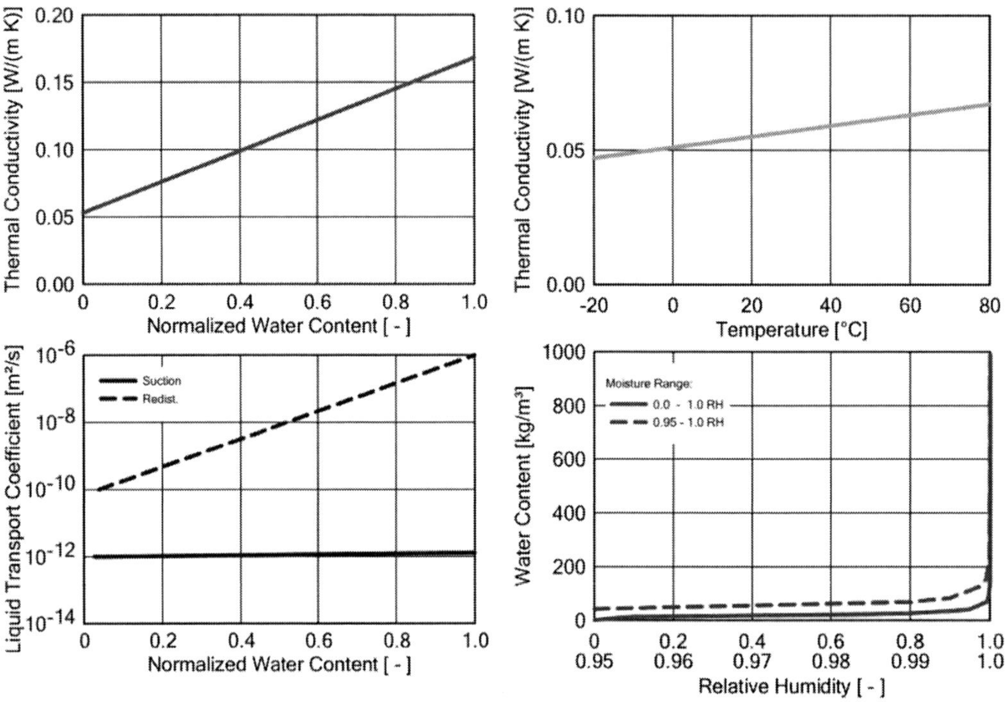

Figure 8. Hygrothermal curves for blowing birch wood wool. Note that both liquid suction and redistribution mechanism contributions to $D_\phi(w)$ are specified.

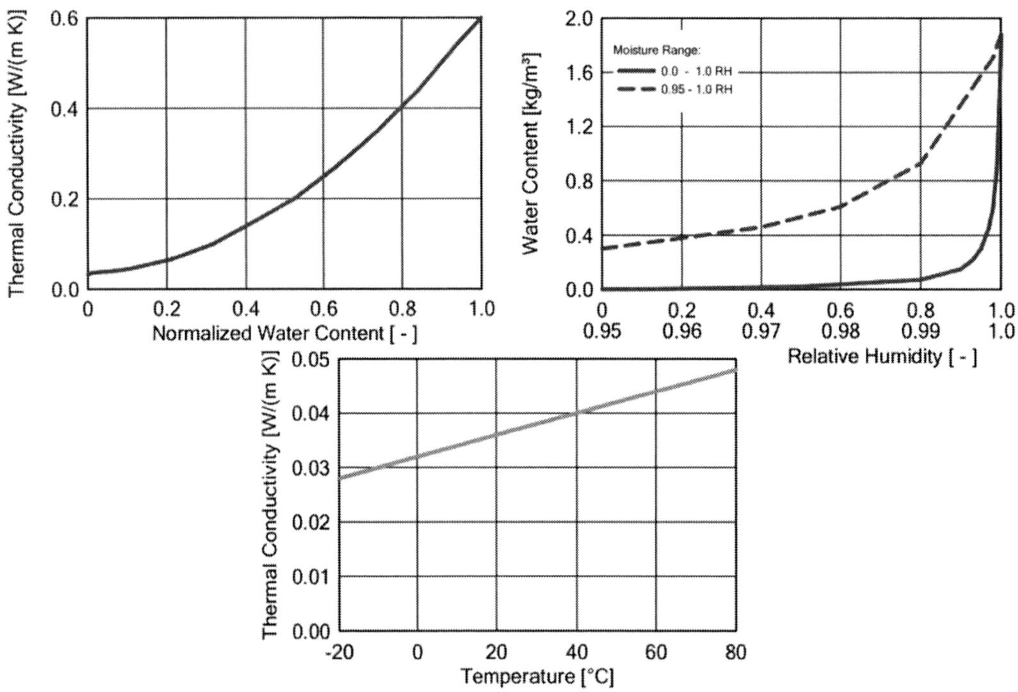

Figure 9. Hygrothermal curves for PAROC eXtra.

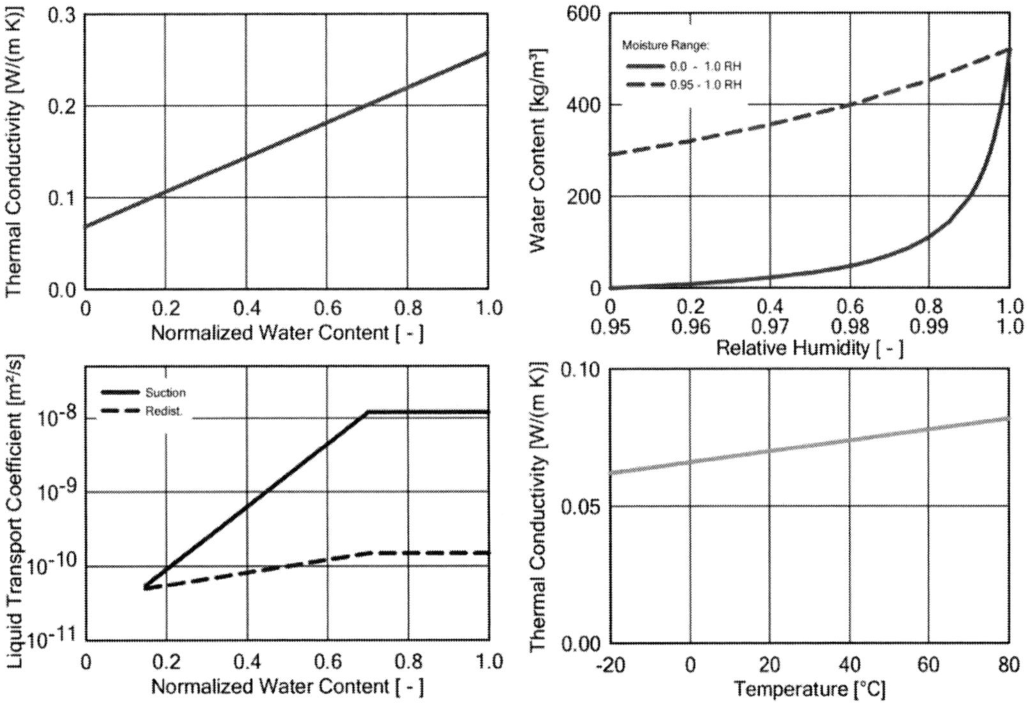

Figure 10. Hygrothermal curves for fibrolite board.

5. Conclusions and outlook

Numerical models have been established for all buildings within the Botanical garden (University of Latvia) test site. Many of these models showed very good agreement with experimental data in terms of RH dynamics, especially for PLY, AER, Ceiling and Floor insulating envelopes. In most cases temperature time series were in excellent agreement. It was found that both numerical models and processed sensor data correctly predict mould growth risks and growth onset most of the time, but especially so for PLY and Floor assemblies. The fact that three independent mould growth risk assessment methods agree very well leads to the conclusion that our numerical models are on the right track.

As a result of cross-verification, many of the previously unknown material parameters and dependencies, as well as initial conditions were established (see section available on demand). The applicability of LIM critical curve models under local climate conditions was ascertained. Numerical models for CER, EXP and LOG envelopes currently show quite strong deviations from the experiment, indicating that respective material models are thus far incorrect. It was found that assemblies are much more sensitive to indoor conditions than initially anticipated. For this reason, sensor readings for interior temperature and RH will be incorporated into current models. Outdoor climate dataset with no 1-year cycle averaging will also be used instead of the current boundary condition.

Acknowledgments

This study was conducted with the financial support of European Regional Development Fund, as part of the project "Development, optimization and sustainability evaluation of smart solutions for nearly zero energy buildings in real climate conditions" (1.1.1.1/16/A/192).

References

[1] 2009 *Buildings, and Climate Change. Summary for Decision-Makers. United Nations Environment Program Sustainable Buildings and Climate Initiative* ISBN 9789280700000

[2] *European Union 2010 Directive 2010/31/EU of the European parliament and council on the energy performance of buildings*

[3] Liyana Othman N, Jaafar M, Harun W and Ibrahim F 2015 *Procedia - Social and Behavioral Sciences* **170** 27–36

[4] Holme J, Geving S and Jenssen J 2008 *Moisture and Mould Damage in Norwegian Houses*

[5] Xu C, Mudunuru M and Nakshatrala K 2015 *Continuum Mechanics and Thermodynamics* **28** DOI 10.1007/s00161-016-0511-4

[6] Viitanen H 2011 *Moisture and Bio-Deterioration Risk of Building Materials and Structures* ISBN 978-953-307-636-2

[7] Boukhattem L, Mustapha B, Hamdi H, Benhamou B and AITNOUH F 2017 *Construction and Building Materials* **148** 811–823

[8] Carll C G and Highley T L 1999 *Journal of Testing and Evaluation, JTEVA* **27(2)** 150–158

[9] Irbe I and Andersone I 2007 *Wood decay fungi in Latvian buildings including cultural monuments*

[10] Haas D, Habib J, Galler H, Buzina W, Schlacher R, Marth E and Reinthaler F 2007 *Atmospheric Environment* **41** 5192–5201

[11] Andersson M, Nikulin M, Kõljalg U, Andersson M, Rainey F, Reijula K, Hintikka E L and Salkinoja-Salonen M 1997 *Applied and environmental microbiology* **63** 387–93

[12] Afshari A, Ross Anderson H, Cohen A, De Oliveira Fernandes E, Douwes J, Górny R, Hirvonen M R, Jaakkola J, Kirchner S, Kurnitski J, Levin H, Mendell M, Mølhave L, Morawska L, Nevalainen A, Richardson M, Rudnai P, Schleibinger H, Schwarze P and Krzyzanowski M 2009 *WHO guidelines for indoor air quality: dampness and mould* ISBN 7989289041683

[13] Bornehag C, Blomquist G, Gyntelberg F, Järvholm B, Malmberg P, Nordvall L, Nielsen A, Pershagen G and Sundell J 2001 *Indoor air* **11** 72–86 DOI 10.1034/j.1600-0668.2001.110202.x

[14] Melton P and Yost P 2014 *Environmental Building News* **23(4)** 1–10

[15] Vinha J 2019 Analysis method to determine sufficient water vapour retarder for timber-framed walls

[16] Cabeza L F, Castell A, Medrano M, Martorell I, Perez G and Fernandez A 2010 *Energy and Buildings* **42** 630–636 DOI 10.1016/j.enbuild.2009.10.033

[17] Miljan M J and Miljan J 2019 *Hygrothermal Performance of Timber External Walls Insulated with Natural and Industrial Materials: Sustainable Buildings in Cold Climates* pp 957–967 ISBN 978-3-030-00661-7

[18] Beinarts I, Grunde U and Jakovics A 2014 *Electrical, Control and Communication Engineering* **7** DOI 10.1515/ecce-2014-0022

[19] Greitans M, Grunde U, Jakovics A and Gendelis S 2013 pp 553–556 ISBN 978-1-4799-1419-7

[20] Ratnieks J, Jakovics A and Gendelis S 2017 *Energy Procedia* **132** 441–446 DOI 10.1016/j.egypro.2017.09.654

[21] Ratnieks J, Jakovics A and Gendelis S 2018 *Energy Procedia* **147** 207–213 DOI 10.1016/j.egypro.2018.07.061

[22] Ratnieks J, Jakovics A and Gendelis S 2015 *Energy Procedia* **78** 2905–2910 DOI 10.1016/j.egypro.2015.11.664

[23] Gendelis S, Jakovics A, Ratnieks J and Bandeniece L 2017 *IOP Conference Series: Materials Science and Engineering* **251** 012053 DOI 10.1088/1757-899X/251/1/012053

[24] Geza V, Jakovics A, Gendelis S and Usilonoks I 2017 DOI 10.22364/mmp2017.19

[25] Sedlbauer K 2001 Prediction of mould fungus formation on the surface of and inside building components

[26] Hofbauer W, Krueger N, Breuer K, Sedlbauer K and Schoch T 2008 *Mould resistance assessment of building materials - material specific isopleth-systems for practical application* ISBN 978-87-7877-270-1

[27] Jakovics A, Gendelis S, Ozolins A, Ratnieks J, Dzerins J, Klavins J and Certoks S 2014 *Hygrothermal Performance of Timber External Walls Insulated with Natural and Industrial Materials: Sustainable Buildings in Cold Climates* ISBN 978-9934-517-15-0

[28] Kunzel H 1995 Simultaneous heat and moisture transport in building components. one- and two-dimensional calculation using simple parameters

Experimental investigation of the influence of the relative humidity of air on the drying process of porous building materials

Ł Cieślikiewicz[1], P Łapka[1], M Kubiś[1], M Wasik[1], K Pietrak[1], P Furmański[1], M Seredyński[1] and T Wiśniewski[1]

[1]Warsaw University of Technology, Faculty of Power and Aeronautical Engineering, Institute of Heat Engineering, 21/25 Nowowiejska St., 00-665 Warsaw, Poland

E-mail: lukasz.cieslikiewicz@itc.pw.edu.pl

Abstract. The paper presents the experimental analysis of influence of the relative humidity of drying air on the drying process of porous building materials. Investigations were conducted by applying the stand which worked in a closed-loop flow and equipped with the following elements: the cooler (condenser), fan with variable rotation speed, throttle, humidifier, heater and control and acquisition systems. During experiments the temperature and velocity of air were equal to 40°C and 3 m/s, respectively, while its relative humidity was varied in the range from 16 to 40%. Two different materials were investigated, i.e., the red brick and mortar. Specimens were placed in the duct and had the top surface exposed to flowing air. Variations of the moisture content and temperature in considered building materials were measured by system of two force meters and thermocouples, respectively. The relative humidity of drying air significantly influenced on the drying process of building materials. The increase of the relative humidity of drying air resulted in the rise of the drying duration, the decrease of the drying rate and the increase of the level of temperature plateau (i.e., equilibrium temperature) which resulted from evaporation in the medium. Moreover, for the mortar drying rates were found much slower than for the red brick.

1. Introduction

Many experimental investigations related to moisture transport in porous materials or drying of these materials were reported in literature [1-8]. These studies were concerned on either the analysis of influence of different factors on the moisture transfer/drying process, generation of data required for the validation of mathematical and numerical models or indirect measurements of hygro-thermal properties of porous materials. Different porous materials were investigated, different drying techniques were studied, and different measurements techniques and concepts of the experimental stand were applied. This brief review is concentrated only on measurements related to moisture transport in porous building materials and its applications.

Many studies of the influence of different factors on hygro-thermal transport phenomena in building materials were carried out to find out hygro-thermal behavior of these materials. They are especially important in the case of new building materials, e.g., hemp concrete, because allow for finding of their potential application fields and benefits of their usage. For example, Colinart et al. [1] performed numerical and experimental analysis of the hygrothermal behavior of a coated hemp concrete wall. A special facility was designed to measure mass variations as well as the temperature and relative humidity fields within coated or multilayered samples on the material and wall scale. In the next paper Chikhi et al. [2] studied the behavior of a block of the cement mortar, subjected to variable external temperature and humidity conditions. The aim of the study was to understand the hygro-thermal behavior of construction walls, to make an adequate design according to the climatic parameters and thus to improve the control of the energy used for heating. In turn, Karagiannis et al. [3] performed experimental and numerical investigations drying kinetics of building materials (e.g., stones, bricks and mortars) saturated with capillary absorbed water. The performed investigation might be very useful in the understanding the causes of decay of building materials and in saving cost and improving the energy efficiency of constructions.

When new heat and moisture transfer models are proposed their credibility analyses should be carried out. The aim of these analyses is to validate proposed models, i.e., to proof their accuracy. Such investigations were performed in many papers. For example, Colinart et al. [1] developed the coupled 1D heat and moisture transfer model. Results of simulations were compared to experimental ones obtained on a specially designed experimental stand. Karagiannis et al. [3] validated a first order drying kinetics mathematical model applied for the simulation of the drying process of capillary absorbed water of building materials. They generated data for validation using the semi-industrial air dryer which consisted of an air tunnel with square cross-section. Van Belleghem et al. [4] proposed the new 3D heat and mass transfer model in porous building materials which was implemented in the commercial software. The model was validated using convective drying experiment, in which the saturated with water ceramic brick sample was dried by flowing dry air over one side of the sample surface. The temperature and relative humidity measurements at different depths in the sample, moisture distribution profiles and mass loss measurements were compared with simulation results. Recently, Busser et al. [5] carried out the review of studies in which the results obtained with numerical models and experimental data were compared. In their paper influence of the experimental parameters and boundary and initial conditions as well as modeling of physical phenomena on the agreement between the experimental measurement and the numerical results was discussed.

In the case of estimating hygro-thermal material proprieties of building materials many indirect methods were proposed. Among them were methods which based on observed (experimental) data and identification techniques (hygro-thermal inverse problems). In these methods parameters (e.g., the relative humidity, moisture contend, temperature, etc.) registered during specially designed experiments were applied in inverse algorithms to find hygro-thermal properties, e.g., sorption curve or vapour permeability. The experiment design in these methods, e.g., applied boundary and initial conditions during experiments, the type of sensors as well as their quantity and location, is of importance and influence on the accuracy of estimated parameters. The problem of optimal configuration of the experimental method was studied by Berger et al [6]. They discussed the concept of Optimal Experiment Design for the estimation of material properties of porous walls which was based on in-site measurements and identification method. The proposed optimal conditions ensure to provide the maximum accuracy of the identification method and thus the estimated parameters. Busser et al. [7] proposed a dynamic measurements and identification method for the determination of the vapour permeability and moisture sorption curve. They developed the experimental facility which enabled measurements of the relative humidity within the material and the total moisture content at the same time. A novel indirect method for the determination of the water retention curve from drying tests performed on capillary active materials was proposed by Bianchi et al. [8]. The method was based on the lumped parameter analysis, assuming approximately uniform temperature and water content inside the specimen and at the specimen surface. Moreover, they applied non-destructive techniques (i.e., the gravimetric analysis, infrared thermography) for measuring water content, drying rate and surface temperature during time for a set of material samples.

In this paper the previously developed experimental stand [9] with the closed-loop flow is applied for investigation of drying process of two building materials available on the Polish market. The paper is organized in the following way. At first the stand, samples and measurement protocol are briefly described. Then results of measurements, i.e., relative moisture contents and variations of temperature in two locations inside samples are shown. At the end studies are summarized.

2. Experimental investigation

Experimental investigations of influence of the relative humidity of drying air on the drying process of porous building materials were carried out with fixed temperature and velocity of air equal to 40°C and 3 m/s, respectively. The relative humidity of air used for drying was varied in the range from 16 to 40%. Two different porous building materials were investigated, i.e., the red ceramic brick and mortar sample. The specially developed stand [9] was used to experimentally investigate the drying process of these materials.

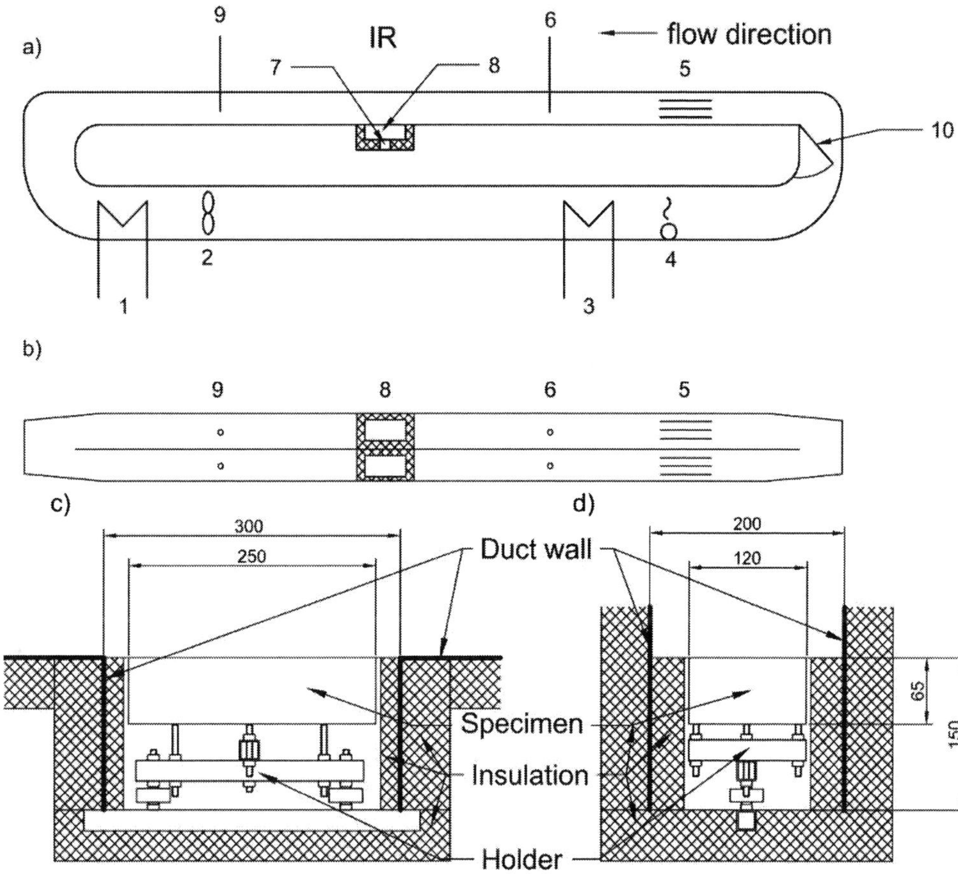

Figure 1. Schematic of the closed-loop flow experimental stand: a) side and b) top view as well as cross-section of the specimen pocket with main dimensions: c) side and d) perpendicular to the flow direction view (where: 1 – cooler/condenser; 2 – fan; 3 – heater; 4 – humidifier; 5 – flow stabilizer; 6 – integrated velocity, temperature and humidity transmitter; 7 – force meter; 8 – sample; 9 – integrated temperature and humidity transmitter; 10 – throttle) [9].

2.1. Experimental stand

The development of the experimental stand used for investigation of influence of the relative humidity of drying air on the drying process of porous building materials was described in detail in [9]. The schematic of the stand is presented on figure 1. Its main features were following:

- Closed-loop air flow,
- Ability to control of the velocity, temperature and relative humidity of drying air in the wide range,

- Ability to measure temporal variations of the mass of the sample and the temperature in several points of the sample,
- Two parallel measurements channels which allow for two measurements at the same time.

The sample was located in the specially designed holder with force meter (see figure 1c) and was thermally insulated and made waterproofed at side and bottom walls. Its top wall was in the same plane as the bottom boundary of the channel. This wall was the only surface in contact with drying air. Therefore, heat and moisture were transported between the sample and drying medium only through the top wall. Temporal variations of temperature in several points of the sample were measured applying 2×0.2 mm T-type wire thermocouples with PFA insulation and external dimensions of 1.3×2.0 mm, located in drillings in the sample. The IR camera might be also applied to measure temperature distribution at the top wall of the sample.

The stand was equipped with the data acquisition and control systems developed based on the measurement and automation devices which provide analog and digital inputs and outputs as well as on the in-house software developed in the LabView environment. The control time step in the control system was 100 ms, while the acquisition time step in the data acquisition system was 6 s. The measurement of drying air parameters took place 1.5 cm above the duct bottom surface and 800 mm before the specimen.

2.2. *Specimens*

Specimens for investigation of influence of the relative humidity of drying air on the drying process were prepared as cuboids of dimensions 250×120×65 mm made of the red ceramic brick and mortar. Ceramic bricks were easily available building materials which came from the Polish market. Mortar specimens were made of market available masonry mortar dry mix and water. They were formed in wooden forms. Next mortar specimens were conditioned in the surroundings for more than one month to obtain the proper mortar binding. In each specimen 5 thermocouples were placed 60 mm from the leading surface, 32.5 mm from the side surface and every 10 mm from the top (drying) surface. Before each measurement specimens were fully submerged in the water bath for 24 h to saturate it with liquid moisture.

Table 1. Air stability indicators for measurements during drying of the ceramic brick.

Mean value of temperature (°C)	Standard deviation of temperature	Mean value of relative humidity (%)	Standard deviation of relative humidity	Mean value of velocity (m/s)	Standard deviation of velocity
40.01	0.21	16.16	2.61	3.08	0.07
39.99	0.18	20.34	1.01	3.07	0.07
40.00	0.17	30.66	0.79	3.15	0.07
40.00	0.29	40.27	0.59	3.09	0.08

2.3. *Drying measurement of the ceramic brick*

In the first experimental series the red ceramic brick was considered. Measurements were carried out with four relative humidity of drying air, i.e.,

- 16% – the lowest operational value of the relative humidity for the stand for drying air at 40°C (for these drying air parameters the dew point temperature was equal to 9.4°C, control of the relative humidity at the stand below 16% was observed difficult due to very low working parameters of the cooler/condenser),
- 20%,
- 30%,
- 40% – the highest practically reasonable relative humidity of drying air at 40°C for the real drying processes in the field condition (for these drying air parameters the dew point temperature was equal to 23.8°C, in hot climates the walls have maximal temperatures up to 20-

25°C, higher values of the relative humidity of drying air might lead to moisture condensation on walls which are drying).

Table 1 presents stability indicators of drying air state, i.e., the mean value and standard deviation of controlled parameters of drying air. The standard deviation for the temperature was approx. 0.2°C which is similar value to standard deviations presented in literature [4]. The standard deviation of the relative humidity was decreasing with the rise in the mean value of the relative humidity which resulted from the increase of the hygric inertia of air with the rise in its moisture content. The standard deviation of the velocity was stable and on the low level, i.e., under 0.1 m/s.

Table 2. Relative moisture content after 24 h and the drying time required to achieve the relative moisture content of 0.3 for the ceramic brick.

Relative humidity of drying air (%)	Relative moisture content after 24 h	Drying time to relative moisture content of 0.3 (h)
16	0.32	25.20
20	0.43	34.37
30	0.44	45.37
40	0.56	62.83

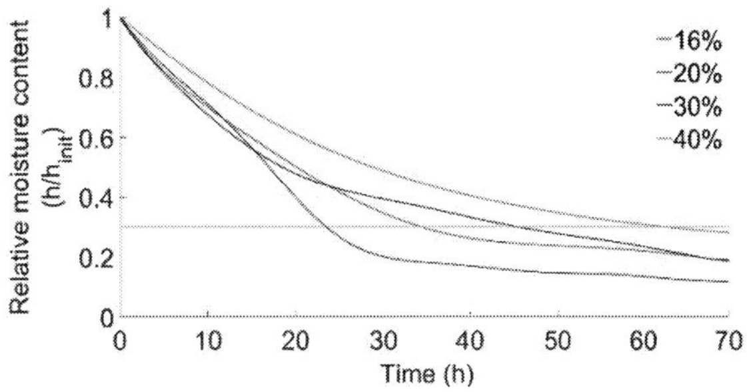

Figure 2. Temporal variations of the relative moisture content for drying of the ceramic brick for different relative humidity of drying air.

Figure 2 presents temporal variations of the relative moisture content for the brick for different relative humidity of drying air, where the relative moisture content was defined as the ratio of the current to the initial moisture content. Table 2 shows relative moisture contents after 24 h of drying process as well as drying times required to achieve the relative moisture content equal to 0.3 for different relative humidity of drying air. In first stage of the drying process drying rates were similar for the all values of the relative humidity of air. This was related to the rapid evaporation from the top surface of the sample. Differences between drying rates for relative humidity of 16, 20 and 30% were on the level of measurement uncertainty but for 40% the drying rate was the smallest which probably was due to water condensation on the surface of the sample. Moreover, for 40% the sample first increase its temperature to the equilibrium temperature in the porous structure (see Fig. 3), which additionally hindered the drying. In the second drying stage it was found that the increase of the relative humidity of air led to the increase of the drying time by 35% between two consecutive measurement points as well as the decrease in the drying rate.

Figure 3 presents temporal variations of temperature of the sample measured in two different points, i.e., 10 and 30 mm from the top surface of the sample. In the first stage of the drying when evaporation from the top surface of the sample was dominant, the level of temperature plateau (i.e., equilibrium temperature in the porous medium) which resulted from evaporation in the porous building material was found to increase with the rise in the relative humidity of drying air. This effect was related to the change of the wet bulb temperature of drying air with the variation of its relative humidity. For the relative humidity lower than 20% the sample temperature in the first drying stage decreased from the initial one to the equilibrium temperature in the porous medium. Then in the second drying stage when the diffusive transport of moisture from the inside of the sample was dominant the temperature was slowly approaching to the asymptotic steady state temperature. For the relative humidity of 30% the initial temperature was close to the equilibrium temperature in the porous medium and then started to increase to the asymptotic steady state temperature. For the relative humidity of 40% the temperature rose to the equilibrium temperature in the porous medium and then again was approaching to the asymptotic steady state temperature.

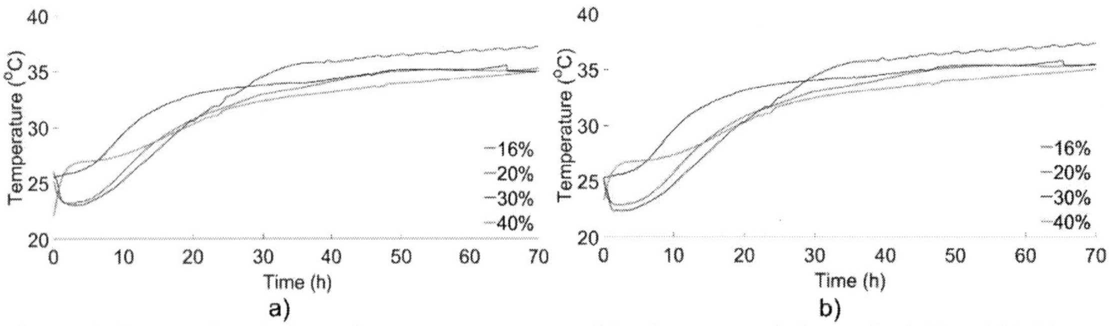

Figure 3. Temporal variations of temperature measured by thermocouple located: a) 10 and b) 30 mm from the drying surface for drying of the ceramic brick for different relative humidity of drying air.

2.4. Drying measurement of the mortar

The second series of measurements was carried out for the mortar sample. Due to low drying rates for this material and long measurement times only two relative humidity of drying air were examined, i.e., 20% and 40 %. Table 3 presents stability indicators of drying air state, i.e., the mean value and standard deviation of controlled parameters of drying air. The standard deviation of the temperature was under 0.2°C which was similar as in the previous measurements for the red brick. Again, the standard deviation of the relative humidity was decreasing with the increase of the mean value of the relative humidity and it was lower than 0.8% which was better than for similar measurements presented in literature [4]. The standard deviation of the velocity again was stable and on the low level, i.e., under 0.1 m/s.

Table 3. Air stability indicators for measurements during drying of the mortar sample.

Mean value of temperature (°C)	Standard deviation of temperature	Mean value of relative humidity (%)	Standard deviation of relative humidity	Mean value of velocity (m/s)	Standard deviation of velocity
39.97	0.13	20.47	0.75	3.22	0.10
40.01	0.18	40.17	0.28	3.17	0.07

Table 4. Relative moisture content after 24 and 192 h of the drying of the mortar sample for different relative humidity of drying air.

Relative humidity of drying air (%)	Relative moisture content after 24 h	Relative moisture content after 192 h
20	0.71	0.54
40	0.73	0.63

Figure 4 presents temporal variations of the relative moisture content of the mortar sample, while table 4 shows the relative moisture content after 24 and 192 h of the drying of the mortar sample with different relative humidity of drying air. During drying process rapid removal of moisture in the first few hours was observed. Liquid moisture at the top surface and close to it was quickly evaporated during this period. Unfortunately, after this stage the drying process rapidly slowed down and after 8 days of drying even the relative moisture content of 0.5 was not obtained. Very low drying rates in the mortar sample as compare to the red brick may be related to different mechanisms of moisture absorption and accumulation in the mortar than in the brick. In ceramic bricks most of moisture is absorbed by action of capillary forces and occupies pores. In the mortar most of moisture is probably chemically absorbed and bound in its structure. This explanation is confirmed by temperature temporal variations presented on figure 5. After rapid evaporation of moisture in the first very short stage of drying, the sample temperature was increasing to the steady state temperature much faster than for the red brick. Moreover, the mortar has a lower vapor permeability than the red brick which additionally hinder moisture transfer from the inside of the sample to the drying medium.

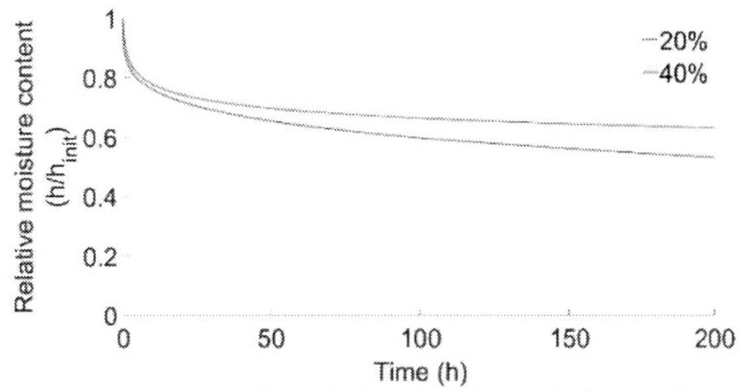

Figure 4. Temporal variations of the relative moisture content for the drying of the mortar sample for different relative humidity of drying air.

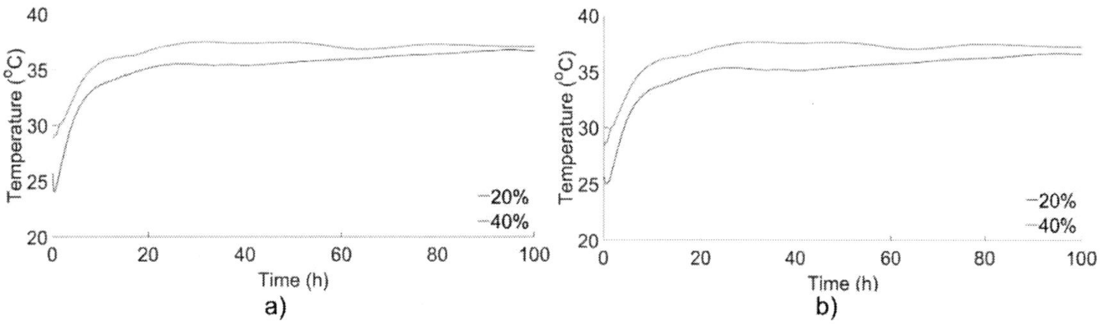

Figure 5. Temporal variations of temperature measured by thermocouple located: a) 10 and b) 30 mm from the drying surface for drying of the mortar sample for different relative humidity of drying air.

3. Conclusion

In this paper experimental investigations of influence of the relative humidity of drying air on the drying process of porous building materials was presented. Two different materials were examined, i.e., the ceramic red brick and mortar. Measurements were carried out for four and two different relative humidity of drying air, respectively. Results of experimental measurements show that the relative humidity of drying air has significant impact on the drying process. For the red brick the increase of the relative humidity of drying air by 10% resulted in the rise of the duration of the drying process by approx. 35%. The relative humidity of drying air affected also the level of temperature plateau in porous building materials. This level increased with increasing of the relative humidity of drying air. For mortar sample the drying rates were found to be much slower than for the red brick due to different mechanism of moisture absorption and accumulation in mortar as well as different hygric parameters than for the ceramic brick.

In the future more materials will be examined, e.g., silicate bricks. Additionally, influence of other parameters of drying air, e.g., the temperature and velocity, will be investigated separately and in superposition. Obtained data of future works will be applied for the validation of mathematical and numerical models of heat and moisture transfer in porous building materials which are currently under development [10, 11] and for optimization of the drying process of masonry walls.

Acknowledgments
This work was supported by the National Centre for Research and Development (Poland) under grant no. POIR.04.01.02-00-0099/16 *Development of innovative technology of drying and moisture sealing of masonry walls, DryWall.*

References
[1] Colinart T, Lelievre D and Glouannec P 2016 Experimental and numerical analysis of the transient hygrothermal behavior of multilayered hemp concrete wall *Energy Build.* **112** 1–11
[2] Chikhi A, Belhamri A, Glouannec P and Magueresse A 2016 Hygro-thermal behaviour of porous building material subjected to different external temperature and humidity conditions *J. Eng. Sci. Technol.* **11** 458–475
[3] Karagiannis N, Karoglou M, Bakolas A, Krokida M and Moropoulou A 2017 Drying kinetics of building materials capillary moisture *Constr. Build. Mater.* **37** 441–449
[4] Van Belleghem M, Steeman M, Janssen H, Janssens A and Paepe M De 2014 Validation of a coupled heat , vapour and liquid moisture transport model for porous materials implemented in CFD *Build. Environ.* **81** 340–53
[5] Busser T, Berger J, Piot A, Pailha M and Woloszyn M 2019 Comparison of model numerical predictions of heat and moisture transfer in porous media with experimental observations at material and wall scales: An analysis of recent trends *Dry. Technol.* **37** 1363–1395.
[6] Berger J, Dutykh D and Mendes N 2017 On the optimal experiment design for heat and moisture parameter estimation *Exp. Therm. Fluid Sci.* **81** 109–122
[7] Busser T, Berger J, Piot A, Pailha M and Woloszyn M 2018 Dynamic experimental method for identification of hygric parameters of a hygroscopic material *Build. Environ.* **131** 197–209
[8] Bianchi J M, Colombo L P M, Ochs F and Feist W 2017 Determination of the water retention curve from drying experiments using infrared thermography: A preliminary study *Int. J. Therm. Sci.* **114** 271–280
[9] Cieślikiewicz Ł, Łapka P, Wasik M, Kubiś M, Pietrak K, Wiśniewski T S, Furmański P and Seredyński M 2018 Development of the experimental stand for investigation of heating and drying phenomena in the porous building materials with one surface of the sample exposed to the flowing air *E3S Web Conf.* **70** 03003
[10] Wasik M, Cieślikiewicz Ł, Łapka P, Furmański P, Kubiś M, Seredyński M, Pietrak K, Wiśniewski T S and Jaworski M 2019 Initial credibility analysis of a numerical model of heat and moisture transfer in porous building materials *AIP Conf. Proc.* **2078** 020106
[11] Łapka P, Wasik M, Furmański P, Seredyński M, Cieślikiewicz Ł, Pietrak K, Kubiś M, Wiśniewski T S and Jaworski M 2018 Preliminary mathematical and numerical transient models of convective heating and drying of a brick *MATEC Web Conf.* **240** 01022

Calculation of initial and long-term heating costs for a nearly zero energy single-family building over 30 years

S Gendelis, M Miķelsons and A Jakovičs

Faculty of Physics, Mathematics, and Optometry, University of Latvia, Jelgavas Str. 3, Rīga, LV-1004, Latvia

E-mail: stanislavs.gendelis@lu.lv

Abstract. In practice, during the design stage of a building, only the initial costs are taken into account. These costs are also discussed with the customer and optimised with the aim of reducing them. It is the global costs over a longer period that show the real difference in the total expenses of living in a particular house, however. In this study, one newly built, nearly zero energy single-family building was studied in detail using a comparative global cost calculation methodology. The most popular insulation materials, construction types, types of windows, and types of floor construction in Latvia are included in the calculation of the costs. The results demonstrate that initially cheaper building means lower global costs during the analysed 30-year operating period due to initial costs are significantly higher than all other costs including energy costs, meaning that changes in construction type and building materials affect the initial cost more than the long-term energy costs. In parallel, different modern heating systems (including heat pumps) were included in the analysis to calculate and compare heating expenses. The results show that almost all the heating systems have the same energy costs regardless of fuel type for the 30-year period, however, the initial costs of these systems vary due to the different used technologies. Therefore, only the sum of the initial and long-term energy costs provides objective information about heating systems that are optimal in the long-term.

1. Introduction

The EU directive on the energy performance of buildings (EPDB) [1] defines the minimum requirements for the energy performance of buildings and their elements and requires that all new buildings must be nearly zero energy buildings (nZEBs) as of 31 December 2020. The low amount of energy that nZEBs require should come mostly from renewable energy sources. This directive demands the calculation of the cost-optimal balance between the primary energy and the total cost during the lifetime of a building through the use of a comparative methodology [2, 3]. The same methodology can also be used for practical purposes, such as choosing the initial building construction or analysing the heating systems over a longer period.

Normally, only the initial construction costs are taken into account during the design stage of a single-family house and these are usually reduced after consultation with the customer. At the same time, the building's operating and maintenance costs lead to a real difference in the total costs during a long operation period (so-called 'cost of ownership'). In the case of nZEBs, this global cost approach becomes especially useful because of the strict requirements in terms of heating and primary energy and the compulsory use of renewable energy sources, which requires more detailed and more comprehensive calculations of a cost-optimal set of solutions.

In this paper, a newly built nZEB single-family house (Fig. 1) with a heated floor area of 154 m^2 is analysed in detail using this global cost calculation methodology. The most popular types of insulation materials in Latvia (loose cellulose wool, mineral wool, EPS/XPS) combined with two external wall types (wooden frame construction, masonry construction), two floor types (slab-on-ground, strip foundation), and windows with wooden and PVC frames, as well as various heating systems are included. The economic calculations of each combination are carried out for a period of 30 years, taking into account the expected energy price increase, discount rate, etc. These calculations show the differences in the total costs and the influence of separate investment (initial, energy, replacement, maintenance, etc.) parts in total balance. Comparing the initial and long-term energy costs for different heating systems allows the optimal system to be chosen during the planning phase, thus eliminating unexpected expenses later on.

As the cold season lasts for a long time in Latvia (around seven months with an average temperature between 0 and 2°C), the highest calculated annual energy consumption in the analysed building is for space heating: 21 kWh/m^2. Energy consumption for other purposes is significantly lower (1 kWh/m^2 for cooling, 4 kWh/m^2 for domestic hot water, 2 kWh/m^2 for mechanical ventilation).

Figure 1. A photo (left) and a cross-section (right) of an analysed nZEB.

2. Methods

The total heating consumption of a whole building is calculated in accordance with Latvian legislation, which is based on the ISO 52016-1 standard [4]. The simplified monthly calculation principle with Latvia's climate data is applied for this methodology, including calculation of the primary energy and CO_2 emissions. The main target of this research is to compare different popular building materials and construction types with the purpose of analysing global and initial costs and to investigate the long-term usage of typical modern heating systems in terms of total costs based on existing building (Figure 1).

Two types of wall constructions (a timber frame filled with insulating material and solid masonry with an external insulation layer) are included in the calculations combined with loose cellulose wool, mineral wool, and EPS insulation materials. Slab-on-ground and strip foundation types are used for the floor construction. A flat roof construction with timber firrings and different insulation materials is viewed. Triple-glazed windows are chosen with either PVC or a more expensive wooden frame. The area of outer walls is 215 m^2, basement/roof area is 101 m^2 and the windows area is 46 m^2. Basic information about the main used materials is given in Table 1. Building is equipped with modern recuperative ventilation system with efficiency 90%.

The combinations of all the materials and construction types used are summarised in Table 2. Combination #1 corresponds to the existing building. The analysed heating systems varied according to energy source and efficiency and included natural gas and wood pellet boilers, four types of heat pumps, and electric heaters.

The next calculation step consisted of financial calculations according to the methodology [2] and defined the total cost of ownership, which consisted of an initial investment (construction costs) and the follow-on costs (energy, maintenance, replacement costs, etc.). The residual value was also taken into account to compare variations with different lifetimes. Initial energy prices corresponding to 2019 in Latvia with a 1% annual increase were assumed as follows: 150 €/MWh for electricity, 41 €/MWh for natural gas, and 35 €/MWh for wood pellets. Other parameters used for the calculation are summarised in Table 3. The most difficult to predict (even for a couple of years) were future energy prices and maintenance costs.

Table 1. Basic parameters of used materials.

Construction	Material	Thermal conductivity λ (W/m/K)	Thickness, cm
Outer walls	Wooden frame	0.2	5*+20*
	Expanded clay	0.245	30
	Perforated ceramic	0.175	25
	Aerated concrete	0.11	30
	Mineral wool	0.35	34/24
	Cellulose wool	0.042	24
	EPS	0.038	10/15
Floor	Concrete	2.0	15
	XPS	0.038	15
	EPS	0.037	25
Roof	Plasterboard	0.25	1
	Mineral wool	0.043	65/53
	Cellulose wool	0.042	35
	EPS	0.039	30

* framing timer with cross section of 5×5 or 20×5 cm

Table 2. Summary of analysed combinations for the nZEB.

#	Floor		Walls			Roof		Window frame
	Type	Insulation	Base	Insulation material	Insulation, cm	Insulation material	Insulation, cm	
1	Slab-on-ground	XPS	Wooden frame	Mineral wool	34	Mineral wool	65	Wood
2	Slab-on-ground	EPS	Wooden frame	Mineral wool	24	Mineral wool	35	Wood
3	Slab-on-ground	EPS	Wooden frame	Cellulose wool	24	Mineral wool	35	Wood
4	Slab-on-ground	XPS	Wooden frame	Cellulose wool	24	Cellulose wool	35	Wood
5	Strip footings	EPS	Expanded clay	Mineral wool	15	Mineral wool	35	Wood
6	Strip footings	EPS	Perforated ceramic	Mineral wool	15	Mineral wool	35	Wood
7	Strip footings	EPS	Aerated concrete	Mineral wool	10	Mineral wool	35	Wood
8	Strip footings	EPS	Aerated concrete	Mineral wool	15	Mineral wool	35	Wood
9	Strip footings	EPS	Aerated concrete	EPS	10	Mineral wool	35	Wood
10	Strip footings	EPS	Aerated concrete	EPS	15	Mineral wool	35	Wood
11	Strip footings	EPS	Aerated concrete	EPS	10	EPS	30	Wood
12	Strip footings	EPS	Aerated concrete	EPS	15	EPS	30	Wood
13	Slab-on-ground	XPS	Wooden frame	Mineral wool	34	Mineral wool	65	PVC
14	Slab-on-ground	EPS	Wooden frame	Mineral wool	24	Mineral wool	35	PVC
15	Slab-on-ground	EPS	Wooden frame	Cellulose wool	24	Mineral wool	35	PVC
16	Slab-on-ground	XPS	Wooden frame	Cellulose wool	24	Cellulose wool	35	PVC
17	Strip footings	EPS	Expanded clay	Mineral wool	15	Mineral wool	35	PVC
18	Strip footings	EPS	Perforated ceramic	Mineral wool	15	Mineral wool	35	PVC
19	Strip footings	EPS	Aerated concrete	Mineral wool	10	Mineral wool	35	PVC
20	Strip footings	EPS	Aerated concrete	Mineral wool	15	Mineral wool	35	PVC
21	Strip footings	EPS	Aerated concrete	EPS	10	Mineral wool	35	PVC
22	Strip footings	EPS	Aerated concrete	EPS	15	Mineral wool	35	PVC
23	Strip footings	EPS	Aerated concrete	EPS	10	EPS	30	PVC
24	Strip footings	EPS	Aerated concrete	EPS	15	EPS	30	PVC

Table 3. Main parameters used in the economic calculations.

Parameter	Value	Estimated lifetime, years
Discount rate, %	5%	-
Maintenance cost - **heat pump**, % annum	7%	15
Maintenance cost - **natural gas boiler**, % annum	3%	15
Maintenance cost - **pellet boiler**, % annum	5%	15
Maintenance cost - **electric heaters**, % annum	3%	30
Maintenance cost - **recuperation system**, % annum	8%	15

3. Results

3.1. Calculation of global and initial costs

The changing the constructions and insulation materials with different thicknesses affects not only the construction costs, but also the *U*-value (see Table 4) and, therefore, total building's heating energy need, which is calculated for each combination. For the windows the weighted average *U*-value is used. Results of the calculated annual heating consumption, as well as all other costs from global cost calculation approach for all the combinations is presented in Table 5. The most important are global and initial costs, as well as energy (mainly heating) costs.

By analysing the obtained results, it can be concluded that the key component in the global cost is the initial costs (including design, purchase of building elements and systems, installation and commissioning costs) the resulting that the lowest construction costs also means the smallest global costs. This can be explained by the fact, that the initial costs are higher than all other costs (Figure 2), leading to significantly less impact of other components. This finding means that significant changes in calculated heating energy consumption (from 21 up to 38 kWh/m^2) due to different construction types and insulation materials does not affect the global costs as much as they might seem to initially. It is important to note that this statement applies only to the 30-year calculation period and to low energy consumption buildings and that for longer time periods and buildings with higher energy consumption the impact of heating costs will increase (assuming that energy prices do not decrease). Figure 3 shows a characteristic distribution of the construction costs for the analysed building (in the case of combination #1).

Table 4. *U*-values (W/m^2/K) for analysed constructions (see Table 2).

Combination	Floor	Walls	Roof	Windows*
1 (13)	0.12	0.10	0.06	
2 (14)	0.19	0.15	0.12	
3 (15)	0.19	0.15	0.12	
4 (16)	0.12	0.15	0.08	
5 (17)	0.18	0.19	0.12	0.77with wooden frame
6 (18)	0.18	0.18	0.12	
7 (19)	0.18	0.18	0.12	
8 (20)	0.18	0.15	0.12	0.80 with PVC frame
9 (21)	0.18	0.18	0.12	
10 (22)	0.18	0.15	0.12	
11 (23)	0.18	0.18	0.12	
12 (24)	0.18	0.15	0.12	

Comparative analysis of the economic calculation results (Table 5) with a view to find the optimal combination for a single-family nZEB shows that there are two options with the lowest initial (and therefore global) costs: #14 and #15. Both consist of a slab-on-ground foundation with 15 cm EPS insulation, a flat roof construction with 35 cm mineral wool insulation, wooden frame outer walls, and windows with PVC frames. The only difference between these combinations is the insulation material in the walls: mineral wool (#14) or loose-fill cellulose (#15), resulting in an ~3% difference in initial and global costs.

Table 5. Summary of main economic calculation for all analysed combinations.

Combination	Annual heating need, kWh/m^2	Global costs, €/m^2	Initial costs, k€	Maintenance costs, k€	Energy costs, k€	Residual value, k€	Replacement costs, k€
1	21.1	907.55	121.52	5.49	6.01	18.61	25.07
2	34.3	875.09	115.01	5.49	8.16	17.22	23.06
3	34.3	898.32	118.58	5.49	8.16	17.22	23.06
4	28.0	909.30	120.05	5.49	7.12	17.42	24.52
5	35.7	1,032.73	144.76	6.94	8.26	44.43	43.21
6	34.5	1,026.71	143.81	6.94	8.05	44.21	43.21
7	34.5	1,042.55	145.57	6.94	8.05	43.36	43.04
8	30.7	1,039.26	145.66	6.94	7.44	43.41	43.10
9	34.5	1,027.25	143.69	6.94	8.05	42.33	41.54
10	30.7	1,032.53	144.83	6.94	7.44	42.96	42.45
11	34.5	1,026.15	144.11	6.94	8.05	41.01	39.62
12	30.7	1,031.42	145.25	6.94	7.44	41.63	40.53
13	22.8	859.19	116.66	5.49	6.16	16.99	20.72
14	36.3	827.17	110.15	5.49	8.38	15.60	18.71
15	36.3	850.40	113.72	5.49	8.38	15.60	18.71
16	29.9	861.20	115.19	5.49	7.32	15.80	20.17
17	37.9	985.13	139.90	6.94	8.54	42.82	38.86
18	36.7	979.07	138.95	6.94	8.32	42.59	38.86
19	36.7	994.91	140.71	6.94	8.32	41.74	38.68
20	32.9	991.52	140.80	6.94	7.69	41.79	38.75
21	36.7	979.62	138.83	6.94	8.32	40.71	37.19
22	32.9	984.79	139.97	6.94	7.69	41.34	38.10
23	36.7	978.51	139.25	6.94	8.32	39.39	35.27
24	32.9	983.68	140.39	6.94	7.69	40.01	36.18

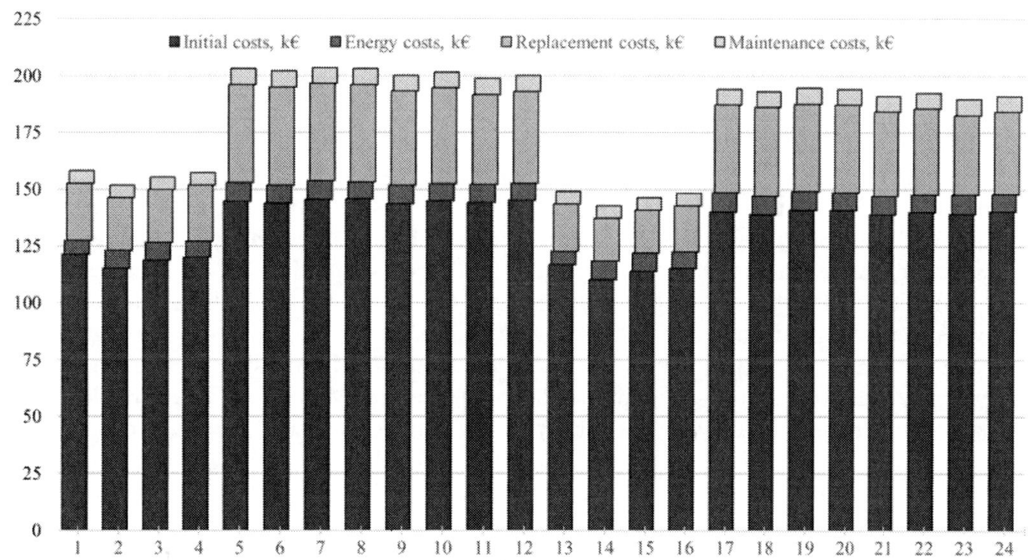

Figure 2. Composition of the global costs for the analysed combinations (see Tables 2 and 5).

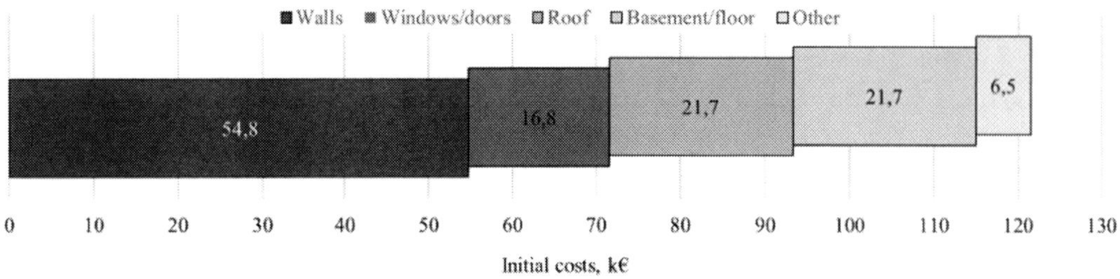

Figure 3. Distribution of the initial construction costs for the analysed nZEB (combination #1).

3.2. *Calculation of heating costs*

The calculation methodology [2] is also very useful for the calculation of heating costs according to the type of heating system and the installation costs. To exclude the influence of any other parameters, the heating systems were varied for the existing building (combination #1) with a calculated annual heating need of 3.2 MWh. The seven most popular and modern heating systems (Table 6) were chosen for the comparative analysis. One half of the analysed heating systems consisted of four heat pumps with two different energy sources: ground (with vertical/borehole or horizontal collectors) or air, and two heat transfer mediums: water and air. Another two heating systems were classical boilers (natural gas and wood pellet) with a required chimney installation and the last one was simple electric heaters connected directly to the power outlet.

Table 6. Summary of analysed heating systems.

Type	Designation	Seasonal efficiency (-)	Initial costs, k€	Heating costs over 30 years, €
Heat pump (borehole/vertical-water)*	HP-V/W	3.7	11.5	3,118
Heat pump (earth/horizontal-water)*	HP-H/W	4.1	9.5	2,814
Heat pump (air-water)	HP-A/W	3.5	10.7	3,296
Heat pump (air-air)	HP-A/A	4.1	7.7	2,814
Natural gas boiler	GAS	0.97	8.1	3,250
Pellet boiler	PEL	0.9	10.0	2,991
Electric heaters	ELE	1	1.9	11,536

* including ground heat exchanger

As the heat pumps produce more thermal energy than the consumed electricity, they are typically characterised by their seasonal efficiency (or seasonal coefficient of performance: SCOP), which describes the average ratio between these energies during a heating season. The difference between these energies comes from the renewable source (ground or air) produced onsite, which is a mandatory requirement for nZEBs. The use of boilers or electric heaters does not satisfy this requirement; therefore, additional renewable energy sources are needed to meet these requirements (e.g. solar energy). It is important to emphasise that the heat pumps included in this research were selected from the top lines of the best manufacturers and have excellent efficiency. In the case of the air-air heat pump, a multi-split system with one outdoor unit and five indoor units was used.

The comparison of the initial (installation) costs and the costs of providing heating energy over 30 years for different heating systems (Figure 4) shows that five heating systems were very close in terms of energy cost regardless of fuel type. The corresponding initial costs varied, however, due to the use of different technologies with specific nuances (borehole drilling, chimney construction, etc.). The electric heaters were significantly different and had low initial costs but high long-term energy costs.

A combination of both initial and long-term energy costs provides objective information about the running costs of different heating systems over a longer period. Analysis of this combination for heating systems (Figure 5) shows that the most cost-effective systems for the calculation period of 30 years were those with lower initial costs (except electrical heaters), i.e. the air-air heat pump and the natural gas heating system. The main reason for this is the very small heating energy consumption of nZEBs, which means that even for a period of 30 years, the energy costs are smaller than the initial costs of the heating systems.

It is important to note that the calculations of the thermal energy produced by the heat pumps are highly dependent on their SCOP, which varies for different soil types, groundwater levels, and outdoor temperatures [5, 6]. A very important factor influencing the heat pump's SCOP is water temperature in the indoor coil as a lower temperature means a smaller temperature difference, which is the most important factor for heat exchange efficiency [7]. As the underfloor heating is designed for heat pumps with water inside coil, the designed SCOP is relatively high; for the radiator heating system efficiency and SCOP will be significantly lower.

The very low initial costs of electric heaters are not optimal in the long run due to high electricity prices, however, this type of heating is economically justifiable in the short term. The choice of initially expensive heating systems is not justifiable from an economic point of view and there should be other reasons, e.g. free summer cooling (use of only the heat exchanger and brine pump) in the case of ground source heat pumps.

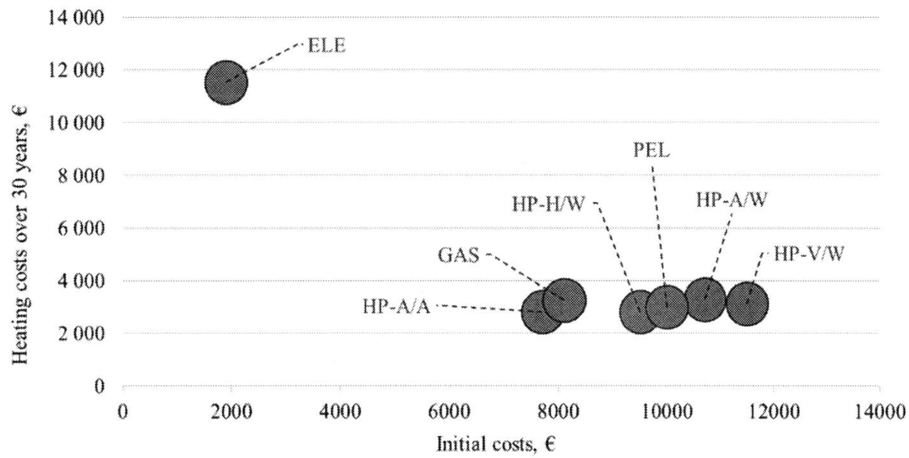

Figure 4. Initial and long-term heating costs for different heating systems (see Table 6).

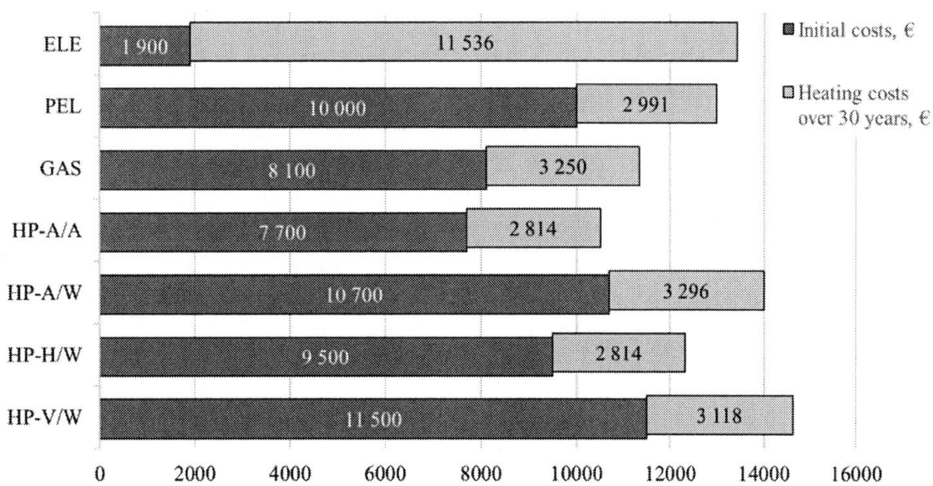

Figure 5. Total running costs over 30 years for different heating systems (see Table 6).

Conclusions

Due to the low energy consumption and relatively high building costs of nZEBs, long-term economic calculations using the global cost methodology are very useful when it comes to choosing materials, construction type, and systems based on long-term operation costs. The most critical assumption used in the calculation methodology that plays a key role in the reliability of the calculated results is the forecast of future energy prices and hard-to-predict maintenance costs.

Long-term calculations show that because constructions costs are significantly higher than all other costs, different construction types and insulation materials affect the initial costs more than the long-term energy costs. It follows that initially cheaper buildings mean lower global costs during the analysed 30-year operating period. At this point, it should be noted that this statement applies only to this time period and to low energy consumption buildings. For longer periods and buildings with higher energy consumption the impact of heating costs will increase (assuming that energy prices do not decrease).

Economic analysis of widely used and modern heating systems (excluding electric heaters) showed that their long-term energy costs were very similar regardless of fuel type, but that their corresponding initial costs varied due to the different costs of the technologies used. The initial costs of electric heaters are very low, but they have high long-term energy costs. An objective indicator of the choice of heating systems is the sum of initial and energy costs, providing information about total costs in the long term.

The most cost-effective systems for the selected building were the air-air heat pump with multiple indoor units or the natural gas heating system. The reason for this is the very small heating energy consumption of the building, which means that even for a period of 30 years the energy costs for such systems are several times smaller than the installation costs. The use of electric heaters is economically justifiable for a shorter period or in case when price difference between electricity and other fuels will increase.

Acknowledgments

This work was supported by the European Regional Development Fund project, "Development, optimization and sustainability evaluation of smart solutions for nearly zero energy buildings in real climate conditions" [1.1.1.1/16/A/192].

References

[1] European Parliament and of the Council 2010 Directive 2010/31/EU of the European Parliament and the Council on the energy performance of buildings *Official Journal of the European Union* **153** 13-35

[2] European Parliament and the Council 2012 Commission Delegated Regulation (EU) No. 244/2012 supplementing Directive 2010/31/EU of the European Parliament and of the Council on the energy performance of buildings by establishing a comparative methodology framework for calculating cost-optimal levels of minimum energy performance requirements for buildings and building elements *Official Journal of the European Union* **81** 129-147

[3] Gendelis S, Miķelsons M and Jakovičs A 2018 Calculations of cost-optimal levels of energy performance requirements for renovated buildings and building elements: a Latvian case *Proc. 18th Int. Multidisciplinary Scientific GeoConference SGEM 2018 (Vienna)* **18(6.4)** 645-652

[4] International Organization for Standardization 2017 ISO 52016-1:2017 Energy performance of buildings -- Energy needs for heating and cooling, internal temperatures and sensible and latent heat loads -- Part 1: calculation procedures

[5] Jun L, Xu Z, Jun G, Jie Y 2009 Evaluation of heat exchange rate of GHE in geothermal heat pump systems *Renew Energy* **34(12)** 2898–904

[6] Busato F, Lazzarin R M, Noro M 2012 Energy and economic analysis of different heat pump systems for space heating *International Journal of Low-Carbon Technologies* **7(2)** 104–112

[7] Verhelst C, Logist F, Van Impe J, Helsen L 2012 Study of the optimal control problem formulation for modulating air-to-water heat pumps connected to a residential floor heating system *Energy and Buildings* **45** 43-53

IOP Conf. Series: Materials Science and Engineering **660** (2019) 012027 doi:10.1088/1757-899X/660/1/012027

Considerations on transients in a water supply distribution network

A Constantin and C S Niţescu

Faculty of Civil Engineering, Ovidius University, 22b Unirii str, Constanta, Romania

Email: aconstantina@univ-ovidius.ro

Abstract. Lipniţa is a village situated in a hilly region, in Constanta County. The water supply system has its own groundwater water source, placed in the middle of the village at a low elevation. Water is conveyed, by a long transmission duct to a storage tank from which it flows by gravity to the consumers. The distribution network has two long mains that intersect in a manifold with isolation valves, situated in the middle of the village. This topology of the system shows high static head for the transmission duct and high elevation differences between the end consumers and those in the centre of the village.
The paper presents an analysis, by numerical simulation, on the transients that occur in two different segments of the system.
The simulation was performed either for the original old steel made network, aiming at identifying the vulnerable cross sections and extreme pressure values. The results showed that maximal pressure values are not dangerous, but cavitation occurs along the pipeline. The same simulation was carried out for the new PEHD pipelines. The change of material and the increase in diameter of the new pipes showed better results regarding pressure variation during transients. Cavitation is avoided.

1. Introduction

In Constanta County, Romania, apart from the large interconnected water supply system that delivers drinking water to the cities and resorts on the Black Sea coast, there are many independent systems which provide water from their own underground sources, especially in the rural area. One of them is the water system of Lipniţa, a village situated in a hilly region, in the South West of the county.

The distribution network is being under modernization process. Therefore an investigation on the safety and reliability of the system is imposed, as a basis for the seizing of the new pipeline and for water quality preservation.

Mechanical failures including pipe breaks, pump's stoppage due to a power failure or abnormal operation generated by wrong valve manoeuvres may put into jeopardy the hydraulic integrity of the pipeline and subsequently water quality may be impaired. Isolation valves play an important role in the system reliability and security by providing a closure function [1]. Transients are usually present in hydraulic systems as a consequence of the demand variation, planned operation and maintenance activities, emergencies, or simple hydrant flushing. Rapid changes in pressure conditions due to surge events can lead to pipe deformation or even collapse, leakage, deformation of the equipment such as pipeline valves, air valves, and protection devices [2]. Sometimes the damage is hidden, but it may cause pipe collapse at a future moment, as transients continue. A clear analysis of the factors influencing hydraulic and physic integrity of the system, and the way the damaged hydraulic integrity affects the water quality is made by [2]. An emphasis is put on low pressure transients which may lead to leakage of contaminants into the pipe, at junctions under sub-atmospheric pressure values. The intruded contaminants can flow with the water downstream from the site of entry as [3] reported, referring to negative pressure values that occur in the distribution system and cause such intrusions. Consequently the admissible concentration values [4] might be overpassed.

These are the main reasons from which we started our study on transients in the water supply system, aiming at preventing possible damages or improper operation.

2. Description of the drinking water supply system

The distribution network is a branched one and the mains follow the direction of the streets (figure 1).

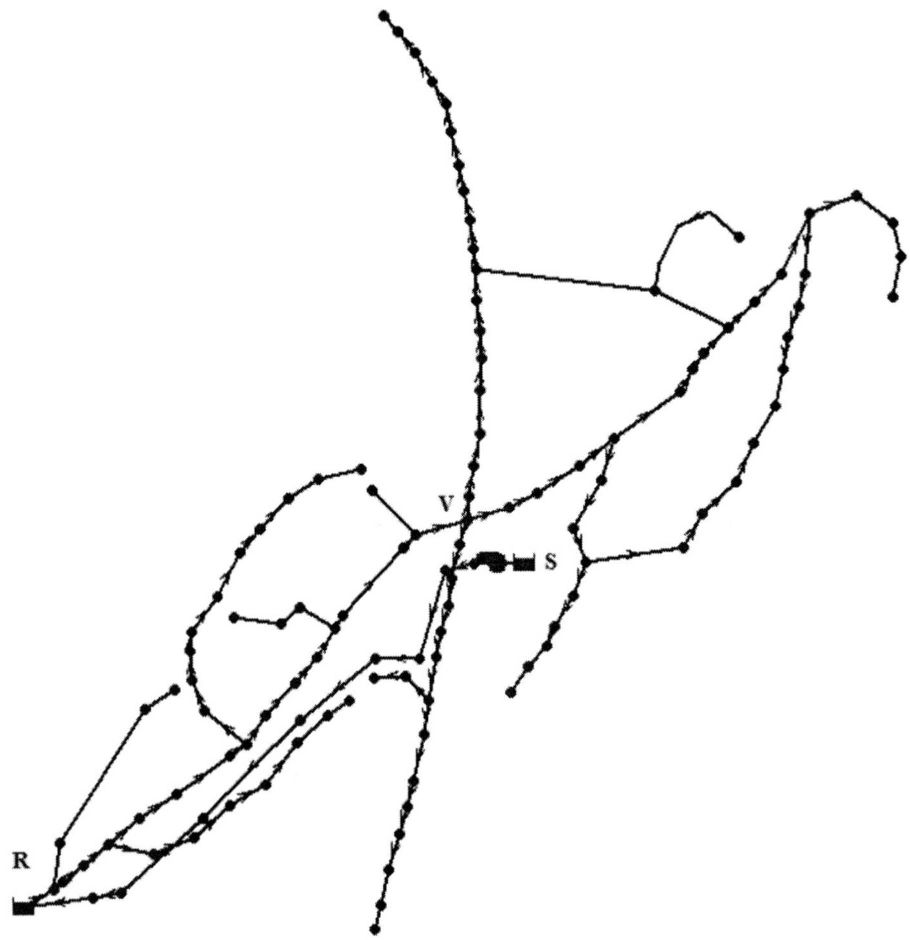

Figure 1.Water supply system in Lipnița.
SR-transmission pipe; RV-network first branch.

The groundwater is pumped from the source S, which is placed in the middle of the village at a low elevation of 63.99m, to a storage tank R, placed at the highest elevation point, of 123.03m, at the end of the main street in the village. As it can be noticed in figure1, there are two long mains that intersect in a manifold situated at 60.83 m of elevation, close to the water facility that comprises the source. The steel made transmission duct, that connects the source to the storage tank, is 850m in length and 150mm in diameter. From the storage tank water flows by gravity to the consumers. The manifold hosts the isolation valves V (figure 1).

This topology of the system shows high static head for the transmission pipe that conveys water from the source to the storage tank. Moreover, there are high elevation differences between the end consumers and the ones in the centre of the village, so pressure, under normal operation, varies in a wide range. The network is a low pressure one [5], [6].

3. Research methods

The investigations on transients were achieved by numerical simulation. We analysed pressure variation on two different segments of the system, considering the most impairing scenarios:

a. a power failure stops the pump, therefore transients occur on the transmission pipe, (SR line in figure1).
b. the isolation valve V (figure 1) is quickly closed, due to an accidental wrong manoeuvring, consequently transients occur on the first network branch, between the storage tank R and the valve V (RV line in figure1);

The software, a non-commercial one, named Hammer, was conceived by the former Institute for Hydraulic Structures in Bucharest, to solve hydraulic shock problems using the characteristics method. The software assumes water flow is one-dimensional, and water is one single phase, [7], [8].

The mathematical model used by this software, composed of the motion equation and the mass conservation equation [7] takes into account either the elasticity of the pipe walls or the water compressibility. The differential equations are linearized. Specific boundary conditions are added, according to the type of the calculus node: constant head reservoir, air chamber, turbine, pump, valve, check valve, node with two or three branches, surge tank etc. [8].

Celerity is considered constant on a pipe section as long as the pipe's diameter or material do not change.

The input data refer to: the length, diameter and wall thickness of the pipes, and the material they are made of. Regarding the nodes, data about the elevation, base demand and the grade line have to be introduced. The head loss are assumed to be concentrated in the nodes, therefore the moduli of hydraulic resistance are also input data. The software calculates the pressure values in the nodes and the results are given in graph of tabular form.

In the first scenario (*a*), water flows by gravity from the storage tank along the pipe, and suddenly the downstream valve V is closed. In this case, the border conditions imposed for the valve obturator refer to the reflexion of the pressure waves at the valve obturator, with no change in sign. So, pressure increases in the first phase of the hydraulic shock. At the other end of the pipe, with constant head storage tank, pressure waves reflect with reversed sign, so in the second phase pressure decreases. The longitudinal profile of the pipe is convex, therefore it is more likely to reach negative pressure values [6] in the nodes of high elevation, as the grade line may fall under the duct axis.

The pipe of 1004,5 m in length was divided in 9 sections, that means 10 nodes.

In the second scenario (*b*), it was assumed that the pump in the well stops as a consequence of a power accidental failure and the check valve undertakes a quick closure. In this case, the border conditions are similar, but due to the fact that water, in the regime case, flows to the constant head tank, pressure decreases in the first phase of the hydraulic shock, in the nodes close to the pump and rises in the second phase.

The pipe of 850 m in length was divided in 7 sections, therefore 8 nodes are considered.

For both scenarios, the simulation was firstly conducted in the case of the original steel made network, and secondly, for the new pipes, to make possible a comparison of the pressure variation during transients. In the first scenario, the new pipe of the first distribution branch is PEHD made, and in the second, the new transmission pipe is a composed one: it is steel made only on the first ascendant section (inside the well), and continues with a horizontal PEHD made pipe.

The old network has been proved to be undersized with respect to the fire flow rate [6]. Therefore mobile pumps were needed to operate at a hydrant, to have enough pressure and capacity for fire fight. Thus, the diameter of the new horizontal PEHD made transmission pipe is enlarged at 160 mm, whereas

the ascendant steel made section is 125 mm in diameter. Likewise, the diameter of the new pipe of the first distribution branch is 110 mm, instead of 80 mm.
The simulation flow rate is 11.52 l/s.

4. Discussion of the simulation results

The replacement of the old pipes with new larger ones modified the velocity field across the distribution network. In Table 1 there are gathered the velocity values for the old and the new pipe that consists of the first branch of the distribution network. The velocity decreases with about 40%.

Table 1.Regime velocity on the network distribution first branch.

Pipe section [m]		N1- N2	N2-N3	N3-N4	N4-N5	N5-N6	N6-N7	N7-N8	N8-N9	N9-N10
Dn 80mm	v [m/s]	2.19	2.19	2.2	2.25	2.4	2.43	2.45	2.58	2.61
Dn 110mm	v [m/s]	1.31	1.31	1.32	1.35	1.43	1.46	1.47	1.55	1.57

Although the new PEHD pipes are smoother than the steel ones, the hydraulic losses decrease insignificantly due to the velocity decrease caused by the diameter enlargement. Therefore, the head is slightly different in the nodes of the same pipe as the material changes.

The extreme pressure values in the most representative nodes of the pipes are given in Table 2, in correlation with their elevation.

Table 2.Extreme pressure values in nodes.

Scenario *a*	Nodes	**N1**	**N2**	**N3**	**N4**	**N5**	**N6**	-
	Elevation [m]	0	63,63	67,19	98,05	113,91	121,65	
Steel	p_{max} [mwc]	176,1	111,7	103,4	59,3	40,2	18,7	
	p_{min} [mwc]	47,0	-10,0	-5,5	-10,0	-7,1	-3,6	
Steel and PEHD	p_{max} [mwc]	133,5	69,5	65,9	32,8	23,4	10,9	
	p_{min} [mwc]	108,4	45,3	45,0	18,1	14,1	6,6	
Scenario *b*	Nodes	**N1**	**N2**	**N3**	**N4**	**N5**	**N7**	**N9**
	Elevation [m]	60,83	69,1	74,16	85,44	93,59	110,08	117,89
Steel	p_{max} [mwc]	98,4	99,7	79,2	66,2	56,2	33,3	16,2
	p_{min} [mwc]	-8,5	-8,7	-2,5	-8,3	-10,0	-10,0	-1,4
PEHD	p_{max} [mwc]	77,6	68,8	63,0	50,9	41,6	22,0	9,0
	p_{min} [mwc]	48,1	40,6	36,4	26,2	18,9	5,0	4,1

Referring to the amplitude of oscillations, we may notice an important decrease when the steel pipes are replaced. In the first scenario it decreases about five times, whereas in the second scenario it is only three times smaller, when the pipes are PEHD made.

The pressure variation over time, in some representative nodes is given in figure 2, in a comparative manner. In figure 2 a, the graphs show pressure variation for the first scenario, in nodes N1, N2 and N 4 of the transmission pipe. The node N1 is close to the pump. The amplitude of the first oscillation is 129mwc, but no harmful values are recorded. In all the nodes along the pipe, except for the first one, N1, minimal pressure is negative. However, in the nodes N2 and N4, cavitation occurs for a few seconds. The oscillations damp rapidly.

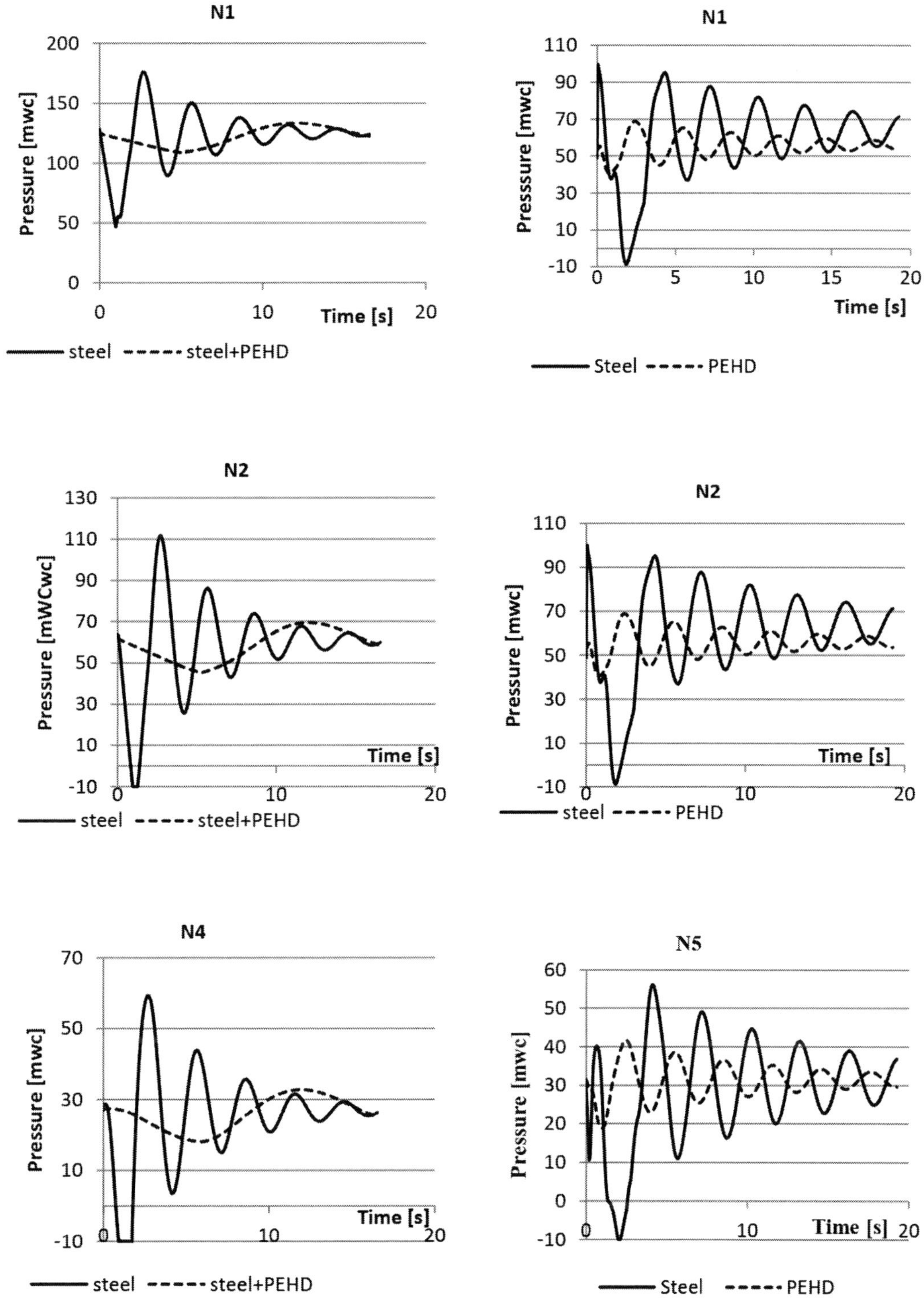

a.Transmission line (first scenario). b.Distribution branch (second scenario).

Figure 2.Pressure variation during transients.

The simulation for the new transmission pipe shows that a sudden stoppage of the pump results in only small oscillations around the regime pressure value, in each node (the dashed line in figure 2a).

The oscillation profile is different for the old and for the new pipe. The amplitude of the first oscillation in node N1 is about 129.1 mwc, and the minimal pressure is 47 mwc, in the case of old steel made pipe. In the other nodes along the pipe the minimal pressure is negative. The damping ratio is 0.025. Similar damping ratio may be noticed at the other nodes of the pipe.

The replacement of the old pipe with a new one, larger in diameter and made of a much more elastic material, results in oscillations with a damping ratio of 0.09. The period increases four times in the case of PEHD made pipe.

In figure 2 b, there are represented the pressure variation graphs in some nodes of the network branch considered in the second scenario. There were selected the nodes N1, N2, and N5, where N1 is close to the valve V. The maximal pressure values are once again harmful, as it happened in the first scenario. The threat comes from the minimal pressure values, which are negative along the duct. On some sections, cavitation is reached. For example, cavitation occurs in the nodes N5 and N7, where the slope of the terrain changes. The amplitude of the first oscillation in node N1 is about 108 mwc, and the minimal pressure is -8.5 mwc. The damping ratio is 0.01.

In the case of the new pipe, larger in diameter and made of PEHD instead of steel, the oscillations have a damping ratio 0.008. The period is comparable for the two cases. The minimal pressure values are positive along the pipe. The amplitude of the first oscillation decreases to 29.5 mwc.

5. Conclusions

The numerical simulation of transients in the old water distribution system Lipnița showed a dangerous decrease of pressure along the pipes in both scenarios, and even the occurrence of cavitation on some segments of the pipes. The hydraulic integrity of the pipes may be affected, and consequently the water quality may be impaired by the intrusion of pollutants through the joints where low negative pressure occurs. There were needed, along the pipe, devices for protection from cavitation.

As the considered water distribution network is a branched one, characterized by long pipe sections with convex longitudinal profile and high difference of elevation between the ends, we recommend a thorough study of transients in similar water distribution networks. Even if maximal pressures are not dangerous, cavitation may occur.

The replacement of the steel pipes with more elastic ones, PEHD made, eliminates the danger of cavitation. The increase of the inner diameter of the pipes, and the subsequent decrease of the velocity contributes to the cavitation danger elimination. Consequently, no protection devices are required.

References

[1] Nekha J and Sumamb K S 2016 Optimal Water Distribution Network Design Accounting for Valve Closure, *Procedia Technology* **24** (2016), pp 332 – 338

[2] National Research Council 2006 Drinking water distribution systems. Assessing and Reducing Risks (Washington DC: the National Academies Press), https://doi.org/10.17226/11728

[3] Friedman M, Kirmeyer G, Pierson G, Harrison S, Martel K, Sandvig A and Hanson A 2005 Development of distribution system water quality optimization plans (Denver, CO: AwwaRF)

[4] STAS 1342-91 1991 *Drink Water. Scope and use field* (Bucharest: The Romanian Institute for Standardization)

[5] STAS 4163-2 -1996 Water supply. Water distributions. Engineering and operation technical directions (Bucharest: The Romanian Institute for Standardization)

[6] NP 133-2013 Regulation regarding the design, execution and exploitation of water distribution systems (Bucharest: MDRAP)

[7] Popescu M 2008 *Hydropower plants and pumping stations. Hydraulic operation during transients* (Bucharest: Editura Universitară)

[8] Popescu M and Arsenie D 1987 *Methodes of hydraulic calculation for hydro-electric plants and pumping stations* (Bucharest, Editura Tehnică)

IOP Conf. Series: Materials Science and Engineering **660** (2019) 012028 doi:10.1088/1757-899X/660/1/012028

Velocity modification at the bridge abutments because scour

B Gjunsburgs[1], **M Vilnitis**[2], **J Parilkova**[3] **and M Bizane**[1]

[1]Riga Technical University, Department of Water Engineering and Technology, Kipsala Str. 6A, LV-1048, Riga, Latvia
[2]Riga Technical University, Department of Construction Management, Kipsala Str. 6A, LV-1048, Riga, Latvia
[3]Brno University of Technology, Department of Water Structures, Veveri 95, 60200, Brno, Czech Republic

E-mail: boriss.gjunsburgs@gmail.com

Abstract. At the corner of the abutment were streamline concentration, sharp water level drop and rapid increase of velocity. Formula for calculation local velocity at the abutments is elaborated. Local velocity is depending on open flow parameters, backwater value, rate of contraction by bridge crossings and depth of scour. Relative local flow modification at the bridge abutment at plain river bed, relative local and critical velocities changes during scour at steady and unsteady flow, uniform bed and at bed layering is studied. Comparison of the local velocity with approach flow and mean velocity under the bridge opening was made. The local velocity V_{lt} is reducing and the critical velocity V0t is increasing because of the scour within the time step at steady or at unsteady flow. When relative local velocity $\beta V_{0t}/V_{lt}$ becomes equal to one the scour stops. For unsteady flow local and critical velocities are different at any steps of hydrograph because of the increased flow discharge and the scour depth. Dependence of the relative local and critical velocities at steady and unsteady flow, development in time and because of the scour, impact layering of the river bed is presented in paper in Figures and compared with tests results.

1. Introduction

The contraction flow by the bridge structures, such as abutments, piers, as well as guide banks, and spur dikes, leads to the flow and velocities modification in the vicinity of the abutment and is one among another reasons for scouring.

Approaching the bridge contraction, the streamlines are bended by an embankment, and the flow goes parallel to it. The flow velocities along the extreme streamline drop almost to minimum values and then gradually increase, which is accompanied by circulation and development of different vortex structures. At the corner of the abutment, is streamline concentration, sharp water level drop, rapid local increase of the velocity and local scour.

Currently in formulas and methods for calculation depth of scour at abutments is used mean velocity of approach flow or Froude number calculated with that velocity.

Scour near bridge structures were studied by Froehlich (1989), Richardson and Davis (1995), Kothuari and Ranga Raju (2001), Coleman et al. (2003), Rahman and Haque (2003), Gjunsburgs and Neilands (2004), Dey and Barbhuiya (2005), Grimaldi et al. (2006), Ghani et al. (2011), Guo (2014), Sheppard at all. (2014) Link at all. (2017) and the others.

The local and critical velocities are changing at the steady and at the unsteady flow because of the scour. The local velocity V_{lt} is reducing and the critical velocity V_{0t} is increasing due to scour within the time step at steady flow. For the next time step local and critical velocities are different because of the unsteadiness of the flow, increased discharge and the scour depths developed in the previous time interval. Relative velocities V_l/V_b, V_{lt}/V_l, $V_{0t}/V_{0,}$ $\beta V_{0t}/V_{lt}$ changes in time at steady and unsteady flow, uniform bed and at bed layering and relative depth of scour h_s/h_f is studied. Relative local velocity ratio $\beta V_{0t}/V_{lt}$ is increasing in time during the scour and when becomes equal to one (critical velocity become equal to local one $\beta V_{0t}=V_{lt}$) the scour stops.

2. Experimental set-up

Tests with fixed bed of the model were made for different contraction of the flow to investigate velocity and water level changes in approach to the embankment, along to it and near the model of the abutment.

With sand bed – to study scour process, velocity changes in time, influence of hydraulic parameters, rate of contraction of the flow, grain size of the bed material and time on the scour. In tests openings of the bridge model were – *50, 80, 120, 200* cm in the flume. Contractions of the flow Q/Q_b, where Q - discharge of the flow, Q_b - discharge of the flow in bridge opening in open flow conditions, were changing from *1.25 to 5.69*, depth of the floodplain was – *7 and 13* cm, Froude numbers were from *0.078 to 0.151*, for steady or unsteady flow. Slope of the first flume – *0.0012.*

Tests with sand bed were made for clear water conditions. Sand was placed 1 m up and down contraction of the flumes. The mean grain size was *0.24 and 0.67* mm in flume with standard deviation with uniform or stratified river bed model. The condition $Fr_R = Fr_f$ were fulfilled, where Fr_R = Froude number for plain river, Fr_f = Froude number in flume. Tests duration in flume was *7 and 14* hours.

3. Method

To calculate the local velocity, we used the Bernoulli equation for two cross sections of an extreme unit streamline. The local velocity at the nose corner of abutments, for the plain river bed, was found from the following formula:

$$V_l = \varphi\sqrt{2g\Delta h} \tag{1}$$

where φ is velocity coefl ficient, Δh is water lever difference at the corner of the abutment

According to tests results velocity coefficient φ in Eq.(1) was depending on flow contraction (Fig. 1).

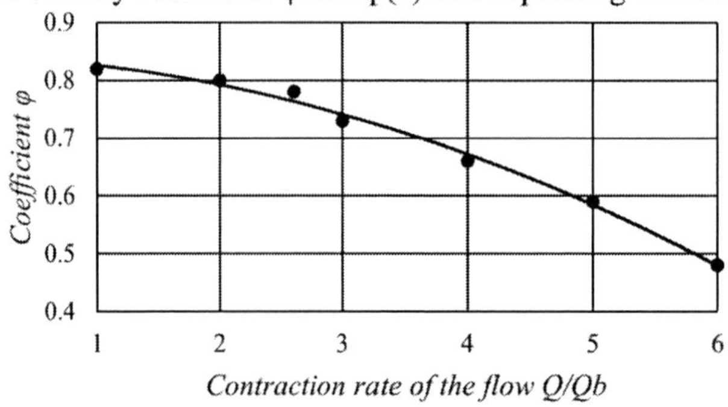

Figure 1. Coefficient φ dependence on flow contraction Q/Q_b

The discharge across the width of a scour hole at the abutment before and after the scour is determined as follows:

$$Q_f = kQ_{sc} \tag{2}$$

where Q_f is a discharge across the width of the scour hole with a plain bed, Q_{sc} is the discharge across the scour hole with a scour depth h_s, and k is a coefficient of changes in discharge because of scour.

Equation (2) can be written as:

$$mh_s h_f V_l = k\left(mh_s h_f + \frac{mh_s}{2} h_s\right) \cdot V_{lt} \tag{3}$$

where mh_s is the width of the scour hole, h_f is a water depth in the floodplain, V_l is local velocity at plain bed near abutment, h_s is the scour depth, and V_{lt} is the local flow velocity at a scour depth h_s. The average depth h_m of scour hole is:

$$h_m = \frac{W}{b} = \frac{mh_s \cdot h_f + mh_s \dfrac{h_s}{2}}{mh_s} = h_f\left(1 + \frac{h_s}{2h_f}\right) \tag{4}$$

where W is the volume of the scour hole, b is the width of the scour hole

From Eq. (3) the local velocity at any depth of scour hole is:

$$V_{lt} = \frac{V_l}{k\left(1 + \dfrac{h_s}{2h_f}\right)} \tag{5}$$

As it was found in tests, the coefficient k is depending on contraction rate of the flow (Fig.2)

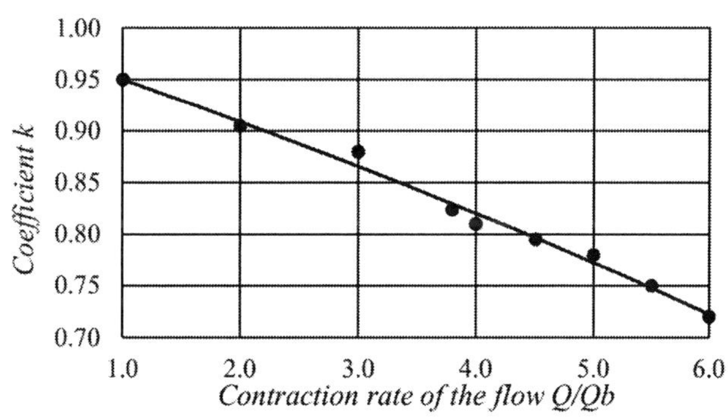

Figure 2. Coefficient k dependence from contraction rate of the flow

The critical velocity for the plane bed V_0 can be determined by the Studenitcnikov's (1964) formula

$$V_0 = 1.15 g^{0.5} (h_f d)^{0.25} \tag{6}$$

With the scour hole development, the corresponding value of the critical velocity V_{0t} for any scour depth h_s can be determined considering the average flow depth $h_f(1 + h_s/(2h_f))$, with coefficient β, the effects arising because the bended flow due to the contraction:

$$V_{0s} = 1.15\beta g^{0.5}\left(h_f d\left(1 + \frac{h_s}{2h_f}\right)\right)^{0.25} = \beta V_0\left(1 + \frac{h_s}{2h_f}\right)^{0.25} \tag{7}$$

β is reduction coefficient of the critical velocity in bended flow

At bed layering, when $h_s>H_{d1}$ (*Hd1* is the thickness of the first layer with d_1) depth of scour is developing by new local and critical velocities which is on the top of the second layer. The local velocity on the surface of the second sand layer with depth H_{d1} is found by the formula:

$$V_{lt2} = \frac{V_{l.}}{k\left(1+\dfrac{H_{d1}}{2h_f}\right)} \tag{8}$$

where H_{d1} is the thickness of the first layer of the river bed with grain size d_1.

The critical velocity on the top of second layer is determined using the average depth of flow $h_m=h_f(1+H_{d1}/2h_f)$, with a scour depth H_{d1} equal to the thickness of the first bed layer:

$$V_{02} = \beta 3.6\cdot d_2^{\,0.25} h_f^{\,0.25}\left(1+\frac{H_{d1}}{2h_f}\right)^{0.25} = V_0\left(1+\frac{H_{d1}}{2h_f}\right)^{0.25} \tag{9}$$

where $V_0=\beta 3.6\, d_2^{0.25} h_f^{0.25}$ is the critical velocity of flow for the grain size d_2, since the layer with exactly this diameter lies on the top of the river bed.

Local and critical velocities are changing in time. The local velocity V_{lt} is reducing and the critical velocity V_{0t} is increasing because of the scour within the time step at steady flow. For the next time step of hydrograph, at unsteady flow, local and critical velocities are different because of the increased discharge of the flow and the scour depth developed in the first step. Is found dependence of the local and critical velocities from unsteadiness of the flow, grain size of the bed material layering river bed.

4. Results

As result of flow modification at the corner of the abutment is local increase in velocities. The difference between average local flow velocity and approach one with increase of the contraction rate of the flow is increasing (Fig.3).

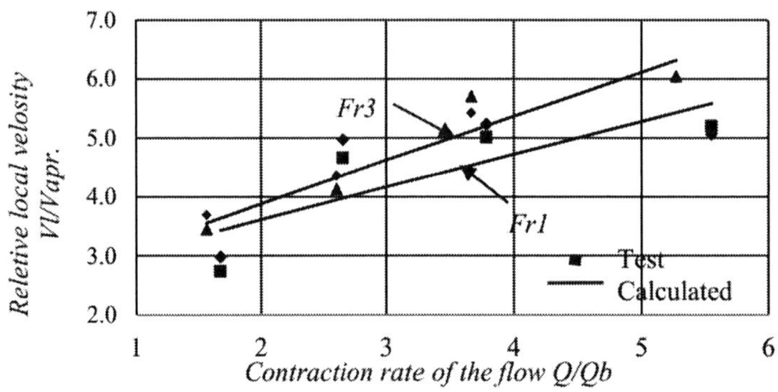

Figure 3. Local and approach velocities ratio at different contraction rate of the flow

Comparison of the flow velocity under the model of the bridge V_b and local velocity at the abutment V_l (Fig.2) show that with increase of the contraction rate of the flow the difference between them is reducing.

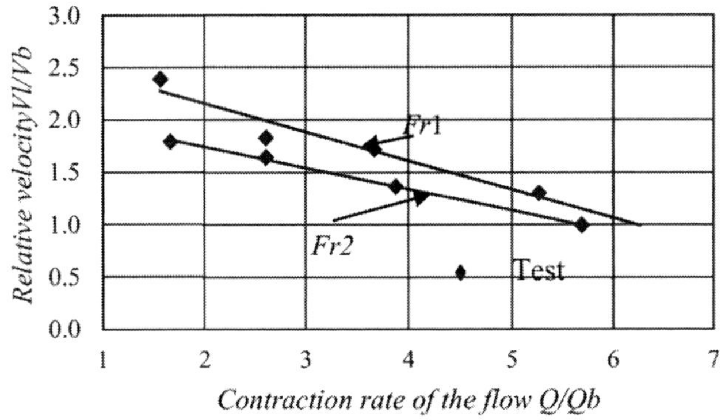

Figure 4. Ratio of the local velocity to velocity in the alingment of the bridge model V_l/V_b to contraction rate of the flow

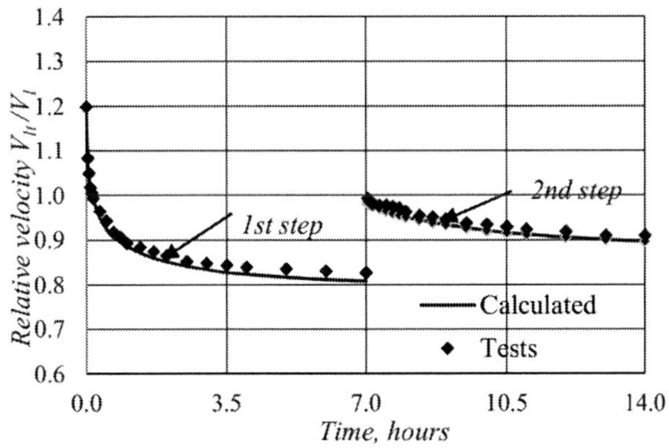

Figure 5. Relative local velocity V_{lt}/V_l development at unsteady flow (Test TL 5)

At steady and unsteady flow, the ration between local velocity at any depth of scour to local velocity at plain bed *Vlt/Vl* is reducing in time as depth of scour increases. Local velocity V_{lt} is reducing in time,than relative velocity V_{lt}/V_l,where V_l is local velocity at plain model bed,is also reducing in time. At the unsteady flow,at the second step of hydrograph,when depth on the floodplain is increacing the the relative local velocity also is reducing (Fig.5).

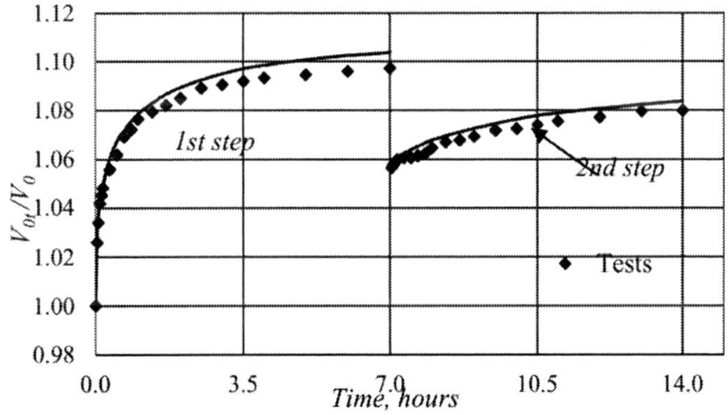

Figure 6. Relative critical velocity *Vot/Vo* development in time at unsteady flow (Test TL 5)

Ratio of the critical velocity at any depth of scour to critical velocity at plain bed in time is increasing because of increses the depth of the flow due to the scour development in time (Fig.6).

Value of the scour depth is depending on ratio local to critical velcities $V_l/\beta V_o$ and it changes with development scour hole in time at steady and unsteady flow conditions.

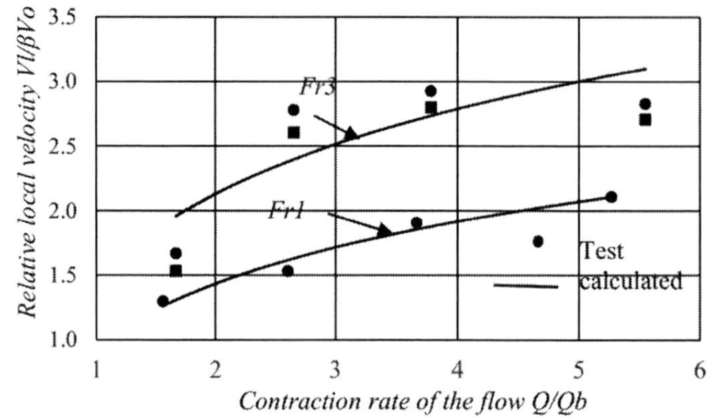

Figure 7. Relative local velocity at plain bed changes with contraction rate of the flow

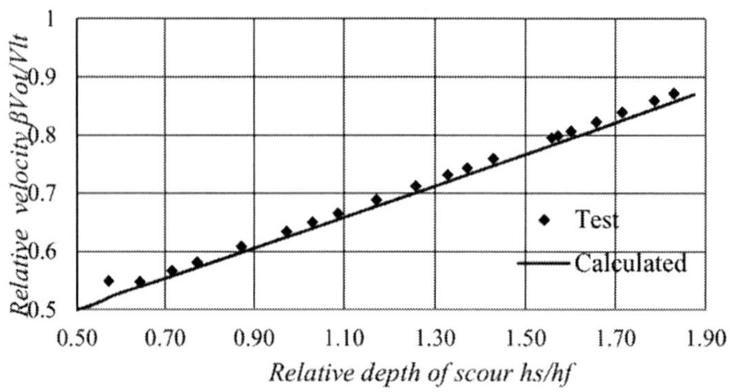

Figure 8. Relative velocity $\beta V_{0t}/V_{lt}$ versus relative depth of scour h_s/h_f at steady flow (Test AL4)

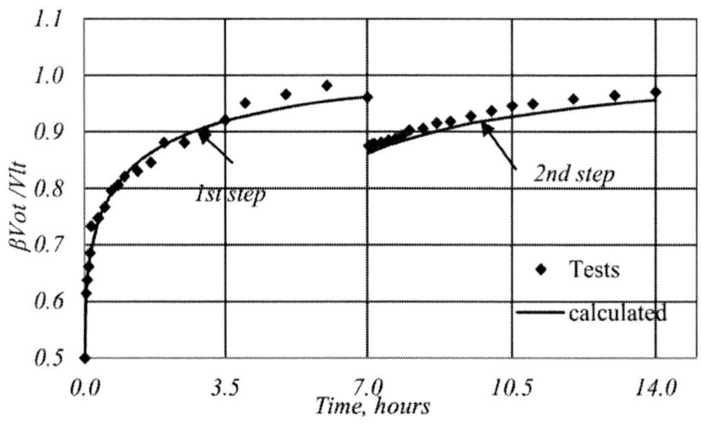

Figure 9. Local velocity versus critical velocity $\beta V_{0t}/V_{lt}$ at unsteady flow in time of scouring

At plain bed,with the increase of the contraction rate of the flow,the difference between local and critical velocities is increasing (Fig.7).

At steady and unsteady flow with increase of the relative depth of scour h_s / h_f the relative local velocity $V_{lt}/\beta V_{ot}$ ratio is reducing (Fig.7) and scour stops, in case, the ration of the velocities becomes equal to 1.

At unsteady flow, with two steps of hydrograph,the ratio of the relative local velocity in changes in time and increased from step to step (Fig.9).

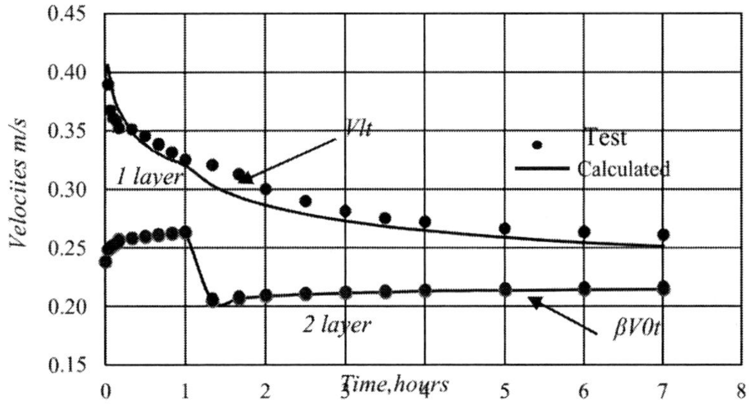

Figure 10. Local velocity V_{lt} and critical velocity V_{0t} changes in time at steady flow and stratified bed conditions (in first layer d_1=*0.67mm* and at the second layer d_2=*0.24mm*,Test AUL5)

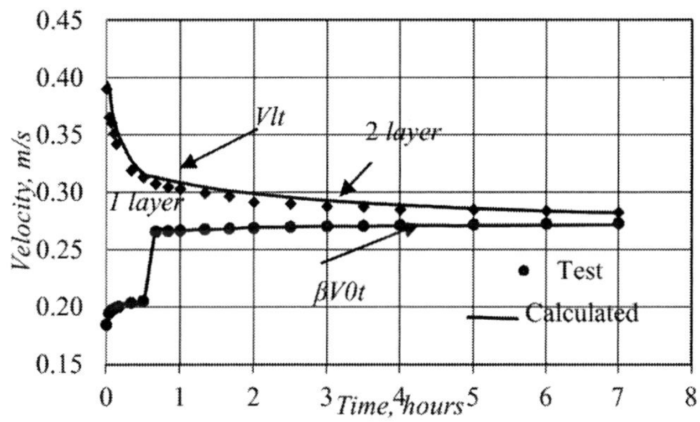

Figure 11. Local velocity V_{lt} and critical velocity V_{0t} changes in time at steady flow and stratified bed conditions (in first layer d_1=*0.24mm* and at the second d_2=*0.67mm*,Test AUL3)

On the top of the next layer with grain size *d* finer that in the first layer critical velocity is reducing, but when first layer is with d_1=*0.67mm* and the second one d_2=*0.24mm* (Fig.10), and increasing when first layer is with d_1=*0.24mm* and the second one d_2=*0.67mm* (Fig.11). When the critical velocity become equal to local one $\beta V_{0t}=V_{lt}$ the scour stops.

5. Conclusion

According to the experimental data, at the corner of the abutment were streamline concentration, sharp water level drop and rapid increase of velocity and local scour hole is developed. Formula for local velocity at plane bed is presented. The local and critical velocities are changing at the steady and at the unsteady flow in time during the scour. Local velocity at plain bed is depending on flow contraction and backwater value.

The local velocity V_{lt} at any stage of scour is determined by Eq. (5) and the critical velocity V_{0t} through the mean depth of $h_m = h_f(1 + h_{equil} / 2h_f)$ near abutments by Eq. (7).

At stratified bed conditions, when $h_s > H_{d1}$, the scour develops in the second layer with d_2. The local and critical velocities on the top of the second or next layer can be determined by Eqs. (8, 9).

The ratio $V_l / V_{ap.}$ (local to approach velocity) is depending on flow contraction and Froude number of the open flow (Fig.3). With increase contraction rate of the flow the difference between local and approach velocity of the flow is increasing and with increase the Froude number is decreasing. Among another factors local velocity, but not approach velocity, is forming the scour hole.

Velocities V_{lt}, V_{0t} and relative ones V_l/V_b, V_{lt}/V_l, V_{0t}/V_0, $\beta V_{0t}/V_{lt}$ are changing in time at steady and unsteady flow and bed layering (Fig.4 - 9). Relative velocity ratio $\beta V_{0t}/V_{lt}$ impact on relative depth of scour hs/h_f, at steady and unsteady flow, is presented (Fig.8,9). With increase of the relative depth of scour h_s/h_f the velocities $\beta V_{0t}/V_{lt}$ is increasing, and in case the ratio of the velocities become equal to 1, the scouring will stop.

River bed layering impact on local and critical velocities change in time during the scour hole development and depending on sequence of the layer with different grain size (Fig.10, 11).

Presented formulas for calculation local and critical velocities at uniform flow and stratified river bed, development relative velocities in time at steady and unsteady flow were confirmed by test result.

References

[1] Froehlich D 1989 Local scour at bridge abutments *In Proceedings of National Conference on* Hydraulic *Engineering*, pp 922–927

[2] Richardson E V and Davis S R 1995 Evaluating Scour at Bridges *HydraulicEngineering Circular*, **No. 18**, Federal Highway Administration, Washington, DC, USA

[3] Kouchakzadeh S and Townsend D 1997 Maximum scour depth at bridge abutments terminating in the floodplain zone *Canadian Journal of Civil Engineering*, **vol. 24**, pp 996–1006

[4] Kothuari U and Ranga Raju K G 2001 Scour around spur dikes and bridge abutments *Journal of Hydraulic Research*, **39(4)**, pp 367-374

[5] Rahman M and Haque M 2003 Local scour estimation at bridge site: Modification and application of Lacey formula *Journal of Sediment Reserch*, **vol 18**, no. 4: pp 8-17

[6] Coleman S, Lauchlan C and Melville B 2003 Clear-water scour development at bridge abutment *Journal of Hydraulic Research*, **41(5)**, pp 521-531

[7] Gjunsburgs B and Neilands R 2004 Local velocity at bridge abutments on plain rivers *Proceedings of River Flow 2004*, Napoli, Italy, in Greco, Carravetta & Della Morte (eds), London: Taylor & Francis Group, **vol.1**, pp 443-448

[8] Dey S and Barbhuiya A 2005 Time variation of scour at abutments Journal *of Hydraulic Engineering*, **131(1)**, pp 11–23, doi: 10.1061/(ASCE)0733-9429(2005)131:1(11)

[9] Grimaldi C, Gaudio R, Cardoso A and Calomino F 2006 Local scouring at bridge piers and abutments: time evolution and equilibrium Ferreira, Alves, Leal & Cardoso eds., *Proc. of Int. Conf. on Fluvial Hydraulics River Flow 2006*, 6-8 September, Lisbon, Portugal, **vol.1**, pp 1657–1664,

[10] Ghani A, Azamathullah H and Mohammadpour R 2011 Estimating time to equilibrium scour at long abutment by using genetic programming, *3rd International Conference on Managing Rivers in the 21st Century: Sustainable Solutions for Global Crisis of Flooding, Pollution and Water Scarcity, Rivers 2011*, 6 – 9 December 2011, Penang, Malaysia, pp 369 – 374

[11] Guo J 2014 Semi-analytical model for temporal clear-water scour at prototype piers *Journal of Hydraulic Research*, **52(3)**, pp 366–374

[12] Sheppard D, Melville B and Demir H 2014 Evaluation of existing equations for local scour at bridge piers *Journal of Hydraulic Engineering*, **140(1)**, pp 14–23

[13] Link O, Pizarro A, Rojas A, Ettemer B, Escauriaza C and Manfreda S 2017 A model of bridge pier scour during flood waves, *Journal of Hydraulic Research*, 2017,**vol.55** ,iss.3, pp 310-323

[14] Studenitcnikov V 1964 Flow erosion capacity and river-bed calculation methods Moscow: Engineering Literature Press.

IOP Conf. Series: Materials Science and Engineering **660** (2019) 012029 doi:10.1088/1757-899X/660/1/012029

Impact flow and bed layering on equilibrium depth of scour at elliptical guide banks

B Gjunsburgs[1], M Vilnitis[2], J Parilkova[3]and M Bizane[1]

[1]Riga Technical University, Department of Water Engineering and Technology, Kipsala Str. 6A, LV-1048, Riga, Latvia
[2]Riga Technical University, Department of Construction Management, Kipsala Str. 6A, LV-1048, Riga, Latvia
[3]Brno University of Technology, Department of Water Structures, Veveri 95, 60200, Brno, Czech Republic

E-mail: boriss.gjunsburgs@rtu.lv

Abstract. In the paper is present study the equilibrium scour at elliptical guide banks at different Fr number of the open flow, contraction, uniform or stratified bed model. A new approach for calculation equilibrium depth of scour at elliptical guide banks is elaborated and method for calculation equilibrium depth of scour bed at elliptical guide banks is presented. Analysis of the test, calculation and computer modelling results is made to estimate impact on relative equilibrium depth of scour: Fr number of the flow, local Froude number- Fr_{VI}, Fr_{Vequil}, flow contraction, relative time, thickness and sequence of the bed layers. Relative equilibrium depth of scour is increasing with increase of Fr number of the open flow, local Fr_{VI} number, $Fr_{V\,equil}$ at equilibrium depth of scour, relative time of scour, flow contraction and with different thickness and sequence of the bed layers. Calculation equilibrium depth of scour with the approach flow velocity and using the grain size only on the top of the bed or for the next layer calculate depth of scour with the same velocity, as it accepted nowadays, and neglecting changes of the flow parameters during the scour, bed layering can lead to wrong results and damages and losses.

1.Introduction

Structures failure due to scour near the foundations are the flow modification in floods and during the scour, flow contraction, layering of the river bed and type, size and shape of the structures.
Equilibrium scour depth at bridge piers, abutments, and spur dikes is studied by Raudkivi and Ettema (1983), Sturm and Janjua (1994), Ahmed and Rajaratnam (2000), Radice et al. (2002), Coleman et al. (2003), Kwak et al. (2004), Joko and Lim (2006), Ballio et al. (2009), Cardoso and Fael (2010), Ghani et al. (2011), Guo (2014), Sheppard et al. (2014), Evangelista et al. (2017) and the others.
Impact on the equilibrium depth of scour at elliptical guide banks open flow Fr number, local Froude number- Fr_{VI}, Froude number at equilibrium stage of scour, densimetric Froude number, flow contraction, relative time and thickness and sequence of the layers is studied.

Relative equilibrium depth of scour is increasing with increase of the open flow Fr number, local Fr_{VI} number, relative time of scour, flow contraction and bed stratification, as changes the thickness and sequence of the bed layers. Calculation equilibrium depth of scour using approach flow velocity and only the grain size on the top of the bed or for the next layer to calculate depth of scour with the same flow velocity as for first layer, as it is excepted nowadays, and neglecting changes of the flow parameters during development of the scour hole in time, can lead to wrong results, damages and losses.

2.Experimental set-up

The tests were carried out in a flume *3.5 m* wide and *21 m* long. The flow distribution between the channel and the floodplain was studied under open-channel flow conditions (Table 1).

The tests with a rigid bed were performed for different flow contractions, in order to investigate the velocity and the water level changes in the vicinity of the elliptical guide banks, along it, and near a modelled abutment.

The aim of the tests with a sand bed was to study the scour processes, the changes in the velocity with time, the effect of hydraulic parameters and the contraction rate of the flow. The opening of the bridge model was *50, 80, 120, and 200 cm*.

The contraction rate of the flow Q/Q_b (Q was the discharge of the flow and Q_b was the discharge of the flow in a bridge opening in open-flow conditions) varied from *1.25 to 5.69* at a depth of floodplain of *7 and 13* cm.

The Froude of the open flow numbers varied from *0.078 to 0.134*. The flume slope was *0.0012*.

The tests with a sand bed were carried out in the conditions of clear water. The sand was placed 1 m up and down the bride model contraction of the flumes.

The mean size of grains was *0.24 and 0.67 mm* with a standard deviation. The scour development in time for the bed material with different grain size thickness or sequence layers was studied. Duration of the tests, with different flow parameters, bed materials and layers thickness and sequence was *7 and 14* hours.

Table 1. Experimental data for open flow conditions in a flume

Test	L cm	h_f cm	V cm/s	Q l/s	Fr	Re_c	Re_f
L1	350	7	6.47	16.60	0.780	7500	4390
L2	350	7	8.58	22.70	0.010	10010	6060
L4	350	7	8.16	20.81	0.098	10270	5590/5660
L5	350	7	9.07	23.48	0.109	11280	6140/6410
L6	350	7	11.10	28.31	0.134	13800	7550/7840
L7	350	13	7.51	35.48	0.067	13700	9740
L8	350	13	8.74	41.38	0.076	16010	11395
L9	350	13	9.90	47.10	0.088	14300	14300

3.Method

The scour depth at elliptical guide banks is equal to the equilibrium depth of scour in the conditions when the local velocity becomes equal to the critical one.

The water discharge for unit width flowing over the scour hole does not change in time, according to tests results, then the scour hole development does not alter the streamlines paths, the depth integrated velocity averaged over the whole scour hole width, assuming free surface and a triangular shape for the scour hole transverse section.

The local velocity at a plain river bed is found by the Bernoulli equation for two cross sections of the extreme unit streamline. The discharge across the width of a scour hole before and after the development of scour is:

$$m \cdot h_{equil} \cdot h_f \cdot V_{lel} = \left(m \cdot h_{equil} \cdot h_f + \frac{m \cdot h_{equil}}{2} \cdot h_{equil} \right) V_{lt} \tag{1}$$

where m is the steepness of the scour hole, h_{equil} is the depth of scour at the equilibrium stage, h_f is the depth of flow at the floodplain, $V_{l\,el}$ is the local velocity, which depends on the flow contraction. The local velocity V_{lt} at an equilibrium stage of scour is determined from Eq. (1):

$$V_{lt} = \frac{V_{l\,el}}{\left(1 + \frac{h_{equil}}{2h_f}\right)} \tag{2}$$

The critical velocity V_{0t} at the equilibrium stage can be determined through the mean depth of flow $h_m = h_f(1+h_{equil}/2h_f)$ near elliptical guide banks at that stage:

$$V_{0t} = 1.15\beta g^{0.5}\left(h_f d_i \left(1 + \frac{h_{equil}}{2h_f}\right)\right)^{0.25} = \beta V_0 \left(1 + \frac{h_{equil}}{2h_f}\right)^{0.25} \tag{3}$$

where β is a coefficient of reduction in the critical velocity due to vortex structures, d_i is the grain size of the bed materials, and $V_0 = 1.15 g^{0.5}(h_f d)^{0.25}$ is the critical velocity at the plain bed (Studenithcnikov, 1964).

The scour at the equilibrium stage stop when the local velocity V_{lt} (Eq. 2) becomes equal to the critical velocity V_{0t} (Eq. 3):

$$\frac{V_{l\,el.}}{\left(1 + \frac{h_{equil}}{2h_f}\right)} = \beta \cdot 3.6 d_i^{0.25} h_f^{0.25} \left(1 + \frac{h_{equil}}{2h_f}\right)^{0.25} \tag{4}$$

From Eq. 4, the equilibrium depth of scour at elliptical guide banks is found:

$$h_{equil} = 2h_f \left[\left(\frac{V_l}{\beta V_0} \right)^{0.8} - 1 \right] \cdot k_\alpha \cdot k_m \tag{5}$$

where k_α is a coefficient depending on the flow crossing angle and k_m is a coefficient depending on the side-wall slope of guide banks.

According to Eq. (5), the equilibrium depth of scour depends on the, contraction rate of flow, backwater value, and the grain size of river bed. With increase in the grain size, the equilibrium depth of scour reduces.

The geology of the river bed is complicated and may be is formed by layers with different thickness and grain sizes. When the depth of scour is less than thickness of the bed layer $h_{equil}<H_{d1}$, equation (5) can be used; however, when the depth of scour is greater than thickness of the bed layer $h_{equil}>H_{d1}$, the scour develops in the second layer with a grain size d_2. If the depth of scour is greater than thickness of the second bed layer $h_{equil}>H_{d1} + H_{d2}$, the scour develops in the third layer with a grain size d_3, and so on.

Then, the equilibrium scour depth is different from that is with one uniform layer. When $h_{equil}>H_{d1}$, the scour develops in the second layer with d_2. Now, to determine the equilibrium depth of scour the local and critical velocities on the top of the second layer must be calculated.

The local velocity on the surface of the second layer is found by the formula:

$$V_{lt1} = \frac{V_{1el}}{\left(1 + \frac{H_{d1}}{2h_f}\right)} \quad (6)$$

where H_{d1} is the thickness of the first layer of the river bed.

The critical velocity is determined from the medium depth of flow $h_{mid}=h_f(1+H_{d1}/2h_f)$ on the floodplain with a scour depth equal to the thickness of the first bed layer:

$$V_{01} = \beta 3.6 \cdot d_2^{\ 0.25} h_f^{\ 0.25} \left(1 + \frac{H_{d1}}{2h_f}\right) \quad (7)$$

where $V_0=\beta 3.6d_2^{0.25}h_f^{0.25}$ is the critical velocity of flow for the grain size d_2, since the layer with exactly this diameter lies on the top of the river bed.

In the second layer the scour hole is forming new velocities with is on the top of the second layer, not velocities at plain bed, as assumed by different authors.

Then, the scour depth in the second layer is determined as:

$$h_{s2} = 2h_f \left[\left(\frac{V_{lt1}}{V_{01}}\right)^{0.8} - 1\right] \cdot k_\alpha \cdot k_m \quad (8)$$

If $h_{s2}>H_{d2}$, but the scour stops in that layer, the equilibrium scour depth is

$$h_{equil} = H_{d1} + h_{s2} \quad (9)$$

4. Results

Local flow modification at the head of the elliptical guide bank lead that local Froude number $Fr_{Vl} = V_l/(g\ h_f)^{0.5}$, local Fr_{Vlt} numbers at the end of test and $Fr_{V\,equil}$ at equilibrium depth of scour are changes, compare with Froude number of the open flow. With increase of the flow contraction the local Froude number Fr_{Vl} increasing (Fig.1) and leads to the increase of the relative equilibrium depth of scour (Fig.2).

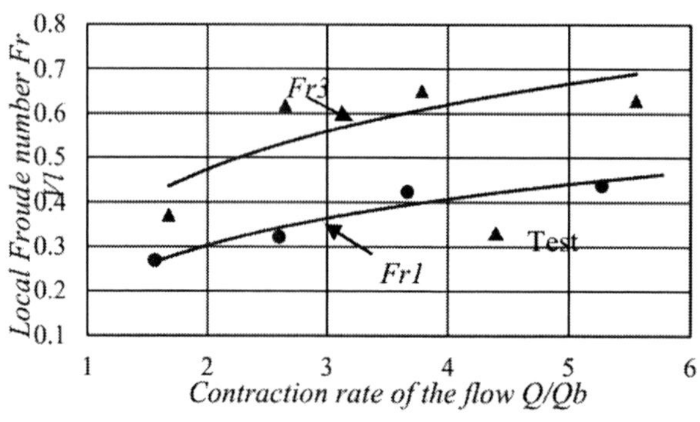

Figure 1. Local Froude number Fr_{Vl} dependence from flow contraction rate Q/Q_b (at different Fr numbers of the open flow)

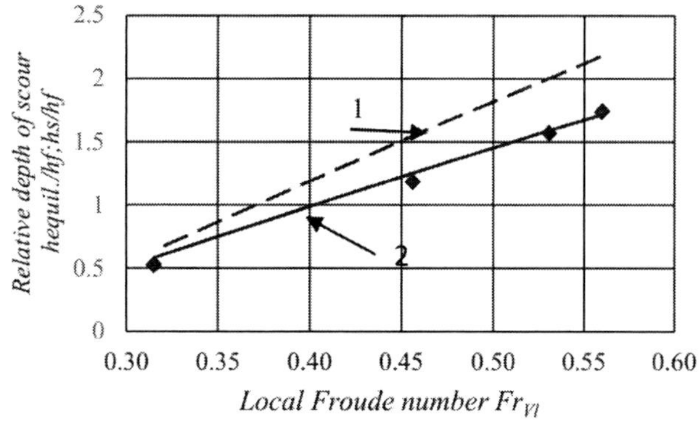

Figure 2. Relative equilibrium depth of scour h_{equil}/h_f *(1)* and relative depth of scour at the end of the test h_s/h_f *(2)* versus local Froude number- Fr_{Vl}

Computer modeled relative equilibrium depth of scour h_{equil}/h_f is increasing with flow contraction as well as relative depth of scour h_s/h_f measured in test. (Fig.3).

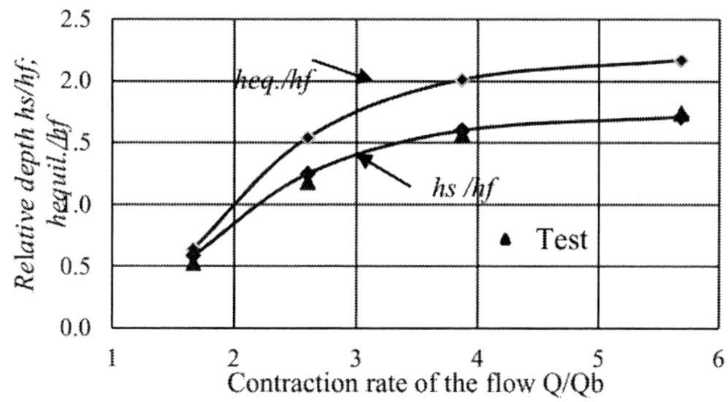

Figure 3. Relative equilibrium depth of scour h_{equil}/h_f and relative depth of scour at the end of the 7 hour long tests hs/h_f versus flow contraction Q/Qb

When the depth of scour is less than thickness of the bed layer $h_{equil}<H_{d1}$, equation (5) can be used; however, when the depth of scour is greater than thickness of the bed layer $h_{equil}>H_{d1}$, the scour develops in the second layer with a grain size d_2. Local velocity on the top of the second layer is changed and can be calculated by equation (6) and new value of the critical velocity by equation (7). If the depth of scour is greater than thickness of the second bed layer $h_{equil}>H_{d1}+H_{d2}$, the scour develops in the third layer with a grain size d_3, and new values of the local and critical velocities and so on.

With increase thickness of the first layer with $d=0.67mm$ the relative depth of scour is reducing. Depth of scour is depending from thickness and sequence of the layers with different grain size, when the depth of scour is greater than thickness of the bed layer $h_{equil}>H_{d1}$, the scour develops in the second layer with a grain size d_2. Sequence of the bed layers and thickness impact on equilibrium depth of scour. The first layer with different thickness H_{d1} is changing from 0 to14 cm is with $d_1=0.67mm$ and the second one is with thickness $H_{d2}=50cm$ $(d_2=0.24)$.

Relative equilibrium depth of scour h_{equil}/h_f is increasing with increase of relative time of scour t_{equil}/t_{test} (Fig.4).

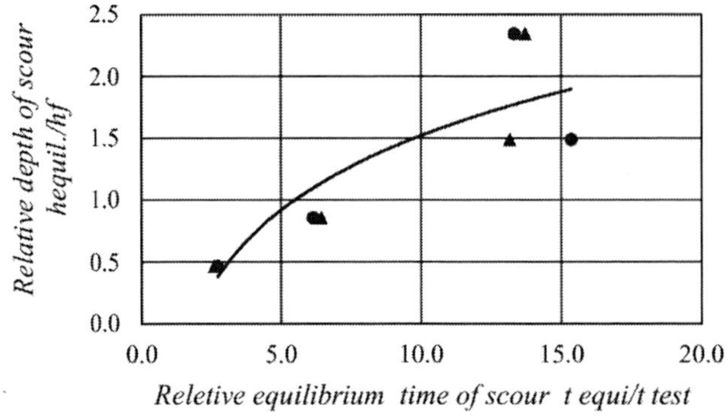

Figure 4. Relative equilibrium depth of scour h_{equil}/h_f versus relative time of scour $t_{equil.}/t_{test}$

In Fig.5 is presented changes of ratio of the relative grain size d/h_s (where d is grain size, h_s depth of scour at the tests at different contraction rate of the flow) at uniform bed.

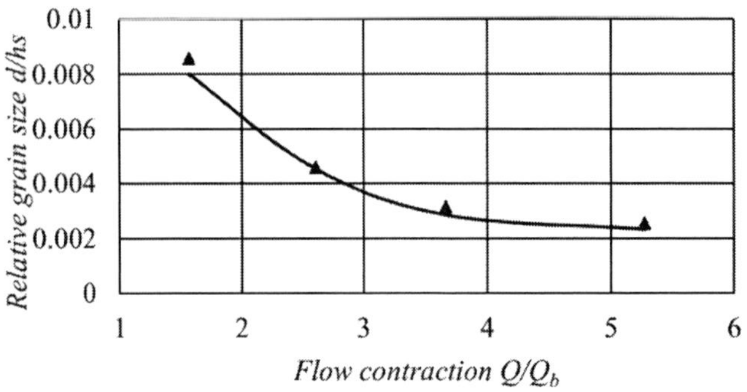

Figure 5. Relative grain size of the bed d/h_s changes at verification of the contraction rate of the flow

Depth of scour is always greater when the fine-sand layer is under the coarse sand layer, and smaller when the fine-sand layer is on the top of the coarse sand layer. According to the results obtained in tests and by the method elaborated, the depth of scour is always greater when a fine-sand layer is under a coarse-sand layer.

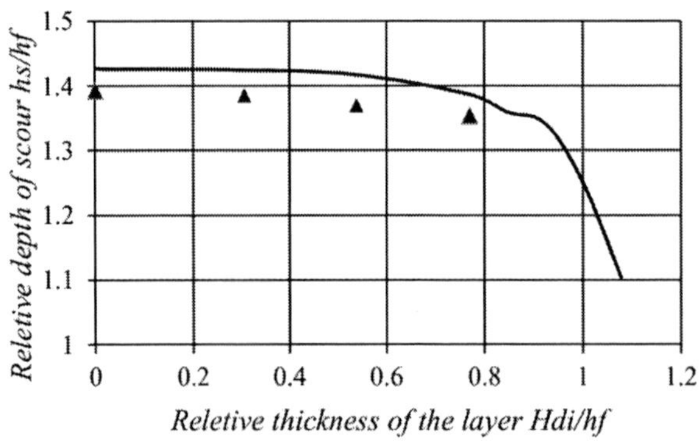

Figure 6. Influence relative thickness of the bed layer *hs/hf* on relative depth of scour *Hdi/hf* (Tests EL23, EL15)

At stratified bed conditions the sequence of the layers has the significant influence on the scour depth value and can be reason of the bridge failure.

With increase of the thickness of the first coarse sand layer ($d_1=0.67mm$) the relative depth of scour is reducing (Fig.6).

5. Conclusion

Method for calculation equilibrium depth of scour at uniform and stratified river bed at elliptical guide banks is presented (Eqs.5, 8).

Analysis of the test, calculation and computer modelling results is made to estimate impact on relative equilibrium depth of scour: Fr number of the flow, local Froude number- Fr_{Vl}, $Fr_{V\ equil}$ at equilibrium depth of scour, flow contraction, relative time, thickness and sequence of the bed layers.

Local flow modification at the head of the elliptical guide bank lead that local Froude number $Fr_{Vl} = V_l/\ (g\ h_f)^{0.5}$, local Fr_{Vlt} numbers at the end of test and $Fr_{V\ equil}$ at equilibrium depth of scour are increasing, compare with Froude number of the open flow. Local Froude number is depending on flow contraction (Fig.1) and with increase of the local Froude number relative equilibrium depth of scour is increasing (Fig.2). Relative equilibrium depth of scour is depending on flow contraction (Fig.3) and relative time of scour (Fig.4). Depth of scour is always greater when the fine-sand layer is under the coarse sand layer, and smaller when the fine-sand layer is on the top of the coarse sand layer.

Relative equilibrium depth of scour is increasing with increase of local *Fr* number, relative time of scour flow contraction and bed stratification with different thickness and sequence of the bed layers. When the depth of scour is less than thickness of the bed layer $h_{equil}<H_{d1}$, equation (5) can be used; however, when the depth of scour is greater than thickness of the bed layer $h_{equil}>H_{d1}$, the scour develops in the second layer with a grain size d_2. Local velocity on the top of the second layer is changed and can be calculated by equation (6) and new value of the critical velocity by equation (7). If the depth of scour is greater than thickness of the second bed layer $h_{equil}>H_{d1} + H_{d2}$, the scour develops in the third layer with a grain size d_3, and new values of the local and critical velocities and so on.

With increase of the thickness of the first coarse sand layer (with $d_1=0.67mm$) the relative depth of scour is reducing (Fig.6).

Calculation equilibrium depth of scour with the approach flow velocity and at river bed layering, using the grain size only on the top of the bed or for the next layer calculate depth of scour with the approach flow velocity, as it excepted nowadays, and neglecting changes of the flow parameters during the scour, river bed layering can lead to wrong results and damages and losses.

References

[1] Raudkivi A and Ettema R 1983 Clear water scour at cylindrical piers *Journal of Hydraulic Engineering,* **No.3 (109)**, pp 338-345

[2] Sturm T W and Janjua N S 1994 Clear-water scour around abutments in floodplains *Journal of Hydraulic Engineering,* **vol. 120(8)**, pp 956–766

[3] Ahmed F and Rajaratnam N 2000 Observations on flow around bridge abutment *Journal of Engineering Mechanics,* **126(1)**, pp 51-59

[4] Radice A, Franzetti S and Ballio F 2002 Local scour at bridge abutments *Procedings. of Int. Conference on Fluvial Hydraulic, River Flow 2002* (Louvian – La – Neuve , Belgium), **vol 2**, pp 1059 – 1068

[5] Coleman S, Lauchlan C and Melville B 2003 Clear-water scour development at bridge abutment *Journal of Hydraulic Research,* **41(5)**, pp 521-531

[6] Kwak K, Lee J H, Park J, and Chung M 2004 Pier scour and erosion characteristics at Choji bridge *Proc. of 2 Inter. Conf. on Scour and Erosion 2004,* **vol.1,** Singapore: Stallion Press, pp 292-300

[7] Gjunsburgs B, Neilands R and Govsha E 2006 Local scour at elliptical guide banks *Procedings. of Int. Conference on Fluvial Hydraulic, River Flow 2006* (Lisbon: Portugal),**vol.2**, pp 1649-1655

[8] Ballio F, Teruzzi A and Radice A 2009 Constriction effects in clear-water scour at abutments *Journal of Hydraulic Engineering,* **vol. 135(2)**, pp 140-145

[9] Cardoso A and Fael C 2010 Time to equilibrium scour at vertical wall bridge abutments *Proc. of ICE-Water Management*, pp 509–513

[10] Ghan A, Azamathullah H and Mohammadpour R 2011 Estimating time to equilibrium scour at long abutment by using genetic programming *Proc. of 3rd Int. Conf. on Managing Rivers in the 21st Century: Sustainable Solutions for Global Crisis of Flooding, Pollution and Water Scarcity, Rivers 2011*, 6-9 December, Penang, Malaysia, pp 369–374

[11] Guo J 2014 Semi-analytical model for temporal clear-water scour at prototype piers *Journal of Hydraulic Research*, **52(3),** pp 366–374

[12] Sheppard D, Melville B and Demir, H 2014 Evaluation of existing equations for local scour at bridge piers *Journal of Hydraulic* Engineering, **140(1**), pp 14–23, doi:10.1061/(ASCE)HY.1943-7900.0000800

[13] Evangelista S, Govša J, Greco M, Gjunsburgs B 2017 Computation of Sediment Discharge in Rivers: The Contributions by Levi and Studenitcnikov Revisited, *Journal of Hydraulic Research*, 2017, **vol. 55**, iss. 4, pp 445-454

[14] Studenitcnikov V 1964 Flow erosion capacity and river-bed calculation methods. Moscow: Engineering Literature Press.

The impact of nano SiO_2 and superplasticizers on hydration and physical mechanical properties of cement

J Malaiskiene[1], V Baneviciene[1] and M Vaiciene[2]

[1]Vilnius Gediminas technical university, Institute of Building Materials, Laboratory of Composite Materials, Linkmenu str. 28, Vilnius, Lithuania,
[2]Vilnius College of Technologies and Design, Civil Engineering Faculty, Antakalnio str. 54, Vilnius, Lithuania

E-mail: jurgita.malaiskiene@vgtu.lt

Abstract. The aim of work is to determine the lowest and effective amount of nano SiO_2 (NS) for cementitious paste and identify the complex impact of NS and chemical admixtures on cement hydration and physical-mechanical properties. In order to achieve the aim, cementitious specimens with different types and amounts of NS were formed. Further, after the selection of NS and its amount, specimens were formed with various superplasticizers. The work analysis the following properties of cementitious mixtures and hardened cement stone: flow diameter, exothermal temperature, density, ultrasound pulse velocity, compressive strength, microstructure and mineralogical composition. It was determined that the greatest strength of cementitious specimens was obtained when 0.02 wt.% of NS was used, and, compared to control specimens, it increased by ~15%. Additionally, the greatest increment in strength was observed for specimens with melamine-based superplasticizer. Compared to control specimens, the strength increased up to 23% and the obtained density and ultrasound pulse velocity were the greatest. XRD analysis showed that all specimens had analogue crystal hydrates formed but their amounts are different. SEM analysis showed that structure of the specimens with NS is denser.

1. Introduction

Currently, an analysis of the impact of various nano- additives (nano-TiO_2, nano-$CaCO_3$, nano-Al_2O_3, carbon nano-tubes, nano silica (NS), etc.) on the properties of cementitious materials gains the popularity. One of the most effective nano- additives is NS which fills the cavities between cement particles and has high reactivity with calcium hydroxide of the cement. Moreover, NS increases the density of cementitious materials, reduces drying shrinkage, capillary absorption and water demand [1-3]. The use of NS makes concrete financially more attractive and reduces the CO_2 footprint of the produced concrete products [4]. Authors [5] suggest using nano particles in small amounts of (0.01-0.005)% by cement weight. Different NS in scientific studies of cementitious materials is used in various amounts – from 0.01% to 0.5% and sometimes more than that [6-9].

Properties of cement mixtures with NS in literature [10-11] show that the strength of all specimens with NS was higher than that of specimens with micro SiO_2. Consistent use of NS in cement mixtures increases compressive strength in early stages by 10–20%, after 28 days – by 16–24%, and after an addition of plasticizers – up to 30% [12]. Other authors [13] state that adding of 0.25% of NS without a plasticizer to the cement paste enhanced the compression strength of the samples after 1-3 days up to 6%, at that, after 7 days cement specimens demonstrated no strength increase, while after 28 days and after 90 days cements specimens even lost some of their strength. Addition of 0.1% of a superplasticizer ensured an increase the compression strength of concrete after 3-7 days by 28-30% and after 28 days – by 10%.

It is determined [14-15] that NS acts as an active pozzolanic additive which accelerates hydration reactions. Additionally, it is proved that NS has an effect on the cement paste structure, and fills the voids between the cement particles [16-17]. It was reported that NS generated more CSH gel in cement stone [18-19]. In addition, the results showed that the suitable content of nano-SiO_2 improve the durability of concrete significantly and such concrete has lower water permeability than the normal concrete [20-21]. In authors' works [22-24], the impact of micro-silica and NS on a durability of cement materials is determined. The results show that NS has to be used together with micro-silica. Moreover, NS has a greater demand for plasticizer and cementing efficiency compared to micro-silica. Authors [25] tested the impact of NS on transporting and mechanical properties of superhydrophobic cement mixtures; therefore, they determined that the nature and amount of superplasticizer are extremely important.

Partial replacement with NS (up to 2.5%) in mortars and cement pastes determines portlandite minerals after 9 hours of hardening. Furthermore, the density reduction effect is observed, i.e. density reduces by 2.4% [26].

Scientists [27] tested NS impact on various properties of ultra light cement composites. Combined expanded glass granules and cenospheres were used as fine aggregates which facilitate the density of ultra light cement composites. The density of these specimens is 1075 kg/m^3 and compressive strength after 28 days of hardening – 20 MPa. Small amount of NS increases a compressive strength of ultra light cement composites up to 27 MPa. Water accessible porosity values of ultra light cement specimens are in the range of 15–19%. It is as well determined that increased viscosity appears due to ultra fine particles in suspension. They detain air in mixtures; therefore, water accessible porosity increases.

Work [28] presents the positive complex impact of nano-clay and NS on compound cement hydration. Chemically bonded water content is analysed and loss on ignition at 900°C after 2 h is determined. It is the main parameter which evaluates the kinetics of Portland cement hydration. In compound cement paste, the content of chemically bonded water increases with the addition of NS/nano-clay amount. It is explained by the hydration speed increment which is caused by the interaction between amorphous alumosilicate NS/nano-clay with free calcium hydroxide to form CSH, CAH and CASH.

Article [29] analysis the impact of three different materials based on silicon dioxide (quartz, ultrafine quartz, and silica fume) on cement hydration, rheology, structure and compressive strength. Calorimetric tests showed that 2.5% of silica fume and NS accelerate the hydration of cement paste. In case of quartz, no effect, compared to control specimen, is observed when a median particle size of 0.6 μm is reached. Other authors [30] analysed the impact of NS size on compressive strength of cement materials. Cement specimens with 40 nm size NS were characterised by a higher compressive strength than those with 12 and 20 nm-sized NS. Such impact is explained by a particles' agglomeration and ineffective dispersion. By changing 3%, 6%, 9% and 12% cement with NS, the optimum NS quantity is 9%.

The amounts of NS used by scientists vary widely, as are the size of particles and their effects. The aim of our work is to select the lowest effective NS amount and to determine its effect on the properties of cement mixtures and hardened specimens using different types of superplasticizers.

2. Materials and Methods

In the research, cement CEM I 42.5 R was used; its chemical composition is presented in Table 1 and physical mechanical properties – in Table 2. Mineral composition of the cement: C_3S – 56.6 %, C_2S – 16.7 %, C_3A – 9.0 %, C_4AF – 10.6 % and 7.1 % others (alkaline sulphates and CaO).

For the research, two types of nano SiO_2 (A and B) were used, their properties are presented in Table 3. The effect of three different superplasticizers (with potassium hydroxide (pH 9.0), based on melamine (pH 9.4) and based polycarboxylate (pH 6.5) on the properties of cementitious mixtures and hardened samples was investigated. Drinking water was used to prepare mixtures. The compositions of cement mixtures are given in Table 4.

Table 1. Chemical composition of cement, %.

CaO	SiO_2	Al_2O_3	Fe_2O_3	MgO	K_2O	Na_2O	SO_3	Cl	L.O.I.
63.2	20.4	4.0	3.6	2.4	0.9	0.2	3.1	0.05	2.15

Table 2. Physical-mechanical properties of cement.

Particle density, g/cm^3	Soundness "Le chatelier", mm	Passing 32 µm, %	Compressive strength after 2 days, MPa	Compressive strength after 28 days, MPa	Vicat initial, min	H_2O, %
3.1	1.0	78.5	30	55	180	27.5

Table 3. Properties of NS.

Mark	Size, nm	Surface area, m^2/g	Purity, %	Relative molecular weight, g/mol	pH (40 g/l)	Relative density, g/cm^3
A	12	205	99.8	60.08	4.2	2.6
B	~10-30 nm	202	99.8	60.08	4.0	2.2

Table 4. The compositions of cement mixtures.

Mark	Cement, g	Nano SiO_2, g	Nano SiO_2, %	Superplasticizer, g	Superplasticizer, %	Water, g
C	3000.0	0.0	0.0	0	0	1050
A1	2997.0	3.0	0.1	0	0	1050
A2	2994.0	6.0	0.2	0	0	1050
A3	2991.0	9.0	0.3	0	0	1050
A01	2999.7	0.3	0.01	0	0	1050
A02	2999.4	0.6	0.02	0	0	1050
A03	2999.1	0.9	0.03	0	0	1050
B2	2999.4	0.6	0.02	0	0	1050
BH	2999.4	0.6	0.02	6	0.2	850
BM	2999.4	0.6	0.02	6	0.2	850
BP	2999.4	0.6	0.02	6	0.2	850

Primarily, only one type of nano SiO_2 (A, particle size 12 nm) was selected for determine optimum amount of this additive. Later, it was checked or the optimum amount (according to the A type SiO_2) of B type SiO_2 would change significantly properties of cement specimens, as the price of A type SiO_2 and B type SiO_2 differs several times (B is cheaper). Since, the results of cement specimens properties with B type SiO_2 were slightly better than with A type, only B type SiO_2 was used in further studies.

When preparing the mixtures, nano SiO_2 was first dispersed in 350 g of water with equipment UZDN-2T. Disperse duration was 15 min at 400 W 22 kHz. The resulting mixture is then mixed with the rest of the water to bring the water temperature to 20°C. If the mixture is prepared with a superplasticizer, the superplasticizer is also poured into the prepared suspension. Mixing was performed in the mixer for 3 min, and 16×4×4 cm samples were formed. The consistence of prepared mortars was established according to the flow diameter established according to EN 1015-3.

The X-ray diffraction (XRD) analysis of the phase composition of materials was carried out upon applying diffractometer DRON-7. The parameters of the tests were following: voltage - 30 kV; current - 12 mA; the range of the diffraction angle – from 4 to 60°, the detector movement step – 0.02°; the duration of the intensity measuring in a step – 0.5 s. Phase identification was carried out by decoding the XRD patterns according to ICDD diffraction databases. The quantitative changes in the XRD patterns were assessed according to the height of the peak of the main diffraction maximum of a mineral.

The microstructure of the materials was explored upon use of scanning electron microscopy (SEM) equipment JEOL JSM-7600F. The parameters of electron microscopy were following: voltage - 10 kV; the distance to the surface of the specimen – from 7 to 10 mm. The peculiarities of the microstructure were identified on examining the split surface of the specimens. The image was formed on registering the signal of secondary electrons. Prior to the test, the cleavage surface was covered with thin golden layer upon applying the method of gold electron vacuum evaporation.

For exothermal tests of mixtures, the specimens were prepared as follows: cement mortar mixtures (the total weight 1500 g) were poured into forms and a thermocouple in a glass pipe was immersed in the form. The prepared mixture with the form was immediately placed into a metal box with internal 50 mm thick foam polymer insulation. Temperature was measured and its values were uninterruptedly fixed until heat emission processes took place in the specimen. The exothermal tests were carried out at the room temperature (20±1) °C, the temperature of the used raw materials and water was the same, i.e. (20±1) °C. This test was provided in semi-adiabatic conditions.

The density of the specimens was established according to EN 196. Ultrasound propagation time is determined using the equipment "Pundit 7" (frequency of converters is 54 kHz) and ultrasound propagation speed is calculated based on the following equation (V, m/s):

$$V = l/\tau; \tag{1}$$

where : l – length of the specimen, m; τ – signal propagation time, s.

The compressive strength of specimens after their hardening in water for 7 and 28 days was established upon using hydraulic press ALPHA3-3000 S according EN 196.

3. Tests results and analysis

Consistence (determined by flow table) results in mixtures with superplasticizers were from 19.7 cm to 20.6 cm. The use of superplasticizers has reduced the amount of water, and it is aimed to ensure that all mixtures do not spread more than 10 mm. In this case, the flow diameter varied from 15.4 cm (BH) to 16.4 cm (BP). The BH superplasticizer also has air-absorbing properties, resulting in lower flow diameter of this mixture.

The results of density and ultrasound pulse velocity after 28 days of specimens hardening are shown in figure 1.

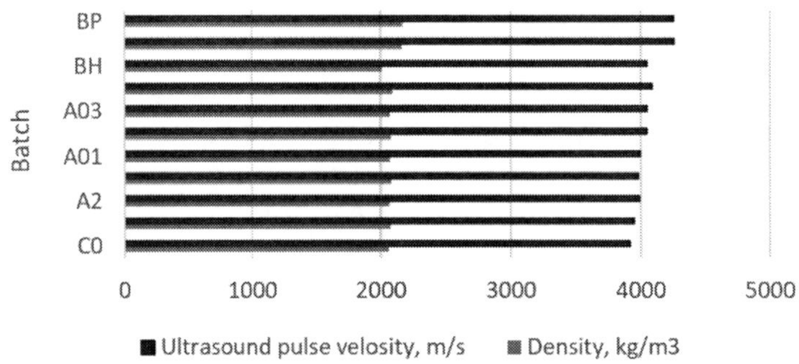

Figure 1. The results of density and ultrasound pulse velocity.

It can be observed that the density values at different amount of NS change only slightly, i.e. up to ~1%. However, a greater density change is seen in addition to the use of plasticizing admixtures. The minimum density is obtained for specimens BH, and compared to control specimen C0, it reduces up to ~3%, while the greatest density is obtained for specimens BM and it increases by ~5%. The difference in density when using different superplasticizers is ~ 8%. Ultrasound pulse velocity (UPV) change after addition of NS is 4%. When superplasticizing admixtures are used, UPV, compared to control specimen, increases by ~9%. The UPV results showed that the NS compacts the structure and especially the positive effect of NS was observed using melamine- and polycarboxylate-based superplasticizers. In specimens with potassium hydroxide-based superplasticizer, more pores are formed. Producers describe this plasticizing admixture as a material which increases water resistance and has air-absorbing properties.

Figure 2 presents the results of the compressive strength of the specimens, which show that the A NS additive when used without superplasticizers increases the compressive strength by ~11% after 7 days of hardening and approx. 16% after 28 days of hardening.

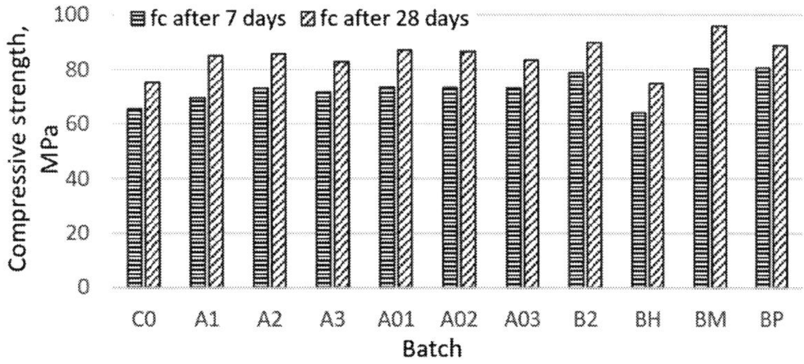

Figure 2. The results of compressive strength (fc).

Interestingly, the changes in the compressive strength of the specimens with A NS depending on the content of admixture varied slightly, i.e. by ~5%. Perhaps this is due to the agglomeration of the particles and the low degree of dispersion. The best results were obtained for Type A NS at 0.02% by cement weight, but this material is more expensive than Type B NS due to its variable size. Therefore, it was tried to mix the best-performing batch with Type B NS at 0.02% by cement weight. The results of compressive strength are even better in this case than in Type A NS. In this case, the compressive strength after 7 days hardening increased by ~17% and after 28 days – by 19%. Similar results were obtained by other scientists who found that 40 nm-sized NS increases the strength more than 12 nm-sized NS [30] and based their results on stronger agglomeration of smaller NS particles. Further studies were conducted with the highest strength having B2 mixtures containing various types of superplasticizers. The results show that by selecting the appropriate superplasticizer, the compressive strength of the specimens increases even more. The best results are obtained when melamine-based superplasticizer is used. The compressive strength of these BM specimens after 7 days compared to the control sample C0 after 7 days hardening increased by 23% and after 28 days – by 27%. Similar tendencies have been found in other authors' works [12]. However, after the addition of a different type superplasticizer (BH), the strength may be reduced by up to ~2%. It should be noted that this chemical admixture is used to improve the frost and water resistances of mixtures.

In the literature, the increase in compressive strength by using the NS additive is justified by its high pozzolanicity, bonding of free calcium hydroxide and water, as well as the formation of a denser structure. In order to determine the effect of our used NS on cement hydration processes, exothermic curves (figure 3) of the most characteristic batches are given and the highest reached temperature and time to reach it are presented in Table 5. Table 5 and figure 3 show that NS accelerates cement binding and raises exothermic temperature. Stronger effects are captured with superplasticizers.

Table 5. The highest temperature and it reaching time based on batches.

Mark	C0	A2	A02	B2	BH	BM	BP
Duration, h	10.97	9.8	10.58	10.72	8.75	9.35	8.20
Max temperature, °C	77.8	84.1	80.1	79.7	84.8	87.5	88.3

The most obvious acceleration of cement binding (2.77 h) is observed when polycarboxylate-based superplasticizer is used, in these specimens obtained the highest exothermic temperature (88.3 °C).

Similar exothermic temperatures have been determined using melamine-based superplasticizer. In this case, cement binding was accelerated by 1.62 h. When comparing the batches of A02 and B2, which used an analogous amount of 0.02% NS, only the characteristics of it were different but the obtained results were similar. The difference in time to reach the maximum temperature is 1.3%, and the maximum difference in temperature is only 0.5%. After adding 0.2% NS, the process began to run faster. Hence, the greater impact on hydration of cement had the content of NS and the composition of the superplasticizer. Literature [31] states that after cement hydration begins, hydrate products diffuse and envelop nanoparticles as kernels, which can promote the cement hydration and make the cement matrix more homogeneous and compact.

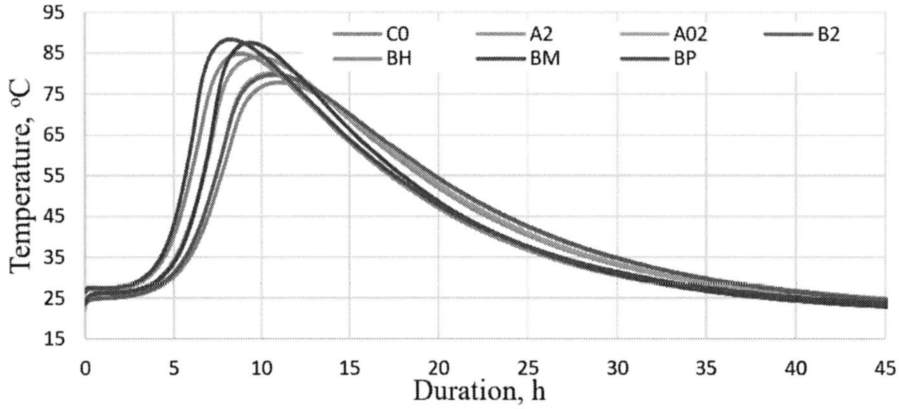

Figure 3. Exothermic curves of characteristic batches.

Figure 4 shows microstructure images of C0 and A2 batch samples after 7 days hardening. It can be observed that the structure of the C0 samples without NS is more porous, with more voids in which the ettringite crystals grow. The structure of samples with NS is denser, less ettringite can be seen while CSH dominates. Therefore, the changes in structure have determined the density and compressive strength.

The results of XRD analysis of the most characteristic specimens are presented in Table 6 and figure 5. Analysis of XRD results by peak intensity shows that NS accelerates CSH formation. It has been found that the use of NS reduces the amount of unreacted minerals alite and belite, reduces the amount of ettringite and increases the amount of CSH and CASH. Likewise, a lower amount of portlandite was obtained from the samples with NS, which is explained by a quicker reaction of NS and calcium hydroxide and formation of CSH and CASH [1, 20]. Such tendencies are even more pronounced with the use of a superplasticizer.

C0 after 7 days A2 after 7 days

Figure 4. Images of microstructure, magnification 5000 times.

Table 6. The results of XRD (intensity) analysis of hydrates formed in compositions.

Mark	Hardening	Ettringite	Calcite	Portlandite	Alite	Belite	C-S-H	CASH
C0	7p	303	187	355	199	178	99	215
A2	7p	286	228	320	195	177	106	214
BM	7p	287	172	264	182	168	108	226

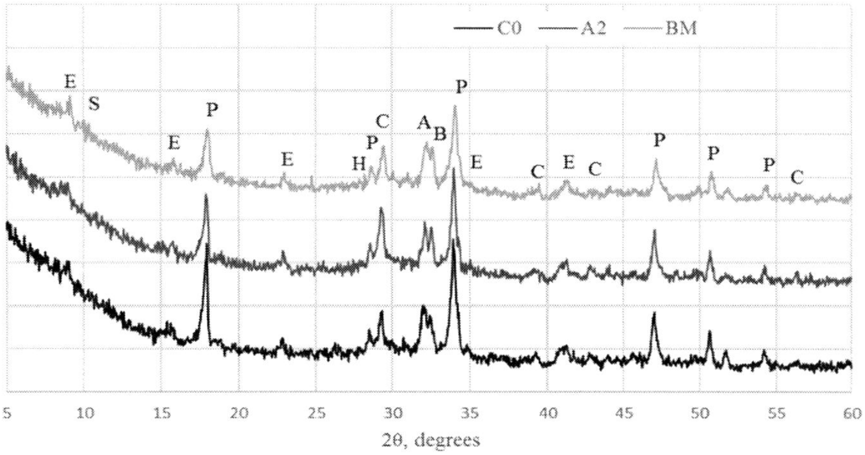

Figure 5. XRD curves of C0 and A2 batches (E - ettringite, S – CASH, P – portlandite, H – CSH, C – calcite, A – alite, B – belite).

Therefore, incorporation of NS in Portland cement materials affects strength development and hydration processes due the following effects [32]: physical effect and pozzolanic activity. Physical effects are mainly attributed to the addition particles filling ability associated to their deposition in the intergranular voids between cement particles, which leads to a denser cement matrix and, therefore, to the mechanical properties improvement. Pozzolanic effect also promotes an increase in the mechanical properties related to the occurrence of the pozzolanic reaction that forms additional strength-providing reaction products i.e., C-S-H and C-A-H [33]. Cement hydration process and physical-mechanical properties highly depends on the type and amount of superplasticizer used. Superplasticizers bind with hydration products crystals and stop or delay the growth of new crystals over the old ones, especially in the initial hydration period. Melamine based plasticizers often retard cement binding, whereas synthetic polycarboxylate resin-based superplasticizers may accelerate hydration. Such different effect of superplasticizers is explained by their action mechanism: polycarboxylate resin-based superplasticizers act under the mechanism of electrostatic and dispersion forces. The dispersive effect of this additive is caused not only by electrostatic repulsion of particles, as it is in the case of melamine superplasticizer, but first of all by spatial repulsion related with long side chains of the ether [34-35].

4. Conclusions

The effect of NS amount on the cement properties was greater than that of NS characteristics. In order to select the minimum effective amount of NS, physical-mechanical properties and best results were obtained using 0.02% NS by cement weight. Then, the maximum density, UPV and compressive strength are obtained. An even greater effect is achieved through the use of superplasticizers. The maximum compressive strength was determined for specimens with 0.02% NS where melamine-based superplasticizer was used. Then, the compression strength was increased by 27% compared to the control samples after 28 days hardening.

It was found that using the NS, the structure of the specimens is denser (macro pores are not visible) and the hydration process is approximately 1 h faster. The most obvious accelerating of cement hydration (2.77 h) is observed when polycarboxylate-based superplasticizer is used; these specimens produced the highest exothermic temperature (88.3 °C). Hydration acceleration was also confirmed by XRD studies, which showed that in specimens with NS due to the intense $Ca(OH)_2$ and SiO_2 reaction, more CSH, CASH, less ettringite and portlandite are produced, lower levels of alite and belite are detected.

References

[1] Rai S, Tiwari S 2018 *Materials Today: Proceedings* 5 9196–9202
[2] Nili M, Ehsani A 2015 *Mater. Des* 75 174–183
[3] Su Y, Li J, Wu C, Wu P, Li Z X 2016 *Compos Eng Part B* 91 595–609
[4] Quercia G, Brouwers H J H 2010 8th fib PhD Symposium in Civil Engineering
[5] Kalasnikov V I, Tarakanov O V, Ibragimov R A, Moroz M N 2014 Betony novogo pokolenija: nano-, mikro- ili mikronanotechnologii? Penza State University of Architecture and Construction, (in Russian)
[6] Wang B 2011 Mater. Sci. Forum 686 432–437
[7] Potapov V V, Kaputin A N Sposoby povyshenija prochnosti betona pri sjatii s izpolzavaniem nanokremnezema, poluchenogo iz gidrotermalnogo rastvora, http://www.findpatent.ru/patent/259/2599739.html (in Russian)
[8] Demjanenko O V, Kopanica N O, Sarkisov J S 2016 Izbrannye doklady II International conference of young scientists and students 193–196
[9] Demjanenko O V, Kopanica N O 2015 II International conference of young scientists 408–411
[10] Subalakshmi P, Saraswathy V, Sivasankaran S K 2012 *Constr Build Mater* 36 648–653
[11] He Z, Liu J, Zhu K 2012 *Energy Procedia* 16 836–841.
[12] Kopanica N O, Sarkisov J S, Demjanenko O V 2016 Herald of Tomsk State University of Architecture and Construction 5 140–150
[13] Potapov V, Kashutin A, Serdan A, Shalaev K, Gorev D 2013 *Nanoindustry* 3(41) 40–49
[14] Jo B W, Kim C H, Tae G H, Park J B 2007 *Constr Build Mater* 21 1351–1355
[15] Ltifi M, Guefrech A, Mounanga P, Khelidj A 2011 *Proc Eng* 10 900–905
[16] Naji G A, Abdul R S, Aziz F N A, Salleh M A M 2010 *Compos. Part B* 41 673–677
[17] Lee H S, Balasubramanian B, Gopalakrishna G V T, Kwon S J, Karthick S P, Saraswathy V 2018 *Constr Build Mater* 159 463–472
[18] Zhu J, Feng C, Yin H, Zhang Z, Shah S P *Constr Build Mater* 2015 101 246–251
[19] Yu R, Spiesz P, Brouwers H J H 2014 *Constr Build Mater* 65 140–150
[20] Wang L, Zheng D, Zhang S, Cui H, Li D, Nann T 2016 *Nanomaterials* 6(12) 241
[21] Ji T 2005 *Cem Concr Res* 35(10) 1943–1947
[22] Li L G, Zhu J, Huang Z H, Kwan A K H, Li L J 2017 *Constr Build Mater* 157 337–347
[23] Li L G, Huang Z H, Zhu J, Kwan A K H, Chen H Y 2017 *Constr Build Mater* 140 229–238
[24] Li L G, Zheng J Y, Zhu J, Kwan A K H 2018 *Constr Build Mater* 168 622–632
[25] Chen H, Feng P, Du Y, Jiang J, Sun W 2018 *Constr Build Mater* 182 620–628
[26] Senff L, Labrincha J A, Ferreira V M, Hotza D, Repettee W L 2009 *Constr Build Mater* 23 2487–2491
[27] Du H 2019 *Constr Build Mater* 199 696–704
[28] Heikal M, Abdel-Gawwad H A, Ababneh F A 2018 *Constr Build Mater* 190 508–516
[29] Flores Y C, Cordeiro G C, Filho R D T, Tavares L M 2017 *Constr Build Mater* 146 524–530
[30] Haruehansapong S, Pulngern T, Chucheepsakul S 2014 *Constr Build Mater* 50 471–477
[31] Jalal M, Mansouri E, Sharifipour M, Pouladkhan A R 2012 *Mater Des* 34 389–400
[32] Cyr M, Lawrence P, Ringot E 2006 *Cem Concr Res* 36(2) 264–277
[33] Costa C, Marques P 2012 IEEE-IAS/PCA 54th Cement Industry Technical Conference 1–9
[34] Shi C, Zhang G, He T, Li Y 2016 *Constr Build Mater* 112 261–26
[35] Wutz K 2001 Proceedings of the Biennial Worldwide Congress on Refractories Unitecr, 85–97.

The Effect of Silica Fume Addition on the Resistance of Concrete to Alkali Silica Reaction

D Nagrockienė[1,2], A Rutkauskas[1], I Pundienė[1] and I Girnienė[1]

[1]Vilnius Gediminas Technical University, Faculty of Civil Engineering, Department of Building Materials, Sauletekio al. 11, LT-10223 Vilnius, Lithuania

E-mail: dzigita.nagrockiene@vgtu.lt

Abstract. Concrete mix made of cement CEM I 42.5 R, silica fume, fine aggregate sand of 0/4 fraction and superplasticizer was tested. Five batches of specimens were made with different silica fume content, where up to 10 % of the binding material was replaced with silica fume. Compressive strength, water absorption, density and resistance to alkali silica reaction were tested in concretes modified with different amounts of silica fume addition. The effect of silica fume addition on the expansion of modified concrete conditioned in 1M NaOH solution for 56 days was tested. Concrete modified with silica fume at 10 % by weight of cement was found to be the most appropriate for structures exposed to alkaline environment. The control specimen without mineral addition had the expansion of 0.113 %, which exceeds the limit value by 0.1 %, whereas the expansion of specimens modified with 2.5 % of silica fume was equal to the limit value, i.e. 0.1 %. The expansion values reduced in specimens modified with 5 % and 7.5 % of mineral addition. After 56 days of testing, the expansion values of these specimens were 0.093 % and 0.082 % respectively. The lowest expansion value was obtained in specimens with the highest content of silica fume, i.e. 10 %. The expansion of these specimens was 0.07 % after 56 days of conditioning in 1M NaOH solution of 80 °C temperature. The tests revealed that concrete modified with silica fume replacing 10 % of cement had better resistance to alkali silica reaction and better durability characteristics and thus can be used as structural concrete.

1. Introduction

Concrete is the most widely used construction material worldwide, with billions of tons produced every year [1]. During the last three decades, great strides have been made in improving the performance of concrete as a construction material [2]. Durability of ordinary Portland cement is of a major issue in civil infrastructure projects [3].

Modern concrete is more than just a mixture of cement, water, and aggregates. Modern concrete more and more often contains mineral components having very specific characteristics that give specific properties to concrete, and also chemical admixtures that have even more specific effects [4].

Concrete consists of aggregate particles, water, and a binding agent (cement and other cementitious materials). The reactions between aggregates and binding agents are extremely complex. Many natural aggregates contain amorphous silica and carbonate, which both react with the alkalis (Na and K) that are released into the mixture from cementitious materials. This interaction is widely known as the alkali-aggregate reaction (AAR). Depending on which minerals are involved, AAR is subdivided into alkali-silica reaction (ASR) and alkali-carbonate reaction (ACR). ASR is generated by the aggregates containing reactive silica minerals [5].

After more than 60 years of research, the knowledge gained on reaction mechanisms has allowed researchers and recommendation committees to provide rules to prevent new ASR cases [6]. Unfortunately, no ways to stop the reaction or to cure an affected structure have been found until now. There are many ASR affected structures around the world which require periodical repairs and preventive maintenance policies creating to ongoing costs [7].

The product of alkali-silica reaction is a gel that surrounds the aggregate in the concrete matrix. This gel can appear as a dark border around the aggregate or a white spot within the aggregate, sometimes the white area extending into the concrete matrix [8]. ASR is a chemical process occurring within the concrete, in which alkali, mostly originating from cement, react with silica originating from the aggregate [9]. Water plays a major role in the degradation of structures affected by ASR [10].

The reaction of amorphous or crystalline SiO_2 fume aggregates with the alkaline pore solution of concrete (pH 13.0–13.5) causes the expansion of concrete. The generated stress may eventually cause cracks in concrete years after the affected structures were built [11]. ASR affected concrete structures exhibit cracking, displacements, structural deformations, pop-outs and reduction in mechanical performance. Service of structures may be severely affected. Gels are usually supposed to be the main cause of the induced swelling and degradation [12].

The formation of the ASR gel alone does not cause cracking; however, when the gel absorbs water, it has a significant potential to swell. The resulting expansion often results in pressures greater than concrete can withstand, which in turn causes cracks in the concrete. The ASR is a multistage process [13-14].

For the design of a concrete mix, it is important to consider the fine particles, as they have an influence on the water demand and, consequently, on the fresh and hardened concrete properties [15].

Different mineral admixtures are often added in concrete to improve the durability, rheology of fresh concrete, and mechanical properties of hardened concrete [16]. Silica fume, a by-product of the ferrosilicon industry, is a highly pozzolanic material that is used as mineral admixture to enhance mechanical and durability properties of concrete. Silica fume (SF) is added to concrete to improve strength and reduce permeability [17]. SF concrete may be appropriate in places where high abrasion resistance and low permeability are of utmost importance or where very high cohesive mixes are required to avoid segregation and bleeding [2]. It may be added directly to concrete as an individual ingredient or in a blend of Portland cement and silica fume [18]. Silica fume can be used in a densified or undensified powder form, a slurry, as a combination at the concrete mixer, or part of a factory-blended cement [19]. Individual particles are extremely small, approximately 1/100th the size of an average cement particle. Because of its fine particles, large surface area, and high SiO_2 content, silica fume is a very reactive pozzolan when used in concrete [20]. Silica induces a pozzolanic reaction that results in the reduction of the amount of calcium hydroxide in concrete, and SF reduces porosity and improves durability [21].

Alkali-aggregate reaction has always been and will always be a concern for the construction industry because there is currently no way to completely cure the problem before it stops by itself, which can take decades to happen after extensive damage has developed [22].

This paper reports the results of research into the effect of liquid suspension of amorphous silica fume on the resistance of concrete containing reactive aggregates to alkali silica reaction. The aim of the paper is to analyse the effect of silica fume addition on alkali silica reaction in concrete mixture containing reactive aggregates.

2. Materials and test methods

Portland cement CEM I 42.5 R and liquid suspension of amorphous SF (as mineral addition) were used for the tests. Chemical compositions of cement and liquid suspension of amorphous silica fume are presented in Table 1 and Table 2. Sand of fraction 0/4 mm was used for the test. Sand specifications are presented in Table 3.

The concrete mix was prepared in accordance with the requirements of standard EN 196-1:2016. 5 batches of concrete mixes were mechanically produced under laboratory conditions for the tests. The batches differed by the type of cement and silica fume content ranging from 0% to 10% by weight of cement (the binder was replaced by silica fume). Compositions of concrete mixtures are presented in Table 4. Concrete mixes were mechanically mixed under laboratory conditions and 40×40×160 mm size specimens were formed in metal casting moulds. Specimens were removed from the moulds after 24 hours of curing and soaked in 20±2°C water for 28 days. After 7 and 28 days of curing in water of 20 ± 2 °C temperature the density of the specimens was measured according to EN 12390-7:2009, the compressive strength was measured according to EN 12390-3:2009/AC:2011. The total water absorption of the specimens was measured after soaking in water for 96 hours. The specimens were soaked in water of 20 ± 2 °C temperature and kept there until the constant mass was achieved. Specimens in the bath must be placed at 15 mm between specimens and the layer of water above the specimens must be at least 20 mm. The constant mass was achieved when the difference in weight every 24 hours was less than 0.1 %. Prior to weighting the specimens were dried with a damp cloth to remove excessive water.

Table 1. Chemical composition of cement.

Chemical composition, %								
SiO_2	Al_2O_3	Fe_2O_3	CaO	K_2O	SO_3	Na_2O	Na_2O	Other
20.76	6.12	3.37	63.5	1	0.8	0.3	0.8	3.35

Table 2. Chemical composition of liquid suspension of amorphous SF.

Chemical composition									
SiO_2	Al_2O_3	Fe_2O_3	CaO	MgO	K_2O	Na_2O	Na_2O	SO_3	Loss on ignition
93.35	0.92	0.71	0.69	0.72	0.86	0.33	0.85	0.13	1.44

Table 3. Mineral and petrographic composition of sand from used quarries.

Mineral and petrographic composition	Sand fraction 0/4
Quartz	62.3 %
Feldspars	13.0 %
Carbonates	14.2 %
Mica	0.1 %
Amphiboles	2.6 %
Crystal rock	7.8 %

Table 4. Compositions of concrete mixes.

SF content, % (by weight of cement)	Cement, %	Aggregate, %	Water, %	Silica fume, %	W/C	Plasticizer, % (by weight of cement)	Slump flow, mm
0	26.88	60.48	12.63	0.00	0.47	0	205
2.5	26.21	60.48	12.63	0.67	0.47	0.20	219
5	25.54	60.48	12.63	1.34	0.47	0.30	215
7.5	24.87	60.48	12.63	2.02	0.47	0.60	205
10	24.19	60.48	12.63	2.69	0.47	0.80	205

The resistance of concrete specimens to ASR was tested in accordance with the standardised methodology Rilem AAR–2. According to AAR-2 method the specimens (40x40x160 mm) are cured in water of 80 °C and further conditioned in 1M NaOH solution of 80 °C for 14 days; afterwards their expansion is measured. The methodology prescribes a longer conditioning time for specimens with aggregates containing slowly reactive rocks. In our case the time of soaking specimens in 1M NaOH solution was extended to 56 days [23].

For ASR tests washed samples are dried at 100 °C–110 °C for 16 hours. The mortar was made of 1 part cement and 2.25 parts of aggregate. 600 g of cement and 1350 g of dry aggregate is required for three prisms (40×40×160 mm). 6 prisms with each aggregate were produced. The water/cement ratio was 0.47. The mortar flow diameter must be from 205 to 220 mm. Polycarboxylate-based superplasticizer was used to obtain the required mortar flow.

Four prisms with each silica fume amount were selected for the test. The prisms were immersed into water of 80 °C. After 24 hours the distance between the inner ends of the metal pins were measured and the specimens were placed into stainless steel containers with 1M NaOH alkali solution at 80 °C. Containers with the specimens were placed into climatic chamber where 80 ± 2 °C temperature was maintained. The elongation of specimens was measured regularly. The specimens were removed after 24 ± 2 hours of hardening in water at 80 °C temperature, and the length, i.e. the zero reading (L_0) was recorded; this is a reference reading for further measuring of specimen expansion. After the specimens were placed into 1M NaOH alkaline solution at 80 °C, the specimen dimensions (L_n) were recorded after 1, 3, 6, 9, 12, 14, 21, 28, 35, 42, 49, 56 days ± 2 hours. The elongation of specimens was measured by a digital dial indicator with the scale interval of 0.001 mm. The indicator was fixed in the frame where the prisms were placed. The frame and indicator axles were inserted into the moulded holes at the ends of the specimen.

The linear expansion of each specimen was obtained by calculating the difference between the length of the specimen at each period of measurement (L_n) and zero measurement (L_o), to the nearest 0.001% of the effective length, as follows:

$$\text{Expansion}, \% = \frac{100 \cdot (L_n - L_0)}{l} \tag{4}$$

Where:

L_o - measurement of the specimen just before the immersion into sodium hydroxide solution (zero reading),

L_n - a reading taken at each period of soaking in sodium hydroxide solution, n being the number of days counted from the zero reading.

l - distance between the inner ends of the metal pins, measured to the nearest of 0.1 mm.

According to RILEM AAR-2 method, aggregates are considered to be reactive when the expansion of concrete specimens conditioned in 1M NaOH solution for 14 days exceeds 0.054 % [24]. According to ASTM C 441 test method, in tests lasting for 56 days the threshold expansion limit is 0.1 %.

3. Test results and analysis

Figure 1 illustrates density and water absorption tests of concrete modified with silica fume.

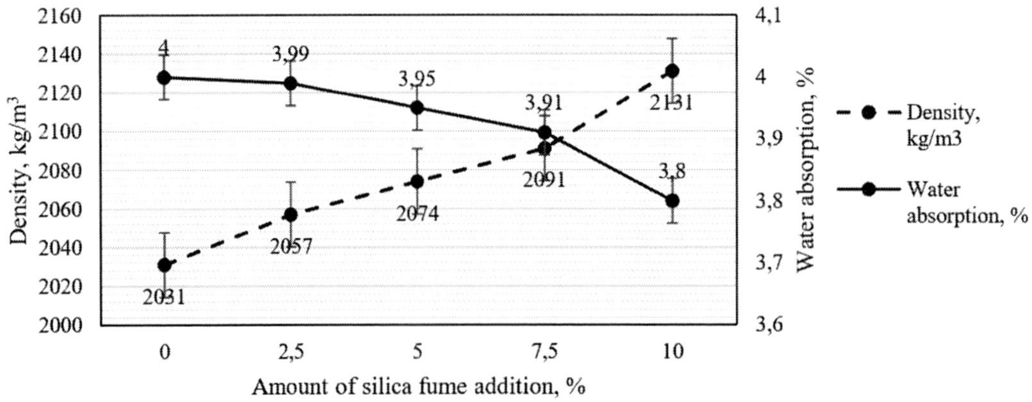

Figure 1. Relationship between density and water absorption of concrete and silica fume content.

The results of tests of concrete modified with silica fume revealed that the highest density of 2131 kg/m^3 was obtained in the batch containing 10 % of silica fume. It was found that the density of specimens was directly related to silica fume content, i.e. the higher amount of silica fume was added to the mix, the higher density of the specimen was obtained. The maximum difference in the density between control specimens and the specimens containing 10 % of silica fume is 100 kg/m^3, i.e. 4.92 %.

Water absorption rate decreases after silica fume is added to concrete mix. A minimum drop of 0.01 % of water absorption value (3.99 %) was recorded in specimens where 2.5 % of the binder was replaced with silica fume. When silica fume content was increased to 5 % and 7.5 %, water absorption values reduced to 3.95 % and 3.91 %. In specimens with the maximum SF content of 10 % the lowest water absorption value was 3.8 %, which is 0.2 % lower than water absorption in control specimens.

Figure 2 illustrates the relationship between the compressive strength of concrete cured for 7 and 28 days and silica fume content.

The mean value of control specimen's compressive strength after 7 days of curing is 46.78 MPa. An increase of 4.05 MPa was observed in specimens where 2.5 % of the binder was replaced with SiO_2. These specimens had the compressive strength of 50.83 MPa. Specimens modified with 5 % of silica fume had the compressive strength of 54.05 MPa, i.e. 7.27 MPa higher compared to the control specimen. When silica fume content was increased up to 7.5 %, the compressive strength reached 56.98 MPa. After 7 days of curing specimens with the maximum 10 % content of silica fume had the compressive strength of 57.39 MPa, which was 22.68 % higher compared to the control specimen containing no silica fume.

The compressive strength values increase with higher SF content after 28 days of curing. When cement is replaced with 2.5 % of silica fume, the compressive strength increases by 3.51 %, i.e. from 52.08 MPa (control specimen) to 53.91 MPa. When 5 % and 7.5 % of the binding material is replaced with silica fume, the compressive strength values increase 9.12 % and 10.06 % and reach 56.83 MPa and 57.32 MPa. The highest compressive strength value (59.06 MPa) was obtained with the highest silica fume content (10 %). The compressive strength of those specimens was 13.4 % higher compared to the control specimen.

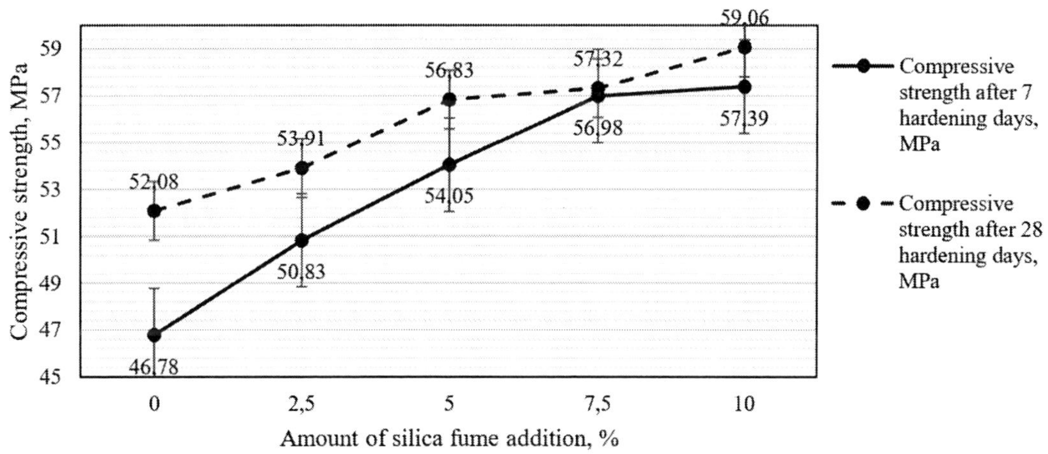

Figure 2. Relationship between the compressive strength of concrete after 7 and 28 days of curing and silica fume content.

Silica fume fills in small voids in concrete structure and increases concrete density, ultrasonic pulse velocity, compressive and flexural strength, but reduces water absorption. Concrete structure becomes more homogeneous, the porosity reduces and the connection between the aggregate and cement matrix becomes stronger.

The highest density, ultrasonic pulse velocity, compressive and flexural strength values, and the lowest water absorption value were obtained when concrete was modified with 10 % of silica fume.

Figure 3 illustrates the relationship between the expansion of silica fume modified concrete specimens made of cement CEM I 42.5 R and silica fume content in the mix after conditioning the specimens in 1M NaOH solution at 80 °C temperature for 56 days. Constant increase of expansion was observed.

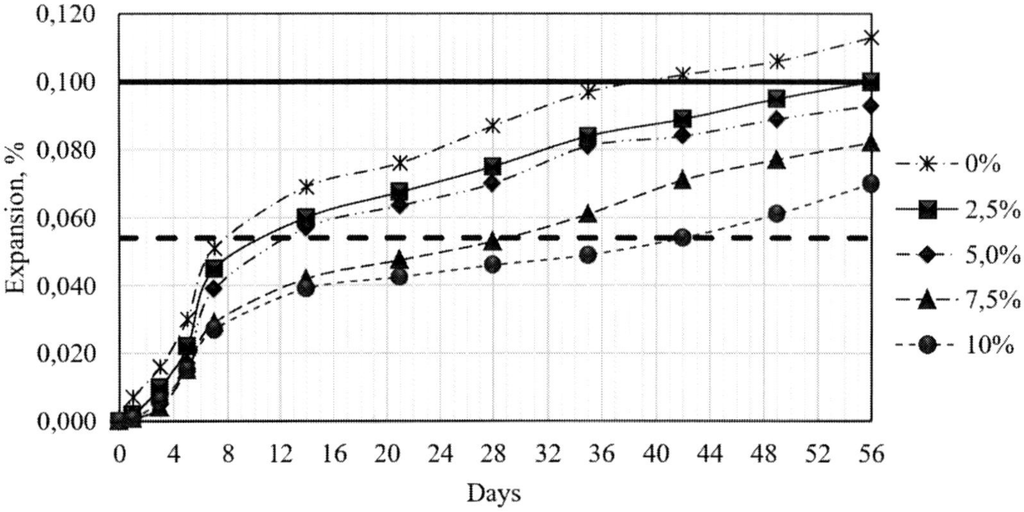

Figure 3. Relationship between the expansion of silica fume modified concrete specimens and silica fume content after conditioning in 1M NaOH solution of 80 °C for 56 days.

The test results showed that in specimens containing 0 %, 2.5 %, and 5.0 % of silica fume addition the average expansion values after 14 days exceeded the allowable expansion limit of 0.054 %. In specimens where 7.5 % and 10.0 % of cement was replaced with silica fume addition the respective expansion values were 0.039 % and 0.042% after 14 days of testing.

After conditioning in the specimens 1M NaOH solution of 80 °C for 56 days, the control specimen without mineral additions had the expansion of 0.113 %, which exceeds the limit value by 0.1 %, whereas the expansion of specimens modified with 2.5 % of silica fume was equal to the limit value, i.e. 0.1 %. The expansion values reduced in specimens modified with 5 % and 7.5 % of mineral addition. After 56 days of testing the expansion values of these specimens were 0.093 % and 0.082 % respectively. The lowest expansion value was obtained in specimens with the highest content of silica fume, i.e. 10 %. The expansion of these specimens was 0.07 % after 56 days of conditioning at 80 °C temperature in 1M NaOH solution.

The test method used revealed that the limit value of silica fume in concrete made of cement CEM I 42,5 R without mineral addition and reactive aggregates is 7.5 %. After 14 and 56 days of testing the expansion of specimens made of concrete mix modified with 7.5 % silica fume did not exceed the threshold expansion limit values of 0.054 % and 0.1 %.

Figure 4 illustrates the relationship between the expansion of specimens conditioned in 1M NaOH solution of 80 °C for 14 and 56 days and silica fume content.

Tests with different amounts of silica fume revealed that the absolute lower expansion values are obtained in specimens where 10 % of the binding material (CEM I 42,5 R) was replaced. SiO_2 was found to have a positive effect on concrete with reactive aggregates in ASR favourable environment.

ASR induced expansion of modified concrete specimens after 14 and 56 days of conditioning in 1M NaOH solution of 80 °C depends on the amount of tested mineral additions and SiO_2 content in them.

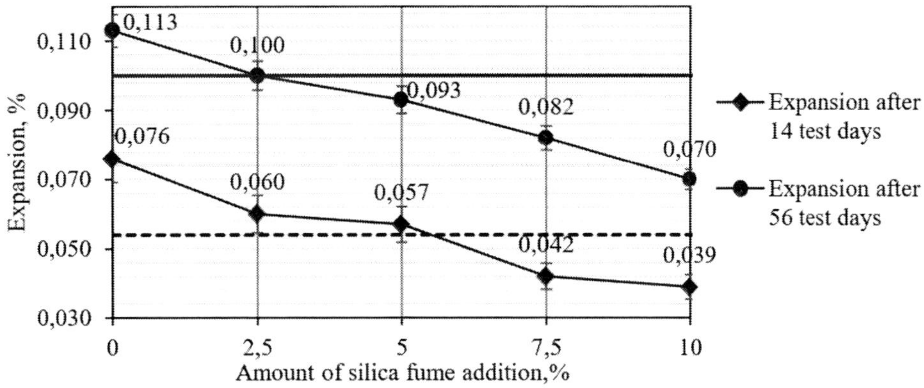

Figure 4. Relationship between the expansions of concrete specimens conditioned in 1M NaOH solution of 80°C temperature for 14 and 56 days and silica fume content in the mix.

A higher content of active mineral addition in concrete mix causes a bigger amount of alkali present in the cement to react and thus reduces the amount of reactive aggregate and alkali reaction product and expansion of concrete.

4. Conclusions

It was found that 10% of silica fume addition by weight of cement increases the density, compressive strengths, and reduces water absorption of concrete. Higher density and lower water absorption hinder corrosion in alkaline environment, i.e. water absorption by the gel and the development of destructive internal stresses are impeded.

Direct influence of silica fume on the corrosion in alkaline environment was determined by the tests. The expansion of specimens decreases with higher silica fume content. The test results showed that the expansion values in specimens made of concrete mix of cement CEM I 42,5 R without silica fume and in specimens made of concrete mix with the same type of cement modified with 10.0 % of silica fume by weight of cement, reduces from 0.076 % to 0.039 % after 14 days and from 0.113 % to 0.070 % after 56 days of testing.

The limit value of silica fume in concrete mix was established. The expansion of specimens does not exceed the threshold limit value of 0.054 % after 14 days of testing and 0.1 % after 56 days of testing when concrete mix is modified with 7.5 % of silica fume by weight of cement.

Amorphous SiO_2 rich silica fume used in concrete manufacture causes a bigger amount of alkali present in the cement to react and thus reduces the amount of reactive aggregate and alkali reaction product and expansion of concrete.

References

[1] Meyer, Ch. 2013. The Greening of the Concrete Industry, The 2013 World Congress on Advances in Structural Engineering and Mechanics (ASEM13) pp. 79-92

[2] Roy, D. K. S.; Sil, A. 2012. Effect of Partial Replacement of Cement by Silica Fume on Hardened Concrete, International Journal of Emerging Technology and Advanced Engineering, Volume 2, pp. 472- 475

[3] Kupwade-Patil, K.; Allouche, E. 2011. Effect of Alkali Silica Reaction (ASR) in Geopolymer Concrete, World of Coal Ash (WOCA) Conference, pp. 1-12

[4] Aitcin, P. C. 2000. Cements of yesterday and today. Concrete of tomorrow. Cement and Concrete Research, 30, 1349-1359

[5] Islam, M. Sh.; Akhtar, S. 2013. A Critical Assessment to the Performance of Alkali-Silica Reaction (ASR) in Concrete, Canadian Chemical Transactions. Volume 1. Issue 4. pp. 253-266

[6] Mather B. 1999. How to make concrete that will not suffer deleterious alkali-silica reaction,

Cement and Concrete Research 29, pp. 1277-1280

[7] Poyet S.; Sellier, A.; Capra, B.; Foray, G.; Torrenti, J.-M.; Cognon, H.; Bourdarot, E. 2007. Chemical modelling of Alkali Silica reaction: Influence of the reactive aggregate size distribution. Materials and Structures, Springer Verlag, 40 (2), pp.229-239

[8] Ballard Z. J.; Caires W. S.; Peters S. R. 2008. Alternate Mitigation Materials for Alkali-Silica Reaction (ASR) in Concrete, Colorado Department of Transportation DTD Applied Research and Innovation Branch. Report No. CDOT-2008-10.

[9] Crucq P. 2005. Development of a method to measure the mechanical behavior of ASR gels. Part I: Literature Study; Alkali-silica reaction, causes, effects and prevention. Microlab Civil Engineering and Geosciences

[10] Comi C.; Kirchmayr B.; Pignatelli R. 2012. Two-phase damage modeling of concrete affected by alkali–silica reaction under variable temperature and humidity conditions, International Journal of Solids and Structures. Volume 49, Issues 23–24, Pages 3367-3380

[11] Leemann A.; Le Saout G.; Winnefeld F.; Rentsch D.; Lothenbach B. 2011. Alkali–Silica Reaction: the Influence of Calcium on Silica Dissolution and the Formation of Reaction Products, Journal of the American Ceramic Society. 94(4). pp. 1243 – 1249

[12] Poyet S.; Sellier, A.; Capra, B.; Foray, G.; Torrenti, J.-M.; Cognon, H.; Bourdarot, E. 2007. Chemical modelling of Alkali Silica reaction: Influence of the reactive aggregate size distribution. Materials and Structures, Springer Verlag, 40 (2), pp.229-239

[13] Latifee, E.R.; Akther, S.; Hasnat, K. 2015. A Critical Review of the Test Methods for Evaluating the ASR Potential of Aggregates. Proceedings of 10th Global Engineering, Science and Technology Conference

[14] Ulm F.-J.; Coussy O; Kefei L.; Larive C. 2000. Thermo-Chemo-Mechanics of ASR Expansion in Concrete Structures, Journal of Engineering Mechanics. pp.233-242

[15] Quercia Bianchi, G.; Lazaro Garcia, A.; Geus, J. W.; Brouwers, H. J. H. 2013. Characterization of morphology and texture of several amorphous nano-silica particles used in concrete, Cement & Concrete Composites. 44 (2013). pp.77–92

[16] Ayub T.; Shafiq N.; Khan S. U. 2013. Durability of Concrete with Different Mineral Admixtures: A Review, International Journal of Civil, Environmental, Structural, Construction and Architectural Engineering. Vol:7, No:8, pp. 601-612

[17] Toutanji H.; McNeil S.; Bayasi Z. 1998. Chloride Permeability and Impact Resistance of Polypropylene-Fiber-Reinforced Silica Fume Concrete, Cement and Concrete Research, Vol. 28, No. 7, pp. 961–968

[18] Guide for the Use of Silica Fume in Concrete. Reported by ACI Committee 234. ACI 234R-06. Reapproved 2012

[19] King D. 2012. The Effect of Silica Fume on the Properties of Concrete as Defined in Concrete Society Report 74, Cementitious Materials. 37th Conference on Our World in Concrete & Structures. Article Online Id: 100037011

[20] Srivastava V.; Agarwal V. C.; Atul; Kumar R.; Mehta P. K. 2013. Silica Fume – An Admixture for High Quality Concrete, J. Environ. Nanotechnol. Volume 2. pp. 53-58

[21] Mondal P.; Shah S. P.; Marks L. D.; Gaitero J.J. 2010. Comparative Study of the Effects of Microsilica and Nanosilica in Concrete, Transportation Research Record Journal of the Transportation Research Board. No. 2141. pp. 6-9

[22] Fournier B.; Bérubé M. A. 2000. Alkali-aggregate reaction in concrete: A review of basic concepts and engineering implications. Canadian Journal of Civil Engineering. Vol. 27. pp. 167-191

[23] RILEM recommended test method: AAR-2. Detection of potential alkali-reactivity. Accelerated mortar-bar test method for aggregates, pp. 1-15

[24] Bumanis G.; Bajare D.; Locs J.; Korjakins A. 2013 Alkali-silica reactivity of foam glass granules in structure of lightweight concrete, Construction and Building Materials Volume 47, 2013, Pages 274-281

IOP Conf. Series: Materials Science and Engineering **660** (2019) 012032 doi:10.1088/1757-899X/660/1/012032

Effect of MWCNT and PCE plasticizer on the properties of cement pastes

G Skripkiunas[1], E Karpova[1], J Bendoraitiene[2] and I Barauskas[3]

[1]Department of Building Materials and Fire Safety, Vilnius Gediminas Technical University, Sauletekio al. 11, LT-10223 Vilnius, Lithuania
[2]Department of Polymer Chemistry and Technology, Kaunas University of Technology, K. Donelaicio g. 73, LT-44249 Kaunas, Lithuania
[3]Department of Silicate Technology, Kaunas University of Technology, K. Donelaicio g. 73, LT-44249 Kaunas, Lithuania

E-mail: ekaterina.karpova@vgtu.lt

Abstract. Polycarboxylate ether (PCE) plasticizer possesses by high water-reduction capability of about 30-40%. Despite the superior water reduction capability, the high dosages of PCE may cause the bleeding in cement systems, which is not suitable for high workability and self-compacting concrete. The current research is devoted to the studying of multi-walled carbon nanotubes (MWCNT) suspension prepared in combination with PCE plasticizer on the rheological properties of cement pastes. The bleeding of cement pastes modified by different dosages of MWCNT was estimated as well. The increase of yield stress in 3.7 and 3.5 times was obtained for cement paste modified by MWCNT suspension in dosage of 0.12% by weight of cement (bwoc) in 5 and 120 min after cement paste mixing, respectively. The increase of plastic viscosity in 2.95 and 1.55 times was obtained for cement paste modified by MWCNT suspension in dosage of 0.12% bwoc in 5 and 120 min after cement paste mixing, respectively. Modification of cement pastes by MWCNT suspension in the dosage of 0.24 % bwoc led to the decrease of bleeding water volume by 21.7% in comparison with cement pastes modified only by PCE plasticizer.

1. Introduction

Application of high workability and self-compacting concrete requires the usage of plasticizing admixtures with high water reduction capability [1]. The last generation of comb-shaped copolymers discovered in the mid-1980s possesses by high water reduction capability and enables to decrease amount of water in cement systems up to 40%. The comb-shaped copolymers consist of backbone presented by carboxylic groups and non-ionic polyether side chains, which are attached to backbone. Mainly, this type of plasticizing admixtures called as polycarboxylate ethers, polycarboxylate esters or polycarboxylates (PCE).

Electrostatic and steric hindrance effects underlie the action mechanism of PCE plasticizers. The negatively charged backbone provides the adsorption of plasticizer onto the positively charged cement particles. Steric hindrance effect is provided by non-adsorbing side chains. The combination of electrostatic and steric hindrance effect makes the action mechanism of PCE plasticizers more efficient in comparison with other types of polymer plasticizers.

The chemical nature of the backbone, length and chemical nature of the side chains, distribution of the side chains along the backbone, anionic charge density and binding between backbone and side chain play an important role in performance of PCE plasticizers [2].

The number of researches are devoted to the investigation of the correlation between molecular structure of PCE plasticizers and fluidity of cement pastes and concrete mixtures [3-5].

Authors of research [3] designed and synthesized the PCE plasticizer with star-shaped topological molecular structure and compared it with PCE of comb-shaped structure. The star-shaped PCE presented the better fluidity and water reduction capability. Researchers [4] established that PCE plasticizers with higher molecular weight, lower side chain densities and shorter side chains have the stronger adsorption properties. Besides the adsorption properties of plasticizers, the critical micelle concentration determines the efficiency of PCE plasticizer in cement systems as well. The addition of PCE plasticizer higher than critical micelle concentration provides the low yield stress and high flowability [5].

The PCE plasticizers require their improvement in terms of avoiding bleeding in cement systems at high dosages of plasticizer. Nowadays, the application of different types of viscosity modifying agent (VMA) such as welan gum, cellulose ethers, polyethylene oxides, polyacrylamide, polysaccharides and some others can be applied for this purpose [6,7].

Nanoadditives with nanometer dimensions of particles, high aspect ratio and unique physical, chemical, electrical properties may be used as a perspective tool to improve the stability of concrete mixture [8,9]. E. Garcia-Taengua et. al. observed the increase of viscosity of fresh mortar modified by 2% of nano-silica [8]. M.R. Du et.al. investigated the effect of MWCNT and methylcellulose on stability of cement pastes [9].

Nowadays, concrete is produced with application of different types of mineral, chemical and nano additives and admixtures simultaneously. For example, the researchers [10, 11] investigated the effect of different surfactants on the properties of MWCNT suspension and cement systems. O.A. Mendoza et.al. highlighted the incompatibility between dosage of surfactant used to obtain an adequate dispersion degree of MWCNT suspension in water and dosage of surfactant necessary to obtain adequate rheological properties of cement system [11]. For this reason the combine effect of additives and admixtures is required to be studied for better understanding of their complex action mechanism and subsequent effect on cement systems.

The present research is focused on the studying of multi-walled carbon nanotubes (MWCNT) suspension prepared in combination with PCE plasticizer on the rheological properties of cement pastes and bleeding of cement pastes modified by different dosages of MWCNT.

2. Materials and Methods

2.1. Materials

The Portland cement without mineral additives CEM I 42.5 R conforming to EN 197-1 with water consumption 26.6% and fineness by Blain of 3552 cm^2/g was used as a binder.

Table 1. Composition of MWCNT suspensions.

Specimen	MWCNT in volume of suspension, %	PCE in volume of suspension, %	Ratio between MWCNT and PCE	Time of ultrasonication, min
D1	0.015	3.8	1:256	6
D2	0.030	3.8	1:128	6
D3	0.060	3.8	1:64	6
D4	0.120	3.8	1:32	6
D5	0.240	3.8	1:16	6
D6	0.048	3.8	1:8	6
D7	0.096	3.8	1:4	6

The MWCNT suspension was prepared from the masterbatch pellets "Graphistrength CW 2-45" produced by the company "Arkema" (France). The masterbatch contains 45 wt. % of MWCNT and 55 wt. % of carboxymethylcellulose (CMC). The MWCNT was characterized by filament length of 100 – 10000 nm and diameter of 15–20 nm.

Distilled water was used as dispersion medium for MWCNT. The polycarboxylate ether of MPEG type (PCE) with specific gravity 1.100 g/cm^3, pH up to 5 at 20 °C and dry content equal to 50.5% was applied in the research as dispersant to distribute the nanotubes in the volume of suspension. The compositions of generated MWCNT suspensions are presented in Table 1.

2.2. Methods

The MWCNT suspension was prepared in the following way. The masterbatch pellets "Graphistrength CW 2-45" were immersed in hot water with temperature about 75 °C and stirred by standard mixer agitation about 1000 rpm during 5 min.

After that the MWCNT suspension was subjected to ultrasonication for a duration of 6 min. The ultrasonication was performed by Bandelin Sonopuls HD 3400 ultrasonic homogenizer (400 W, 20 kHz) with probe VS 200 T (Ø 25mm, amplitude – 82μm).

The prepared MWCNT suspension was characterized by pH, electrical conductivity tests, measurements of surface tension, particle size analysis and measurements of zeta potential.

pH test was performed by pH-meter Mettler-Toledo EL20 with limits of error ± 0.01 pH. The electrode with gel electrolyte was used.

Electrical conductivity was measured by Mettler-Toledo EL30. The measurement range of it is 0.1 to 199.1 mS/cm with limits of error ± 0.5 %. The electrode for medium to high conductivities (10 μS/cm ...500 mS/cm) was applied.

Surface tension of MWCNT suspension was measured based on stalagmometric method by stalagmometer Traube with measurement accuracy 1 mN/m. The stalagmometric method consists in determination of drops number which are formed in the course the flowing of liquid from capillary tube of known volume. The surface tension of studied liquid is calculated based on Equation (1):

$$\sigma_x = \sigma \cdot \frac{N}{N_x} \cdot \frac{d_x}{d} \quad (1)$$

where: σ- surface tension, mN/m; N - number of drops; d - density, g/cm^{3}of water; σ_x, N_x, d_x- the same for studied liquid.

Particle size analysis and zeta potential measurements of MWCNT suspension were performed by particle size and zeta potential analyser Delsa Nano C, Beckman Coulter. Particle size analysis was based on the principle of photon correlation spectroscopy with resolution of measurements placed from 0.6 nm to 7 μm and measurement accuracy of 0.05 nm.

Table 2. Content of materials for 1 m^3 in tested cement pastes.

Specimen	Cement, kg	Water, kg	W/C ratio	MWCNT, % bwoc [a]	PCE, % bwoc [a]
REF	935	280	0.30	0	0
PCE	935	234	0.25	0	1.0
PCE+0.00375	935	234	0.25	0.00375	1.0
PCE+0.0075	935	234	0.25	0.0075	1.0
PCE+0.015	935	234	0.25	0.015	1.0
PCE+0.030	935	234	0.25	0.030	1.0
PCE+0.060	935	234	0.25	0.060	1.0

[a] bwoc -by weight of cement

Zeta potential measurements was based on the electrophoretic light scattering principle with range of measurement from -150 mV to +150 mV and measurement accuracy of 0.005 mV.

The cement pastes with composition presented in Table 2 were prepared for rheological test. The MWCNT suspensions with composition presented in Table 1 were used as mixing liquid and mixed together with cement for 180 s at high speed by standard mixer according to EN 196-1. Rheological properties of cement pastes were tested in different time after pastes mixing: 5, 60 and 120 min. The rheology tests were carried out at temperature 20±2 °C. The rotational rheometer Rheotest RN 4.1 with coaxial cylinders was used for rheological tests. Shear rate ranged from 100.0 s-1 to 0.1 s^{-1} during testing time. Cement pastes after mixing between tests were placed in the plastic containers and carried in the laboratory conditions with temperature 20±2 °C and relative humidity not less than 65 %. The cement pastes were premixed before rheological test by hands in plastic container using metallic scoop for 1 min. The yield stress (τ_0) and plastic viscosity (K) were determined in the course of approximation of flow curve (dependence between shear stress (τ) and shear rate (γ) in the range from 0.1 to 100 s^{-1} based on Herschel-Bulkley model. The Herschel-Bulkley model is more widely used model for description of flow behaviour of cement systems with shear thinning or shear thickening effect, which is expressed by the Equation (2):

$$\tau = \tau_0 + K \cdot \gamma^n \tag{2}$$

where: τ - shear stress, Pa; τ_0- yield stress of the cement paste, Pa; K- plastic viscosity, Pa·s; γ - shear rate, s^{-1}; n - shear thinning or shear thickening index if $n < 1$ or $n > 1$, respectively.

Amount of bleeding water was determined according the following method. 50 g of MWCNT suspension was pour into the 100 ml container, after that 50 g of cement was added for 1 min. Subsequently, MWCNT suspension and cement were mixed continuously for 4 min in the container and were poured into a graduated cylinder of 30 ml. The graduated cylinder with cement paste was placed onto the table and the initial volume of cement paste was written. The graduated cylinder was standing without moving and shaking during the test. The volume of sedimented cement in cement paste after water bleeding was determined after 2 hours since the reference point and every 30 min after it. The observation of water bleeding was performed till the moment when the volume of cement paste did not changed. Water separation (water bleeding) was calculated according the Equation (3):

$$K = \frac{a-b}{a} \cdot 100 \tag{3}$$

where: K –coefficient of water bleeding, %; a - initial volume of cement paste, cm^3; b - volume of sedimented cement, cm^3.

3. Results and Discussion

3.1. Analysis of MWCNT suspension homogenity

The results of particle size analysis of MWCNT suspensions with PCE plasticizer subjected to ultrasonicaion are presented in Fig. 1.

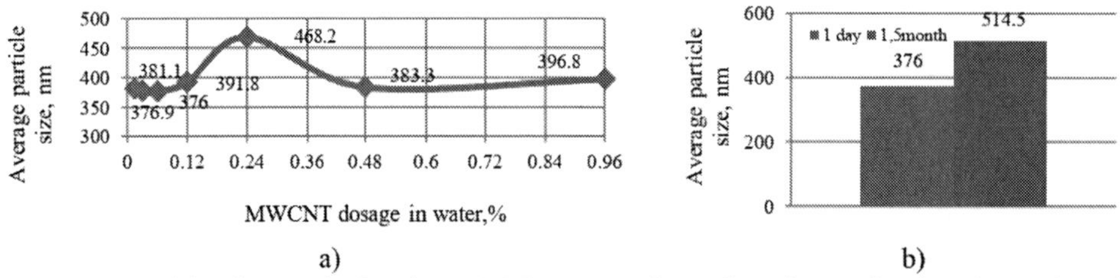

a) b)

Figure 1. Particle size analysis of MWCNT suspension after ultrasonication (a) and particle size analysis of MWCNT suspension with 0.06% of MWCNT in water after 1.5 month of storage (b).

IOP Conf. Series: Materials Science and Engineering **660** (2019) 012032 doi:10.1088/1757-899X/660/1/012032

The graph in Fig.1a presents the particle size analysis of MWCNT suspension after ultrasonication. The average diameters of particles are placed on the same level. The minimum average diameter of particles was achieved of 376 nm for suspension with MWCNT dosage equal to 0.06% in water. The suspension with 0.06% of MWCNT was tested after 1.5 months of storage. It is revealed the increase of average particle size up to 514.5 nm, caused by coagulation processes in MWCNT suspension (Fig.1b).

The MWCNT suspension was characterized by pH, electrical conductivity, zeta potential and surface tension measurements as well.

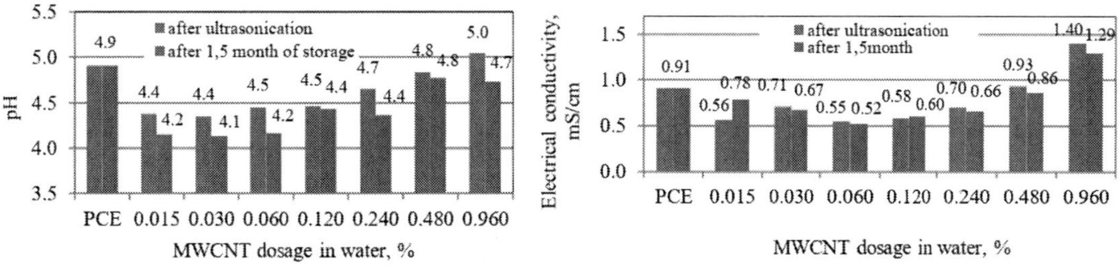

a) b)

Figure 2. pH of MWCNT suspension (a); electrical conductivity of MWCNT suspension (b).

Fig.2a demonstrates the changes in pH level of suspensions with different MWCNT dosage after ultrasonication and 1.5 months of storage. Increase of MWCNT dosage changes the pH level of the suspensions not significant. pH of MWCNT suspensions slightly decreased in the process of storage. This process can be explained by coagulation processes which take part in MWCNT suspensions during the time.

Test of electrical conductivity revealed that inrease of MWCNT dosage to 0.96% increase electrical conductivity from 0.91 to 1.40 mS/cm as it is seen in Fig. 2b. Changes in electrical conductivity during the 1.5 months of storage did not observed. The dimensions of electrical conductivity for suspensions with MWCNT dosage from 0.015 to 0.24% demostrated smaller electrical conductivity in comparison with sample without MWCNT, which can be caused by the electrostatic interactions between MWCNT and plasticizers.

Fig.3a shows that the surface tension of MWCNT suspensions increased with increase of MWCNT dosage in water. Increase of MWCNT dosage up to 0.96% led to the increase of surface tension from 62 to 65 mN/m.

Zeta potential of MWCNT suspension increased with the increase of MWCNT dosage to 0.96% as it seen in Fig.3b. Fig.3c shows that zeta potetial did not changes significantly during 1.5 months of storage.

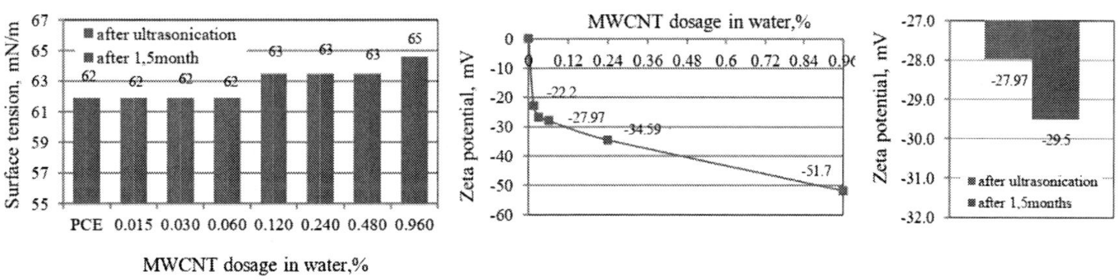

a) b) c)

Figure 3. Surface tension of MWCNT suspension (a); zeta pontetial of MWCNT suspension after ultrasonication (b); zeta potential of MWCNT suspension with MWCNT dosage of 0.06% in water after 1,5 months of storage (c)

The results of MWCNT suspension analysis confirmed that usage of MWCNT suspension is more suitable after ultrasonication process when degree of coagulation is not high.

3.2. Rheological properties and bleeding of cement pastes with MWCNT suspension

The Fig.4, 5 present the changes in rheological behaviour of nanomodified cement pastes with different dosage of MWCNT in 5 and 120 min after cement pastes mixing, respectively.

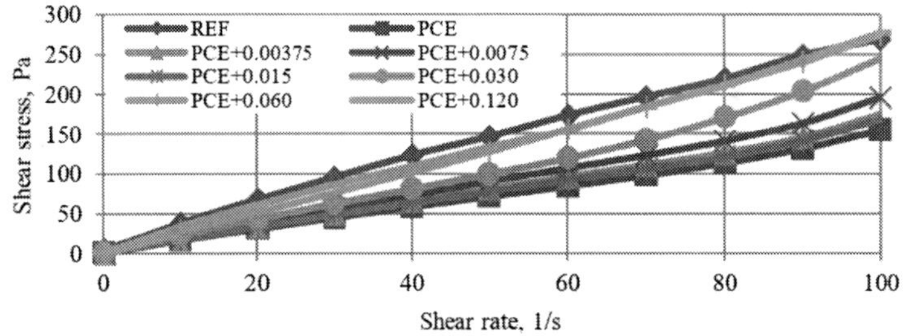

Figure 4. Flow curves of nanomodified cement pastes in 5 min after cement paste mixing.

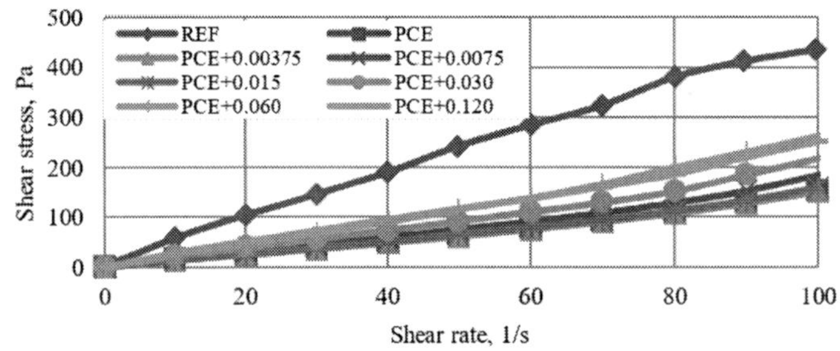

Figure 5. Flow curves of nanomodified cement pastes in 120 min after cement paste mixing.

Rheological behaviour of cement pastes modified by PCE plasticizer changes from the shear thinning to shear thickening. Addition of MWCNT at the dosage 0.060 and 0.012% by weight of cement (bwoc) make the behaviour of cement paste closer to shear thinning. However, this effect did not observed in 120 min after cement paste mixing. Modification by MWCNT suspension more than 0.030% bwoc increases the plastic viscosity of cement pastes in 5 and 120 min after cement pastes mixing. The experimental flow curves were approximated by Herchel-Bulkley model. The obtained equations and thixotropy indexes are available in Table 3.

Table 3. Equations based on Herschel-Bulkley model and thixotropy indexes of cement pastes.

Specimen	Flow curves equations		Thixotropy Indexes	
	After 5 min	After 30 min	After 5 min	After 30 min
REF	τ=4.97+3.470$\gamma^{0.940}$	τ=4.2+3.750$\gamma^{1.050}$	0.940	1.050
PCE	τ=0.37+0.610$\gamma^{1.197}$	τ=0.26+0.600$\gamma^{1.190}$	1.197	1.190
PCE+0.00375	τ=0.30+0.663$\gamma^{1.200}$	τ=0.32+0.600$\gamma^{1.190}$	1.200	1.190
PCE+0.0075	τ=0.32+0.750$\gamma^{1.200}$	τ=0.38+0.698$\gamma^{1.193}$	1.200	1.193
PCE+0.015	τ=0.29+0.650$\gamma^{1.230}$	τ=0.212+0.610$\gamma^{1.190}$	1.230	1.190
PCE+0.030	τ=0.34+0.900$\gamma^{1.200}$	τ=0.45+0.780$\gamma^{1.190}$	1.200	1.200
PCE+0.060	τ=0.43+1.163$\gamma^{1.100}$	τ=0.55+0.895$\gamma^{1.230}$	1.100	1.230
PCE+0.120	τ=1.37+1.178$\gamma^{1.090}$	τ=0.92+0.930$\gamma^{1.210}$	1.090	1.210

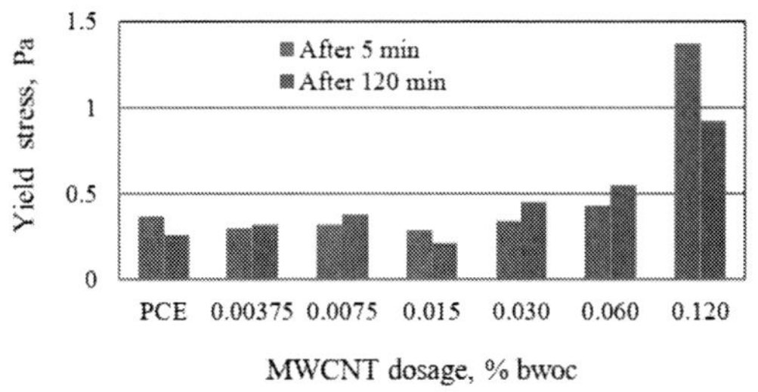

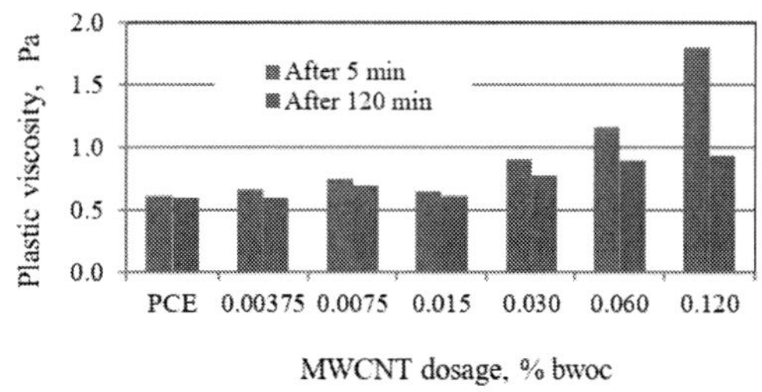

a) b)

Figure 6. Yield stress (a) and plastic viscosity (b) of cement pastes with differenet MWCNT dosage.

The changes of yield stress and plastic viscosity for cement pastes with different dosage of MWCNT in 5 and 120 min after cement pastes mixing are presented in Fig. 6a and Fig. 6b, respectively. Addition of MWCNT at the dosage more than 0.03% bwoc increases the yield stress. The maximum increase was observed for cement paste modified by 0.120% bwoc. The yield stress was increased by 3.7 and 3.5 times in 5 and 120 min after cement paste mixing, respectively.

Fig. 6b demonstrates the changes of plastic viscosity for cement pastes modified by different MWCNT dosage. The MWCNT dosage more than 0.030% bwoc increases the plastic viscosity. The maximum increase of plastic viscosity was established for cement paste modified by 0.120% bwoc. The plastic viscosity increased by 2.95 and 1.55 times in 5 and 120 min after cement pastes mixing, respectively.

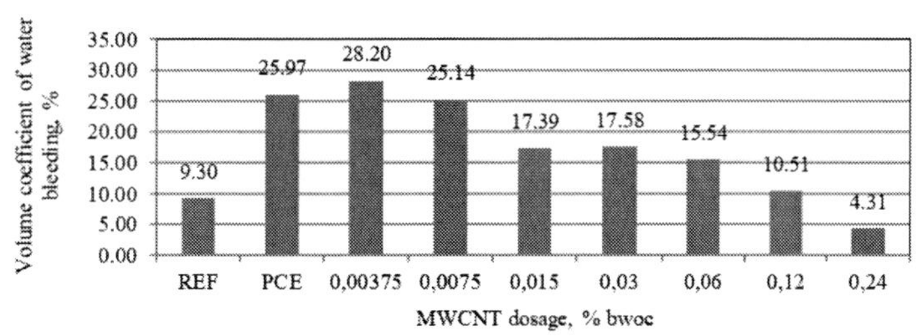

Figure 7. Water bleeding of nanomodified cement pastes.

The water bleeding test of nanomodified cement pastes revealed the decrease of bleeding water with modification of cement pastes by MWCNT as it is seen in Fig.7. The maximum decrease by 21.7% was obtained for cement paste modified by MWCNT of 0.24% bwoc in comparison with sample modified only by PCE plasticizer.

4. Conclusions

The following conclusions are drawn in this study:

1. The MWCNT suspension with average diameter of particles of 376 nm was established. The results of MWCNT suspension analysis confirmed that usage of MWCNT suspension is more suitable after ultrasonication process when degree of coagulation is not high.

2. The flow behavior of cement pastes modified by MWCNT suspension with PCE plasticizer changed from shear thinning to shear thickening in comparison with reference cement paste without admixtures.

3. The increase of yield stress in 3.7 and 3.5 times was obtained for cement paste modified by MWCNT suspension in dosage of 0.12% bwoc in 5 and 120 min after cement paste mixing, respectively. The increase of plastic viscosity in 2.95 and 1.55 times was obtained for cement paste modified by MWCNT suspension in dosage of 0.12% bwoc in 5 and 120 min after cement paste mixing, respectively. The increase of yield stress and plastic viscosity during time for cement pastes

modified by MWCNT suspensions with different ultrasonication treatment may be caused by Van-der-Waals, electrostatic and steric forces, which appear between cement particles in the course of modification by plasticizer and MWCNT.

5. Modification of cement pastes by MWCNT suspension in the dosage of 0.24 % bwoc led to the decrease of bleeding water volume by 21.7% in comparison with cement pastes modified only by PCE plasticizer.

6. Effect of MWCNT suspensions on yield stress, plastic viscosity and bleeding may give the opportunity to use them as effective viscosity modifying agent for high workability and self-compacting concrete.

References

[1] Ha Thanh Le, Muller M, Siewert K, Ludwig Horst-Michael: The mix design for self-compacting high performance concrete containing various mineral admixtures. *Materials and Design* 2015, 72: 51-62.

[2] Gelardi G, Mantellato S, Marchon D, Palacios M, Eberhardt AB, Flatt RJ: 9 - Chemistry of chemical admixtures, in: Aitcin PC, Flatt RJ (Eds.) *Sci. Technol. Concr. Admix*, Woodhead Publishing 2016: 149-218.

[3] Liu X, Guan J, Lai G, Wang Z, Zhu J, Cui S, Lan M, Li H: Performances and working mechanism of a novel polycarboxylate superplasticizer synthesized through changing molecular topological structure. *Journal of Colloid and Interface Science* 2017 (504): 12-24.

[4] Winnefeld F, Becker S, Pakusch J, Gotz T: Effects of the molecular architecture of comb-shaped superplasticizers on their performance in cementitious systems. *Cement and Concrete Composites* 2007 (29): 251-262.

[5] Qian Y, Lesage K, Cheikh KE, Schutter GD: Effect of polycarboxylate ether superplasticizer (PCE) on dynamic yield stress, thixotropy and flocculation state of fresh cement pastes in consideration of the Critical Micelle Concentration (CMC). *Cement and Concrete Research* 2018 (107): 75-84.

[6] Pei R, Liu J, Wang S: Use of bacterial cell walls as a viscosity-modifying admixture of concrete. *Cement and Concrete Composites* 2015 (55): 186-195.

[7] Khayat KH: Viscosity-Enhancing Admixtures for Cement-Based Materials - An Overview. *Cement and Concrete Composites* 1998 (20): 171-188.

[8] Meng W, Khayat KH: Effect of graphite nanoplatelets and carbon nanofibers on rheology, hydration, shrinkage, mechanical properties, and microstructure of UHPC. *Cement and Concrete Research* 2018 (105): 64-71.

[9] Garcia-Taengua E, Sonebi M, Hossain KMA, Lachemi M, Khatib J: Effects of the addition of nanosilica on the rheology, hydration and development of the compressive strength of cement mortars. *Composites Part B* 2015 (81): 120-129.

[10] Nadiv R, Vasilyev G, Stein M, Peled A, Zussman E, Regev O: The multiple roles of a dispersant in nanocomposite systems. *Composites Science and Technology* 2016 (133): 192-199.

[11] Mendoza Reales OA, Arias Jaramillo YP, Ochoa Botero JC, Delgado CA, Quintero JH, Toledo Filho RD: Influence of MWCNT/surfactant dispersions on the rheology of Portland cement pastes. *Cement and Concrete Research* 2018 (107): 101-109.

Waste glass as a replacement for fine fraction in high-performance concrete

D Mariaková[1,2], Z Jirkalová[1,2] and L Laiblová[1,2]

[1]University Centre for Energy Efficient Buildings of CTU in Prague, Buštěhrad, Czech Republic
[2]Czech Technical University, Prague, Czech Republic

E-mail: diana.mariakova@fsv.cvut.cz

Abstract. Research project is focused on the experimental verification of high-performance concrete (HPC) with full replacement of fine fraction. This paper deals with HPC mixtures containing recycled glass materials (waste glass powder from municipal waste, waste glass powder from photovoltaic panels, grinding waste glass from jewellery). The research is based on the results of previous student grant. A recipe that was made during 2017 is used and optimized according to the waste glass powder properties. The influence of different local sources is monitored. The HPC mixtures are tested for mechanical properties (flexural strength, compressive strength) and compared with the results from the previous research. Flowability is tested and all the mixtures are workable. The mixture with grinding glass (GG) showed the highest results in flexural and compressive strength compared to the reference sample, the use as a substitute for fine fraction in HPC is conceivably attitude. The municipal waste glass mixture and the mixture with the waste glass from photovoltaic panels is not usable as HPC.

1. Introduction

Last few years, things like decreasing amount of non-renewable materials (and sources) belong to the main issues to deal with. Meanwhile, the high energy consumption to obtain these materials has negative impact on the environment – it happens during the mining, processing and subsequent using the materials [1]. It is probably time to think about what the real issue is – the decreasing number of non-renewable materials or the threat it poses to the environment?

Despite the factors mentioned above, utilization of non-renewable materials continues. With the use of non-renewable materials, a large amount of waste is produced – in fact, it concerns more or less all new materials. Almost 40% of waste comes from civil engineering. According to the latest available data (2014) it was almost 18 million tons [2]. The thing is, that this waste, once produced, needs to be stored somewhere, or better – recycled.

After the water, concrete is the second most used material in the world. About 7-8 % of total CO_2 production is from cement fabrication. Therefore, in recent years, is an effort to reduce this material or to lower environmental impacts by using new types of optimized or recycled concrete [3].

Recycling in its original meaning is returning to the process in which the waste originated (original purpose or same system). In a broader sense, recycling can be considered as any reuse of material – it doesn't have to be necessary the utilization in the same system as it was before. However, any return of the waste to the process is considered as success – at least for our environment.

In recent years is increasing tendency to use special kinds of concrete with greater strength, resistance or durability. Such concretes are usually classified in group of High Performance Concretes (HPC) or

Ultra High Performance Concrete (UHPC). HPC or UHPC have very small fraction of used aggregates compared with ordinary concrete and they also have very small water/cement ratio due to the use of superplasticizers. Unfortunately, these concretes have almost always a large amount of cement. One of the sustainable approaches is to replace some components with more environmentally friendly materials. .

One of the appropriate materials is glass, which is one of the most recyclable materials (almost 100%) [4]. On one hand, glass is chemically resistant to external influences, have excellent heat-insulation properties. On the other hand, it is non-decomposable material, therefore it is necessary to landfill waste glass – but due to the transporting and landfilling waste glass, impurities may occur. Utilization of waste glass in construction industry is convenient, because the demands of glass purity are decreasing.

There are several studies that deal with replacing some of the concrete components by glass. Two main ways to use waste glass in concrete are possible – aggregate replacement or cement replacement. There are also researches that combine these attitudes.

Hongjian Du and col. investigated the influence of replacing cement by glass powder in high volume up to 60% to mechanical and durability properties of concrete [5]. Topcu and col. investigated the tensile bending strength of concrete with waste glass as coarse aggregate [6]. Waste glass powder was used as a pozzolanic material in Shayan and Xu research – they replace up to 30% of cement in concrete mix [7]. Another research groups were investigated possibility of replacing aggregate with waste glass in concrete [8–10]. Based on them, it could be told, it is more suitable to use waste glass as a fine aggregate in concrete. Schwarz and col. compared the properties of concrete containing waste glass powder with concrete containing fly ash [11]. There are several studies to investigate the amount of waste glass powder in concrete [10,12,13]. There are different results, replacement range varied from 10% to 40% - ideal amount should be around 10% - 20% [10]. Soliman and Tagnit-Hamou developed ultra-high-performance glass concrete with a compressive strength up to 220 MPa. However, this is the only one study dealing with ultra-high-performance concrete (UHPC) [14].

Waste glass utilization is appropriate in HPC (or UHPC) because less water is used there than in regular concrete. With growing amount of water, the risk of alkali-silica reaction (ASR) increases. There may be a lot of situation which may cause degradation of concrete, when considering replacing silica powder by waste glass powder, one of the most important causes is ASR.

It is important to notice that on basis of alkali-silica reaction are significant only aggregates with amorphous silica [15]. Beside amount of waste glass powder in concrete, there is question about size of particles. There were few researches made on this topic [7,16,17]. Research by Z. Bažant found that particles smaller than 0,25mm caused no expansion. On the other hand, particle size around 1,5mm caused excessive expansion [16]. Based on this researches, it is possible to say, that the waste glass powder is suitable for utilization in high performance concrete as partial replacement as fine aggregate in concrete. The potential risk is lower compared to coarse aggregate - it is known that particles smaller than 1mm are reactive too, but the pressure is so small that there are no cracks at the surface. More detailed research or experiments are not part of this work.

According to Carsana and col. research that investigated comparison between waste glass and other supplementary cementitious materials, mortars with waste glass showed long term durability – the samples did not show any degradation after seven years' immersion in water and compressive strength was higher [18].

There are many types of waste glass, which could be possibly useful for this issue. This paper investigates utilization of grinding glass, waste glass from municipal waste and from photovoltaic panels as a replacement for fine grains. The research is partially based on the results from previous work and should verify the results and the effect of changing local resources – especially cement [19].

2. Materials

First experiments were made based on the original basic recipe optimized during our research program in 2017. The reference mixture of HPC (REF_1), the mixture with grinding glass (GG_1) and the mixture with waste glass powder from municipal waste (MWG_1) were made and tested to verify the strength, freeze-thaw resistance and other properties.[19] HPC was made following the basic recipe (**Table 1.**).

IOP Conf. Series: Materials Science and Engineering **660** (2019) 012033 doi:10.1088/1757-899X/660/1/012033

This approach was chosen to verify the effect of changing local raw materials and sources and the age of waste glass powder on the mechanical properties of high performance concrete. In 2018, new mixtures followed the basic recipe were made (**Table 1.**). Cement I 42.5R, technical silica sand, silica fume, superplasticizers and water were used. Silica powder was added in the reference sample (REF_2). After that, the basic recipe was changed – silica powder was replaced with grinding glass (GG_2), then with the milled waste glass from municipal waste (MWG_2). The weight of fine fraction stayed the same (325kg/m^3).
Same thing was done with powder from photovoltaic panels (PP_{2A}). However, these samples cracked and it was not possible to test them.

Table 1. Basic recipe used in 2017, repeated in 2018 – Reference sample($REF_{1,2}$), grinding glass($GG_{1,2}$) and milled waste glass($MWG_{1,2}$).

Component (kg/m^3)	$REF_{1,2}$	$GG_{1,2}$	$MWG_{1,2}$	PP_{2A}
Cement I 42.5R	680	680	680	680
Technical silica sand	960	960	960	960
Silica powder (ground quartz)	**325**	0	0	0
Grinding glass	0	**325**	0	0
Milled waste glass	0	0	**325**	0
Photovoltaic panels powder	0	0	0	**325**
Silica fume (microsilica)	175	175	175	175
Superplasticizers	29	29	29	29
Water	171	171	171	171

It follows that two HPC samples remained (GG_2, MWG_2) to examine and compare with reference sample (REF_2). The differences were evident in the structure of concrete.
The next step was to optimize the recipe containing photovoltaic panels powder. The possibility of an unwanted reaction that led to the cracks in the concrete structure was eliminated by excluding silica fume (microsilica) from the mixture (**Table 2.**). Water-cement ratio was increased to 0,27. These steps proved mixture usable, cracks were successfully eliminated.

Table 2. Optimized recipe 2018 – Photovoltaic panels (PP_{2B}).

Component	PP_{2B}
Cement I 42.5R	650
Technical silica sand	1200
Photovoltaic panels powder (ground quartz)	**240**
Silica fume (microsilica)	0
Superplasticizers	30
Water	180

IOP Conf. Series: Materials Science and Engineering **660** (2019) 012033 doi:10.1088/1757-899X/660/1/012033

3. Experiments

For now, samples made according to the basic recipe (**Table 1.**) and optimized recipe (**Table 2.**) have been tested – REF_2, GG_2, MWG_2, PP_{2B}. Prepared samples were examined to determine:

- Flow table test
- Tensile bending strength
- Compressive strength

3.1. Flow TableTest

The mortar flow is a relative measure of workability similar to slump. This method is suitable for cement mortars and fine-aggregate concrete. Mortar flow test is described in the Czech standard CSN EN 1015-3 (2000). Mortar flow is most sensitive to water and air content. This test determines how much a mortar sample flows when it is unconfined and consolidated. Mortar is placed inside conical form. When the form is removed, the mortar is vibrated as the flow table rises and drops 15 times. The mortar changes from a conical shape. Mortar flow is reported as the final diameter of the mortar on the flow table.

3.2. Tensile Bending Strength

Tensile bending strength is described in Czech standard CSN EN 12390-5. The test is realized on block specimens with dimension 40 x 40 x 160mm. Flexural strength is defined by three-point press according to CSN EN 1015-11. The test is over after the sample breaks. The highest force registered during the test is used for calculation tensile bending strength. It is always necessary to test at least three samples.

3.3. Compressive Strength

Compressive strength is verified on fragments of samples after flexural strength test. The tested surface A_c is 40 x 40 mm. There are certain conditions to keep – about samples (EN 12350-1, EN 12390-1, EN 12390-2 or EN 12504-1), machine (EN 12390-4) and procedure. The test is done when the sample breaks. The maximum measured force is recorded. The whole method is described in CSN EN 12390-3. All results are evaluated by methods mathematical statistics.

Tensile bending strength and compressive strength tests were made after 28 days.

4. Results and Discussion

The samples were tested to determine the properties of concrete. The following chapter describes the results and discussions of all tests.

4.1. Flow Table Test

The effects of waste glass powder on the workability of fresh concrete were tested. Until now, the dependence between the spill rate and the strength has not been found [20]. The workability of fresh concrete with the waste glass powder replacement was different, but not inferior (**Figure 1., Figure 2.**). As a result of the flow table test on the workability of the fresh concrete, the slump value was determined to be 23.5 – 28.2 cm (**Table 3.**)

Figure 1. Flow table test REF_2 (left), GG_2 (right).

Figure 2. Flow table test MWG_2 (left), PP_{2B} (right).

Table 3. Results of flow table test.

Type of sample	Spill (cm)
REF_2	25
GG_2	23.5
WGP_2	26.75
PP_{2B}	28.2

4.2. Tensile Bending Strength

In previous research was proven, that the longer hardening time has positive impact to the tensile bending strength [19]. Mixture with grinding glass (GG_1) and mixture with milled waste glass (MWG_1) were compared with the reference sample contains silica powder (REF_1) in 2017. It was the first basic research on this topic and the results were usable. However, it was necessary to repeat the experiments to provide the workability of the mixture and to investigate the influence of the use of different local sources in the mixture. Influence of the higher age of waste glass powder was monitored as well.

Mixtures and samples (based on the original recipe from 2017) were made in 2018 – reference sample (REF_2), mixture with grinding glass (GG_2), mixture with milled waste glass (MWG_2) and new mixture with powder from photovoltaic panels (PP_{2B}).

Figure 3. compares the results from 2017 (REF_1, GG_1, MWG_1) with the one from 2018 (REF_2, GG_2, MWG_2). Values from the mixture PP_{2B} are new, so they have no comparison. In 2018, the values of tensile bending strength mixture with grinding glass (GG_2) were 15% lower compared with the reference sample (REF_2). The similar results were provided in 2017 (20%). Milled waste glass sample (MWG_2) values are even lower than in 2017. Additional optimization of current MWG_2 recipe will be required before next experiments, otherwise this mixture could not be used as HPC. It seems that higher age of milled waste glass and freshness of cement or its slight differences can affect partially the resulting mechanical properties of HPC.

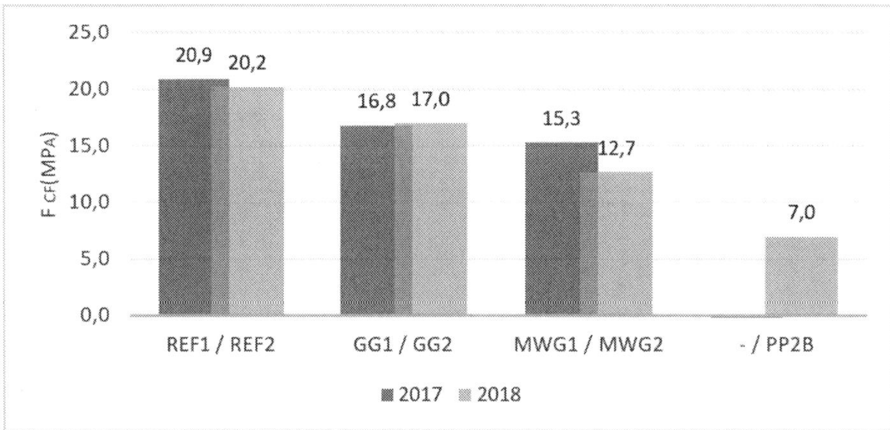

Figure 3. Tensile bending strength results comparison – years 2017, 2018 and values of new recipe from 2018.

4.3. Compressive strength

The compressive strength test was performed on sample fragments. The differences in the structure are visible – samples with waste glass powder have more pores in the structure (**Figure 4.** and **Figure 5.**).

Different particle size may also have impact to the results. Another factor could be chemical composition. Due to these facts, the compressive strength is lower. However, mixture with grinding glass provide usable results. Mixture with waste glass from photovoltaic panels has the lowest results (**Figure 6.**). The use of different age of local sources have influence on the results – especially freshness of cement. The decrease of the compressive strength in mixture with waste glass powder from municipal waste can be caused by the chemical reaction created by cement with waste glass powder.

Figure 4. Fragments of samples after flexural strength test REF_2 (left), GG_2 (right).

Figure 5. Fragments of samples after flexural strength test MWG_2 (left), PP_{2B} (right).

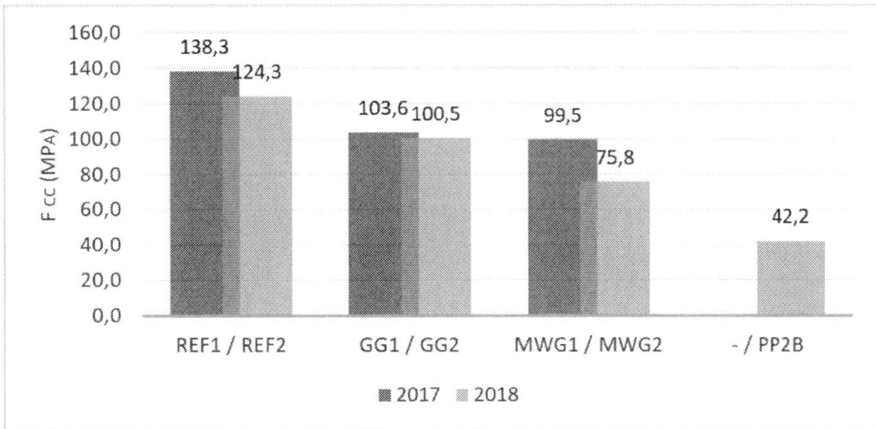

Figure 6. Compressive strength results comparison – years 2017, 2018 and values of new recipe from 2018.

5. Conclusion

Waste glass powder, obtained from different sources, was used as a full replacement of silica powder in high performance concrete. The samples were examined in several tests. This research builds on previous study from 2017. This paper was made to compared the results from 2018 with the results from 2017, monitored the influence of the use of different local materials and freshness (especially cement) and the influence of age of the waste glass.

Glass powder from grinding jewellery, milling municipal waste and photovoltaic panels was used. These sources are only a part of the waste that can be reused as a replacement in concrete.

During the first round of experiments, basic recipe optimized during 2017 was used. The utilization of waste glass powder from photovoltaic panels has proven to be inappropriate in this case. The cracks appeared and the samples had to be excluded from further testing. The recipe was optimized during this research and the samples with waste powder from photovoltaic panels were tested as well.

The tests to find out tensile bending strength and compressive strength were made on samples with grinding glass, milled waste glass and glass from photovoltaic panels. The results of new mixtures containing the waste glass were compared with the reference sample. A comparison with the results of previous research (2017) was made as well.

Especially the mixture with waste glass powder from grinding glass achieved usable numbers. The flow table test ends up with spill range 23.5 cm. However, the workability was the worst of the tested samples. The tensile bending strength was 84,2% compared with the reference sample (considered as 100%). The compressive strength was 80,8% compared with the reference sample (considered as 100%)

The mixture containing the glass powder from milling municipal waste had convenient workability. However, the results of tensile bending strength test and compressive strength test were slightly worse, compared with reference sample and mixture with grinding glass. The recipe should be optimized before further testing, because the mixture does not have any high performance parameters yet.

Photovoltaic panels powder used in a HPC mixture that was made according to the recipe from 2017 was unsuitable, the cracks appeared. The recipe was optimized; flow table test provides workability of the mixture. The tensile bending strength test and the compressive strength test was made as well – the results were low compared to the reference sample and compared to other mixtures with waste glass. We suppose that even after further optimizing the recipe, this mixture will not be usable in term of high-performance concrete. It is probably due to a different chemical composition of the glass from photovoltaic panels.

The risk of alkali-silica reaction must be dealt with for these concrete mixtures. Tests will be conducted in the following research. A long-term experiment should verify the low ASR risk. The samples will be visually checked and tested every year. Up to this point, the potential risk of ASR was based on other researches work.

Based on research, the use of waste glass powder in high performance concrete is a suitable way to reduce the need for primary raw materials used to produce concrete. Waste glass properties are conveniently advantageous for use in concrete (pozzolanic properties, similar chemical composition etc.). Due to the increasing tendency to use special concrete with higher strength or durability, it is necessary to find sustainable approach to replace non-ecological components like cement. In view of the deteriorating condition of our planet, this approach is necessary and will continue to increase in the next years. The resulting strength values demonstrate the potential to use grinding glass powder as a replacement for the fine fraction in HPC.

Acknowledgements

This work has been supported by the Ministry of Education, Youth and Sports within National Sustainability Programme I, project No. LO1605 – University Centre for Energy Efficient Buildings – Sustainability Phase and MPO TRIO FV 10397 – RENCO Recycled Environmental Concrete for Building Construction. This work was also supported by the Grant Agency of the Czech Technical University in Prague, grant No. SGS19/093/OHK1/2T/1 - Use of waste materials in HPC in design applications and grant No. SGS18/108/OHK1/2T/11 – Environmental aspects of high performance cement composites and concrete with recycled aggregate including their durability and service life.

References

[1] Mlčochová V 2006 New findings about recycled concrete *Recycling 2006 - Možnosti a perspektivy recyklace stavebních odpadů jako zdroje plnohodnotných surovin* (Brno)

[2] GREEN Solution s.r.o. 2017 Program předcházení vzniku odpadu

[3] Hájek P, Fiala C and Kynčlová M 2011 Life cycle assessments of concrete structures - a step towards environmental savings *Structural Concrete* **12** 13–22

[4] Vacková J 2013 *Problematika třídění odpadů u různých věkových kategorií obyvatel*

[5] Du H and Tan K H 2017 Properties of high volume glass powder concrete *Cement and Concrete Composites* **75** 22–9

[6] Topçu İ B and Canbaz M 2004 Properties of concrete containing waste glass *Cement and Concrete Research* **34** 267–74

[7] Shayan A and Xu A 2004 Value-added utilisation of waste glass in concrete *Cement and Concrete Research* **34** 81–9

[8] Mohajerani A, Vajna J, Cheung T H H, Kurmus H, Arulrajah A and Horpibulsuk S 2017 Practical recycling applications of crushed waste glass in construction materials: A review *Construction and Building Materials* **156** 443–67

[9] Ismail Z Z and AL-Hashmi E A 2009 Recycling of waste glass as a partial replacement for fine aggregate in concrete *Waste Management* **29** 655–9

[10] Khatib J M, Negim E M, Sohl H S and Chileshe N Glass Powder Utilization in Concrete Production *European Journal of Applied Science 4*
[11] Schwarz N, Cam H and Neithalath N 2008 Influence of a fine glass powder on the durability characteristics of concrete and its comparison to fly ash *Cement and Concrete Composites* **30** 486–96
[12] Shayan A and Xu A 2006 Performance of glass powder as a pozzolanic material in concrete: A field trial on concrete slabs *Cement and Concrete Research* **36** 457–68
[13] Pattengil M and Shutt T Use of ground glass as pozzolan
[14] Soliman N and Tagnit-Hamou A 2016 Development of ultra-high-performance concrete using glass powder – Towards ecofriendly concrete *Construction and Building Materials* **125** 600–12
[15] Breitenbücher R Alkalicko-křemičitá reakce – důsledky pro cementobetonové kryt *Strasse + Autobahn* 2006 pp 205–9
[16] Bažant Z *Fracture mechanics of concrete structures*
[17] Jin W, Meyer C and Baxter S 2000 "Glascrete" - Concrete with Glass Aggregate *ACI materials journal* 208–13
[18] Carsana M, Frassoni M and Bertolini L 2014 Comparison of ground waste glass with other supplementary cementitious materials *Cement and Concrete Composites* **45** 39–45
[19] Mariaková D, Vlach T and Pavlů T 2019 Glass waste powder utilization in high performance concrete *Acta Polytechnica CTU Proceedings* **21** 24–7
[20] Domone P 1998 The Slump Flow Test for High-Workability Concrete 11Communicated by C.D. Pomeroy. *Cement and Concrete Research* **28** 177–82

IOP Conf. Series: Materials Science and Engineering **660** (2019) 012034 doi:10.1088/1757-899X/660/1/012034

Preliminary research for long lasting self-healing effect of bacteria-based concrete with lightweight aggregates

L Dembovska[1], D Bajare[1], A Korjakins[1], D Toma[1] and E Jakubovica[1]

[1]Riga Technical University, Department of Building Materials and Products, Institute of Materials and Structures, Kalku str. 1, LV-1658, Riga, Latvia

E-mail: laura.dembovska@rtu.lv

Abstract. A typical phenomenon in concrete structures is crack formation. Shrinkage cracks may appear in early stage of concrete. In late age cracks may appear due to process of serviceability and external weather conditions. Larger cracks interfere structural integrity, micro-cracks may result in durability problems as increase of water absorption, which results in concrete degradation and reinforcement corrosion. Manual maintenance of the concrete structure is costly and labour consuming; therefore, functionality of self-healing phenomenon of concrete structures has been inspected.
In this study self-healing phenomenon of bacteria-based concrete has been investigated. Calcium lactate and yeast medium with spore-forming alkaliphilic bacteria Bacillus Lentus have been chosen. Bacillus Lentus spores with medium were integrated (by using vacuum technique) into expanded clay granules (LECA) with a size of 2-4 mm. Direct addition of bacteria (without immobilization) in concrete limits the survival of bacteria from 1 to 2 months. In this study, to increase viability of bacteria at high pH media in concrete and thus the time-related functionality of self-healing agents, bacteria were immobilized in LECA.
The mechanism of crack healing in bacteria-based concrete presumably occurs through metabolic conversion of calcium lactate to calcium carbonate. The influence of bacteria on the mechanical and physical properties of samples was estimated after 28 days and after 6 years. From the results obtained, it can be concluded that concrete samples with Bacillus Lentus showed an increase of 22% in compressive strength compared to the reference sample.
The main task of this scientific research is to identify, if self-healing phenomenon happens for the samples, which are 28 days old and for samples which are 6 years old.
Results from the optical investigation showed that self-healing phenomenon (after 14 days of healing) has occurred for all samples and cracks in concrete were partly closed with calcium carbonate.

1. Introduction

Concrete is the most essential and popular material for building structures, due to high serviceability, compressive strength, and availability of raw materials.[1] However, the mechanical strength and durability of concrete structures can be seriously deteriorated by micro or macro-cracks.[2] Typically, penetrations of water and chloride through the cracks drastically affect the durability of concrete structures and cause the corrosion of steel bars, which consequently lead to failure issues.[3] Therefore, it is vital to heal or repair these cracks to maintain the serviceability of concrete structures.[4] Cracks can be repaired manually in rare situations, because most of these repairing operations are restricted by accessibility, location, cost, and environment problems.[5] Hence, self-healing of cracks has become a necessity for concrete structures, and attracts rising attention of researchers. On the other hand, the autogenous repairing phenomenon of concrete structures has been observed in natural environment for many years. It is accomplished by formation of calcium carbonate and continuous hydration of anhydrated cementitious materials, when water and carbon dioxide are available.[6] However, without further modification or improvement, the efficiency of those healing activities is relatively low. Therefore, various strategies have been developed to improve self-healing behavior,[7] typically though inserting kinds of additions into the cementitious materials, such as crystalline admixture[8], polymers[9], hollow fiber[10], mineral admixtures[11], encapsulations[12], nanoparticles[13], and microorganisms[14].

In this study, in order to prolong the service life of concrete structures the self-healing concrete with alkaliphilic bacteria from the genus of Bacillus and organic compounds was developed. Production of the self-healing concrete mix does not differ from the ordinary cement mix production of the lightweight concrete contained cement, sand, lightweight aggregates and water. Before mixing lightweight aggregates were impregnated with the bacteria Bacillus Lentus and the organic mineral compound (calcium lactate and yeast) in order to encapsulate self-healing agent and prevent it from the degradation at high pH of concrete. Bacteria spores, calcium lactate and yeast extract were dissolved in water and solution was used for impregnation of bacteria into the pores of the expanded clay granules (LECA) by using vacuum treatment. After impregnation and drying the bacteria spores localize inside of the lightweight aggregates, but they are not active. They become active when cracking of concrete appears and presence of water and oxygen is possible. After activation the bacteria start to degrade the calcium lactate and produce calcium carbonate that is sealing the open cracks.

The main goal of this research is to estimate the Bacillus Lentus activity after pre-cracking of 28 days and 6 years old lightweight concrete samples. The crack self-sealing phenomenon was estimated by optical microscope. Physical and mechanical properties of bacteria-based concrete were determined.

2. Materials and methods

2.1. Preparation of the healing agent

The bacteria-based healing agent consisted of spores derived from alkaliphilic bacteria of the genus Bacillus and organic mineral compounds. Calcium lactate and yeast medium with spore-forming alkaliphilic bacteria Bacillus Lentus have been chosen to produce bacteria-based concrete. Calcium lactate is a salt that consists of two lactate anions for each calcium cation (Ca2+). It is prepared commercially by the neutralization of lactic acid with calcium carbonate or calcium hydroxide. Calcium lactate is also found in daily dietary supplements as a source of calcium. It is also available in various hydrate forms, where calcium lactate pentahydrate is the most common.

The healing agent was incorporated in lightweight expanded clay aggregate (LECA) via solution absorption during 24 hours, 72 hours and an impregnation under vacuum with calcium lactate, yeast extract and bacteria spores solution.

Two solutions were prepared for impregnation into LECA. Firstly, water with calcium lactate (1.5% added of total liquid medium) and yeast (0.03% of total liquid medium). Calcium lactate and yeast extract were heated at 40 °C for about 1 h to dissolve in water. Then the same steps were taken as for the LECA impregnation with water and bacteria spores.

Maximal amount of bacteria and calcium lactate and yeast extract medium which is possible to impregnate in the LECA was calculated.

2.1.1. Absorption test for 24 and 72 hours. For 24 hour and 72 hour absorption test 500g of LECA was taken and immersed in 3 solutions. To avoid aggregates flowing to the top, LECA was put inside the fabric cloth and pushed to the bottom of the container with 0.5mm sieve. After 24 and 72 hours, LECA was taken out of the solutions, dried with wet cloth and mass was measured. Then LECA was placed in an oven at 20°C until it reaches a constant weight.

After 24 hour absorption test it was concluded, that solution uptake for LECA is in a range of 18-22%. Calcium lactate amount inside the LECA aggregates ranges from 0.27-0.33%, yeast extract ranges from 0.0054-0.0066%, bacteria spores $2x10^8$.

After 72 hour absorption test it was concluded, that solution uptake for LECA is in a range of 24-25%. Calcium lactate amount inside the LECA aggregates ranges from 0.36-0.375%, yeast extract ranges from 0.0072-0.0075%, bacteria spores $2x10^8$.

2.1.2. Impregnation under vacuum. Initially LECA was dried at 20°C for 24 h to remove moisture content. The weighted amount (500 g) of the LECA was added in the vacuum desiccator at −0.95 bar by employing a suction pump. The LECA was spread in between two metallic sieves with a mesh size 0.125 mm to enhance surface contact between the aggregate and the water. Liquid medium was injected from the top through a funnel to flow under gravity past the LECA through sieves. The medium was kept injected inside the desiccator until the porous aggregate was fully submerged. The top sieve is installed to prevent LECA floating over the water surface when a vacuum is applied. The bottom sieve is installed to prevent settling of the heavier LECA mix when the vacuum is removed. The vacuum was applied for 10 min. The medium is assumed to keep impregnation into LECA as long as air bubbles keep appearing under vacuum.

After impregnation test it was concluded, that solution uptake for LECA is in a range of 55-57%. Calcium lactate amount inside the LECA aggregates ranges from 0.825-0.855%, yeast extract ranges from 0.0165-0.0171%, bacteria spores $2x10^8$.

Following the impregnation, the LECA were dried for approximately 5-6 days at standard temperature (20±2°C), until a constant weight was achieved. During drying procedure, the spores can remain in the dormant state, since the pH of the environment is not adequately high for them to activate and start germinating. It was found, that after drying, the impregnated LECA did not increase their initial weight (only 0.05%) due to the addition healing agent.

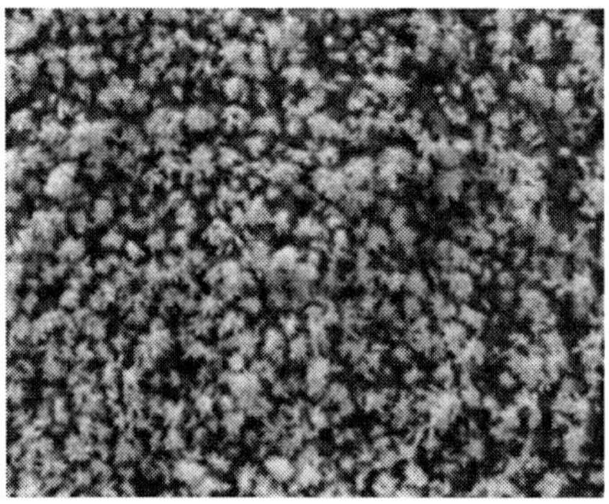

Figure 1. Calcium lactate is crystallized after five days of drying at 37 °C.

As from the Figure 1 it is seen, after drying of LECA calcium lactate crystallized on the surface of LECA. In order to ensure, that the crystallized medium do not have negative aspect on the hardened concrete and their mechanical properties, it was concluded to prepare cubes 10x10x10 cm and test concrete compressive strength.

2.1.3. Samples for mechanical properties. The aim of this study was to determine how the self-healing medium, such as calcium lactate, yeast extract and bacterial spores, affect the compressive strength of concrete matrix. Sample compositions are given in Table 1.

Main ingredients for production of the bacteria-based concrete were: Portland Cement CEM I 42.5N, locally available sand (combination of 2 fractions and quartz powder), lightweight expanded clay aggregates and water. In this study lightweight expanded clay aggregates (LECA) from Ltd "Fibo" were used with size from 2 to 4 mm, density - 250 kg/m^3, porosity – 82%.

Table 1. Sample compositions.

Components	Weight, g	REF	SH1	SH2	SH3
Lightweight expanded clay aggregates	292. 0	x			
Lightweight expanded clay aggregates (bacteria)	292.0		x		
Lightweight expanded clay aggregates (calcium lactate, yeast extract)	292.0			x	
Lightweight expanded clay aggregates (medium and bacteria)	292.0				x
0.5-1 mm sand			397.0		
0.25-0.5 mm sand			346.0		
0.125-0.25 sand			186.0		
CEM I 42.5 N			384.0		
Water			192.0		
W/C			0.5		

Compressive strength of bacteria-based lightweight concrete was determined according to LVS EN 1015-11 using specimens with dimensions 10x10x10 cm. The density of samples was measured in accordance with EN 1097-7. An optical microscope "AmScope" was used to evaluate self-healing of the concrete cracks. Total porosity for LECA was obtained from specific gravity obtained by Le Chatelier flask (ASTM C188).

2.1.4. Samples for self-healing test. LECA was impregnated with self-healing substances which consisted of bacterial medium - calcium lactate, yeast extract and bacterial spores. Impregnation process consisted is described before in 2.1.3.

For self-healing testing ability, SH3 samples with dimensions of 4x4x16 cm were prepared (Table 1). After filling the concrete mass into moulds, the samples were vibrated for 3 min on the vibration table (40 Hz) to ensure that the concrete mass was evenly compacted. The samples were hardened in the moulds for 3 days, covered with polyethylene film. Afterwards the specimens were cured in water for 2 days, then in final stage - the curing continued in a chamber with air temperature of + 20 ± 5 °C and relative humidity of air ≥95%.

2.2. *Cracking of mortar prisms*

After removing the samples from the curing chamber, the wet areas were wiped with a dry cloth. The dimensions of the samples were determined using the gauge.

Three-point-bending (with a span of 100 mm) was used for the crack induction on 28-days-old samples and 6-years-old samples. A single crack was created in each specimen. A vertical load was applied in the middle of the specimens until the formation a stable crack. While loading, the crack opening increased constantly by 0.0005 mm/s until it reached approximately 0.4 mm. The crack width was monitored via two Linear Variable Differential Transducers attached on the front and the back of the side of the specimens. When the crack width reached 0.4 mm the specimens were slowly unloaded. After unloading, the crack width reduced to approximately 0.27-0.36 mm. The depth of the crack was not monitored during bending, but it was observed through microscopic images.

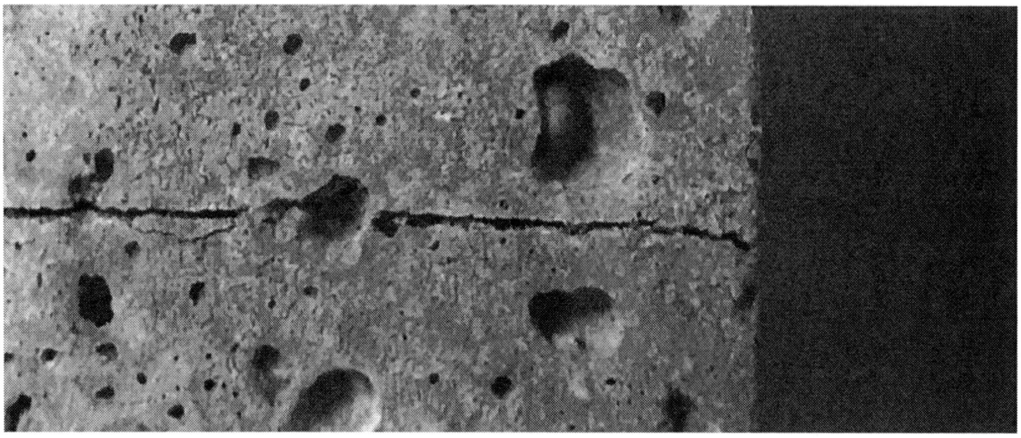

Figure 2. Pre-cracking for 6-years old sample.

2.3. *Healing treatment*
Following the crack creation specimens were placed horizontally in a plastic container filled with tap water for crack healing. Samples were completely immersed in water while placed on the spacers. The container was kept open to the atmosphere at standard room temperature (20±2 °C) with (60±10) % RH. Samples at the age of 28 days and 6-years-old samples were left to heal for 14 days. Extra water was added, to keep a constant liquid-to-solid ratio.

2.4. *Crack inspection*
Crack inspection was used to primary evaluate the sealing efficiency of the bio-based healing agent. The inspection was conducted in two steps, i.e. right after crack creation and after healing treatment.

3. Results
3.1. *Mechanical and physical properties of the concrete matrix*
The density of the samples ranged from 1736 to 1777 kg/m^3. Scientific literature describes that calcium lactate, a bacterial nutrient, has a positive effect on the compressive strength of concrete, improving it by 10%.[15] In addition to calcium lactate, a yeast extract that functions as a bacterial nutritional supplement was added. According to literature yeast extract may significantly reduce the compressive strength of concrete.[16]

Table 2. Physical and mechanical properties of the 28-days old 10x10x10 cm samples.

No	Composition	W/C ratio	Age of concrete	Density, kg/m^3	Compressive strength, MPa	Concrete class
1.	REF	0.5	28 days	1736	21	C20/25
2.	SH1	0.5	28 days	1767	22	C20/25
3.	SH2	0.5	28 days	1777	27	C20/25

After 28 days of hardening, the concrete samples were tested. The compressive strength of the REF sample was 21 MPa, the sample with medium impregnated in LECA (SH2) - 27 MPa and the sample with Bacillus Lentus bacteria spores impregnated in LECA (SH1) - 22 MPa.

From the results obtained, it can be concluded that both of the concrete samples with self-healing reagents showed better compressive strength results compared to the reference sample. The sample with bacterial nutrients impregnated in LECA showed 22% increase in compressive strength compared to the reference sample.

It is concluded that adding calcium lactate (1.5% of the total liquid medium) and a small amount of yeast extract (0.03% of total liquid medium), compressive strength increases and the medium gives a positive effect on bacterial growth and development.

3.2. Results from the self-healing test

The specimens were loaded until the crack opening reached 0.4 mm. However after unloading the crack width was < 0.3 mm. Specimens (28-days old and 6-years old) were submerged in water for 14 days for self-healing to occur.

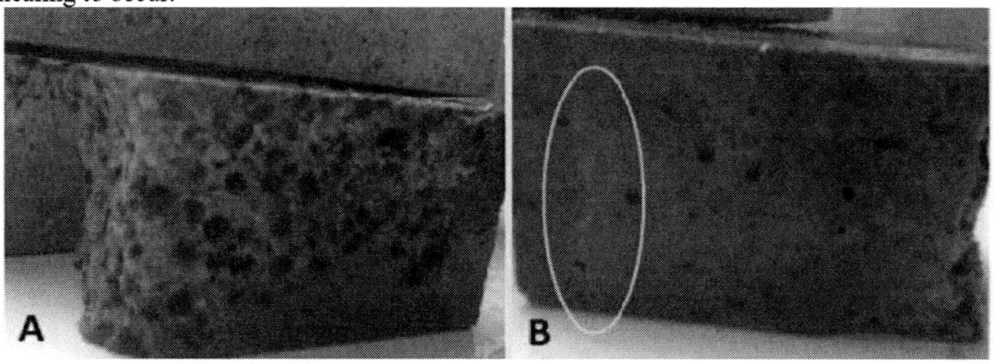

Figure 3. 28-days old sample after cracking (A), self-healing of the cracks for 28-days old sample after 2 weeks in water (B).

Figure 4. Calcium carbonate percipitation in the cracks of 28-days old SH3 sample.

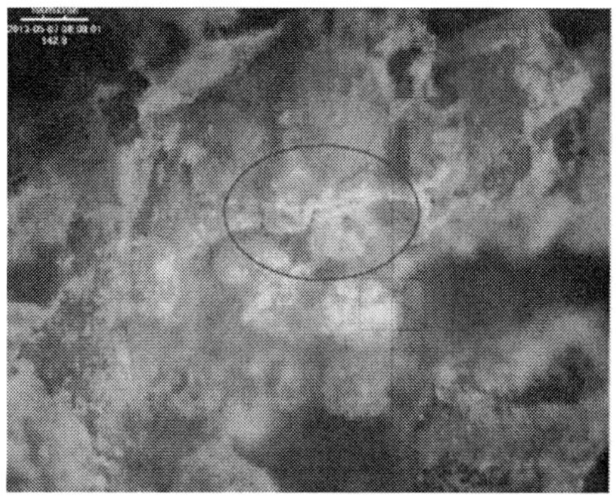

Figure 5. Calcium carbonate percipitation in the cracks of 28-days old SH3 sample.

In Figure 4 and 5 cracks in the concrete matrix at 40x magnification using an optical microscope are shown. White crystalline material precipitates on both sides of the crack. The cracks are not completely healed, but precipitation is observed throughout the concrete matrix which had direct water contact. Crack size varies up to 0.3 mm.

The self-healing results from the 28-day-old samples were observed after 2 weeks (Figure 4 and Figure 5). It is concluded, that self-healing processes are happening and this medium and Bacillus Lentus are successful in ensuring self-healing ability of concrete. In the scientific literature it is mentioned, that the crack healing process with bacteria can last from 2 weeks to several years.[14]

In the beginning of the experiment, it was considered, that all the healing agent inside the LECA will be released when LECA will be intersected by crack and will convert to $CaCO_3$. In reality, only part of the healing agent is released, while the rest remains in the lightweight aggregate (it was observed in the samples after 14 days of healing).

Autogenous healing processes, that occur during the healing treatment in water for 28-days old samples were not considered.

6-years old samples were pre-cracked and tested on self-healing ability. All the conditions for the test experiment were ensured the same as for the 28-days old samples in year 2013. Samples were loaded with 50kN large force, crack openings were up to 0.3 mm.

In Figure 6 the self-healing process has happened in the range of 60% to 90%, depending on the size of cracks in the concrete matrix. Precipitation of calcium carbonate was determined.

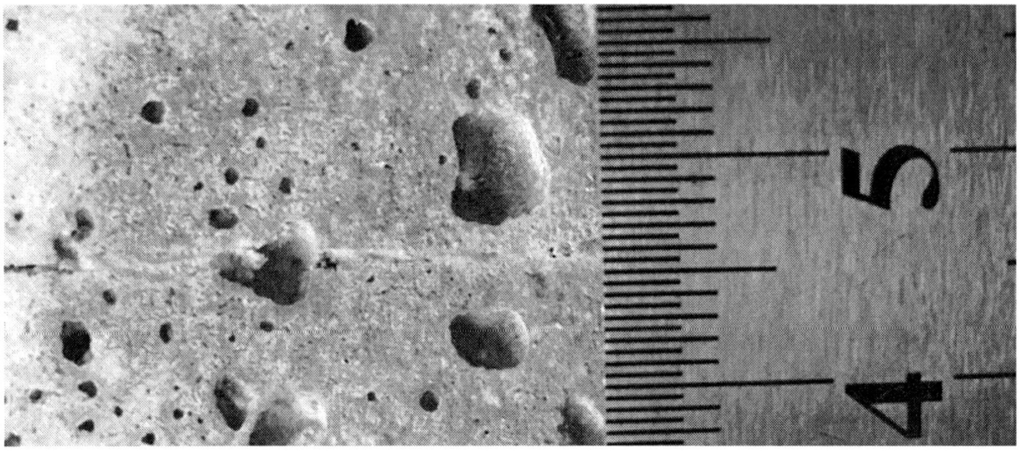

Figure 6. Calcium carbonate precipitation after 2 weeks for 6-years old SH3 sample.

It is possible to exclude the fact that these calcium carbonate precipitations for 6-years old samples are from hydration processes (Figure 6), because concrete samples were made and maintained under appropriate conditions to allow complete hydration of the concrete, and the water cement ratio (0.5) of the samples was not high.

It can be stated, that crack closure percentage increases for the specimens immersed in water for longer time period (specimens were monitored each day during time period of 14 days). During the monitoring process of the cracks it was concluded, that the narrower the crack the more efficient is the sealing.

4. Conclusions

It was concluded, that reactions, that take place inside the cracks are complex and depend on several factors, such as local crack width, the presence of oxygen, the duration of healing treatment etc. Experimental approaches might overestimate the crack closure due to the hypothesis that the crack filling is constant along the crack length, but it is not.

It is concluded that addition of calcium lactate (15% of total liquid medium) and a small amount of yeast extract (0.03% of total liquid medium) increase the compressive strength and the medium gives a positive effect on bacterial growth and development. Addition of calcium lactate (15% of total liquid medium) to a concrete mix improved compressive strength by 22%. The yeast extract in low dose (0.03% of total liquid medium) had no negative effect on the compressive strength of the concrete. Addition of bacterial spores to the concrete did not have a negative effect on compressive strength, compared to the reference samples, the compressive strength increased by 2.31%.

After 2 weeks in the water for 28-days samples cracks were partially healed, in the range of 10% to 90%, depending on the size of the crack in the concrete matrix.

Bacteria immobilization in LECA ensures bacteria viability after many years. After 2 weeks in water for 6-years-old samples crack healing was observed, in the range of 60% to 90%, depending on the size of crack. We can conclude, that Bacillus Lentus spores have survived 6 years old dry period and calcium carbonate precipitation in the cracks is obvious. Self-healing of 6-years old samples have happened.

IOP Conf. Series: Materials Science and Engineering **660** (2019) 012034 doi:10.1088/1757-899X/660/1/012034

References

[1] Mehta P K 1986 Concrete: Structure, properties and materials *New Jersey, USA: Prentice-Hall*

[2] Muhammad NZ, Shafaghat A, Keyvanfar A, Majid MZA 2016 Tests andmethods of evaluating the self-healing efficiency of concrete: A review. *Constr Build Mater.* 112:1123–1132

[3] Basheer L, Kropp J, Cleland DJ 2001 Assessment of the durability of concrete fromits permeation properties: A review. *Constr Build Mater.* 15:93–103

[4] Yıldırım G, Keskin ÖK, Keskin SB, Şahmaran M, Lachemi M. 2015 A reviewof intrinsic self-healing capability of engineered cementitious composites:Recovery of transport and mechanical properties. *Constr Build Mater.* 101:10–21

[5] Vijay K, Murmu M, Deo SV 2017 Bacteria based self-healing concrete–Areview. *Constr Build Mater.*152:1008–1014

[6] Jacobsen S, Sellevold EJ 1996 Self healing of high strength concrete after deteri-oration by freeze/thaw. *Cem Concr Res.* 26:55–62

[7] De Belie N, Gruyaert E, Al-Tabbaa A, et al. 2018 A review of self-healing con-crete for damage management of structures. *Advanced Materials Interfaces* 1800074:1–28

[8] Ferrara L, Krelani V, Moretti F 2016 On the use of crystalline admixtures incement based construction materials: From porosity reducers to promotersof self healing. *Smart Mater Struct.* 25:084002

[9] White SR, Sottos NR, Geubelle PH, et al. 2001 Autonomic healing of polymercomposites *Nature* 409:794–797

[10] Dry C 1994 Matrix cracking repair and filling using active and passive modesfor smart timed release of chemicals from fibers into cement matrices *Smart Mater Struct.* 3:118–123

[11] Tittelboom KV, Gruyaert E, Rahier H, De Belie N 2012 Influence of mix com-position on the extent of autogenous crack healing by continued hydrationor calcium carbonate formation *Constr Build Mater.* 37:349–359

[12] Zemskov SV, Jonkers HM, Vermolen FJ. 2011 Two analytical models for theprobability characteristics of a crack hitting encapsulated particles: Appli-cation to self-healing materials *Comput Mater Sci.* 50:3323–3333

[13] Perez G, Erkizia E, Gaitero JJ, Kaltzakorta I, Jiménez I 2015 Synthesis and char-acterization of epoxy encapsulating silica microcapsules and amine functio-nalized silica nanoparticles for development of an innovative self-healingconcrete. *Mater Chem Phys.* 165:39–48

[14] Khaliq W, Ehsan MB. 2016 Crack healing in concrete using various bio influ-enced self-healing techniques. *Constr Build Mater.*102:349–357

[15] Tittelboom KV, Wang J, Araújo M, et al. 2016 Comparison of different approaches for self-healing concrete in a large-scale lab test. *Constr BuildMater.* 107:125–137

[16] Thao TDP, Johnson TJS, Tong QS, Dai PS 2009 Implementation of self-healingin concrete–Proof of concept. *IES J Part A Civ Struct Eng.* 2:116–125

Lightweight concrete with expanded aggregate from siliceous rocks

L V Zakrevskaya[1], I A Gandelsman[1], A A Gavrilenko[1] and P A Lubin[1]

[1]Department of Construction Industry, Vladimir State University, Vladimir, 600000, Russia

E-mail: lvzak@mail.ru

Abstract. The purpose of this study is to develop technology for the production of a new building composite made of silicon-containing rocks. An aggregate is foamed granules obtained on the basis of diatomite with the addition of caustic soda, dolomite, water and liquid glass. The material composition, materials processing, physical and mechanical properties have been observed. The problem of alkali-silicate interaction typical for traditional Portland cement and glass phase is solved through the use of magnesia binder. Samples on Portland cement and magnesia binder were obtained, the aggregate content was varied. Some aggregate fractions were used to obtain high packing density in the composite. Concrete was estimated according to density, compressive strength and tensile strength, water absorption and thermal conductivity. Analysis of the material structure was made. The results of the study of the microstructure of synthesized concrete indicate the presence of a cement matrix in which granules of various sizes are tightly packed. The results of X-ray phase analysis of synthesized concrete indicates that the cement matrix consists of forsterite crystals. Approximate savings in the use of concrete with the developed aggregate is about 10-20%. Such savings are possible due to the fact the surface occurrence of siliceous rocks allows them to be mined in an open way with minimal costs compared to other rocks, with a lower calcination temperature during material creation.

1. Introduction

The increased need for lightweight concrete sets the task of expanding its range on the basis of natural resources with improved thermal and structural properties.

The demand for high-quality building materials is growing every year. The most important performance indicators of building materials are strength, durability, environmental friendliness, reliability and cost.

In accordance with the Federal Law # 261-FZ "On Energy Saving and Increasing Energy Efficiency and Amendments to Certain Legislative Acts of the Russian Federation" special attention should be paid to thermal insulation of buildings.

In recent years monolithic construction in which lightweight concrete structures perform heat-insulating and enclosing functions is gaining popularity.

In modern literature this topic is represented by some studies that confirm the effectiveness of the use of diatomite rocks to create lightweight aggregates for concrete. Despite all the existing studies the potential of this rock is not fully disclosed and needs further study [1,2].

The main objectives of the study are: study of physical and mechanical properties of granular thermal insulation material, obtained on the basis of silica raw materials, and development of concrete compositions with high performance indicators.

IOP Conf. Series: Materials Science and Engineering **660** (2019) 012035 doi:10.1088/1757-899X/660/1/012035

Many technical and operational characteristics of concrete depend on the type and properties of the aggregate. Expanded clay is "the main field of activity" which worked out new approaches to the design of lightweight concrete mixtures on porous aggregates over the past decades [3].

Some scientific papers on this topic confirm the fact that the strength of concrete directly correlates with the strength of porous aggregate granules. However studies of the properties of the "aggregate-cement matrix" zone show that the nature of the bond between the aggregate and the cement matrix strongly affects both the strength and other properties of lightweight concrete [4,5,6].

Siliceous opal-cristobalite rocks, the reserves of which Russia ranks first in the world, belong to multi-purpose raw materials. The use of these rocks in construction as well as their origin and distribution has been studied in many papers.

Siliceous rocks are known to be light fine-pored powders composed of the smallest opal fragments of diatoms and cristobalite, as well as clay minerals [5,7].

The main and most common diatomite rocks are: diatomite and tripoli (Figure 1, 2) [7].

Figure 1. Diatomite.

Figure 2. Tripoli.

Diatomites and tripoli (celite, rock flour, diatomaceous earth, infusorial earth) are light fine-porous rocks. Diatomites are composed of the remains of diatomite algae (60-80%) and have a typically organogenic structure. They are loosely cemented, characterized by grayish-yellow color, very high porosity (20-84%) and low density (0.35-1.6 g/cm3). Sometimes they have a microlayered structure. The age of diatomites is the Paleogene-Neogene and Quaternary. In Russia large deposits of diatomites are found in the Volga region, the Urals and Transcaucasia. There is the possibility of using diatomite rocks as reactive amorphous silica in the Vladimir region.

Tripolis are similar to diatomites externally and in composition. In the wet state tripolis are easily rubbed with fingers. Opal (75-80%) (Figure 3), clayey matter, quartz grains, glauconite, carbonates and occasionally residues of diatoms are detected under a microscope. The content of fractions is smaller than 0.01 mm - 36.2 - 46.2%. The specific gravity is equal to 2.4 kPa; the density varies from 0.45 to 1.4 g/cm3; the porosity varies from 41 to 81%. High porosity (especially closed porosity) leads to low thermal conductivity. Diatomite rocks are used as hydraulic additives for cements, raw materials for the production of bricks and light materials for backfilling [8, 9, 10].

Table 1 shows the most important properties of diatomite rocks.

Table 1. The physical properties of diatomite rocks.

Density ρ_0, kg / m^3	True density ρ_{true}, kg / m^3	Porosity,%	Compressive strength, MPa	Mohs Hardness Scale	Thermal conductivity, W/(m •°C)
350-1600	2000-2500	20-84	2-15	1-4	0,09-0,5

Some authors [9, 10, 11] state that the chemical composition of diatomites and tripolis is almost the same, but the microstructure is different. This is due to the difference in the geological age of the deposits. The chemical composition of iatomites is shown in Table 2.

Table 2. The chemical composition of diatomites and tripolis in the Vladimir region.

	SiO_2, mass%	Al_2O_3, mass%	Fe_2O_3, mass%	CaO, mass%	MgO, mass%	K_2O, mass%
Diatomites	70 – 96	5 – 15	2 – 5	0,5 – 5	0,5 – 3	0,1 – 1,45
Tripoli	66,2 – 84,7	2,7 – 18,6	0,2 – 5,6	0,24 – 22,1	0,1 – 1,8	0,4 – 1,5

The phase composition of siliceous rocks is represented by three minerals: opal, cristobalite and quartz. The predominant mineral is opal, the content of which is usually 56-98%. In all siliceous rocks there is quartz, the content of which can vary from 5 to 35%. Cristobalite is almost absent in diatomites. In other varieties of siliceous rocks its content can reach 20%. Opal consists of extremely small particles or porous aggregates with the developed inner surface [12-15]. Scanning electron microscopy of the opal sample is shown in Figure 3 [16].

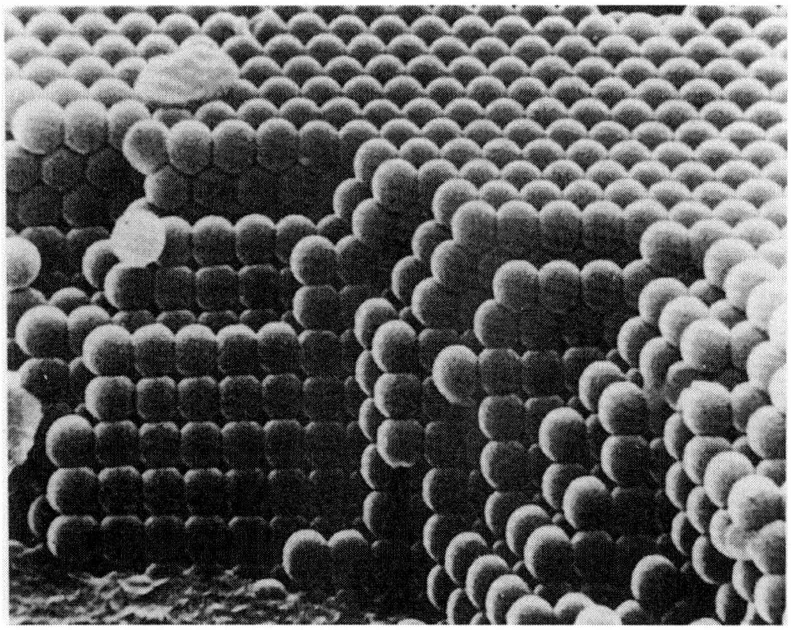

Figure 3. Scanning electron microscopy of the opal sample.

The aim of this study is to develop technologies for the production of a new building composite made of silicon-containing rocks. Low thermal conductivity and density with sufficient strength are the main advantages of the obtained aggregate. The problem of alkali-silicate interaction typical for traditional Portland cement and glass phase is solved through the use of magnesia binder.

2. Experemental

The following raw materials were used for the study:

- diatomite
- causticsoda
- water glass
- dolomite

• water

The obtained material received the working title "Diapen".

"Diapen" is a building material of a new generation, synthesized from silica rocks with a porous, chemically stable structure and a raw material component for lightweight concrete.

Table 3 shows the composition of the raw material mixtures to obtain "Diapen".

Table 3. Composition of lightweight siliceous aggregates.

Compositio n	Diatomite, mass%	Causticsoda , mass%	Waterglass, mass%	Water, mass%	Dolomite, mass%
C1	57	20	12	5	6
C2	60	20	12	4	4
C3	65	17	9	3	6
C4	70	14	6	6	4
C5	72	13	5	6	4

Preparation of raw materials and production of granules consisted of the following process steps:

• grinding diatomite
• dosage of main components
• adding caustic soda, water glass and water
• mixing ingredients
• production of granules by extruder
• swelling in the rotary kiln at temperature of 880 degrees

The appearance of the obtained aggregate granules is shown in Figure 4.

Figure 4. "Diapen", obtained from siliceous rocks.

For the further research were chosen the composition C2 because of it has the less density nd thermal conductivity in compare with the other ones.

Figure 5 shows the microstructure of "Diapen" granules, obtained by scanning electron microscopy. This microstructure allows to conclude that "Diapen" is a material with closed pores on the outer surface and developed porosity inside the granule.

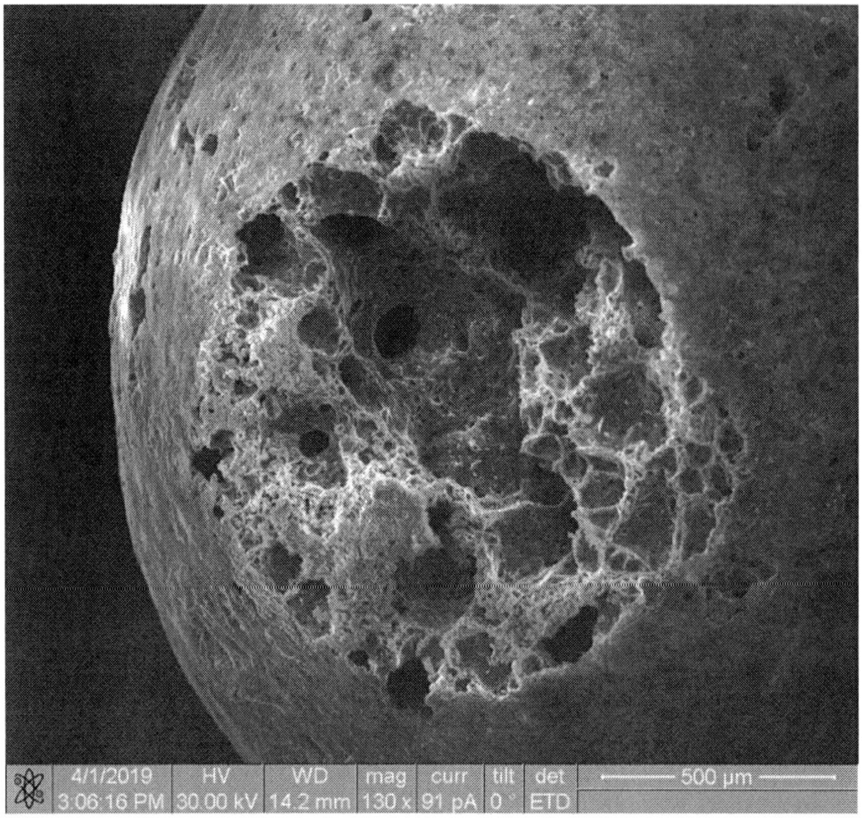

Figure 5. The surface and internal structure of "Diapen" granules.

Mechanical and operational characteristics of "Diapen" were studied to assess the quality of the obtained material. The study of density, compressive strength in the cylinder and water absorption was carried out in accordance with GOST 9758-2012. Table 4 shows the measurement results of physical and mechanical properties of the porous aggregate.

Table 4. Properties of the porous aggregate.

	Fraction of the aggregate, mm	
	1 – 5	5 – 20
Bulk density, kg/m3	295,10	191,90
Compressive strength, MPa	3,77	2,66
Water absorption, mass%	4,2	6,3
Water absorption, volume%	6,2	6,5
Thermal conductivity W/(m*C)	0,085	0,070

It is recommended to use magnesia binders because the obtained aggregate has a high content of silicates in its composition, representing the danger of alkali-silicate interaction in concrete with this aggregate[17-20]. To test the theoretical assumptions experimental samples of lightweight concrete were made on traditional Portland cement and magnesite cement. Table 5 shows the composition of lightweight concrete with different binders.

Table 5. Composition of lightweight concrete.

	Magnesia cement, mass%	Aggregate, mass%	Sand, mass%	Grouting fluid, mass%
LB-C/M-1	13,43	23,58	52,25	10,74
LB-C/M-2	16,44	24,36	48,35	10,85
LB-C/M-3	19,49	25,12	44,50	10,89

	Portland cement, mass%	Aggregate, mass%	Sand, mass%	Water, mass%
LB-C/P-1	11,82	23,58	38,48	26,12
LB-C/P-2	13,50	24,36	40,00	22,14
LB-C/P-3	14,20	25,12	42,50	18,18

Concrete samples were taken from specially prepared laboratory mixes of concrete mix and tested. Samples were solidified for 28 dys at the air temperature of 20^{0}C and relative humidity of 95% in accordance with GOST 18105-2010. Samples were tested and stored after hardening at the air temperature of 20^{0}C and relative humidity of 55%.

The sand used in to produce concrete was middle-sized with a density of 1670 kg/m^3. The binder was portland cement B42,5.

The density of concrete was determined in the air-dry state and according to GOST 12730.1. Compressive strength, flexural strength and tensile strength of samples were determined according to GOST 10180-2012. Determination of water absorption was carried out in accordance with GOST 12730.3-78. Determination of thermal conductivity was carried out in accordance with GOST 30256-94. Samples were tested using the MIT-1 device under normal conditions.

3. Results and discussion

Comparative characteristics of concrete on Portland cement and magnesia cement are shown in the Table 6.

Table 6. Results of property testing.

	LB-C/M-1	LB-C/M-2	LB-C/M-3	LB-C/P-1	LB-C/P-2	LB-C/P-3
Density, kg/m3	692	690	690	664,60	660,5	656,0
Compressive strength, MPa	8,76	9,00	9,00	5,15	5,20	5,15
Water absorption, volume%	10,8	10,0	10,5	11,3	8,7	8,9
Thermal conductivity W/(m*C)	0,315	0,320	0,328	0,291	0,25	0,24

The sample LB-C/M-2 was selected for the study of the structure and phase analysis. The results of the study of the microstructure of synthesized concrete shown in Figure 6, 7 indicate the presence of a cement matrix in which granules of various sizes are tightly packed. X-ray phase analysis of the sample is shown in Figure 8.

Figure 6. Cement matrix with forsterite crystals.

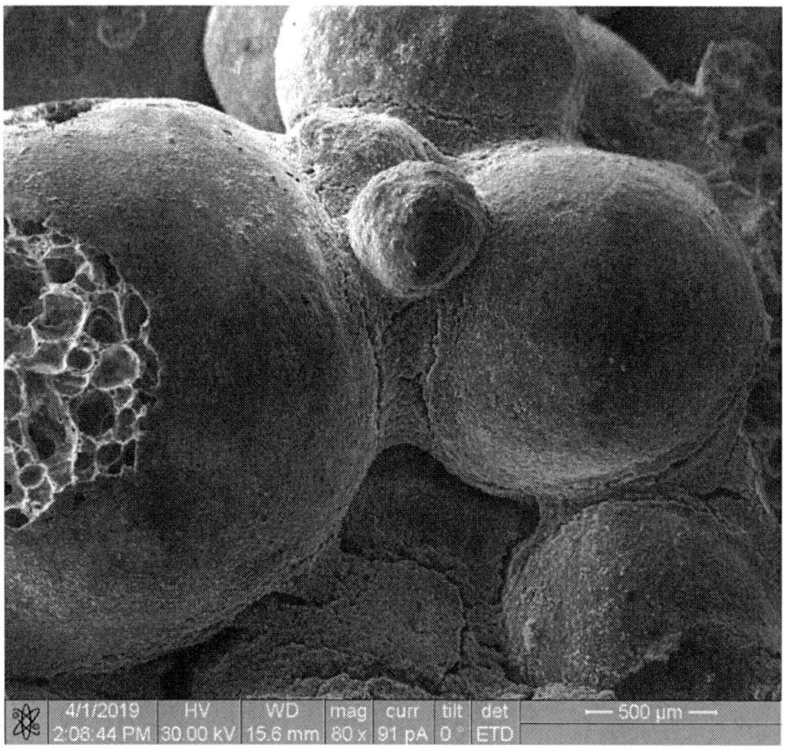

Figure 7. Small and large fractions, the densest packing.

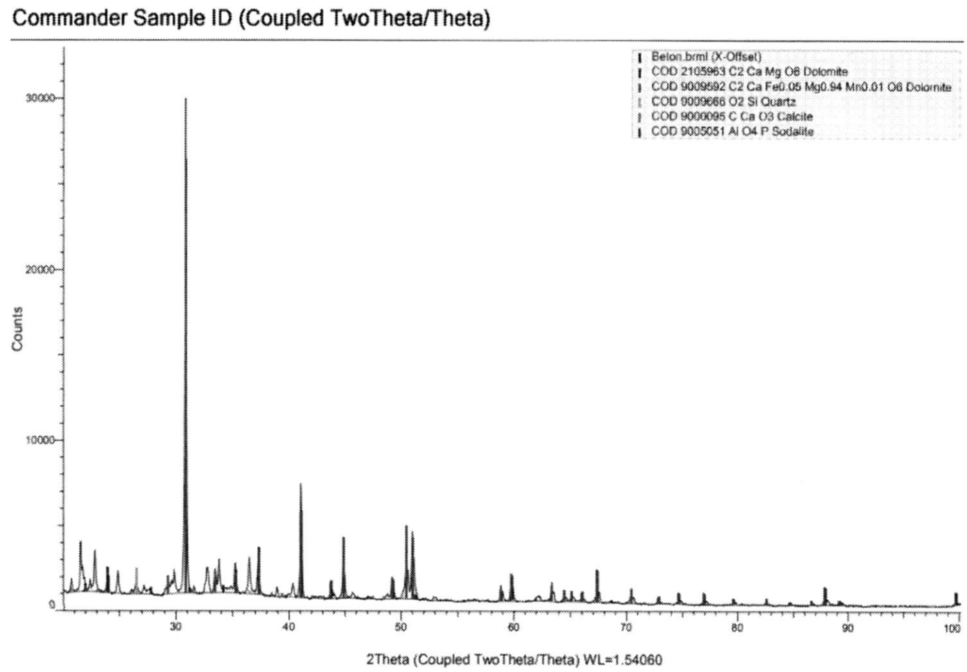

Figure 8. X-ray phase analysis of the sample.

4. Conclusion

One of the reasons for the high mechanical strength of "Diapen" and concrete with "Diapen" is the formation of needle-like crystals that create reinforcement cages throughout the concrete volume. We can see it in the study of the structure using scanning electron microscopy and X-ray phase analysis.

Test results convincingly showed that the developed material has good thermal insulation and strength properties that allow to produce high-quality material for enclosing and thermal insulation structures, especially for actively developing monolithic-frame and low-rise construction, as well as using the developed aggregate as loose-fill insulation.

It should be noted that the extraction of siliceous rocks in the Vladimir region is about 2 times cheaper than the extraction of raw materials for the production of expanded clay. Its cost (coupled with the lower firing temperature in the process of creating) is lower than the cost of production of expanded clay. The surface occurrence of siliceous rocks allows to be mined in an open way with minimal costs compared to other rocks used for similar purposes. Approximate savings in the use of concrete with the developed aggregate will be about 10-20%.

The evaluation of concrete with "Diapen" properties according to GOST 9758-86 and their comparison with analogues shows that the obtained material is not inferior to traditional building materials and can be recommended as a highly effective thermal insulation and thermal insulation structural material.

References

[1] Qian T, Li J, Ma H and Yang J 2016 *Adjustable thermal property of polyethylene glycol/diatomite shape-stabilized composite phase change material* Polym. Compos. **37**, 854–860

[2] Karaman S, Karaipekli A, Sarı A and Biçer A 2011 *Polyethylene glycol (PEG) /diatomite composite as a novel shape-stabilized phase change material for thermal energy*

storage Sol. Energy Mater. Sol. Cells **95** 1647–1653

[3] Zhang D, Zhou J, Wu K and Li Z 2005 *Granular phase change composites for thermal energy storage* Sol. Energy **78** 351–480

[4] Yu S, Wang X and Wu D 2014 *Microencapsulation of n-octadecane phase change material with calcium carbonate shell for enhancement of thermal conductivity and serving durability: synthesis, microstructure, and performance evaluation* Appl. Energy **114** 632–643

[5] Ling T and Poon C 2013 *Use of phase change materials for thermal energy storage in concrete: an overview* Constr. Build. Mater. **46** 55–62

[6] Al-Degs Y, Khraisheh M A M and Tutunji M 2001 *Sorption of lead ions on diatomite and manganese oxides modified diatomite* Water Res. **35** 3724–3728

[7] Hadjar H 2008 *Elaboration and characterisation of new mesoporous materials from diatomite and charcoal* Micropor. Mesopor. Mat. **107** 219–226

[8] Garderen N, Clemens F, Mezzomo M, Bergmann C and Graule T 2011 *Investigation of clay content and sintering temperature on attrition resistance of highly porous diatomite based material* Appl. Clay Sci. **52** 115–121

[9] Li M, Kao H, Wu Z and Tan J 2011 *Study on preparation and thermal property of binary fatty acid and the binary fatty acids/diatomite composite phase change materials* Appl. Energy **88** 1606–1612

[10] Xu B and Li Z 2013 *Paraffin/diatomite composite phase change material incorporated cement-based composite for thermal energy storage* Appl. Energy **105** 229–237

[11] Wen R, Zhang X, Huang Z, Fang M, Liu Y, Wu X, Min X, Gao W and Huang S 2018 Preparation and Thermal Properties of Fatty acid/diatomite Form-Stable Composite Phase Change Material for Thermal Energy Storage *Solar Energy Materials and Solar Cells* **178** 273-279

[12] Letelier V, Tarela E, Muñoz P and G Moriconi 2016 Assessment of the Mechanical Properties of a Concrete made by Reusing both: Brewery Spent Diatomite and Recycled *Aggregates Construction and Building Materials* **114** 492-498

[13] Wang R, Ren M, Gao X and Qin L 2018 Preparation and Properties of Fatty Acids Based Thermal Energy Storage Aggregate Concrete *Construction and Building Materials* **165** 1-10

[14] Qian T and J 2018 Octadecane/C-Decorated Diatomite Composite Phase Change Material with Enhanced Thermal Conductivity as Aggregate for Developing structural–functional Integrated Cement for Thermal Energy Storage. *Energy* **142** 234-249

[15] Zhang G 2013 *Microstructural modification of diatomite by acid treatment, high-speed shear, and ultrasound* Micropor. Mesopor. Mater. **165**, 106–112

[16] Jung K, Jang D and Ahn K 2014 *A novel approach for improvement of purity and porosity in diatomite (diatomaceous earth) by applying an electric field* Int. J. Miner. Process **131** 7–11

[17] Sun Z, Yang X, Zhang G, Zheng S and Frost R 2013 *A novel method for purification of low grade diatomite powders in centrifugal fields* Int. J. Miner. Process **125** 18–26

[18] Zhang J, Ping Q W, Niu M, Shi H and Li N 2013 *Kinetics and equilibrium studies from the methylene blue adsorption on diatomite treated with sodium hydroxide* Appl. Clay Sci. **83–84** 12–16

[19] Qin Y 2015 *Sodium sulfate–diatomite composite materials for high temperature thermal energy storage* Powder Technol. **282** 37–42

[20] Cao L, Tang Y and Fang G 2015 Preparation and properties of shape-stabilized phase change materials based on fatty acid eutectics and cellulose composites for thermal energy storage *Energy* **80** 98-103

IOP Conf. Series: Materials Science and Engineering **660** (2019) 012036 doi:10.1088/1757-899X/660/1/012036

Drying Shrinkage Deformation Comparison Between Foam Concrete, Geopolymer Concrete, Disintegrated, and Non-disintegrated Cement Mortar

R Gailitis[1], A Sprince[1], L Pakrastins[1], G Sahmenko[1] and T Kozlovskis[1]

[1] Faculty of Civil Engineering, Riga Technical University, 1 Kalku Street, LV-1658, Riga, Latvia.

E-mail: rihards.gailitis@edu.rtu.lv

Abstract. Foamed concrete has been known as a building material for nearly 100 years. In the beginning, it was used as an insulation material with very low density. Since then there have been attempts to make this material more load-bearing and structural. In present-day foamed concrete is being used in soil reinforcement, building blocks and in other sorts of building applications [1]. Another innovative material - the geopolymer concrete has been around only for 40 years. It is being used in buildings and infrastructures objects such as railroads, reservoirs, and houses and others. The main benefit of the geopolymer is that it is green material that is partially made by utilizing waste products. The geopolymer manufacturing carbon footprint is 2 times less than the Portland cement carbon footprint. Another way to reduce Portland cement carbon footprint is to reuse old cement. In the past few decades, there has been a considerable amount of researches regarding the partial replacement of cement using disintegrated cement in cement mortar or concrete. As it is known to obtain powder mineral filler material planetary ball milling is applied, but it is ineffective. It has been discovered that grinding by collision is a more effective method for refining brittle material. One of the ways to refine is to disintegrate with disintegrator. This raises the question of whether old cement disintegration together with sand can improve its long-term properties and what differences do these different cement and alkaline activated compounds have. The aim of this article is to determine the difference of shrinkage deformation for foamed concrete and disintegrated cement mortar which is Portland cement based cement composites and geopolymer concrete which represents alkali-activated cement composites. The size of all shrinkage specimens was 46mm in the diameter and 190mm in height. The shrinkage deformations of the specimen were determined by consistently measuring specimen deformation displacement. Shrinkage deformation values for foamed concrete in the 81st day did reach 11.85mm $*10^{-2}$, disintegrated old cement mortar 4.88mm $*10^{-2}$, non-disintegrated new cement mortar 5.02mm $*10^{-2}$, non-disintegrated old cement mortar 4.33mm $*10^{-2}$, but for geopolymer concrete, on the 81st day, it was 3.73mm $*10^{-2}$.

1. Introduction

One of the dominant technologies in mining industries for the production of minerals and in material treatment is grinding by the ball mills. However, grinding by collision is a more effective method for refining brittle material and one of the few machines that do collision grinding is disintegrator. The grindability of different hardness mineral materials using milling by collision in disintegrator does not differ in particle size after multiple milling stages while the materials with higher hardness have shown a decrease in the mean particle size after one step milling about 20% [2].

Foamed concrete is a cellular cementitious material obtained by introducing foam in the cement matrix. This process gains development of air voids within the specimen microstructure, and thus contributes to obtaining properties such as low self-weight, thermal insulating characteristics, acoustic absorption, increased fire resistance and better workability. It also reduces the strength of the material [3,4]. Conventionally foamed concrete has a density of 280-1800 kg/m3. The density of the cellular (foamed) concrete can be controlled by adding a calculated amount of proper foam into the slurry of water and cement, with and without the addition of aggregate. Foam stability is one of the most important factors for foamed concrete. Unstable foam can cause segregation and uneven density through the material. However, it is relatively difficult to control foam stability because it is affected by many aspects such as foam agent, foam achieving technology, water/cement ratio, and others [4–8]. However, foamed concrete also shows disadvantages, such as low strength, high drying shrinkage, and water absorption. It has been determined that there is a close relationship between the pore structure of foamed concrete and the properties of foamed concrete such as strength, water absorption and drying shrinkage [9].

Another novel material is geopolymer. Geopolymer belongs to a group of novel three-dimensional inorganic materials. This novel material got multiple beneficial properties such as low density, low cost, environmentally friendly nature, and high mechanical performance [10]. As a composite material geopolymer concrete is two or more constituent material arrangement. A continuously called matrix and the dispersed phase or phases, either fibers or particulates, in order to develop another material with the desired combination of properties [11].

Geopolymer is comparable in performance to ordinary Portland cement [12]. Geopolymer concrete main advantage is its contribution to the environment. It is estimated that carbon footprint made by geopolymer concrete manufacturing is 26 to 46% less than Portland cement concrete if in concrete mix Portland cement is replaced completely [11,12]. It is reckoned that production of 1 ton of caolin geopolymer contributes to 0.180 tons of CO_2, that is 6 times less than the manufacturing of Portland cement concrete [13].

According to statistics the restrained tensile stress rather than imposed load is the main factor in introducing cracking in concrete structures, especially throughout the construction period. The volume of concrete continues to decrease after hardening without external load; this phenomenon is called shrinkage. The shrinkage stresses are caused by restrained effect produced by reinforcement in order to maintain compatibility of concrete and reinforcement strains along with the self-balanced equilibrium conditions. According to capillary tension theory, the main driving force of concrete shrinkage is the capillary pressure in the concrete pore walls when water is absorbed in the pores. The capillary pressure will vary as the pore structure changes, which results in different shrinkage characteristics of the concrete [14,15].

In contrast to Portland cement concrete, that shrinks primarily due to negative capillary pressure that leads to a volume contraction of the cement paste during water loss, the shrinkage of geopolymer occurs not only due to water loss over time during reaction and evaporation but also because the pore structure depends on several critical factors, such as alkaline activator, water content, binder materials, and curing condition. The pores in geopolymer are generated during reaction, mainly from small disordered zeolites, which makes it fundamental to understand the correlation between mix design parameters and pore structure as part of the mechanism of geopolymer shrinkage behavior [16]. However, there are also drawbacks in the metakaolin-based geopolymers. The high specific surface area of the platy metakaolin particles leads to excessive mixing water demand and high yield stress. Moreover, the high water/binder ratio will have an apparent deleterious effect on the pore structure, durability, and efflorescence of the geopolymer products. On the other hand, fly ash is made from spherical particles that could work as a "ball bearing" and reduce the viscosity of the paste, therefore improving particle packing to refine pore structure of the binder [17].

In terms of sustainable raw material management, it is crucial to recycle industrial waste as much as possible and also to develop new technologies that not only reduce industrial waste landfills but also produce materials with new added value [18].

This study shows how does shrinkage affect materials due to their microstructure that is achieved with different binders and different fillers.

2. Material and methods

Foam concrete mix was produced in Warmhouse Ltd. industrial turbulence mixer with the effect of cavitation. The foamed concrete mix was prepared according to the composition shown in Table 1.

Table 1. Foamed concrete mix.

Ingredients	Units	Mass proportion
Cement CEM I 42.5N	kg/m^3	1.0000
Quartz sand 0/0.5mm	kg/m^3	1.2500
Foam agent (Tukums) mixed together with 0.35l H_2O	kg/m^3	0.0044
PP fibers	kg/m^3	0.0013
Micro silica Elkem 971 U	kg/m^3	0.0333
Water	kg/m^3	0.3333
Plastificator Stachema	kg/m^3	0.0129
W/C	-	0.3333

Cement mortar mixes were produced in laboratory planetary mixer Hobart, using fine-graded mineral aggregate (quartz sand) and specially prepared white (Aalborg) cement. Mortars with disintegrated and non-disintegrated cement were prepared according to the composition shown in Table 2.

Table 2. Disintegrated and non-disintegrated cement mortar mix.

Ingredients	Units	Mass proportion
Cement CEM I 42.5N	kg/m^3	1.0000
Quartz sand 0.4/1mm	kg/m^3	1.7500
Quartz sand 0/0.5mm	kg/m^3	1.2000
Water	kg/m^3	0.5000
Plastificator Stachema	kg/m^3	0.0040
W/C	-	0.5000

Cement amount in the composition is the same for mortars with old (stored in a laboratory for 10 years) and fresh Aalborg cement. In figure 1 is shown how different is the fresh and old cement.

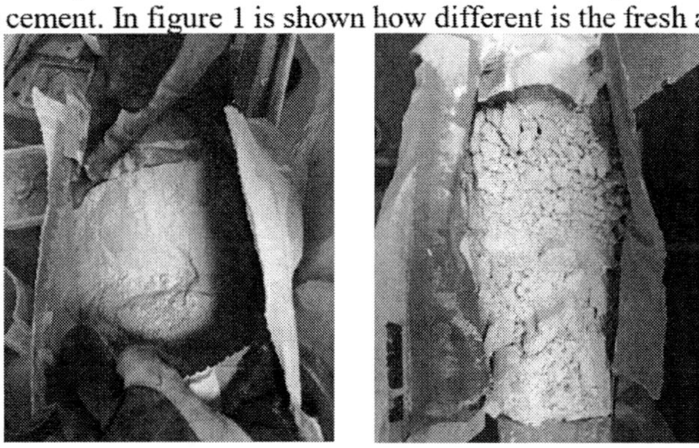

Figure 1. Fresh and old cement.

Dry cement and sand mixes are prepared using disintegrator DESI-16C in order to recover cement binding properties. Sand and old cement were disintegrated together in disintegrator for 5 minutes as it is shown in figure 2 and figure 3.

Figure 2. Disintegration process.

Figure 3. Disintegrated and non-disintegrated old cement with sand.

Geopolymer cylindrical specimen matrix was based on fly ash sourced from the local power plant in Skawina city (Poland). This kind of fly ash is suitable for geopolymers because of its physical properties and chemical composition. The fly ash contains spherical aluminosilicate particles as well as it is rich with oxides such as SiO_2 (47.81%), Al_2O_3 (22.80%). The high value of SiO_2 and Al_2O_3 gives advantages for geopolymerization.

Geopolymer specimens were prepared using sodium promoter, fly ash, sand (ratio sand and fly ash – 1:1). The process of activation has been made by 12M NaOH solution combined with the sodium silicate solution (at a ratio of 1:2.5). In order to make the composite the technical NaOH in flakes was used and water solution of sodium silicate R–145. Tap water was used instead of the distilled one. The alkaline solution was prepared by means of pouring the aqueous solution of sodium silicate and water over solid sodium hydroxide. The solution was mixed and left thru the night until its temperature is stabilized and the concentrations equalized. The fly ash, sand, and alkaline solution were mixed for about 15 minutes by using a low-speed mixing machine (to receive the homogenous paste). Next, the mix was poured into the plastic molds. The specimens were hand-formed and then the air bubbles were removed with vibrations. Molds were heated in the laboratory dryer for 24h at 75 °C. Then, the specimens were unmolded [19,20].

All prepared specimens had the following dimensions: Ø 46 x 190 mm.

For shrinkage deformation tests, 6 aluminum plates (10 x 15 mm) were glued to each specimen in pairs. Afterward, strain gauges were attached to those plates.

Shrinkage deformations were monitored for the first two weeks every day, afterward every two days.

Figure 4. Prepared shrinkage test specimens.

3. Results and discussion

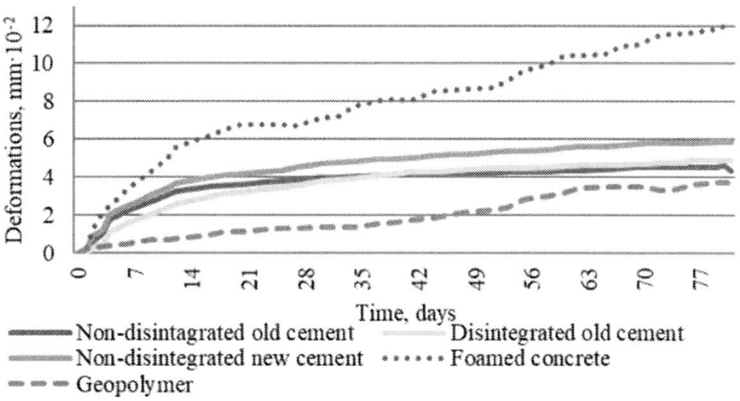

Figure 5. Test sample shrinkage deformations.

From figure 5 it appears that foamed concrete is the most exposed compound to shrinkage deformations. In contrast, the disintegrated and non-disintegrated compounds show relatively significant deformations in the first two weeks, but afterward, the deformation change is relatively small. Geopolymer concrete, in contrast, shows a constant increase throughout the testing time.

Because of the different relative humidity levels thru the specimen testing graphs in figure 6 show the shrinkage deformations in context with the factor that causes them – the change of the relative humidity (delta value that represents the change of the relative humidity throughout the test).

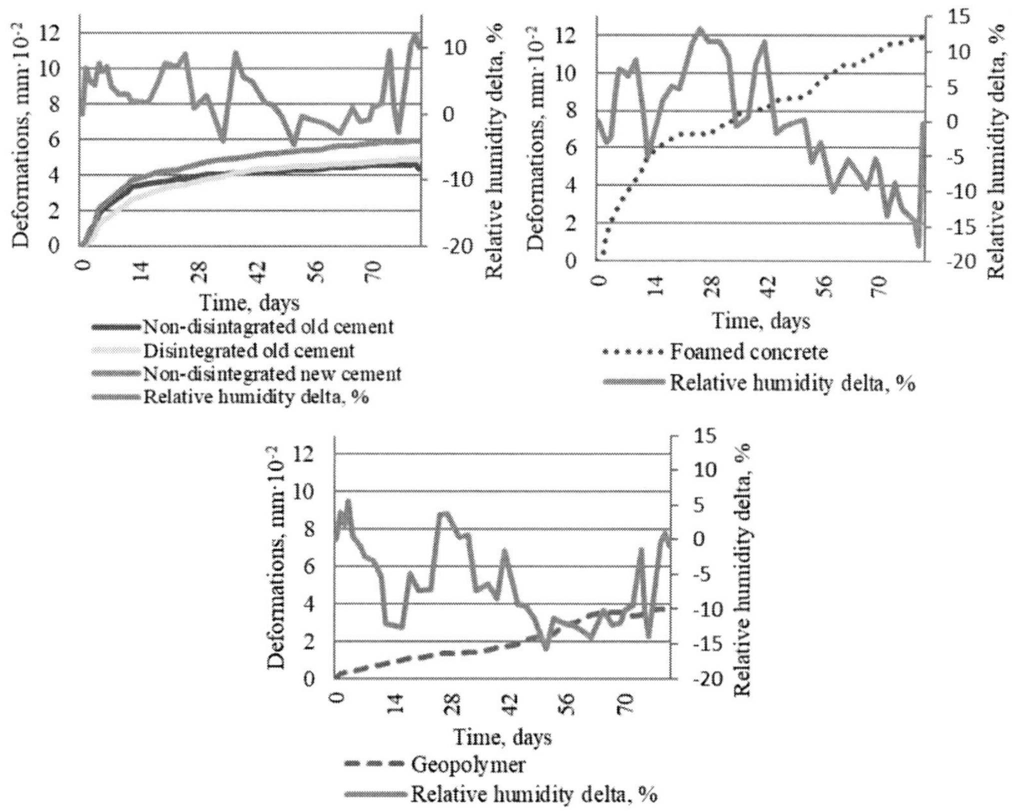

Figure 6. Disintegrated non-disintegrated cement mortar, geopolymer and foamed concrete test sample shrinkage deformations with relative humidity.

From figure 6 it can be deduced that all the tested compounds have got similar shrinkage deformation reaction time to the change of relative humidity level. The average response in shrinkage deformations from drop or increase to changes in relative humidity level is 1.5 days.

Due to the previous statement regarding the link between relative humidity and shrinkage, it is necessary to get a factor that would represent material willingness to shrink. So, by dividing shrinkage deformation with relative humidity variation in time (delta) the shrinkage deformation factor is achieved.

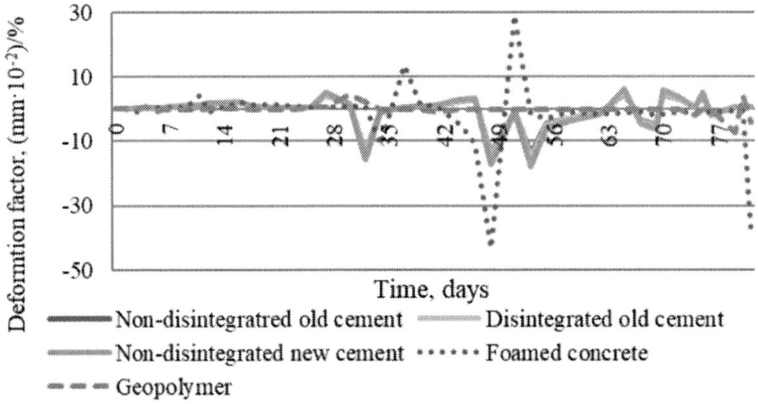

Figure 7. Disintegrated non-disintegrated cement mortar, geopolymer, and foamed concrete test sample shrinkage deformation factor.

From figure 7, it appears that the most affected compound is the foamed concrete that is followed by the disintegrated and non-disintegrated compounds and the most resistant is geopolymer.

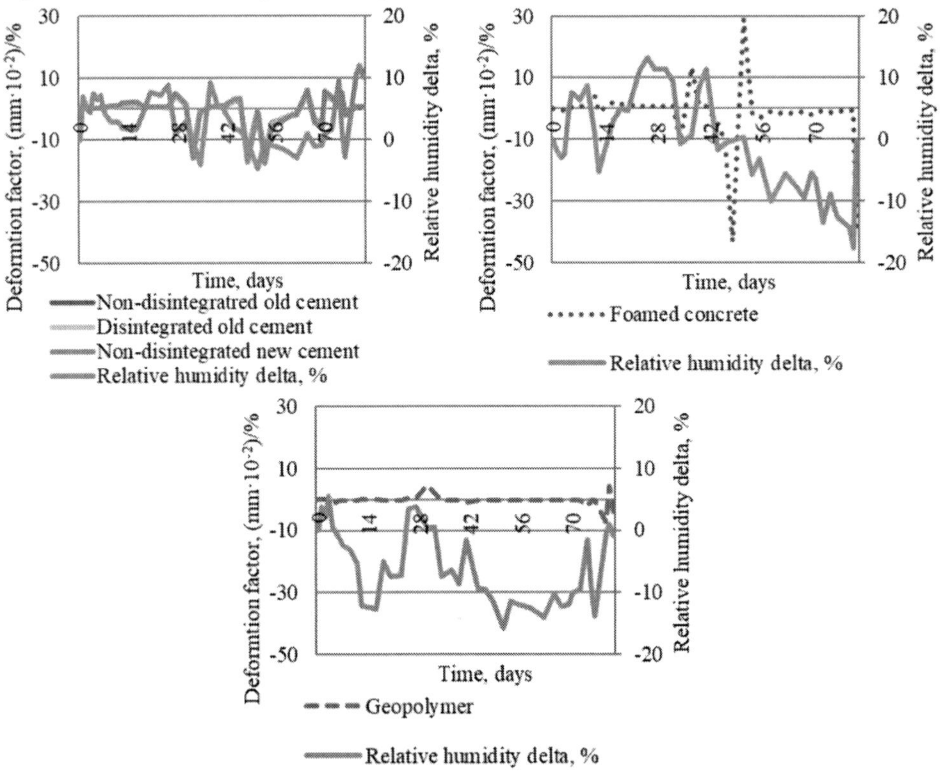

Figure 8. Disintegrated non-disintegrated cement mortar, geopolymer and foamed concrete test sample shrinkage deformations with relative humidity.

Relative humidity delta value that is shown in figure 6 and figure 8 is the relative humidity value that throughout the test differs from the relative humidity value that was at the beginning of the test. The relative humidity level at the beginning of the tests for disintegrated and non-disintegrated cement mortars was 17.9%, for geopolymer 29.0% and f or foamed concrete it was 46.0%. Using these values the relative humidity delta value is calculated using equation (1):

$$RH_{\Delta} = RH_{t=0} - RH_{t=n} \tag{1}$$

where:
RH_{Δ} – relative humidity delta value;
$RH_{t=0}$ – relative humidity at the beginning of the test;
$RH_{t=n}$ – relative humidity value n days after the test began.

To determine the effect of the humidity level change on to shrinkage factor change figure 8 was made.

The shrinkage deformation factor is calculated by dividing shrinkage deformation with relative humidity delta value. Every single shrinkage deformation reading is divided with relevant humidity delta value.

From the deformation factor graphs, it appears that geopolymer is the most resistant to the moisture level variations from all the tested compounds. The most affected seem to be foamed concrete and the disintegrated and non-disintegrated cement compounds.

4. Conclusions

Regarding shrinkage, the most sensitive from the tested compounds was foamed concrete. Much more resistant to shrinkage is regular cement mortar and disintegrated cement mortar. The lowest shrinkage shows geopolymer concrete. The shrinkage deformations for disintegrated old cement mortar is 23%, non-disintegrated new cement mortar 25%, non-disintegrated old cement mortar 15% and foamed concrete 68% greater than geopolymer concrete.

From figure 8 it is also determined that geopolymer concrete is less likely to gain or lose shrinkage deformations due to rapid increases or decreases of relative humidity in the environment. In this case, foamed concrete shows the higher possibility to gain or lose deformation due to change in humidity gains or lose in the environment.

This means that if all the tested compounds are used as a structural material then there is a bigger possibility for foamed concrete to develop cracks in structure due to variations in relative humidity levels in an environment that could lead to loss of its structural properties. These conclusions also mean that due to higher sensitivity to humidity changes, especially, rapid ones, it is recommended to somehow shield foamed concrete structures from direct moisture interactions that could lead to high structure saturation with water.

Acknowledgments

This work has been supported by the Latvian Council of Science within the scope of the project 'Long term properties of innovative cement composites in various stress-strain conditions' No. lzp-2018/2-0249.

References

[1] Mugahed Amran Y H, Farzadnia N and Abang Ali A A 2015 Properties and applications of foamed concrete; a review. *Construction and Building Materials* **101**, pp 990-1005

[2] Goljandin D, Kulu P, Käerdi H, Bruwier A. Disintegrator as device for milling of mineral ores. *Material Science* **11**, pp 398-402

[3] Falliano D, De Domenico D, Ricciardi G, Gugliandolo E 2019 Compressive and flexural strength of fiber-reinforced foamed concrete: Effect of fiber content, curing conditions and dry density. *Construction and building materials* **198**, pp 479-493

[4] Suleyman Gokce H, Hatungimana D and Ramyar K 2019 Effect of fly ash and silica fume on hardened properties of foam concrete. *Construction and building materials* **194**, pp 1-11

[5] Nguyen TT, Bui HH, Ngo TD, Nguyen GD, Kreher MU, Darve F 2019 A micromechanical investigation for the effects of pore size and its distribution on geopolymer foam concrete under uniaxial ceompression. *Engineering Fracture Mechanics* **209**, pp 228-244

[6] Ghorbani S, Ghorbani S, Tao Z, de Brito J, Tavakkolizadeh M. 2019 Effect of magnetized water on foam stability and compressive strength of foam concrete. *Constr Build Mater* **197**, pp 280-290

[7] Li T, Wang Z, Zhou T, He Y, Huang F 2019 Preparation and properties of magnesium phosphate cement foam concrete with H2O2 as foaming agent. *Constr Build Mater* **205**, pp 566-573

[8] Vatin NI, Barabanshchikov YG, Komarinskiy M V., Smirnov SI. 2015 Modification of the cast concrete mixture by air-entraining agents. *Magazine of Civil Engineering* **56**, pp 1-10

[9] Sun C, Zhu Y, Guo J, Zhang Y, Sun G 2018 Effects of foaming agent type on the workability, drying shrinkage, frost resistance and pore distribution of foamed concrete. *Constr Build Mater* **186**, pp 833-839

[10] Yan S, He P, Jia D, Wang J, Duan X, Yang Z, et al. 2017 Effects of high-temperature heat treatment on the microstructure and mechanical performance of hybrid C_f-SiC_f-$(AL_2O_{3_p})$ reinforced geopolymer composites. *Composites Part B* **114**, pp 289-298

[11] Lokuge W, Wilson A, Gunasekara C, Law DW, Setunge S 2018 Design of fly ash geopolymer concrete mix proportions using multivariate adaptive regression spline model. *Constr Build Mater* **166**, 472-481

[12] Luna-Galiano Y, Leiva C, Villegas R, Arroyo F, Vilches L and Fernández-Pereira 2018 Carbon fiber waste incorporation in blast furnace slag geopolymer- composites. *Materials Letters* **233**, pp 1-3

[13] Nguyen KT, Le TA and Lee K 2018 Evaluation of the mechanical properties of sea sand-based geopolymer concrete and the corrosion of embedded steel bar. *Constr Build Mater* **169**, pp 462-472

[14] Huang L, Hua J, Kang M, Zhou F and Luo Q 2019 Capillary tension theory for predicting shrinkage of concrete restrained by reinforcement bar in early age. *Constr Build Mater* **210**, pp 63-70

[15] Baleviius R, Augonis M, Bistrickait R and Diliūnas S 2018 Shrinkage effect on Cracking Resistance of Flexural Reinforced Concrete Members. *Mechanica* **24**, pp 11-18

[16] Ling Y, Wang K and Fu C 2019 Shrinkage behavior of fly ash based geopolymer paste with and without shrinkage reducing admixture. *Cement and Concrete Composites* **98**, pp 74-82.

[17] Yang T, Zhu H and Zhang Z 2017 Influence of fly ash on the pore structure and shrinkage characteristics of metakaolin-based geopolymer pastes and mortars. *Constr Build Mater* **153**, pp 284-293.

[18] Mucsi G, Szenczi Á and Nagy S 2018 Fiber reinforced geopolymer from synergetic utilization of fly ash and waste tire. *Journal of Cleaner Production* **178**, pp 429-440

[19] Korniejenko K 2018 Geopolymers for increasing durability for marine infrastructure. American Concrete Institute, ACI Special Publication, Vol. 2018-June, Issue SP 326, pp 20.1-20-10

[20] Łach M, Mikuła J and Hebda M 2016 Thermal analysis of the by-products of waste combustion. *Journal of Thermal Analysis and Calorimetry* **125**, pp 1035-1045

Effect of sodium lignosulfonate superplasticizer on the early hydration of cement with different contents of cubic C_3A

A Mezhov [1] and K Kovler [1]

[1] National Building Research Institute - Faculty of Civil and Environmental Engineering, Technion - Israel Institute of Technology, 3200003, Haifa, Israel

e-mail: alexander.m@technion.ac.il

Abstract. In this research, the influence of the sodium lignosulfonate superplasticizer (LS) on the early hydration of two types of cements containing different amount of cubic C_3A was studied. It is shown that at 0.1% addition of LS has no significant influence on the hydration, while at higher dosages LS contributes to the additional formation of ettringite. This phenomenon is observed for both types of cement. The higher ettringite formation was found for the cement containing higher C_3A amount. However, LS addition retards the hydration, which is indicated by the reduction of the portlandite amount formed.

1. Introduction

Lignosulfonates (LS) are one of the most popular superplasticizers (SP) in the market, since their first introduction in 1930s [1]. LSs are the by-product of the pulping industry. Thus, LS's have a low price, and despite relatively low efficiency in comparison to polycarboxylate ethers, LS's still remains highly demanded superplasticizers in concrete production [2].

LS is based on lignin – natural polymer presented in wood. Initially lignin is a water insoluble matter which undergoes sulfonating process in presence of sulfite or bisulfite salts containing various counter-ions (Na^{+2}, Ca^{+2}, Mg^{+2} etc.). Depending on its chemical composition LS differently influences on the properties of the fresh and hardened concrete [3]. Among others sodium and calcium lignosulfonates are the most common ones being used in the concrete industry.

Regardless of the type of the superplasticizers, most of them retard cement hydration. This shortcoming is a consequence of their performance. When introduced into water-cement paste SP is adsorbed on the cement particles. Then, depending on the SP's nature, plasticizing effect is reached - either through the increment of the electrostatic forces or due to steric hindrance. Therefore, adsorption is a key interaction between SP and cement. Nonetheless, it is adsorption that causes retardation of the cement hydration due to the inhibiting the contact between the cement particles and water. It is known that cement minerals have a different affinity to the SP's; the highest attraction exhibit C_3A and C_4AF phases [4]. Since C_3A hydrates way earlier than C_4AF, its content has a significant influence on the early-age hydration of cement.

Hydration of cement is an exothermic process, the typical heat flow curve obtained for the cement used in the current research is presented in figure 1. The retardation effect caused by SP's application is usually associated with an increment of the stage II – induction period [5]. Since stage III is associated with the precipitation of C-S-H and portlandite, and the stage IV is associated with an intensive secondary ettringite formation caused be renewed C_3A dissolution [6], the influence of the LS introduction is appropriate to study right after these stages. Moreover, it was pointed out that C_3A polymorphs significantly affect the performance of polycarboxylate ethers (PCE) and polynaphthalene sulfonate (PNS) superplasticizers [7].

The current study is to elucidate the effect of cubic C_3A polymorph on the early-age hydration of the cement paste in the presence of sodium lignosulfonate. For this purpose, the hydration of two cements containing different amount of cubic C_3A was stopped by the solvent replacement method and the hydration product was investigated by XRD.

2. Materials and methods

2.1 Materials

CEM I 52.5N supplied by Nesher Cement, Israel is used for this study. Its mineral composition is presented in table 1.

Table 1. Phase composition of cement (wt.-%), obtained by XRD-Rietveld analysis.

Chemical composition	Cem_A (CEM I 52.5N)	Cem_B (model cement)
Alite	61.02	58.31
Belite	12.95	12.37
C_3A cubic	5.86	8.81
C_3A orthorhombic	-	-
C_4AF	12.42	11.87
Bassanite ($CaSO_4{\cdot}0.5H_2O$)	1.87	2.97
Anhydrite	0.28	0.27
Portlandite	1.83	1.75
Calcite	2.25	2.15
Arcanite	0.97	0.93
Aphthitalite	0.55	0.53

Model cement was prepared by mixing the industrials cement (table 1), laboratory made C_3A cubic and calcium sulfate hemihydrate, which was added in order to compensate the acceleration of the setting time initiated by the C_3A addition. The amount of hemihydrate added was chosen to keep the initial C_3A to hemihydrate ratio as 73/27 by weight [8].

The cubic modification of C_3A was synthesized from compacted powders of calcium carbonate (Spectrum) and alumina (Camag). The powders were mixed in plastic vials (50 ml) in a vortex mixer (Vortex-Genie 2 – Scientific Industries, Inc) and then compacted into pellets of about 3.5 g and 13 mm in diameter using a hydraulic press. The pellets were calcined at 900 °C for 1 hour and sintered at 1450 °C for 2 hours in Pt-crucibles in a high-temperature chamber (Nabertherm LHT 02/17 LB, Nabertherm, Germany) and finally quenched with a fan to a room temperature within minutes. The pellets were ground in a mortar and sieved. The particles passed 45 μm sieve was retained. The obtained powder comprised 98.7 wt.-% of C_3A and 1.3 wt.-% of free lime. The chemical composition of model cement (Cem_B) is presented in table 1.

Superplasticizer was commercially available sodium lignosulfonate (Sodyum Lignosülfonat) produced by Likit Kimya, Turkey.

All cement pastes were prepared at a water to cement ratio w/c = 0.4. Sodium lignosulfonate superplasticizer was preliminarily dissolved in the mixing water and then added to cement.

2.2 Methods

2.2.1 X-ray diffraction (XRD)

The XRD analysis was performed using a Malvern PANAlytical EMPYREAN X-ray diffractometer. The XRD scans were analyzed using the HighScore Plus software. Analyzes were performed using the external standard method. As a standard was used α-Al_2O_3 with 100 % crystallinity. Commercial name is CR1 produced by Baikowski, France.

Sample preparation: for each XRD sample, the hydration of LS-cement suspensions was stopped using the solvent replacement method [9]. After 12 hours of hydration, cement paste (5 g) prepared at w/c = 0.4 were gently ground by mortar and immersed in iso-propylalcohol (IPA) for 5 min and then moved in a chamber for 24 hours, 35 °C, low vacuum 500 mbar. The dried powder was milled and sieved, particles passed 75 μm were retained and analyzed by XRD.

2.2.2 Isothermal calorimetry

Calorimetric measurements were performed using the Isothermal Calorimeter TAM Air model (TA instruments). Measurement precision of the heat flow was ±4 μW at a 95% confidence interval. Fresh pastes consisting of cement (5 g) and deionized water (2 g) were prepared and mixed by vibration at 60 rpm for 10 s. The samples were then introduced into the calorimeter, which measured the heat flow for 48 hours, providing 6 measurements per minute. All the tests were conducted at 21°C with three independent mixes for each LS dosage in the cement paste.

2.2.3 Particles size distribution

The analysis was performed by means of Particle Size Analyzer Malvern Mastersizer 2000 in iso-propylalcohol (IPA).

3. Results and discussion

The particle size distribution of the cements, gypsum hemihydrate and cubic C_3A is shown in figure 2. It can be seen that the model cement (Cem_B) is almost similar to the reference Cem_A. The only difference is a small "tail" of the particles bigger than 100 μm, that is evidently caused by hemihydrate addition.

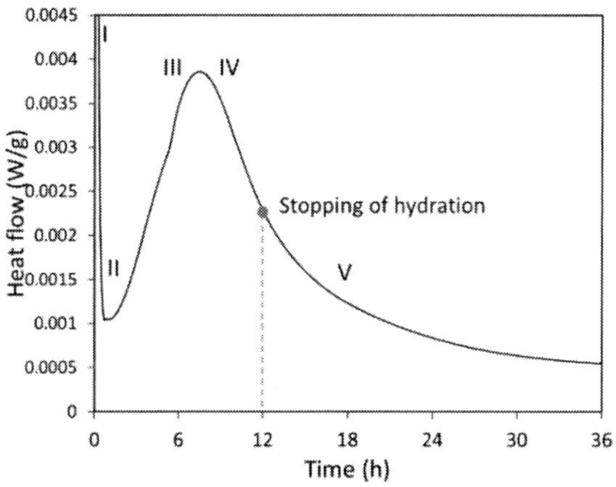

Figure 1. Heat flow of OPC (Cem_A).

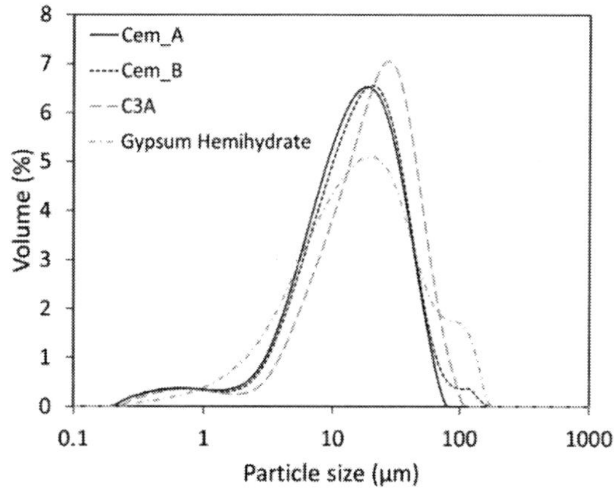

Figure 2. Particle size distribution.

The hydration of both cements was stopped after 12 hours of hydration. The XRD scans of the cement prepared with different LS dosages are presented in figure 3 (a), (b). The quantitative composition of the obtained phases is presented in figure 4.

The addition of higher LS dosage significantly oppressed the hydration of alite and consequently the formation of portlandite, that naturally induces an increment of the induction period (stage II in figure 1), i.e. the more alite remains, the less portlandite is formed, and the longer retardation of hydration.

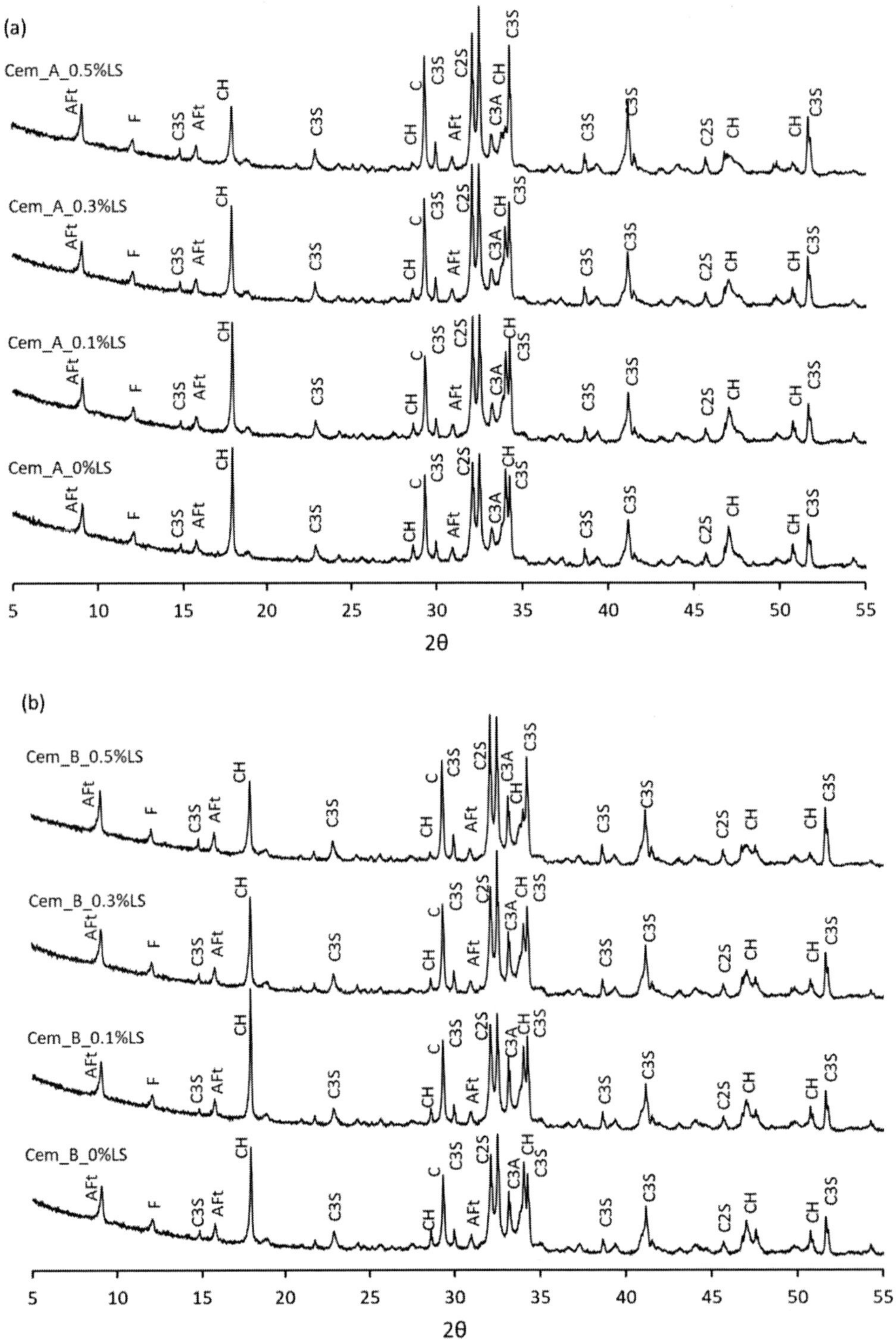

Figure 3. XRD curves from 2θ = 5° to 55°; (a) Cem_A; (b) Cem_B. Hydration was stopped after 12 hours. Displayed peaks: AFt - ettringite; CH – portlandite; C3S – alite; C2S – belite; C3A – aluminate cubic; F – aluminoferrite; C – calcite.

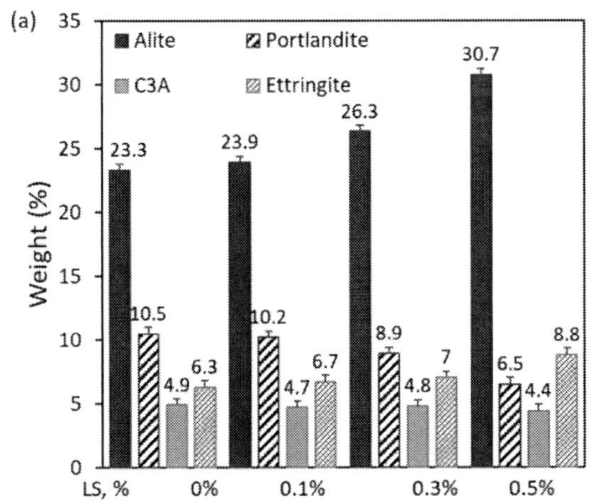

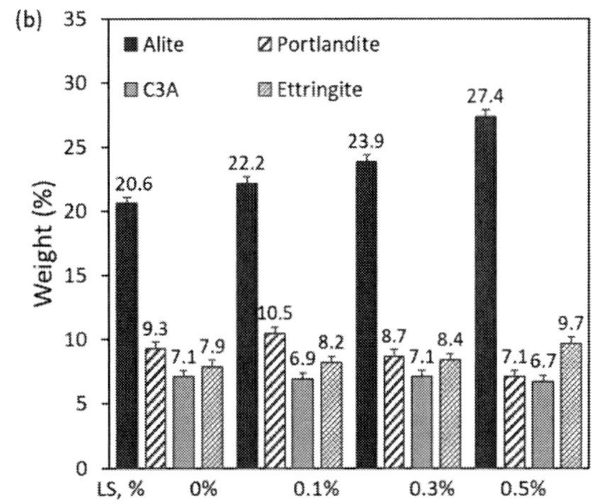

Figure 4. The phase composition of (a) Cem_A; (b) Cem_B prepared with various LS dosages obtained analysis after 12 hours of hydration with respect to the LS dosage. All the rest cement phases, i.e. belite, C_4AF etc. have not been significantly changed.

Simultaneously, the amount of cubic C_3A is slightly reduced. Interestingly, that irrespective to the cement type the amount of ettringite formed is increased with a LS dosage. Evidently, an increment of LS dosage promotes the intensive formation of ettringite.

Recently this phenomenon was pointed out for the ettringite formation in the presence of calcium lignosulfonate [10], [11]. Obviously, this phenomenon repeats for the cement paste prepared with sodium lignosulfonate. The increasing of the C_3A and gypsum contents naturally boost the amount of ettringite formed. At the same time, the difference in alite amount (figure 4 (a), (b)) is related to the initial difference between the cement used (table 1). It is worth noting that up to 0.1% LS addition the amount of porlandite and alite remains the same for both types of cement. At the higher dosages the portlandite amount is decreased meaningfully, that evidently extends the induction period of hydration. Since the C_3A has a higher affinity to the LS than other phases, it is expected that an increment of C_3A amount reduces the retardation effect.

4. Conclusions

In the current research the influence of sodium lignosulfonate on the early hydration products was investigated. The hydration of two type of cements with similar mineral composition, but different amounts of C_3A and calcium sulfate hemihydrate, were stopped after 12 hours of hydration. It was found that the influence of 0.1% LS addition on the development of the hydration products appears to be negligible. At higer dosages LS induces acceleration of the ettringate formation that was detected for both cement types. The highest increment of ettringite was found for the LS dosage of 0.5%. At the same time, LS retards the hydration of cement by extending the induction period. The retardation is embodied in reduced portlandite formation. Evidently, LS is consumed by alite that inhibits its hydration. The increasing of cubic C_3A (Cem_B) content results in higher ettringite formation and at the same time does not significantly affect the portlandite formation up to the LS dosage 0.3%. Therefore, it can be concluded that LS presumably is adsorbed by the cubic C_3A phase.

References

[1] Gelardi G, Mantellato S, Marchon D, Palacios M, Eberhardt A B and Flatt R J 2016 *Sc. and Tech. of Concr. Adm.* **9** 149
[2] Flatt R and Schober I 2012 *Und. the Rheol. of Concr.* **7** 144
[3] Arel H Ş and Aydin E 2017 *Constr. Build. Mater.* **157** 1084
[4] Yoshioka K, Tazawa E I, Kawai K and Enohata T 2002 *Cem. Concr. Res.* **32** 1507

[5] Marchon D and Flatt R J 2016 *Sc. and Tech. of Concr. Adm.* **12** 279
[6] Jansen D, Goetz-Neunhoeffer F, Stabler C and Neubauer J 2011 *Cem. Concr. Res.* **41** 602
[7] Alonso M M and Puertas F 2015 *Constr. Build. Mater.* **78** 324
[8] Quennoz A and Scrivener K L 2013 *Cem. Concr. Res.* **44** 46
[9] Zhang J and Scherer G W 2011 *Cem. Concr. Res.* **41** 1024
[10] Wang X, Pang Y, Lou H, Deng Y and Qiu X 2012 *Cem. Concr. Res.* **42** 1549
[11] Colombo A, Geiker M, Justnes H, Lauten R A and De Weerdt K 2018 *Cem. Concr. Res.* **107** 188

Properties of thin wall cement composites reinforced with AR glass, carbon and PVA fibres

R Moceikis[1], A Kičaitė[1], G Sahmenko[2] and A Korjakins[2]

[1]Vilnius Gediminas Technical University, Department of Building Materials and Fire Safety, Saulėtekio al. 11, LT-10223 Vilnius
[2]Department of Building Materials and Products, Institute of Materials and Structures, Faculty of Civil Engineering, Riga Technical University, Kalku Str. 1, Riga, LV 1658, Latvia

E-mail: rimvydas.moceikis@vgtu.lt

Abstract. Thin wall cement composites find a wide range of special applications for making light weight façade exterior wall and urban decorative elements, as well as a part of load bearing constructions. AR glass fibre is the most popular reinforcing material for producing glass fibre reinforced concrete (GRC). Possibilities for use alternative Carbon and PVA fibres are analysed in this work. The study is focused on investigation of technological and mechanical properties of thin wall (15 mm) cement composite plates reinforced by tree types of fibres.

1. Introduction

Conventional manufacturing innovations offer wide range of production methods for fibre reinforced concrete, such as 3D- Molding and 3D Printing. Using these methods, different unique products can be made from fibre reinforced cementitious composites- from flower and trash pots to stunning exterior precast concrete wall panels [1-4].

In many GRC manufacturing processes, self- consolidating concrete (SCC) is preferred due to good workability and final product quality. This is very important in case of thin mold casting process, when concrete mix must fill narrow spaces- 10 mm to 20 mm, depending on product thickness. Studies showed, that 12 mm long glass fibres are a good choice for self-consolidating fibre reinforced premix, because most of the filaments still stay in bundles even after very intense mixing speeds after 1 minute up to 1000 RPM.

The interest of modern researchers and manufactures of fiber reinforced decorative elements started in UK. First known application of glass fibres in construction was The Credit Lyonnais Building, built in London in 1956 (Figure 1). Even though glass fibres used for reinforcement of exterior panels were from regular glass, this architectural masterpiece is still standing today and demonstrating all- round performance capability of the product [5].

Typical E- glass (fiberglass) was invented in 1932 by R.G. Slayter which is still widely used for fiber- reinforced polymer composites. Fibre glass chemical corrosion in alkali environment is well understood and investigated process and for this reason, E- Glass should never be used in cementitious composites [6-9]. First generation alkali resistant (AR) glass was invented in 1970's and it's chemical resistance to hydroxyl ion attack drastically improved [10]. Nevertheless, GRC composite loses significant portion of it's initial strength under natural weathering conditions due to complex hydration processes which induce micro surface flaws on the surface of glass filaments and increases brittle behavior of the composite. For this reason, alternative fibres for GRC such as carbon or PVA (Polyvinyl Alcohol) should be also investigated.

Figure 1. The Credit Lyonnais Building, built in London in 1956, exterior panels made from early developments of glass fibre reinforced concrete (GRC).

Pictures source: https://twitter.com/royalfineart/status/770968500852649984

One research showed, that carbon fibres significantly reduced concrete slump and decreased compressive strength by up to 58%. On the other hand, tensile splitting strength increased by 17% and flexural fracture energy increased by 41%, when 0,25% of fibres were added from mass of cement. Scanning electron micrographs of the fractured concrete specimens revealed that carbon fibres were distributed in SCC without any fibre clumping or fibre balling [11].

In another research it was found that CFRSCC mixtures with carbon fibres content up to 0.75% satisfactorily passed the requirements of SCC. Slump flow time was increased with the increase in carbon fibres content because the inclusion of fibres slowed the flow of CFRSCC mixture by making it more viscous [12].

PVA fibres has high contribution to post- peak flexural behavior with single-crack strain–hardening response. Studies show that short polyvinyl alcohol (PVA) fibres can significantly improve mechanical characteristics as well as long term behaviour of cement-based composites [13,14]. Flexural displacement (ductility) at ultimate load increased by 50-150 times and ultimate flexural strength by 50-250% than plain concrete [15,16].

2. Materials and Methods

Rapid hardening Portland cement CEM I 52,5R was used as a binder. Quartz sand was chosen as matrix filler (Figure 1). Polycarboxilic ether- based superplasticizer was used as a water reducing agent. For matrix reinforcement, three different fibres were used- AR glass, carbon and polyvinyl alcohol (PVA) with the same length- 12mm. Properties of fibres are compared in Table 1.

Table 1. Properties of fibres used in this investigation.

Properties	AR Glass fibres	Carbon fibres	PVA fibres
Tensile strength, MPa	1400	3000	800
Modulus of elasticity, GPa	74	350	40
Thermal resistance, °C	-50…+350	300	100
Density, kg/m³	2500	1570	1300
Fibre length, mm	12	12	12
Filament diameter, μm	18	5	660
Filaments per bundle, pcs	200	3000	1

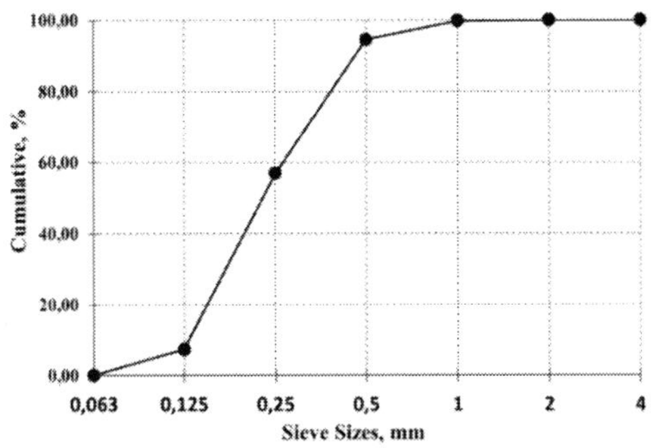

Figure 2. Cumulative curve of quartz sand filler.

Mix compositions were designed according EN 15191 and common self- compacting GRC practice and are given in Table 2. Binder to filler ratio was 1:1, water-cement ratio (W/C) – 0.36, superplasticizer – 1.1% from mass of cement. Fiber content is calculated by mass from the whole weight of the matrix. Mass proportions of PVA and glass fibres were calculated so that volumetric concentrations of fibres were the similar (compositions GF and PV). Significantly lower contents of carbon were chosen due to high negative impact to mix workability, because high shear mixer with up to 800 rpm was used for all tested compositions.

Table 2. Compositions of concrete mixtures (proportions by mass).

Mix Id.	CEM I 52,5R	*PCE*[a]	W/C	*Quartz sand*	*Glass fibres*	*Carbon fibres*	*PVA fibres*
REF	1	1.1%	0.36	1	0	0	0
GF	1	1.1%	0.36	1	2.9%	0	0
C1	1	1.1%	0.36	1	0	0.7%	0
C2	1	1.1%	0.36	1	0	0.2%	0
PV	1	1.1%	0.36	1	0	0	1.5%

[a]*Polycarboxilate ether based superplasticiser*

Workability of fresh concrete was tested according EN 1170-1. Compressive strength was tested according EN 196-1 with prisms 40x40x160, after 28days curing in natural conditions. Fracture characteristics was determined according EN 1170-4, by casting concrete boards 525x525x15mm and cutting them into specimen plates 262x50x15 mm, which were oriented in perpendicular direction. Flectural test was carried out in accordance with 4-point loading scheme. Distance between support 224 mm and distance between loading points 75 mm was provided. Plates were tested after 28 days and bending curves plotted.

Figure 3. Glass fibres (on the left) and PVA (on the right).

3. Results and Discussion

3.1. Workability

Best mix workability was achieved with PVA and glass fibres while the use of carbon fibres had significant impact on the mix workability. Considerable decreasing of mix slump may be due to a high dispersity and specific distributioof carbon fibres in cement matrix. Slump test results are given in Table 3 and Figure 3.

Glass fibres consist of 200 individual filaments, which are kept in a bundle due to high electrostatic forces between them. During high shear mixing process most of the glass filaments do not disintegrate from the bundles (Figure 4) and there was no significant slump difference between reference composition REF without fibres and GF with glass fibre.

Carbon fibres showed very weak bond between the filaments and completely disintegrating from the bundles, resulting in severe slump modification. Even with as little as 0,2% of carbon fibre, concrete mix by no means can be considered as self- compacting and requires intense vibration to cast the moulds.

On the other hand, 660 μm diameter PVA fibres showed almost no impact to the mix workability, very similar to REF composition without fibres. This fact makes PVA good alternative matrix reinforcement, when mix workability is discussed.

Table 3. Slump test results.

Mix Id.	Slump cm
REF	19
GF	18
C1	0
C2	0
PV	18

a) b) c)

Figure 4. Slump test photos a) Reference specimen REF, without fibres, b) GF with glass fibre, c) C1 with carbon fibres.

Figure 5. SEM micrograph with glass fibre bundle in hardened matrix.

3.2. Flexural resistance

Flexural tests with thin plates 15x50x262mm were done and load- deflection curves made to describe fracture characteristic with different types of fibres. Results showed that glass and PVA fibres significantly increase ductility of the matrix.

Flexural strength values at different deflections are given in Table 4 and flexural resistance curves in Figure 5. Unreinforced reference samples REF showed usual brittle behaviour with average flexural capacity 6.8MPa. Carbon significantly increased flexural capacity of the matrix- up to 10.7 MPa. But fracture behaviour of carbon reinforced matrix was brittle, and specimens ruptured below 2 mm deflection. Glass fibre specimens ruptured below 5 mm deflection and maintained 40% of maximum flexural capacity at 3 mm deflection. PVA fibred showed highest matrix ductility with 62% of maximum capacity at 6 mm deflection and 84% at 3 mm.

Table 4. Flexural stress at 0 mm, 3 mm and 6 mm stroke (plates 15x50x262, 28 days).

Composition	Flexural stress after 28 days, **MPa** at 0,25 mm deflection	at 3mm deflection	at 6 mm deflection
REF	6,8	0	0
GF	10,9	4,3	0
C1	9,1	0	0
C2	10,7	0	0
PV	8,9	7,5	5,5

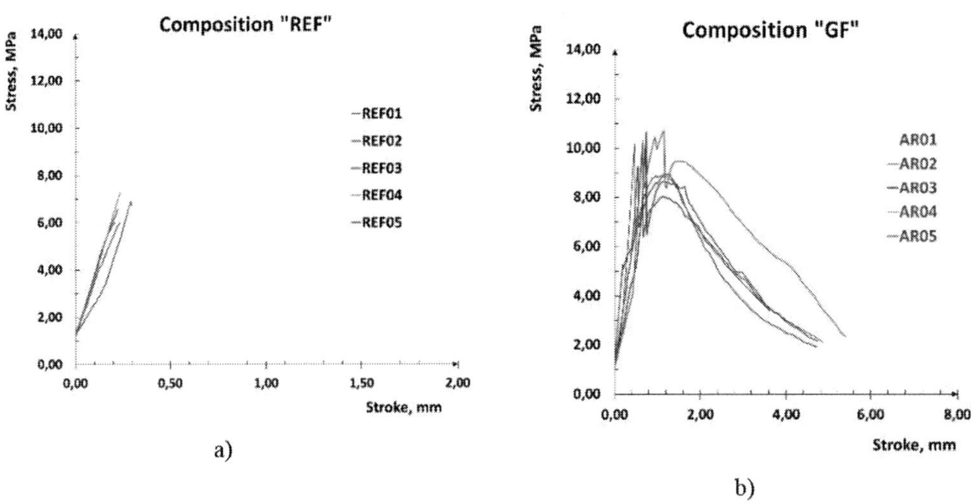

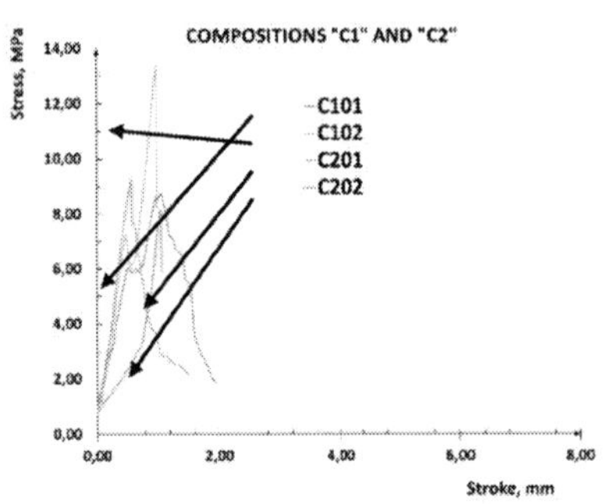

c)

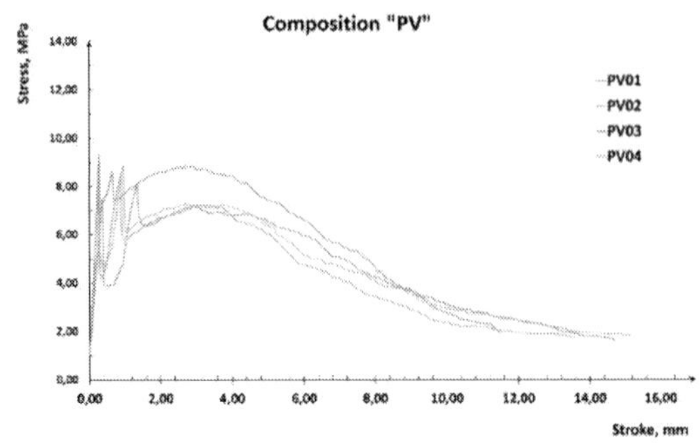

d)

Figure 6. Flexural resistance curves: a) Reference specimens REF, without fibres, b) GF with glass fibre, c) C1,C2 with carbon fibres, d) Composition PV with PVA fibres.

3.3. Concrete densities and water absorption

Hardened concrete densities are given in Figure 6 and water absorption in Figure 7. Results show, that refence sample REF had densest matrix and lowest water absorption. Added glass and carbon fibres decreased matrix density by 3,4%, PVA- by 1,4%. Added glass fibres increased matrix water absorbtion by 10%, carbon fibres- by 1,4%, PVA fibres did not have any significant influence.

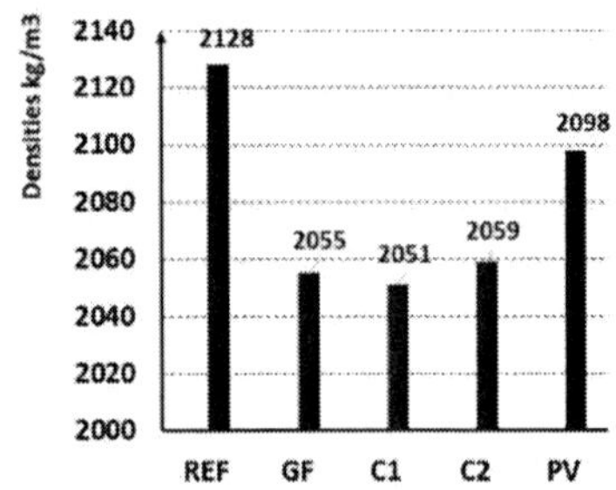

Figure 7. Hardenned concrete densities.

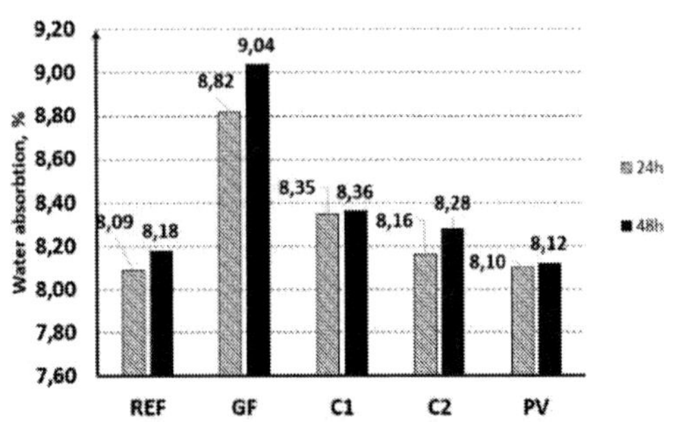

Figure 8. Water absorption after 24 and 48 hours.

4. Conclusions

Results of this study demonstrate possibility to use polyvinyl alcohol (PVA) fibres as alternative to AR glass in ductile GRC composites. Introduction of carbon fibre to cementitious matrix significantly increases flexural capacity, but the composite still ruptures in a brittle behavior. Similar volume of PVA fibres as AR glass in the same Portland cement matrix significantly increased composite ductility.

Flexural capacity with PVA was by 18% lower than typical glass fibre reinforced composite (GRC). Workability tests according to EN 1170-1 slump method showed, that PVA fibres has minor impact on slump, when compared with reference matrix sample without fibres. On the other hand, carbon filaments easily disintegrate from fibre bundles during high shear mixing process and mix workability goes far beyond any self-compacting limits and needs intense vibration to cast the molds. Glass fibres had the highest effect on matrix water absorption, increasing it by 10% from the reference sample.

Acknowledgement

"The financial support of European Regional Development Fund project Nr.1.1.1.1/16/A/007 „A New Concept for Sustainable and Nearly Zero-Energy Buildings" is acknowledged".

References

[1] Gregory T, 2011 The Next Generation – Future Manufacturing Developments *GRC Symposium, Istanbul,* 2011
[2] Stein C, Keynote: GRC for the Restoration of Historic Buildings *GRC Symposium, Istanbul,* 2011
[3] Henriksen T, Lo S and Knaack U, 2016 The impact of a new mould system as part of a novel manufacturing process for complex geometry thin-walled GFRC *Architectural Engineering and Design Management, vol. 12, no.3, pp. 231-249, 2016.*
[4] Ma Y and Che Y, 2015 A brief introduction to 3D printing technology *GRC Symposium, Dubai,* 2015.
[5] Rickard C, 2015 GRC developments in Australia *GRC Symposium, Dubai, 2015*
[6] Diamond S, 1985 Durability of Glass Fibre Reinforced Concrete *Proceedings in PCI symposium, Illinois, USA, 1985*
[7] Paul A, 1977 Chemical durability of glasses: a thermodynamic approach *Journal of Materials Science 12: 2246-2268*
[8] Majumdar A J, West J M, Larner L J, 1977 Properties of glass fibres in cement environment *Journal of Materials Science 12: 927-936*
[9] Orlowsky J, Raupach M, Cuypers H, Wastiels J, 2005 Durability modeling of glass fibre reinforcement in cementitious environment *Materials and Structures 38: 155-162*
[10] Ali M A, Majumdar A J and Sigh B, 1975 Properties of glass fibre cement- the effect of fibre length and content *Journal of materials science, vol. 10, pp. 1732-1740, 1975*
[11] Safiudin Md, Yakhlaf M and Soudki K A, 2018 Key mechanical properties and microstructure of carbon fibre reinforced self-consolidating concrete *Construction and building materials, vol. 164, pp. 477-488*
[12] Safiudin Md, Yakhlaf M and Soudki K A, 2013 Properties of freshly mixed carbon fibre reinforced self-consolidating concrete *Construction and building materials, vol. 46, pp. 224-231*
[13] Sprince A, Korjakins A, Pakrastinsh L, 2013 Time-dependent Behavior of High Performance Fiber-reinforced Concrete. *Advanced Materials Research Vol. 705PP. 75-80. Trans Tech Publications, Switzerland*
[14] Sprince A, Pakrastinsh L, Vatin N, 2016 Crack formation in cement-based composites *IOP Conference Series: Materials Science and Engineering. Vol. 123, Issue 1*
[15] Arisoy B, Wu H C, 2008 Material characteristics of high-performance lightweight concrete reinforced with PVA *Construction and building materials, vol. 22, pp. 635-645*
[16] Toutanji H, Xu B, Gilbert J, Lavin T, 2010 Properties of polyvinyl alcohol fiber reinforced high-performance organic aggregate cementitious material: Converting brittle to plastic *Construction and building materials, vol. 24, pp. 1-10*

Investigations on the foam concrete production techniques suitable for 3D-printing with foam concrete

V Markin[1], G Sahmenko[2], V N Nerella[1], M Näther[3], V Mechtcherine[1]

[1] Institute for Construction Materials, Technische Universität Dresden, 01602 Dresden, Germany
[2] Department of Building Materials and Products, Institute of Materials and Structures, Riga Technical University, Latvia
[3] Endowed Chair of Construction Machines, Technische Universität Dresden, Institute for Construction Materials, 01602 Dresden, Germany

E-mail: mechtcherine@tu-dresden.de

Abstract. With high thermal insulating property and low density, foam concrete has high potential in 3D-printing applications such as residential buildings. The synergy of the foam concrete and 3D-printing process pre-requisites research and optimization of foam concrete production. This paper presents investigations on the appropriateness of two different methods for production of foam concrete: 1) mixed foaming method in turbulence (colloidal) mixer and 2) mixed foaming method in cavitation disintegrator. Performance of synthetic and protein based foaming agents were examined. The dosage of the foaming agent was varied from 0.7% to 1.2% by weight of cement to produce foam concretes with densities ranging between 800 kg/m^3 and 1500 kg/m^3. The study shows that foaming ability of the designed concrete compositions depends primarily on their w-b ratio. Moreover, foaming with lesser dosage of foaming agent is possible when turbulence mixer is used. Furthermore, mechanical and physical properties of foam concrete, including compressive strength, bending strength, and water absorption were reported.

1. Introduction

Developments in the application of the 3D-concrete-printing (3DCP) technology in the construction industry over the past few years are providing the construction industry a chance to increase their productivity and cost efficiency [1, 2]. In the recent years sophisticated methods and theoretical approaches in conjunction with material development were proposed [3–5]. Nevertheless, the development of the 3DCP for construction industry is in the early stage. The high degree of automation with utilization of robotics impose new requirements concerning quality control, safety, and compliance with building standards [6]. In addition, the rheological requirements on the concretes suitable for the placement and extrusion during printing are primary challenges associated with implementation of 3DPC. Due to the complex requirements and novel nature of printable concretes, their optimal production techniques are still to be developed and adopted for on-site application. The research at the hand investigates methods of the foam concrete (FC) production and assesses their applicability for the 3D-foam-concrete-printing (3DFCP).

Initial studies and practical use of cement based materials for digital concrete construction are mostly related to usage of ordinary normal weight concrete or mortar (bulk density 2000-2500 kg/m^3) [1, 2]. Although, the normal weight concrete is irreplaceable for the construction of numerous infrastructural projects, there is a significant subdivision of applications where novel materials could be used. In recent years, various non-traditional building materials, such as clay (also with addition of straw as fibre reinforcement) and geopolymers were applied in 3DPC [7–10]. Sustainability and efficiency are some of the primary motives for the development of 3D-printing technologies, this also includes reduction of material consumption. Therefore, it is necessary to develop economical and sustainable concrete compositions for 3D-printing. A promising way to achieve this target is the use of foam concrete in 3D-printing. The significantly lower density of foam concrete allows to decrease construction self-weight and improve thermal insulation properties. In the context of 3D-printing, lower self-weight makes it possible to decrease pressure on the bottom layers thus improving stability of printed structures in the fresh state. Upon this background, 3D-printable foam concrete was recently introduced as a versatile material with excellent physical and sufficient mechanical properties [7, 11, 12].

The primary reason to regard foam concretes as economical and ecological materials is the presence of air cells up to 80% of their total volume. The large volume of air cells in foam concrete are often introduced by mechanical aeration of cement mortar using foaming agents [13]. Two primary methods of foam concrete production are: pre-foaming and mixed-foaming [14]. In the former, separately generated/prepared foam and cement matrix (paste/mortar) are mixed together to form foam concrete. In mixed-foaming method, foaming agent is added directly to the cement matrix mixer; this means cement matrix, foam, and subsequently foam concrete are generated in one mixer. The research at hand focuses on mixed-foaming method using intensive turbulence mixer, and cavitation disintegrator. It is known that intensive mixing improves dispersion of agglomerated cement and micro-filler particles and promotes accelerated hydration processes in foam [15]. Various concepts for delivering of the foam concrete to the print head are presented in Figure 1. All presented concepts presume mixed-foaming production method. In the Concept 1 after mixing, foam concrete is placed manually to the reservoir of the print head; see Figure 1a.

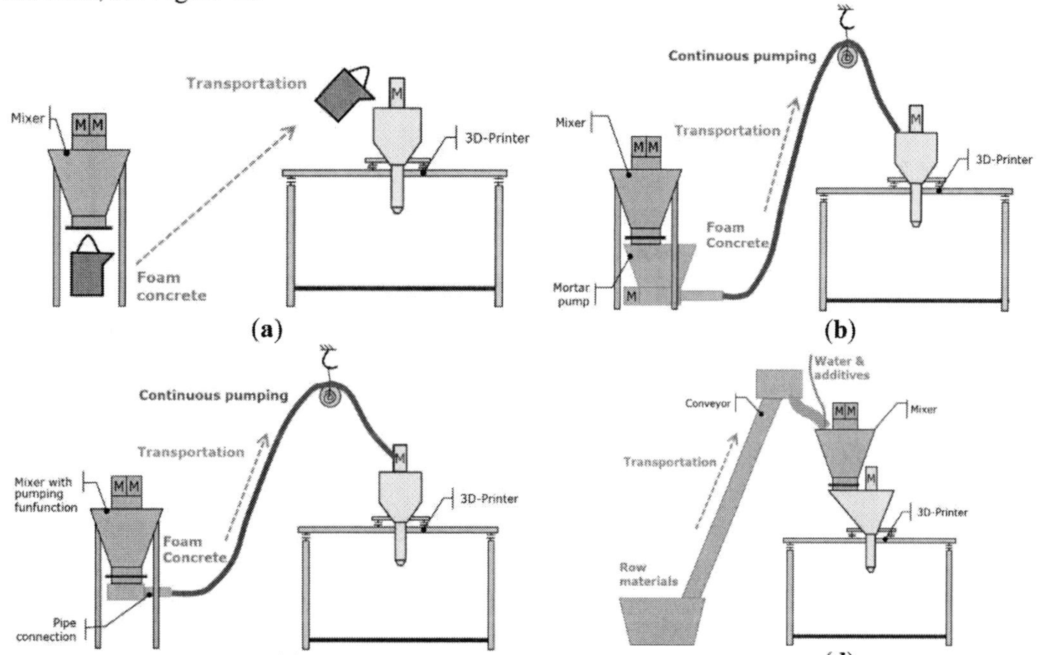

Figure 1. Different concepts of feeding system for continues foam concrete printing: (**a**) Concept 1: manual filling; (**b**) Concept 2: mixing and pumping; (**c**) Concept 3: pumping to integrated mixing system; (**d**) Concept 4: convening and fully integrated mixing system.

This concept prerequisite additional labour involvement and slows down construction rate. In Concept 2 foam concrete mixer is directly connected to a pump, which continuously delivers the foam

concrete to the print head reservoir; see Figure 1b. In comparison to the Concept 1, the Concept 2 eliminates time required for manual transportation and refiling of the print head reservoir.
In the Concept 3, the foam concrete mixer is directly connected to a print head. Technical particularities of the mixer used in Concept 3 allows delivery of the foam concrete to the print head reservoir. The Concept 3 appears as very suitable for 3DFCP and the technical particularities of chosen mixers in this study match overall technical requirements of this concept; see Section 2.2. In Concepts 4 mixer should be installed on the print head. The Concept 4 is similar to Concept 3, except position of the mixer and that supply of raw materials is automated. Raw materials in predefined quantities are transported and mixed on the printhead. One of the considerable advantages of Concepts 4 is that stability of the foam concrete is not effected by the pumping procedure. However, technical realisation of the Concept 4 is challenging. Existing solutions for inline mixing of dry constituents with water and admixtures must be adopted for mixing of the foam concrete on a concrete printhead [16]. It is worth noting that presented concepts in Figure 1 are also applicable to the normal weight printable concretes and alternative building materials used in 3DPC [8–10]. In such cases, instead of foaming agent, other chemical additives could be added.

2. Materials and methods

2.1. Materials

This study was conducted with a type II Portland composite cement CEM II/A-M (S-LL) 52.5 R (OPTERRA Zement GmbH, Werk Karsdorf, Germany). Hard coal fly ash Steament H-4 (STEAG Power Minerals GmbH, Dinslaken, Germany) was chosen as a secondary cementitious material. Details of chemical composition are provided in Table 1. A polycarboxylate ether (PCE) based Superplasticizer (SP) (MasterGlenium SKY 593, BASF Construction Solutions GmbH, Trostberg, Germany) was used in the cement-based matrix to adjust the workability at reduced water contents. The SP is characterized by density of 1050 kg/m^3 and water content of 77% by mass. A tenside based foaming agent (Centripor SK155, MC-Bauchemie GmbH & Co. KG, Bottrop, Germany) and a protein based foaming agent (Oxal PLB6, MC-Bauchemie GmbH & Co. KG, Bottrop, Germany) were used for production of the FC.

Table 1. Chemical composition of cement and fly ash.

Material	Density (kg/m^3)	Chemical composition (%)											
		Residue	SiO2	Al_2O_3	Fe_2O_3	CaO	MgO	SO_3	Ka_2O	Na_2O	Loss on ign.	CO_2	CL
CEM II/A-M(S-LL) 52.5 R	3.120	0.74	20.63	5.35	2.82	60.94	2.14	3.52	1.05	0.22	3.47	2.87	0.07
Fly ash H4	2.220					3.6		0.6		2.9	1.8		< 0.01

2.2. Overview of the mixers used

The details of two different mixers used for production of the foam concrete in this study are presented below:
(a) Cavitation disintegrator (CD). Initially CD was designed for production of the stable fuel mixtures and water-fuel emulsions; to improve heavy fuel oil by dispersing asphalt-resin components and to increase the efficiency of input additives. In addition, CD may be applied for preparation and activation of other emulsion and dispersion systems [17]. A novel hydrodynamic dispersion method for fine particles was developed and patented by Polakovs et al. [18]. Mechanical activation is associated with cavitation effect, which is achieved by high-speed rotation of toothed disks (up to 7000 rpm); see Figure 2. It must be noted that the application of CD for the preparation of FC is an innovative idea [15]. Since usage of CD is neither well known nor widespread in the construction industry, its technical components are worth noting. In CD, a set of toothed disks and impeller, fixed on a motorized shaft, form the rotor with frustoconical shape. The fixed body of the disintegrator is made of a conical lid and rectangular

grooves corresponding to toothed disks of the rotor. The body contains an inlet branch pipe and an outlet branch pipe. The electric motor is protected from liquid by a ferrule, installed in the intermediate disk, fastened together with the body on the motor flange. If the ferrule leaks, the liquid flows out on the leakage opening (ferrule operating monitoring). To discharge air from system, a plug is set in the upper part of the body. At first cement-based matrix with addition of the foaming agent are filled in the supply reservoir. During operation of the CD, the mixture cycles numerous times from supply reservoir through the tube to the disintegrator. CD used in this study had an output capacity of 10 m³/h, power 5.5 kW and was equipped with a frequency converter to control the rotational speed of the mixer working shaft in the range of 50 rpm to 7000 rpm.

(b) Laboratory turbulence mixer (TM). The TM used in this study allows to produce foam concrete in both techniques: mixed-foaming and pre-foaming method. Additionally, cement based slurries could be also produced and pumped under application of the air pressure. The mixer consists of conical mixing tank, electrical motor, bearing, pressure compensated coupling and vertical shaft, see Figure 2. The mixer enables mixing under pressure (up to 0.7 bar), which allows producing low density FC as well as unloading and pumping of the foam concrete mixture through excess pressure. The mixer is equipped with a frequency converter to control the rotational speed of the mixer working shaft in the range of 20 rpm up to 1000 rpm. Particular advantage by application of this type of the mixer in conjunction with 3DPC is integrated possibility of pumping of the foam concrete. Consequently, foam concrete could be pumped directly to the print head eliminating necessity of a separate screw pump. According to the technical data sheet of the TM-Mixer, the produced foam concrete under applied air pressure could be pumped up to 100 m distance. However, the producer doesn't specify spatial arrangement of the pipe by the given distance.

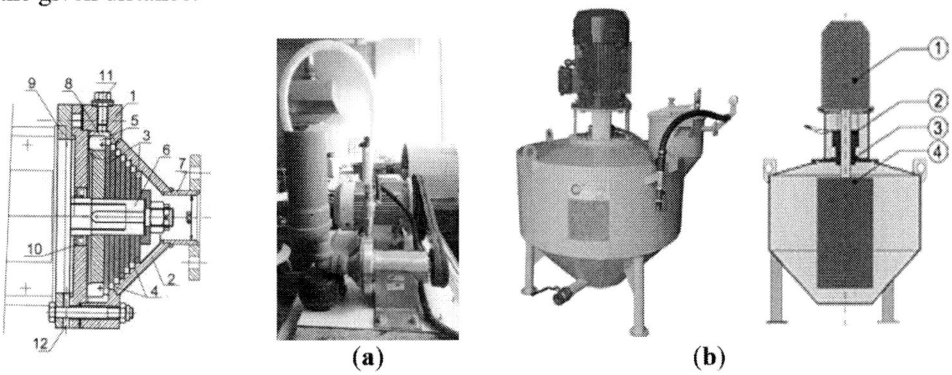

Figure 2. Overview of used mixers: (a) cavitation disintegrator (CD): 1. body; 2. conical lid; 3. toothed disks; 4. rectangular recess; 5. impeller; 6. shaft; 7. inlet branch pipe; 8. outlet branch pipe; 9. electric motor; 10. ferrule; 11. plug; 12. leakage opening, and (b) laboratory turbulence mixer (TM): 1. electrical engine; 2. bearing; 3. pressure compensated coupling; 4. vertical shaft.

2.3. Experimental procedures

The concrete mixing procedure was organized in the following way: all dry materials were premixed with addition of water and SP for approximately 2 min using brick trowel. At the end of premixing, foaming agent was added and mixed for additional 15 s. Then the mixture was filled into reservoir of cavitation disintegrator or turbulence mixer, respectively. In total, six foam concrete compositions were designed, see Table 2. Operating principles as well as technical procedures of used mixers are different, therefore different mixing protocols were used for the mixers; see Table 3.

In the fresh state, the plastic density of the mixtures was measured right after foam concrete production. The workability of the samples was assessed by visual observation of the homogeneity of performed FC. The compressive and flexural strength of foam concrete mixture was measured by casting of prism specimens with dimensions 160x40x40 mm^3 following the procedure provided by DIN EN 1015-11 (2007-05) [19]. At an age of 24 hours, produced specimens were wrapped into the polyethylene foil and stored under approx. constant temperature until testing day. Water absorption of the designed foam concrete compositions was measured in accordance with DIN EN 772-11 [20].

Table 2. Mixture compositions per m^3.

Constituents		Designed composition of mixture (per m^3)					
		M-1	M-2	M-3	M-4	M-5	M-6
Cement	(kg)	405	405	405	405	405	405
Fly ash (H-4)	(kg)	192	192	192	192	192	192
Tap water	(kg)	189	189	243	189	189	192
SP SKY 593	(%)*	0.5	0.7	0.1	0.5	0.5	0.6
FA SK-155	(%)**	1.2	1.2	1.2	-	0.7	0.7
FA Oxal PLB6	(%)**	-	-	-	1.2	-	-
$(w/c)_{eq}$	-	0.39	0.39	0.5	0.39	0.39	0.4
Used Mixer		CD	CD	CD	CD	TM	TM

* in the mass percentage of the binder, i.e. the total amount of cement and fly ash
** in the mass percentage of the cement

Table 3. Foam concrete mixing procedure.

Cavitation disintegrator (CD)	Turbulence mixer (TM)
0 min-2.0 min: 2100 rpm	0 min-2.0 min: 1500 rpm
2.0 min-4.0 min: 2400 rpm	2.0 min-4.5 min: 3000 rpm
4.0 min-6.0 min: 3600 rpm	4.5 min-6.0 min: 4800 rpm
6.0 min: Conveying of the FC with settled ca. 2100 rpm	6.0 min: Setting of 1 bar air pressure for conveying of the foam concrete

3. Results and discussion

3.1. Verification of the density

Figure 3 gives an overview of measured plastic (fresh state) densities of all tested mixers and corresponding drying loss after accelerated drying in a drying chamber at a temperature of 105 ± 5 °C until no change of mass is identified. As shown in the Figure 3, it was possible to achieve different densities of the foam concrete by varying of the constituencies and mixing methods. Drying loss is corresponding to the achieved plastic density and increases with reduction of the density [21]. Considering plastic density of 1712 kg/m^3 of composition M-4-CD it could be concluded that addition of protein based foaming agent didn't reveal sufficient foaming by mixed-foaming technique using CD-mixer. Based on performance of protein based foaming agent using CD-Mixer it could be expected that similar inadequate performance would be achieved by use of TM-Mixer.

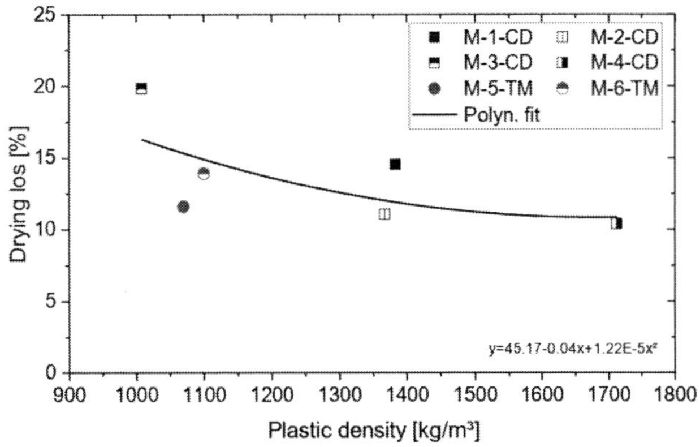

Figure 3. Drying loss depending on density of the foam concrete.

3.2. Effectiveness of the dosage of the foaming agent

Figure 4 presents the relationship between dosage of foaming agent and plastic density of foam concrete. Major factor influencing plastic density are dosage of the foaming agent and initial followability of the cement-based matrix that is influenced by w-b ratio and amount of the SP.

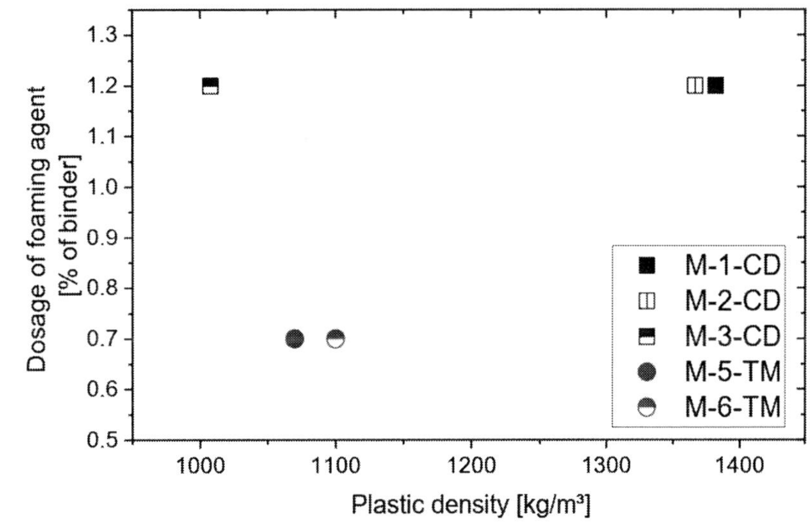

Figure 4. Effectiveness of the dosage of the foaming agent SK155 in different mixture compositions.

In case of CD-Mixer, increase of the flowability of the cement-based matrix by increasing the SP dosage from 0.5% to 0.7% (in the percent by mass of the binder, %bmob) yielded in achieving lower density. Reduction of the amount of SP while increasing w-b ratio in the composition M-3-CD yielded in achieving lower density in comparison to the compositions M-1-CD and M-2-CD. The influence of the flowability of the cement-based matrix on the foam concrete density is also evident in compositions M-5-TM and M-6-TM. Similar to the CD-Mixer, increase of SP-dosage yielded in reduction of plastic density when TM-Mixer was used.

Since compositions M-1-CD and M-5-TM distinguish only amount of foaming agent and type of used mixer, the results of these mixtures could be used to compare the performance of tested mixers. Using TM-Mixer with 0.7 %bmob of the foaming agent, foam concrete with plastic density of 1100 kg/m³ could be produced. In contrast, using CD-Mixer with a higher (1.2 %bmob) dosage of the foaming agent foam concrete with plastic density of 1382 kg/m³ could be produced. It is remarkable that, in case of TM-Mixer, higher foam volume i.e. lower foam concrete density could be achieved even with lower dosage of foaming agent.

3.3. Hardened state Properties of the foam concrete

The compressive and flexural strengths of tested foam concretes at various ages are presented in Table **4**. Mechanical properties of the foam concretes correlated with their density at the particular age. As shown in Figure 5, density at the testing age could vary from measured plastic density. The plastic density of the composition M-6-TM was 1100 kg/m³ and 28-day density was ca. 1230 kg/m³. This discrepancy can be imposed by pressure applied on the foam concrete during the pumping through the extruder and pipe (approx. 1.5 m with bend) to the casting molds. Presented results in the Figure 5 underline significant influence of the density of the foam concretes on their mechanical properties.

Table 4. Mechanical properties of the foam concrete.

Mix	Mixer	Plastic density [kg/m³]	Compressive strength [N/mm²]			Flexural strength [N/mm²]		
			7d	28d	>120d	7d	28d	>120d
M-1-CD	Cavitation disintegrator	1383	20.2	25.3	30.0	3.3	3.2	3.2
M-2-CD		1367	20.2	28.0	31.0	2.9	1.7	3.3
M-3-CD		1008	6.1	9.4	10.9	1.9	1.2	1.4
M-4-CD		1712	35.0	47.4	53.0	4.1	1.3	3.0
M-5-TM	Turbulence mixer	1070	14.2	17.4	19.3	2.4	1.6	1.4
M-6-TM		1100	4.4	5.8	4.9	1.7	1.3	0.9

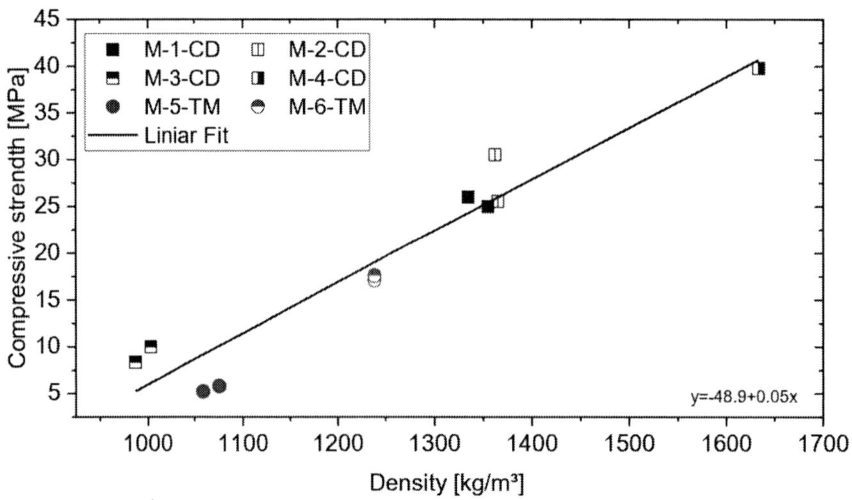

Figure 5. Relationship density – 28 d compressive strength for foam concretes prepared in 2 types of the mixers.

The results of the compressive tests show that, all tested foam concrete compositions increased their compressive strength with age; see Figure 6. Similar to normal weight concrete, the rate of the strength development was greater within first 7 days and decreases gradually afterwards; see Figure 6. Foaming of the composition M-4-CD was affected by ineffectiveness of the protein based agent, and as result of that, volumetric proportion of the cement was considerably higher than other compositions. This resulted in higher rate of the strength development compared to other compositions containing more entrapped air and corresponding lower volumetric proportion of the cements.

In contrast to compressive strengths, certain rate of the degradation of the flexural strength of all tested specimens over age is evident; see Table 5. In the course of aging over time, the water in the foam concrete evaporates due to drying. This leads to drying shrinkage among other adverse effects. Due to these and the highly porous structure of foam concrete, cracks may form in the inner pore microstructure [22, 23]. As a consequence, tensile and flexural behavior of foam concrete decreases, as confirmed by the advancement of flexural strengths of the measured foam concretes. It must be noted here that, all tested specimens have identical storing conditions, and density on the testing day vary in admissible deviation of 2-5 %.

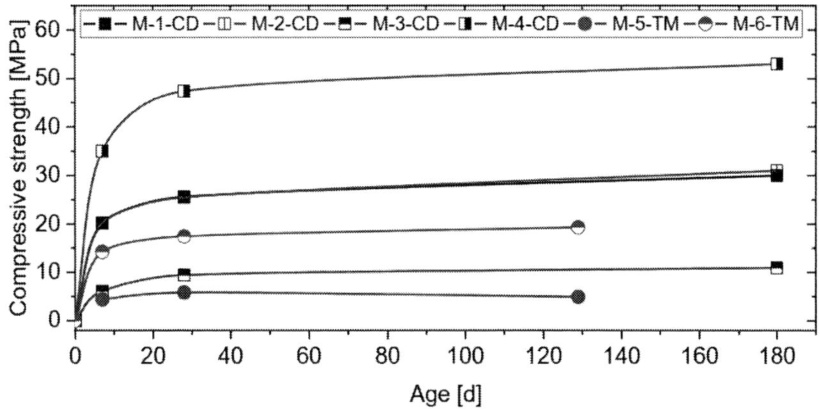

Figure 6. Long term compressive strength development.

3.4 Water absorption

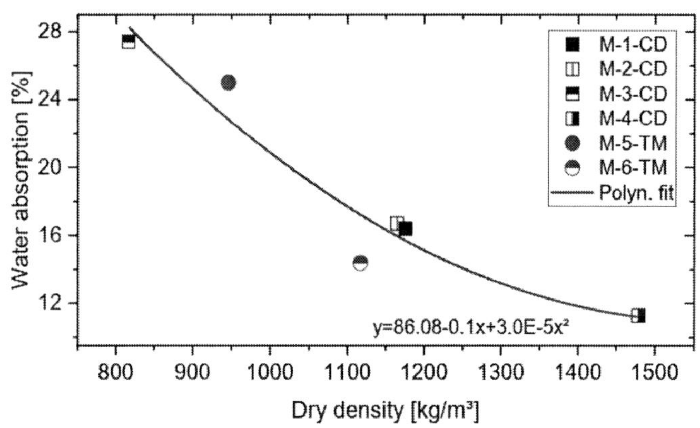

Figure 7. Relationship dry density – water absorption, for foam concretes prepared in 2 types of mixers.

Density of the foam concrete influence rate of the water absorption, which is evident in Figure 7. For instance, composition M-3-CD with the lowest dry density exhibited highest water absorption of ca. 27% by volume. Results of the composition M-4-CD shows that decrease of the porosity and corresponding increase of the density implies reduction of the water absorption. This results underlines that production technique has also influence on the rate of the water absorption. It could be seen that the composition M-7-TM with lower density than compositions M-1-CD and M-2-CD has lower water absorption. These results are based on quality of pores effected by the production of the foam concrete using two different types of mixers. Thus, formation of pores by use of TM-Mixer has better influence on resistance against water absorption and associated durability of produced foam concrete.

Conclusion

This study has shown that turbulence (colloidal) mixer and cavitation disintegrator can be applied for the production of the foam concretes with varying mechanical and physical properties. Obtained range of the densities – 800 kg/m³ to 1500 kg/m³ – correspond to the 28 day compressive strength from 5 MPa to 45 MPa. The flexural strength of tested foam concretes tends to decrease with their age, potentially due to effects of drying shrinkage. This finding needs to be further examined by microscopic investigations with aim to describe and prevent the origins. The successful production of foam concrete following mixed-foaming method using CD- and TM-Mixer made it feasible to deliver foam concrete directly to the printhead of a 3D-printer. In addition, analyses of the achieved densities revealed beneficial influence of the flowability of the cement-based matrix on foam concrete properties. Considering this fact, subsequent addition of the accelerator after foaming of a flowable cement-based matrix need to be explored to meet requirement on the consistency of printable foam concretes presented in [7]. Water absorption measurements revealed the influence of used mixers and corresponding different foaming techniques on microstructure and pore distribution of foam concretes.

Acknowledgments

The authors express their sincere gratitude to the German Federal Ministry for the Environment, Nature Conservation, Building and Nuclear Safety (BMUB) for funding this project in the framework of the research initiative Zukunft Bau of the Federal Institute for Research on Building, Urban Affairs and Spatial Development (BBSR). We also thank our industrial partners OPTERRA Zement GmbH, MC Bauchemie Müller GmbH & Co. KG, Kniele GmbH, and BAM Deutschland AG.

References

[1] Schutter G de, Lesage K, Mechtcherine V, Nerella V N, Habert G and Agusti-Juan I 2018 Vision of 3D printing with concrete — Technical, economic and environmental potentials *Cement and Concrete Research*

[2] Buswell R A, Leal de Silva W R, Jones S Z and Dirrenberger J 2018 3D printing using concrete extrusion: A roadmap for research *Cement and Concrete Research* **112** 37–49

[3] Roussel N 2018 Rheological requirements for printable concretes *Cement and Concrete Research* **112** 76–85

[4] Wangler T *et al* 2016 Digital Concrete: Opportunities and Challenges *RILEM Tech Lett* **1** 67

[5] Perrot A, Rangeard D and Pierre A 2016 Structural built-up of cement-based materials used for 3D-printing extrusion techniques *Mater Struct* **49** 1213–20

[6] Krause M, Otto J, Bulgakov A and Sayfeddine D Strategic optimization of 3D concrete printing using the method of CONPrint3D® *35th International Symposium on Automation and Robotics in Construction (ISARC 2018)*

[7] Markin V, Nerella V N, Schröfl C, Guseynova G and Mechtcherine V 2019 Material Design and Performance Evaluation of Foam Concrete for Digital Fabrication *Materials*

[8] Perrot A, Rangeard D and Courteille E 2018 3D printing of earth-based materials: Processing aspects *Construction and Building Materials* **172** 670–6

[9] Panda B, Paul S C, Hui L J, Tay Y W D and Tan M J 2017 Additive manufacturing of geopolymer for sustainable built environment *Journal of Cleaner Production* **167** 281–8

[10] Nematollahi B, Vijay P, Sanjayan J, Nazari A, Xia M, Naidu Nerella V and Mechtcherine V 2018 Effect of Polypropylene Fibre Addition on Properties of Geopolymers Made by 3D Printing for Digital Construction *Materials (Basel, Switzerland)* **11**

[11] Falliano D, Gugliandolo E, Domenico D d and Ricciardi G 2019 Experimental Investigation on the Mechanical Strength and Thermal Conductivity of Extrudable Foamed Concrete and Preliminary Views on Its Potential Application in 3D Printed Multilayer Insulating Panels *1st International Conference on Concrete and Digital Fabrication* **19**

[12] Lublasser E, Adams T, Vollpracht A and Brell-Cokcan S 2018 Robotic application of foam concrete onto bare wall elements - Analysis, concept and robotic experiments *Automation in Construction* **89** 299–306

[13] Ramamurthy K, Kunhanandan Nambiar E K and Indu Siva Ranjani G 2009 A classification of studies on properties of foam concrete *Cement and Concrete Composites* **31** 388–96

[14] Aldridge D Introduction to foamed concrete: what, why, how?

[15] Namsone E, Šahmenko G and Korjakins A 2017 Durability Properties of High Performance Foamed Concrete *Procedia Engineering* **172** 760–7
[16] Wilson Ricardo Leal da Silva 2017 3D Concrete Printing - from material design to extrusion. [Presentation Slides] Retrieved from Danish Technological Institute website: https://www.dti.dk/projects/3d-printed-buildings/36993
[17] Justs J, Shakhmenko G and Mironovs, V., Kara, P. Cavitation Treatment of Nano and Micro Filler and Its Effect on the Properties of UHPC
[18] Mironovs V., Polakovs A. and Korjakins A. 2011 A Method and apparatus for suspension preparation *Patent LV14364B Int.cl. C01B33/00 RTU*
[19] DIN EN 1015-3 Prüfverfahren für Mörtel für Mauerwerk – Teil 3: Bestimmung der Konsistenz von Frischmörtel (mit Ausbreittisch); Deutsche Fassung EN 1015-3:1999+A1:2004+A2:2006
[20] Deutsches Institut für Normung e.V. and European Committee for Standardization. Prüfverfahren für Mauersteine – Teil 11: Bestimmung der kapillaren Wasseraufnahme von Mauersteinen aus Beton, Porenbetonsteinen, Betonwerksteinen und Natursteinen sowie der anfänglichen Wasseraufnahme von Mauerziegeln; Deutsche Fassung EN 772-11:2011 (accessed 17 Apr 2019)
[21] Amran Y M, Farzadnia N and Abang Ali A A 2015 Properties and applications of foamed concrete; a review *Construction and Building Materials* **101** 990–1005
[22] Jones M, Zheng L and Mohammad M 2016 Bubble Structure, Stability and Rheology of Foamed Concrete *The 9th International Concrete Conference 2016: Environment, Efficiency and Economic Challenges for* 682–94
[23] Jones M R, Ozlutas K and Zheng L 2017 High-volume, ultra-low-density fly ash foamed concrete *Magazine of Concrete Research* **69** 1146–56

IOP Conf. Series: Materials Science and Engineering **660** (2019) 012040 doi:10.1088/1757-899X/660/1/012040

Correlation between ultrasound velocity, density and strength in metal-ceramic composites with added hollow spheres

A Tatarinov[1], A Shishkin[1] and V Mironovs[1]

[1] Laboratory of Powder Materials, Riga Technical University, Kipsalas Street 6A-331, LV-1048, Riga, Latvia

E-mail: aleksejs.tatarinovs@rtu.lv

Abstract. Ultrasound velocity is known as a non-destructive predictor of strength in some construction materials like concrete. In multi-phase materials, where the physical properties of components differ by orders, the strength prediction becomes more ambiguous due to more complicated interactions between the components. As an example of such material, sintered metal-ceramic composites with added hollow spheres were taken. The matrix was a mixture of powdered iron and clay, where hollow admixtures were cenospheres or microsphreres of fly ash. By varying the iron-to-clay ratio, sintering temperature and cenospheres content, 17.4% variation of the bulk density ρ and 35.3% variation of the compression strength σ were achieved. Ultrasound velocity C was measured in cylindrical specimens and showed 8.6% variation. All mentioned parameters positively correlated with each other. The relationship between C and σ had expressed non-linear character. Normalization of the relationship by ρ helped improving the C-σ correlation. The non-linear monotonous function $C = \rho^{-2}\cdot\sigma^{-4}$ provided the closest correlation with the Spearman rank correlation coefficient 0.942 and the Pearson linear correlation coefficient 0.943. The hollow spheres content was the main determinant of density ρ in this material.

1. Introduction

Innovative materials offer growth of application opportunities in machinery, construction, transport and other fields. During the recent decades, promising expectations have been assigned to a diversity of types of syntactic foams [1]. Metal matrix composites and metal/ceramic composites are designed to reduce weight and costs of products maintaining high strength properties [2]. Addition of hollow particles such as fly ash cenospheres into metal matrix composites and syntactic foams may significantly reduce the total weight of the composite and increase its impact absorption properties [3]. On the other hand, the addition of hollow particles results in decrease of the mechanical strength as the effect similar to increased porosity [4]. The trade-off between the achieved lightweight, low costs and the necessary strength is an important issue in designing of the syntactic foams with hollow fillers.

Non-destructive evaluation of the mechanical properties of syntactic foams and strength prediction can be valuable components of the material development and testing. Ultrasonic examination based on measuring of quantitative parameters of ultrasound propagation, the most commonly, ultrasound velocity also called as ultrasound pulse velocity and speed of sound has proven itself in different testing applications. The most widely the method has been used for examination of cementitious materials, particularly, concretes [5]. Correlations between ultrasound velocity and compression strength of concretes are well established and are put into testing standard and serve as a basis for ultrasonic testing [6]. Usually, this correlation is of a non-linear type that is explained by both non-linear dependences of ultrasound velocity on the elasticity modulus (as follows from the theory of elasticity and experiment, the velocity is proportional to the square root of the elasticity modulus [7] and of the elasticity modulus versus the strength [8]. Various authors offer empirical dependences of the strength on the velocity, the most commonly described by the equation: $\sigma = a*\exp(b*C)$, where σ is strength, C is ultrasound velocity; a and b are empirical coefficients for different compositions and conditions, such as dry or wet [9].

There were attempts to apply ultrasonic measurements for properties characterization in syntactic foams. Influence of volume fraction of different components, particularly, micro-balloons on ultrasound longitudinal and shear wave velocities and attenuation has been demonstrated [10]. However, the creation of a generalized strength prediction model based on the ultrasonic parameters is difficult due to complex influences of the composition, structure, and interaction between components in the multi-phase material. The inclusion of cenospheres - hollow gas-filled particles in a high concentration may create unexpected effects.

The purpose of the study was to test the applicability of ultrasound velocity measurements for the strength prediction in metal-ceramic composites with hollow particles (cenospheres) and investigate the empirical correlation. The variables that affected density and strength of the material were the metal-ceramic ratio, the presence or absence of the cenospheres content, and the sintering temperature, i.e. the factors related to composition, structure and manufacturing technology.

2. Materials and methods

2.1. Composite specimens

The purpose of material selection was to provide such range of specimens, where the variation of strength was determined by influence of three independent factors: composition of the metal-ceramic base, bonds between the metal and ceramic components and presence of hollow particles as a low density filler. Previous attempts have demonstrated the feasibility of fabrication of such metal-ceramic syntactic foams with cenospheres with compression strength varying in the range 150-340 MPa [11]. In the present study, two variations of the metal-ceramic compositions were laid – with the weight ratios of raw iron powder to clay 1:1 and 2:3. The ceramic materials was natural homogenised illite red clay from Liepa deposition (Lode JSC, Liepa, Latvia). The homogenised clay was dried at 105 °C for 24 h, milled with a RETSCH PM 400 (Retsch, Haan, Germany) ball mill for 15 min, and sieved to limit the particle size to <100 μm. The bonds within the ceramic component and between the ceramic and metal components were affected by gradual variation of sintering temperature. Four grades of sintering temperature were applied – 1050, 1100, 1150 and 1180 ^{0}C, stopped at the maximum above which burning of clay was possible. The hollow particles were cenospheres – waste or byproduct of coal burning at thermal power plants presenting hollow air-filled spheres obtained from Biotecha Latvia Ltd (Riga, Latvia). The cenospheres are typically sized 0.1 – 0.3 mm in diameter with a silica wall 10 – 30 microns thick and have bulk density in the range of 0.4 – 0.8 g/cm^3. Chemical and grading compositions for the applied clay and cenospehers were described in details in the previous studies of the authors [11, 12]. Cenospheres were uniformly mixed to one part of specimens with the weight ratio to the rest metal-ceramic content of 2:5, while another part of specimens contained no cenospheres. Introduction of cenospheres had the purpose to reduce total bulk density of the composite making it more lightweight, but compromising the material strength to some extent. The groups of specimens were designated according to the ratios of bulk weights of three components in raw mixtures before sintering: M, C and S, where M – is metal (iron) powder; C – is clay or ceramic component; and S –is cenospheres content. Four basic groups were formed: MCS 4-6-4; MCS 5-5-4; MCS 4-6-0; and MCS 5-5-0 with corresponding ratios of M, C and S components. Two first groups contained cenospheres and two last groups were composed only of metal and ceramics. Each group was divided into subgroups by 4 mentioned grades of sintering temperature. Each subgroup contained 5 identical specimens. Thus, the total number of specimens reached 4x4x5 = 80. The sintered specimens had cylindrical shape with a diameter 19 mm and a height in the range of 16-20 mm depending on the shrinkage during sintering.

2.2. Ultrasonic testing

Ultrasound velocity was determined by conventional pulse time-of-flight measurements using through transmission. The experiment layout is presented in figure 1. To generate and receive ultrasonic signals, a pair of miniature custom-made piezoelectric transducers was used. The excitation pulse was a two-period tone-burst sine enveloped by the Gauss function. The working frequency of 500 kHz coincided with one of resonant frequencies of the transducers. It was selected by the following reasons. Due to the presence of hollow particles and pores in the volume, the material exhibited high attenuation of ultrasound at frequencies approaching to 1 MHz and higher. It caused weak detection of the first arrival of ultrasonic signals masked by noises and being under the level of resolution. By decrease the frequency much lower 500 kHz, the ultrasonic wavelength becomes comparable with the specimen's size that may cause undesirable manifestations of geometrical dispersion of the velocity. Ultrasonic wavelength at 500 kHz was in the range of 4-8 mm that is smaller than the specimens' size but several times larger than the elements of internal structure – sintered metal-ceramic clusters and cenospheres – allowing to consider the medium as a continuous one.

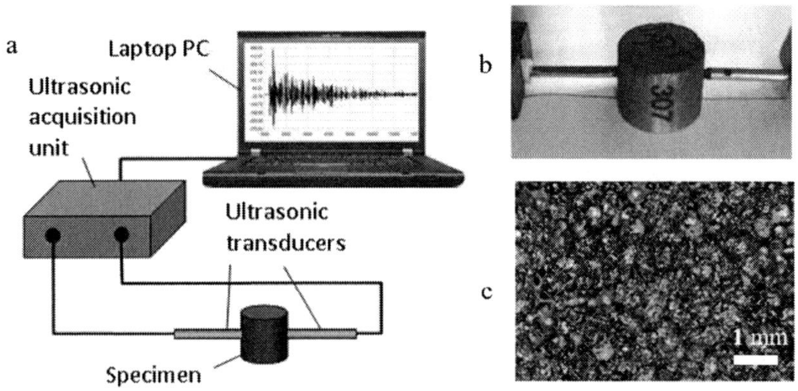

Figure 1. Illustrations of experiment: a – layout of ultrasonic testing; b – transducers application to MCS specimen; c – microphotograph of MCS internal structure at magnification x200.

Ultrasound velocity was calculated as C = D/(t – dt), where D was the specimen's diameter, t – time-of-flight, and dt – time delay in the transducers measured at zero base. Pulse repetition time of 1 millisecond was chosen sufficient to ensure that all reverberation from multiple reflections was damped. Time-of-flight t was measured by the first arrival of the pulse and related to the fastest propagating longitudinal wave mode. Other parameters of the testing device were the following: excitation voltage 140 V peak-to-peak, sampling rate 30 MS/s, ADC 10-bit, averaging number 32, and tuneable amplification. Measurements were repeated 4 times in each specimen with replacing the transducers. The average reproducibility of ultrasound velocity was 25 m/s in specimens without cenospheres and 35 m/s in specimens containing cenospheres.

2.3. Density and strength testing

Because the specimens' shape was a regular cylinder, it allowed determination of the bulk density by simple weighting specimens in the air and measurement of the diameter and height at several locations. After passing ultrasonic testing, the specimens were statically loaded by compression to the maximum force, after reaching of which the destruction occurred. For compression tests, a universal testing machine of Instron, model 8801, was used. Strain rate was 0.01kN/s. Compression strength was determined as the ratio of the maximum force to the specimen's cross-section.

3. Results and discussion

The summarized data in all specimens' dataset, including average values of ultrasound velocity, compression strength and bulk density with some statistical estimates are presented in Table 1. The data shows the variance of all parameters, where compression strength has the highest relative variation and ultrasound velocity has the lowest one. According to the skewness estimations, the data distribution is approximately symmetric for ultrasound velocity and density (<0.5 in absolute value), but strength has a moderate positive skewness with upper values more dispersed than lower ones. Low values of kurtosis (<3 in absolute value) indicate that the values distribution is uniform for all parameters with no outlier data problems. Variation coefficients of the parameters in subgroups of specimens of the similar composition and sintering temperature (n=5) are the measures of data deviation in conditionally identical specimens. On the average in the subgroups, it was 1.99% for ultrasound velocity, 6.25% for compression strength and 1.39% for bulk density. The ratios of variation coefficients in the entire dataset of specimens to the same in subgroups was 4.3 for ultrasound velocity, 5.7 for compression strength and 12.5 for bulk density. Therefore, ultrasound velocity and compression strength had close relative variability caused by the factors of composition and sintering temperature, but it was 2-3 times higher for density.

Table 1. Ultrasound velocity, compression strength and bulk density in specimens' dataset (n=80).

Parameter, measurement units	Average value	Standard deviation	Variation coefficient (%)	Skewness	Kurtosis
Ultrasound velocity (m s^{-1})	3270	± 280	± 8.6	0.05	– 0.91
Compression strength (MPa)	183	± 64	± 35.3	0.56	– 0.50
Bulk density (g cm^{-3})	3.82	± 0.66	± 17.4	0.31	– 1.01

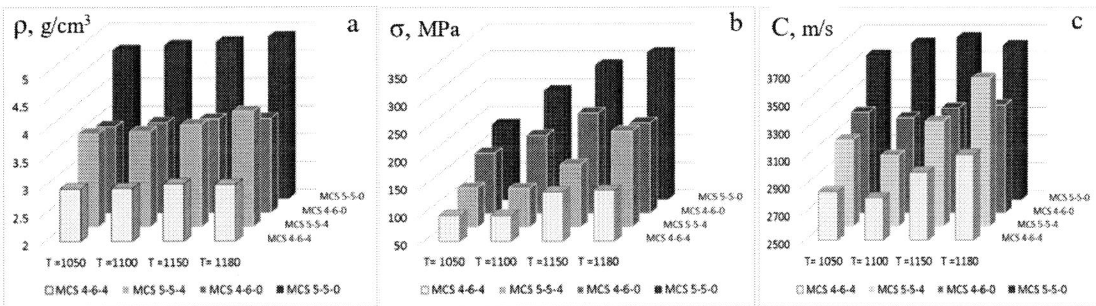

Figure 2. Bulk density ρ (a), compression strength σ (b) and ultrasound velocity C (c) in groups of MCS specimens' groups differing by their composition and sintering temperature T.

Influence of the varied factors – composition and sintering temperature on three studied parameters – ultrasound velocity, compression strength and bulk density is illustrated in figure 3 (a-c). The following common trends were revealed. All the parameters increased by increase of sintering temperature, reaching maximum at the highest temperature of 1180^{0}C. The explanation of this fact is strengthening of bonds in the sintered components till achievement of the maximum strength and stiffness of the ceramic component. Simultaneous compaction of the material caused the increase of its density. In the relation to the specimens' composition, all the examined parameters were higher in the specimen's groups of higher iron content due to a higher strength, stiffness and density of iron comparing to ceramics. Addition of cenospheres in the composites had a similar effect to the introduced porosity explained by the hollow structure and low strength of cenospheres. It caused the compromised strength of the composite in the whole, as well as lower ultrasound velocity and lower density. Despite the common trends to increase with sintering temperature and increase of iron content, the changes of each parameter had their individual peculiarities according to changes of sintering temperature and composition. It was caused, firstly, by a complex nature of the mutual dependences, and secondly, by unaccounted factors related to measurement such as non-uniformity of the specimens by the volume. Additional effects on strength can be caused by a complex interference between the components with sharply differing properties – metal powder particles, clay and cenospheres. Such effects were found for metal-cement composites with additives that manifested in voids and tiny cracks on the boundaries between the components [13].

The strength of correlation between the parameters was quantified by the Pearson product-moment correlation coefficient as the measure of linear dependence and by the Spearman's rank correlation coefficient (Spearman's rho), the measure of the relationship described by a monotonic function, including a nonlinear one. Cross-correlations between the examined parameters presented in Table 2 show that all the parameters are positively correlated.

Table 2. Cross-correlations between ultrasound velocity (C), compression strength (σ) and bulk density (ρ) by Pearson and Spearman.

Type of correlation	Pearson		Spearman	
Parameter	σ	ρ	σ	ρ
C	0.875	0.886	0.894	0.867
σ	-	0.816	-	0.781

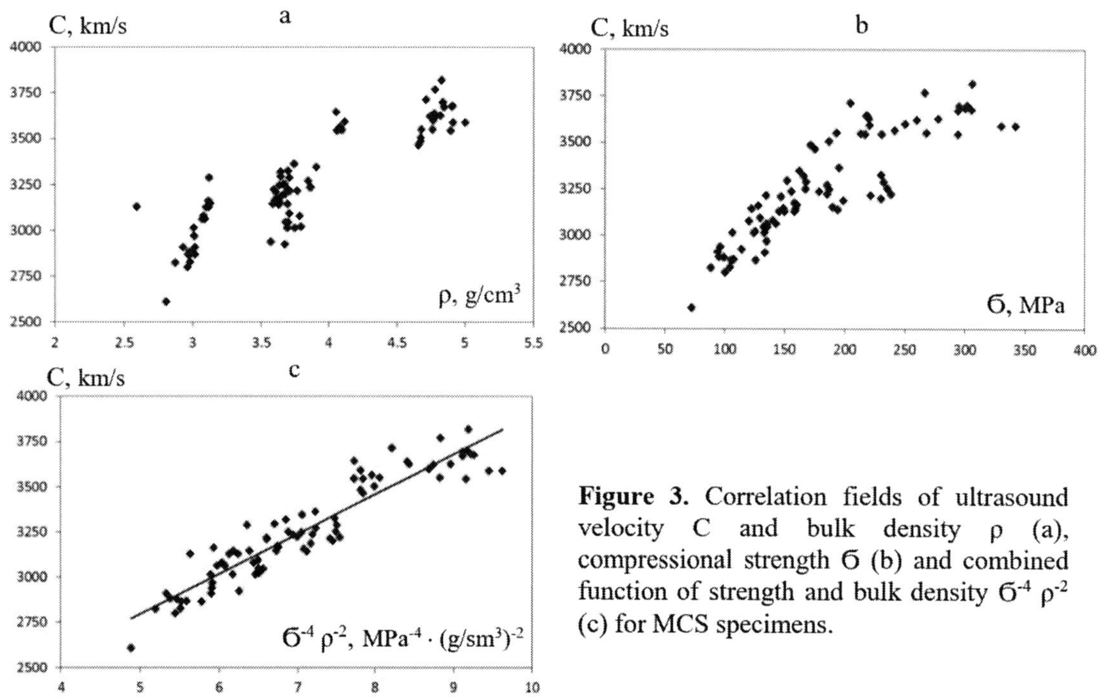

Figure 3. Correlation fields of ultrasound velocity C and bulk density ρ (a), compressional strength σ (b) and combined function of strength and bulk density $\sigma^{-4}\rho^{-2}$ (c) for MCS specimens.

The Pearson and Spearman correlation coefficients ranging between 0.8 and 0.9 show close dependences between the parameters. Increase of ultrasound velocity is due to both increase of strength and density. The discrepancy with the theory of elasticity teaching that ultrasound velocity is inversely proportional to the square root of the density [7] is explained by the positive correlation of the density with the strength. The specimens containing cenospheres are less dense and mechanically weaker than the specimens without it. The same relates to the specimens sintered at lower temperature in comparison to those sintered at higher temperature. In the majority of construction materials such as concrete, plastics and metal alloys, the strength positively correlates with the elasticity modulus. According to the theory of elasticity and the practice of ultrasonic testing, speed of sound is directly proportional to the square root of the elasticity modulus [7]. Hence, the increase of ultrasound velocity accompanying the increase of density occurs due to associated effects of the increment of the strength and elasticity modulus rather than the increase of density.

Correlations fields of ultrasound velocity C and bulk density ρ (a) and compressional strength σ (b) are presented in figure 3 (a, b). In order to find closer correlations and bring the dependences to a linear character that is more convenient for the strength prediction, varying degrees of strength σ and combinations of products σ with bulk density ρ at varying degrees were tried to correlate with C. The results of these trials are presented in Table 3. By increasing the negative degree of σ and applying its product with ρ, the Pearson linear correlation coefficient increased reaching its maximum of 0.943 for the dependence C = σ-4 ρ-2. The results demonstrated the possibility of creation of a strength prediction model for metal-ceramic composites with hollow inclusions increasing the inner porosity based on measurements of ultrasound velocity related to the longitudinal wave and bulk density of the material. At a constant ratio of metal and ceramic contents, the material density will be determined by the content of hollow inclusions.

Table 3. Correlation between ultrasound velocity (C) and mathematical expressions of compression strength (σ) and bulk density (ρ) by Pearson and Spearman.

Type of correlation	Expressions of σ and ρ					
	σ	σ^{-2}	σ^{-4}	$\rho\,\sigma$	$\rho\,\sigma^{-2}$	$\rho^{-2}\,\sigma^{-4}$
Pearson	0.875	0.895	0.902	0.907	0.934	0.943
Spearman	0.894	0.904	0.904	0.942	0.938	0.938

4. Conclusion

The study has demonstrated a non-linear character of dependence of ultrasound velocity on strength and density in sintered metal-ceramic composites with hollow fillers, where sintering temperature and the material composition were variables. The strength prediction model for such materials based on measurement of velocity of the longitudinal ultrasonic wave has to account bulk density in order to provide the closest correlation.

Acknowledgement

Authors acknowledge Prof. Irina Hussainova from the Tallinn University of Technology for her help in producing experimental material. The study was supported by the project 7045.6 of the Riga Technical University.

References

[1] Duarte I and Ferreira J M F 2016 *Materials* **79** 1
[2] Karakulov V V and Smolin I Y 2016 *International Conference on Physics of Cancer Proceedings (Tomsk)* vol 1760 (AIP Publishing LLC) 020024
[3] Santa Maria J A, Schultz B F, Ferguson J B, Gupta N and Rohatgi P K 2014 *J Mater Sci* **49** 1267
[4] Orbulov I N and Ginsztler J 2012 *Acta Polytechnica Hungarica* **9** 43
[5] Malhotra V M and Carino N J 2004 Handbook on nondestructive testing of concrete (Boca Raton: CRC press) p 384
[6] British Standards Institution 2004 EN 12504-4 Testing concrete – determination of ultrasonic pulse velocity
[7] Krautkrämer J and Krautkrämer H 1990 Ultrasonic testing of materials (Berlin: Springer-Verlag) p 497

[8] Jurowski K and Grzeszczyk S 2015 *Procedia Engineering* **108** 584
[9] Panzera T H, Christoforo A L, Cota F P, Borges P H R and Bowen C R 2010 Ultrasonic pulse velocity evaluation of cementitious materials *Advances in Composite Materials - Analysis of Natural and Man-Made Materials* ed P Tesinova (Rijeka: Intech) chapter 17 pp 411-436
[10] Mylavarapu P and Woldesenbet E 2008 *J. Cellular Plastics* **44** 203
[11] Rugele K, Lehmhus D, Hussainova I, Peculevica J, Lisnanskis M and Shishkin A 2017 *Materials* **10** 828
[12] Shishkin A, Mironovs V, Zemchenkov V, Antonov M and Hussainova I 2016 *Key Engineering Materials* **674** 35
[13] Mironov V, Pundiene I, Tatarinov A and Baroninsh J 2015 *Construction Science* (Riga: Riga Technical University) **16** 16

Sustainable construction success indicators

M Vilnītis[1], **VA Lapsa**[2] **and A Veinbergs**[3]

[1] PhD Professor, Faculty of Civil Engineering, Riga Technical University, Riga, Latvia
[2] PhD Leading researcher, Faculty of Civil Engineering, Riga Technical University, Riga, Latvia
[3] PhD student, Faculty of Civil Engineering, Riga Technical University, Riga, Latvia

E-mail: martins.vilnitis@rtu.lv

Abstract. The proposed paper examines the success indicators of sustainable construction projects, and its relationship to building codes with specific focus on safety, quality and budget aspects in terms of their implications and impacts prior, during and after construction. Such a research will be very beneficial to the construction industry and the government of all countries in terms of significant money savings and preservation of national wealth and resources through sustainable construction approaches. Consideration of sustainability at the design stage is emphasized. Building safety and the roles of building codes in prevention of structural failures are also covered in the paper together with factors affecting building failures and methods for their prevention in early design phases. Construction process is introduced in the paper from the perspective of modern total quality management. Concepts of quality management, quality control, quality assurance and six sigma and how they relate to building sustainability are discussed with examples. At least the importance of appropriate maintenance of building during expected life time are partly covered in the paper. It was necessary, in an initial stage, to define the key indicators that should be considered in order to assess the main aspects of sustainable construction. This paper intends to present the analysis that was performed in order to select some of quality evaluation indicators.

1. Introduction

Sustainable building is a concept that began its development nearly two decades ago. With sustainable buildings, it is intended to establish building practices that allow minimizing the buildings impacts, build and manage buildings with an adequate balance between environment, society and economy. Several studies have been developed with the goal of understand the reason of these weak implementation. They have concluded that one of the main reasons is because the building stakeholders still consider sustainable practices more expensive [5] [9] [11] [13].

In order to perform a cost-benefit analysis of sustainable construction solutions, it was necessary, in an initial stage, to define the key indicators that should be considered in order to assess the main aspects of sustainable construction. The assessment process could be simplified by using a ready-made wide set of individual sustainability impact indicators for construction products and processes, sensitive to the parameters that determine them to a great extent [24] [2] [16].

2. Safety, Quality and Productivity in Construction

A recent industry survey indicates that there is synergy between lean production and safety management practices in construction [3]. Lean production is based on the Toyota Production System (TPS) to achieve best quality, lowest cost, shortest lead time, best safety and high morale. Lean production tools include Just-In-Time (delivering right items at right time and amount); autonomation ("jidoka" - never letting defective items go to next station; stopping workers and machines when defects are significant); production leveling (reducing variability and inconsistency); and continuous improvement (there is no best; there is always an opportunity to improve processes). These tools can be applied to improving safety management practices such as management commitment, staffing for safety, planning for safety, safety education/training, worker involvement, evaluation and recognition, subcontractor involvement, and accident investigation. Since the lean approach combines quality with productivity, and it can improve safety, one can conclude that safety, quality and productivity can be collectively improved [18]. The question of how safety is related to productivity often comes up in decisions concerning how much attention and investment must be directed toward safety programs in the construction industry, especially in smaller and mid-sized firms.

Construction projects are expected to simultaneously achieve multiple goals concerning cost, schedule, quality, productivity and safety. These factors are related, while they may be in apparent conflict with each other. Because of pressures on cost, schedule, and productivity, workers may engage in unsafe behavior, believing that this enables them to be more productive On the other hand, there is strong belief in the industry that no trade off exists between safety and productivity; and in fact they might be synergistic [12] [17] [23].

There is some evidence to support the notion of safety being negatively correlated with productivity. However, there is stronger support for the opposite; that is, safety and productivity are positively correlated, meaning there would be no loss of productivity due to taking safety measures or executing safety programs in a construction project. On the negative correlation side, there are reports in the safety literature that workers may ignore safety due to productivity concerns. In a recent study, safety audits conducted at 200 residential sites showed 59 % non-compliance, meaning widespread disregard for safety. It was found that this came about as a result of concern over taking time off work to learn new safety measures; fear of cost of compliance; and lack of knowledge (inadequate training). In another study, it was observed that managers believed there was not enough time to work safely and safety practices decreased productivity. Research on the impacts of schedule pressures on safety indeed showed increases in injury rate when such pressures are present. Another research study found that workers cut corners on safety to be more productive for fear of losing their jobs. In a study on roofers, it was observed that productivity dramatically decreased when fall arrest systems were used. Workers lost a lot of time adjusting their lanyards on steep roofs resulting in loss of productivity. In a more balanced study, it was concluded that there was a strong negative correlation between safety and productivity when new safety strategies were implemented; however, the correlation turned positive after some time elapsed in the project.

In support of positive correlation between safety and productivity, a U.K. study indicated that productivity losses were higher with safety violations than with preventive safety practices. Similar evidence in a different study suggested that safety management is a positive influence on productivity because of declining task performance by the entire crew in case of injuries. According to this study, damaged equipment and materials; time for additional record keeping, accident investigation and training also slow down production. According to Hinze's [12] distraction theory, workers are more productive when they perceive hazards to be minimal, than working with the perception of being in a high hazard environment. Productivity is compromised when distractions due to hazards turn the worker's attention to mitigating safety risks. A recent study on manufacturing productivity vs. safety concluded that management should view safety and operating performance measures as being in concert with each other rather than competing entities. There is more scrap, rework and less employee involvement when safety performance is poor. Based on 18 case analyses, a study by Maudgalya, [17] revealed that several organizations realized a payback of their monetary investment in safety initiatives in less than a year. Following the implementation of workplace safety initiatives, organizations observed an average increase of 66% in productivity, 44% in quality, and 71 % in cost benefits. The study concluded that as a business case, safety can assist an organization in achieving the long-term benefit of

IOP Conf. Series: Materials Science and Engineering **660** (2019) 012041 doi:10.1088/1757-899X/660/1/012041

operational sustainability. Finally, a study by McGraw [18] suggests that contractors are experiencing positive business outcomes from safety programs in terms of shortened schedules, budget/cost savings, increases in project return on investment (ROI), improved reputation, improved market share, and improved project quality.

3. Sustainable Construction Key Indicators

In order to study the sustainable construction it was very important to select the proper indicators. These indicators should include the main building impacts and assess the particular aspects of the socioeconomic context. Generally, building sustainability assessment (BSA) tools evaluate the buildings sustainability level through the aggregation of the building performance in a group of sustainability indicators, such as environmental, social and economic indicators. The first developed BSA tool was BREEAM - Building Research Establishment Environmental Assessment Methodology [6], in 1990. After this, other BSA tools were developed such as: LEED - Leadership in energy and environmental design [15], developed in the United States of America, SBTool – Sustainable Building Tool [12] developed by International Initiative for a Sustainable Built Environment (iiSBE), CASBEE - Comprehensive Assessment System for Building Environmental Efficiency [7] developed in Japan, among many other [20].

In order to define the sustainable building key indicators, four European initiatives have been analysed, the Sustainable Building Challenge 2011 [21] and 2013 [20] key indicators, the Sustainable Building Alliance, the SuPerBuildings [22] and the OPEN HOUSE project [19]. These initiatives are chosen because they are European initiatives that have performed recent work with the goal of define sustainable construction key indicators. The selection of indicators was performed considering the following aspects [4]:

1. All indicators with a subjective assessment were excluded;
2. All indicators whose performance cannot be changed by design options or whose assessment goes beyond the building boundaries were excluded;
3. All indicators whose performance was difficult to translate into economic terms were excluded.

Additionally, it was intended that the key indicators list is broad enough to include the mainly sustainable buildings impacts, but also concise enough in order to make the assessment practicable (see Table 1).

Table 1. Indicators selected by at least three initiatives [4].

DIMENSION	CATEGORIES	SELECTED INDICATORS
ENVIRONMENT	Energy and Emissions	Non-renewable primary energy
		Renewable primary energy
		Green House Gases emissions
	Water	Water consumption
	Materials and Waste	Materials embodied energy
		Ozone depletion potential
		Acidification potential
		Eutrophication potential
		Photochemical oxidation potential
		Reused and recycled materials
		Responsible sourcing materials
		Waste production
SOCIETY	Users health and comfort	Indoor air quality
		Lightning
		Thermal comfort
		Acoustic comfort
	Process quality	Integrated design project
		Commissioning
ECONOMY	Economy	Life Cycle Costs

Through the revision of the results obtained in some European initiatives, developed with the goal of select sustainability key indicators, was possible to observe that despite all the efforts, there are still some differences between the lists of key indicators developed [1] [8] [10].

4. Proposed Methodology of Quality Evaluation

The topicality of the problem in the world is confirmed by Directive 2014/24/EU of the European Parliament and of the Council in Article 49 in which the "best" proportion of price and quality is named as the main criterion of the award of contract. In order to clarify, it is recommended to claim a "smaller" proportion, i.e., a cheaper offer for higher quality. However, the word "quality" remains comprehensive, abstract, non-weighted by one number and therefore a too freely usable concept here, as well, which will lead to unfair competition and sustainable disregard of construction principles.

Economic profitability – economic efficiency or benefit gained by a customer, may be mentioned as the second criterion. In order to assess the economic efficiency correctly, not only costs should be taken into account, but also the following factors:

1. Contract performance terms;
2. Operating costs;
3. The quality of construction works;
4. Aesthetic characteristics;
5. Functional specification;
6. Environmental compliance;
7. Technical advantages;
8. The availability of spare parts;
9. Security of supply process;
10. Other contract-related factors

The aim of this research is to show that it is completely enough with the price lists available in the market, quality documents of materials, specifications of construction details, tariff rates of construction workers, time norms and cost estimates, to calculate the economic efficiency of materials, construction and the whole building by a single and calculated efficiency price of the unit of measurement. However, in order to determine this, it is not enough with only the use of market units of measurement, but the standardized efficiency (which is also quality) units of measurement of both; the material and construction must definitely be used in the calculations. The offered quality assessment methodology contains the following components:

1. Quality concept assumed as the materials, structures, as well as all construction utility functions (strength, load bearing capacity, heat resistance, frost resistance and many others) indicators that client are looking for and willing to pay for them.
2. Each utility function has its own specific normative measurement, which may differ from the units used in the market.
3. Utility functions could be found in the project, both as materials and structures specifications as well as all the building life time parameters.
4. The price for each component of the utility function could be calculated dividing the construction materials or structures market price to the utility normative measurement specified in the Technical project. With similar approach can evaluate the usefulness of the project as a whole. In that case as the utility function should be chosen expected building life time, specified in the contract or project, less than operating expenses during building life time.
5. As the criteria of competitiveness for certain project or building could be set the utility function for the all components usefulness or the total price for all project usefulness.
6. Utility functions evaluation, comparison of prices and how to achieve designed quality for the lowest price (optimum choice) will be the responsibility of designer and this prove his competitiveness.

The offered methodology does not consider such competitiveness criteria as guarantee scope (the time may be considered), all compulsory or other types of insurance, however, it is also a cost which may be easily added to the calculated quality price. An example of the practical application of the offered quality assessment methodology – a thermal house construction and technology is included in the next chapter.

5. Experimental Results and Discussion

This practical example shows another positive effect of the application of the offered methodology – minimization of the building costs and the social effect obtained from it. By applying the offered methodology it is possible to build a low energy consumption house in a much cheaper way than from traditional construction and technologies.

Let us deal with the price dependence of concrete efficiency – its compression strength on the compression strength classes as the first which is clearly shown graphically (see Figure 1).

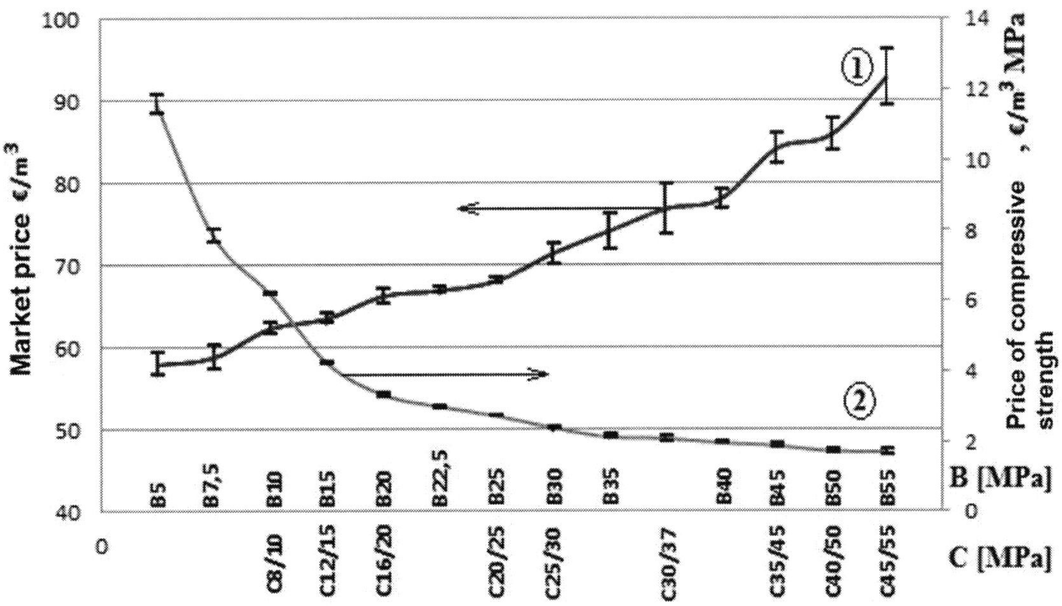

Figure 1. Concrete market price (1) and a price of its strength (2) depending on the concrete class.

Upon increasing concrete strength, its market price of 1m3 may certainly grow. But what is the economic efficiency of 1m3 of concrete? No one gains any benefit from the fact that the supporting construction occupies a large capacity in the building and is heavy. It means that the market unit of measurement 1m3 may not be an efficient unit of measurement of concrete.

By investigating the curve it may be established that with the increase of concrete strength, the price of its construction efficiency function, compression strength, rapidly decreases. It may also be seen that upon exceeding the concrete strength of 45 MPa, this relation seemingly becomes stable. However, the technical and economic benefit to be gained behind this border is easily explained – it is the further decrease of concrete consumption, own weight and load of the construction on the lower constructions. A similar relation may be observed when evaluating the efficiency of another building construction material, steel. Its market unit of measurement is tons, from which no one gains any benefit. With the increasing strength of steel, the market price of 1t certainly increases, as well, while the price of efficiency unit of strength measurement [€/t·MPa] falls. Steel with high strength is used as cold-rolled, light thin walls and rational profiles.

Other building properties standardizing the quality of building material may be evaluated in the same way by cost, for example, chemical resistance or frost resistance: the cost of one cycle of frost resistance? The efficiency criterion of heat insulation materials will certainly be their heat resistance $1/\lambda$ by the unit of measurement mK/W.

The economic efficiency price of the roof covering efficiency may be evaluated similarly by dividing its market price €/m2 by the efficiency criterion – service life in years. Then, upon choosing the material with this price [€/m2·years] a customer and designer will see that the most expensive material in the market will be the cheapest one in the long term.

By applying the offered methodology the building construction itself may also be evaluated. It will be the load carrying capacity price for the supporting constructions –for columns €/kN, beams €/kN m, while for plates €/kN m2. The application of the methodology is comparatively simple because load carrying capacity may be found in both the design, and in the specifications of built-up constructions, while the prices are indicated in the price lists of the supplier or manufacturers of materials.

It may easily be seen that the efficiency function of the separating constructions – external walls and their elements, windows, is their heat resistance RT= 1/U with the unit of measurement m2K/W. It should have monolith construction included in the design, while market prices should be indicated in the cost estimate together with all construction costs. The heat resistance (or its opposite unit W/m2K) for built-up constructions, panels, windows may be found in the specifications or in the manufacturer's certifications, but a competent customer or designer may choose its market price appropriate to a minimum and guaranteed efficiency price.

For economically efficient optimum selection of building materials a scheme should be used which is clearly shown in the table (see Table 2). Different external wall materials are included here for comparison – those which include both functions at the same time – load carrying capacity + heat insulation, as well as specialized ones – provided for only load carrying, or to be used only for heat insulation of the building with the highest values of both these efficiency functions (strength and heat resistance).

Table 2. Market prices of construction building materials and their efficiency prices, according to RTU, Construction Technology department survey.

Material	Density kg/m^3	Price €/m^3	Heat resistance 1/λ K·m/W	Price of heat resistance €·W/m^4·K	Compressive strength MPa	Price of compressive strength €/m^3 MPa
Concrete C35/45	2400	84,00	0,49	144,90	45,00*	1,87
Ceramic blocks	650	113,00	5,65	20,00	12,5	9,04
AAC blocks	400	67,00	9,10	7,36	3,00	22,33
Wooden structure (pine)	600	165,00	8,20	20,1	8,5·16,0	19,4÷103
Foam polystyrene (EPS60)	15	57,23	30,30	1,89	0,060**	953,8

As it is seen, the highest concrete compression strength class found in the construction market is 45 MPa (150 x 150x 150 mm for cubes*), but it has the least heat resistance. On the other hand, polystyrene foam EPS 60 has the highest heat resistance and it is 30.3 m.K/W, but its compression strength (at 10% of deformation**) is insignificant – only 0.06 MPa. However, unlike mineral wool, it is completely sufficient in the technological sense for the remaining heat insulation formwork. Upon comparing the prices of efficiency functions of both these specialized building materials with the efficiency prices of traditional materials, it may be seen that the cost indicators of both efficiency functions – strength and heat resistance of both, that ceramics, aerated concrete blocks and wood are more expensive. It may be easy to make a conclusion on the basis of these numbers that it is not profitable technically nor economically to try to combine implementation of both aforementioned efficiency functions in one external wall material, because both will be too expensive as a result. It is also obvious that it is most profitable in the technical and economic sense to make a separating construction from two separate and high-specialized materials – so that the load carrying material performs its function with the lowest strength price, as well as with the least consumption and mass, while the heat insulation material should be chosen with the minimum (necessary only for assembly) durability and minimum heat resistance $1/\lambda$ price.

How does such a division of efficiency functions of building materials, together with the cost minimization of these functions, determine the building properties of the construction and how will the improvement of its efficiency functions affect construction costs? We will deal with the building of the external wall construction of thermal houses as an example of the offered evaluation methodology of the efficiency factors (see Figure 2 and Figure 3).

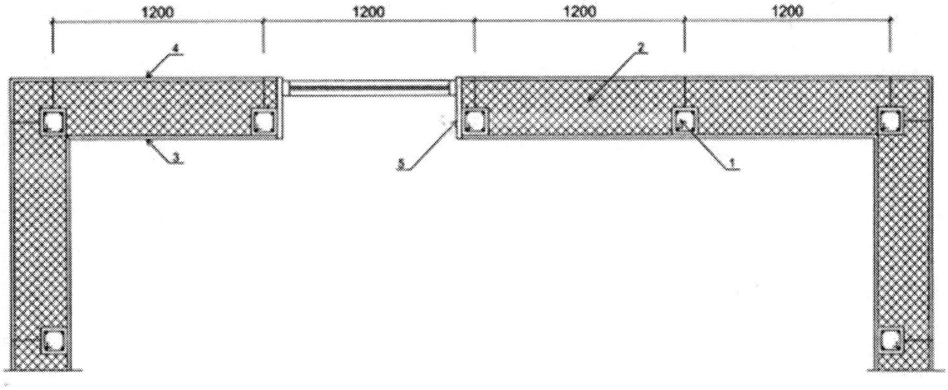

Figure 2. An external wall construction of a thermal house in the remaining formwork. 1- concreted columns, 2- heat insulation formwork, 3- hard internal finish, 4- external finish of the wall, 5- heat insulation window classes.

Figure 3. Concreting of the external wall of a thermal house.

The economic and technical indices of the external wall construction of the thermal house are divided here in two parts (see Table 3). The first part – from p.1.1. to p.1.8., shows that upon making an optimum division of the efficiency functions in the external wall construction together with their cost minimization (Column 5) by comparing with constructions made of traditional materials with both combined efficiency functions (Column 3 and 4) is obtained by a building company. These are considerably decreased construction costs. The second part (p.2.1 – p. 2.4) shows what else is obtained by a customer apart from the cheapest price within the long-term process of the use of the building. The thermal external wall heat resistance is 8.27 m2.K/W, which far exceeds the minimum 5.555... m2.K/W required in the Latvian Construction Standard LBN 002-15, is also the cheapest. Upon a customer's wish, there are no difficulties to increase the heat resistance of thermal walls even more by the offered technology. Besides, the 1m2 weight of the external wall is considerably less, which allows to make the constructions with smaller volume and mass, as well as cheaper respectively.

Table 3. Technical – economical comparison of outside wall.

Indicators of the economical and technical benefits for outside wall	Unit of measurement	Type of outside wall		
		Ceramic blocks, 40 cm λ=0,177 W/m·K	AAC blocks, 40 cm λ=0,11 W/m·K	Expanded polystyrene blocks 40 cm as formwork filled with concrete core
1	2	3	4	5
1. **Economical indicators**	€/m²	34,49	36,18	31,00
1.1 Materials (with VAT)				
1.2 Labour costs	€/m²	14,74	11,37	5,14
1.3 Social security taxes 23,59 %	€/m²	3,48	2,68	1,21
1.4. Mechanisms and hand tools	€/m²	1,47	1,14	0,51
1.5 Transport costs	€/m²	2,76	2,89	3,03
1.6 Contractor profit	€/m²	5,69	5,13	4,09
1.7 VAT 21%	€/m²	13,15	11,84	9,45
1.8. Total costs	€/m²	75,79	71,23	54,43
Indicators of the economical and technical benefits for outside wall	Unit of measurement	Type of outside wall		
		Ceramic blocks, 40 cm λ=0,177 W/m·K	AAC blocks, 40 cm λ=0,11 W/m·K	Expanded polystyrene blocks 40 cm as formwork filled with concrete core
1	2	3	4	5
2.**Technical indicators**				
2.1. Weight	kg/m²	360,00	160,00	78,00
2.2. Heat resistance	m²K/W	2,26	3,810	8,27
2.3. Price of heat resistance	$\frac{€W}{m^4K}$	33,54	18,70	6,58
2.4. Heat loses for heating season	kWh/m²	43,12	25,57	11,78
	€/m²	3,04	1,81	0,83

The offered methodology may also be easily applied for profitability assessment of the whole building, for example, by dividing its price (or construction costs) with the service life indicated in the design. Such an assessment might not match the efficiency assessment of particular constructions in terms of cost, because one covering panel may serve (carry load) a much longer than the life of the building. On the other hand, if the service life of the building is determined, some of its maintenance indicators may be used as a competitiveness criterion, for example, the forecast heating costs in the unit of time. Upon multiplying them with the service life of the building, the costs of the determined efficiency, in this example heating costs, are obtained within the entire service life of the building. They will certainly be important only to the customer, but they must determine the cooperation of both, designer and builder, upon developing a sustainable building product. The assessment methodology of the offered efficiency functions may also be used concurrently – for materials, construction and for the whole building

6. Conclusions

The work presented in the paper selects a set of indicators that will be used in the cost-benefit analysis to sustainable construction. Due to the goal of the work, some specificities were applied in the selection of indicators. However, when selecting a set of indicators to assess sustainability, it is important to analyse the most significant buildings impacts and the socio-cultural context. As the result, we could make several statements:

1. All the most important sustainability criteria can be measured by money, either evaluating materials and structures or construction in general.
2. Construction costs of the Building could be reduced by maximization of sustainable construction criteria, and the minimization of their costs. In such a way, we could increase economic benefits which are important for client as well as for construction companies.
3. Using proposed methodology of quality evaluation as a result the low energy buildings could be built cheaper, from economic life time prospective, than average (traditional) buildings.

References

[1] Alwaer H and Clements-Croome D J, 2010 Key performance indicators (KPIs) and priority setting in using the multi-attribute approach for assessing sustainable intelligent buildings *Building and Environment* 45(4): 799-807

[2] Alyami S H and Rezgui Y 2012 Sustainable building assessment tool development approach *Sustainable Cities and Society* 5(0): 52-62

[3] Antillon E I, Alarcón L F, Hollowell M R and Molenaar K R 2012 A Research synthesis on the interface between lean construction and safety management *Proc. Construction Research Congress* ASCE West Lafayette

[4] Araújo C, Bragança L, Almeida M 2013 Sustainable Construction Key Indicators *Building Sustainability Assessment Tools* 505-512

[5] Bragança L et al 2010 Building Sustainability Assessment *Sustainability* 2(7): 2010-2023

[6] BREEAM 2012 BREEAM - Building Research and Consultancy's Environmental Assessment Method, Retrieved 27-12-2012, 2012, from http://www.breeam.org.

[7] CASBEE 2012 CASBEE - Comprehensive Assessment System for Building Environmental Efficiency, Retrieved 27-12-2012, 2012, from http://www.ibec.or.jp/CASBEE/english/overviewE.htm.

[8] Chen Y et al 2010 Sustainable performance criteria for construction method selection in concrete buildings *Automation in Construction* 19(2): 235-244

[9] Cole R J 1999 Building environmental assessment methods: clarifying intentions *Building Research & Information* 27(4-5): 230-246

[10] Ding G K C 2008 Sustainable construction - The role of environmental assessment tools *Journal of Environmental Management* 86(3): 451-464

[11] Freyd A-C 2012 Annual Report 2012 - Sustainable Building Alliance, *Sustainable Building Alliance*

[12] Hinze J 1996 The distraction theory of accident causation *Proc., Int. Conf. On Implementation of Safety and Health on Constr.* Sites, CIB Working Commission W99: Safety and Health on Construction Sites, L. M. Alvez Diaz and R. J. Coble, eds., Balkema, Rotterdam, The Netherlands, 357–384

[13] Huovila P 2012 Sustainability and performance assessment and benchmarking of buildings - Final report, Finland, VTT Technical Research Centre of Finland

[14] iiSBE 2012 SBTool 2012, Retrieved 21-01-2013, 2012, from http://iisbe.org/sbtool-2012

[15] LEED 2013 LEED - in Energy & Environmental Design, Retrieved 27-12-2012, 2012, from http://www.leedonline.com

[16] Mateus R and L Bragança 2011 Sustainability assessment and rating of buildings: Developing the methodology SBToolPT-H *Building and Environment* 46(10): 1962-1971

[17] Maudgalya T, Genaidy A and Shell R 2008 Productivity—Quality—Costs—Safety: A sustained approach to competitive advantage—A systematic review of the National Safety Council's case studies in safety and productivity *Hum. Factors Ergon. Manuf. Serv. Ind.* 18 (2), 152–179

[18] McGraw Hill Construction 2013 Safety management in the construction industry: Identifying risks and reducing accidents to improve site productivity and project ROI *SmartMarket report*

[19] OPEN HOUSE 2010 OPEN HOUSE - Seven Framework Programme Home Page, Retrieved 06/06/2013, 2013, from http://www.openhouse-fp7.eu/
[20] SB Conferences 2013 Sustainable Building 2013-2014 Conferences Series - SB Challenge, Retrieved 06/06/2013, 2013, from http://sbconferences.org/concursos/
[21] SBChallenge11 2011 Sustainable Building Chalenge 2011, Retrieved 06/06/2013, 2013, from http://sbchallenge.iisbe.org/
[22] SuPerBuildings, 2012 SuPerBuildings - Sustainability and Performance Assessment and Benchmarking of Buildings, Retrieved 08/06/2013, 2013, from http://cic.vtt.fi/superbuildings/
[23] Veltri A, Pagell M, Behm M and Das A 2007 A data-based evaluation of the relationship between occupational safety and operating performance *J. of Safety, Health and Env. Research*, Vol.4. No.1
[24] Zavrl M Š et al 2010 D1.3. Definition of indicators, sustainability performance levels and procedures to evaluate them, OPEN HOUSE

Life cycle assessment for masonry exterior wall assemblies made of traditional building materials

Z Zimele[1], M Sinka[1], D Bajare[1] and A Jakovics[2]

[1]Institute of Materials and Structures, Faculty of Civil Engineering, Riga Technical University, Kalku iela1, Riga, LV-1658, Latvia
[2]Faculty of Physics and Mathematics, University of Latvia, Zellu 23, Riga LV-1002, Latvia

E-mail: zinta.zimele@rtu.lv

Abstract. Building materials contribute a significant amount of CO_2 to overall greenhouse gas emissions and this environmental impact should be reduced to improve the sustainability of the construction industry. Although new and low-carbon materials are emerging, the majority of materials used are still traditionally produced and are highly carbon intensive, such as masonry blocks and insulation slabs – rock wool or expanded polystyrene. To assess which of the traditionally used building materials are more sustainable, in this paper, wall models of the most widely used masonry blocks and insulation materials were analysed using life cycle assessment. The wall models were created to fulfil the requirements of nearly-zero energy buildings. The assessment showed that the lowest impact on the environment is from aerated concrete blocks and expanded polystyrene insulation, which is mostly due to low weight and raw materials consumption compared to other materials. On the other hand, expanded polystyrene insulation poses more danger to humans and the environment in its use phase than other materials due to emissions during fire and degradation, thus should be used with caution.

1. Introduction

The building sector contributes up to 30% of global annual greenhouse gas emissions and consumes up to 40% of all energy [1,2]. The European Union has a number of official documents and guidelines aimed at reducing CO_2 emissions by up to 20% by 2020 [3], up to 40% by 2030 [4] and the energy performance of the Buildings Directive requires all new buildings to be nearly zero-energy by the end of 2020 [4]. In this study traditionally used material assemblies for exterior wall construction are analyzed using Life cycle assessment (LCA) to assess their impact on the environment, mostly focusing on greenhouse gas (GHG) emissions.

This study examined outer wall construction models consisting of a traditional masonry wall with a thermal insulation layer according to nearly zero-energy building requirements. Aerated concrete, hollow ceramic blocks, and expanded clay concrete were chosen as constructive materials, as they are equally suitable for use in exterior walls because of their physical properties – lower thermal conductivity than concrete and greater compressive strength amongst other types of lightweight concrete. To reach the requirements defined for nearly zero-energy buildings, rock wool and polystyrene foam slabs were chosen as a thermal insulation material.

In this study, the main focus was to analyze the impact on the environment by calculating LCA of the selected wall models using SimaPro software to find a solution for construction with the least impact on the environment.

2. Methodology

2.1. Methods: Life cycle assessment

All buildings have an impact on the environment at all stages of their life cycle - materials must be quarried, mined or harvested, transported to factories and manufactured. The building products have to be transported to the site, lifted into place and fixed into position. The finished buildings have to be operated, heated and cooled. All of those listed stages have an impact on the environment that can be calculated and analyzed using the life cycle assessment software SimaPro. As the factors of building operation, in the scope of this study, are considered identical for all chosen materials, the "cradle to gate" system is analyzed.

The main value for comparing different outer wall constructions is the thermal transfer coefficient which is defined as U=0.105 (W/m^2K). The examined area (unit) of the outer walls made of different constructions is $1m^2$. The main environmental impact category which is compared is the *Global Warming Potential (GWP)*, which expresses climate change by the emission of greenhouse gases like carbon dioxide (CO_2) and methane (CH_4) and it is measured in *CO_2 equivalents*. Additionally other impact category are considered - acidification potential (views gases that cause acid deposition, measured in kg SO_2 eq), abiotic depletion potential (the consumption of non-biological and biological resources , MJ and kg Sb eq), human toxicity potential (calculated index of a chemical released in the environment such as arsenic, sodium dichromate, and hydrogen fluoride, which are dangerous to human health, kg 1,4-DB eq), ozone layer depletion potential (the damage of various gases into stratospheric compounds reduce ability to prevent ultraviolet light entering the atmosphere, kg CFC-11 eq), photochemical oxidation (pollution of photochemical ozone, kg C2H4 eq), eutrophication (chemical nutrient concentration in ecosystems which lead to abnormal productivity, kg PO_4^{3-} eq), eco-toxicity (tolerable concentrations for ecosystems caused by heavy metals, kg 1,4-DB eq), that is divided into fresh-water aquatic ecosystems, marine ecosystems and terrestrial ecosystems [5,6].

2.2. Materials

The outer wall constructions of the buildings are designed with different constructive materials - lightweight concrete blocks (200 mm thickness), hollow ceramic blocks (250 mm thickness), and aerated concrete blocks (300 mm thickness), with variations of thermal insulation material - rock wool or polystyrene slab of different thickness. The outer wall constructions are designed to achieve the thermal transfer coefficient U=0.105 (W/m^2K).

2.2.1. Lightweight concrete blocks. The lightweight concrete blocks contain expanded clay aggregates mixed with cement. Lightweight expanded clay aggregates are artificial, and their manufacturing involves heating clay particles at a temperature of 1150 -1250 °C in a rotary kiln. After heat treatment, the clay particles expand and become about four to five times larger compared to their original size and take the shape of pellets. A hard-sintered crust is formed on the outer surface, while quite light and highly durable aggregates with a porous clinker-like structure may be produced inside it [7]. The lightweight concrete block manufactured by *Weber* (as hereinafter *Fibo*) with a density of 740 kg/m^3 and thermal conductivity is 0.22 W/mK [8] is used.

2.2.2. Hollow ceramic blocks. The hollow ceramic blocks are produced using clay - forming, then drying at a temperature of 50-150°C and sintering at 900-1000 °C [9]. The hollows in the structure of ceramic blocks increase their thermal performance. Hollow ceramic blocks are lighter compared with traditional clay bricks, easier to handle due to their dimensions and usually, single-block wall constructions are made. The density of the hollow ceramic blocks "*Keraterm*" manufactured by *Lode* (as hereinafter *Keraterm*) is 783 kg/m^3 and thermal conductivity is 0.13 W/mK [10].

2.2.3. Aerated concrete blocks. The aerated concrete blocks are produced using cement, lime, fly ash, and a foam formation agent (aluminum powder). It is a high-efficiency constructive thermal insulating material made through the molding and steam curing processes. The apparent density of the aerated concrete manufactured by *Bauroc* (as hereinafter *Bauroc*) is low (375±25 kg/m^3) and the thermal conductivity of it is even lower than that of the hollow ceramic blocks - 0.09 W/mK [11].

2.2.4. Rock wool. Two types of rock wool are examined in this study - Paroc Linio 15 and Paroc Linio 80. The difference between Paroc Linio 15 and Paroc Linio 80 is density and thermal conductivity. Paroc Linio 15 density is 120 kg/m^3 and thermal conductivity is 0.037 W/mK while Paroc Linio 80 density is 85 kg/m^3 and thermal conductivity is 0.04 W/mK [12].

2.2.5. Polystyrene foam slab. Polystyrene foam slab *Tenapors NEO EPS 100* (as hereinafter *EPS*) with a density of 20 kg/m^3 and thermal conductivity of 0.03 W/mK [13] is used. The main component of a polystyrene foam slab is styrene (C_8H_8), which is derived from petroleum or natural gas and formed by a reaction between ethylene (C_2H_4) and benzene (C_6H_6), benzene is produced from coal or synthesized from petroleum. Styrene is polymerized either by heat or by an initiator such as benzoyl peroxide [14].

2.3. Exterior wall models.
The outer wall construction models design are of Fibo and Paroc Linio 15 (as hereinafter *Fibo-Paroc 15*); Fibo and EPS (*Fibo-EPS*); Keraterm and Paroc Linio 80 (*Keratem-Paroc 80*); Keraterm and EPS (*Keratem-EPS*); Bauroc and Paroc Linio 80 (*Bauroc-Paroc 80*) and Bauroc and EPS (*Bauroc-EPS*) Table 1. For all models the same indoor and outdoor finishing is used – stucco (2 mm thick stucco for indoors and 10 mm thick stucco for outdoors).

Table 1. Exterior wall models.

	Fibo-Paroc15, mm	**Fibo-EPS, mm**	**Keraterm-Paroc 80, mm**	**Keraterm-EPS, mm**	**Bauroc-Paroc 80, mm**	**Bauroc-EPS, mm**
Stucco, 1750 kg/m^3 [15]	12	12	12	12	12	12
Fibo, 740 kg/m^3	200	200	-	-	-	-
Keraterm, 783 kg/m^3	-	-	250	250	-	-
Bauroc, 400 kg/m^3	-	-	-	-	300	300
Cement mortar for Fibo and Keraterm, 1800 kg/m^3 (16,6 kg/m^2)	2	2	2	2	-	-
Cement mortar for Bauroc, 1400 kg/m^3 (9 kg/m^2)	-	-	-	-	2	2
Adhesive mortar, 1670 kg/m^3 [15]	2	2	2	2	2	2
Paroc Linio 15, 120 kg/m^3	350	-	-	-	-	-
Paroc Linio 80, 85 kg/m^3	-	-	340	-	300	-
EPS, 20 kg/m^3	-	300	-	280	-	240
Thickness of wall, mm	**564**	**514**	**604**	**542**	**614**	**554**

3. Results

The LCA calculation is made using SimaPro8 software. Outer wall models with an area of $1m^2$ are analyzed taking into account all manufacturing stages of the building materials used. For LCA the environmental impact is calculated using the CML-IA baseline V3.04 method. This CML method was created by the University of Leiden in the Netherlands in 2001 and contains more than 1700 different flows. The CML-IA baseline being the most common impact categories used in LCA [5]. The outer wall's component emissions impact on the environment was analyzed separately in GWP impact category Table 2.

Table 2. GWP by $1m^2$ of outer wall.

	Fibo-Paroc15, kg CO_2 eq	**Fibo-EPS, kg CO_2 eq**	**Keraterm-Paroc 80, kg CO_2 eq**	**Keraterm-EPS, kg CO_2 eq**	**Bauroc-Paroc 80, kg CO_2 eq**	**Bauroc-EPS, kg CO_2 eq**
Stucco	1.47	1.47	1.47	1.47	1.47	1.47
Fibo	65.06	65.06	-	-	-	-
Keraterm	-	-	57.16	57.16	-	-
Bauroc	-	-	-	-	53.20	53.20
Cement mortar for Fibo and Keraterm	11.76	11.76	11.76	11.76	-	-
Cement mortar for Bauroc	-	-	-	-	1.57	1.57
Adhesive mortar	7.98	7.98	7.98	7.98	7.98	7.98
Paroc Linio 15	60.43	-	-	-	-	-
Paroc Linio 80	-	-	39.59	-	36.69	-
EPS	-	24.78	-	23.13	-	19.42
Wall kg CO_2 eq	**146.71**	**111.06**	**117.96**	**101.50**	**100.91**	**83.41**

The results in Table 2 show that *Fibo-Paroc 15* (area 1 m^2) produces 146.71 kg CO_2 eq, the biggest impact of emissions is made by the Fibo masonry wall –65.06 kg CO_2 eq, the insulation material contributes - kg 60.43 CO_2 eq. The constructive model made of *Fibo-EPS* (1 m^2) gives 111.06 kg CO_2 eq in GWP. The highest impact comes from the Fibo masonry wall, but the thermal insulation layer gives only 24.78 kg CO_2 eq. The CO_2 emissions of *Fibo-EPS* are about 24% less compared with the same structural wall construction made with Paroc Linio 15.

1 m^2 of *Keraterm-Paroc 80* wall in the GWP impact category makes 117.96 kg CO_2 eq, the biggest impact is made by Keraterm masonry wall -57.16 kg CO_2 eq, which is about 12% less than from Fibo. The GWP from thermal insulation Paroc Linio, 80 is 39.59 kg CO_2 eq, which is 33% of all CO_2 emissions. Comparing *Fibo-Paroc 15* and *Keraterm-Paroc 80*, the results of CO_2 emissions differs depending on the insulation, the PAROC Linio 15 gives a bigger impact than Paroc Linio 80 because of its density, Paroc Linio 15 density is 120 kg/m^3 but Paroc Linio 80 is only 85 kg/m^3 [12]. Facade rock wool CO_2 emissions from 1 kg of material are 1.44 kg CO_2 eq by Paroc Linio 15 and 1,37 kg CO_2 eq by Paroc Linio 80. The GWP of the rock wool production is dominated by the direct emissions during the fabrication of the rock wool, which contribute over 60 % to the total greenhouse gas emissions amount to 1.01 kg CO_2-eq/kg rock wool, other material emissions are from briquette fabrication, hard coal and phenol in the supply chain [16].

The wall of *Keraterm-EPS* makes 101.50 kg CO_2 eq of GWP emissions, the main impact is made by Keraterm, while EPS makes 23.13 CO_2 kg eq, which is 23% of all emissions.

According to the GWP impact category the *Bauroc-Paroc 80* wall contributes 100.91 kg CO_2 eq, where the highest impact is by Bauroc masonry wall – 53.20 kg CO_2 eq, which is about 7% less compared to Keraterm (1 m^2 Keraterm of 250 mm thickness makes 57.16 kg CO_2 eq, while 1 m^2 Bauroc of 300 mm thickness makes 53.20 kg CO_2 eq). The GWP emissions for Paroc Linio 80 is 36.69 kg CO_2 eq, which is 37% of total CO_2 emissions. In this case, comparing models of *Keraterm-Paroc 80* and *Bauroc-Paroc 80*, obtained results of CO_2 emissions are similar because the same insulation material (Paroc Linio 80) is used.

The wall with lowest greenhouse gas emissions is *Bauroc-EPS*, it contributes only 83.41 kg CO_2 eq, where the highest impact is made by Bauroc masonry wall -53.20 kg CO_2 eq, while thermal insulation EPS makes 19.42 kg CO_2 eq, which is 23% of total CO_2 emissions.

The choice of insulation material plays a major role in CO_2 emissions, as the LCA calculations show that reducing the thickness of the polystyrene foam slab by 40 mm reduces CO_2 emissions by 16% (3.71 kg CO_2 eq), which is better compared with rock wool (7% or 2,9 kg CO_2 eq).

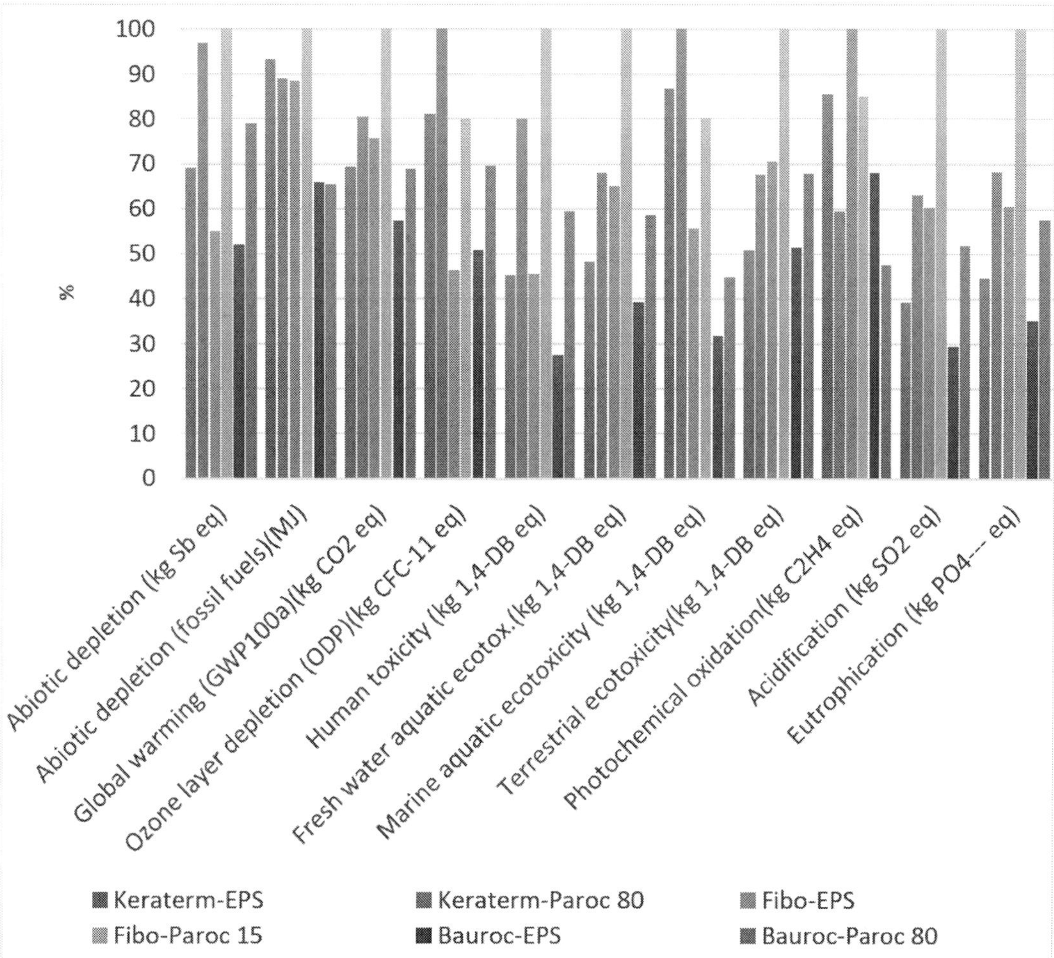

Figure 1. Impact of masonry walls ($1m^2$) according to the method: CML-IA baseline V3.04.

From the results showed in Figure 1. the biggest impact is made by the wall model *Fibo-Paroc 15*, except in impact category ozone layer depletion and marine aquatic ecotoxicity, where the biggest impact is made by *Keraterm-Paroc 80,* in photochemical oxidation the biggest impact is by *Fibo-EPS.* In photochemical oxidation the biggest emissions come from EPS which is 62% of total emissions, while Fibo makes 31% of all emissions. The results of *Bauroc-EPS* show lower impact in almost all of the impact category, except in the category of photochemical oxidation, where the highest impact is given by EPS (approximately 75% from total emissions).

4. Discussions

As shown in the results, masonry wall construction *Bauroc-EPS* has a lower impact on global warming emissions (CO_2 emissions - 83.31 kg CO_2 eq) compared with the other wall models. The highest emissions are by *Fibo-Paroc 15* - 146.71 kg CO_2 eq, which is 43% more compared to *Bauroc-EPS*. The total CO_2 emissions are higher for all models consisting of a rock wool insulation layer than with a polystyrene foam slab. The reason is the fact that the insulation layer used has a higher density – the density for rock wool is 85-120 kg/m^3, but for polystyrene foam slabs - 20 kg/m^3.

It should be noted, that the highest impact of abiotic depletion (fossil fuels), ozone layer depletion and human toxicity is caused by construction materials, not by thermal insulation materials. The highest impact of the photochemical oxidation emissions comes from masonry walls with EPS, it is because of the polystyrene foam slabs used in the models.

This study is based on the LCA report of different building exterior wall models to find a suitable wall model with less impact on GWP according to nearly zero-energy building conditions. The achieved results of 1m^2 of external walls with the lowest impact are summarized in Table 4.

Table 3. Outer wall results according to the Global warming potential (GWP100a), kg CO_2 eq for 1m^2.

External wall	**Global warming potential (GWP100a), kg CO_2 eq for 1m^2**	**Essential Impact contributor**
Keraterm-EPS	101.5	Hollow ceramic block.
Keraterm-Paroc 80	117.96	Hollow ceramic block.
Fibo-EPS	111.06	Lightweight concrete block.
Fibo-Paroc 15	146.71	Lightweight concrete block and rock wool (Paroc Linio 15) – equally makes the impact. Paroc Linio 15 makes more CO_2 emissions than Paroc Linio 80.
Bauroc-EPS	83.31	Aerated concrete block.
Bauroc-Paroc 80	100.91	Aerated concrete block makes the biggest impact (53% from all emissions), and 37% emissions are from used rock wool (Paroc Linio 80).

In Table 4 it can be seen that there are wall models that can be improved – for the masonry wall like *Fibo-Paroc 15* - thermal insulation replacement to Paroc Linio 80 can reduce CO_2 emissions.

The best results comparing masonry wall constructive solutions by GWP according to nearly zero-energy building is *Bauroc-EPS*, which is also the most popular masonry wall form in a single-family house building in Latvia, because the aerated concrete block has better thermal insulation properties than the hollow ceramic block and lightweight concrete block, as a result, there is a need for a thinner layer of thermal insulation.

5. Conclusions

From the results obtained from the LCA it can be concluded that the masonry wall construction *Bauroc-EPS* has the lowest impact on global warming by CO_2 emissions - 83.31 kg CO_2 eq which is 10-43% lower compared to other wall models. From masonry wall constructive materials Bauroc showed the lowest impact in GWP, yet much of the raw materials used in aerated concrete blocks production may consist of recycled materials, including copper mine tailings and fly ash, a byproduct of coal-fired power plants, thus reducing CO_2 emissions [17].

All examined wall models with polystyrene foam slab (EPS) thermal insulation showed lower CO_2 emissions than walls with rock wool, contrary to what was expected, as EPS is considered to have the highest environmental impact of traditional insulation materials. If the reference functional unit in LCA calculation would be 1kg of material instead of 1 m^2, then the claim that EPS has higher environmental impact would be justified, because 1 kg of EPS material contributes 4.1 kg CO_2 eq/kg, while from1 kg of Paroc is 1.37 – 1.44 kg CO_2 eq/kg (by SimaPro calculation). Obviously, EPS is a better thermal insulation according to CO_2 emissions in the stage of production, but this material is a petroleum-based non-biodegradable foam material, which when for long periods is exposed to UV rays photodegrades. By this process EPS can to get into the food chain causing pollution in fauna (for animal it causes starvation or death) [14]. Biodegradation of EPS does occur but at a very slow rate in the natural environment - EPS biodegradation in mixed microbial soil (including silt loam, cow manure, activated sludge, decaying plastics, etc.) ranged from 0.04 to 0.57% after 11 weeks [18]. By biodegradation EPS persists for long periods of time as a solid waste increasing the amount of plastic waste [19], because of its high resistance to water, mould, fungi, and bacteria. The burning of EPS is more toxic than rock wool [20]. EPS consist of three precursors - benzene, pentane and styrene - that impact on human health through various means of exposure, both in the production and the use of polystyrene foam products (packaging and disposable tableware) [14]. The benzene is carcinogenic to humans, with long-term exposure increasing the risk of leukemia. The pentane in EPS is in low levels, is a highly flammable gas which may evaporate from EPS during processing and storage, thus creating a fire hazard when the product is exposed to ignition sources or high heat. Styrene in EPS is reasonably anticipated to be a human carcinogen [14]. The environmental impact of the product listed above is a reason for reducing the production and use of EPS products, despite the fact that this insulation material is better by comparing wall models according CO_2 emissions in LCA calculation of material production stage.

Acknowledgement
The financial support of the European Regional Development Fund Project No. 1.1.1.1/16/A/192 "Development, optimization and sustainability analysis of smart solutions for nearly zero- energy buildings under natural climate condition" is acknowledged.

References
[1] UNEP SBCI, Buildings, and Climate Change. Summary for Decision-Makers, United Nations Environment Programme Sustainable Buildings and Climate Initiative (2009), pp 62
[2] Wolf C De, Pomponi F and Moncaster A 2017 Measuring embodied carbon dioxide equivalent of buildings //*A review and critique of current industry practice Energy Build* 140 pp 68–80
[3] European Union 2010 Directive 2010/31/EU of the European parliament and council on the energy performance of buildings pp 13–35
[4] Liobikiene Genovaite and Mindaugas B 2017 The European Union possibilities to achieve targets of Europe 2020 and Paris agreement climate policy // *Renew. Energy*, vol. 106 pp 298–309
[5] Pré, Mark Goedkoop, Michiel Oele, Jorrit Leijting, Tommie Posioen, Ellen Meijer Introduction to LAC with SimaPro // Pré, January 2016
[6] Aitor P Acero, Cristina Rodríguez, Andreas Ciroth 2017 LCIA methods Impact assessment methods in Life Cycle Assessment and their impact categories // Greendelta, pp 23
[7] Vijayalakshmi R and Ramanagopal S 2018 Structural Concrete using Expanded Clay Aggregate: A Review // *Indian Journal of Science and Technology* Vol 11(16), DOI: 10.17485/ijst/2018/v11i16/121888, April 2018
[8] Weber 2017 Fibo blocks [ONLINE] Available at: https://www.lv.weber/search-content/content_type/product/activities/fibor-bloki-59 [Accessed 16 April 2019]
[9] Balázs Nagy, Elek Tóth 2016 Hydrothermal behaviour of hollow and filled ceramic masonry block// International RILEM Conference on Materials, Systems and Structures in Civil Engineering Conference segment on Moisture in Materials and Structures, pp 279-289
[10] Lode 2019 Keraterm 44 [ONLINE] Available at: https://lode.lv/produkts/keraterm-44-s [Accessed 16 April 2019]

[11] Bauroc 2019 Bloki Universal [ONLINE] Available at: https:// bauroc.lv/product/universal [Accessed 16 April 2019]
[12] Paroc. 2019 PAROC Izolācija un citi izstrādājumi Būvniecība [ONLINE] Available at: https://www.paroc.lv/izstradajumi/konstrukcijas [Accessed 16 April 2019]
[13] Tenapors 2019 Tenapors Neo EPS [ONLINE] Available at: http://www.tenapors.lv/lv/siltumizolacijas-plaksnes/grieztas-putu-polistirola-plaksnes-eku-siltumizolacijai/tenapors-neo-eps [Accessed 16 April 2019]
[14] Manu Chandra, Colin Kohn, Jennifer Pawlitz 2016 Grant Powell Real cost of Styrofoam// Saint Louis University pp 46
[15] Sakret 2018 Sakret produktu katalogs [ONLINE] Available at: http://www.sakret.lv/2018/09/26/sakret-jaunais-produkta-katalogs/ [Accessed 16 April 2019]
[16] Karin Flury, Rolf Frischknecht 2012 Life Cycle Assessment of Rock Wool Insulation// ESU-services Ltd., fair consulting in sustainability, Uster, pp 27, Juni 2012
[17] Neha Bhatia 2017 Life Cycle Assessment as a tool for Material Selection - A comparison of Autoclaved Aerated Concrete and VSBK Brick Wall Assembly// 30th International Plea Conference, Ahmedabad pp 8
[18] Shanshan Yang, Anja Malawi Brandon, Defeng Xing, J Yang, J W Pang, C S Criddle, N Q Ren, Weimin Wu 2018 Progresses in Polystyrene Biodegradation and Prospects for Solutions to Plastic Waste Pollution// IOP Conference Series Earth and Environmental Science 150(1):012005, May 2018
[19] Ba Thanh Ho, Timothy K. Roberts, Steven Lucas An overview on biodegradation of polystyrene and modified polystyrene: the microbial approach// Critical Reviews in Biotechnology, 38:2, pp 308-320, 2017
[20] Anna A Stec, T Richard Hull 2011 Assessment of the fire toxicity of building insulation materials// *Energy and Buildings* 43 (2-3), pp. 498-506

IOP Conf. Series: Materials Science and Engineering **660** (2019) 012043 doi:10.1088/1757-899X/660/1/012043

Identification and Monitoring of Degraded Real Estate in the Context of Urban Environmental Analysis and Development Planning in the City of Riga

I Amoliņa[1], M Zīra[2] and I Geipe[1]
1 Institute of Civil Engineering and Real Estate Economics, Riga Technical University, Latvia, 6 Kalnciema Str. Riga, LV- 1048
2 College of Law, Latvia, Kr. Valdemara 1c, Riga, LV-1010

E-mail:iveta.pukite@rtu.lv

Abstract. Nowadays real estate plays an important role in the life of a person: it is one of the important elements of the quality of life (dwelling, workplace), as well as of public spaces, especially the outdoor space, i.e., the elements that make up the image of the city and the aesthetic enjoyment of the city. The topical issue under consideration is directly related to the physically deteriorated real estate that falls into the category of degraded or even dangerous buildings and structures. The article aims at investigating the problems of identifying, assessing and monitoring the degraded real estate in the context of urban development. The topicality of degraded objects is also observed in the main national and municipal development planning documents. The quality of real estate is assessed by different institutions according to the specificity and necessity of their activity, for example, planning and monitoring of territory development, application of real estate tax, as well as determination of the value of real estate. However, according to the research conducted within the article, information on real estate is not always up to date, available or linked between different holders of information sources. Different institutions evaluate the same thing at different levels of detail, but there is currently no fast, efficient and accurate exchange of information. This can be explained by disorganised administrative and legal issues, as well as the inefficiency of mutual communication.

1. Introduction

Real estate is one of the types of property that provides benefits to its owner and often directly affects the environment, processes and people. Not only is the safety and quality of the property important, but also the scenic and aesthetic enjoyment it provides to the passers-by, tourists or its users, which in turn promotes economic development. The importance of high-quality, architecturally successful and energy-efficient real estate is particularly emphasised, even with the granting of various awards, while problematic or degraded property is financially and/or administratively restricted with the aim of restoring it. Great responsibility for property quality, design and availability lies directly on the owner of the real estate because, according to the national legislation, responsibility is not only for its internal content and quality, but also for surrounding people.

Degradation includes both environmental degradation associated with loss of biodiversity, soil degradation, ozone depletion, reduction in water quality [1], [11], lower energy efficiency, higher maintenance costs [5], and the territory with known or anticipated pollution, abandoned, unused or under-utilised property where past activities have caused environmental pollution but have the potential for restoration or other economically viable use.

In the urban environment, the degraded state of the property (structure, territory) is more used in the context of its socio-economic activity or use; however, separate attention is also paid to the soil. Soil chemical degradation [2] or pollution is one of the types of soil degradation that can occur in water bodies (surface, underground, as well as in sludge). In the urban environment, it is possible to distinguish physical forms of soil degradation, such as its covering with artificial coating with limited permeability (asphalt, cement), compression and erosion [2].

The European network of experts defined brownfields as sites affected by past use of the site and adjacent areas, abandoned or unused, contaminated or considered to be contaminated, mainly located in a developed urban environment, as well as sites that can be returned to useful use as a result of interference [12].

The aim of the article is to investigate the problems of identifying, assessing and monitoring the degraded real estate in the city of Riga.

To achieve the aim, the following tasks are defined:

1) to study scientific conclusions on the identification of real estate degradation;
2) to investigate real estate identification and monitoring activities;
3) to study statistics of degraded property;
4) to identify the possibilities of classification of degraded structures in the city of Riga.

Limitations of the research are as follows:

1) the research covers the period of 2011–2015;
2) the research covers the identification methods of real estate degradation and methods of degradation category determination in Riga.

Within the framework of the research, the authors have used the following methods: a descriptive method, summarising the information and taking into account the scientific literature review, the statistical and analytical methods.

2. Property Identification and Monitoring

At present, there are no common criteria and methodology for identifying and assessing brownfields in Latvia.

The identification and monitoring of degraded objects and brownfields in Riga are carried out in a fragmentary way, such as built-up territories are regularly surveyed, while regular surveys of unbuilt territories do not currently exist, and surveys also depend on the ownership of property (Figure 1).

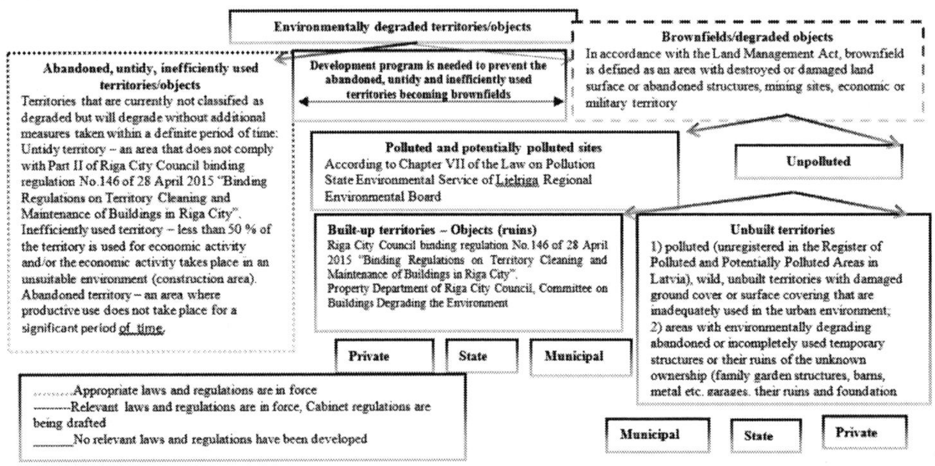

Figure. 1. The administrative and organisational structure for the identification and monitoring of degraded objects and brownfields in Riga [12].

In Latvia, the concept of brownfield/degraded object is used differently. The definition of soil degradation is explained as a result of changes occurring or taking place under the influence of natural processes and human activities, due to which the possibility to use the soil for economic, environmental and cultural functions is reduced. In contrast, brownfield is defined as an area with destroyed or damaged land surface or abandoned structures, mining sites, economic or military territory [14]. Degraded real estate has formed and continues to form due to various international and local socio-economic factors. Here, problems of financial and/or legal nature are often observed.

In Riga, one can observe a significant number of real estate properties at different quality levels, including degraded/non-degraded objects. The vastness of the city, the large number of real estate properties and the changing environment indicate a large amount of information that should be processed, surveyed, evaluated, administered, monitored, etc. However, there is still no single information system that contains information on each building to the extent necessary to make quality decisions, such as the purchase of real estate.

The degradation level of structure elements refers to the degradation of the performance of structures in the context of the decrease in performance of building envelope elements and energy supply systems. Declining performance is affected by wear and tear, inappropriate management, inadequate maintenance [5], as well as the degradation of mechanical properties of building structural materials [13]. Maintenance rather than wear and tear is given a greater role or influence on the occurrence of degradation. As a result, the energy efficiency of the building is reduced and the costs of the building lifecycle increase [5]. This essential condition already indicates the importance of the owner's administration or maintenance, which is often even a daily matter. Consequently, the area of quality real estate maintenance and management should be particularly highlighted and supported as a long-term policy. In the long run, this would affect the quality assurance of the building fund, its impact on cost reduction and sustainable management.

In the context of urban planning in the municipality of the city of Riga, surveys are carried out to identify the total quality or degradation of objects and territories. In 2004, within the framework of the development of Riga Spatial Plan for 2006–2018, criteria for the identification of degraded territories were developed (based on the experience of the European Union in the management of brownfields, incl. the Contaminated Land Rehabilitation Network for Environmental Technologies in Europe – CLARINET and available materials within the European expert network Concerted Action on Brownfield and Economic Regeneration Network – CABERNET), land and building resources were identified that were previously used economically, but were no longer actively managed or abandoned or contaminated during the research. In Riga, 185 environmentally degrading objects (buildings) and 142 brownfields were identified [3]. In Riga, there are currently over 600 degraded buildings and 200 brownfields [4], which have a negative impact on the city's ecosystem, public safety and economy.

Brownfields also include data from the Latvian Environment, Geology and Meteorology Centre on polluted and potentially polluted areas. At present, there are polluted areas of 104.1 ha and potentially polluted areas of 283.8 ha in Riga, which make up the largest concentration of polluted and potentially polluted areas in Latvia [10].

The Soviet industrial heritage also includes contaminated water objects [15]. Likewise, modern entrepreneurship has a negative impact on the quality of the environment, for example, despite the port contribution to the economy of the state and Riga, residents in the vicinity of the port regularly suffer from odours, noise, vibration and landscape degradation. Therefore, in the Sustainable Development Strategy of Riga, neighbourhoods of Riga port (Mangalsala, Vecmilgravis, Milgravis, Sarkandaugava, Bolderaja and Daugavgriva) are identified as priority areas for improving the quality of the public outdoor environment [6].

The problem of degraded objects is highlighted in the development planning documents of the municipality of Riga: the Sustainable Development Strategy of Riga until 2030 and Development

Program of Riga for 2014–2020 [7]. Political incentives foresee the priority of revitalising degraded objects rather than building new ones, as well as providing support for business development in brownfields [6]. Referring to the above-mentioned planning documents, the European Union funding is attracted for the revitalisation of certain objects/territories.

3. Results and Discussion

Urban environment surveys and information are highly valued as a tool for long-term policy definition, such as spatial planning, support from European Union programs, and as a valuable monitoring indicator. However, this is a general type of study or a general visual assessment, and it does not assess different elements of each structure and their physical assessment. Figure 2 shows a scheme for identifying and monitoring a degraded object in the context of urban analysis and development planning in the municipality of the city of Riga.

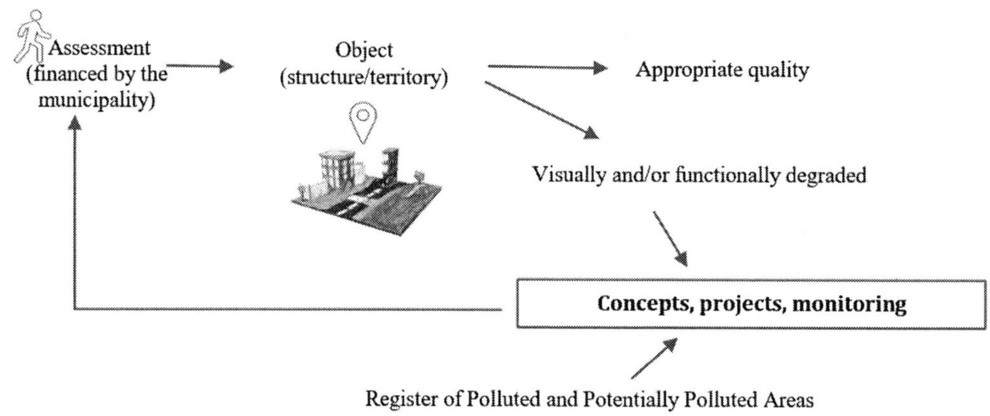

Figure 2. Identification and monitoring of a degraded object in the context of urban analysis and development planning.

This type of degraded object survey and monitoring takes place on average every seven years, related to the drawing up of territorial development planning documents. It is appreciated that territories are also evaluated in these surveys because at other stages of the quality determination of real estate, only fragmentary attention is devoted to territories. In Riga, brownfields have been divided into natural, residential, business and mixed-use areas. Part of the land is degraded while being at the state of reservation, such as red line corridors reserved for future street infrastructure development, but it is currently not being properly utilised, resulting in degradation and/or contamination.

Riga City Council's Committee on Buildings Degrading the Environment determines whether the buildings (structures) qualify as environmentally degrading, collapsed or posing threat to human health and safety [8]. The Committee classifies buildings or structures that are degraded, collapsed or pose a threat to human health and safety not only for the purpose of administering a real estate tax, but also consider applications of building owners claiming to postpone the classification of a structure, if it is manifested that a structure will be restored, and evaluates the change in classification assigned to the structure according to its actual visual and technical state, as well as considers other related issues. Committee's meetings are convened as necessary but at least once a month [16].

It is the responsibility of the owner or the actual possessor of the real estate to maintain the facade of the buildings and other external structures in a technical condition and appearance that does not degrade the environment or ruin the surrounding landscape. Buildings that do not comply with the requirements of urban and landscape conservation are classified according to their impact on human safety in categories A, B or C.

Category A structure/ruin is **completely or partially destroyed, and it is a structure in a dangerous condition posing threat to human safety**. Category B structure/ruin is **a structure in a potentially dangerous technical condition posing threat to human safety**. Category C structure/ruin is **an environmentally degrading structure** [9].

According to the statistical information, in Riga there are mainly Category B structures, which account for 51 % of the total degraded structures, followed by Category A structures (34 %), and Category C structures (15 %). From the point of view of property ownership, structures that are environmentally degrading, destroyed or pose threat to human security are mainly privately owned (87 %), state owned (7 %), municipality owned (3 %) and of unknown ownership (3 %) [17].

The register of environmentally degrading buildings includes environmentally degrading objects and they are subject to an increased real estate tax rate of 3 %. The total number of objects in Riga reached 617 on 4 April 2019. The register also includes objects to be monitored, i.e., objects that are maintained in accordance with the requirements of regulatory enactments and currently their number has reached 161, as well as objects with environmentally degrading features that are planned to be examined by the Committee on Buildings Degrading the Environment, and their number is currently 5103 [17]. Figure 3 illustrates the main steps of the classification of environmentally degrading buildings in the context of real estate tax administration in the municipality of the city of Riga.

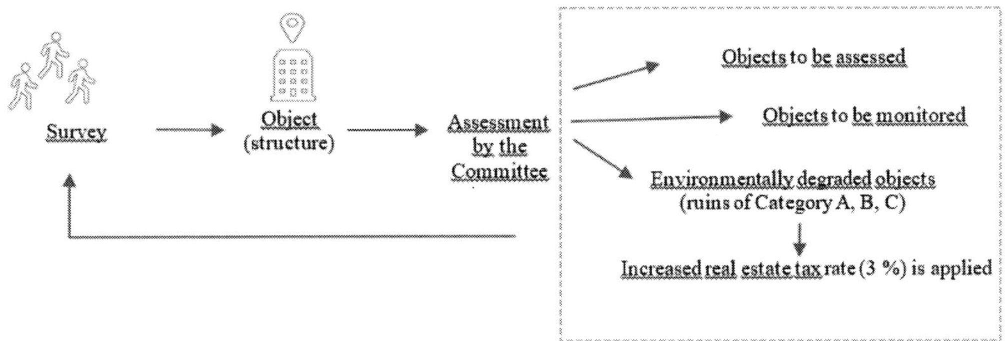

Figure 3. Classification of environmentally degrading buildings in the context of real estate tax administration.

In the case of this survey, surveys are carried out by several persons who can theoretically interpret objects in different ways. Although a large number of the Committee representatives provide a comprehensive view of building evaluation, it should be borne in mind that in this case only structures are assessed, so the analysis of brownfields remains outside the assessment.

The physical deterioration of a building is much more detailed, which is often the basis for various decisions, but this information is not always relevant to the current situation without updating it.

The indicator of physical deterioration of a building affects the cadastral value of the object, which in turn affects the amount of the real estate tax and the value of the property. Information on building deterioration or physical condition is included in the State Information System of Real Estate Cadastre under the section "Basic Data of Structures".

As an example, the city of Riga is considered, which is dominated by residential buildings put into operation in the periods of 1900–1914, 1925–1940 and 1960–1975; their number is 3769, 3542 and 5772, respectively. According to the assessment of physical deterioration, most buildings fall into categories of physical deterioration ranging from 21 % to 60 % (see Figure 4).

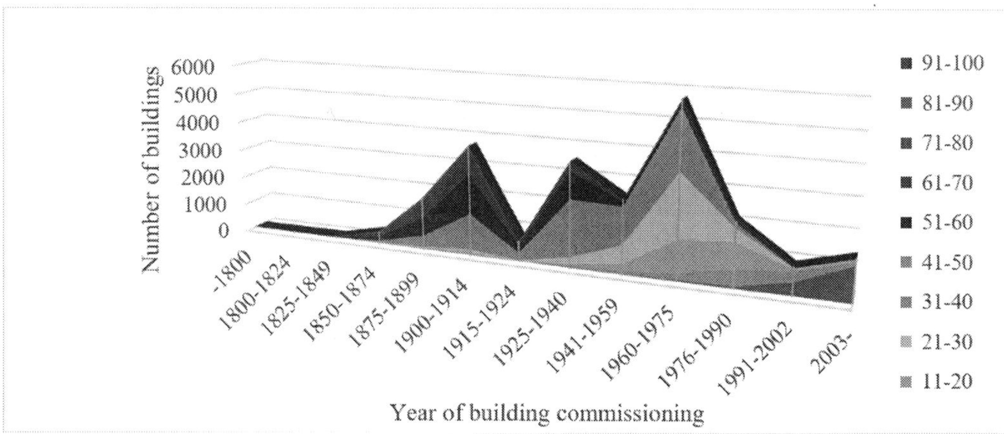

Figure 4. Buildings by the year of commissioning and physical deterioration, % [18, calculations made by the authors].

A high proportion of buildings in poor technical condition with deterioration above 50 % indicates poor quality of living environment. Approximately 27 % of residential buildings in Riga have a depreciation rate above 50 % and these are mainly buildings put into operation in the periods of 1875–1899, 1900–1914 and 1925–1940.

Within the Building Deterioration Detection Service, buildings are surveyed locally, determining the ratio (percentwise) of physical deterioration of the main structural elements (foundation, walls, intermediate floors and roof). After the survey, information is updated in the State Information System of Real Estate Cadastre (at the same time, the cadastral value of the building and assessment of real estate are updated). The determination of building deterioration is shown schematically in Figure 5.

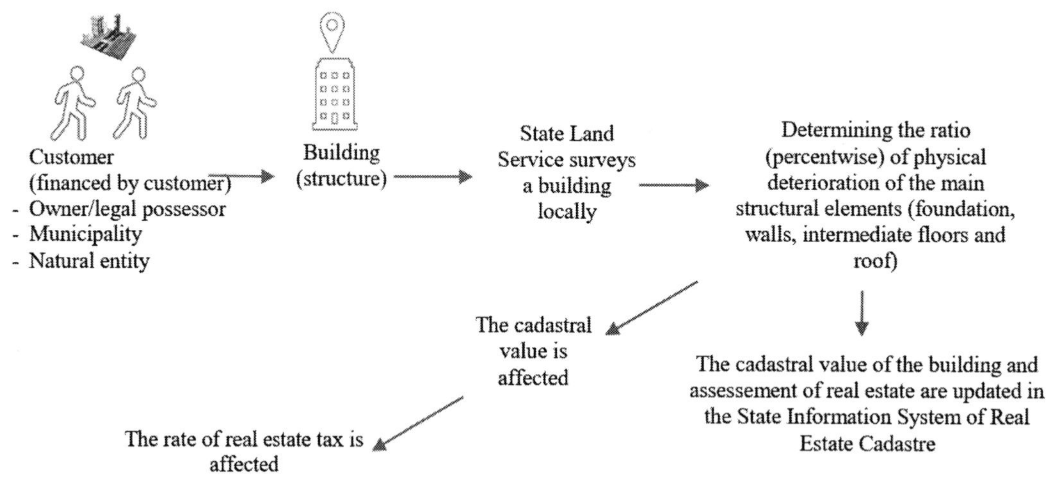

Figure 5. Determination of building deterioration.

Determining building deterioration is a very important process in the entire real estate context, which should be given special attention as it affects both value and cost in the short and long term. Therefore, it would be important to carry out such assessments on a regular basis and possibly on request by the state or municipal authorities to have information on the condition of the entire building stock. In addition, as the technical opinions of buildings demonstrate, the information here also depends on the knowledge of the performer and the work done.

Supervision of the operation of structures is ensured by Riga City Construction Board, which makes decisions on the technical examination of the structure or built-in construction products, gives an instruction to eliminate the identified hazards, if the danger of the structure has been detected, — prohibits its operation until the prevention of danger.

Despite different approaches, state and municipal institutions are constantly looking for and applying solutions to force and encourage owners to restore real estate in the city themselves, especially the degraded property. According to the statistics, in recent years the activity of restoring the degraded property has become more active in Riga. If 16 such objects were restored in 2011, then in 2017 their number already reached 205 [19].

According to the regulatory framework, each owner is responsible for keeping their property in good condition. Legislation prescribes a number of conditions to maintain real estate quality and physical preservation throughout their lifetime. They include cleaning conditions, energy efficiency requirements, microclimate, service availability, structure and equipment inspection and maintenance, financial organisation and other management issues.

Reviewing scientific opinions and political documents of the city, the authors set out the criteria that should be taken into account when determining the level of degradation of objects (see Table 1). To perform a comparative analysis at an urban level, for example, when deciding on priority objects to be revitalized, impact weights should be developed and incorporated in the proposed criteria. Moreover, additional information should be added, such as ownership, encumbrances, the cultural and historical significance of the object, the range of persons affected, as well as the potential environmental and socio-economic benefits gained after restoring the object as opposed to the loss if the object is not restored. It is important to evaluate not only the short-term and long-term income losses, but also the maintenance costs and increased real estate tax.

Table 1. Criteria for Detection of Real Estate Degradation [developed by the authors].

No.	Criterion	Notes
1.	Environmental pollution caused by the object	CO2 emissions, thermal areas, soil and water pollution, waste. The larger the affected area size (and/or depth), the greater the weight to be assigned.
2.	Deterioration	Physical and moral assessment of buildings and their elements, utilities, engineering structures.
3.	Visual appearance	Visual appearance of facades and other elements of buildings, as well as architectural integrity in a common building space.
4.	Safety	Compliance with relevant operational requirements.
5.	Intensity of technical maintenance	Regular and qualitative assessment of the necessity of restoration work, frequency and quality of the works to be performed.
6.	Functionality	Intensity of real estate use and functionality.
7.	Financial sustainability	The operation of an object on the basis of sustainable solutions and/or materials. Assessed in the context of Indicator 1 (such as fossil fuels, heat loss) and Indicator 6, such as partial or incomplete use of property (empty/abandoned property that is gradually degrading).

In the case of degraded property, dismantling, reconstruction or renovation are identified as voluntary restoration measures. They are also influenced by building regulations of a respective municipality, in the case of Riga, these are the regulations for the use and construction of the territory of Riga, which set the requirements for the use of land plots and structures in the territory of the municipality. In Riga, it is recommended to demolish an environmentally degrading building, as it degrades the environment and to construct a new building in place of the degrading buildings in accordance with the regulations for the use and construction of the territory of Riga for new buildings [9].

Conclusion

Inappropriately administering or abandoning a real estate object, it loses its value, attractiveness both morally and physically, and starts degrading over time.

In municipality of the city of Riga, real estate appraisal takes place at several levels, such as survey in the context of urban development and policy making, survey, analysis and evaluation of degraded buildings for application of a real estate tax rate, technical survey findings upon owner's or manager's request, various surveys on forced restoration, determination of building deterioration in the context of cadastral information, as well as surveys in the nature performing real estate valuation. There are several evaluations of each building that are not included in a single information system, but are used to make specific decisions.

When the owner or property manager detects deviations from the norm or identifies various deficiencies when deciding on the restoration of the property and starting the restoration, usually the costs are also lower in the long run, and the property is not classified as a degraded one by the supervising authorities, thus the real estate tax rate is not increased and the property is not prevented from being exploited.

In cases where the real estate does not conform to the quality stipulated in the regulatory enactments and its owners do not take the appropriate measures for maintaining the building in good technical condition, several administratively motivating mechanisms are imposed, for example, penalties, increased real estate tax rate, or ultimately Riga City Council's decision is taken on the forced restoration of the object.

Despite the rapid dynamics of real estate restoration, there is still a need for complex solutions to implement the ongoing process of property restoration, as well as to prevent the emergence of new degraded objects. At present, there is no clear definition of a set of preventing measures in the context of empty buildings in order to prevent degradation of the property.

References

[1] Capelle-Blancard G. and Laguna M A 2010 How does the stock market respond to chemical disasters? *Jornal of Environmental Economics and Management. Elsevier* vol 59 (2), pp 192–205

[2] Carla S S Ferreira, Rory P D Walsh and António J D Ferreira 2018 Degradation in urban areas *Elsevier*- Nr.5 pp 19 - 25

[3] Degradēto teritoriju izpēte Rīgas pilsētā 2004 Rīga: *SIA "Grupa 93" pp* 59 – 66

[4] Par aktualizētā Rīgas attīstības programmas 2014.–2020.gadam Rīcības plāna un Investīciju plāna 2014.–2016.gadam apstiprināšanu: Rīgas domes lēmums Nr.2491 (21.04.2015.)

[5] Eleftheriadis G and Hamdy M 2017 Impact of building envelope and mechanical component degradation on the whole building performance: a review paper *Elsevier vol* 132 pp 321 - 326

[6] Rīgas ilgtspējīgas attīstības stratēģijas līdz 2030. gadam un Rīgas attīstības programmas 2014. – 2020. gadam ieviešanas uzraudzības pārskats par periodu no 2014. līdz 2016. gadam Rīga: Rīgas domes Pilsētas attīstības departaments, 2017 pp 57

[7] Rīgas attīstības programma 2014-2020.gadam Rīga: Rīgas domes Pilsētas attīstības departaments, 2014 pp 82.

IOP Conf. Series: Materials Science and Engineering **660** (2019) 012043 doi:10.1088/1757-899X/660/1/012043

[8] Rīgas domes Vidi degradējošu būvju komisijas nolikums: Rīgas domes nolikums Nr.58 (2014, 11.februāris)

[9] Rīgas pilsētas teritorijas kopšanas un būvju uzturēšanas saistošie noteikumi: Rīgas domes saistošie noteikumi Nr.146 (pieņemts 28.04.2015.) 13.punkts// "Latvijas Vēstnesis". Nr.84 (5402) (2015.30.aprīlis)

[10] Latvijas Vides, ģeoloģijas un metroloģijas centrs https://www.meteo.lv:8443/lapas/vide/piesarnoto-un-potenciali-piesarnoto-vietu-registrs/piesarnoto-un-potenciali-piesarnoto-vietu-registrs?id=1527&nid=373

[11] Proikakia M, Nikolaoub I, Jonesc N, Malesiosd C, Panayiotis Dimitrakopoulose D and Evangelinos K 2018 Community perceptions of local enterprises in environmentally degraded areas. *Journal of Behavioral and Experimental Economics* vol 73 pp 116–124 https://doi.org/10.1016/j.socec.2018.01.007

[12] Metodikas izstrāde degradēto objektu un teritoriju identificēšanai. Projekts "Integrēta plānošana un sadarbības modelis degradēto teritoriju revitalizācijai (Baltic Urban Lab)" Rīga: SIA "Eiroprojekts", 2017 pp 4

[13] Topolář L, Štefková D and M Hoduláková 2017 The Assessment of Cement Mortars after Thermal Degradation by Acoustic Non-destructive Methods. *IOP Conference Series: Materials Science and Engineering. IOP Conf. Series: Materials Science and Engineering* vol 251

[14] Zemes pārvaldības likums (pieņemts 30.10.2014.) 1.pants// "Latvijas Vēstnesis". Nr.228 (5288) (2014, 15.novembris)

[15] Pārskats par Kīleveina grāvja piesārņojuma izpēti.- Rīga: SIA „Vides Konsultāciju Birojs", 2015.- 25.lpp

[16] Vidi degradējošo būvju komisija.- http://grausti.riga.lv/lv/vidi_degradejoso_buvju_komisija/.- (Accessed 15.09.2018.)

[17] Statistiska.- http://grausti.riga.lv/lv/statistika/.- (Accessed 04.04.2019.)

[18] http://www.vzd.gov.lv/lv/pakalpojumi/buvju-un-dzivoklu-ipasniekiem/datu-aktualizacija/ekas-nolietojuma-noteiksana/#Par-pakalpojumu.- (Accessed 11.12.2017.)

[19] Vidi degradējošās būves Rīgā.- http://grausti.riga.lv/lv/statistika/.- (Accessed 22.03.2019.)

IOP Conf. Series: Materials Science and Engineering **660** (2019) 012044 doi:10.1088/1757-899X/660/1/012044

Competence of local government Building Board in elimination of the consequences of arbitrary construction

I Luce[1] and I Pukite[2]

[1]College of Law, Latvia, Kr. Valdemara Str. 1c, Riga, LV-1010, Latvia
[2] Riga Technical University, Latvia, 6 Kalnciema Str. Riga, LV- 1048, Latvia

E-mail: intra@jk.lv

Abstract. In accordance with the provisions of regulatory enactments of the Republic of Latvia, the law obliges the local governments to establish a Building Board within their administrative territory or, in co-operation with other local governments to develop joint Building Boards and ensure the resources required for the operation of a Building Board, thus providing for the legality of construction process in their administrative territory. Building Board is an institution of special status, established at the local government and it is one of the municipal structures that has been delegated to control the legality of the building process in the administrative territory of local government, therefore, the Building Board is an intermediate administration institution that is bound by external regulatory enactments that are binding on derivative public entity, and, therefore the administrative procedure is binding on them as well.The objective of the article is to study the competence of Building Board in the elimination of the consequences of arbitrary building by determining the principal functions of the Building Board, signs of the detection of the arbitrary building and the competence of the personnel of the Building Board and the competences of the Building Board personnel in the review of arbitrary construction cases. The article evaluates the administrative acts of the Building Board that impose the duty of the addressee of the administrative act to eliminate the consequences of arbitrary building.

1. Introduction

Building is the sector of economy that falls within the competence of the Ministry of Economics. To make the planning, control, organisation and monitoring of the construction processes easier and more efficient, the government has delegated these rights to local governments. In each administrative territory a local government has been established, building boards are established under its authority and regulate the building processes in the particular municipality.

Many individuals in contemporary developed world own real estate that they wish to transform and make more attractive by reconstructing the existing buildings or renewing them, as well as constructing new buildings on their land plots. Foreign investors in Latvia invest their financial resources in construction and implementation of different new projects. Real estate companies have been established in the country that are involved in the transactions in this area. Some of them are implementing new projects by constructing residential villages, multiple apartment homes or commercial objects.

The importance of building is determined by significant proportion thereof in the development of gross domestic product. Building is relatively labour-intensive process and, on national scale, it attracts approximately one half of all capital investments. Building is associated with any sector of economy, since it creates basis for that – in the form of buildings and constructions. Therefore, development of viable building and construction industry is important for every country.

Building board is a special status institution that has been developed at each local government and is one of municipal structures that are authorised to control building processes in the administrative territory of the local government.

Building are occupied by people and would seem reasonable to expect that the way they are designed and constructed protects people's health and well-being, and the effectiveness of this would be verified in practice. However, the construction sector is unusual in that is does not routinely evaluate the performance of its artefacts [12].

Construction project success depends on the multi – firm project organizations involved working together satisfactorily. The success of construction projects is a fundamental issue for most governments, users and communities [11].

To ensure that the process of building is successful, the following factors must be observed: safety and quality; past performance; environment; management and technical aspects; organization; experience; size/type of previous projects; finance [4].

The objective of the article – to determine the competence of building board in the prevention of arbitrary construction. Due to this, the authors study and analyse the following:

1) The legal aspect of the development of building boards;
2) The competence of the building board and implementation thereof;
3) What the meaning of arbitrary building is;
4) The competence of the building board in the elimination of arbitrary consequences.

Building board is an intermediate administration institution bound by external regulatory enactments that are binding upon derivative public entity. The hierarchy that must be observed, while making decisions – the Constitution of the Republic of Latvia, laws, Regulations of the Cabinet of Ministers, Binding municipal regulations.

In accordance with the regulatory enactments, the administrative act issued by the Building Board or actual action can be contested by submitting a complaint to the Building Board. The decision of the City Council regarding the contested administrative act or actual action can be appealed at the Administrative District Court. [6]

2. Materials and methods

The scientific research methods used in the article of the author:

1) descriptive method that will help to determine the institutes of administrative law, for instance, Building Board and its place in the system of administrative institutions;
2) the dogmatic method will help correctly understand administrative norms;
3) the analytical method will help to determine and to assess the content of legal concepts and legal principles.

All of the aforementioned methods enable to obtain research results and to develop theoretical conclusions in the area of elimination of consequences of arbitrary building.

Significant number of the regulatory enactments of the Republic of Latvia is required to reach the set target. Internet resources of specialised government institution websites were used to obtain the aforementioned regulatory enactments, international data bases were used. At the same time, legal practice was examined during the development of the article.

3. Results and discussion

3.1. Building Board as the institution developed by a derivative public entity

Safety of building is the practice of designing, constructing, operating, maintaining and removing buildings in ways that no one has deteriorated health, suffered injuries or died due to the use of the building [7].

State government institution structure of Latvia consists not only of direct administration (state) institutions, but also derivative public entities, including local governments. The existence of local governments arises from the concept that direct administration entities cannot perform all administrative tasks in a centralised an effective manner in the entire territory of the country. There are many administrative tasks, the effective implementation of which is possible only with the knowledge of the local conditions, including the needs of the residents of respective administrative territory [5].

The local government that is responsible for the operation of the respective building board in its territory is a specific form of state administration, where the residents living at the places of residence of the country are involved in the implementation of the state administrative operations. Three differences between local governments and other derivative public entities have been mentioned in scientific literature: the local government is operating in a particular territory, the local government is operating in the interests of the residents, the municipal government is independent in its actions.

Construction Law defines the concept – building board as a municipal institution or a structural unit, or an institution established by several local governments. The function of building board can be performed by several structural units of the local government, each of these institutions has certain competence that arises from this law [9]. This means that the Building Board is an institution established by the local government and is subordinate to the chairman of the City Council. The head of the Building Board manages and organises the work of the Building Board. The head of the Building Board is appointed and removed from office by the City Council. Internal and external regulatory enactments determine the competences of the Building Board.

To ensure that personnel working at the Building Board act within the limits of the provisions of regulatory enactments and their competences, the Construction Law, which came in effect on 1 October 2014 and series of subordinated regulations of the Cabinet of Ministers must be known. Special emphasis in these regulatory enactments was placed on the process of organising construction works and responsibilities of the involved parties, simultaneously reducing the number of bureaucratic procedures.

Success in construction projects is dependent on the effective organization of multiple, specialized teams, each of which brings its own ability, experience, knowledge and skill towards completing the joint project, but which also bring their own objectives, goals and management styles, which may not be entirely complimentary [6].

Existing buildings in operation, if properly operated, are considered to be safe for people [3]. However, practice proves the fact that, upon the inspection of the buildings on site, reconstruction works, which are not approved by the Building Board and fail to conform to construction norms, are performed.

Considering the development of contemporary technology, changes in the regulatory enactments, more and more attention is paid to safety requirements, while being in the premises [8]

Construction Law describes the duties performed by the Building Board:

1) controls the construction process and conformity thereof to the requirements of this Law and other regulatory enactments that govern construction;
2) provides information on the conditions of territory use and construction, as well as the engineering networks in the territory;
3) informs on the legal basis of the construction in progress and provides the information on the building;
4) reviews applications and makes decisions on the construction idea, as well as checks the execution of the conditions included in the decisions and the content of submitted documents in the amount provided for by the regulatory enactments;
5) reviews the plan of construction supervision;
6) appoints construction inspector for the control of construction at the construction object and determines the schedule of mandatory visits to the construction site. In the event, where the building needs construction supervision, the aforementioned schedule will be drawn up by evaluating the main phases of construction works determined by the construction supervision plan;
7) upon the receipt of information on the non-compliance of the construction product at the construction site with the requirements of regulatory enactments, decides on the need to ask the submitter of the construction idea to assess the effect of incompatibilities on significant requirements set for the building;

8) reviews alternative technical solutions for the provision of the accessibility of environment, if required, by requesting the opinion of the experts of the non-governmental organisations of the respective sector in the cases, where the observation of the technical requirements of construction normative regulations;
9) commissions constructions;
10) reviews applications and makes decisions on the change of the type of use of a building or a part thereof after reconstruction;
11) consults on the procedure of building process;
12) performs other actions associated with the construction process and conformity thereof with the requirements of regulatory enactments;
13) registers construction permits issued by other institutions;
14) consults on the possibilities of building in the respective territory;
15) co-operates with the bureau and institutions that perform the functions of the Building Board in accordance with the procedure provided by general construction normative regulations [9].

The aforementioned normative enactment allows to conclude that the competence of Building Boards is clearly indicated in the Construction Law.

The objective of the Building Board is the development of high quality city space and provision, implementation and control of the legality of the construction process in the administrative territory, in accordance with the spatial plan of the respective territory, construction regulations and other regulatory enactments.

Building Board regulations and Construction Law allow to come to the conclusion that one of the main objectives of the Building Board is to provide the legality of the building process, thus, the Building Board has the duty to set an objective of making arbitrary construction "uncomfortable".

To ensure that the Building Board acts within the limits of its competence, it requires employees with the education in engineering sciences, construction, architecture, design, spatial planning.

However, basically, to perform their office duties, they must have the appropriated education in the area of construction, as well as building inspectors must comply with the requirements of Cabinet Regulation No. 499 *Regulations Regarding Building Inspectors*, adopted on 19 August 2014. The requirements for building inspectors have been indicated in the regulation, the procedures for their inclusion or exclusion from the register of building inspectors, as well as the procedure for supervision of professional activities have also been provided.

The employees of Building Board, who adopt regulatory enactments that are binding on natural persons and issue administrative acts on behalf of the Building Board, are officials in accordance with the law *On Prevention of Conflict of Interest in Activities of Public Officials*. Part Two, Section 4 of the law provides – Also persons who in fulfilling official duties in institutions of a public person have the following rights in accordance with laws and regulations, shall be considered to be public officials: to issue administrative acts; to fulfil supervisory, control, inquiry or punishment functions in relation to persons who are not directly or indirectly subordinate to them.

Definitely, Section 2 of *Administrative Procedure Law* is also binding on employees of Building Board, as it defines the principal targets of administrative procedure: 1) to ensure the observance of basic democratic, law-governed state principles, especially human rights, in specific public legal relations between the State and a private person; 2) to subject actions of executive power relating to specific public legal relations between the State and a private person to the control of an independent, impartial and competent judicial power; 3) to ensure just, accurate and effective application of the norms of law in public legal relations [9].

When performing their work duties, Building Board employees may not have subjective opinions during decision making process and they must act in accordance with regulatory enactments, therefore, the employees of Building Boards, who make decisions on behalf of the Building Board, are public officials, who must observe the law *On Prevention of Conflict of Interest in Activities of Public Officials* in order to prevent adoption of subjective decisions.

When Building Board performs its functions and makes decisions that are binding on private entities – natural person, legal entity of private law or a union of such persons [10], the compliance with Administrative Procedure Law must be ensured, while making decisions on the merits or implementing actual actions. Thus, considering the above, the Building Board, its structural unit or its officials are the institution that implements functions provided for by Construction Law.

The Building board, in accordance with the Administrative Procedure Law, initiates administrative cases on the basis of applications, at its own initiative and based on an order of a superior institution or a report of another institution. The institution may also implement actual action. Actual action of the institution - an action in the area of public rights that does not manifest as a legal act and is targeted at the creation of actual consequences, if the private person is entitled for this action, or infringement of subjective rights or legal interests of the person has arisen or may arise as a result of such action. Actions of institution that, independently of the intention of the institution, generate such actual consequences, as a result of which the private person has faced or can face significant infringement of rights shall also constitute an actual action. Procedural actions of an institution [10].

Actual action can manifest in the following cases:

a) Building Board receives an application, which indicates that the addresses wishes to receive a certificate that certifies the absence of the construction on site. Action of the Building Board - the building inspector drives to inspect the object and issues a certificate on the absence of the building on site.

b) A case, where the application with the request to issue a certificate on the status of the new construction has been submitted. Action of the Building Board - the building inspector inspects the object and compares with the construction project approved at the Building Board, if no deviations from the building project are detected at the object, the building inspector shall issue the certificate on the status of new construction.

When the Building Board receives an application, where signs of arbitrary construction are detected, it is referred in the form of a resolution to the Building Administration division, which evaluates on the merits, whether the submitter must provide answer on the merits. The chief building inspector of the Building Board evaluates the facts indicated in the application and performs visual surveying of the object, as well as assesses the information at the disposal of the Building Board. Section 134 of Cabinet Regulation No. 500 *General Construction Regulations*, adopted on 19 August 2014 provides that a construction inspector must prepare opinion regarding each inspection.

The construction inspector shall prepare an opinion on the visual inspection of the object, where the actual situation at the object is identified, as well as the opinion on the presence or absence of signs of arbitrary building at the object is drawn up. If the construction inspector detects signs of arbitrary building, the administrative procedure for the elimination of the consequences of arbitrary building shall be always initiated at the Building Board.

Simultaneously, a building inspector must evaluate the time period, when arbitrary construction was performed, because Section 239 of Latvian Administrative Violations Code indicates the cases, where administrative violation case cannot be initiated. Most frequently, administrative violation case of arbitrary building cannot be initiated, because the periods indicated in Section 37 of Latvian Administrative Violations Code have expired. Latvian Administrative Violations Code provides that administrative violations case can be initiated no later than within six months since the date, when the violation was committed, but, if the violation was lasting — since the date, when the violation was ceased [13].

The building inspector can also initiate an administrative violation case, if they have received an application containing signs of arbitrary building,– upon the receipt of information that point to potential commission of an administrative violation, they shall decide on the initiation of administrative violation case or on the refusal to initiate an administrative violation case.

Thus, a building inspector has three options of initiating an administrative violation case:

1) Upon the receipt of an application – in the form of resolution on the application;
2) In the form of a decision;

3) In the form of an opinion.

Considering the above, the conclusion can be made that, depending on legal situation, two parallel procedural actions may occur: within the framework of an administrative procedure and within the framework of an administrative violation case.

4. Concept of Arbitrary Building

Arbitrary building is the term that refers to construction works that are performed without the construction permit or prior to the moment, when a note regarding the fact that certain conditions have been met is made in the construction permit, confirmation card or explanatory note in the cases, where the respective decision is required in accordance with the regulatory enactments, as well as building works that fail to conform to the construction project and the requirements of the regulatory enactments are classified as arbitrary building. Term arbitrary building also refers to the use of the building or part thereof that does not conform to the type of use indicated in the project, or building works that have been commenced without the respective project documentation, if such is required in accordance with regulatory enactments [9].

The definition of arbitrary building indicated above allows the conclusion that the fact of arbitrary construction can be detected, if particular conditions have occurred. Therefore, arbitrary building – building works:

1) That have been commenced without construction permit or respective project documentation, if such is required in accordance with regulatory enactments;
2) That have been commenced or are performed prior to the date when a note certifying that particular conditions have been met is made in the construction permit, confirmation card or explanatory note, if the respective decisions are required in accordance with the requirements of regulatory enactments.
3) That fail to conform to the building project and requirements of regulatory enactments.

The definition of arbitrary building also indicates that the term arbitrary building refers to the use of the building or part thereof in a way that is different from the type of use indicated in the project. The information above allows the conclusion that the term arbitrary building refers not only to building works that are not approved in accordance with the requirements of regulatory enactments, but, potentially, also to the use of the building in a way that differs from the approved type of use.

In essence, the regulation of the Construction Law also refers to the cases, where arbitrary building is not detected during the process of building works, but later – during the use of illegally built (or non-commissioned) building. In these cases, the legislator has delegated certain competences to the building inspector, upon the detection of the fact of arbitrary building, to suspend further operations at such building site until the decision of the Building Board on the restoration of the previous condition or continuation of construction is made [1]. Therefore, based on the decision, conclusions can be made that arbitrary building is not only construction works performed during the building process, but also buildings or constructions that have already been built or rebuilt during use, which are not commissioned.

5. Implementation of Building Board Competence, while Eliminating the Consequences of Arbitrary Building

The elimination of the consequences of arbitrary building is commenced at the moment, when the Building Board, upon the performance of an inspection, detects the signs of arbitrary building, or a person that has performed arbitrary building (or purchased the real estate, where arbitrary building had been performed) voluntarily submits the respective application to the Building Board.

The construction permit shall not be issued, if the construction idea fails to conform with the municipal spatial plan, local plan and detailed plan, except for the cases, where building idea refers to the object of national interest, if the building idea has not been approved by the owner of the land plot or the owner of the building in the cases, where construction works are planned in the existing building, or where the owner of the land is not informed in the cases, when it is required by regulatory enactments, if, the building project has not been developed in minimum version in accordance with the requirements of regulatory enactments that govern building sector, except for the cases, where the drafting of the respective building project is not required, if the initial evaluation of the planned building or the environmental impact assessment has not been drawn up in the cases, when it is required by regulatory enactments. When the Building Board has evaluated the submitted building idea in minimal version thereof and detects any of the aforementioned cases, it declines building permit and issues a negative administrative act to the submitter of the building idea, but, if the Building Board fails to detect any of the aforementioned cases, it issues the building permit.

Thus, considering the above, it can be concluded that construction works can be commenced, if the Building Board has made a note in the construction permit certifying that all design conditions included in the permit have been met, the conditions for the commencement of building works have been met and the construction permit has become non-contestable." [9]. At the same time, it must be considered that prior to the issue of a construction permit by the Building Bard, the construction inspector shall inspect the site of construction idea implementation to verify that no arbitrary building has occurred at the site.

The note on the conformity with design conditions is one of the pre-requisites for the validity of the building permit in terms of material-law. Meanwhile, since the moment of the promulgation of the building permit, it will take effect in procedural meaning. The note that design conditions have been met is a procedural decision, whereby the institution confirms, whether the construction project confirms to the conditions of the building permit and that the construction works can really be commenced [15].

Issue of an administrative act. Prior to the issue of an administrative act to an addressee, the Building Board, in accordance with Section 66 of the Administrative Procedure Law, must assess the content of necessity considerations and decide on:

1) the necessity of the administrative act for the reaching of a legal (legitimate) objective;
2) the suitability of an administrative act for the reaching of the respective objective;
3) the need for the administrative act, i.e. on whether this objective can be reached by means that are less limiting to the rights or legal interests of the administrative procedure participants;
4) the appropriateness of the administrative act, after comparing the violation of the rights of natural person, benefit for public interests and considering the fact that considerable infringement of the rights of a natural person can be justified only by significant public benefit [10].

When the Building Board has evaluated the information at the disposal of the Building Board and clarified the conditions of the case, as well as heard opinion of the administrative procedure participant, it shall immediately evaluate the conditions of the case and issue:

1) mandatory administrative act, if the applicable legal norm provides that the administrative act must be issued;
2) free administrative act, if the institution has the freedom to act and the issue of an administrative act is useful;
3) administrative act, whereby the issue of favourable administrative act to the submitter is refused, due to the fact that the issue of the administrative act is unjustified or it is not useful;
4) the decision on the termination of the case due to the lack of facts or uselessness, if the case was initiated at the initiative of the institution, including - on the basis of information (complaint) submitted by another natural person [10].

If the institution recognises that arbitrary building occurs at the object, the institution shall decide on the restoration of the previous state or the permission to continue construction. The administrative act of the institution, considering the final conclusion of the institution on the presence or absence of arbitrary building, the duty can be imposed on the person to prepare the building documentation required for the execution of the decision. Thus, the opinion of the building inspector is an intermediate decision" [15].

In the cases, where the building inspector detects arbitrary building, they will suspend the works and draw up the respective opinion, while the institution will adopt one of the following decisions:

1) on the restoration of the previous state, if the building of the particular object in the respective territory is not permitted in accordance with the regulatory enactments or building works were commenced prior to the note in the building permit on the conformity with design conditions – irrespectively of the conditions due to which the respective action was not implemented;

2) on the permission to perform construction works after the compliance with the requirements of the regulatory enactments that govern building, but in the cases, where harm to environment was inflicted as a result of arbitrary building, — after the elimination of the harm in accordance with the procedures of the regulatory enactments that govern environmental protection. If the decision on the permission to perform building works was not implemented within the period indicated by the Building Board, the Building Board may adopt a decision on the restoration to the previous state.

To reduce administrative load, the Building Boards commenced the practice, where the opinion on the inspection of the building includes the decision on the elimination of the consequences of arbitrary building, the Building Board entitles the building inspector to make decisions on the elimination of the consequences of arbitrary building.

Upon making the decision on the elimination of the consequences caused by arbitrary building, the Building Board must evaluate the possibility of deciding on lesser infringement of the rights of the person. Therefore, upon the evaluation of usefulness considerations and commensurability, the conclusion must be made that the most commensurable decision made with the purpose of elimination of the particular arbitrary building is to provide the addressee with the possibility of developing construction documentation for building works at the real estate within the limits of particular procedural period, or to restore the previous legal status of the real estate by granting freedom of action to the addressee to select most appropriate means for the purposes of eliminating the consequences of the particular arbitrary building [16].

Simultaneously, the opinion on the inspection of the building includes the warning on the mandatory execution of the administrative act.

For instance, Building Board of Riga City Council has indicated in its management report that 588 administrative acts regarding the elimination of consequences of arbitrary building were issued by Arbitrary Building Assessment Department (Legal Support Department) of Building Control Division of the Building Board in 2015, 1,038 (opinions as decisions on the elimination of the consequences of arbitrary construction or termination of an administrative case) administrative acts regarding the elimination of consequences of arbitrary building were issued in 2016, wile, 1,078 administrative acts regarding the elimination of consequences of arbitrary building (opinions as decisions on the elimination of the consequences of arbitrary construction or termination of an administrative case) were issued in 2017. The number of issued administrative acts demonstrates that the Building Council has increased the rate of combating arbitrary building in the administrative territory of the city of Riga.

Voluntary execution of the administrative act. The addressee of the administrative act shall execute the act voluntarily. Therefore, the Building Board, upon evaluating the use fullness considerations and commensurability, by the administrative act, gives the addressee a possibility of choosing how to eliminate the arbitrary building consequences at their real estate. Thus, the addressee selects whether they will develop building documentation or restore the previous state. If the addressee of the administrative act chooses to return the previous state at the real estate, they will return the previous state at the real estate and inform the Building Board thereof.

Mandatory execution of the administrative act. If the addressee fails o perform voluntary execution, the administrative act can be executed according to mandatory procedure [10]. The Building Board issues an administrative act regarding the elimination of consequences of arbitrary building and immediately includes the warning on mandatory execution of the administrative act. Mandatory execution of the administrative act shall be performed by the institution that has issued the administrative act, namely - the Building Board.

At the same time, attention must be paid to the fact that the costs of mandatory execution of the administrative act shall be enforced on the addressee of the administrative act.

To commence mandatory execution of the administrative act, the Building Board must establish that the administrative act was not executed within the established period of time, as well as that it was not contested and therefore it became irrevocable and has to be executed.

Mandatory execution is implemented in accordance with the conditions of the warning, which could include - the use of mandatory payment or direct force. The executive institution, based on the external regulatory enactment and considering the considerations of usefulness, can select measures for mandatory execution, by changing them until the result is reached. If the administrative act imposes a duty on addressee to perform certain actions or to refrain from certain actions and they fail to perform this duty, mandatory payment may be enforced from the addressee.

Considering the aforementioned, the Building Board must establish that the Building Board possesses insufficient information to issue an executive order and all pre-requisites for imposing mandatory payment on the addressee for the failure to voluntarily execute the administrative act have set in. At the same time, the Building Board invites the addressee to execute the administrative act and eliminate the consequences of arbitrary building at the object, since the payment of mandatory payments imposed by an executive order does not release the addressee from the execution of the administrative act.

The application of mandatory payment sis not an objective in itself, but a means for the provision of the execution of the administrative act. Thus, in every particular case, the usefulness of application thereof must be evaluated. Generally, the fact that the person has commenced the voluntary execution of an administrative act and that this execution cannot be completed due to reasons independent of the intention of such person, could be taken into consideration, when making the decision on the need to impose the mandatory payment and the amount thereof" [2].

Conclusion

The objective of the Building Board is the development of high quality city space and provision, implementation and control of the legality of the construction process in the administrative territory of the local government, in accordance with the spatial plan of the respective territory, construction regulations and other regulatory enactments. Building Board has set objective – to make arbitrary building "inconvenient".

At the same time, it must be considered that arbitrary building is not only detected during construction works, but also buildings or constructions that have already been built or rebuilt during use, which are not commissioned.

The elimination of the consequences of arbitrary building is commenced at the moment, when the Building Board, upon the performance of an inspection, detects the signs of arbitrary building, or a person that has performed arbitrary building (or purchased the real estate, where arbitrary building had been performed) voluntarily submits the respective application to the Building Board.

Persons, who are not employed in building sector cannot be certain that, upon the purchase of real estate, there are no signs of arbitrary building at the property, therefore the authors would suggest that a note regarding arbitrary building should be included in the land register. Thus, the new owners of the real estate would be informed on the situation at the real estate.

Analysis of the requirements and duties established for the employees of the Building Board, the most important ones are – 1. Competence; 2. objectivity and independence; 3. professional qualification; 4. honesty and ethics; 5. responsibility.

This article is the first part of a series of articles, the second part will analyse the five aforementioned requirements and duties of employees working for Building Boards. More in depth analysis will be given in the part II at article

References

[1] Administrative court of the Latvian Republic 2017 *Decision of the administrative court of 9 February 2017 on the case No. A420141217*, Administrative court of the Latvian Republic

[2] Administrative court of the Latvian Republic 2017 *Administrative District Court decree of 21 March 2017* No. A420158117, Administrative court of the Latvian Republic
[3] Abdelaziz Elsorady D 2014 Assessment of the compatibility of new uses for heritage buildings, *Journal of Cultural Heritage,* 15, 511–521. DOI: 10.1016/j.culher.2013.10.011
[4] Alzahrani J.I., Emsley M.W 2013 The impact of contractors' attributes on construction project success: a post construction evaluation. *International Journal of Project Management,* 31 (2): 313–322
[5] Briede J., Danovska E., Kovaļevska A 2016 *Administrative Rights: Textbook.* Riga: Courthouse Agency, 2016, – pp. 148
[6] City of Riga. *Building Board regulation of the City of Riga* (adopted on 21 July 2008. No. 117). Retrieved March 6, 2019, from http://www.rpbv.lv/nolikums
[7] Chen W. T., Chen T.-T., Lu Ch. Sh., Liu Sh.-Sh 2012 Analysing relationships among success variables of construction partnering using structural equation modelling: a case study of Taiwan's construction industry, *Journal of Civil Engineering and Management*, 18 (6): 783–794. DOI:10.3846/13923730.2012.735062
[8] Druķis P.,Gaile L., Pakrastiņš L. 2017 Inspection of public buildings based on risk assessment. *Procedia Engineering*. 172, 247 – 255. DOI: 10.1016/j.proeng.2017.02.106
[9] Ferrari S., Zagarella F. 2015 Costs assessment for building renovation cost-optimal analysis. *Energy Procedia,* 78, 2378 –2384. DOI: 10.1016/j.egypro.2015.11.193
[10] Latvian Republic Saeima 2013 Construction Law, *Latvian Journal*
[11] Latvian Republic Saeima 2001 Latvian Republic Supreme Council Administrative Procedure Law. *Latvian Journal*
[12] Neringa G., Audrius B., Nerija B., Jorge, L 2013 Development of a Conceptual Critical Success Factors Model for Construction Projects: A Case of Lithuania. *1th International Conference on Modern Building Materials, Structures and Techniques, MBMST 2013.* Pages 392-397. DOI: 10.1016/j.proeng.2013.04.051
[13] Sharpe T. 2019 Ethical issues in domestic building performance evaluation studies. *Journal Building Research & Information.* Volume 47, Pages 318-329. DOI: 10.1080/09613218.2018.1471868
[14] Supreme Council and Government Reporter 1984 Latvian Administrative Violations Code. *Supreme Council*
[15] Supreme Court of the Republic of Latvia 2017 The decision of the Administrative Cases Department of the Supreme Court of the Republic of Latvia of 8 June 2017 on the case No. 670003717, SKA-958/2017, *Supreme Court of the Republic of Latvia*
[16] Supreme Court of the Republic of Latvia 2017 The decision of the Administrative Cases Department of the Supreme Court of the Republic of Latvia of 22 December 2017 on the case No. 680021117 , SKA- 1488/2017, *Supreme Court of the Republic of Latvia*
[17] Supreme Court of the Republic of Latvia 2005 Objective investigation principle - interpretation and application, Retrieved March 01, 2019, from *http://www.at.gov.lv/files/uploads/files/6_Judikatura/Tiesu_prakses_apkopojumi/2016/1-Obj_izm_princips_visparinajums-2005.do*

Evaluation of Bridge and Landscape Interactions

I Rozentale[1] , A Paeglitis[1] and J Smirnovs[1]

[1]Department of Road and Bridges, Riga Technical University, Kalku Street 1, Riga, Latvia, LV 1658

E-mail: ilze.rozentale@rtu.lv

Abstract., The community has become more and more focused on preserving the natural landscape. That is an essential part of the quality of human life. Every member of society wants to enjoy a high-quality landscape. That is evidenced by the documents adopted in recent years, which set out how to evaluate the interaction between structures and landscape. The European Landscape Convention covers all areas with both natural and human-changed landscape. These paper analyses methods and criteria for the assessment of bridges as a component of the landscape, focusing on the possibilities and experiences of applying multi-criteria methods. Particular attention dedicated to the interaction between the bridge and the environment, the criteria and methods for its assessment, which is one of the stages of the visual (technical and visual) and aesthetic (separate elements and overall design) of the bridges, which includes both landscape and bridge impact assessment. Paper analysis the possible application of criteria and methods for evaluation of existing and new bridges as a part of the landscape.

1. Introduction

Society estimate the landscape is an essential part of the quality of his life and want to enjoy a high-quality landscape. Evidence of this is the European Landscape Convention, which covers natural, rural, urban and peri-urban areas, including inland water and marine areas. The Convention covers both the landscapes which considered as excellent and degraded landscapes. Convention aims to promote protection, management, and planning of the landscape as well as to organise cooperation of landscape issues [1].

One of the elements of the landscape is the bridge transport infrastructure element, which must be both functional and elegant. Often, due to its size, bridges are highly visible in the neighbouring area and could impact the environment positively or negatively. The aesthetic impression created by the bridge depends on how the bridge looks in compositional harmony with the surrounding environment. Interaction between the bridge structural solutions and environment is one of the leading architectural and aesthetic problems of bridge construction. Relationships between the bridge design aesthetic qualities and its impact on environment condition will be model with a mathematical model. Authors in paper [2] gives a general scheme for the evaluation of the aesthetic and visual qualities of bridges (Figure 1).

The purpose of this paper is to find out the criteria for the assessment of bridge and landscape interaction.

2. Landscape assessment

European Landscape Convention [1] defines a landscape as the territory within the meaning that people perceive it, and which is the natural or result of human activity and interaction: in urban and rural areas, degraded or cared areas, gorgeous or casual places. The landscape, according to individual type, can be assessed by a value of specific criteria. That describes the visually aesthetic landscape quality, cultural and historical significance as a landscape ecosystem quality [3].

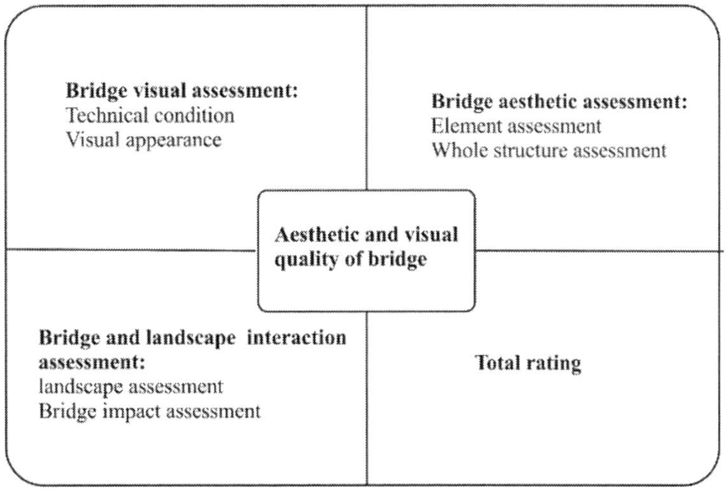

Figure 1. A proposed general scheme for bridge assessment.

Studies of aesthetic quality in Latvia described for different landscape levels and types: natural elements as primary aesthetic sources; natural and rural landscapes; urbanized landscapes. Landscape researchers, geographers Olgert Nicodemus and Aija Melluma [4] have made a significant contribution to the study of the aesthetic quality of Latvian natural landscapes, emphasizing the importance of the subjective perception of the man in the interpretation of landscape aesthetics. Research on landscape aesthetic quality using the aesthetic quality assessment methodology was performed by Daiga Zigmunde [5], developing an evaluation method with criteria of aesthetic quality based on the basic principles of composition and evaluation criteria of aesthetic quality based on composition. Figure 2 shows the criteria for the aesthetic quality of the landscape and their assessment based on D Zigmunde's study [5]. Literature review shows that for landscape identification and classification there are many different methods and systems, that are widely discussed in the paper [6]. The author indicates 54 methods for landscape assessment used worldwide and indicates the simplest one. These methods identify the landscape quality mostly with natural components (geology, soil, etc.) and components characterising human intervention (land use, settlements, etc.). They also are suitable for landscape inventory and land use planning. In the paper [7], the use of the term "landscape" is limited to visual qualities. The methodology used by the authors [7] consists of 5 parts. First two parts refer to the criteria given in Figure 2. The evaluation of the landscape quality performs by the observer survey and evaluation by using an assessment of the intensity of landscape elements. The visual quality of the landscape consists of the evaluation of each component in the visual perception and relatively uniform landscapes classified by the geographic information system (GIS). In this study observers rated the photos, giving rating points, such as "best" +4 points, "worst" -4 points. As a result, an average visual quality index (AVQ index) determined for each landscape. Authors developed a scale of measurement of landscape attributes and elements, who will be used for the landscape aesthetic evaluation and based on the Regulations No.240 issued by Cabinet of Ministers of the Republic of Latvia "General Regulations for Planning, Use and Building of the Territory" [8] . The results in the form of criteria and scoring of aesthetic quality shown in Figure 2.

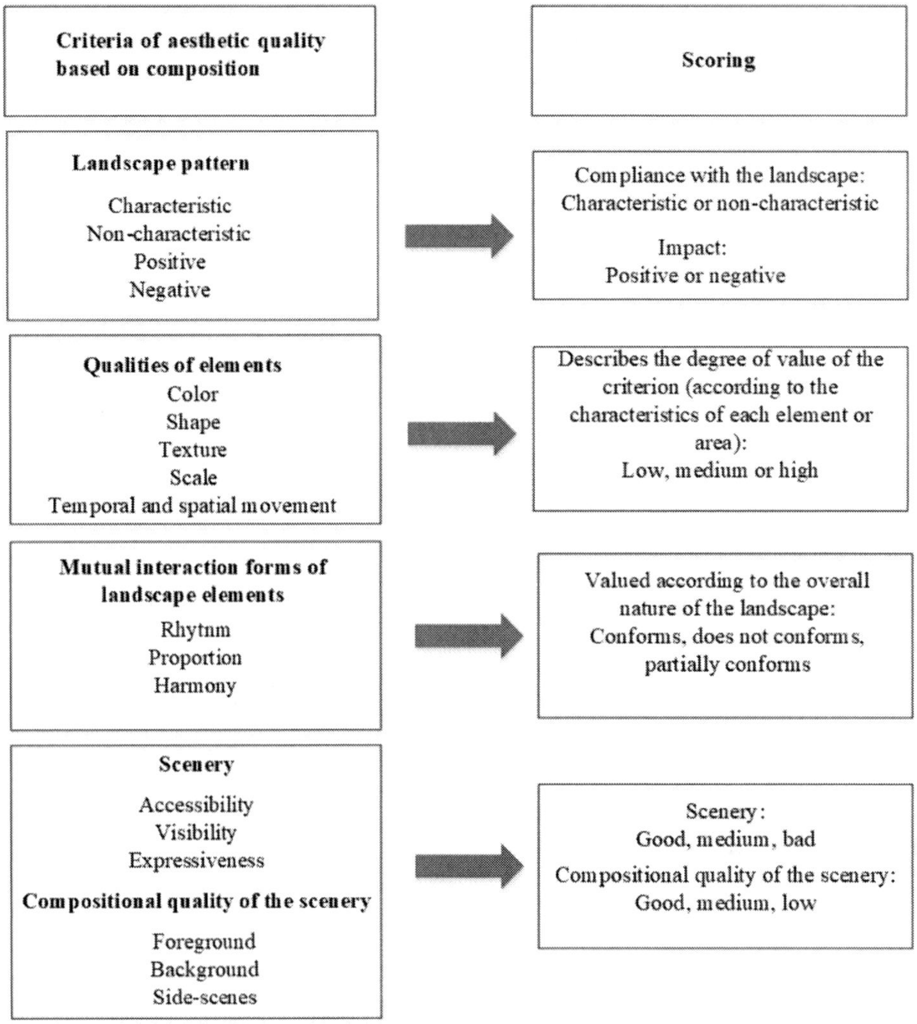

Figure 2. Criteria and scoring of aesthetic quality.

The evaluation of the landscape and bridge interaction is relatively complicated and based on different criteria. Flexible well-structured method with multiple decision-making criteria (MDMC) will solv this task. It has been developed many MDMC methods with different applications. An overview of the application of the MDMC methods for the period from 2000 to 2014 is found in [9]. Also other authors research of the MDMC methods [10], [11], [12], etc., shows that they have used similar approaches. This paper does not try to analyse in detail and compare MCDM methods, but looks at the possibility of its practical application in landscape evaluation. The MCDM method for each criteria include the "weight" that show the importance of criteria. Paper [7] describe one of the “weight" detection methods. In table 1 is showed the measurement scale from [7] with the “weight” corrections.

In table 1 and 2 given evaluations and the scale of measurements is not enough accurate because it is affected by the human subjective assessment and the large number of criteria that complicate the use of MDMC. To assess human action impact on the landscape, it should be assessed not only the landscape but also a bridge and landscape interaction.

Table 1. Scale of measurement of landscape attributes and elements [7] with corrections.

Criteria	Importance ("weight") of criteria	Scoring
Presence of water	0.01	No water= 0; River= 1; Lake= 2; Dam=3: Large areas of water =4
Water movement	0.01	No movement= 0; Movement= 1
Type of land use	0.03	Non built-up territory = 1; Built-up territory = 2
Land use - non built-up territory	0.05	Forest area = 1; Agricultural use = 2; Public outdoor space = 3
Percentage of land covered by vegetation	0.03	0-25%= 0; 25-50%= 1; 50-75%= 2; 75-100%= 3
Type of vegetation	0.03	Meadow plants or cereal crops = 1; Mix vegetation (bushes and trees)= 2; Trees= 3
Horizon	0.01	Hidden= 0; Almost flat =1; Some small hills= 2
Presence of man-made elements with high influence (roads, bridges, objects with cultural heritage)	0.02	None= 0; One element= 1; Two elements= 2; Three or more elements= 3
Presence uf building environment	0.02	No buildings= 0; Low-storey buildings (<3 storeys) = 1 ; Multi storey buildings (> 3 storey) = 2; Mixed buildings = 3
Building density	0.03	No buildings= 0; Individual buildings = 1;Low building density = 2; High building density = 3
Land use - built-up territory - residental buildings	0.05	No buildings = 0; Private house building = 1; Garden house building = 2; Country estate building = 3; Terraced house building = 4; Multi-apartment house building= 5
Land use - built-up territory - public building	0.05	No buildings = 0; Office buildings = 1; Building of business = 2;Building of tourism and recreational establishments, Building of cultural institutions = 3; Sport buildings = 4; Building of defence and security institutions= 5; Building of educational and scientific institutions, Building of health protection and social care institutions = 6
Land use - built-up territory - industrial building	0.05	No building = 0; Building of light industry undertakings = 1; Building of heavy industry and primary processing undertakings = 2; Building of agricultural production undertakings = 3; Extraction of mineral resources = 4; Building of waste management and recovery undertakings = 5
Land use - built-up territory - technical building	0.03	No building = 0; Building of light industry undertakings = 1; Building of heavy industry and primary processing undertakings = 2; Building of agricultural production undertakings = 3; Extraction of mineral resources = 4; Building of waste management and recovery undertakings = 5
Number of colours	0.05	One colour= 1; Two colours= 2; Three or more colours= 3
Internal contrast	0.05	Weak colour contrast= 0; Clear colour contrast= 1
Presence of alignments	0.03	None= 0; Presence of alignments= 1
Scale effect	0.05	No element presents scale effect= 0; Presence of scale effect= 1
Focal view	0.05	No focal view = 0; Focal view = 1
Mutual interaction forms of landscape elements (rhytm, proportion, harmony)	0.15	Conforms to the overall nature of landscape= 2; Partially conforms = 1; Does not conforms= -1
Scenery (accessibility, visibility, expressiveness)	0.1	Good = 2; Medium = 1; Bad = -2
Compositional quality of the scenery	0.1	Good = 2; Medium = 1; Low = -1

3. Bridge impact assessment

The question about the interaction of bridge and landscape already viewed in the Roman architect Vitruvius 10-volume essay "The Architecture" [13]. Vitruvius underline that architecture includes both - structures and their placement. Henry Grattan Tyrrell, in a paper published in 1912 [14], defining requirements for beautiful bridges mentioned compatibility with the environment. Also, Gottemoeller F. [15], Leonhardt F [16] in terms of aesthetically pleasing bridges mentions both integrations into the environment and the landscape. As mentioned [2] then directly Christian Menn emphasizes bridge compatibility with the environment associated with the spatial components – scale, a character of the landscape and topography.

Evaluation of bridge and environmental interactions closely linked to the visual quality of the landscape, characterized by three characteristics - vividness, intactness, unity, and harmony between all elements of the landscape, including the bridge. These characteristics discussed in the article [2]. Principles of Bridge Impact Assessment for Landscape based on the Principal Steps of the Visual Impact Assessment for Highway Projects (VIAHP) Handbook [17] are: 1. determination of the affected environment, 2. determination of main viewpoints, 3. assessment of bridge visual impact, including changes of visual environment and reaction of viewers such us impact on the landscape and vegetation, impact from bridge integration degree into street network or landscape, impact on the spatial urban or landscape context, impact on the surrounding character.

Selection of the viewing point/points is essential for determining the interaction between the bridge and the landscape. This issue extensively discussed in the article [18]. The authors note that there are no clear criteria for determining the point of view, but given the type of criteria types, as summarized in Table 2. Using images created with Google Map, Table 3 attempts to simulate the impact of distance on bridge perception based on Table [18].

Table 2. Type of Viewpoint selection criteria [18].

Division		Viewpoint selection criteria
Division by center	Landscape resources	The place where see excellent landscape resources
	Users	The place where density is high
Division by Location of View	Inside View Point	A central point inside of the area
	Outside View Point	A central point surrounding the area
	Distance View/Middle Distance/Close-range	Prediction point of landscape change by distance
Division by View Point Use	Reputational	A standard viewpoint to protect a good landscape
	Formational	A standard point to form a good landscape
	Management type	A standard point to manage a poor landscape

The evaluation methodology for determining the visual impact of the bridge to the environment/landscapes can be used methodology given in [19], which includes determining of the impact levels of the visual resources depending on the percentage of visual characteristics occupied the basic viewing area from the primary viewshed.

Estimation of the bridges from the aesthetic and visual point of view can be based on the 4 class system used in Finland for evaluation of bridge site [20] supplemented by Class V without requirements [2] and assigning a rating of 5 - Class 1 to 1 Class 5.

Table 3. The appearance of the bridge depending on the distance of viewing point.

Divide	Distance	Visible landscape and bridge	Target area/distance (scheme and satellite view)
Close -range (a)	Located in radius about of 200m of bridge		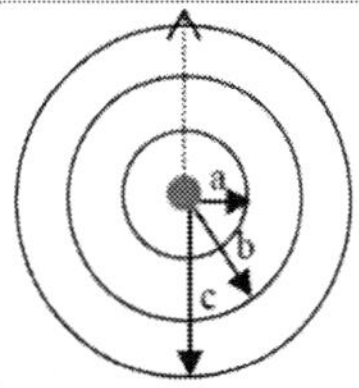 
Middle distance (b)	Located in radius about of 500m of bridge		
Distance view (c)	Located in radius about of 1km of bridge		

A comparative assessment of landscape and bridge effects has been carried out using the methodology and evaluation system from [20], information of Table 1, and MCDM. The simple assessment scheme is shown in Figure 3. The results summarized in Table 4.

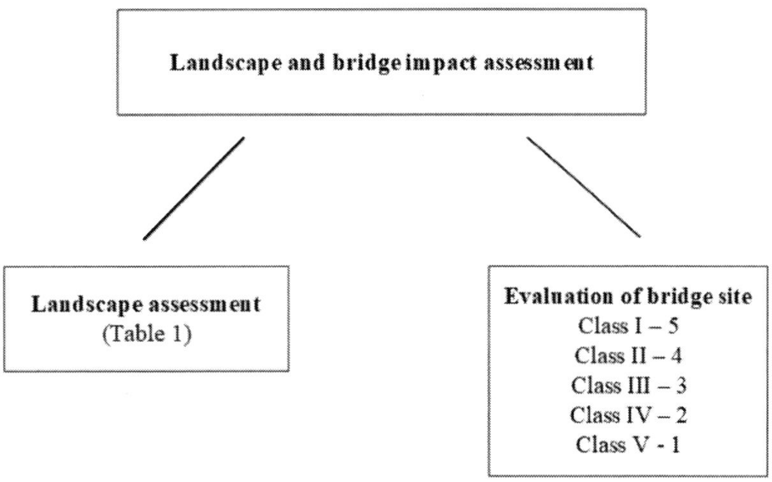

Figure 3. Simple assessment scheme.

Table 4. The results of the assessment.

Bridge over Lorupe ravine (photo R.Salcēvičs)

Landscape assessment -1.83
Bridge impact – 8.15 (max 10)
Bridge site – Class 1 – 5

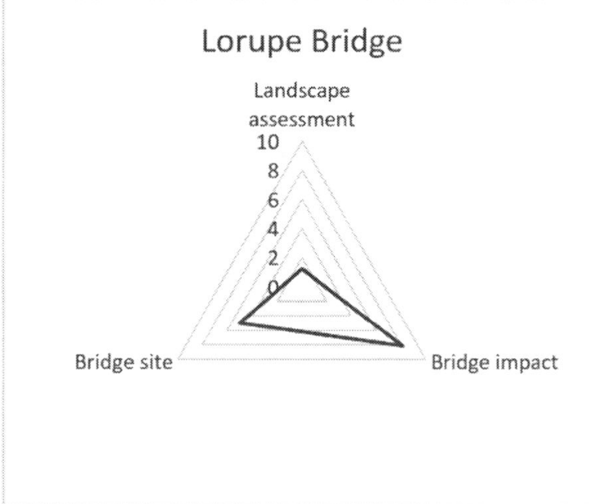

4. Conclusion

For measuring of industrial changed landscape quality could be used the MCDM method. For real use of MCDM method the number of criteria must be reduced and the criteria should be redefined. For the general evaluation of the bridge could be obtained a single evaluation scale, that could could be used as the basis for the methodology for assessing of the visual, aesthetic and technical quality of the bridges in Latvia.

References

[1] "European Landscape Convention," *European Treaty Series Nr.176,* p. 7, 2000

[2] Rozentale I, Paeglitis A, "Criteria for assessment of bridge aesthetic and visual quality," in *The 3rd International Conference on Innovative Materials, Structures and Technologies (IMST 2017)*, Riga, 2017

[3] Matisovs I, "Pilsētvides ainavas kvalitate un to ietekmējošie faktori Latgalē," in *Environment. Technology. Resources. Proceedings of the International Scientific and Practical Conference*, Rezekne, 2005

[4] Nikodemus, O., Rasa, I., "Gaujas Nacionālā parka ainavu estētiskais vērtējums:," 2005. [Online]. Available: https://www.yumpu.com/lv/document/view/27262474/gaujas-nacionala-parka-ainavu-estetiskais-vertejums-dabas-

[5] Zigmunde D, "Evaluation criteria of protected landscape aesthetic quality," in *Research for rural development 2007. International scientific conference proceedings*, Jelgava, 2007

[6] Simensen T, Halvorsen R, Erikstad L , "Methods for landscape characterisation and mapping: A systematic review," *Land Use policy,* vol. 75, pp. 557-569, 201

[7] Arriaza M, Cañas-Ortega F , Cañas-Madueño, J A, Ruiz-Avilés P, "Assessing the visual quality of rural landscapes," *Landscape and Urban Planning,* vol. 69, no. 1, pp. 115-125, 2004

[8] Cabinet of Ministers , *Regulation No. 240 General Regulations for the Planning, Use and Building of the Territory,* Riga: Latvijas Vēstnesis, 2013

[9] Abbas Mardania, Ahmad Jusoh, Khalil MD Nor, Zainab Khalifah, Norhayati Zakwan, Valipour, "Multiple criteria decision-making techniques and their applications – Multiple criteria decision-making techniques and their applications –," *Economic Research-Ekonomska Istraživanja* , vol. 28, no. 1, 2015

[10] Vyas Gayatri S, Misal Chetan S, "Comparative Study of Different Multi-criteria Decision-making Methods," *International Journal of Advanced Computer Theory and Engineering (IJACTE),* vol. 2, no. 4, pp. 09-12, 2013

[11] Mota, Pedro Jorge Gomes, "Comparative Analysis of Multicriteria Decision Making Methods," Pedro Jorge Gomes Mota, Faculdade de Ciências e Tecnologia, Universidade Nova, Lisboa., 2013

[12] Patrick T H., Velasquez M, "An Analysis of Multi-Criteria Decision Making Methods," *International Journal of Operational Research,* vol. 10, no. 2, pp. 56-66, 2013

[13] Vitruv, De Architectura Libri Decem, Marix, 2012

[14] Tyrrell Henry Grattan, Artistic bridge design : a systematic treatise on the design of modern bridges according to aesthetic principles, Chicago: The Myron C.Clark Publishing Co, 1912

[15] Gottemoeller Frederick, Bridgescape: the art of designing bridges, Hoboken, New Jersey: John Wiley& Sons, Inc., 2004, p. 316

[16] .Leonhardt F, Briicken - Asthetik und Gestaltung /Bridges -Aesthetics and Design, Stuttgart: Deutsche Verlags-Anstalt GmbH, 1982, p. 308

[17] U.S. Department of Transportation , "Guidelines for the Visual Impact Assessment of Highway Projects," U.S. Department of Transportation , Washington, D.C., 2015

[18] Yoon Jung Sik, Bur-Deul Yoon, Ki Hun Kim, Chang Hoon, "Comparative Study of View Point Types on Landscape Evaluation," *International Journal of Civil and Environmental Engineering,* vol. 6, no. 11, 2012

[19] INTEGRAL, landscape Architecture&Visual Planning, "Continuation of Boggabri Coal Mine, visual impact assessment," Integral, Scarborough, 2010

[20] Salokangas L, "ETSI Project - Bridge Life Cycle Optimisation Stage 3," Aalto University, Helsinki, 2013

[21] L. Salokangas, "ETSI PROJECT (Stage 2) Bridge Life Cycle Optimisation," Helsinki University of Technology, Helsinki, 2009

Asphalt Recycling Technologies: A Review on Limitations and Benefits

M Rathore[1,2], M Zaumanis[2] and V Haritonovs[2]

[1]"Celu Eksperts", Ikskile, Latvia,
[2] Faculty of Civil Engineering, Riga Technical University, Riga, Latvia

E-mail: mukul.rathore@edu.rtu.lv

Abstract. Recycling of asphalt pavements gained popularity in 1970s due to the rapidly increasing cost and reducing supply of oil and hence - bitumen. However, presently there are other reasons to develop this technology, the main being the ability to utilize non-renewable resources. Sustainability of asphalt pavement construction can only be ensured if we are able to utilize the available reclaimed asphalt pavement (RAP) material completely in the new construction. Another potential advantage is the mutual benefit of using RAP in conjunction with warm mix asphalt (WMA). This is due to compensation of hardened binder from RAP with softer binder of WMA. For such mixes, significant energy savings can be realized due to the lower mixing and compaction temperature as well as due to lesser efforts required for compaction. This paper summarises several approaches that can be adopted to maximize the amount of RAP material in asphalt mixtures. In addition, the current recycling practices implemented in the industry are discussed to identify the main reason behind the lacking confidence in using high RAP content in mixes. The environmental and economical benefits of high rate of recycling are also discussed. Finally, the possible research directions that will allow to incorporate high proportion of RAP without compromising the performance of pavement are discussed.

1. Introduction

The rapidly increasing cost of paving materials, and increased awareness towards environmental damage have lead researchers to come with a more economical and environment friendly pavement construction technology. One of the approaches followed to address above issues was to recycle the materials used in road construction. Recycling of pavement reduces the consumption of virgin material, which in turn would cut down the cost and energy associated with construction of pavement. Moreover, by recycling we will be able to save the valuable landfill spaces that are required for dumping the materials of these demolished pavements. During the past few years, a number of government and private organisations were encouraged to check the feasibility of recycled materials in pavement construction and conduct research on improving the quality of these materials.

As per EAPA 2017 report [1], the total production of hot and warm mix asphalt combined in European countries was around 300 million tonne as shown in Figure 1 and the trend shows that demand is likely to increase in future years. From this data, we get an estimate about the enormous quantity of material that is required for the road construction every year. By replacing a significant proportion of virgin materials with the recycled material, can save huge amount of taxpayer's money and at the same time mitigating the environmental damage due to pavement construction. The basic principle behind asphalt recycling is that hot mix asphalt (HMA) laid pavements retain materials of considerable value

IOP Conf. Series: Materials Science and Engineering **660** (2019) 012046 doi:10.1088/1757-899X/660/1/012046

even after reaching end of their usable service lives [2]. By milling the top layers of such pavements, reclaimed asphalt pavement (RAP) material can be extracted which consists of aged asphalt and crushed stones. This RAP material can be fractionated, separated or screened in order to obtain the desired gradation and incorporate maximum amount of RAP in the mix [3].

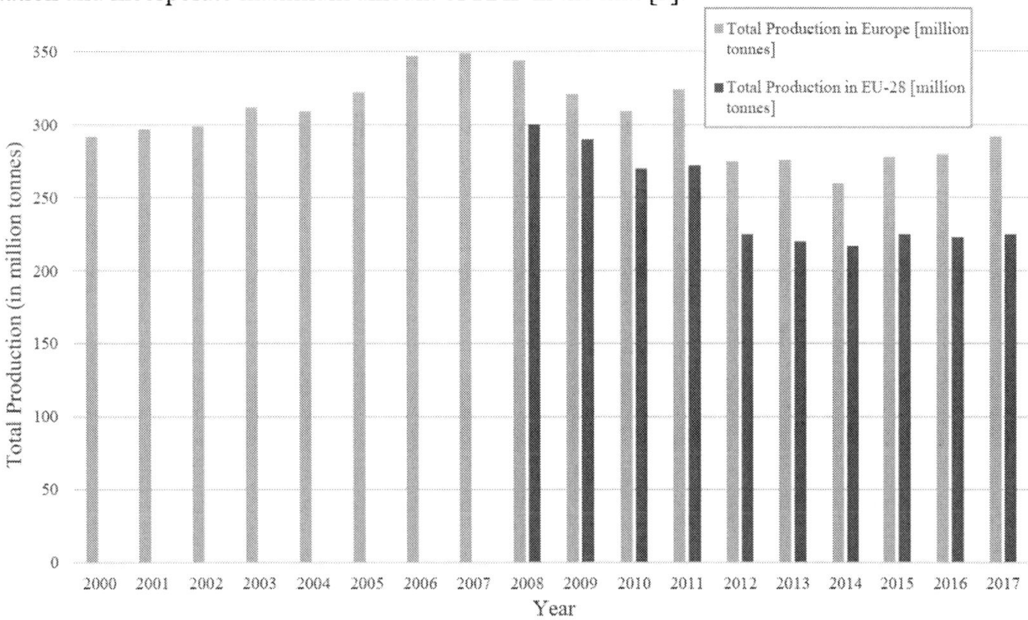

Figure 1. Total production of hot and warm mix asphalt in Europe from 2000 to 2017 (EAPA 2017).

Since, the emergence of this technology, the industry has been constantly looking for different ways to improve the quality of recycled mixes and utilize higher proportion of RAP material in asphalt mixes. United States currently uses 99% of asphalt mixture reclaimed from old asphalt pavements by putting back into new pavements [4]. The number of states with producers reporting average RAP percentages 20% or greater has increased significantly, rising from 10 states in 2009 to 24 states in 2017, however the average RAP percentage is still around 20-35% [4]. Recent researches have established that RAP replacement at proportions above 50% is feasible to produce new HMA mixtures, obtaining satisfactory results in the mechanical properties [5]–[8]. Zaumanis et al. 2014 [9] states, "100% recycling can provide true sustainability by closing the materials cycle and allows use of reclaimed asphalt in the same high value application as that of conventional asphalt".

The RAP content is mainly limited by practical issues such as quality of production and hence the performance of the mixtures in the asphalt plant. RAP materials gets aged at high temperatures and using high proportion can potentially lead to early cracking of the roads [10]. Studies have shown that increased ageing will reduce the stress relaxation capacity of the binder which means it decreases the cracking resistance of the mixture [11]. Another reason is that due to the milling operation, huge amount of fine particles are generated as a result of crushing of aggregates. These finer particles limit the maximum content of RAP usage as it limits the gradation requirements for the mix design at high RAP content. Apart from this, the plants also do not allow the higher proportion of RAP as in batch plants typical range of RAP incorporation is 10–20% and drum plants can accommodate upto 50% [10]. To restore the original properties of RAP material, it is required to add some rejuvenating agents. However, the actual mechanism that occurs during the addition of a rejuvenator is not very well known. This makes the performance of RAP mixes to be unpredictable and hence poses a restriction on increasing the RAP content. Lack of any proper method for characterisation for RAP material quality and little knowledge about blending of virgin and RAP binder is also major reason for being unpredictable about performance of recycled mix and hence conservative on using higher RAP contents.

2. Evolution of asphalt recycling

The recycling of asphalt pavements started gaining popularity in 1970s in United States during the Arab oil embargo when the cost of crude oil was increasing significantly [12]. The construction practices started evolving leading to extensive study on utilization high proportion of RAP material in bituminous pavements. In 1979, a field demonstration project No. 39 by FHWA was carried out in New Jersey with incorporation of around 50% RAP [13]. The use of RAP became very popular in State transport departments until the implementation of Superior Performing Asphalt Pavements (Superpave®) mixture design method in late 1990s. The Strategic Highway Research Program did not provide guidance for the use of RAP in HMA. Thus, many departments stopped allowing use of high content of RAP [12]. Finally, at the end of 20th century, FHWA's Superpave® Mixtures ETG developed interim guidelines for the use of RAP in the Superpave®. Since then, recycling of pavement has become very popular and huge amount of asphalt material is recycled every year. In United States, asphalt material is recyled at higher rate than any other material which helps to lower overall material costs, allowing road owners to achieve more roadway maintenance and construction activities within limited budgets [4]. Now due to the advancement of hot mix plants, it has become possible to incorporate higher percentages of RAP as well as rejuvenating agent effectively during the asphalt production process. As per 2017 survey, the European countries, Belgium, Finland, Great Britain, Hungary and Slovakia were recycling more than the 90% of the available RAP in hot and warm mix asphalt production for surface layers [1].

3. Benefits of asphalt recycling

3.1 Economical Benefits

Using reclaimed asphalt materials have significant economical advantages, but these come true only if similar performance to conventional asphalt is ensured. The most expensive and economically variable material in an asphalt mixture is the asphalt binder [12]. By using the RAP material in surface and intermediate layers, can make the pavement construction very economical as the expensive virgin binder will be replaced by less expensive binder from RAP. The consumption of natural aggregates is also reduced by using aggregates obtained from RAP. This will also result in huge economical saving considering the reduction in aggregate requirement and cost of transporting the material. Although there is an additional cost associated with the milling operation and rejuvenators. Studies have suggested that total cost can be reduced upto 35% by using RAP content at 50% [14]. The cost benefits of using high RAP mixtures depend on a numbers of factors which can be material cost, plant efficiency, government incentives etc. Thus it becomes very important to carry out the economic reassessment before choosing the recycling methodology. It will also be very important to lay down standard protocols for reclamation, transportation, storage, mixing and compaction of RAP material so that it is economically feasible.

3.2 Environmental Benefits

With concerns about global warming and energy consumption, the asphalt industry has been looking for ways to lower its carbon footprint. Recycling of old asphalt pavement materials is a sustainable option of road construction which can result in considerable reduction in consumption of natural resources. It also addresses the disposal issue of RAP material and thereby saving huge landfill spaces. Another advantage is the synergy between RAP material and warm mix asphalt, which can give combined environmental benefits. However, there are also some emissions associated with the plant production of high RAP mixtures. But recently the advances in asphalt production plant technology have allowed shielding of RAP from contact with the flame [15] and modern air pollution control technologies can be used to filter the emissions before releasing them into the atmosphere [10].

4. Limitations of RAP use

To understand the limitations of high RAP usage in plant production, NAPA [4] conducted a voluntary survey on asphalt mixture producers, representing 238 companies with 1,158 production plants in the United States, the most common inputs from the groups are shown in Figure 2.

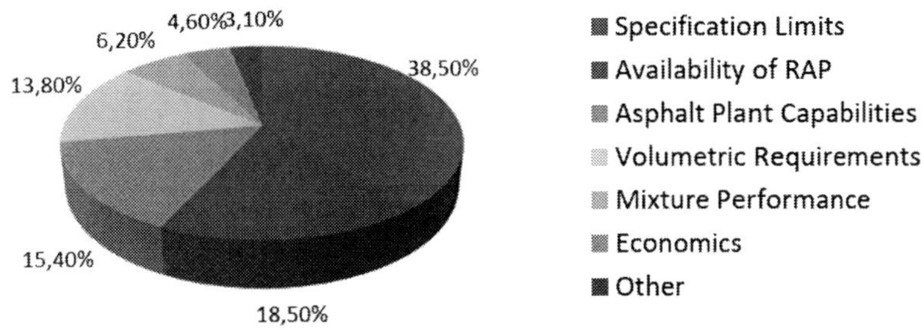

Figure 2. Reported factors limiting the use of RAP (NAPA 2017).

4.1 RAP aggregate fractions

To ensure that the RAP mixtures perform equally to that of conventional mixtures, the basic requirement is that same fractions of RAP aggregates are used as that of virgin aggregates. However, due to the milling operation, excessive fines are generated and it becomes very difficult to fulfil the gradation requirements. This limits the maximum content of RAP material to which it can be used. The inhomogeneity of RAP has been reported as a problem, but recent survey of RAP variability by [16] shows that RAP gradation is generally more consistent than that of virgin aggregates. Zaumanis et al. 2018 [17] showed that the variability of cored samples allowed 20% RAP in mixture design, whereas the milling increased the maximum RAP content to 30%, and stockpile mixing further increased the allowed RAP content to 40%. Thus, it is very important to manage the RAP material at very early stages to allow higher percentages of RAP incorporation.

4.2 Blending

For production of a RAP mixture, it is required that aged asphalt from RAP material interacts with virgin binder and with the rejuvenator also if used and form a homogeneous film thickness over the aggregate. However, the studies have shown there is a partial blending and thus it is practically not possible to extract the aged bitumen completely from the RAP material [18]. Thus, quantification of blending is very important to estimate the black rock portion of the RAP that does not blends completely. Several researchers have developed the methods to estimate the blending [19]–[21]. Assumption of complete binder extraction may lead to underrated binder requirement that may lead to cracking, ravelling or moisture damage. However, assuming low blending may overrate the binder quantity required and can ultimately lead to plastic deformation of the mix [22]. Zaumanis and Mallick 2013 [21], observed that the diffusion rate can be significantly enhanced with increasing temperature and time of mixing.

4.3 Fatigue Cracking

Cracking is an important concern when dealing with high content RAP mixtures. The aged binder in RAP material may lead to fatigue cracking in the resultant mixtures. The increase in RAP proportion in pavements escalates the potential of such cracking that is one of the main reasons for reluctance for government agencies to allow very high RAP content. However, rejuvenators are used to reduce the cracking susceptibility of mixtures due to aged binder. Studies have shown that with rejuvenators, the fatigue resistance of recycled mixtures was found to be higher than that of virgin mixture [23]. Some studies suggested use of bio binders or modifiers to reduce the fatigue susceptibility of high RAP mixes [24]–[26]. The effect of long-term ageing did not show a significant change on fatigue characteristics of high RAP mixture with and without rejuvenators [6]. Hence, it becomes very important to develop new method for charactering the actual cracking resistance of high RAP mixtures containing rejuvenators.

4.4 Rutting
Due to the aged binder, the mixtures containing high amount of RAP are generally less prone to plastic deformations. However, when a softer grade binder or a rejuvenator is used it can increase the rutting susceptibility of mixing due to over softening of mix. Hence, proper care must be taken to investigate the rutting potential when softer grade binders or rejuvenators are incorporated [16]. Mogawer 2013 [27] observed that the rejuvenators increased the rutting and moisture susceptibility of the 40% RAP and 5% RAS mixes. Im et al. 2016 [28] found that the rutting performance of all rejuvenated binder samples and asphalt mixtures were decreased as compared to the virgin materials. These results clearly indicate the need of careful selection of the rejuvenator dosage or virgin binder to avoid rutting due to excessive softening of binder.

5. Maximizing the RAP content

5.1 Use of Rejuvenators to maximize RAP content
When RAP is incorporated into virgin materials, the resultant mix become very stiff and the compaction of the mix becomes very difficult. Therefore, to compensate the effect of hardened binder from highly oxidised RAP material, softer grade virgin binders or recycling agents are generally used. Rejuvenators are recycling agents which are suitable either for highly oxidized mixture or mixture containing a large percentage of RAP [29]. Rejuvenation of bitumen is replacement of the oils lost during aging process and balancing the bitumen proportions such that it no longer stays brittle in nature [29]. To be able to use higher rates of reclaimed asphalt in the asphalt mixture production rejuvenators can play an important role. However, the selection of rejuvenator and dosage is very important issue, as by using an inappropriate rejuvenator type or dosage, the mix either can have no effect on its properties or can become very soft causing the plastic deformation in the mix. Hence, rejuvenator is not encouraged or even not allowed for recycling in some states in US due to the uncertainty of the rutting properties of recycled mixtures containing a rejuvenator [29]. Therefore, it is always recommended to determine the dosage of rejuvenator in laboratory based on the properties of RAP material instead of using a preselected content. The decision, whether the use of a softer binder or a rejuvenator is technically advised, depends on e.g. the quality and stiffness of the binder of the reclaimed asphalt, the amount of reclaimed asphalt to be added, and the characteristics of the mixture to be produced. As RAP percentage is increased (>50 %), recycling agents offer many unique benefits as compared to the use of softer binders [23]. The dosage of rejuvenator is very important parameter for performance of recycled mix. Im et al. 2016 [28] developed a step by step new mix design method in which they firstly determine the range of rejuvenator dosage on the basis of characterisation of blends. Further, with the optimum binder content calculated performance tests, they determine the optimum rejuvenator content based on performance of recycled asphalt mixes produced as a counterpart to control mixes. Zaumanis et al. 2014 [9] utilized six different rejuvenators for developing a method of determination of optimum rejuvenator content according to Superpave design asphalt binder specifications in which they determined the minimum and maximum rejuvenator content to satisfy high and low PG temperature, and then an optimum dosage can be determined based on statistical evaluation.

5.2 Use of WMA with RAP
Using WMA technology, higher percentages of RAP can be incorporated into asphalt mixes because of compensation of the stiffer asphalt binder in RAP with the softened binder achieved through certain

warm mix additives. The resulting mix can be properly compacted in the field due to the reduced viscosity. Lowering production and paving temperature for WMA can cause considerable changes in the properties of bitumen hardening in the production process. Less ageing of the asphalt binder during the production and pavement processes increases the service life of the pavement with fewer occurrences of cracking [30]. The general opinion is that less ageing during the production and paving processes tends to improve the flexibility of the pavement, which reduces susceptibility to fatigue and temperature cracking resulting in improvement of pavement longevity [31]. Since WMA mixtures are more susceptible to rutting, replacing a part of virgin aggregates with RAP material will improve the rutting performance of the mix. Laboratory tests have indicated that WMA mixtures with higher percentage of RAP exhibit higher resistance to rutting, better resistance to moisture damage, and better fatigue performance [32]. Thus uses of warm mix additives with or without reducing the temperature can be a good option to incorporate the high proportion of RAP. However, as the binder activation will be lower, more studies are required to evaluate the performance of warm recycled mixes.

5.3 RAP management to maximize usage

By following the management practices that ensure the quality of RAP is not degraded starting from its milling to the laying of pavement will allow to maximize the content of RAP. The milled material obtained can be of high quality such that it may not require any processing. However in some cases it is desirable to screen out some oversize particles to maximize the amount of RAP. The RAP particles can also be separated into coarse and fine stockpiles that may be used wherever required. This separation of RAP on the basis of size increases the control over quality and reduces the variability. The basic goals of processing RAP are [16]:

- Creating a uniform stockpile of material from a collection of different RAP materials from various sources.
- Separating or breaking apart large agglomerations of RAP particles to a size that can be efficiently heated and broken apart during mixing with the virgin aggregates.
- Reducing the maximum aggregate particle size in the RAP so that the RAP can be used in surface mixes (or other small nominal maximum aggregate size mixtures).

In most cases, processed RAP will be moved from the location it is screened or crushed to another location for making it convenient to feed into the asphalt plant. However, there is another opportunity that the material is remixed to improve its consistency. It is done most commonly to prevent or limit the segregation. Arc-shaped, uniformly layered stockpiles are preferred for storing milled or unprocessed RAP material (i.e., material of various sizes). As with virgin aggregate, conical stockpiles or small, low-sloped piles are preferred for storing processed RAP material [12].

6. Summary and discussion

This paper summarizes some practices adopted to maximize the amount of RAP material in asphalt mixtures. Several studies were discussed to identify the main reasons for limiting the RAP content and understand the benefits of high RAP mix. Based on this study, following points can be drawn out:

i. The proportion of RAP content has a very high impact on economics of pavement construction. Hence, life cycle cost analysis studies of RAP production can be an important tool to understand the benefits of RAP use and can motivate asphalt producers to use higher percentages of RAP in the mix.
ii. The limits laid down by specifications including aggregate gradation, fine content, binder requirements and mix performance were one of the main reasons for low utilization of RAP material. A performance based methodology which allows use of very high RAP contents can make the RAP utilization more efficient.
iii. The innacurate assessment of degree of blending between aged asphalt and virgin asphalt results in unpredictable performance of RAP mixes. Studies should focus on extensive characterisation of blending to improve the mix design parameters.

iv. It is evident that the rejuvenators play a very important role when high RAP contents are used. However, the chemical process that occurs in rejuvenation of bitumen is not very clear. Studies carried out in this direction to will help to understand the rejuvenator diffusion mechanism.
v. Studies show that WMA allows tó incorporate higher percentages of RAP due to compensating effect. However, maximum proportions of RAP will also depend on the type of warm mix technology which gives scope to study various types of WMA technologies and their impact on high RAP mixes.
vi. Maximum RAP content significantly depends on quality of RAP material. By laying out standard procedures for milling, storing, and preserving the RAP material will allow to maintain the homogeneity and quality of RAP material and thus having a better performing mix.

Acknowledgements
This study is part of SAFERUP!, an Innovative Training Network devoted to "Safe, Accessible and Urban pavements". SAFERUP! project has received funding from the European Union's Horizon 2020 research and innovation programme under the Marie Skłodowska-Curie grant agreement No 765057.

References

[1] EAPA, "Asphalt in Figures 2017," *Eur. Asph. Pavement Assoc.*, 2018
[2] REBECCA MCDANIEL; R. MICHAEL ANDERSON, *Recommended Use of Reclaimed Asphalt Pavement in the Superpave Mix Design Method: Technician's Manual.* 2001
[3] Shrum E D 2010 "Evaluation of Moisture Damage in Warm Mix Asphalt Containing Recycled Asphalt Pavement *Masters Thesis*
[4] Brett T C R, Williams A, Audrey Copeland 2018 Asphalt Pavement Industry Survey on Recycled Materials and Warm-Mix Asphalt Usage: 2017 *Natl. Asph. Pavement Assoc.*, vol. 8th Annual, no. Information Series 138 (8th edition)
[5] Zaumanis M, Cavalli M C and Poulikakos L D 2018 "How not to design 100% recycled asphalt mixture using performance-based tests," *Road Mater. Pavement Des.*, vol. 0, no. 0, pp. 1–13
[6] Mogawer Walaa S, Austerman Alexander, Roque Reynaldo, Underwood Shane, Mohammad Louay and Zou Jian 2015 "Aging and rejuvenators: Evaluating their impact on high RAP mixtures fatigue cracking characteristics using advanced mechanistic models and testing methods," *Road Mater. Pavement Des.*, vol. 84, pp. 1–42
[7] Carrión A J del Barco, Presti D Lo and Airey G D 2015 "Binder design of high RAP content hot and warm asphalt mixture wearing courses," *Road Mater. Pavement Des.*, vol. 16, pp. 460–474
[8] Tran N, Taylor A, Turner P, Holmes C and Porot L 2016 "Effect of rejuvenator on performance characteristics of high RAP mixture," *Asph. Paving Technol. Assoc. Asph. Paving Technol. Tech. Sess.*, vol. 85, pp. 283–313
[9] Zaumanis M 2014 "100 % Recycled Hot Mix Asphalt and the Use of Rejuvenators," Worcester Polytechnic Institute
[10] Zaumanis M and Mallick R B 2015 "Review of very high-content reclaimed asphalt use in plant-produced pavements: State of the art," *Int. J. Pavement Eng.*, vol. 16, no. 1, pp. 39–55
[11] Molenaar A A A, Hagos E T and M F C van de Ven 2010 "Effects of Aging on the Mechanical Characteristics of Bituminous Binders in PAC," *J. Mater. Civ. Eng.*, vol. 22, no. 8, pp. 779–787
[12] Copeland A 2011 "Reclaimed Asphalt Pavement in Asphalt Mixtures : State of the Practice," *Fed. Highw. Adm.*, no. FHWA-HRt-11-021
[13] Hellriegel E J 1980 "Recycling Asphalt Pavements.," *Fed. Highw. Adm.*, vol. 1, pp. 325–342
[14] Willis J R, Turner P, Julian G, Taylor A J, Tran N and F de G Padula 2012 "Effect of Changing Virgin Binder Grade and Content on RAP Mixture Properties."
[15] Kent Hansen M C, Dave Newcomb 2011 "Asphalt Tops the Charts for Environmental

Stewardship-Again," *National Asphalt Pavement Association*, vol. 16, no. 5
[16] West R C 2015 "Best Practices for RAP and RAS Management," *Natl. Asph. Pavement Assoc.*, p. 44
[17] Zaumanis M, Oga J and Haritonovs V 2018 "How to reduce reclaimed asphalt variability: A full-scale study," *Constr. Build. Mater.*, vol. 188, pp. 546–554
[18] Imad Q A, Al-Qadi L, Samuel, Carpenter H, Geoff Roberts and Hasan Ozer 2009 "Determination of Usable Residual Asphalt Binder in RAP," *Fed. Highw. Adm.*, no. 09
[19] Gottumukkala B, Kusam S R, Tandon V and Muppireddy A R 2018 "Estimation of Blending of Rap Binder in a Recycled Asphalt Pavement Mix," *J. Mater. Civ. Eng.*, vol. 30, no. 8, p. 04018181
[20] Stimilli A., Virgili A and Canestrari F 2015 "New method to estimate the 're-activated' binder amount in recycled hot-mix asphalt," *Road Mater. Pavement Des.*, vol. 16, pp. 442–459
[21] Zaumanis M and Mallick R B 2013 "Finite Element Modeling of Rejuvenator Diffusion in RAP Binder Film – Simulation of Plant Mixing Process," *Multi-Scale Model. Charact. Infrastruct. Mater.*, pp. 407–419
[22] Al-Qadi I L, Elseifi M and Carpenter S H 2007 "Reclaimed asphalt pavement - A literature review," *Fed. Highw. Adm.*, vol. 07, no. FHWA-ICT-07-001, p. 23
[23] Martins Zaumanis R F, Rajib B. Mallick 2014 "Evaluation of different recycling agents for restoring aged asphalt binder and performance of 100 % recycled asphalt," *Mater. Struct.*
[24] Fini E H, Hosseinnezhad S, Oldham D J, Chailleux E and Gaudefroy V 2017 "Source dependency of rheological and surface characteristics of bio-modified asphalts," *Road Mater. Pavement Des.*, vol. 18, no. 2, pp. 408–424
[25] Fini E H, Al-Qadi I L, You Z, Zada B and Mills-Beale J 2012 "Partial replacement of asphalt binder with bio-binder: Characterisation and modification," *Int. J. Pavement Eng.*, vol. 13, no. 6, pp. 515–522
[26] Cavalli M C, Zaumanis M, Mazza E, Partl M N and Poulikakos L D 2018 "Aging effect on rheology and cracking behaviour of reclaimed binder with bio-based rejuvenators," *J. Clean. Prod.*, vol. 189, pp. 88–97, Jul.2018.
[27] Mogawer W S, Booshehrian A, Vahidi S and Austerman A J 2013 "Evaluating the effect of rejuvenators on the degree of blending and performance of high RAP, RAS, and RAP/RAS mixtures," *Road Mater. Pavement Des.*, vol. 14, pp. 193–213
[28] Im S, Karki P and Zhou F 2016 "Development of new mix design method for asphalt mixtures containing RAP and rejuvenators," *Constr. Build. Mater.*, vol. 115, pp. 727–734, Jul.
[29] Mhlongo S M, Abiola O S, Ndambuki J M and Kupolati W K 2014 "Use of Recycled Asphalt Materials for Sustainable Construction and Rehabilitation of Roads."
[30] Zaumanis M and Smirnovs J 2011 "Analysis of Possibilities for Use of Warm Mix Asphalt in Latvia," in *3rd International Conference CIVIL ENGINEERING`11 Proceedings*, pp. 57–64
[31] Perkins S W 2009 "Synthesis of Warm Mix Asphalt Paving Strategies for use in Montana Highway Construction," *Fed. Highw. Adm.*, no. November, 2009
[32] Zhao S, Huang B, Shu X, Jia X and Woods M 2012 "Laboratory Performance Evaluation of Warm-Mix Asphalt Containing High Percentages of Reclaimed Asphalt Pavement," *Transp. Res. Rec. J. Transp. Res. Board*, vol. 2294, no. 1, pp. 98–105

Evaluation of adhesion between bitumen and aggregate with the digital image processing method

A Riekstins[1,2], V Haritonovs[1] and A Balodis[2]

[1]Riga Technical University, Department of Roads and Bridges, Riga, Latvia
[2]SJSC Latvian State Roads, Riga, Latvia

E-mail: riekstins.arturs@gmail.com

Abstract. Although adhesion in asphalt between bitumen and aggregates is one of the essential properties, it is still evaluated visually. European Standard EN 12697-11 provides 3 test methods for the preparation of samples for evaluation - rolling-bottle method, static method and boiling water stripping method. The method used in Latvia is different from EN 12697-11, but with similarities. Bitumen coverage is evaluated visually for all options. The result is the average value of two parallel operators. This principle of evaluation is much criticized as inaccurate and subjective. Today's technologies are evolving, and in many fields, they are more reliable than a human. This study aims to determine whether a visual assessment method can be replaced by a digital image processing method. To evaluate the accuracy of the image processing method, four different mineral materials - granite, dolomite, quartz diorite, gravel, were selected. Samples obtained during the test procedure were tested visually and using an image processing method. For image evaluation, a mini photo studio was created, where pictures were taken using a professional camera and a budget smartphone. The smartphone was used to determine if it is possible to get high-quality images and results using budget smartphones in our pockets.

1. Introduction

The basic property for quality asphalt concrete is strong adhesion between bitumen and aggregates. [1, 2, 3]. Adhesive properties are known to be affected by the chemical composition, surface tension and viscosity of bitumen, as well as oil used in the production of bitumen. [4, 5] Likewise, adhesion is influenced by the properties of the aggregates, such as chemical composition, surface roughness, shape, amount of cut surface and porosity. [4, 5] To improve the adhesion properties, the bitumen is modified by the addition of a variety of improving additives.[1]

To test the adhesion properties between bitumen and aggregates, European Standard EN 12697-11 provides three different methods: rolling-bottle method, static method and boiling water stripping method [6]. In Latvia, the method used to evaluate adhesion has been remained from GOST standard and is described in Road Specifications 2019. [7] All of the methods mentioned above of sample evaluation are done visually by two independent operators. Coverage is rated as a percentage with an accuracy of 5 %. In figure 1 can see a visual example of the evaluation of bitumen coating. The method of visual assessment has caused many disputes.

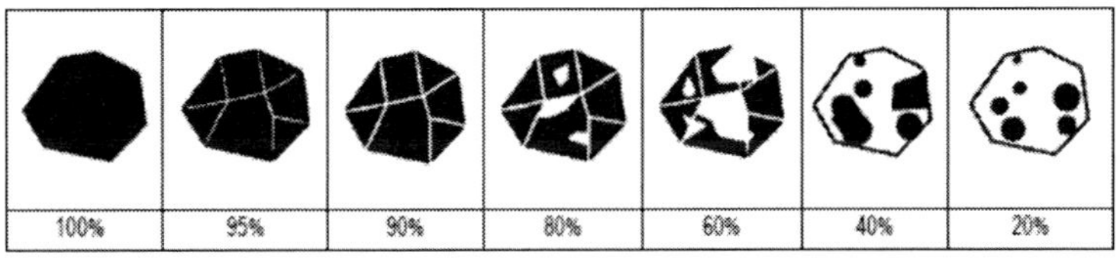

Figure 1. A visual example of evaluation of bitumen coverage [6].

The main factors that influence the results of the visual assessment test are operator skills, lighting, colour of mineral material. The darker the mineral material, the more difficult it is to distinguish it from bitumen. [8] In practice, there are often cases where the difference between the results of operators are significant, which can affect the long-term quality of the road. It may also be a reason for imposing unjustified sanctions on the builder and just because everyone sees it differently.

The human eye can distinguish about 10 million colours. [9] However, colour perception is an extremely complex psychological process in which the subjective attribute of human colour vision has a huge impact. Light has a great effect on colour perception. Therefore it is necessary to improve the objectivity of the test.

To improve the accuracy of the test, it is necessary to reduce the impact of the human factor. One of the possible solutions is to use a digital image processing method instead of visual assessment. [8, 10, 11] The resulting images are processed in an image processing computer program, trying to highlight colour contrasts. In this way, it is possible to determine the ratio of covered areas to the total aggregates area. The main difficulty from previous studies to obtain accurate results was the colour of the aggregates, the falling shadows and the light reflection. [8, 10, 11]. If the colour of the mineral material in the sample is different, it is difficult to evaluate the exact coverage of the bitumen.

This study aimed to compare the visual assessment with the digital image processing method. Two different photo devices were used for the digital image processing method - a professional camera and a budget smartphone.

2. Experimental section

2.1. Raw materials

In Latvia, aggregates of different colours and shapes are used for the production of asphalt concrete. Therefore, four visually different mineral materials were selected to verify the accuracy of the image processing method: dolomite, granite, quartz crystal, gravel.

Bitumen 70/100 was used as a binder.

2.2. Preparation of samples

According to the requirements of LVS EN 12697-11 [6], for each aggregate, a density was determined according to EN 1097-6 [12]. The amount of mineral material in all samples was constant 510 grams. The necessary amount of bitumen was calculated from the density of the mineral material. The results can be seen in table 1.

α- Necessary amount of bitumen;
ρ_m- Density of mineral material;

$$\alpha = \frac{2650}{\rho_m} * 16 \tag{1}$$

Table 1. 4 different mineral material densities (at 25°C).

Material	Density Mg/m3	Bitumen amount, g
Dolomite	2.789	15.20
Granite	2.780	15.25
Quartz diorite	2.779	15.26
Gravel	2.733	15.52

The materials were heated at 150 °C for at least 3 hours until the material is all over warmed up. All the steps for mixing and testing are described in Road Specifications 2019 [7]. In the next day operator from sample choose at least 30 particles which are fully covered by bitumen. These particles are placed on a metal mesh so that they do not interfere with each other. Next step is to place a metal mesh with particles in boiling water for 10 minutes. During the water boiling process, laboratory assistant must remove excess bitumen from the surface of the water. After boiling, the sample should be cooled down, and after that, it is possible to do the last step – visual assessment.

2.3. Obtaining of images

A photo studio was set up to get high-quality images. To create a studio, it is advisable to choose a dark room so that the surrounding lights do not create shadows on the sample. Necessary equipment for studio creation: two stands, petri dish, two adjustable light sources, a green paper sheet. In figure 2 a) and b) can see a drawing of the photo studio and the picture of the created studio.

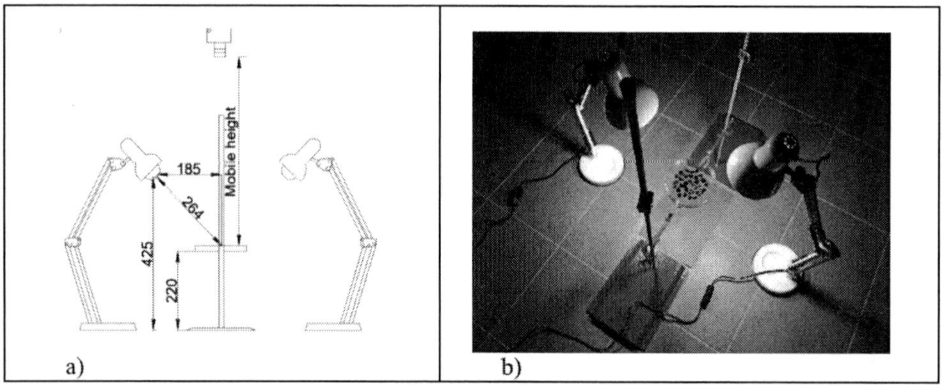

Figure 2. a) drawing of a photo studio; b) photo studio.

When designing distances and angles, the aim was to minimise falling shadows and reflections of the light on the sample. The bitumen surface has a fine structure, so it reflects light. A petri dish filled with water was used to make a less reflection on bitumen. Water diffuse the light rays [13], so the reflection on the bitumen is limited.

Similarly, attention should be paid to the height of the petri dish from the ground, as shadows may be formed on the green background sheet. Adjustable table lamps with LED bulbs were used as light elements. It is advisable to use LED bulbs as other types of light bulbs can create light waves. The light sources were placed in front of each other. The height of the camera was not fixed to allow adjustment as needed. Considering the procedures which were described above, it was possible to start taking photos.

First, 30 stones were placed in a petri dish filled with water. The appropriate shooting height was found, and a picture was taken. Two devices were used to create photos. The first one was a professional photo camera - Canon EOS 600D with 18 megapixels APS-C CMOS sensor. The second one was a budget smartphone Huawei P9 lite with a camera specification of 13 MP, f / 2.0, 1/3 ", 1.12μm, AF. Then, manually, all the bituminous stones were inverted and photographed again. Then the sample was mixed again and repeated the steps described above two times. As a result, six different images with a camera and six different images with a smartphone were obtained for each type of mineral material.

2.4. Image processing

An important role in photo processing was to choose an image processing program. It was found that the ImageJ photo processing program was used to perform similar work [11]. This program is simple, and it is possible to process the image in a concise time. There is a good reasons to use this program because the key is to get accurate results in a short time. The use of the program is also free. Here are steps to process the image:

1. To draw the contour around the sample in the petri dish;
2. Delete background;
3. Do action: analyse – measure;
4. To open Image-Adjust-Colour threshold;
5. Take action with Hue, Saturation, Brightness adjustment. Get pictures from which is possible to measure the stone area and then the bitumen coverage area (see Figure 3).

Bitumen coverage was calculated by using the following formulas:
A_1 - Cropped image area;
A_2 - Green background area;
A_3 - Total area of green background and uncovered aggregate particle area;
A_{bit} - Calculated by the formula:

$$A_{bit} = A_1 - A_3 \quad (2)$$

A_{akm} - Calculated by the formula:

$$A_{akm} = A_1 - A_2 \quad (3)$$

Bitumen coverage is calculated according to the formula:

$$Bp = (A_{bit}/A_{akm}) * 100\% \quad (4)$$

where:
A_{bit} – Area with bitumen coverage;
A_{akm} – Total aggregate particle area;
B – Bitumen coverage on aggregates.

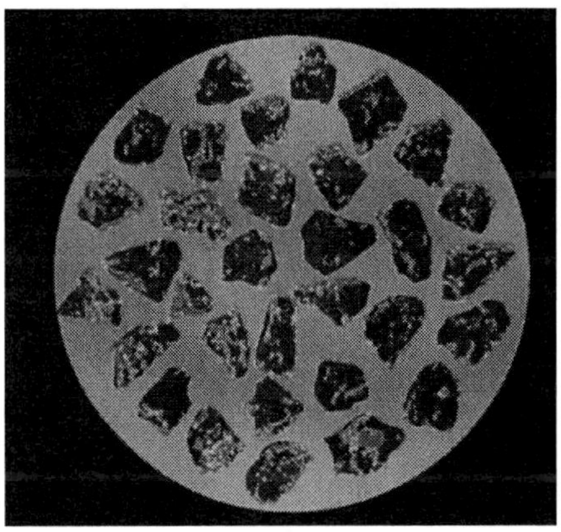

Figure 3. Processed images with highlighted bitumen coverage.

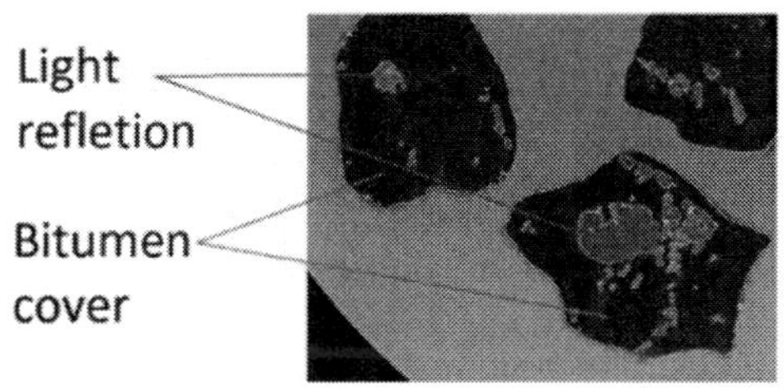

Figure 4. Reflection areas on aggregate surface.

2.5. Inaccuracy caused by reflection

After viewing the photos, it was found that a small reflection was formed on the sample (see Figure 4). A decision was made to evaluate the effect of reflections on the final result. It was assumed that the untested sample had a 100% bitumen coverage. Each sample was photographed four times, mixing the sample each time. A total of 16 images were obtained. Table 2 shows the results obtained after the photos were processed.

Table 2. Average reflection size.

Dolomite	Bitumen coverage, %	Granite	Bitumen coverage, %	Quartz diorite	Bitumen coverage, %	Gravel	Bitumen coverage, %
Photo no. 1	95	Photo no. 1	97	Photo no. 1	97	Photo no. 1	94
Photo no. 2	97	Photo no. 2	98	Photo no. 2	97	Photo no. 2	96
Photo no. 3	96	Photo no. 3	97	Photo no. 3	97	Photo no. 3	95
Photo no. 4	96	Photo no. 4	96	Photo no. 4	96	Photo no. 4	94
Average	96	Average	97	Average	97	Average	95

The effect of reflection on different materials is variable. The shape of the surface can explain it. It was assumed that an average reflection is 4% if the material is 100% covered. However, if the sample coverage is 75% after boiling, only 3/4 of the total sample area can produce a reflection. In this case, the reflection effect is only ~ 3%. The formula for calculating bitumen coverage was improved as follows:

$$R=B/100 \quad (5)$$

$$B_{gal}=B+(R*4), \text{ where} \quad (6)$$

B - calculated bitumen coverage before correction, %;

R – reflection correction coefficient, %.

3. Results

3.1 Results from visual assessment

The Visual evaluation was carried out by five independent operators. The results showed significant differences (see Table 3). The biggest difference between evaluations was from the sample with the smallest bituminous coverage. The smallest difference between the results of the evaluation was given by a gravel sample.

Table 3. Test results from a visual assessment.

Material	Dolomite	Granite	Quartz diorite	Gravel
Operator 1	43 %	84 %	72 %	81 %
Operator 2	52 %	80 %	71 %	80 %
Operator 3	56 %	80 %	72 %	85 %
Operator 4	53 %	75 %	63 %	81 %
Operator 5	46 %	77 %	69 %	78 %
Average	**50 %**	**79 %**	**70 %**	**81 %**
Amplitude between results	13 %	9 %	9 %	7 %

3.2 Evaluation using a photo processing program

Dolomite is a monochrome and light material, but when images were processing, it was found that the difference between Canon EOS 600D and Huawei P9 lite was 8% and it is the most significant difference between all the samples.

Table 4. Test results with dolomite when photo processing method was used.

Dolomite	Canon EOS 600D		Huawei P9 lite	
	Side 1	Side 2	Side 1	Side 2
	Mixed			
Photo evaluation no. 1	49 %	43 %	54 %	58 %
	Mixed			
Photo evaluation no. 2	45 %	45 %	52 %	53 %
	Mixed			
Photo evaluation no. 3	46 %	47 %	52 %	56 %
Average	**46 %**		**54 %**	

The granite selected was various colours and gave the impression that it will be difficult to process it using a computer program. In the end photo processing for granite took more time. Result difference between Canon EOS 600D and Huawei P9 lite was the smallest 5%.

Table 5. Test results with granite when photo processing method was used.

Granite	Canon EOS 600D		Huawei P9 lite	
	Side 1	Side 2	Side 1	Side 2
	Mixed			
Photo evaluation no. 1	85 %	86 %	81 %	81 %
	Mixed			
Photo evaluation no. 2	82 %	85 %	75 %	82 %
	Mixed			
Photo evaluation no. 3	88 %	82 %	76 %	83 %
Average	**85 %**		**80 %**	

Gravel material typically consists of different types of mineral materials. Their colour may vary. In this sample, it was hard to make good repeatability. Every time the sample was mixed the different result was possible to get. Therefore, image processing may be problematic with gravel type material. The difference in results between Canon EOS 600D and Huawei P9 lite was 7%.

Table 6. Test results with gravel when photo processing method was used.

Gravel	Canon EOS 600D		Huawei P9 lite	
	Side 1	Side 2	Side 1	Side 2
	Mixed			
Photo evaluation no. 1	80 %	85 %	81 %	74 %
	Mixed			
Photo evaluation no. 2	83 %	73 %	72 %	82 %
	Mixed			
Photo evaluation no. 3	83 %	81 %	68 %	67 %
Average	**81 %**		**74 %**	

The chosen quartz diorite surface was with a homogeneous colour. The difference in results between Canon EOS 600D and P9 lite was 4%.

Table 7. Test results with quartz diorite when photo processing method was used.

Quartz diorite	Canon EOS 600D		Huawei P9 lite	
	Side 1	Side 2	Side 1	Side 2
	Mixed			
Photo evaluation no. 1	63 %	63 %	64 %	68 %
	Mixed			
Photo evaluation no. 2	60 %	56 %	61 %	65 %
	Mixed			
Photo evaluation no. 3	63 %	66 %	71 %	66 %
Average	**62 %**		**66 %**	

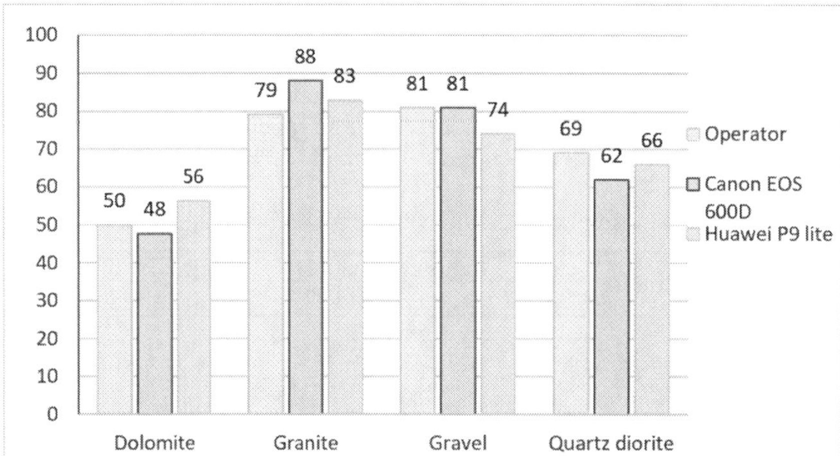

Figure 5. Comparison of the results of visual and image processing method.

Conclusions

1. Images taken with a high resolution professional camera are easier to process, the more precise the contours of the bitumen coverage. Images taken with the smartphone were more difficult and inaccurate to process. The difference between the two electronic devices averaged 4-8%.
2. Reflection error evaluation improved the accuracy of the results.
3. The smaller the bitumen coverage, the more inaccurately results are evaluated by the human.
4. The image processing method shows the smallest scattering when is used a professional camera to capture images.
5. For dark mineral materials, it is significantly more difficult to distinguish the bituminous coverage from the surface of the material.
6. If the mineral material contains stones of different colours (gravel), then it is not recommended to use the image processing method.

Further work

As a future option, we see the use of computer learning to estimate bitumen coverage. It could improve accuracy because the material could be heterogeneous in colour. However, to do this, it would be necessary to improve the accuracy of the image processing method. It is possible to do it by using a higher resolution camera and developing the photo studio lighting.

References

[1] Oliviero R C, Teltayev B and Angelico R 2017 Adhesion Promoters in Bituminous Road Materials: A Review *J. Appl. Sci.* **7** (5) 524

[2] Wistuba M P, Grothe H, Grönninger J and Handle F 2012 5th Eurasphalt&Eurobitume Congress (Istanbul) P 5 EE–197

[3] Pasandín A R and Pérez I 2015 The influence of the mineral filler on the adhesion between aggregates and bitumen I. *J. of Adhesion and Adhesives* **58** 53–8

[4] Twagirimana E 2014 Evaluation of adhesion properties in bitumen-aggregate systems for winter surfacing seals using the bitumen bond strength test (Stellenbosch University)

[5] Khan A, Redelius P and Kringos N 2016 Evaluation of adhesive properties of mineral-bitumen interfaces in cold asphalt mixtures *J. Construction and Building Materials* **125** 1005–21

[6] LVS EN 12697-11 "Bituminous mixtures. Test methods for hot mix asphalt. - Part 11: Determination of the affinity between aggregate and bitumen"

[7] SJSC Latvian State Roads 2019 Road Specifications 2019 Riga, Latvia

[8] Mulsow C 2012 Determination of the degree of gravel aggregate-bitumen-coverage by multi-directional reflectance measurements *International Archives of the Photogrammetry, Remote Sensing and Spatial Information Sciences*, XXXIX-B5, 39–43

[9] Wyszecki, Gunter 2006 Chicago. Color *World Book Inc.* **824**

[10] Kallen H, Heyden A and Astrom K Measurement of Bitumen Coverage of Stones for Road Building, Based on Digital Image Analysis *IEEE Workshop on the Applications of Computer Vision* (WACV), 2012

[11] Lamperti R, Lantieri C, Sangiorgi C, Bitelli G and Simone A Semi-Automatic Evaluation of the Degree of Bitumen Coverage on Bitumen-Coated Aggregates *RILEM Bookseries 8th RILEM International Symposium on Testing and Characterization of Sustainable and Innovative Bituminous Materials*, 2015 pp. 15–24

[12] LVS EN 1097-6 "Tests for mechanical and physical properties of aggregates. - Part 6: Tests for mechanical and physical properties of aggregates."

[13] Anthoni J F 2000-2005 Water and light in underwater photography. Available: http://www.seafriends.org.nz/phgraph/water.htm

Effect of wetting/drying on the properties of OSB/3 and birch plywood

R Kask[1], H Lille[1], M Kiviste[1], R Tamm[1] and J O Lääne[2]

[1] Institute of Forestry and Rural Engineering, Estonian University of Life Sciences, Tartu, Estonia,
[2] Institute of Technology, Estonian University of Life Sciences, Tartu, Estonia

E-mail: mihkel.kiviste@emu.ee

Abstract. In this study the thickness swelling (TS), restore dimensions (RD), density (D) and modulus of rupture (MOR), after ten cycles of wetting over/drying, of commercial boards, i.e. 3-layered OSB/3 ("SWISS KRONO", thickness 12mm) and 9-layered birch (*betula sp*) plywood with phenol-formaldehyde (UPM-Kymmene, Otepää, Estonia), were investigated according to EN standards. Test samples (dimensions of 290×50×12 mm³) for determination of a change in the investigated parameters, according to the standard EVS EN 326-1:2002, were cut from the boards in the parallel (‖ - major axis) direction to the face (outer layer) strand. The cutting forehead and the facet edges of the test samples were coated with the water repellent mastic *Kiilto Fibergum* ("KIILTO OY"). The specimens were altered by soaking in a tank with water (22±1°C) for 24 hours and dried for 48 hours in a ventilated drying box at 65 ± 1°C. Before testing, the samples were conditioned in a climatic chamber at a relative humidity of 65% at 21°C. The surfaces and cross-sections of the initial dry samples after the first wetting/oven-drying cycles were studied using the X-ray technique. The analytical exponent- and linear-fractional equations were used to approximate the obtained experimental data depending on the number of cycles.

1. Introduction

Wood-based panel products such as plywood, fibreboard, particleboard and oriented strand board (OSB) provide a substitute for solid wood while retaining the requisite structural properties with favourable performance and low cost. Plywood is the dominant wood-based panel type, accounting for 42 % of all wood-based panel production in 2016. The production of OSB accounted for 7.2% in 2016 [1]. However, the global OSB market is expected to grow approximately 28% by 2022. OSB is taking an increasingly higher share in the structural panel products market in applications such as walls and roof sheathing, I-beam webs and single layer flooring [2]. Thus plywood and OSB are the largest growing wood-based panels in the global market.

Plywood is a material manufactured from thin layers or "plies" of wood veneer which are glued together with adjacent layers having their wood grain rotated up to 90 degrees to one another. This cross-graining has several important benefits: it reduces the tendency of wood to split when nailed in at the edges; reduces expansion and shrinkage, providing improved dimensional stability; and makes the strength of the panel consistent across all directions. There are usually an odd number of plies, so that the sheet is balanced, which reduces warping. The face veneers of a typical plywood panel have higher grade than the veneers. The principal function of the core layers is to increase separation between the outer layers where bending stresses are the highest, thus increasing the modulus of rupture (MOR) of plywood.

Oriented strand board (OSB) is a material made from large, flat strands of wood, where the outer layers' strands are oriented parallel to the long edge of the board or to the production line, and the core layer is often formed of smaller strands oriented at right angles to the outer layers [3]. The outer layers of OSB are compressed to attain higher density, ensuring the basic physical and mechanical properties of the boards. The moisture transfer properties of OSB differ from those of both plywood and construction lumber [4]. OSB is a hygroscopic material and its dimensional stability and physico-mechanical properties are dependent on many factors including adhesive content, water sorption, number of soaking-oven-drying periods and ambient temperature.

The objective of the current study was to find out and compare the effect of wetting and oven-drying on the physico-mechanical properties of OSB and plywood. For this, commercially available 3-layered OSB/3 and 9-layered birch (betula sp) plywood were investigated. The investigated physical and mechanical properties of OSB/3 and plywood were thickness swelling (TS), restore dimensions (RD), density (D) and modulus of rupture (MOR), respectively.

2. Experimental methods

Twelve OSB/3 and plywood test samples (dimensions 290×50×12 mm3) per each wetting/oven-drying cycle were prepared for determination of a change in the investigated parameters according to standard EVS EN 326-1:2002 [5]. The test samples were cut in parallel (‖ - major axis) to the face (outer layer) strand from two plywood boards and from one OSB/3. The cutting forehead and the facet edges of the test samples were coated with the water repellent mastic Kiilto Fibergum ("KIILTO OY"). The samples were altered by soaking in a tank with water (22±1) °C for 24 hours and dried after that (for 48 hours) in a ventilated drying box at 65 ± 1°C. Before testing the samples were conditioned in a climatic chamber at 21°C and relative humidity 65%.

All investigated parameters of the test samples were determined immediately after they were taken out of the water tank or the drying box and were arranged in accordance with the standards: TS and RD (EN 317:1996) [6], MC (EN 322:2002) [7] and D (EN 323: 1993) [8]. Also, the modulus of rupture (MOR) of OSB/3 and that of plywood were compared. MOR was determined by the three-point bending test performed in accordance with EN 310:1993b [9], using the computer-controlled mechanically actuated universal testing machine Instron 3369. Deflection for calculating the modulus of elasticity was measured by an optical gauge (Advanced Video Extensometer 2663-821). Density was determined as the ratio of the mean mass of twelve test samples to their volume, in our case for dry samples [10, 11].

The surfaces and cross-sections of the dry samples at the initial time point and after the first wetting/oven-drying cycle were studied using the X-ray technique (Yxlon FF35 CT).

The following assumptions were made for the mathematical expressions (values of the investigated properties vs number of wetting/oven-drying cycles): the approximation curve, first, cuts ordinate or starts from zero; and second, enables to determine limit values, in which the studied properties stabilize.

The following analytical linear-fractional and exponent equations were used to approximate the obtained experimental data depending on the number of cycles [12]:

$$Y(x) = (b(Y_i - Y_f)/(ax + b)) + Y_f; \tag{1}$$

$$Y(x) = Y_f(1 - e^{-ax}), \tag{2}$$

where Y_i, Y_f are the calculated initial ($x = 0$) and final ($x = \infty$) values of the investigated properties, x is the number of wetting/oven-drying cycles and a and b are constants.

The equations allowed to predict the investigated parameters of the specimens to a certain extent when their values after applying a small number of soaking/oven-drying cycles were known.

The initial and final values of the properties and constants should be determined so that the measured experimental data are approximated in the best way by minimizing the square of error (least squares regression). This problem was solved by using the program *Mathcad 15* with the regression function *genfit(vx,vy,vg,F)*. The TS and RD trend line of plywood were found by using the program *MS Excel 2016* with the regression analysis function.

It is usually described by a parameter that defines the range within which the true values of the quantity to be measured is estimated to fall within the given confidence range 95%. Calculation of the expanded uncertainty of measured properties is carried out according to the standard EVS-EN 326-1: 2002 [5]. In our study the mean values of the experimental data were used for approximation.

3. Results and discussion

The average MC of the test samples after the first 24 hours of soaking (at 22 ± 1°C) were 25.3% and 43.5% (latter is Wet) for plywood and OSB/3, respectively. After 48 hours of drying in a ventilated box at 65 ± 1°C the average MC was 7.7% and 12.0% (latter is Dry) for plywood and OSB/3, respectively. The percentages of TS and RD of OSB/3 board and plywood, compared to their initial values, are presented in Figure 1 and Figure 2. It should be noted that the values of the calculated percentages are taken from the approximation curves.

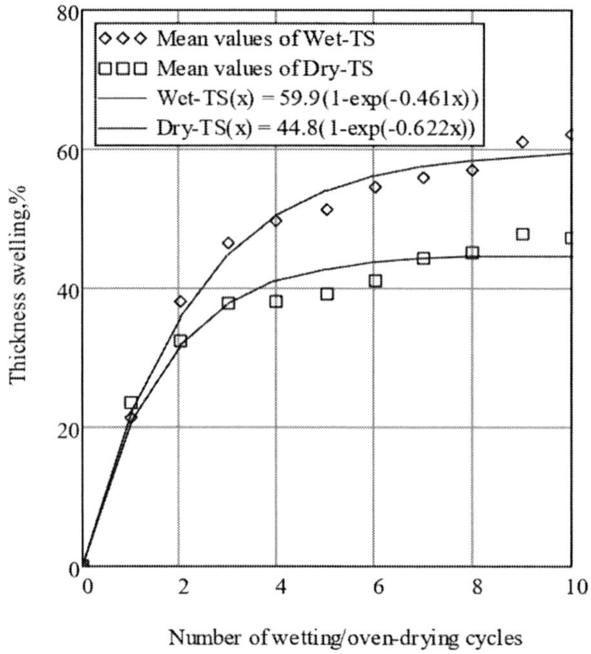

Figure 1. Dependence of Wet–thickness swelling and Dry – thickness swelling of OSB/3 on the number of soaking/oven-drying cycles: mean values of the experimental data and the curves of approximation with the determined quantity according to equation (2).

The mean thickness of OSB/3 board after the first wetting/oven-drying cycle was 20.7% and after the first soaking cycle 22.1% of the initial values, which indicates that plastic deformations of the resin matrix are not restored.

The thickness 5.15% of plywood after the first wetting was markedly smaller than that of OSB/3 board and decreased even down to initial thickness after drying and its RD was - 1.35%. It is evident that for plywood, the values of TS remained constant within the experimental error irrespective of the number of wetting/oven-drying cycles (Figure 2).

The mean values of density depending on the number of wetting/oven-drying cycles in Figure 3 for OSB/3 were determined after drying. A notable reduction in the density values was found after the first soaking/oven-drying cycle and the obtained $D_m(1)$ was 87.8% of $D_m(0)$. Based on the limits of experimental error, the density values did not decrease significantly after three soaking/oven-drying cycles and accounted for 74 % of their initial values. Thus, the OSB/3 panel had lost one quarter of its initial density while its air permeability had consequently increased.

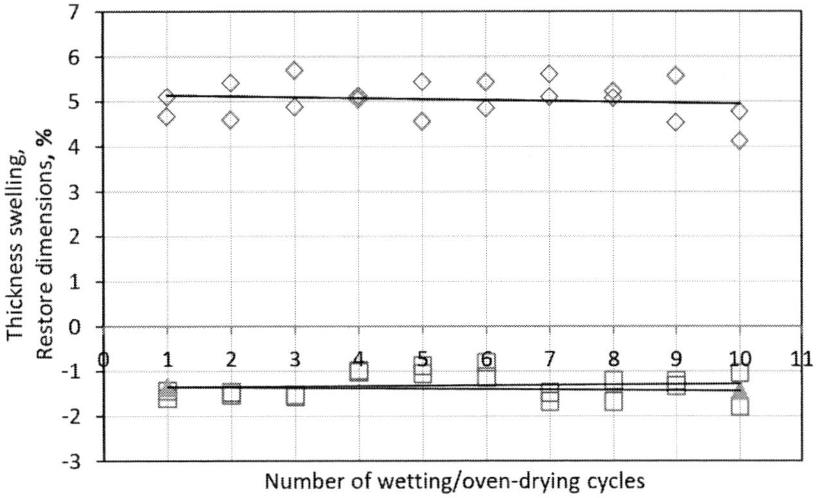

Figure 2. Dependence of Wet –TS and Dry – TS of plywood on the number of soaking/oven-drying cycles: mean values of the trend lines are 5.15% and -1.35%, respectively.

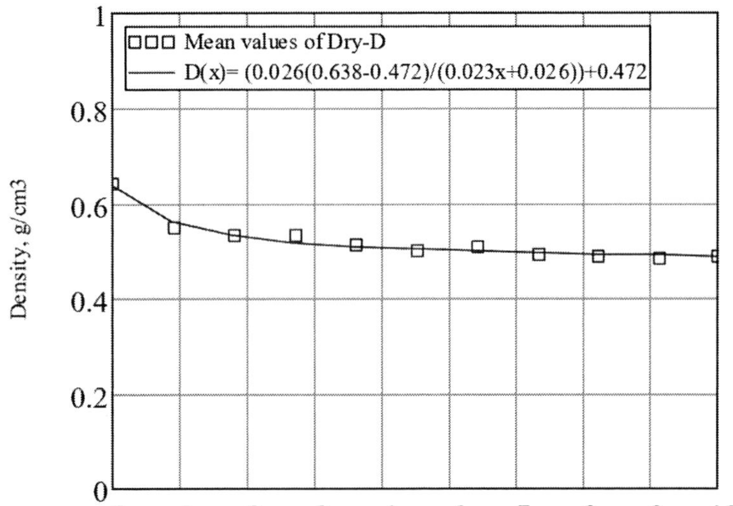

Figure 3. Dependence of the density of OSB on the number of soaking/oven-drying cycles: mean values of the experimental data and the formula and curve of approximation with the determined quantity according to equation (1).

The retention values vs initial values of MOR recorded after the first wetting/oven-drying cycle, were 79.8% of Dry-MOR, but only 44.2% of Wet-MOR for OSB/3. The following wetting/oven-drying cycles had less impact on Dry-MOR compared to Wet-MOR, see Figure 4.

The retention values vs initial values of MOR, recorded after the first wetting/oven-drying cycle, were: Dry-MOR 80.0%, but Wet-MOR 52.8% for plywood. The MOR values of the both wet and dry samples did not decrease considerably after the first cycle (Figure 5). On the other hand, the decrease of Wet- MOR values for both plywood (almost 53%) and OSB/3 (about 44%) was highly significant.

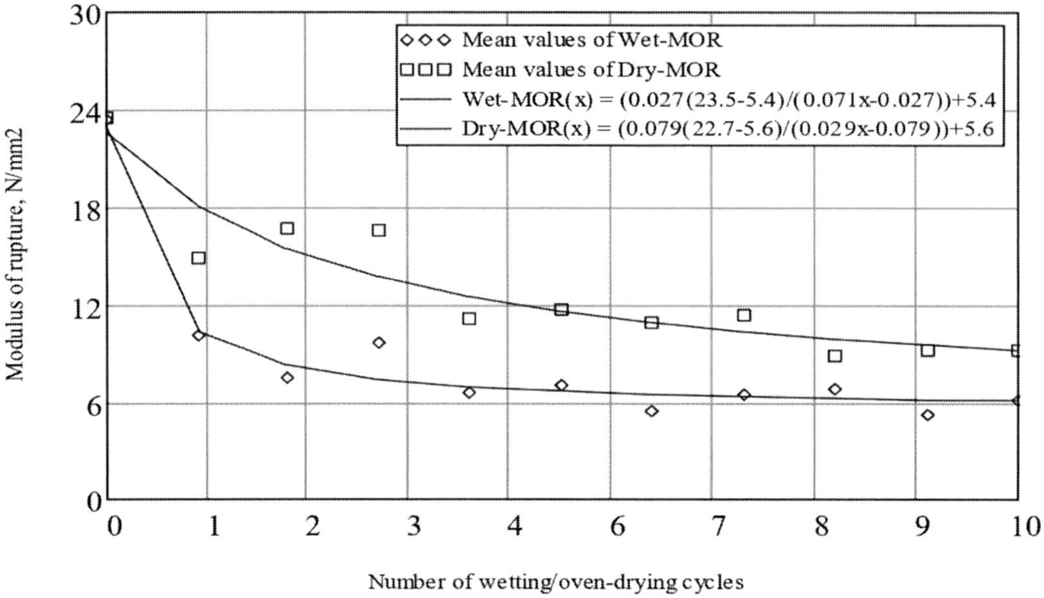

Figure 4. Dependence of Wet–MOR and Dry-MOR of OSB/3 on the number of soaking/oven-drying cycles: mean values of the experimental data and the curves of approximation with the determined quantity according to equation (1).

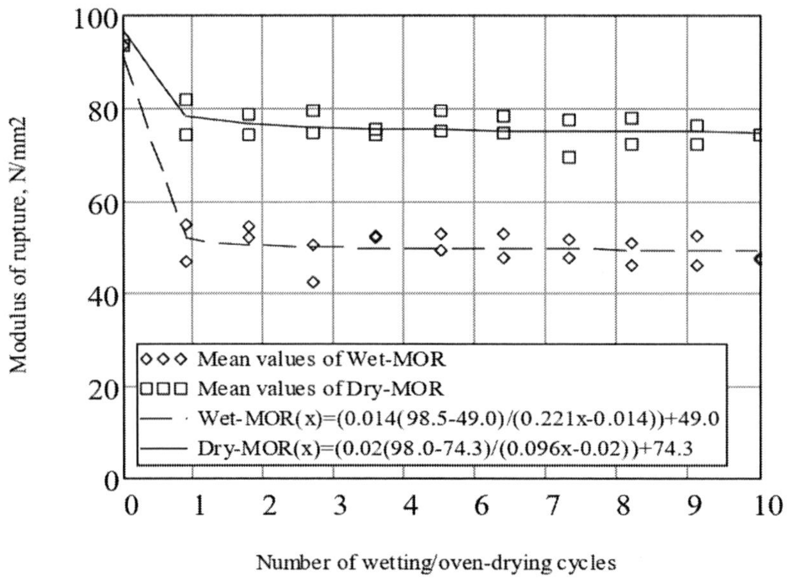

Figure 5. Dependence of Wet – MOR and Dry-MOR of plywood on the number of soaking/oven-drying cycles: mean values of the experimental data and the curves of approximation with the determined quantity according to equation (1).

This means, that both plywood and OSB/3 failed to meet the requirements for the ultimate limit state (e.g. as a sheathing board ensuring rigidity) in the case of 24 h of soaking in water.

X-ray investigations conclude with a manipulation of OSB/3 board of large increase in TS and decrease in density and MOR even after the first wetting/oven-drying cycle (Figure 6). The X-ray investigations allowed to make the following conclusion. A small amount of phenol formaldehyde

IOP Conf. Series: Materials Science and Engineering **660** (2019) 012048 doi:10.1088/1757-899X/660/1/012048

resin (matrix) tended to segregate along the wood strands boundaries forming an amorphous matrix in which the individual strands are embedded. Wood strands swelled during wetting and their increasing volume created stresses in the embedded boundaries of the resin matrix, resulting in plastic deformations. Next, as the volume of wetted strands decreased in oven-drying, the strands consequently (partially or fully) ripped off from the embedded boundaries of the matrix or fractured parallel or perpendicular to the grain. Because of the voids generated between the fracture surfaces of the strands and the embedded boundaries of their plastically deformed resin matrix, the dimensions of the samples did not restore after drying. Also, comparison of the TS of the wetted and dry samples indicated that the plastic deformations of the matrix did not recover, being 20.7% of Dry-TS and 22.1% of Wet-TS after the first wetting/oven-drying cycle and after the first wetting, respectively. Also, many surface strands fractured in volume and the MOR of OSB/3 decreased considerably after the first cycle (Figure 4).

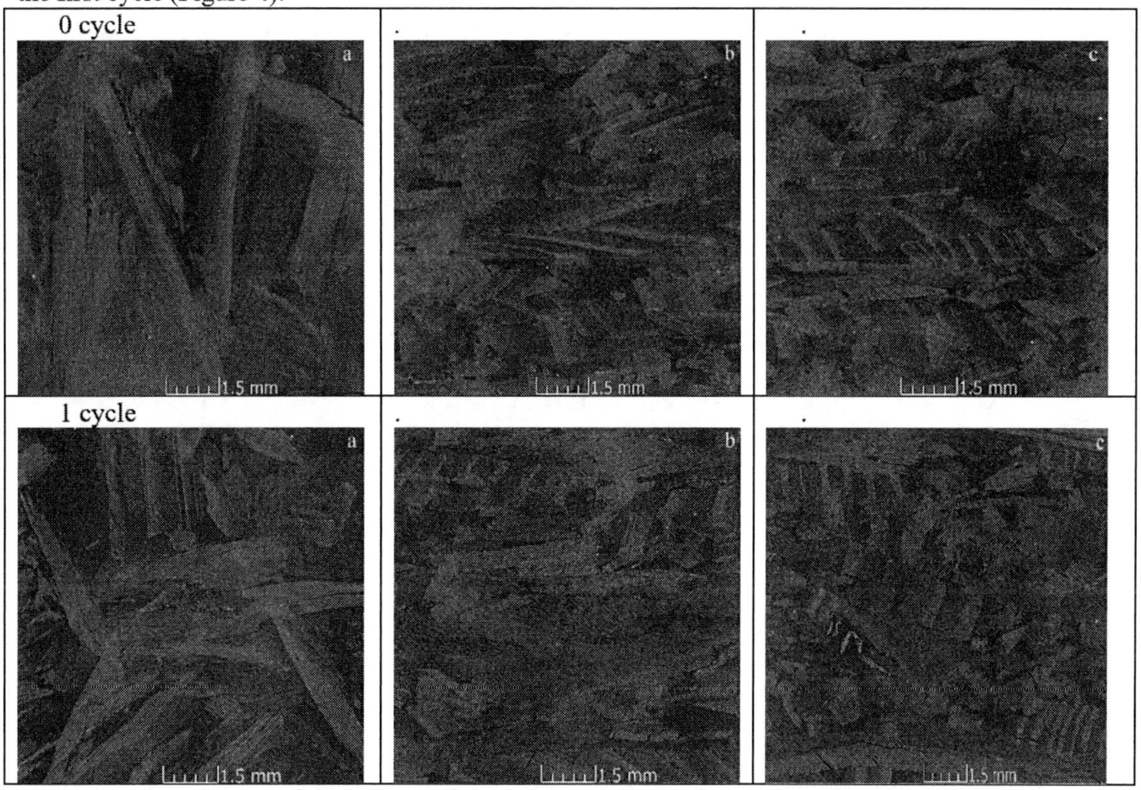

Figure 6. X-ray image of the texture of the 3-layered OSB/3 board from the: a)face (outer) layer, b) middle cross-section, in parallel with II axis, c) in parallel with T axis; purveyance dry (0 cycle), and after the first wetting/oven-drying cycle.

Owing to frequent precipitation in the practice of construction sites, wetting of OSB/3 boards before assembly or when they are used as a sheathing material (before installation of external cladding) results in generation of air voids even after consecutive drying. These air voids also significantly increase the air permeability of OSB/3 boards [13], as a consequence of which they do not satisfy the requirements for nearly zero-energy buildings. The present analysis is limited to the methodology used and to the data obtained from the above described experiments of the investigated panels.

4. Conclusions

After the first wetting the mean values of the wetted thickness swelling of OSB/3 board and plywood were 22.1% and 5.15%; the wetted modulus of rupture was 44.2% and 52.8% of the initial value, respectively. Thus, 24 h of wetting resulted in a significant decline in the modulus of rupture of both OSB/3 and plywood boards, which failed to meet the requirements for the ultimate limit state as a part of the load bearing structure (e.g. as a sheathing board ensuring rigidity) of a building.

After the first wetting/oven-drying cycle the mean values of dry thickness swelling (restore dimension) of OSB/3 and plywood were 20.7% and -1.4%; the dry modulus of rupture accounted for 79.8% and 80.0% of its initial value, respectively. The density of dry OSB/3 board was 87.8% of its initial value.

The X-ray investigations of the OSB/3 samples revealed a significant change in the studied parameters after the first wetting and drying cycles as a result of non-restored plastic deformations of the resin matrix, in which embedded strands were jammed after wetting. For the same reason, the strands ripped off during oven-drying, resulting in the generation of air voids.

The proposed analytical functions approximated satisfactorily to the experimental data of thickness swelling, density and modulus of rupture depending on the number of soaking/oven-drying cycles.

Acknowledgement
The X-ray investigations of the current study were supported by the European Regional Fund, project No. 2014-2020.4.01.16-0183, Smart Industry Centre (SmartIC).

References

[1] Food and agriculture organization of United Nations 2016 Global Forest Products: Facts and Figures. Available from: http://www.fao.org/3/I7034EN/i7034en.pdf
[2] Grand View Research 2018 Wood Based Panel Market Size, Share & Trends Analysis Report by Product (Plywood MDF Particleboard Softboard Hardboard) by Application (Furniture Construction Packaging) and Segment Forecasts 2018 – 2025 p 141
[3] Irle M A et al 2012 *Handbook of Wood Chemistry and Wood Composites* 2nd, ed R M Rowell (Boca Raton: CRC Press, Taylor & Francis Group) pp 321-412
[4] Carll C and Wiedenhoeft A 2009 *Moisture Control in Buildings: the Key Factor in Mold Preventions* 2nd ed ASTM International, MNL 18
[5] *Estonian Centre for Standardization*, 2002. EVS-EN 326-1: 2002 Wood-based panels. Sampling, cutting and inspection Part 1: Sampling and cutting of test pieces and expression of results (in Estonian)
[6] *Estonian Centre for Standardization* 2000 EVS-EN 317: 2000 Particleboards and fibreboards – Determination of swelling in thickness after immersion in water. (in Estonian). Available from: https://www.evs.ee/tooted/evs-en-317-2000
[7] *European Standard* 1993 EN 322: 1993 Wood-based panels – Determination of moisture content
[8] *European Standard* 1993 EN 323: 1993 Wood-based panels – Determination of Density
[9] *European Standard* 1993 EN 310: 1993b. Wood-based panels - Determination of modulus of elasticity in bending and of bending strength
[10] Tamm R 2018 Physical and Mechanical Characteristics of OSB/3-Boards after Repetitive Soaking and Oven-Drying Unpublished thesis (MScs) Estonian University of Life Sciences Tartu: Estonia (in Estonian) https://dspace.emu.ee/xmlui/browse?type=author&value=Tamm%2C+Renee
[11] Kruus S 2016 Changes in Physical and Mechanical Properties of Moisture Resistent Birch Plywood after Soaking and Oven-Drying Cycles Unpublished thesis (MScs) Estonian University of Life Sciences Tartu: Estonia (in Estonian) https://dspace.emu.ee/xmlui/browse?type=author&value=Kruus%2C+Silver
[12] Sooru M, Kasepuu K, Kask R and Lille H 2015 *Proc. 2nd International Conference on Innovative Materials, Structures and Technologies* vol 96 (Riga: Latvia IOP Publishing Ltd) 10.1088/1757-899X/96/1/012075.
[13] Hodoušek M, Böhm M, Součková, A and Hýsek Š 2018 *BioResources* **3** 4856

IOP Conf. Series: Materials Science and Engineering **660** (2019) 012049 doi:10.1088/1757-899X/660/1/012049

Damping Identification and Prediction for Laminated Composite Plates

M Wesolowski[1], **E Barkanov**[2]
[1]Department of Structural Mechanics, Koszalin University of Technology, Koszalin, Poland
[2]Institute of Materials and Structures, Riga Technical University, Riga, Latvia
E-mail: miroslaw.wesolowski@tu.koszalin.pl

Abstract. Laminate composite structures are widely used in different sectors of industry due to their remarkable stiffness and damping performance. The present study focuses on the damping identification and prediction for laminated composite plates. The damping model based on the strain energy method was used for damping coefficients identification and for modal loss factor prediction. The study concerns with the damping coefficients identification for a single layer of an unidirectionally laminated plate. The reliability of the proposed identification method is validated by numerical modal loss factor prediction for a plate made of the layers in question but having different geometric characteristics. Further, the predictions of the modal loss factors were also validated by a vibration test. All the vibration tests necessary for the current study were performed in a non contact manner using laser vibrometer. The results delivered from the identification method were proven to be credible for modal loss factor prediction.

1. Introduction

Laminate composite structures are widely used in the form of plates, beams, or cylindrical shells. Besides their exceptional stiffness properties, the composites structures exhibit highly preferable damping characteristics. The damping identification as well as damping prediction have been intensively studied considering laminated composite plates and beams [1, 2, 3, 4, 5]. As these two procedures are mutually dependent on each other, the identification procedure seems to be more crucial in this relation. It is so, because the identification procedure delivers the necessary information on the damping coefficients which are further used to predict the modal damping of structures composed of a material in question. Among the methods for damping coefficients identification, the inverse technique based on dynamic response has been appreciated the most [6, 7, 8, 9]. The inverse technique combines the experimental and numerical analyses. In the numerical analysis, a proper selection of the applied damping model is the core issue, since it is used for the both, identification and prediction of damping.

The study concerns with the damping coefficients identification of a single layer of an unidirectionally laminated plate. In the present work the damping model based on the strain energy method [2, 10, 11] was used for damping coefficients identification and for modal loss factor prediction. The damping model was combined with the finite element solutions. The identification was performed using an inverse technique which was based on the design of experiments and the response surface methodology [12]. The reliability of the identification method was validated by numerical modal loss factors predictions for a plate made of the same material but having different geometric characteristics. Further the predictions of the modal loss factors ware also validated by vibration test. All the vibration tests necessary for the current study were performed in a non-contact manner using a laser vibrometer.

2. Materials and samples

Damping properties of two laminated plates were investigated in the present study. One plate with a rectangular shape was used for the damping coefficients identification purpose. The second plate, square in shape, was used for predictions of modal loss factors based on the identified damping coefficients. Both plates were built up of the same carbon/epoxy prepreg material by means of the vacuum bagging. The plates had length a, width b, ply thickness t, and total hight h (Table 1). The plates were composed of 16 layers of plies having material properties as listed in Table 2. A global Cartesian co-ordinate system (x, y, z) was set-up for the plates, with z axis downward and normal to the reference surface (Figure 1a). For each layer a lamina co-ordinate system $(1,2,3)$ was defined with the direction 1 along the lamina fibres, direction 2 transverse to the fibres direction, and 3 - through the thickness direction. A lamination angle (Φ) of a single layer was defined between the x-axis and the 1-axis.

3. Inverse technique

The inverse technique based on experiment design and the response surface methodology is used in the present study for damping coefficients identification. The whole procedure is given schematically on Figure 1b and can be summarised as follows. In the first step a plan of experiments (DoE) is developed depending on the number of identification parameters l and number of experiments p. Then, the finite element analyses are performed in the reference points of the experiment design and plate's eigenfrequencies and corresponding eigenvectors are calculated numerically. Based on these data the modal loss factors are then calculated for each reference point. The results of these calculations are then taken to determine simple functions using the response surface methodology. In parallel, a vibration test is carried out with the purpose to extract frequency response function (FRF) which is used for experimental modal loss factor assessment. An identification of damping coefficients is performed in

Table 1. Geometric characteristics of the laminated plates used in the investigation.

Sample	a [mm]	b [mm]	h [mm]	t [mm]	Layers	Φ [deg]
Plate 1	290.0	200.0	2.6	0.1625	16	0
Plate 2	300.0	300.0	2.6	0.1625	16	0

Table 2. Properties of the constituent plies of the two laminated plates used in the investigation.

Material	E_1 [GPa]	E_2 [GPa]	G_{12} [GPa]	ν_{12}	ρ [kg/m^3]
Carbon/Epoxy	116.80	8.85	4.10	0.36	1546.60

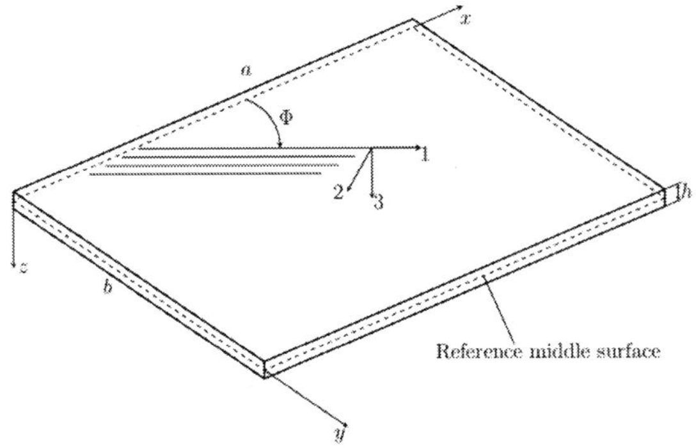

(a)

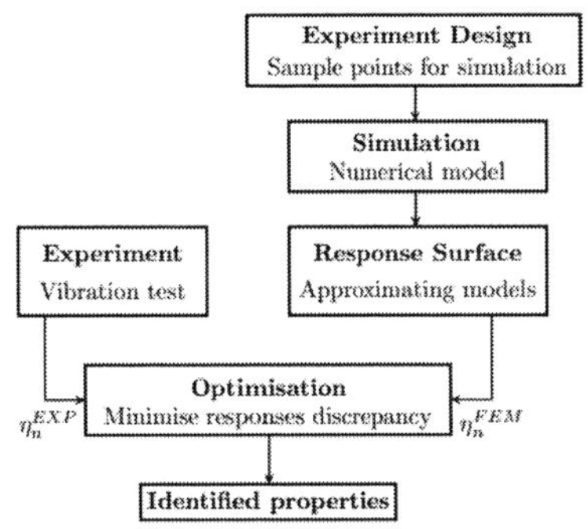

(b)

Figure 1. a) Geometry of laminated plates. b) Inverse technique procedure.

the final step minimising the error functional between experimental and numerical data.

3.1. Response surface methodology

Response surface methodology (RSM) is a mathemathical procedure for empirical model building [13]. By carefully generated DoE, the aim of RSM is to optimize an output variable (a response) which is affected by several independent input variables (identification parameters). The DoE is a set of runs (experiments), in which changes are made in the input variables in order to observe the changes in the response. The input variables are denoted as $x_1, x_2, ..., x_l$ and the response is denoted as y. The relationship between the response and the input variables can be represented as:

$$y = f(x_1, x_2, ..., x_l) \quad (1)$$

The RSM starts with finding an approximation to the true functional relationship between y and the input variables. For the most cases, the approximations are taken in the form of first or second-order polynomials. The domain of these polynomials is enclosed in the region of the input variables.

3.2. Identification parameters

The parameters to be identified were specific damping capacity coefficients (SDC coefficients) of constituent layers of the laminated rectangular composite plate (Plate 1). The unidirectional layer of the plate was considered homogeneous and transversely isotropic with respect to the fibre direction. In the current investigation a thin plate is considered with a linear elastic material and the plane stress state. For such a case, a vector of three SDC coefficients corresponding to the in plane lamina directions is required to be identified:

$$\mathbf{x} = [\mathbf{x}_1, \mathbf{x}_2, ...\mathbf{x}_l] = [\psi_{11}, \psi_{12}, \psi_{22}] \quad (2)$$

3.3. Design of experiments

The main part of the RSM is the design of experiments (DoE). The objective of the DoE is the selection of the sample points (reference points) in which the response should be calculated. A location of a single reference point is defined by the value and number of the identification parameters. The initial informations for developing the design are therefore a number of identification parameters l and the number of experiments p. The reference points shall be distributed as regularly as possible in the narrow domain of the identification parameters.

3.4. Objective function

The objective function is expressed as an error estimator describing the discrepancy between the experimental modal loss factors and the corresponding modal loss factors calculated from RMS approximation polynomials:

$$\Theta(\mathbf{x}) = \sum_{n=1}^{I} \mathrm{w}_n \cdot \left[\frac{\left(\eta_n^{EXP}\right)^2 - \left(\eta_n^{FEM}(\mathbf{x})\right)^2}{\left(\eta_n^{EXP}\right)^2} \right]^2 \tag{3}$$

where η_n^{EXP} are the experimentally determined modal loss factors; η_n^{FEM} are the modal loss factors calculated from the approximation polynomials; $\mathbf{x}$ is the vector of unknown parameters; I is the total number of the the modal loss factors considered in the analysis, w_n is the integer having value 1 or 0 and was used for the selection of the modal loss factors which entered the objective function.

3.5. Optimisation method

In order to find the minimum of the objective function, the following optimisation problem was solved

$$min\Theta(\mathbf{x}) \quad \text{Subject to constraints:} \quad \mathbf{x}_i^L \leq \mathbf{x}_i \leq \mathbf{x}_i^U,\ i = 1,2,\ldots,l \tag{4}$$

where $\mathbf{x} = [\mathbf{x}_1, \mathbf{x}_2, \ldots \mathbf{x}_l]$; $\mathbf{x}_i^L$ and $\mathbf{x}_i^U$ are lower and upper bounds of the identification parameters, respectively. The Hooke-Jeeves Direct Search method was used to find the optimal unknown vector which minimise the objective function.

4. Numerical Model

The numerical models of the laminated plates were developed using the Finite Element Method (FEM). The finite element code SIMULIA/ABAQUS was used to model the laminated plates with the use of SR4 layered shell elements. The elements' displacement field was described according to the First Order Shear Deformation Plate Theory. The linear elastic material model was applied for the plates. For the need of the current study the eigenvalue problem of undamped free vibrations was represented as:

$$\left(\mathbf{K} - \omega_n^2 \mathbf{M}\right) \phi_{\mathbf{n}} = 0 \tag{5}$$

where $\mathbf{K}$ and $\mathbf{M}$ are the stiffness and mass matrices of the plate, respectively; $\phi_{\mathbf{n}}$ are the eigenvectors (mode shapes) of the corresponding eigenvalues $\omega_n = 2\pi f_n$, where f_n are eigenfrequencies. The Linear perturbation Frequency analysis was solved with the Lanczos mode-extraction method. The eigenvalues and the corresponding eigenvectors of the FFFF plate (all edges free) were then extracted. The graphical representation of the FE model is given on Figure 2.

5. Damping model

An applied damping model for the laminted plates was based on the modal strain energy method [3, 10]. According to the method, the specific damping capacity (SDC) Ψ_n of a plate is defined as the ratio of the total dissipated energy ΔU_n to the maximum strain energy U_n stored in the plate during a stress cycle at n^{th} mode of vibration:

$$\Psi_n = \frac{\Delta U_n}{U_n} = 2\pi\eta_n \tag{6}$$

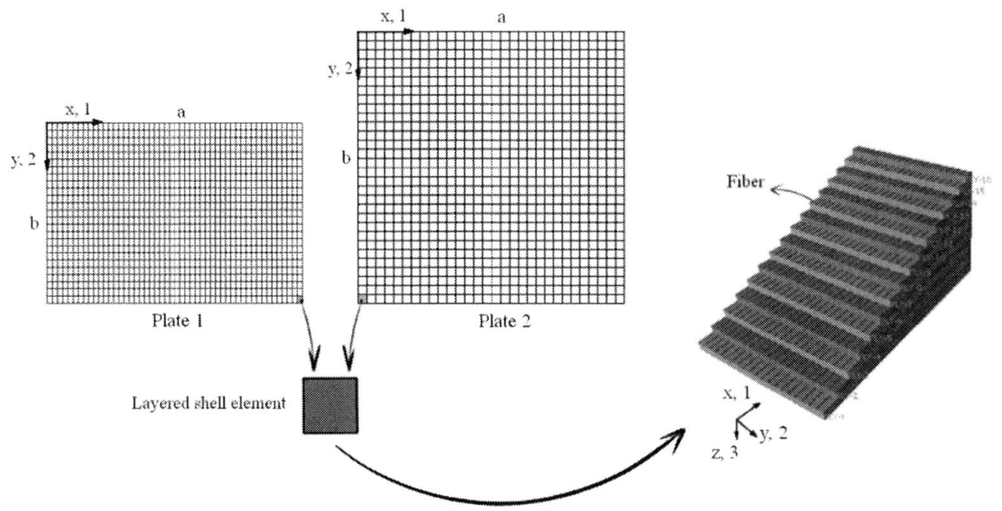

Figure 2. Finite Element Models of laminated plates.

where η_n is the modal loss factor. The maximum strain energy is calculated using the mode shapes of the laminated plate obtained from the solution of the undamped vibration as stated in [14].
In regard to the strain energy and the finite element model, the selected elements of the strain and stress tensors are related to each other as follows:

$$U^e_{ij,k} = \frac{1}{2}\int_{V^e_k} \varepsilon^e_{ij,k} \cdot \sigma^e_{ij,k} \cdot dV^e_k \tag{7}$$

where $U^e_{ij,k}$ is strain energy, $\varepsilon^e_{ij,k}$ are the strain components, $\sigma^e_{ij,k}$ are the stress components, V^e_k is the volume of the layer k of the finite element e, $ij = 11, 12, 22$ are directions in the lamina coordinate system. For the laminated plate composed of the same layers, the total strain energy of the plate is given as:

$$U_n = (U_{11})_n + (U_{12})_n + (U_{22})_n \tag{8}$$

where the each U_{ij} component is the sum of the strain energies stored in all layers of the plate's elements on a given ij direction. Likewise, the dissipated energy is given as:

$$\Delta U_n = \psi_{11}(U_{11})_n + \psi_{12}(U_{12})_n + \psi_{22}(U_{22})_n \tag{9}$$

where the ψ_{ij} terms are the SDC coefficients responsible for energy dissipation on a given ij direction. Having calculated the strain energy U_n and the dissipated energy ΔU_n the modal loss factor can be calculated according to the Eq. 6.

6. Experimental set-up

In the current research a non-contact method for vibration sensing is applied using the scanning laser vibrometer POLYTEC PSV-400-B. The PSV-400-B system is built of a scanning head, a controller, a junction box, a PC station, and a power amplifier (Figure 3). The PSV-400-B system has the ability

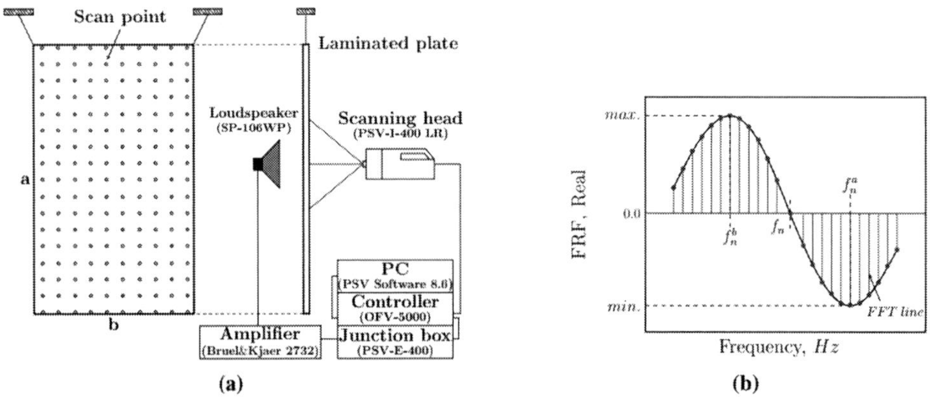

Figure 3. a) Experimental set up with the laser vibrometer. b) Modal loss factor estimation method.

of scanning procedure (multipoint measurements). This gives the possibility for storing an average Frequency Response Function (FRF) and provides a high quality of operational mode shapes of a specimen. The scanning procedure is as follows. A specimen is driven to vibrate by means of a loudspeaker which is provided with an input signal generated by the internal generator. As a result, the specimen starts vibrating within the frequency range of the input signal. The vibrometer controller automatically moves the measurement laser beam (using scanner mirrors) to the each point of the defined scanning grid and senses its back-scattered light. The photo-detector (highly sensitive digital decoder VD-07) measures the time depended vibration velocity and validates measurements with the signal-to-noise ratio. After the measurements have been performed at each point, the averaged time response is transformed to the frequency domain using Fast Fourier Transform. As a result, the FRF of the specimen is obtained. The estimation of the modal loss factors based on the method proposed in [15], and it makes use of the Real part of the mobility FRF:

$$\eta_n^{EXP} = \frac{1-\left(f_n^b/f_n^a\right)^2}{1+\left(f_n^b/f_n^a\right)^2} \tag{10}$$

where η_n^{EXP} is the experimentally measured modal loss factor, n stands for a mode number; f_n^a and f_n^b are frequencies estimated from the Real part of the FRF (Figure 3b). A more comprehensive description of the experimental set-up can be found in [16].

7. Identification and prediction procedure

7.1. Identification of SDC coefficients

A unidirectional laminated composite plate (Plate 1) was used for the identification procedure. The identification procedure was built up with Isight software and is given schematically on Figure 4. The finite element analysis was performed at first on the numerical model of the laminated plate. The selected eigenfrequencies and eigenvectors were then extracted - 8 flexural modes in the range of 0 ÷ 600 Hz (Table 3). The strain and stress components corresponding to the extracted eigenvectors were stored in the Excel 1 file. To do so, each shell element was decomposed to separate layers and the strain ε_{ij} and stress σ_{ij} components of the layers were extracted. In addition the volumes of the layers of each

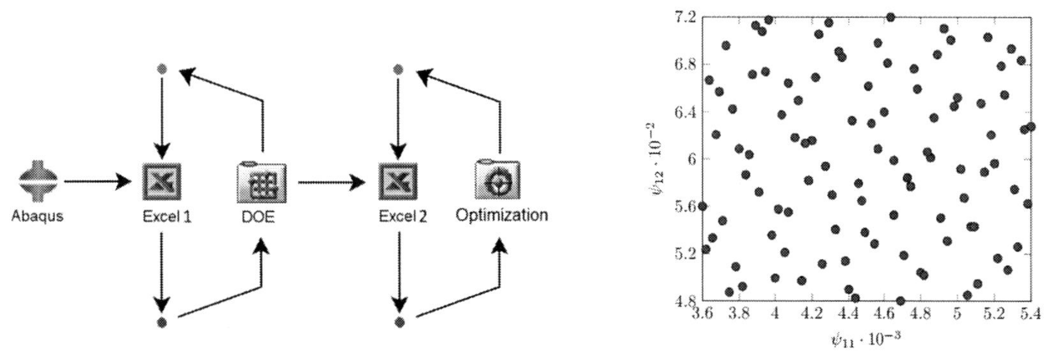

(a) (b)

Figure 4. a) Isight optimization flow. b) Example of the DoE distribution.

elements were also extracted from the finite element model. The obtained volumes were multiplied by the strain and stress components giving the strain energy of the single layer of an element. Summing up all the layers gave the strain energy stored in the single element. The strain energy of the laminated plate for a particular mode shape was obtained by summing up all the energies of the individual elements. Then, the initial values of the SDC coefficients of the layers were used to calculate the dissipated energy components. Finally, the SDC of the whole plate (Ψ_n) could be calculated according to Eq. 6 as well as the modal loss factor (η_n^{FEM}) according to Eq. 6. The procedure was repeated for all extracted mode shapes.

The DoE was performed next. Three SDC coefficients were chosen as the identification parameters. The Latin Hypercube Technique was used for developing of the design of experiments. The design domain for the SDC coefficients was defined as: $\psi_{11} = 0.0038 \pm 20\%$, $\psi_{12} = 0.040 \pm 20\%$, $\psi_{22} = 0.040 \pm 20\%$. What follows, the number of the input variables was set to $l = 3$ and the number of the reference points was set to $p = 100$. An example of the generated DoE is given on Figure 4b. Calculations of the SDC (Ψ_n) of the plate for each point of the DoE were performed using the data of the mode shapes stored in Excel 1 file. After having calculated the Ψ_n, the modal loss factors (η_n^{FEM}) were calculated using Eq. 6. Then, the first order approximation polynomials were developed by applying response surface methodology on the calculated η_n^{FEM}. These polynomials were then exported to the Excel 2 file and served for a rapid calculation of $\eta_n^{FEM}(\mathbf{x})$ in terms of SDC coefficients ψ_{11}, ψ_{12}, and ψ_{22}. These $\eta_n^{FEM}(\mathbf{x})$ values were then used to build the objective function given by Eq. 3.

Next, the vibration tests were performed. The excitation range was set up to $0 \div 600$ Hz, and the frequency response functions (FRFs) were stored. The Real parts of the FRFs were then used to asses the modal loss factors. From the obtained FRFs, 8 resonances were captured. However, the mode shape type (1,2) was of a poor quality and was not used for the estimation of the modal loss factor. On that reason 7 resonances were used for the modal loss factors η_n^{EXP} calculations. The calculations were performed according to the Eq. 10 and the results were stored in Excel 2 file.

In the last step, the identification problem was defined. The objective function was defined as given in Eq. 3. The identification parameters were subjected to the following constraints: $0.0036 \leqslant \psi_{11} \leqslant 0.0054$, $0.048 \leqslant \psi_{12} \leqslant 0.072$, $0.040 \leqslant \psi_{22} \leqslant 0.060$. In order to eliminate the mode shape (1,2) from the objective function, the integer w_n was set to 0 as follows: $w_3 = 0$. The optimisation was performed with the Hooke-Jeeves Direct Search method with the process termination after 40 iteration and linear scaling of the identification parameters.

Table 3. Experimental results.

Resonance (n)	Mode shape $(u,v)^*$	f_n^{EXP} $[Hz]$	$\eta_n^{EXP} \cdot 10^{-3}$
1	1,1	75.2	10.36
2	0,2	157.6	8.99
3	1,2	225.1	-
4	2,0	278.6	0.74
5	2,1	317.7	2.74
6	0,3	445.5	6.97
7	2,2	465.1	6.75
8	1,3	504.9	7.66

* u - nodal line perpendicular to the x -axis; v - nodal line perpendicular to the y -axis

7.2. Loss factor prediction

Having obtained the vector of the desired damping coefficients given as $\mathbf{x} = [\psi_{11}, \psi_{12}, \psi_{22}] = [0.42; 6.23; 4.78] \cdot 10^{-2}$, the modal loss factors of the square laminated composite plate (Plate 2) were predicted numerically. To do so, the proposed damping model was applied and the procedure given in Section 7.1 was repeated omitting the steps required for identification purpose. In order to validate the predicted modal loss factor values, the vibration test was performed on the Plate 2 in a manner as described in Section 6. The experimental mode shapes used for damping prediction, the corresponding FEM eigenvectors, and modal loss factors are given in Table 5.

8. Results

The inverse method was applied for the identification of the SDC coefficients for the rectangular laminated plate - Plate 1. The experimentally measured modal loss factors, corresponding resonances and the mode shape types are listed in Table 3. The results of the identification procedure are given in Table 4. The reliability of the applied identification method was verified by the numerical predictions of the modal loss factors for the square laminated plate - Plate 2. For this purpose the identified SDC coefficients were used. In order to validate the numerical predictions, the experimental modal loss factors of the Plate 2 were also extracted from the vibration test. As given in Table 5 a good convergence was found between the numerical predictions and the experimental results. The predicted values of the modal loss factors differs less than 10 % in most cases. As a matter of fact, the discrepancy of the results can be further reduced by application of the vacuum chamber for vibration testing. Besides, the frequency resolution of the extracted FRFs shall be set up to be lower what might have an influence on the results convergence.

Table 4. Identification results.

SDC coefficients	$\psi_{11} \cdot 10^{-2}$	$\psi_{12} \cdot 10^{-2}$	$\psi_{22} \cdot 10^{-2}$
	0.42	6.23	4.78
$min\Theta(\mathbf{x})$	0.057		

Table 5. Mode shapes and their corresponding modal loss factors - experimentally measured (EXP) and numerically predicted (FEM ABAQUS).

$(u,v)^*$	EXP	FEM ABAQUS	f^a, Hz	f^b, Hz	η^{EXP} 10^{-3}	$(\eta^{FEM})^{**}$ 10^{-3}	Difference %
(1,1)	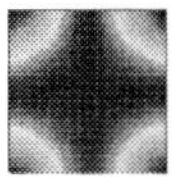	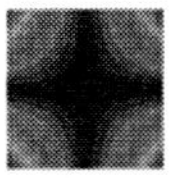	45.45	45.00	9.95	9.82	1.27
(0,2)			69.75	69.25	7.19	7.66	6.42
(1,2)	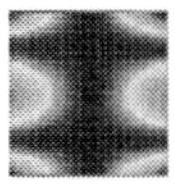		120.75	119.75	8.32	9.05	8.87
(0,3)			194.00	192.50	7.76	7.64	1.63
(2,1)	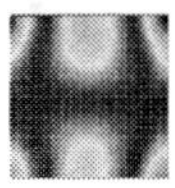	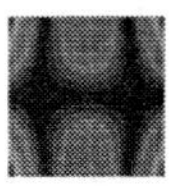	263.30	262.75	2.09	1.88	10.11
(2,2)	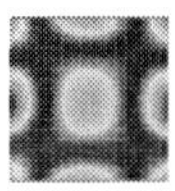		327.00	325.75	4.90	4.31	11.96
(0,4)			387.75	385.00	7.12	7.60	6.79
(2,3)		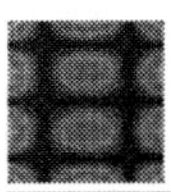	444.75	441.75	6.77	6.20	8.32

*u - nodal line perpendicular to the x - axis; v - nodal line perpendicular to the y - axis
**Predicted with ABAQUS using the identified SDC coefficients vector $\mathbf{x} = [0.42; 6.23; 4.78] \cdot 10^{-2}$

9. Conclusions

Two laminated plates were investigated in the present work. One plate, rectangular in shape, was used to utilise the inverse method for the identification of the SDC coefficients. The second plate made of the same material but rectangular in shape, was used for modal loss factor prediction based on the identified SDC coefficients. By doing so, the reliability of the proposed inverse method was verified and was proven to be credible. Moreover, the numerical predictions were also validated by experimental vibration test and a good results convergence was found. However, further research shall be focused on the SDC coefficients identification for plates being in vacuum conditions.

Acknowledgements
The research leading to these results has received the funding from the Faculty of Civil Engineering Environmental and Geodetic Sciences, Koszalin University of Technology (grant no. 504-01-03).

References

[1] Rikards R, Chate A and Korjakin A 1994 Damping analysis of laminated composite plates by finite element method *Mech. Compos. Mater.* **30** 68

[2] Berthelot J-M and Sefrani Y. 2004 Damping analysis of unidirectional glass and Kevlar fibre composites *Compos. Sci. Technol.* **64** 1261

[3] Berthelot J-M and Sefrani Y 2007 Longitudinal and transverse damping of unidirectional fibre composites *Compos. Struct.* **79** 423

[4] Yang J, Xiong J, Ma L, Wang B, Zhang G and Wu L 2013 Vibration and damping characteristics of hybrid carbon fiber composite pyramidal truss sandwich panels with viscoelastic layers *Compos. Struct.* **106** 570

[5] Diveyev B 2007 Loss factor prediction for laminated *J. Achiev. Mater. Manuf. Eng.* **25** 41

[6] De Visscher J, Sol H, De Vilde W P and Vantomme J 1997 Identification of the damping properties of orthotropic composite materials using mixed numerical experimental method *Appl. Compos. Mater.* **4** 13

[7] Ledi K S, Hamdaoui M, Robin G and Daya E M 2018 An identification method for frequency dependent material properties of viscoelastic sandwich structures *J. Sound Vib.* **428** 13

[8] Wei Sun W, Wang Z, Yan X and Zhu M 2018 Inverse identification of the frequency-dependent mechanical parameters of viscoelastic materials based on the measured FRFs *Mech. Syst. Signal Process.* **98** 816

[9] Wesolowski M and Barkanov E 2012 Model errors influence on identified composite material properties *Compos. Struct.* **94** 2716

[10] Adams R D and Bacon D G C 1973 Measurement of the flexural damping capacity and dynamic Young's modulus of metals and reinforced plastics *J. Phys. D: Appl. Phys.* **6** 27

[11] Maheri M R, Adams R D and Hugon J 2008 Vibration damping in sandwich panels *J. Mater. Sci.* **43** 6604

[12] Kovalovs A and Rucevskis S 2011 Identification of elastic properties of composite plate *IOP Conf. Series: Materials Science and Engineering* **23** 012034, DOI: 10.1088/1757-899X/23/1/012034

[13] Toropov V V, Filatov A A and Polynkin A A 1993 Multiparameter structural optimization using FEM and multipoint explicit approximations *Struc. Optim.* **6** 7

[14] Rikards R and Chate A 1997 Vibration and damping analysis of laminated composite and sandwich shells *Mech. Compos. Mater.* **4** 209

[15] Barkanov E and Gassan J 1994 Frequency response analysis of laminated composite beams *Mech. Compos. Mater.* **30** 484

[16] Janeliukstis R, Rucevskis S, Wesolowski M, Kovalovs A and Chate A 2016 Damage Identification in Polymer Composite Beams Based on Spatial Continuous Wavelet Transform *IOP Conf. Series: Materials Science and Engineering* **111** 012005

Behaviour of Timber-Concrete Joints in Hybrid Members Subjected to Flexure

R Vasiljevs, D Serdjuks, J Gerasimova, K Buka-Vaivade and A Orhun Erüz

Institute of Structural Engineering and Reconstruction, Faculty of Civil Engineering, Riga Technical University, Kipsalas st. 6A/6B, LV-1048, Riga, Latvia

E-mail: karina.buka-vaivade@rtu.lv

Abstract. Possibility to develop rigid timber-concrete joint for hybrid timber-concrete specimens was stated. Behaviour of timber-concrete rigid and compliant joints were evaluated numerically and by experiment for the members subjected to flexure. Small scale hybrid timber-concrete specimens with the length, width and thickness equal to 400, 95 and 43 mm were investigated for case of three points bending. The small-scale hybrid timber-concrete specimens consists from the layers of cement base finishing mass Sacret BAM and timber boards of strength class C24 with thicknesses equal to 25 and 18 mm, correspondingly. The rigid timber-concrete joint was provided by the pieces of crushed granite, which were strengthened on the surface of the timber boards by epoxy glue Sica Dur 330. Dimensions of the crushed granite pieces changes within the limits from 2 to 25 mm. The compliant timber-concrete joint was provided by the screws with diameter and length equal to 4 and 40 mm, correspondingly. The screws were placed under the angles equal to 45 and 90 degrees relatively to the direction of fibres of the timber layers. Load-carrying capacity and maximum vertical displacements were gotten for groups of small-scale hybrid timber-concrete specimens with rigid and compliant timber-concrete joints. It was shown that providing of rigid timber-concrete joint enables to decrease 1.86 - 3.50 times the maximum vertical displacements of hybrid timber-concrete specimens. Load-carrying capacity of the specimens grows by 16.5 – 50.0% at the same moment.

1. Introduction

Combination of timber and concrete becomes enough popular especially during the last years [1-7]. Timber is used in combination with the concrete for producing prefabricated and cast in place load-carrying hybrid composite members subjected to flexure mainly. It can be beams or panels of floors and roofs mainly (figure 1).

Combination of timber and concrete enables to increase load-carrying capacity of structural members, it fire resistance and sound insulation. Load-carrying capacity of structural members grows so as protective or finishing layers of concrete or another cement-based composite are involved into common work with the timber members and take up a part of the applied load. The fire resistance grows so as concrete layer play a role of protective covering in the case [8]. Sound absorption of timber and concrete hybrid members grows so as timber has approximately three-time higher sound absorption coefficient in comparison with the concrete.

Figure 1. Timber-concrete hybrid members subjected to flexure [2].

At the present moments exist two major types of timber to concrete joints. This are compliant and rigid ones. The compliant joints usually are realized by the mechanical fasteners but the rigid ones by the adhesives. Investigations of behaviour of timber to concrete joints indicates, that the hybrid structural members realized by the compliant and rigid timber to concrete joints, are characterized by the comparable load-carrying capacity [4-8]. But hybrid structural members realized by the rigid timber to concrete joints, are characterized by the increased rigidity. But theoretically hybrid structural member realized by the rigid timber to concrete joints, possess a potential for increase of it load-carrying capacity in comparison with the analogous members realized by the compliant joints due to the more rational structural materials use in the first case [4]. More rational materials use and probable increase of the load-carrying capacity is joined with the normal stresses distribution between the joined layers caused by the different compliances of the joints in the both mentioned above cases [4]. Information regarding behaviour of the hybrid timber-concrete panels with rigid timber to concrete joints, which are solved by the adhesives, is enough miserable at the present moment. Comparison of behaviour of the hybrid timber-concrete panels created by the mechanical fasteners and adhesives is another question, which cause a certain interest. So, aim of the current study is to evaluate experimentally possibility to increase load-carrying capacity of the hybrid timber-concrete panels subjected to flexure with the rigid timber to concrete joint in comparison with the analogous members with compliant once. Behaviour of timber to concrete joints in hybrid timber-concrete panels subjected to flexure should be investigated for the purpose.

2. Choice of specimen's parameters

2.1. Choice of type of mechanical fasteners for compliant timber to concrete joint

Several types of mechanical fasteners so as cuttings with the different shapes are widely used for compliant timber to concrete joints. The major variants of compliant timber to concrete joints, realized by the mechanical fasteners, are shown on figure 2 [4].

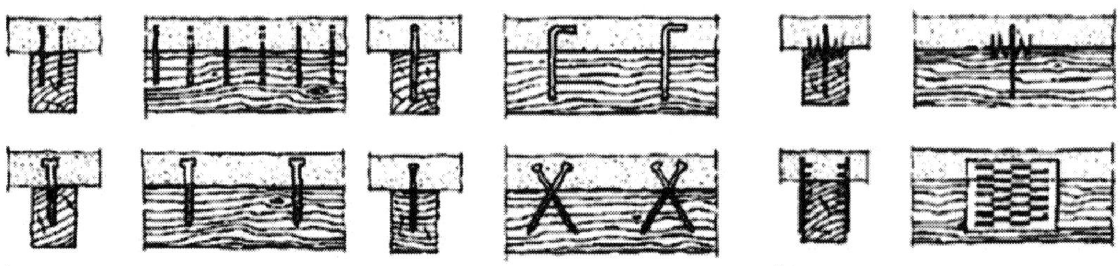

(a) (b)

Figure 2. Timber to concrete joints by the mechanical fasteners: (a) steel bars and screws; (b) tooth steel plates and punched steel plates [4].

So, straight and curved steel bars, screws, placed vertically and under the angles so as tooth steel plates and punched steel plates can be mentioned as the major types of mechanical fasteners, used for compliant timber to concrete joints. The major shapes of cuttings used in combinations with the mechanical fasteners for compliant timber to concrete joints are shown on the figure 3 [4].

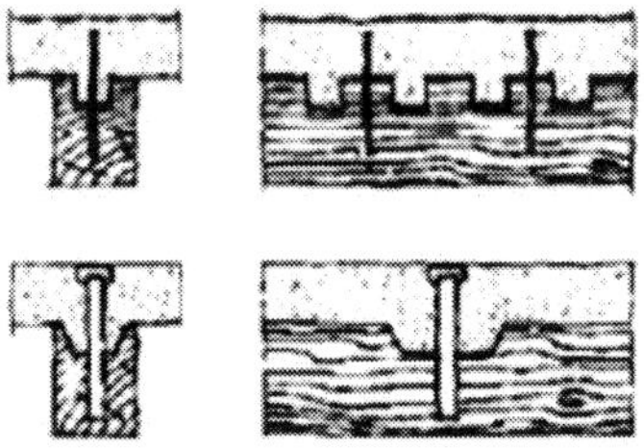
(a)
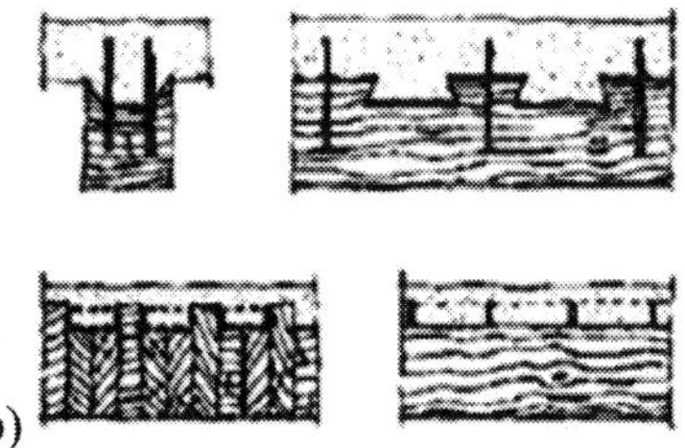
(b)

Figure 3. Timber to concrete joints by the cuttings and mechanical fasteners: (a) round and trapezium cuttings; (b) trapezium cuttings and timber boards joined by the nails with steel plates [4].

So, specimens where compliant timber to concrete joint will be provided by the screws are chosen for the investigations. Three variants of the screw's placement were considered. The screws were placed under the angles 90° and 45 ° relatively to longitudinal axis of the specimens, correspondingly, for the two first variants. Two screws were placed in each point under the angle 45 ° relatively to longitudinal axis of the specimens, for the third variant [9].

So, third variant of the screw's placement is characterized by the two times increased amount of screws in comparison with the two first variants.

Twelve small scale hybrid timber-concrete specimens were prepared with compliant timber to concrete joint provided by the screws with length and diameter equal to 40 and 4 mm, correspondingly. Three sub-groups by the four specimens in the each were prepared with the screw's placement by the variants, which are shown in figure 4. Specimens have the length, width and thickness equal to 400, 95 and 43 mm, correspondingly. The small-scale hybrid timber-concrete specimens consists from the layers of cement base finishing mass Sacret BAM and timber boards of strength class C24 with thicknesses equal to 25 and 18 mm, correspondingly. Thickness of cement base finishing mass layer was equal to 25 mm so as it is a maximum size of the crushed granite pieces used for the specimens with the rigid timber to concrete joint. The cement base finishing mass Sacret BAM has modulus of elasticity and density equal to 30000 MPa and 20 kN/m3, correspondingly. Its strength class is C20 C.

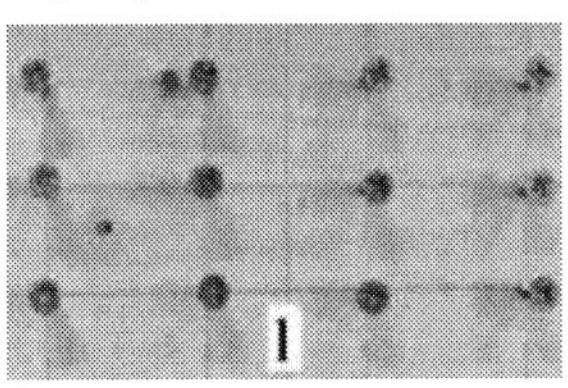

(a)
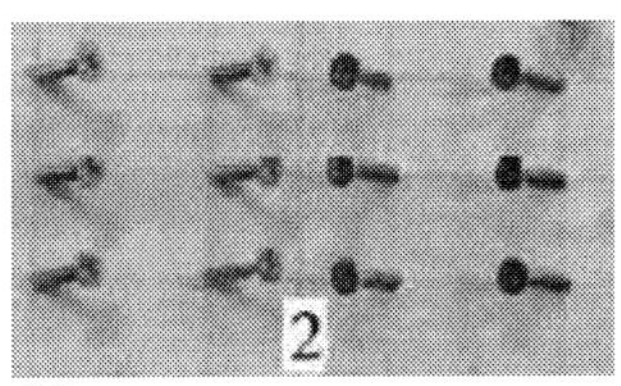

(b)
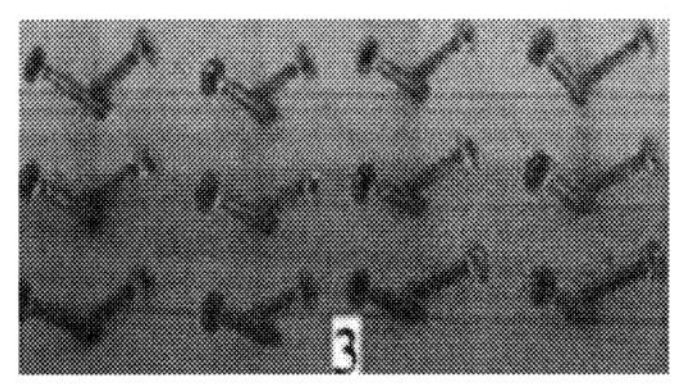

(c)

Figure 4. Considered variants of screws placement: (a) screws were placed under the angle 90° relatively to longitudinal axis of the specimens; (b) screws were placed under the angles 45°; (c) two screws were placed in each point under the angle 45 ° [9].

2.2. Description of specimens with rigid timber to concrete joint

The rigid timber to concrete joint can be provided by the glue and different elements, which can be used to provide additional mechanical bonds between the timber and concrete. Possibility to obtain rigid timber to concrete connection by the glue only was investigated before [10]. Possibility to develop a hybrid timber to concrete member by the gluing together cross-laminated timber (CLT) panel and prefabricated panel from the cement base finishing mass Sacret BAM by the epoxy glue Sica Dur 330 was considered [10]. CLT panel consisting from three layers of C 24 strength class timber with thickness in 20 mm each and total thickness in 60 mm was glued together with the prefabricated panel from the cement base finishing mass Sacret BAM with thickness in 30 mm. Length of the panels was equal to 2 m. The panels were freely supported with the span equal to 1.8 m and loaded by the three-point bending scheme, as it is shown on the figure 5 c). Six such hybrid timber to concrete specimens were prepared and tested for getting it real load-carrying capacities (figure 5). The maximum vertical displacements of the specimens were predicted by the transformed section method as for the members with the rigid timber to concrete joint. The difference between the results obtained by the transformed section method and by the experiment does not exceeds 7 %. It enables to conclude, that rigid timber to concrete joint was obtained. But during the test and after failure of the prepared specimens it was stated, that the contact between the CLT panel and prefabricated panel from the cement base finishing mass was not obtained by all the surface. So, variant, when the bonds between the timber and concrete will be provided by the pieces of crushed granite joined with the surface of timber by the glue and playing the role of keys s is considered in the current investigation. Existence of such keys should increase the effectiveness of the timber to concrete joint and load-carrying capacity of the hybrid member subjected to flexure.

(a) (b) (c) (d)

Figure 5. Preparation and testing of hybrid timber to concrete specimens with the rigid joint provided by the glue: (a) CLT panel with the surface covered by the epoxy glue; (b) prefabricated panel from the cement base finishing mass Sacret BAM in the moulds; (c) hybrid timber to concrete specimen during the testing; (d) look on the side of the hybrid timber to concrete specimens after it collapse [10].

Twelve small scale hybrid timber-concrete specimens were prepared with rigid timber to concrete joint. The rigid timber-concrete joint was provided by the pieces of crushed granite, which were strengthened on the surface of the timber boards by epoxy glue Sica Dur 330. The epoxy glue Sica Dur 330 has modules of elasticity and shear strength equal to 12800 and 15 MPa, correspondingly. Dimensions of the crushed granite pieces changes within the limits from 2 to 25 mm. Three sub-groups by the four specimens in the each were prepared with the different dimensions of the pieces of crushed granite, which are shown in figure 6.

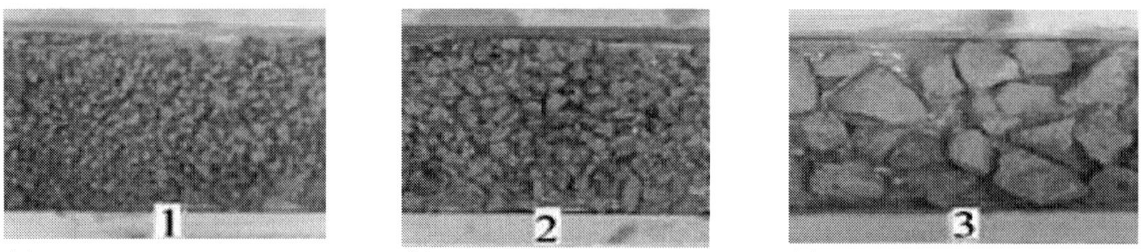

(a) (b) (c)

Figure 6. Considered variants of crushed granite pieces dimensions: (a) 2-5 mm; (b) 5-8 mm; (c) 16-25 mm [9].

The pieces of crushed granite were joined with the surface of the timber by the glue, as it is shown on figure 6. The hardening of the glue was provided in course of two days and then the cement base finishing mass was casted into the moulds. So, the pieces of crushed granite work as a dowel, which are involved into the layer of the cement composite. Geometrical parameters of hybrid timber-concrete specimens with rigid timber to concrete joint are the same as for the specimens with the compliant joints, which were explained above [9].

3. Description of laboratorian experiment

Twenty-four small scale hybrid timber-concrete specimens were tested by the scheme of three point bending at the action of static loading (figure 7). All of the twenty four small scale hybrid timber-concrete specimens of both considered groups have the same dimensions of cross-sections, what is necessary to evaluate behaviour of compliant and rigid timber to concrete joints and to have possibility for comparison of results. Values of the lengths and spans for all the specimens were also the same and equal to 400 and 300 mm, correspondingly (figure 7 (a)). The specimens were tested after twenty-eight days from the specimen's formation.

Intensity of the concentrated force applied in the middle of the specimens span and maximum vertical displacements are two parameters, which were taken under the control during the experiment. The expected load-carrying capacities of the specimens were previously determined by the using of transformed section method [11]. The maximum vertical displacements were measured by two mechanical indicators placed in the middle of the span (figure 7 (a)).

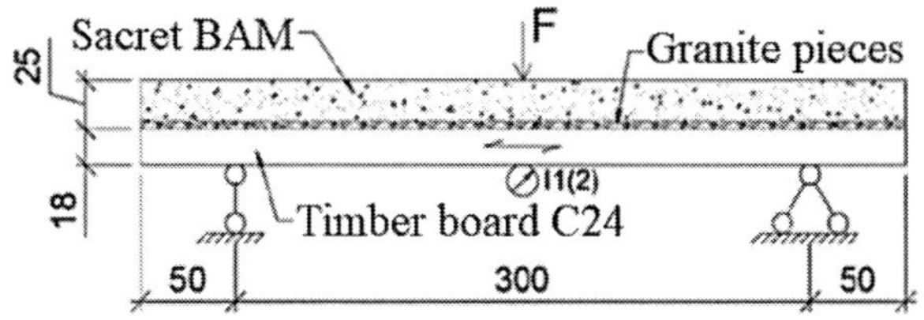

(a)

(b)

Figure 7. Testing of hybrid timber-concrete specimens: a) scheme of the specimens testing; b) failure mode of the specimen with the rigid timber to concrete joint and granite pieces dimensions changing within the limits from 16 to 25 mm [9].

Precision of the maximum vertical displacements measurements was equal to 0.01.mm. All the specimens were loaded till the failure to determine their load-carrying capacities. The vertical load was applied with the steps equal to 2 kN with the speed equal to 2 mm/min [9]. All the specimens were loaded till the failure.

4. Behaviour of timber-concrete joints in hybrid members subjected to flexure

The experiment was carried out for two groups of specimens which are divided into three sub-groups each. Each from the six sub-groups consists from four specimens. The difference between the specimens of the sub-groups for rigid timber to concrete joint is joined with the dimensions of the granite pieces dimensions (figure 6). The difference between the specimens of the sub-groups for compliant timber to concrete joint is joined with the screw's placement (figure 4) [9].

The dependences of the maximum vertical displacements of the hybrid timber-concrete specimens subjected to flexure on the intensity of the vertical applied load were obtained together with the load-carrying capacities for all the specimens. The maximum vertical displacements as a function from the vertical load for the all sub-groups of tested specimens which are differed by the screw's placement and crushed granite pieces dimensions are shown on figure 8. The final point for each graph corresponds to the mean load-carrying capacity of the each sub-group of specimens. Each point on the graphical dependences was obtained as the mean value obtained for all specimens of the sub-group [9].

The values of scatters for the specimens with the compliant concrete to timber joints were observed within the limits from ± 0 to ± 0.07 mm. The scatters for the sub-groups of specimens where the screws were placed under the angles 90° and 45 ° relatively to longitudinal axises, changes within the limits from ± 0 to ± 0.06 mm and from ± 0.03 to ± 0.07 mm, correspondingly. The scatter for the sub-groups of specimens where two screws were placed in each point under the angle 45 ° relatively to longitudinal axis, changes from ± 0.03 to ± 0.06 mm [9].

The values of scatters for the specimens with the rigid concrete to timber joints were observed within the limits from ± 0.01 to ± 0.06 mm. The scatters for the sub-groups of specimens where the dimensions of the pieces of crushed granite, changes within the limits from 2 to 5mm, 5 to 8 and 16 to 25 mm, were within the limits from ± 0.02 to ± 0.03 mm, ± 0.02 to ± 0.06 mm and from ± 0.01 to ± 0.03mm, correspondingly [9].

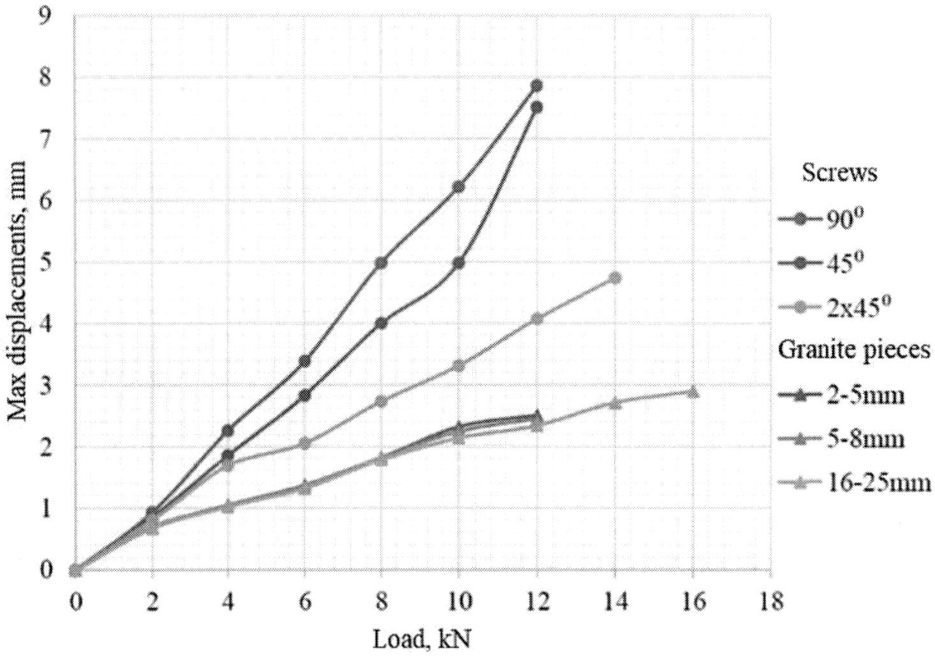

Figure 8. The maximum vertical displacements as a function from the vertical load for the all sub-groups of tested specimens [9].

Comparison of the dependences shown on the figure 8 enables to conclude, that the most rigid specimens are sub-group with the rigid timber to concrete joint with the dimensions of the crushed granite pieces changes within the limits from 16 to 25 mm. The maximum mean vertical displacement for the specimens of this sub-group under the action of vertical concentrated force equal to 12 kN is equal to 2.34 mm. It is 5.13 and 7.26% less, than the vertical displacements of specimen's subgroups, where dimensions of crushed stone pieces changes within the limits from 5 to 8 mm and from 2 to 5 mm, correspondingly. So, influence of the dimensions of the crushed stone pieces on the maximum vertical displacements of the specimens subjected to flexure is insufficient. But it can be stated, that using of the rigid timber to concrete joint instead of compliant ones, enables to decrease maximum vertical displacements of the specimens subjected to flexure from 62.55% to 3.39 times. These results confirm the conclusion regarding increased stiffness of the hybrid timber to concrete members with the rigid timber to concrete joint in comparison with the analogous members, where the joint is compliant [9].

The maximum load-carrying capacities obtained for the all sub-groups of specimens are shown on the figure 9. Each from the six load-carrying capacities was determined as the mean value obtained for the four specimens of the each sub-group. The values of load-carrying capacities for the sub-groups of specimens where the screws were placed under the angles 90° and 45 ° relatively to longitudinal axises, changes from 8.40 to 12.60kN and from 9.40 to 11.80kN, correspondingly. The mean values of load-carrying capacities for the sub-groups are equal to 10.8 and 11kN, correspondingly. The load carrying capacity for the sub-group of specimens where two screws were placed in each point under the angle 45 ° relatively to longitudinal axis, changes from 13.60 to 14.00kN, correspondingly. The mean value of load-carrying capacities for the sub-group was equal to 13.9kN (figure 9) [9].

The values of load-carrying capacities for the sub-groups of specimens where the dimensions of the pieces of crushed granite, changes within the limits from 2 to 5mm, 5 to 8 and 16 to 25 mm, were within the limits from 11.20 to 12.80kN, from 12.00 to 13.40kN and from 15 to 17.8kN, correspondingly. The mean values of load-carrying capacities for the sub-groups are equal to 12.10kN, 12.90 kN and 16.20kN, correspondingly (figure 9) [9].

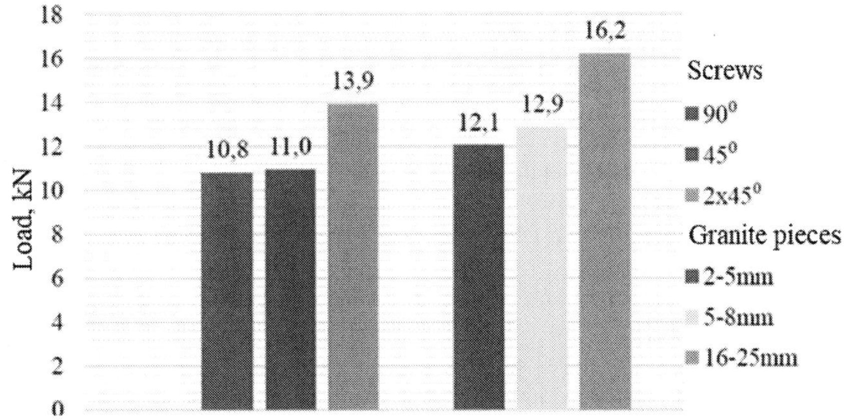

Figure 9. Maximum load-carrying capacities obtained for the all sub-groups of specimens [9].

The obtained results enable to conclude, that using of the rigid timber to concrete joint instead of compliant ones, enables to increase the load-carrying capacity of the hybrid timber to concrete specimens subjected to flexure from 16.55 to 50%. Increase of the crushed granite pieces from 2-5 to 16-25 mm enables to increase load-carrying capacity of considered specimens by 33.88%. So, possibility to increase load-carrying capacity of the hybrid timber-concrete members subjected to flexure with the rigid timber to concrete joint in comparison with the analogous members with compliant once was checked and confirmed experimentally.

5. Conclusions

Possibility to increase load-carrying capacity of the hybrid timber-concrete members subjected to flexure with the rigid timber to concrete joint in comparison with the analogous members with compliant once was checked and confirmed experimentally. It was stated, that using of the rigid timber to concrete joint instead of compliant ones, enables to increase the load-carrying capacity of the hybrid timber to concrete specimens subjected to flexure from 16.55 to 50%. Increase of the crushed granite pieces from 2-5 to 16-25 mm enables to increase load-carrying capacity of considered specimens by 33.88%.

It was stated, that using of the rigid timber to concrete joint instead of compliant ones, enables to increase the rigidity of the hybrid timber to concrete specimens subjected to flexure from 62.55% to 3.39 times. Increase of the crushed granite pieces from 2-5 to 16-25 mm enables to decrease maximum vertical displacements of considered specimens by 7.26 %.

Acknowledgement

The research leading to these results has received the funding from Riga Technical University, Faculty of Building and Civil Engineering Grant "DOK.BIF/18".

References

[1] Goremikins V, Serdjuks D, Buka-Vaivade K, Pakrastins L and Vatin N I 2017 Prediction of behaviour of prestressed suspension bridge with timber deck panels *The Baltic Journal of Road and Bridge Engineering.* **12(4)** 234–240
[2] KLH Massivholz GmbH 2018 *Timber concrete composites* p 12
[3] Al Ali M, Bajzecerova V and Kvocak V 2017 Design methods of timber-concrete composite ceiling structure *Magazine of Civil Engineering* **5(73)** 88–95
[4] Kanocz J and Bajzecerova V 2015 Timber – concrete composite elements with various composite

connections. Part 3: adhesive connection *Wood Research* **60** 939–52
[5] Kanocz J, Bajzecerova V and Steller S 2013 Timber – concrete composite elements with various composite connections. Part 1: screwed connection *Wood Research* **58** 555–570
[6] Miotto J L and Dias A A 2015 Structural efficiency of full-scale timber–concrete composite beams strengthened with fiberglass reinforced polymer *Composite Structures* **128** 145–154
[7] Ceccotti A 2002 Composite concrete-timber structures *Progress in structural engineering and materials* **4** 264–275
[8] Nedryshkin O, Gravit M and Grabovyy K 2018 Modeling fires in structures with an atrium in the FDS field model *MATEC Web of Conferences* **193** 03023
[9] Vasiljev R 2018 *Load-carrying Capacity Analyse for Hybrid Composite Member* (Riga: Riga Technical university) p 114
[10] Gerasimova J 2017 *Testing of design method for hybrid elements subjected to flexure* (Riga: Riga Technical university) p 87
[11] Buka-Vaivade K, Serdjuks D, Goremikins V, Vilguts A and Pakrastins L 2017 Experimental Verification of design procedure for elements from cross-laminated timber *Procedia Engineering* **172** 1212–9

Natural vibration of stepped nanoplate with crack on an elastic foundation

M Hossain and J Lellep

Institute of Mathematics and Statistics, University of Tartu, Tartu, Estonia

E-mail: mainul.hossain@ut.ee

Abstract. The small scale effect on the vibrational characteristic of isotropic, rectangular nanoplate embedded in an elastic medium is investigated. The formulation is based on the plate theory on aggregate with the nonlocal elasticity theory. The solution procedure is derived using the governing differential equations of physical phase that are converted into set of linear algebraic equations. Latter these are solved by computer code. The effects of aspect ratio, step, crack and rotatory inertia on the different modal vibrations of nanoplate are explored. The results show the significant effect of different physical and geometrical parameters on the vibration of nanoplate.

1. Introduction

Nanoplate is a two dimensional structural form of nanomaterial which is commonly used as a component in nano-electromechanical systems [1]. At nanoscale, the laws of physics operate in an unfamiliar way because of two important reasons: high surface-to-volume ratio and atomic attraction. When the size of building blocks gets smaller, the surface area of the material increases by six orders of magnitude [2]. To encounter these effects non local theory of elasticity [3] is very popular among the researchers.

It is obvious, cracks or cracks like defects form on the plate during the processes of manufacturing. The cracks are modelled by a massless rotational springs that connect two neighbouring segments together. On the other hand, steps are the sudden changes in geometry or height of the nanoplate. Sometime designer fabricates stepped plate for using material effectively. In addition, rotatory inertia is the ability to resist the change of rotational speed on a specific axis. The effect of rotatory inertia is small on natural frequency. When the natural frequency is very large, the effect of rotatory inertia is significant. It is reasonable to assume that the elastic foundation is formed by means of identical, independent, closely spaced, discrete and linearly elastic springs. Elastic foundation model is developed by Winkler. According to his assumption, the vertical displacement of a point of the elastic foundation is proportional to the pressure at that point and does not depend on the pressure at the adjacent points.

There are only a few of studies on the vibration of nanoplate on an elastic foundation considering the effect of steps, cracks and rotatory inertia. Whereas, some researchers showed their interest on nanoplate on elastic foundation. Daneshmehr et al.[4], Ansari et al.[5], Sobhy[6] investigated the free vibrations of the functionally graded nanoplate on an elastic foundation. They showed the frequency ratio decreases with the increase in size scale parameter value. They also analyzed elastic foundation stiffness, plate aspect ratio and side-to-thickness ratio on the behavior of nanoplates. In addition, Pouresmaeeli et al. [7], Murmu and Adhikari [8], T Natsuki and J Natsuki [9] analyzed the vibration of double nanoplates on an elastic foundation. They presented analytical scale-based nonlocal approach which is useful for n-nanoplates system of grapheme based nanocomposites. The influences of small scale coefficient, stiffness of the external and internal mediums and aspect ratio on the frequencies of the double nanoplates were also elucidated. Furthermore, Ruocco and Mallardo [10] investigated the buckling and vibration behaviour of imperfect nanoplates via nonlocal Mindlin plate theory. Their numerical results showed that the influence of the imperfection on the buckling load depends on the boundary conditions. Similarly, Ebrahimi et al. [11] presented magneto-electro-elastic heterogeneous porous material plates resting on elastic foundations. They discussed the effect of material graduation exponent, porosity volume fraction, magnetic potential, electric voltage, various boundary conditions, elastic foundation parameters and plate side-to-thickness ratio on natural frequencies.

In this paper, a modified Euler-Bernoulli plate theory and Eringen nonlocal theory of elasticity are used. The main purpose of this paper is to discuss the effect of aspect ratio, steps, cracks and rotatory inertia on the vibration of nanoplate which is placed on an elastic foundation with different support systems. Elastic foundation is considered as Winkler one parameter model. It is also assumed that the nanoplate has stepped cross section and that the plate is weakened with crack located along the line of step. The influence of the crack is modelled as simple elastic spring that connects two adjacent segments of the nanoplate. Most of the authors negated the rotatory inertia of the nanoplate for simplifying their calculation. However, our concern is to measure the effect of rotatory inertia on the vibration of nanoplate. The simply supported and clamped supported nanoplates are considered and analysed in greater detail.

2. Method and Analysis

Consider a linearly elastic, isotropic, stepped plate resting on elastic foundation, undergoing linear vibrations. Euler-Bernoulli modified plate theory and Eringen nonlocal theory are used to derive the governing differential equations of motion. Exact solution techniques are used to solve these governing equations. The aim of the study is to determine the eigenfrequencies of natural vibrations of the nanoplate and to clarify the sensitivity of eigenfrequencies on the geometrical parameters of the plate and the physical parameters of the material.

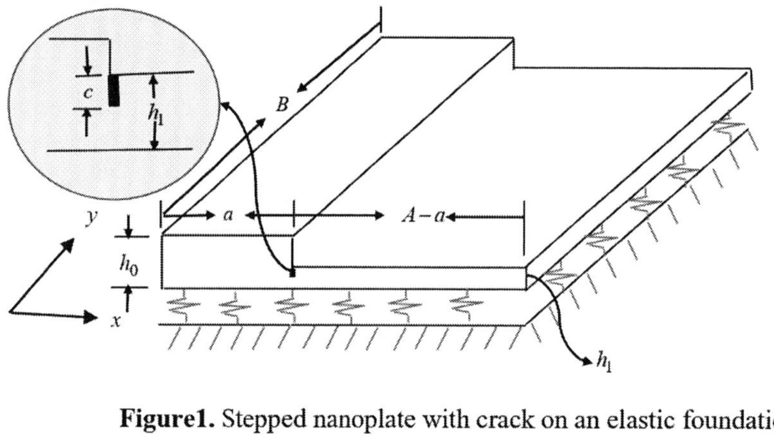

Figure1. Stepped nanoplate with crack on an elastic foundation.

Let us consider a nanoplate of length A along x axis and width B along y axis where the edges of the plate are placed at $x=0, x=A, y=0$ and $y=B$ respectively. Simply supported and fully clamped nanoplates are studied separately in different cases. The coordinate axis $0x$ and $0y$ coincide with the axis of corresponding plate and the origin of coordinates is located at the left-hand end of the plate. It is assumed that the nanoplate has rectangular cross sections with the height

$$h=\begin{cases} h_0, & x\in(0,a) \\ h_1, & x\in(a,A) \end{cases} \tag{1}$$

In Eq. (1) the quantities h_0, h_1 and a are considered as given numbers. The nanoplate is weakened with a crack of length c at $x=a$. The crack length is expected to be constant. Thus the crack area is

$$S_c = cB \tag{2}$$

According to Eringen's nonlocal theory, the stresses in the nanoplate can be written as

$$\left(1-(e_0a)^2\nabla^2\right)\sigma_{ij} = \sigma_{ij}^c \tag{3}$$

where $\sigma_{ij}, \sigma_{ij}^c$ and ∇^2 are the stress tensor of the nonlocal elasticity, the classical local stress tensor and the Laplace operator respectively. The quantity e_0a is a nano length scale where a is an internal characteristic length (e.g. C-C bond length, granular distance, lattice parameter) and e_0 is a calibration constant appropriate to each material. The magnitude of e_0 is determined experimentally using the atomic dispersion curves of the plane. A conservative estimate of the scale coefficient $e_0a < 2.0$ nm for a single-walled carbon nanoelement has been supported by many researchers. The nonlocal constitutive relations (Eq. (3)) for two-dimensional case can be written as:

$$\sigma(X,Y) - (e_0a)^2\nabla^2\sigma(X,Y) = D\delta(X,Y) \tag{4}$$

where $D = \frac{EI}{1-\nu^2}$, E is the Young's modulus and $\delta(X,Y)$ is the local strain. Eq. (4) can be expressed in terms of moment and deflection as below

$$M\left(1-(e_0a)^2\nabla^2\right) = -\left(D\frac{\partial^4 W}{\partial X^4} + 2D\frac{\partial^4 W}{\partial X^2\partial Y^2} + D\frac{\partial^4 W}{\partial Y^4}\right) \tag{5}$$

In the absence of any applied transverse mechanical and thermal loads, dynamic equation of plate can be written as

$$D\frac{\partial^4 W}{\partial X^4} + 2D\frac{\partial^4 W}{\partial X^2\partial Y^2} + D\frac{\partial^4 W}{\partial Y^4} = -I_0\frac{\partial^2 W}{\partial t^2} + I_2\left(\frac{\partial^4 W}{\partial X^2\partial t^2} + \frac{\partial^4 W}{\partial Y^2\partial t^2}\right) - KW \tag{6}$$

where K stands for the elasticity modulus of the foundation. Assume that D is constant. Substituting Eq. (5) into Eq. (6), we obtain the following equations:

$$\left(D\frac{\partial^4 W}{\partial X^4} + 2D\frac{\partial^4 W}{\partial X^2\partial Y^2} + D\frac{\partial^4 W}{\partial Y^4}\right) + \left(1-(e_0a)^2\nabla^2\right)\left[I_0\frac{\partial^2 W}{\partial t^2} - I_2\left(\frac{\partial^4 W}{\partial X^2\partial t^2} + \frac{\partial^4 W}{\partial Y^2\partial t^2}\right) + KW\right] = 0 \tag{7}$$

For free vibration, the transverse displacement is assumed to be of the form $W(X,Y,t) = W(X)Sin\left(\frac{Y}{B}\right)e^{i\omega_c t}$, where ω_c is the angular frequency. We can write (7) as below

$$D\left(\frac{\partial^4 W}{\partial X^4} - 2\frac{1}{B^2}\frac{\partial^2 W}{\partial X^2} + \frac{1}{B^4}W\right) - \omega_c^2 I_0 W + KW + I_2\left(\omega_c^2\frac{\partial^2 W}{\partial X^2} - \omega_c^2\frac{1}{B^2}W\right) + (e_0a)^2\left(\omega_c^2 I_0\frac{\partial^2 W}{\partial X^2} - I_0\omega_c^2\frac{1}{B^2}\right)$$
$$-(e_0a)^2 I_2\left(\omega_c^2\frac{\partial^4 W}{\partial X^4} - 2\omega_c^2\frac{1}{B^2}\frac{\partial^2 W}{\partial X^2} + \omega_c^2\frac{1}{B^4}W\right) - (e_0a)^2 K\left(\frac{\partial^2 W}{\partial X^2} - \frac{1}{B^2}W\right) = 0 \tag{8}$$

The dimensionless variables are defined as follows:

$$x = \frac{X}{A}; \zeta = \frac{A}{B}; \varepsilon = \frac{e_0a}{A}; w = \frac{W}{A}; \omega = \omega_c A^2\sqrt{\frac{\rho h_0}{D}}; I_0 = \rho h_0; I_2 = \frac{1}{12}\rho h_0^3; r_0 = \frac{1}{12}\left(\frac{h_0}{A}\right)^2; k = \frac{KA^4}{D}; \gamma = \frac{h_1}{h_0}; \omega_1 = \gamma\omega \tag{9}$$

Using the dimensionless variable, Eq (8) can be written as below

$$\frac{\partial^4 w}{\partial x^4}\left(1-\varepsilon^2 r_0\omega^2\right) + \frac{\partial^2 w}{\partial x^2}\left(-2\zeta^2 + r_0\omega^2 + \varepsilon^2\omega^2 + 2\zeta^2\varepsilon^2 r_0\omega^2 - \varepsilon^2 k\right) -$$
$$w\left(\omega^2 - \zeta^4 - k + \varepsilon^2\omega^2\zeta^2 + r_0\omega^2\zeta^2 + \varepsilon^2 r_0\omega^2\zeta^4 - \varepsilon^2 k\zeta^2\right) = 0 \tag{10}$$

The solution of Eq. (10) has the exponential form:

$$w(x) = Ae^{i\lambda x} \tag{11}$$

Inserting Eq. (11) into Eq. (10) and dividing through by $Ae^{i\lambda x}$, we obtain the characteristic equation:

$$\lambda^4\left(1-\varepsilon^2 r_0\omega^2\right)+\lambda^2\left(2\zeta^2-r_0\omega^2-\varepsilon^2\omega^2-2\zeta^2\varepsilon^2 r_0\omega^2+\varepsilon^2 k\right) \\ -\left(\omega^2-\zeta^4-k+\varepsilon^2\omega^2\zeta^2+r_0\omega^2\zeta^2+\varepsilon^2 r_0\omega^2\zeta^4-\varepsilon^2 k\zeta^2\right)=0 \tag{12}$$

Let

$$P=1-\varepsilon^2 r_0\omega^2$$
$$Q=2\zeta^2-r_0\omega^2-\varepsilon^2\omega^2-2\zeta^2\varepsilon^2 r_0\omega^2+\varepsilon^2 k$$
$$R=\omega^2-\zeta^4-k+\varepsilon^2\omega^2\zeta^2+r_0\omega^2\zeta^2+\varepsilon^2 r_0\omega^2\zeta^4-\varepsilon^2 k\zeta^2$$

The equation (12) has the solutions:

$$\lambda_{1,2}=\pm i\sqrt{\frac{\sqrt{Q^2+4PR}-Q}{2P}}=\pm i\alpha \qquad \lambda_{3,4}=\pm\sqrt{\frac{\sqrt{Q^2+4PR}-Q}{2P}}=\pm\beta$$

Thus, the solution can be presented for the two different parts of the plate separately

For $0\le x\le a$

$$w(x)=C_1 Sinh(\alpha_0 x)+C_2 Cosh(\alpha_0 x)+C_3 Sin(\beta_0 x)+C_4 Cos(\beta_0 x) \tag{13}$$

For $a\le x\le A$

$$w(x)=C_5 Sinh(\alpha_1 x)+C_6 Cosh(\alpha_1 x)+C_7 Sin(\beta_1 x)+C_8 Cos(\beta_1 x) \tag{14}$$

Boundary conditions for simply supported edges

$$w_i(0)=0, w_i''(0)=0, w_i(A)=0, w_i''(A)=0 \tag{15}$$

Boundary conditions for clamped supported edges

$$w_i(0)=0, w_i'(0)=0, w_i(A)=0, w_i'(A)=0 \tag{16}$$

Intermediate conditions for steps are

$$\begin{gathered} w_i(x_i)=w_{i+1}(x_i) \\ w_i''(x_i)=w_{i+1}''(x_i) \\ w_{i+1}'(x_i)-w_i'(x_i)=K_c w_i''(x_i) \\ w_i'''(x_i)+\psi_i w_i'(x_i)=w_{i+1}'''(x_i)+\psi_{i+1}w_{i+1}'(x_i) \\ \psi_i=\varepsilon^2\omega_i^2 \end{gathered} \tag{17}$$

For crack, the following relations are used

$$K=\frac{72\pi\left(1-v^2\right)}{Ebh_0^2}f(s) \tag{18}$$

$$\text{where } f(s)=\int_0^z zF^2(z)dz$$

$$F(s)=1.93-3.07(s)+14.53(s)^2-25.11(s)^3+25.80(s)^4, \quad s=\frac{c}{h_0}$$

A linear system of algebraic equations is formed considering the solutions of differential equations (13), (14) and the total set of boundary conditions with respect to unknown constants $C_1 ... C_8$. Since it is a linear homogeneous system, the non-trivial solution exists under the condition where the determinant of the system vanishes. This essential condition leads to the system for determination of natural frequencies. This set of linear homogeneous equations is solved by determining the eigenvalues using computer code developed in Maple, various classical boundary conditions are also applied within the Maple code. Thus, the natural frequencies for different physical and geometrical parameters of the plate are evaluated.

3. Results and Discussion

In this section, the practical applicability of the presented method are demonstrated by various tabular and graphical examples. Our results show decent correlation with multiple cracks in nanobeam [12], stepped nanobeam [13] and other available results in the open literature. However, the literature on the investigation of the effects of rotatory inertia and crack on the vibration of nanoplate on an elastic foundation is very limited. The key objective is to study the influences of nonlocal parameter, crack and step location, step height and rotatory inertia on the natural frequency of the nanoplate on an elastic foundation.

Table 1. Natural frequency versus aspect ratio for different nonlocal parameter and frequency mode.

	$\varepsilon=0$			$\varepsilon=1$			$\varepsilon=0$			$\varepsilon=1$		
Mode	1	2	3	1	2	3	1	2	3	1	2	3
ζ	S-S						C-C					
0	9.869	39.476	88.826	2.993	6.205	9.372	22.372	61.675	120.908	6.057	8.896	12.454
0.6	10.229	39.840	89.181	3.052	6.233	9.390	22.574	61.941	121.192	6.036	8.902	12.443
1	10.869	40.476	89.824	3.155	6.285	9.424	22.938	62.424	121.723	6.004	8.916	12.428
2	13.870	43.480	92.827	3.596	6.519	9.583	24.789	64.711	124.189	5.930	8.985	12.397
3	18.868	48.475	97.831	4.233	6.891	9.840	28.298	68.635	128.358	5.990	9.120	12.421

Table 2. Natural frequency versus aspect ratio and height length ratio for different nonlocal parameter and frequency mode.

		$\varepsilon=0.5$			$\varepsilon=1$			$\varepsilon=0.5$			$\varepsilon=1$		
Mode		1	2	3	1	2	3	1	2	3	1	2	3
h/A	ζ	S-S						C-C					
0.1	0	5.278	11.781	17.792	2.981	6.105	9.043	10.842	16.734	22.926	5.964	8.613	11.714
0.2	0	5.215	11.256	16.196	2.945	5.833	8.232	10.428	15.383	19.829	5.710	7.903	10.100
0.3	0	5.114	10.518	14.284	2.888	5.450	7.260	9.833	13.710	16.639	5.349	7.031	8.453
0.1	0.5	5.363	11.821	17.817	3.022	6.124	9.055	10.832	16.745	22.923	5.950	8.618	11.709
0.2	0.5	5.297	11.291	16.214	2.985	5.849	8.240	10.419	15.392	19.826	5.697	7.906	10.097
0.3	0.5	5.192	10.546	14.297	2.925	5.464	7.266	9.825	13.717	16.638	5.338	7.033	8.451
0.1	1	5.611	11.939	17.889	3.140	6.182	9.090	10.808	16.782	22.915	5.913	8.631	11.695
0.2	1	5.537	11.394	16.270	3.099	5.899	8.268	10.398	15.418	19.820	5.664	7.917	10.087
0.3	1	5.421	10.631	14.336	3.033	5.504	7.285	9.806	13.737	16.636	5.312	7.041	8.446

The natural frequencies of stepped nanoplate with a crack on an elastic foundation are Presented in Table 1-2 and Figure 2 -9, where $A=1nm, E=180GPa, v=0.3$ are considered. Table 1 describes the natural frequency in different modes versus aspect ratio, nonlocal parameter and support system. Table 1 shows natural frequency increases with the increase of aspect ratio and decreases with the increase of nonlocal parameter. Table 2 presents the natural frequency versus aspect ratio and height length ratio for different values of nonlocal parameter, frequency mode and support system. Where natural frequency decreases with the increase of height length ratio. The height length ratio or rotatory inertia is more effective in the higher mode of frequency. At the same time, it is less effective in the higher aspect ratio.

In Figure 2, 3 the relationship between the natural frequency and the elastic foundation coefficient is shown for the nanoplate without step and crack in simply supported, clamped supported plate respectively. The natural frequency increases with the increase of elastic foundation coefficient. In Figure 4, 5 illustrate the relationship between natural frequency and step height ratio in different support systems. It is very obvious that natural frequency increases with the increase of step height ratio. The effect of aspect ratio is more effective in simply supported plate than claimed supported plate.

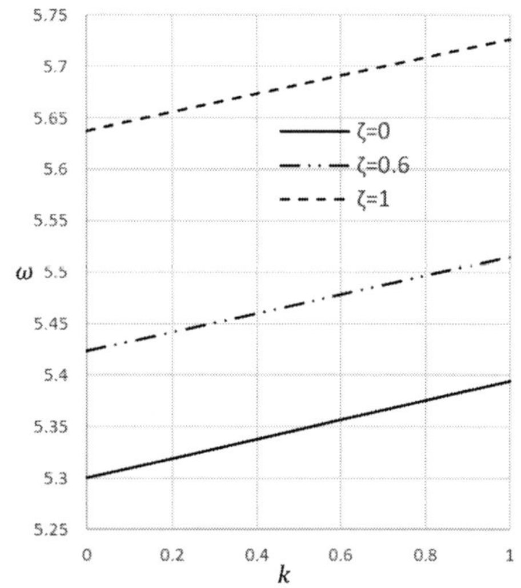

Figure 2. Natural frequency versus elastic foundation coefficient for different aspect ratio (S-S).

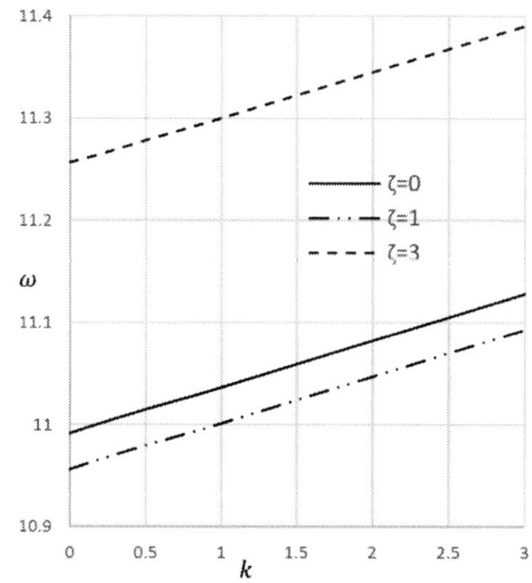

Figure 3. Natural frequency versus elastic foundation coefficient for different aspect ratio (C-C).

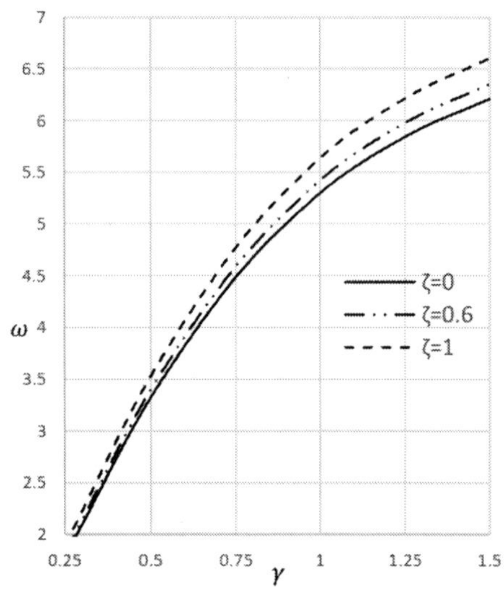

Figure 4. Natural frequency versus step heights ratio for different aspect ratio (S-S).

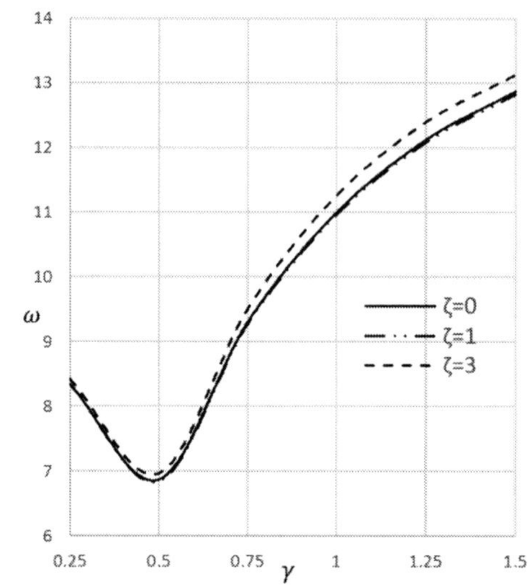

Figure 5. Natural frequency versus step heights ratio for different aspect ratio (C-C).

Figure 6, 7 represent the natural frequency versus crack depth ratio for different aspect ratios and support systems where natural frequency decreases with the increase of crack depth ratio. Finally, in Figure 8, 9 reveal the relationship between natural frequency and step location for different aspect ratio and different support systems. When step location is $a=0$, it represents thinner plate and its natural frequency is lower. On the other hand, when step location is $a=1$, it represents thicker plate and its natural frequency is higher.

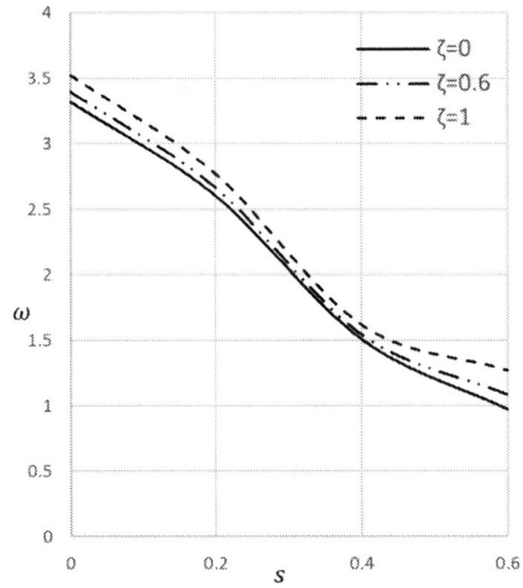

Figure 6. Natural frequency versus crack depth ratio for different aspect ratio (S-S).

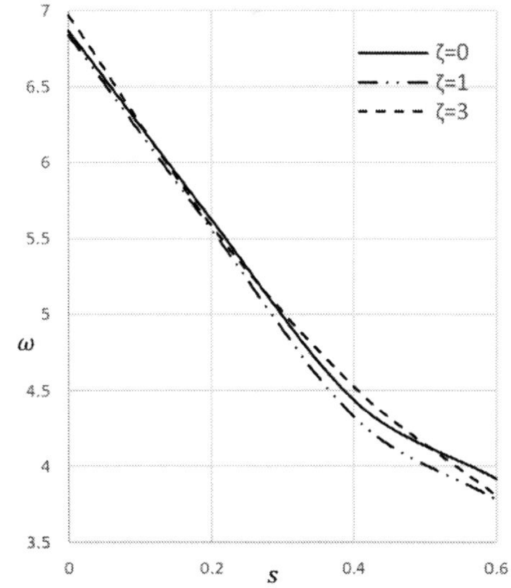

Figure 7. Natural frequency versus crack depth ratio for different aspect ratio (C-C).

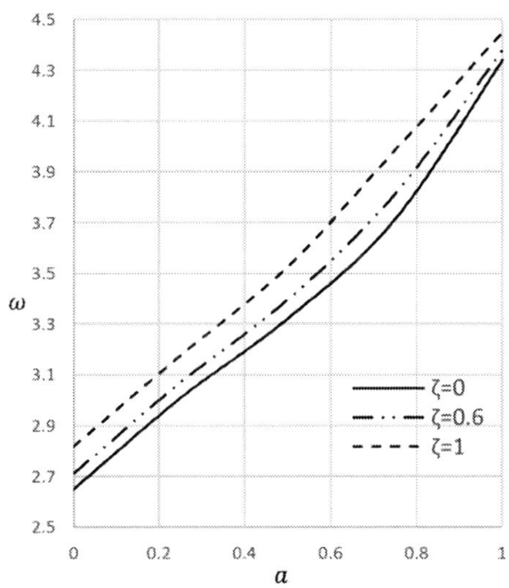

Figure 8. Natural frequency versus step location for different aspect ratio (S-S).

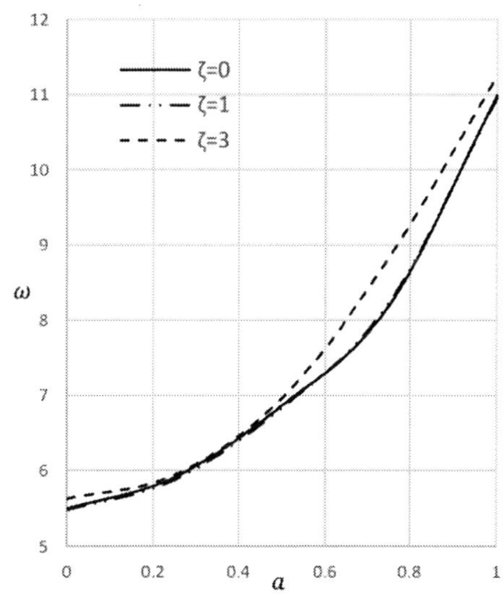

Figure 9. Natural frequency versus step location for different aspect ratio (C-C).

It can be summarized that the effects of different geometrical parameters such as aspect ratio, step, crack and rotatory inertia are noticeable in dynamic nanoplate. Similarly, the different physical parameters such as nonlocal coefficient and spring constant have significant effect on the vibration of nanoplate.

IOP Conf. Series: Materials Science and Engineering **660** (2019) 012051 doi:10.1088/1757-899X/660/1/012051

4. Conclusion

The nonlocal effect on the vibration of isotropic, rectangular, single layer nanoplate embedded on an elastic foundation has been illustrated. The solution technique is based on the classical plate theory combined with nonlocal elasticity theory for encountering the small scale effect. The elastic foundation is considered according to Winkler one parameter model. The governing equations are solved using exact solution technique where matrix of coefficients is formed and resolved with the help of computer code. The effects of physical parameter such as nonlocal parameter, elastic medium parameter and geometrical parameter such as step, crack, rotatory inertia on the vibration of nanoplate have been investigated. It is shown that increasing the aspect ratio, coefficient of elastic foundation and step height ratio, the natural frequency of nanoplate increases. On the other hand, increasing the rotatory inertia and crack depth ratio, the natural frequency of nanoplate decreases. The clamped supported plate is less influenced than simply supported plate due to the effect of different parameters. This solution can be used for designing nanoelectromechanical system effectively.

Acknowledgements

The partial support from the Institutional Research Funding IUT2057 of Estonian Ministry of Education and Research is gratefully acknowledged. Also, the support from Estonian Doctoral School in Mathematics is acknowledged.

References

[1] Tantra R 2016 Nanomaterial Characterization, An Introduction *John Wiley & Sons, Inc.*

[2] Kumar N and Kumbhat S 2016 Essentials in nanoscience and nanotechnology *John Wiley & Sons, Inc.*

[3] Chakraverty S and Behera 2017 Static and Dynamic Problems of Nanobeams and Nanoplates *World Scientific Publishing Co. Pte. Ltd.*

[4] Daneshmehr A, Rajabpoor A and Hadi A 2015 Size dependent free vibration analysis of nanoplates madeof functionally graded materials based on nonlocal elasticity theory with high order theories *International Journal of Engineering Science* **95** 23–35.

[5] Ansari R N, Shahabodini A and Shojaei M F 2016 Nonlocal three-dimensional theory of elasticity with application to free vibration of functionally graded nanoplates on elastic foundations *Physica E* **76** 70–81.

[6] Sobhy M 2015 A comprehensive study on FGM nanoplates embedded in an elastic medium *Composite Structures* **134** 966–980.

[7] Natsuki T and Natsuki J 2018 Transverse impact analysis of double-layered graphene sheets on an elastic foundation *International Journal of Engineering Science* **124** 41–48.

[8] Murmu T and Adhikari S 2011 Nonlocal vibration of bonded double-nanoplate-systems *Composites: Part B* **42** 1901–1911.

[9] Pouresmaeeli S, Fazelzadeh S A and Ghavanloo E 2012 Exact solution for nonlocal vibration of double-orthotropic nanoplates embedded in elastic medium *Composites: Part B* **43** 3384–3390.

[10] Ruocco E and Mallardo V 2019 Buckling and vibration analysis nanoplates with imperfections *Applied Mathematics and Computation* **357** 282–296.

[11] Ebrahimi F, Jafari A and Barati M R 2017 Vibration analysis of magneto-electro-elastic heterogeneous porous material plates resting on elastic foundations *Thin-Walled Structures* **119** 33–46.

[12] Roostai H and Haghpanahi M 2014 vibration of nanobeams of different boundary conditions with multiple cracks based on nonlocal elasticity theory *Appl. Math. Modell.* **38** 1159-1169.

[13] Lellep J and Lenbaum A 2018 Free vibrations of stepped nanobeams *Int. J. Comp. Meth. Exp. Meas.* **6 (4)** 716-715.

Efficient heat-insulating material based on technogenic anhydrite

G I Yakovlev[1], D A Kalabina[1], G N Pervushin[1], R Drochytka[2], K A Bazhenov[1], A F Gordina[1] and J N Ginchitskaya[1]

[1] Kalashnikov Izhevsk State Technical University, 426069 Izhevsk, Russia
[2] Brno University of Technology, 601 90 Brno, Czech Republic

E-mail: 4450539@gmail.com

Abstract. Physical and mechanical properties of the heat-insulating composition based on high-strength anhydrite binder with expanded perlite sand as a lightweight aggregate have been studied. The study has evaluated the influence of two poring components, air-entraining additives and aluminum powder suspension, on the main characteristics of the composition: the compressive strength of the sample, its average density, thermal conductivity, and water absorption. The studies have shown that adding an air-entraining agent in an amount of 1.2% by the weight of dry fluoroanhydrite significantly influenced the decrease in the average density (up to 37%) of the material. The developed composition with a lightweight aggregate based on expanded perlite sand and an air-entraining additive can be used in construction as an efficient heat-insulating material, including for filling cavity walls during frame construction.

1. Introduction

The main task of modern building materials science is to increase the use of recycled materials in the production of building materials to improve the environmental situation in places of industrial dumping and to reduce the cost of the finished product. A cheap replacement of natural gypsum is waste from various industries, such as phosphogypsum [1-4], red gypsum [5, 6], gypsum formed during flue gas desulphurisation [7, 8], and fluoroanhydrite produced during the production of hydrofluoric acid. Despite the environmental friendliness of gypsum, its use in construction remains quite narrow. Most often it is used in the form of internal plasters, gypsum boards, and blocks for partitions. However, the potential use of gypsum is much wider. Gypsum blocks can also be used as external bearing elements with proper hydrophobization [9, 10]. Another promising method of application is the development of porous gypsum compositions, which can be obtained as a foamed or expanded gypsum solution, in which porous structure is formed by chemical additives [11, 12] or using lightweight aggregates [13, 14].

To achieve the set objective, while studying foamed gypsum mixtures [11], aluminum sulphate was used as a gas-forming additive. The average density was 694 kg/m^3, and the compressive strength 0.63 MPa. The expanded gypsum with ammonium bicarbonate had a bulk density of 756 kg/m^3 and a compressive strength of 0.35 MPa.

Study [12] provides information on the development of porous gypsum materials with an average density of 200 to 700 kg/m^3. The composition with a density of 500 kg/m^3 provided a thermal conductivity of 0.16 W/m·K and a compressive strength of 0.7 MPa

Article [15] provides a solution for the production of lightweight gypsum composites based on a high-density matrix with expanded perlite as an ultralight aggregate. The binder was made on the basis of gypsum waste formed during flue gas desulphurisation. A liquid acid agent obtained from aluminum sulfate and dissolved in water citric acid was used as the acid gas-forming component [16]. 20% perlite being added to the mass of gypsum, the average density of the material decreased to 300 kg/m^3, the thermal conductivity coefficient to 0.08 W/mK, while the compressive strength decreased to 0.3 MPa. 5% perlite being added to the mass of gypsum, the average density of the material was 547 kg/m^3 , the strength was 2 MPa and the thermal conductivity was 0.12 W/mK.

The objective of this study was to reduce the average density and thermal conductivity of the fluoroanhydrite composition with expanded perlite sand as a lightweight aggregate to the values that meet the requirements for thermal insulation materials. Earlier studies of fluoroanhydrite compositions with expanded perlite sand content from 45% to 85% by the volume of dry matter showed that it is impossible to reduce the average density without using additional components - when the content of perlite is 85%, the average density decreases only to 660 kg/m^3 while the compressive strength of the sample at the age of 7 days falls to 0.75 MPa. It has been experimentally proved that a further increase in the content of expanded perlite is impractical since the samples are destroyed during demolding. Therefore, in order to further reduce the density of the material, poring agents, aluminum powder and an air entraining agent, were added to the composition.

2. Materials and methods

The produced compositions were based of high-strength anhydrite binder prepared in accordance with work [17], 75% expanded perlite sand being added by the volume of dry matter as an ultra-light aggregate.

For producing a high-strength binder, powdered fluoroanhydrite was used, which is a waste product of hydrofluoric acid produced by "HaloPolymer". To activate the processes of structure formation of fluoroanhydrite, 3% aqueous solution of sodium phosphate Na_3PO_4 was used.

The chemical composition of fluoranhydrite [18] is given in Table 1.

Table 1. Chemical composition of fluoranhydrite, %.

CaO	SO_3	CaF_2	SiO_2	Al_2O_3	Fe_2O_3
35.0-36.5	not less than 45	2.2-5	2.6-3.4	0.5-0.7	0.2-0.95

The conducted dispersion analysis of ground fluoroanhydrite using a SALD-7500nano Shimadzu laser analyzer showed that the average particle diameter is 10.5 μm (Figure 1). It is necessary to note the presence of a nanodispersed component in the composition of fluoroanhydrite with the average particle size of 140 nm.

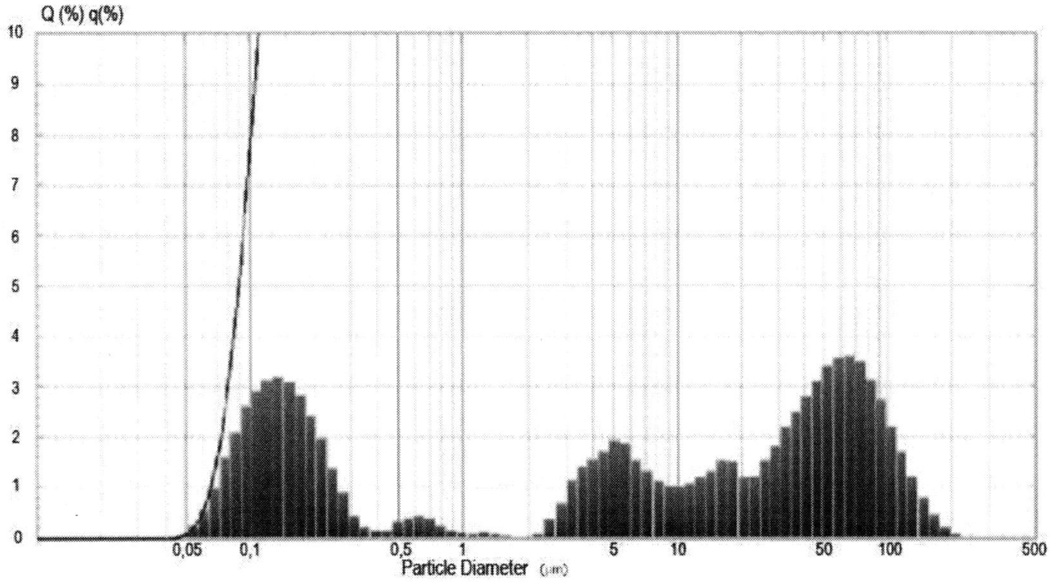

Figure 1. Dispersion analysis of ground fluoroanhydrite.

Expanded perlite sand (GOST 10832-2009) with a bulk density of 98.9 kg/m^3 was used in the experiment. The chemical composition of expanded perlite sand is given in Table 2.

Table 2. Chemical composition of expanded perlite sand, %.

SiO_2	Al_2O_3	K_2O	TiO_2	$CaO + Mg_2O + Fe_2O_3$
73	15	4.7	5.0	2.3

Dispersion analysis (Figure 2) showed that the average particle size of the expanded perlite sand is 65 μm, up to 72% of the particles of the expanded sand have sizes up to 100 μm.

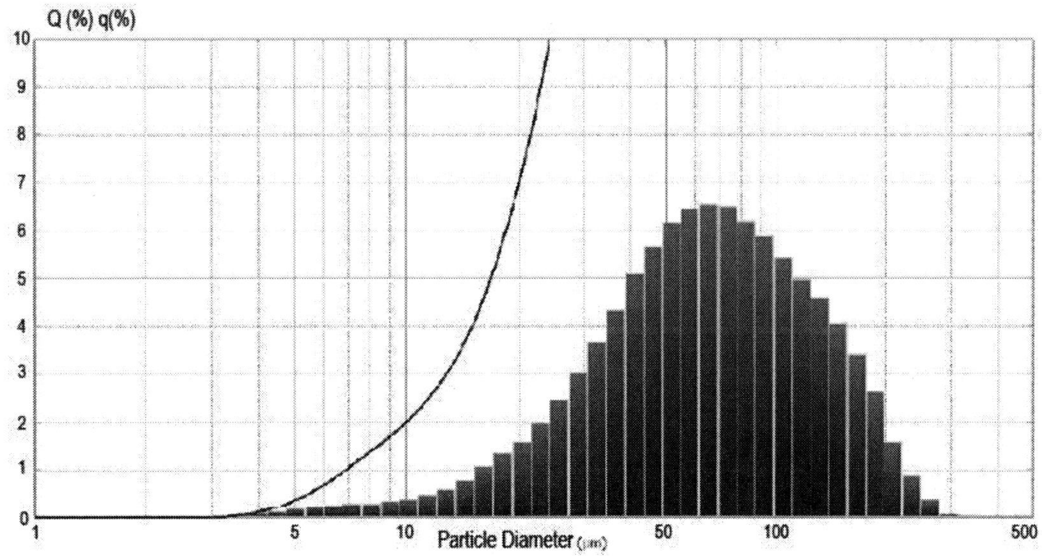

Figure 2. Dispersion analysis of expanded perlite.

Aluminum powder suspension and Poralan STA air-entraining additive produced by the Czech company Stachema were used as poring agents. In order to ensure uniformity of the mixture and uniform distribution of the additive in the composition, the aluminum powder suspension was prepared on the basis of surfactant in a ratio of 1 part of powder to 10 parts of surfactant.

Mass proportions of the main components:

-water 48%

-fluoroanhydrite 41%

-expanded perlite sand 11%

Poring components were added to the composition consisting of high-strength fluoroanhydrite binder and expanded perlite sand. In order to ensure better adhesion, expanded perlite sand was added to a previously prepared solution based on fluoroanhydrite mixed with a 3% aqueous solution of sodium phosphate Na_3PO_4. Such a sequence makes it possible to harden the composition due to the physicochemical interaction between the binder matrix and the surface of the expanded perlite particles, which is confirmed by analyzing the microstructure of the sample with a perlite content of 45% by the volume of dry matter performed on the MIRA3 TESCAN microscope at the AdMaS Technical University of Brno (Figure 3a). Figure 3b shows chemical interaction of the binder at the interphase boundary between perlite and the fluoroanhydrite matrix with the appearance of needle-like new formations [19].

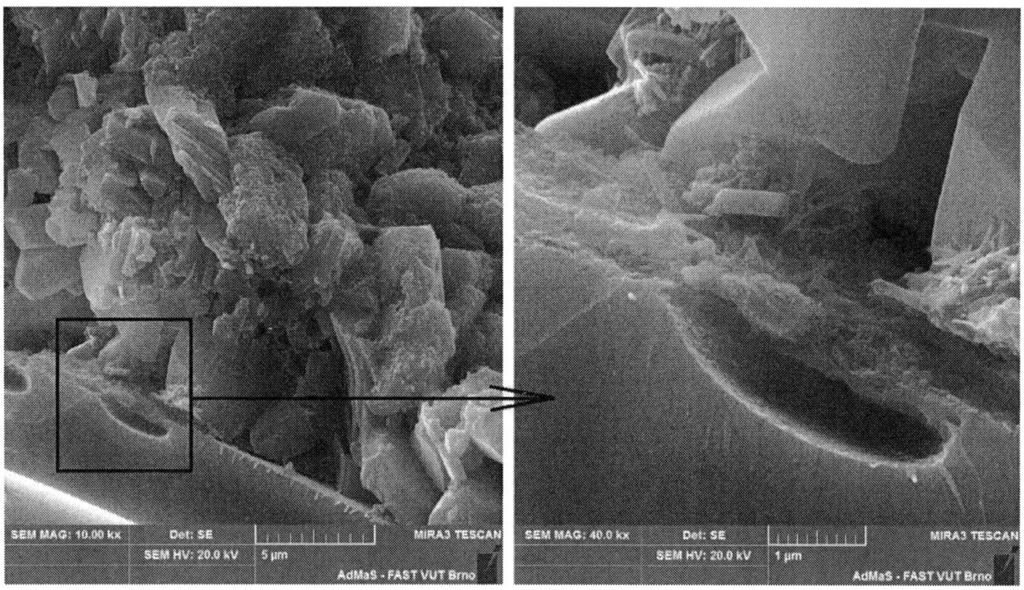

a) b)

Figure 3. a) Microstructure of the interface at 10000-fold magnification b) Physical and chemical interaction of the binder matrix with the surface of the expanded perlite at 40000-fold magnification.

The resulting mixture was put in molds of 40x40x160 mm and hardened in the air. In the process of hardening, the structure of the composition was formed due to the hydration of fluoroanhydrite with the formation of a binding matrix based on calcium sulfate dihydrate. Test samples of the compositions were carried out in accordance with the requirements of GOST 23789-2018.

3. Results and discussions

To study the physical and mechanical properties, the following samples were prepared:

1 - composition with 75% expanded perlite, without additives;

2 - composition with 75% expanded perlite with aluminum powder suspension in an amount of 0.1% by the weight of dry fluoroanhydrite;
3 - composition with 75% expanded perlite with an air entraining additive in the amount of 1.2% by the weight of dry fluoroanhydrite.

The results of physical and mechanical tests are given in Table 3.

Table 3. Physical and mechanical properties of anhydrite composition.

Composition	Flexural strength, MPa		Compressive strength, MPa		Density kg/м³	Thermal conductivity W/mK.	Softening coefficient C_s	Water absorption %
	7 days	28 days	7 days	28 days				
1	0.553	1,417	1.325	2,732	873	0.228	0.42	36.4
2	0.534	*	0.711	*	587	0.131	0.44	82.2
3	0.597	*	0.912	*	546	0.127	0.50	96.5

*Indicators were not determined for technical reasons.

As seen from Table 3, adding a suspension of aluminum powder allowed to reduce the average density of the composition by 33%, the tensile strength of the sample practically being not changed, and the compressive strength of the sample decreased by 46%. Using air-entraining additives allowed to reduce the average density of the composition by 37%, and the compressive strength of the sample at the age of 7 days decreased by 31%, which is not critical for a heat-insulation material.

Analysis of the microstructure of the samples with an electronic USB microscope (Fig. 4) made it possible to establish the presence of large pores (up to 2 mm) in the composition without additives. Aluminum powder being added, large pores are not observed, and there is an increase in the number of small pores (0.2-0.5 mm), which leads to the material shrinkage during the hardening process. This causes microcracks and reduced strength. Air-entraining additive being used, the number of small pores also increased. In this case, the pores are distributed more evenly than with aluminum powder, microcracks are absent.

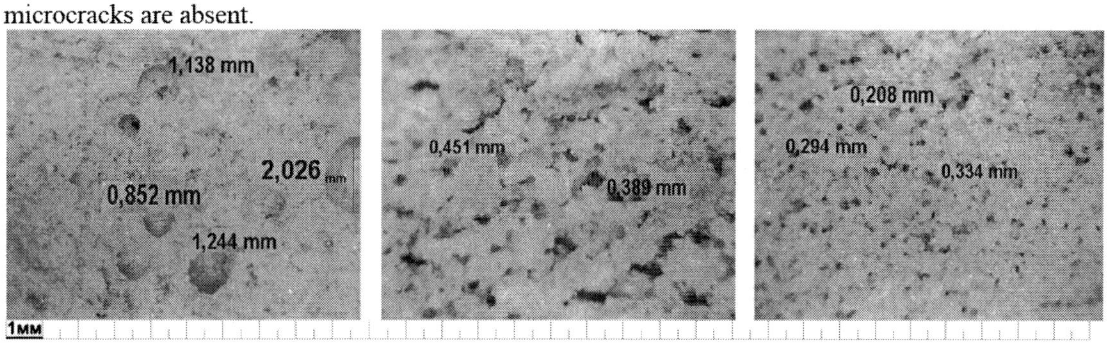

a) b) c)

Figure 4. Microstructure of samples at 50-fold magnification a) the sample without additives; b) with aluminium powder; c) with an air-entraining additive.

Thus, we can conclude that the most effective solution to the problem of reducing the average density of the fluoroanhydrite composition with expanded perlite sand was the use of Poralan STA air-entraining agent. Its use allows to obtain a heat-insulating composition with the following parameters: the average density is 546 kg/m³, the thermal conductivity coefficient is 0.127 W/mK, the compressive strength on the 7th day is 0.912 MPa.

Conclusion

The developed fluoroanhydrite composition with a lightweight aggregate based on expanded perlite sand and Poralan STA air-entraining additive can be used in construction as an effective heat-insulating material, including for filling cavity walls during frame construction. It fully meets the requirements for the values of physico-mechanical parameters imposed on heat-insulating materials and complies with the task of expanding the field of application of industrial wastes in building materials science.

Acknowledgements

The study was carried out within the framework of a state task on the order of the Ministry of Education of the Russian Federation [project No. 16.7823.2017 / 7.8], with financial support from Izhevsk State Technical University named after M.T. Kalashnikov [08.06.01 / 18YGI].

References

[1] Degirmenci, N. Utilization of phosphogypsum as raw and calcined material in manufacturing of building products // *Construction and Building Materials.* 2008. Vol. **22**, pp. 1857–1862

[2] Grinevich A.V., Kiselev A.A., Kuznetsov E.M., Bur'yanov A.F., Ryashko A.I. Plaster knitting from α-Saso4.0.5n2o - withdrawal of production of extraction phosphoric acid. *Stroitel'nye Materialy.* 2014. № 7. pp. 4-8. (In Russian)

[3] Garg M, Jain N, Singh M. Development of alpha plaster from phosphogypsum for cementitious binders // *Construction and Building Materials.* 2009. Vol. **23**(10), pp. 3138-3143

[4] Oreshkin D.V. Environmental problems of complex development of a subsoil at large-scale utilization of technogenic mineral resources and waste in production of construction materials. *Stroitel'nye Materialy.* 2017. № **8**. pp. 55-63. (In Russian)

[5] Hughes P., Glendinning S. Deep dry mix ground improvement of a soft peaty clay using blast furnace slag and red gypsum // *Quarterly Journal of Engineering Geology and Hydrogeology.* 2004. Vol. **37**(3), pp. 205-216

[6] Fauziah I., Zauyah S., Jamal T. Characterization and land application of red gypsum: a waste product from the titanium dioxide industry // *Science of Total Environment.* 1996. Vol. **188**, pp. 243–251

[7] Tesárek P., Drchalová J., Kolísko J., Rovnaníková P., Cˇerny´ R. Flue gas desulfurization gypsum: study of basic mechanical, hydric and thermal properties // *Construction and Building Materials.* 2007. Vol. **21**: pp. 1500–1509

[8] Guo XL, Shi HS. Thermal treatment and utilization of flue gas desulphurization gypsum as an admixture in cement and concrete // *Construction and Building Materials.* 2008. Vol. **22**, pp. 1471–1476

[9] TESÁREK, P.; ROVNANÍKOVÁ, P.; KOLÍSKO, J.; ČERNÝ, R. Properties of hydrophobized FGD gypsum // *Cement-wapno-beton.* 2005. Vol. **5**, pp. 255-264

[10] Gordina A.F., Yakovlev G.I., Polyanskikh I.S., Kerene Ya., Fisher Kh.B., Rakhimova N.R., Bur'yanov A.F. Plaster compositions with complex modifiers of structure. *Stroitel'nye Materialy.* 2016. № **1-2**. pp. 90-95. (In Russian)

[11] Colak A. Density and strength characteristics of foamed gypsum // *Cement & Concrete Composites.* 2000. Vol. **22**, pp. 193–200

[12] Skujans J., Vulans A., Iljins U., Aboltins A. Measurements of heat transfer of multi-layered wall construction with foam gypsum // *Applied Thermal Engineering.* 2007. Vol. **27**, pp. 1219–1224

[13] Sayil B., Gurdal E. The physical properties of polystyrene aggregated gypsum blocks. *In: Lacasse MA, Vainer DJ, editors. Proceedings of 8th international conference on durability of building materials and components (8dbmc).* Canada: Vancouver. 1999. pp. 496–504

[14] Santos AG. PPF-reinforced, ESP-lightened gypsum plaster // *Materiales de Construccion.* 2009. Vol. **59**, pp. 105–124

[15] A. Vimmrová, M. Keppert, L. Svoboda, Ro. Cˇernyˊ. Lightweight gypsum composites: Design strategies for multi-functionality // Cement & Concrete Composites. 2011. Vol. **33**, pp. 84–89
[16] Svoboda L., Vimmrová A. Acid agent for preparation of foamed gypsum // *Czech utility design no. 17844,* Prague; 2007
[17] Yakovlev G.I.,Tulegenova A.V., Pervushin G.N., Keriene J., Gordina A.F., Bazhenov K.A., Ali Elsayed Elrefaei. Multifunctional Admixture Used for Activating Fluoroanhydrite. *20th International Building Materials Conference «Ibausil»*, 12-14 September 2018, Weimar, Germany. Band 2, pp. 559-568
[18] *Ftorangidrit. Sertifikat kachestva* [Fluoroanhydrite. Quality certificate] № 921/2/124398. HaloPolymer Company, Perm, 2017. (in Russian)
[19] G. I. Yakovlev, G. N. Pervushin, V. P. Grakhov, D. A. Kalabina, A. F. Gordina, Yu. N. Ginchitskaya, K. A. Bazhenov, V. V. Troshkova, R. Drokhitka, V. G. Khozin. Constructional and thermal insulation material based on high-strength anhydrite binder. *Intellektual'nye sistemy v proizvodstve.* 2019. T. 17, № **1**, pp. 144-151. (in Russian)

Lamina properties non-destructive characterisation of asymmetric carbon fiber reinforced laminates

E Namsone[1] and D Ermakov[2]

[1]Institute of Materials and Structures, Riga Technical University, Kalku str. 1, LV-1658, Riga, Latvia
[2]Department of Mechanics of Composite Materials and Structures, Perm National Research Polytechnic University, Perm, Russia

E-mail: endija.namsone@inbox.lv

Abstract. A non-destructive technique – numerical-experimental method, based on vibration tests is applied for the characterization of lamina properties of asymmetric carbon fiber reinforced laminates. The main idea of this procedure is the determination of simple mathematical models (response surfaces), using the finite element analysis at the reference points of experimental design. Therefore, a significant reduction of computational efforts can be attained. For the purpose of characterization, two asymmetric carbon fiber laminated specimens were prepared, using the prepreg manual layout technology. To evaluate the elastic material properties obtained, the numerical resonant frequency, determinated by using identified material properties, were successfully compared with experimentally measured and high accuracy with average relative error less than 1% was reached.

1. Introduction

Carbon fiber reinforced plastics are widely used now in different lightweight constructions requiring high stiffness-to-weight and strength-to-weight ratios. Unfortunately, the material data provided by manufacturers very often do not contain all necessary information for the analyses [1], design [2] and optimization [3-5] of advanced composite structures. For this reason, the technical data of such materials are estimated in the research laboratories by using conventional fracture methods [6], ultrasonics [7] or inverse technique based on low-frequency vibrations [8-10]. Very often the conventional fracture and ultrasonic methods do not give a desired accuracy of the obtained results or do not allow to find all the required parameters of composite laminates with arbitrary lamina angles. In this case mixed numerical-experimental technique could be successfully applied.

The general idea of the inverse technique based on vibration tests is minimization of the error functional between experimental and numerical parameters of structural responses. [11-13] This approach is especially useful for a characterization of laminated composites since it has a non-destructive character that enables many repeated tests on the same specimen, and provides more flexibility and less restriction in a choice of the test specimen geometry avoiding by this way a damage introduction in the test specimen. Moreover, an influence of experimental and numerical errors on the identified material properties have been intensively investigated [14-16] that allows to speak about high accuracy and reliability of the mixed numerical-experimental technique based on vibration tests.

In most cases, an identification of material properties from vibration tests is carried out using unidirectional or symmetric laminates. The present investigations demonstrate that asymmetric carbon fiber reinforced laminates could be used also for an accurate characterization of lamina mechanical properties by using the inverse technique based on low-frequency vibrations.

2. Inverse technique

The inverse technique in the present work is realised by using vibration tests and includes physical experiments, numerical modelling and identification procedure, consisting of planning of experiments, response surface methodology and minimization of error functional (Figure 1). First of all, the plan of experiments is created in dependence on the number of experiments and number of identified parameters. Then the finite element code is used for the modelling of reinforced composite laminates. To obtain dynamic parameters, the finite element analysis is applied in the reference points of experimental design. In the next step, the acquired numerical data are used for the determination of approximation functions, using the response surface method. The resonant frequencies are determinated from the experimental data of vibration tests. In the final stage, the required parameters of material are identified by minimizing the error functional that describes the residuals between the experimental and numerical data.

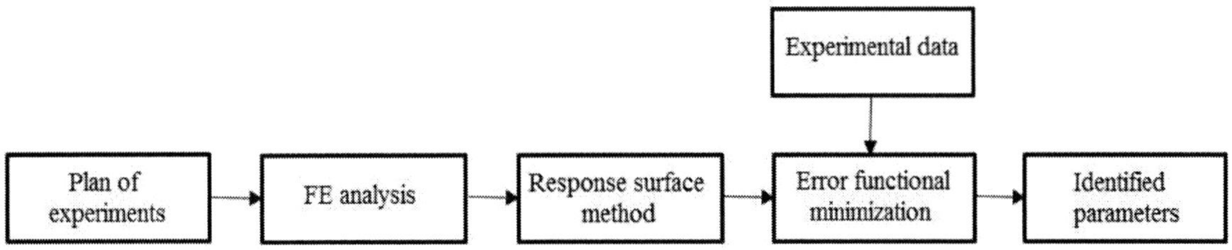

Figure 1. The inverse technique.

2.1. Vibration tests

The vibration testing is fast and low-cost method for an identification of elastic properties. To study the behavior of specimens, one of the common used procedure - experimental modal analysis - is applied. During this analysis, consequent modal parameters, like modal frequency and shape of mode, are determinated. The frequency response functions of the system are provided by using the known forces and appropriate output vibrational responses. Usually, the forces are applied at a number of locations and input signals are measured in many points.

2.2. Numerical modelling

In the present inverse procedure, the finite element method is used for the modelling and dynamic analysis of laminated composite specimens. The finite element code ANSYS 19.2 is applied for modelling the composite specimens using the first order shear deformation theory. In this case the displacements can be expressed:

$$u = u_0 + z\gamma_x \quad , \quad v = v_0 + z\gamma_y \quad , \quad w = w_0 \tag{1}$$

where u_0, v_0, w_0 are the displacements in the reference plane, z is the coordinate of the point of interest in the reference plane, and γ_x, γ_y are the rotations according to the transverse shear deformations. Then the following eigenvalue problem is solved:

$$\mathbf{K}u = \omega^2 \mathbf{M}u, \tag{2}$$

where $\mathbf{K}$ is the stiffness matrix, $\mathbf{M}$ is the mass matrix, u is the displacement vector and ω is the resonant frequency. The eigenvalue problem was solved by the Lanczos method.

2.3. Material identification procedure

The material identification procedure is based on vibration tests and response surface methodology. The main idea of this procedure is the determination of simple mathematical models (response surfaces) using only the finite element analysis in the reference points of experimental design. Minimizing the error functional, that describes the residuals between the experimentally and the numerically calculated frequencies, the identified parameters are obtained.

2.3.1. Planning of experiments

There are different methods for the planning of experiments, for example, D-Optimal, Central Composite, Factorial, Uniform Design, Latin Hypercube, Orthogonal Array, Minimal MSD, etc. In the present work, the points of experiments are distributed as uniformly as possible and there is a physical relevance to the minimum of potential energy of repulsive forces for a set of points of unit mass. For this reason, the consequent criterion is used:

$$\Phi = \sum_{i=1}^{k} \sum_{j=i+1}^{k} \frac{1}{l_{ij}^2} \Rightarrow \min \tag{3}$$

where l_{ij} is the distance between the points with numbers i and j ($i \neq j$). The matrix B_{ij} describes the plan of experiment, where the domain of factors and design space is determined as $x_j \in [x_j^{\min}, x_j^{\max}]$. The points of experiments are calculated by a consequent expression:

$$x_j^{(i)} = x_j^{\min} + \frac{1}{k-1}\left(x_j^{\max} - x_j^{\min}\right)\cdot\left(B_{ij} - 1\right), \tag{4}$$

where $i = 1,2,...,k$, $j = 1,2,..n$. The basic information for the development of the plan is the number of experiments k and number of the design parameters n. It is possible to determine a plan of experiment for each k and each n, but in that case, it enlarges computational efforts. Thereby, each plan of experiment is determined only once and can be applied for different design situations. The minimal number of experiments is obtained as follows:

$$k = (n+1)\cdot(n+2) \tag{5}$$

2.3.2. Response Surface Technique

For the approximation of the numerical results, the response surface methodology is used. There are various methods of the approximation that can be applied for the development of the plan of experiments – Polynomial, Rational Functions, Kriging, Splines, Neural Networks etc. Approximations can be completed by the second order polynomials using the least square method:

$$\overline{F}(x) = \beta_0 + \sum_{i=1}^{K} \beta_i x_i + \sum_{i=1}^{K} \sum_{j=i}^{K} \beta_{ij} x_i x_j \tag{6}$$

The error of approximation is determined using the following equations:

$$\sigma = \sqrt{\frac{\sum_{i=1}^{N}\left(F(x^i) - \bar{F}(x^i)\right)^2}{N}}\,; \quad \sigma_0 = \sqrt{\frac{\sum_{i=1}^{N}\left(F(x^i) - \bar{F}(x^i)\right)^2}{N-L}}\,, \tag{7}$$

where σ is the squared error, σ_0 is the standard deviation, $F(x^i)$ and $\bar{F}(x^i)$ are the values of original and approximating functions in the sample points x^i of experiments, and N is the number of sampling points.

2.3.3. *Minimization of error functional*

In order to identify the required elastic material properties, the error functional, that describes the residuals between the numerical and experimental parameters of structural responses, should be formulated:

$$\Phi(x) = \sum_{i=1}^{I}\left(\frac{\left(f_i^{EXP}\right)^2 - \left(f_i^{FEM}(x)\right)^2}{\left(f_i^{EXP}\right)^2}\right)^2 \Rightarrow \min \tag{8}$$

where f_i^{EXP} are experimentally measured frequencies; f_i^{FEM} are resonant frequencies of regression equations (determined by the response surface method using numerical calculations); I is the number of resonant frequencies used in the identification method. A constrained nonlinear optimization problem for the present minimization should be solved:

$$\begin{aligned} &\min \Phi(x), H_i(x) \geq 0, G_j(x) = 0 \\ &i = 1,2,\ldots,I,\; j = 1,2,\ldots,J, \end{aligned} \tag{9}$$

where I and J are the numbers of equality and inequality constraints. This minimization problem is replaced by an unconstrained problem; therefore, the constraints are taken into account with the penalty functions. In order to solve the optimization problem, a random search method is used.

3. Samples preparation

While solving the inverse problem of determining the elastic characteristics, multilayer asymmetric carbon fiber samples were manufactured. The specimens consist of four anisotropic layers with the following reinforcement schemes [0°/90°/0°/90°] and [-45°/45°/-45°/45°]. Selected thicknesses and reinforcement schemes are applied in the manufacture of power shells in many aircraft structures. Dimensions for both specimens are given in Table 1.

Table 1. Dimensions of composite specimens.

Dimension parameters of specimens	Specimen 1 [0°/90°/0°/90°]	Specimen 2 [-45°/45°/-45°/45°]
Mass m, g	22.9	25.4
Width b, mm	44.9	45.0
Length L, mm	400.0	400.0
Thickness t, mm	0.8	0.9
Density ρ, kg/m^3	1593.5	1422.6

For manufacturing samples, the prepreg manual layout technology was chosen. The carbon fabric uniformly impregnated with a polymer binder was used. The prepreg technology allows to obtain complex shape monolithic products with minimal tooling. The technology of prepreg molding has the following characteristics: low complexity of laminate laying, clean process, high quality composite material, increased requirements for tooling and support materials (high polymerization temperature). The essence of the method is the following: cutting sheets on a single-sided matrix of the required shape to obtain the required thickness, vacuuming under the polyethylene film, curing in an autoclave at high temperature and pressure. CM-PregF-T27 200/1250 CP0041 45 prepreg carbon fiber was used as the samples material.

The process of samples manufacturing can be divided into the following steps:

1) Prepreg patterns marking;
2) Cutting of the prepreg on the plotter (Figure 2 (a));
3) Laying the cut prepreg on a one-sided matrix of the required shape to obtain the required thickness in accordance with the reinforcement scheme;
4) Strengthening the prepreg over the entire area of adhesion (Figure 2 (c));
5) Laying of auxiliary layers on the prepreg surface: perforated film and absorbent layer (Figure 2 (d));
6) Sealing the mold with a vacuum film and providing a vacuum (Figure 2 (b), (e));
7) Placing the laid prepreg into the autoclave and setting the required curing parameters (Figure 2 (f));
8) Removing the cured laminate from the autoclave and removing the vacuum film after the curing process is complete (Figure 2 (g), (h));
9) Mechanical processing of the cured laminate to obtain the samples of the required size.

(a)
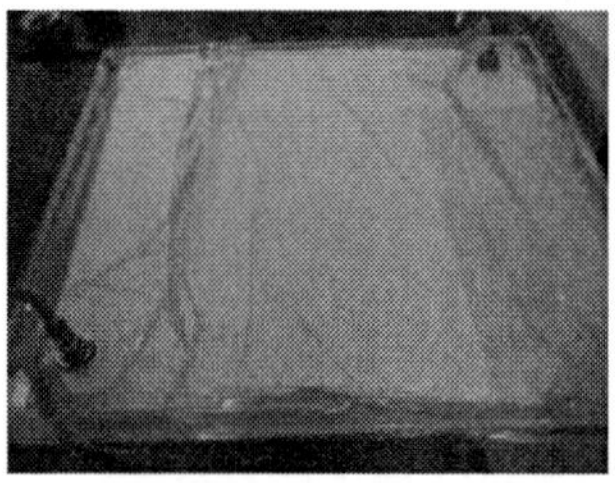
(b)

(c)
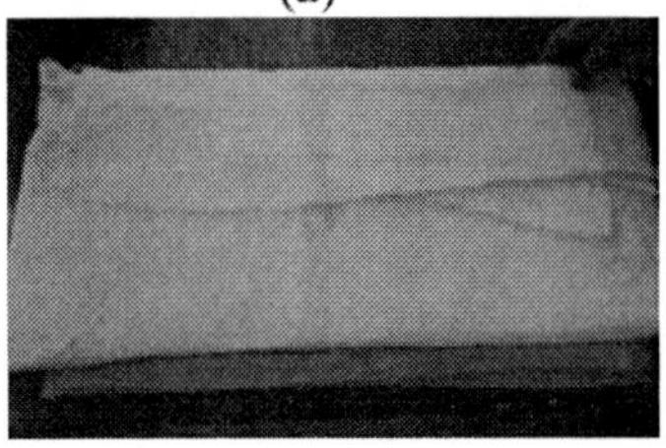
(d)

(e)

(f)

(g)

(h)

Figure 2. Stages of carbon fiber prepreg samples manufacturing by the method of manual layout.

According to the results of the realized technological experiments, the production modes were selected for the method of prepreg manual layout with autoclave molding and sample cutting, which made it possible to manufacture defect-free samples for further testing.

4. Identification of material properties

Two samples of asymmetric carbon fiber reinforced laminates are tested for the identification of following elastic constants:

- two Young's moduli: E_1, $E_2 = E_3$;
- shear moduli: $G_{12} = G_{13}$;
- Poisson's ratio: $\nu_{12} = \nu_{13}$.

It should be noted that not all of the elastic constants are sensitive to frequencies, so some of them are assumed as fixed values [11]:

- shear modulus: $G_{23} = 3.5$ GPa.
- Poisson's ratio: $\nu_{23} = 0.3$.

For the vibration tests, the experimental set-up, based on Scanning Laser Vibrometer POLYTEC, is used for the samples with free-free boundary conditions. For this boundary condition, samples are hanged on thin threads, fixed on the top of samples (Figure 3).

The measurements are done at specified scanning points. The signal analyser converts the average values of input and output signals into the frequency domain, using the Fourier transformation. From the response frequency function the structural resonant frequencies and shapes of mode can be easily obtained. After that, the resonant frequencies are calculated by the function transformation to the modal analysis program. During the physical experiments, 9 first resonant frequencies are measured for the specimen 1 and 12 first resonant frequencies – for the specimen 2.

The plan of experiments for the composite specimens is formulated for 4 design parameters and 30 experiments. The limits for the maximum and the minimum values of elastic parameters are given in Table 2.

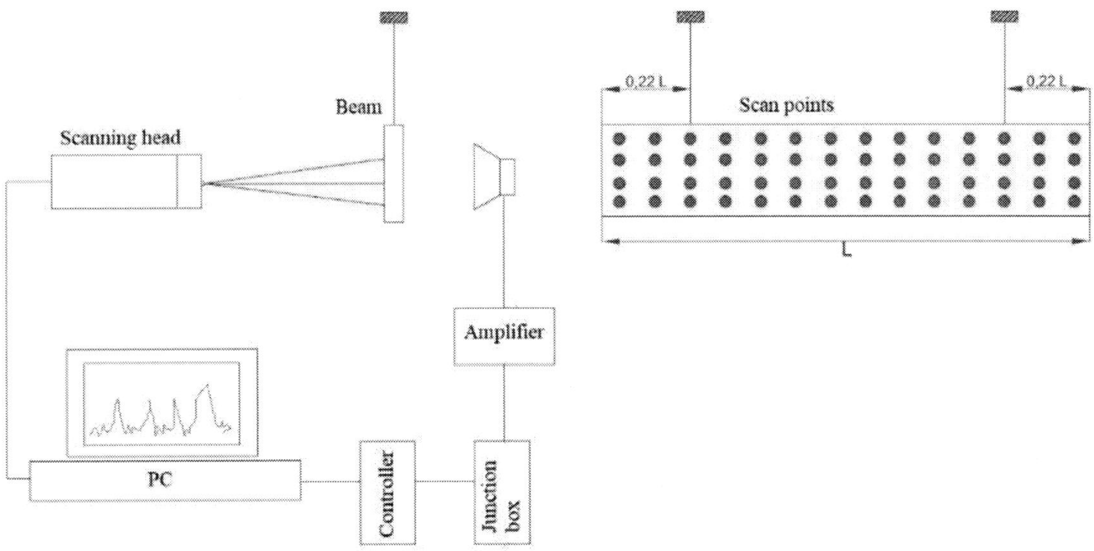

Figure 3. Experimental set-up.

The finite element model of samples is built by linear structural shell elements SHELL 181. The element SHELL 181 is a four-node element and it is suitable for the analysis of thin shells. The finite element has six degrees of freedom in each node – 3 displacements in x, y and z directions and 3 rotations about x, y and z axes. In order to achieve the necessary accuracy for the 16 first frequencies of specimen

1, 125x14 regular finite element mesh is developed, but for the 13 first frequencies of specimen 2 – 107x12 regular mesh.

Table 2. Boundaries of identified parameters for composite specimens.

Parameters of identification	Specimen 1		Specimen 2	
	Minimum value	Maximum value	Minimum value	Maximum value
Young's modulus E_1, GPa	110	145	75	105
Young's modulus $E_2 = E_3$, GPa	12	14.5	5	8
Shear modulus $G_{12} = G_{13}$, GPa	4.5	6	2.5	4
Poisson's ratio $\nu_{12} = \nu_{13}$	0.2	0.4	0.2	0.4

For the approximation and optimization of results, the program EDAOPT was used. As a result of approximation procedure, the equations of regressions were obtained for each resonant frequency. To determine the elastic constants of both specimens, the error functional was minimized. The identified elastic constants for both specimens are presented in Table 3.

Table 3. Identified elastic constants of composite specimens.

Dimension parameters of specimens	Specimen 1	Specimen 2
Young's modulus E_1, GPa	137.0	91.8
Young's modulus $E_2 = E_3$, GPa	13.6	5.75
Shear modulus $G_{12} = G_{13}$, GPa	5.3	3.2
Shear modulus G_{23}, GPa – *fixed value*	3.5	3.5
Poisson's ratio $\nu_{12} = \nu_{13}$	0.33	0.34
Poisson's ratio ν_{23} – *fixed value*	0.30	0.30

To verify the results, the experimentally measured frequencies were compared to the numerical resonant frequencies, obtained by using the identified material properties. The accuracy of eigenvalues was determinated by calculating the relative errors:

$$\Delta_i = \frac{\left| f_i^{FEM} - f_i^{EXP} \right|}{f_i^{EXP}} \cdot 100\%, \tag{10}$$

The obtained results are presented in Table 4 and it is seen that for both specimens the average relative errors are less than 1%.

5. Conclusions

An inverse technique, based on vibration tests and response surface methodology, was successfully applied for an identification of the elastic material properties of asymmetric carbon fiber reinforced composite specimens. The main advantage of the identification procedure, based on non-direct optimization methodology, was the reduction of computational efforts.

Table 4. Frequencies and relative errors for composite specimens.

	Specimen 1			**Specimen 2**		
Mode	f_i^{EXP} (Hz)	f_i^{FEM} (Hz)	Δ_i (%)	f_i^{EXP} (Hz)	f_i^{FEM} (Hz)	Δ_i (%)
1	33.250	33.146	0.31	18.125	18.149	0.13
2	81.875	82.024	0.18	52.250	52.099	0.29
3	91.750	91.383	0.40	106.875	106.717	0.15
4	-	169.878	-	183.750	183.795	0.02
5	179.375	179.191	0.10	204.500	204.483	0.01
6	-	268.810	-	284.000	284.354	0.12
7	296.625	296.308	0.11	407.875	408.764	0.22
8	-	383.332	-	414.000	410.411	0.87
9	443.000	442.815	0.04	556.375	557.063	0.12
10	-	517.288	-	617.500	619.223	0.28
11	619.250	618.673	0.08	727.500	729.118	0.22
12	824.250	824.298	0.01	-	792.879	-
13	-	855.398	-	829.625	832.352	0.33
14	1058.375	1059.459	0.10	-	-	-
15	-	1063.677	-	-	-	-
16	-	1299.833	-	-	-	-
AVERAGE	-	-	0.13	-	-	0.23

For the purpose of identification, two asymmetric carbon fiber samples were prepared. To manufacture defect-free samples, the method of prepreg manual layout with autoclave molding and sample cutting, was applied.

The results, obtained during the identification procedure and the dynamic analysis, were sufficiently accurate for both samples as the values of average relative errors were less than 1%. It is necessary to note that out-of-plate material properties could be identified also using the present methodology. However, in this case the thickness of tested samples should be increased (not less than 4 mm).

Acknowledgements
This work has been partly supported by the Government Perm Krai, research project No. C-26/797.

References

[1] Barkanov E and Gassan J 1994 Frequency Response Analysis of Laminated Composite Beams *Mech Compos Mater* **30** 664-674
[2] Hufenbach W, Kroll L, Holste C, Tager O and Barkanov E 2001 Design of Dynamically Loaded Fiber-Reinforced Structures Considering Their Vibro-Acoustic Behaviour *Mech Compos Mater* **37** 145-152
[3] Barkanov E, Ozoliņš O, Eglītis E, Almeida F, Bowering M.C. and Watson G 2014 Optimal

Design of Composite Lateral Wing Upper Covers. Part I: Linear Buckling Analysis *Aerosp Sci Technol* **38** 1-8
[4] Barkanov E, Eglītis E, Almeida F, Bowering M.C. and Watson G 2016 Optimal Design of Lateral Wing Upper Covers. Part II: Nonlinear Buckling Analysis *Aerosp Sci Technol* **51** 87-95
[5] Barkanov E, Eglītis E, Almeida F, Bowering M.C. and Watson G 2016 Weight Optimal Design of Lateral Wing Upper Covers Made of Composite Materials *Eng Optimiz* **48** 1618-1637
[6] Hearmon R.F.S 1961 *Introduction to Applied Anisotropic Elasticity* (Oxford: Oxford University Press)
[7] Filipczynski L, Pawlowski Z, Wehr J 1966 *Ultrasonic Method of Testing Materials* (London: Butterwoths)
[8] Frederiksen P.S. 1997 Experimental Procedure and Results for the Identification of Elastic Constants of Thick Orthotropic Plates *J Compos Mater* **31** 360-382
[9] Rikards R and Chate A 1998 Identification of Elastic properties of Composites by Method of Planning of Experiments *Compos Struct* **42** 257-263
[10] Araujo A.L, Mota Soares C.M, Moreira de Freitas M.J, Pedersen P and Herskovits J 2000 Combined Numerical-Experimental Model for the Identification of Mechanical Properties of Laminated Structures *Compos Struct* **50** 363-372
[11] Rikards R, Abramovich H, Green T, Auzins J and Chate A 2003 Identification of Elastic Properties of Composite Laminates *Mech Adv Mater Struc.* **10** 335-352
[12] Barkanov E, Chate A, Ručevskis S and Skukis E 2007 Characterisation of Composite Material Properties by an Inverse Technique *Key Eng Mater* **345-346** 1319-1322
[13] Barkanov E, Wesolowski M, Akishin P and Mihovski M 2018 *Techniques for Non-Destructive Material Properties Characterisation.* In:Non-Destructive Testing and Repair of Pipelines (Springer International Publishing) pp 191-207
[14] Wesolowski M and Barkanov E 2012 Model Errors Influence on Identified Composite Material Properties *Compos Struc* **94** 2716-2723
[15] Barkanov E, Wesolowski M, Hufenbach W and Dannemann M 2015 An Effectivness Improvement of the Inverse Technique Based on Vibration Tests *Compos Struc.* **146** 152-162
[16] Wesolowski M and Barkanov E 2016 Air Damping Influence on Dynamic Parameters of Laminated Composite Plates *Measurement* **85** 239-248

Numerical Comparison of HPFRC and HPC Ribbed Slabs

K Buka-Vaivade[1], J Sliseris[1], D Serdjuks[1], G Sahmenko[2] and L Pakrastins[1]

[1]Institute of Structural Engineering and Reconstruction, Faculty of Civil Engineering, Riga Technical University, Kipsalas st. 6A/6B, LV-1048, Riga, Latvia
[2]Institute of Materials and Structures, Faculty of Civil Engineering, Riga Technical University, Kipsalas st. 6A/6B, LV-1048, Riga, Latvia

E-mail: karina.buka-vaivade@rtu.lv

Abstract. The problem of limited resources is actual nowadays. One of the ways for solving this problem can be the use of high-performance fiber reinforced concrete as a structural material for elements subjected to the flexure. The dispersive reinforcement improves the properties of the concrete and allows reducing the required dimensions of cross-section and ratio of longitudinal reinforcement for members subjected to flexure. This research includes the numerical comparison of ultimate limit state and serviceability limit state of ribbed slabs from high-performance concrete (HPC) and high-performance fiber reinforced concrete (HPFRC) behavior, considering the different deformative properties of these materials. The effective use of these materials has been studied using a nonlinear finite element model. This model is created by using a beam type element with a multilayer cross-section, where each layer has a non-linear stress-strain relationship, according to the layer materials such as HPC, HPFRC, or steel bars. The finite element model used for the limited plates is validated. Results of experiment showed, that used calculation model describes with sufficient accuracy the behaviour of the ribbed slabs. The difference between design and experimentally determined breaking load is less than 6% for ribbed slab. The use of HPFRC for ribbed slabs with the spans from 6 to 12 m is justified so as using of HPFRC instead of HPC enables to increase 42-46% the intensity uniformly distributed load for the ribbed slabs with the same cross-sections and allowed deflections.

1. Introduction

As the amount of people in the world is growing, the problem of availability of non-renewable resources is actual. It is important to use resources efficiently. High-performance concrete (HPC) allows to design thin-walled load-bearing structures, thus reducing the self-weighing of structure and saving resources.

High-performance concrete is brittleness material, which is dangerous characteristic for construction material. The use of fibers insignificantly affects the strength of the concrete peak, but it reduces the fragility of high-performance concrete and significantly improve the deformative properties of the concrete in post-peak stage by fiber bridging effect [1-6].

Probably, cross-section dimensions and amount of additional longitudinal reinforcement for elements subjected to the flexure can be decreased by adding fibers in high-performance concrete composition. Thus, reducing cost of labor and delivery of additional longitudinal reinforcement. Thereby, the aim of this study is numerical comparison of behavior of high-performance fibre reinforced concrete (HPFRC) and HPC ribbed slabs to determine possibility of decreasing of slabs cross-section and ratio of longitudinal reinforcement. The obtained results can be used for the initial design stage to select the used material.

2. Methods

2.1. Object of investigation
The elements subjected to the flexure are selected with the configuration of the ribbed slabs as it is shown in figure 1, with the longitudinal reinforcement bars embedded in the ribs of the slab. This configuration makes it possible to effectively integrate different solutions, such as communications and phase-change materials into the slab structure without increasing the overall height of the structure and maintaining the height of the story.

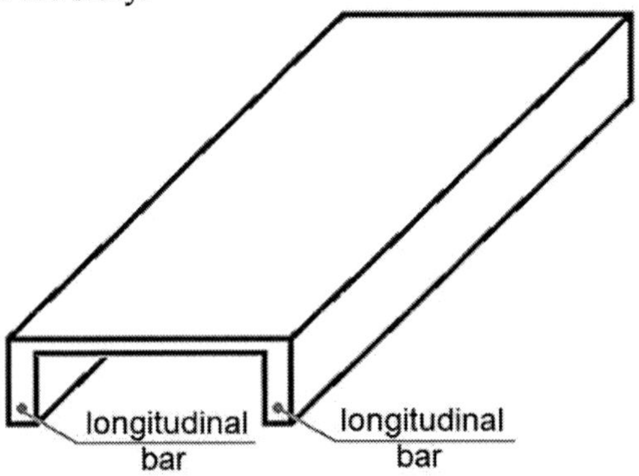

Figure 1. Configuration of ribbed slab.

For comparison of behavior of HPC and HPFRC with steel fiber dosage 25-35 kg/m3, high strength concrete of C80/95 class [7] and high-strength steel of longitudinal reinforcement with yield strength of 1100 MPa are used. Used stress-strain curves of the HPC and HPFRC are shown in figure 2.

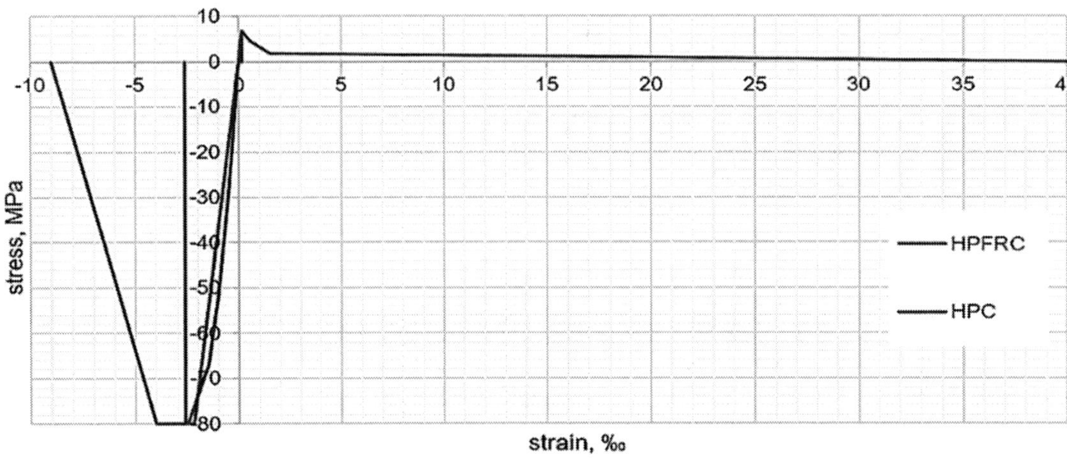

Figure 2. Stress-strain curves of HPC and HPFRC.

The free supported ribbed slabs with width equal to 1 m and span from 6 to 12 meters are analyzed by using uniformly distributed permanent load and imposed load with characteristic value 5 kN/m2.

2.2. Design model
The developed numerical model of the slab allows appreciating the improvement of deformative properties of concrete from the use of fibers by defining the stress-strain curve of the respective

material. The developed non-linear design module based on the MATLAB has been used for creating slab calculation models. The structure is modelled by multi-layer beam-type finite elements, where for each of the layers is assigned its own non-linear stress-strain curve. This module allows designing a structure from HPC and HPFRC with an additional longitudinal reinforcement as it is shown in figure 3.

Figure 3. Cross-sections of calculation model (1 - high strength concrete with or without dispersed reinforcement; 2 - longitudinal reinforcement bars from steel).

The non-linear behavior of HPC and HPFRC materials is characterized by degradation of modulus of elasticity cause of cracking by special damage parameter of the material [8]. The mechanical approach of continuous damage is used for each layer. The non-linear mathematical problem is addressed using the modified Newton-Raphson approach, where a stiffness matrix is recalculated in each load step.

3. Model validation

The developed numerical model which considered non-linear behavior of high-performance fiber reinforced concrete material is validated by the experimental data for free supported specimen of ribbed slab with span 1.23 m in tree-point bending (figure 4 *a)*). The geometric dimensions of the slab prototype cross-section are shown in figure 4 *b)*. Each rib has B500 steel longitudinal bar with a diameter of 6 mm. The bars are located 10 mm from the outer edge. The protective layer of the left bar is 7 mm from the bottom and of the right bar - 11.5 mm.

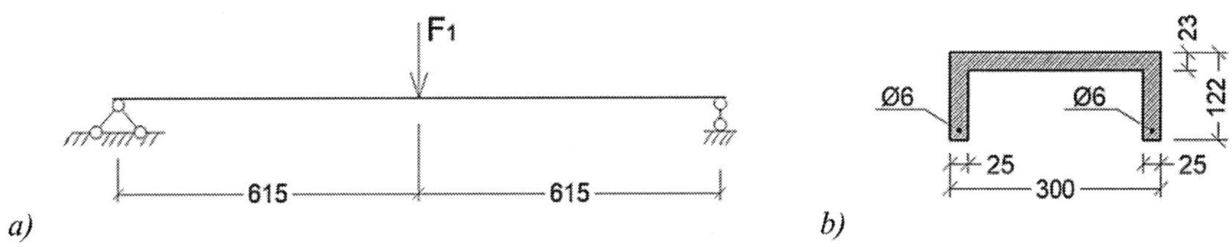

Figure 4. Ribbed slab specimen *a)* design scheme; *b)* geometrical dimensions of cross-section.

Ribbed slab from high-performance synthetic macro fiber reinforced concrete is considered as object of study. Polypropylene (PP) Strux 90/40 fibers were used. The amount of fiber is 0.5% of the concrete volume. For 40 mm long fibers with an aspect ratio of 90, the tensile strength is 620 MPa, the modulus of elasticity - 9.5 GPa and the elongation - 10%.

Compressive strength of cube specimen with dimensions 10x10x10 cm from high-performance synthetic macro fibers reinforced concrete, according to compressed cube tests, is 92.5 MPa. Consequently, $f_{cm} = 78\,MPa$ has been accepted for the compressive strength of the high-performance concrete, 41 GPa for the modulus of elasticity and 4.6 MPa for tensile strength.

A 10 x 10 x 10 cm cube (figure 5) has been modelled to determine the stress-strain curve of the high-performance synthetic macro fibers reinforced concrete, defining separately parameters of high-performance concrete and PP Strux 90/40. A model with 20 mm long fibers and twice the amount of fibers - 1612 fibers is selected because of the low stiffness of synthetic fibers at its addition time and, because fibers in the concrete structure do not retain their original straight shape.

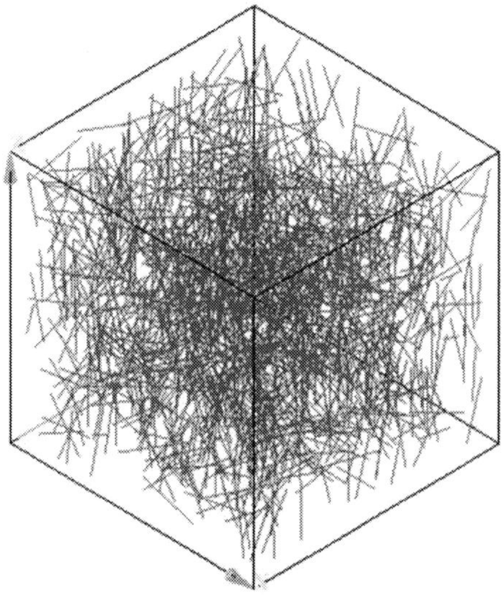

Figure 5. Random placement of fibers (taking into account the effect of form walls) in the cube with 20 mm long 1612 polypropylene fibers.

Tensile stress-strain curve for high-performance PP Strux 90/40 fibers reinforced concrete obtained by the modelling of cube sample in tension is shown in figure 6.

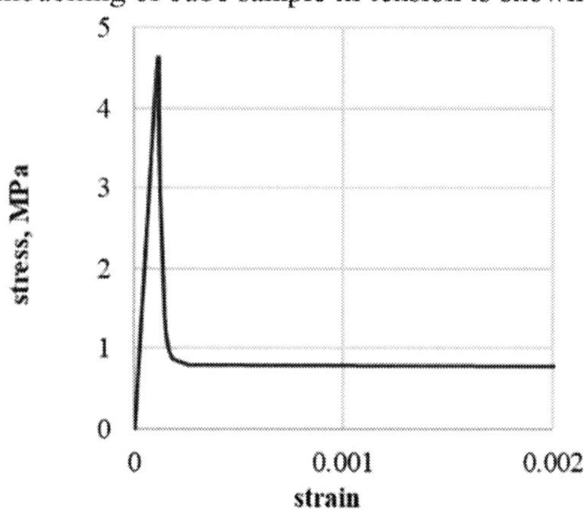

Figure 6. Tensile stress-strain curve for high-performance PP Strux 90/40 fibers reinforced concrete.

Calculation criterion of the slab: is a 100-fold reduction of the modulus of elasticity, corresponding to the damage factor [8] $D = 1 - E/E_0 = 0.99$. The diagram of the slab specimen damage factor $(1 - E/E_0)$ in the three-point bending is shown in figure 7.

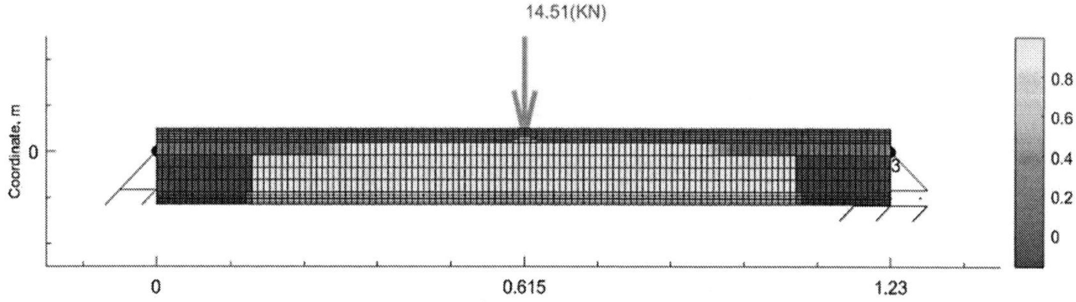

Figure 7. Scheme of the slab specimen damage in three-point bending.

The hydraulic loading device and the displacement measuring devices used in ribbed slab specimen three-point bend test were connected to the computer. Measurements from these devices were fixed automatically by computer. Experimental load-displacement curve of ribbed slab specimen in three-point bending was obtained. Process of the laboratory testing is shown in figure 8.

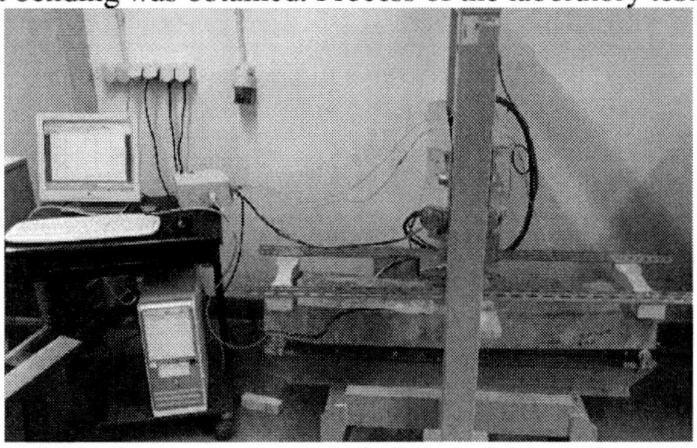

Figure 8. Laboratory testing of the ribbed slab specimen in three-point bending.

Load-displacement curves of ribbed slab specimen in three-point loading, according to laboratory testing and numerical model calculations, are summarized in figure 9.

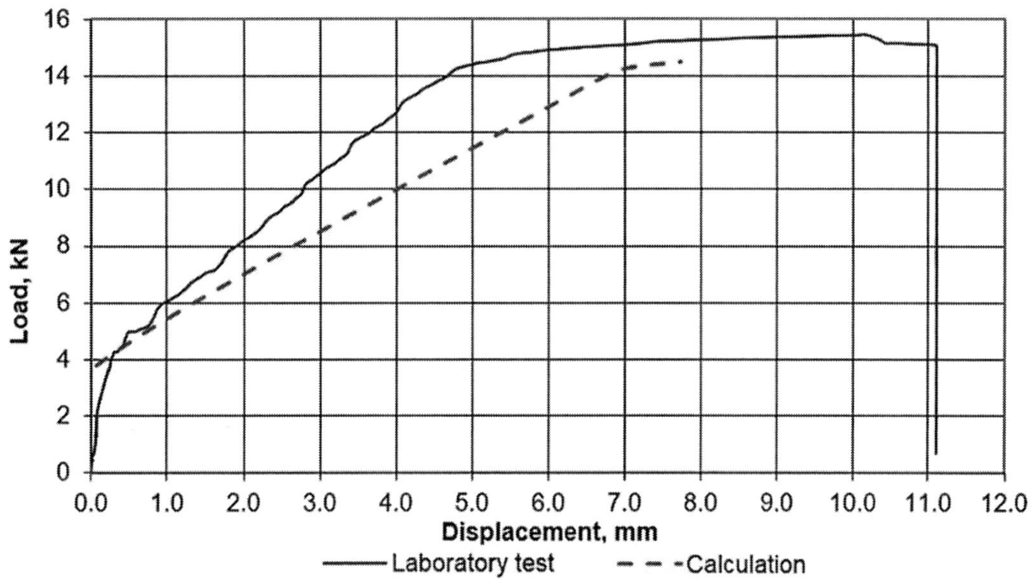

Figure 9. Load-displacement curves for ribbed slab specimen.

As a result of the non-linear calculation, by modelling the ribbed panel with the multi-layered final elements, the destructive load of the panel was 14.51 kN, at the moment when the experimental destructive load for the ribbed panel was 15.42 kN in the three-point bending test. Consequently, the created calculation model has shown a precision of more than 94% with regard to the destructive load.

4. Results and discussion

The maximum bending moment value in the middle of the ribbed slab shelf as a function from the thickness of the shelf for the high-performance concrete and high-performance fiber reinforced concrete and comparison of required shelf thickness at concentrated load F=20 kN in the middle of the shelf span is summarized in table 1.

Table 1. The dependence of the maximum bending moment value in the middle of the slab shelf.

Material	Calculation equation for bending moment M	Required shelf thickness t (mm)
HSC	$M = 0.001\ t^2 - 0.007\ t + 0.0618$	55
HPFRC	$M = 0.0013\ t^2 + 0.0104\ t - 0.3669$	45

The calculations show that the shelf of the ribbed slab from high-performance fiber reinforced concrete can take up a bending moment value of 38-41% higher than the shelf from the high-performance concrete with the same thickness. For example, taking into account the technological limitations of manufacturing and assuming a step of 5 mm of the thickness of the slab shelf, the thickness of the shelf, which will provide a load-bearing capacity of F = 20 kN at the slab width of 1 m, which causes about 2.5 kNm bending moment, is 45 mm for HPFRC, while for the HPC it is 55 mm, which is by 22% higher.

According to the obtained results from the calculations of the maximum moment in the middle of the free supported ribbed slabs with width equal to 1 m and uniformly distributed permanent load and imposed load with characteristic value 5 kN/m2 for slabs with various heights and bars diameters of the ribs and spans 6, 8, 10 and 12 meters, differences, between the load bearing capacities of the slabs of HPC and HPFRC with additional longitudinal reinforcement, are small. The load bearing capacities of the HPFRC slabs are 5-10% higher than that of HPC slabs.

The calculations of the serviceability limit state are the determining factor for ribbed slabs as structures subjected to the flexure. The maximum bending of the slab in the middle of the span is determined by the slab self-weight and distributed imposed load. A 1/250 part of the slab span has been accepted as a maximum available value of the vertical displacements.

The curves for the HPC and HPFRC ribbed slabs with 4 different cross-sections, describing the value of distributed load q, at which vertical displacements in the middle of the slab span is equal with L/250 for slabs with span from 6 to 12 meters, is summarized in figure 10.

As it can be seen from figure 10, the slabs from high-performance fiber reinforced concrete can take up distributed load, which is 42-46% higher, than that load for the slabs made of high-performance concrete with the same cross-section parameters.

As a result, it can be concluded, that the use of HPFRC can reduce the material consumption of the high-performance concrete and longitudinal reinforcement. For example, comparison of the material consumption of ribbed slab with span 10 m, which needs cross-section of HPC H340-D22 or HPFRC H320-D16 (according to the figure 10) and taking into account that width of the rib of the slab is rounded up by 10 mm and is equal to sum of 3 diameters of longitudinal bar and 20 mm, is summarized in table 2.

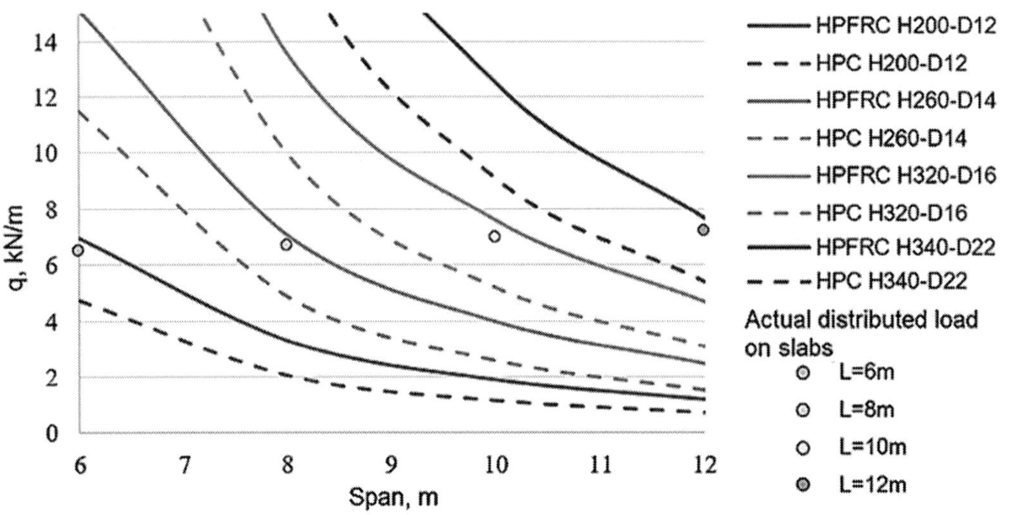

Figure 10. Uniformly distributed load, which satisfy the serviceability of the slabs on their spans for different slab height and reinforcement bars diameters for HPC and HPFRC. HPFRC – high-performance fiber reinforced concrete; HPC – high-performance concrete, H - slab height, D – bar diameter.

Table 2. The comparison of material consumption of the 1 m wide ribbed slab with span 10 m.

Material	Shelf thickness, mm	Width of the rib, mm	Slab height, mm	Diameter of bar, mm	Area of concrete, mm^2	Δ, %	Area or bars, mm^2	Δ, %
HSC	55	90	340	22	106 300	-	760.27	-
HPFRC	45	70	320	16	83 500	21.4	402.12	47.1

5. Conclusions

Possibility to increase load-bearing capacity of the high-performance concrete ribbed slabs subjected to flexure by adding fibers was checked and confirmed by calculations. The use of high-performance fiber concrete as ribbed slab material is effective for slabs with span from 6 to 12 meters in comparison with high-performance concrete slabs with the same cross-sections. The shelf of HPFRC slab takes up a bending moment, which is 38 ... 41% higher than for the shelf of HPC slab. The HPFRC slab can take up 42 ... 46% higher distributed load than HPC slab with the same parameters of cross-section and allowed vertical displacement.

The difference in the ultimate load between the calculated with developed numerical model and experimental results is less than 6.0%. The specimen behavior of the numerical model is similar to the experimentally obtained of the slab, which leads to the conclusion that the developed constitutive model of material behavior with sufficient precision describes the behavior of the ribbed slab subjected to the flexure.

A performance analysis and cross-comparison of elements from HPC and HPFRC leads to the conclusion that the use of HPFRC for slabs with span 6-12 m allows reducing of the cross-section dimensions of the ribbed slab and the additional amount of longitudinal reinforcement more than 20%. Considering the reduction of materials to be used, the logical conclusion is that the costs of transporting and labor cost will also be reduced. In addition, dispersible reinforcement increases safety of the structure thanks to the beneficial effects of fibers on the plastic properties of concrete. The obtained results of calculations allow selecting of the used material and initial parameters of the slab cross-section for the first stage of design process.

Acknowledgement

The financial support of European Regional Development Fund project Nr.1.1.1.1/16/A/007 "A New Concept for Sustainable and Nearly Zero-Energy Buildings" and the funding from Riga Technical University, Faculty of Building and Civil Engineering Grant "DOK.BIF/18" are acknowledged.

References

[1] Olivito R S and Zuccarello F A 2010 An experimental study on the tensile strength of steel fiber reinforned concrete *Composites: Part B* **41** 246–255
[2] Singh H 2017 *Steel fiber reinforced concrete - behavior, modelling and design* (Ludhiana: Springer) p 172
[3] Song P S snd Hwang S 2004 Mechanical properties of high-strength steel fiber-reinforced concrete *Construction and Building Materials* **18** 669–673
[4] Vougioukas E and Papadatou M 2017 A model for the prediction of the tensile strength of fiber-updated concrete members, before and after cracking *Fibers* **5(3)** 27
[5] Kazemi M T, Golsorkhtabar H, Beygi M H A and Gholamitabar M 2017 Fracture properties of steel fiber reinformed high strength concrete work of fracture and size effect methods *Construction and Building Materials* **142** 482–489
[6] Aydin S 2013 Effects of fiber strength on fracture characteristics of normal and high strength concrete Periodica *Polytechnica Civil Engineering* **57(2)** 191–200
[7] EN 1992-1-1 "Eurocode 2: Design of concrete structures - Part 1-1: General rules and rules for buildings"

IOP Conf. Series: Materials Science and Engineering **660** (2019) 012054 doi:10.1088/1757-899X/660/1/012054

[8] Sliseris J 2018 Numerical analysis of reinformed concrete structures with oriented steel fibers and re-pack *Engineering Fracture Mechanics* **194** 337–349.

IOP Conf. Series: Materials Science and Engineering **660** (2019) 012055 doi:10.1088/1757-899X/660/1/012055

Axial strengthening of RC columns by steel encasement with direct fastening connections

R K L Su[1] and Z W Shan[2]

[1] Associate Professor, The University of Hong Kong, Hong Kong, China
[2] PhD student, The University of Hong Kong, Hong Kong, China

E-mail: klsu@hku.hk

Abstract. Reinforced concrete (RC) columns are crucial structural members that sustain vertical loads. Their strength may deteriorate due to fire, corrosion of the reinforcements or design errors. In response, a new strengthening method that uses a steel encasement to increase the axial capacity and ductility of RC columns is proposed in this paper. This method is quick and convenient because direct fastening is used to assemble the system. Four columns including one control column and three strengthened columns have been tested to examine the reliability and effectiveness of the proposed method. Although direct fastening plays a crucial role in the structural performance of the proposed method, current design guidelines have not incorporated this new connection technique. Hence, extensive connection tests have been conducted to evaluate the key factors (e.g. fastener type, fastener spacing, protuberance, etc.) that affect the strength of the connections. New design equations are subsequently proposed based on the test results for shear connections that use direct fastening.

1. Introduction

Reinforced concrete (RC) columns are the main type of support used in structures to resist vertical loads [1]. They play a critical role in protecting structures, in particular, moment resisting frames, from collapse. However, the axial strength of RC columns could deteriorate over time due to three possible reasons [1-4]: (i) an increase in service load, (ii) deterioration of the construction materials, and (iii) insufficient design capacity. To mitigate the strength deficiency problem, various strengthening methods have been proposed that enhance the structural safety of RC columns. The merits and limitations of these strengthening methods will be discussed herein.

The section enlargement technique, which is also called reinforced concrete jacketing, can effectively enhance the strength and ductility of RC columns. However, the available usable floor area will be reduced due to the enlarged column sections which can cause inconvenience during usage, especially in metropolitan cities like Hong Kong where property prices are premium so residential units are smaller in size. Furthermore, this method is time and labor consuming [1-3]. To overcome these shortcomings, alternative strengthening methods such as fiber reinforced polymer (FRP) jacketing and steel jacketing have been developed.

FRP jacketing is a newer technique with a fast and easy installation process, high strength and superior durability, and has been extensively studied worldwide [5-9]. Passive confinement can occur with transverse dilation, which results in the tri-axial compressive stress state in concrete and increases in the strength and ductility of the strengthened column. However, this method is only effective in strengthening circular or square columns with round but not sharp corners [10]. Square columns with sharp corners have to be trimmed into round corners to facilitate the application of FRP jacketing.

Steel jacketing is the use of steel cages or steel encasements for strengthening purposes. It is more preferred over FRP jacketing for strengthening rectilinear RC columns. Steel cages are four steel angles welded together with a steel batten [11-12]. Nevertheless, the welding of vertical weld seams is extremely challenging and requires a high level of skill and on-site quality control. Moreover, welding can cause residual stress which could lead to premature failure. To mitigate the negative effects from welding, bolts rather than weld connections are often used in steel encasement. Furthermore, to eliminate the stress lagging between the existing columns and new steel plates, the use of post-compressed plates (PCPs) has been proposed in some of the previous studies [13-15]. However, the use of through bolts to connect the concrete column and steel jacket with the PCP method can damage the concrete. In addition, precision positioning of the bolt holes at the steel plate and concrete is required to allow the bolts to pass through. This is not easy to implement on-site.

To address the aforementioned issues existing in the previous strengthening methods, a nonconventional encasement that comprises four steel plates fixed with steel angles by using the direct fastening method is proposed in this study. Direct fastening is an effective method to quickly and easily connect two steel components. Unique hardened fasteners are driven into steel material by using a powder-actuated gun or battery-actuated gun [16]. In this approach, additional axial capacity is provided by exterior steel plates. Furthermore, the steel encasement is joined by using direct fastening which can provide some confinement to the concrete core. The shear behavior of connections that are joined by direct fastening governs the degree of confinement and hence the peak strength and the post-peak behavior of the strengthened column. When compared to other strengthening methods, the time for the erection of the steel encasement by the proposed method is extremely short. Therefore, the cost of labor can be substantially reduced.

Direct fastening is carried out by driving unique hardened fasteners into steel materials with a powder- or battery-actuated fastening tool. This connection method has been widely used for connecting non-structural components. However, the application of this joining method to structural components has not been fully explored in the literature, not to mention the availability of relevant design guides. Lu et al. [17] conducted an experimental study and proposed design equations for screwed connections, but these are not applicable to connections joined by direct fastening owing to the different anchoring mechanisms. Lu et al. [18] further examined the behavior of connections joined by powder-actuated fastening in cold-formed steel sheeting at ambient and elevated temperatures. However, their study only involved joining very thin sheets with a single knurled fastener. Hence, to facilitate the application of direct fastening, a comprehensive study that takes the effects of more variables into consideration should be conducted.

In this study, the shear behavior of connections secured by direct fastening are experimentally investigated and design expressions will be proposed. Factors such as type, number and arrangement of the fasteners, thickness of the base steel plate and fastener spacing will be considered. Furthermore, four columns are tested to validate the effectiveness of the steel encasement connected by direct fastening in improving the axial strength and displacement ductility of rectangular RC columns (with lateral dimensions of 500 mm or smaller) which are commonly found in low-rise RC buildings.

2. Connection tests

The connections between the steel plates and angles in RC columns that are strengthened by using a steel encasement system with direct fastening connections should resist significant shear force incurred from the transverse expansion due to the Poisson's effect. The load transfer process of these connections is similar to single lap joints subjected to tensile force at the ends. Therefore, single lap joints shown in figure 1 were designed and tested to study the failure process of the connections. In total, 100 samples were tested. To differentiate the different connection samples, they were properly labeled. Take for example, S275-A-4-5-6(30)-L: S275 denotes that the nominal yield strength of the steel material is 275 MPa; 'A' denotes that the sample is tested at room temperature; '4' represents the nominal diameter of the fastener; '5' signifies the thickness of the base steel plate; '6(30)' means that there are 6 fasteners with a fastener spacing of 30 mm; and 'L' means that the arrangement of these 6 fasteners is parallel to the load direction or Type 5 as illustrated in figure 1. The thickness, width and length of connected steel plate are 3 mm, 100 mm and 200 mm respectively. The thickness of the base steel plate varies from 3 mm to 6 mm while the width and length are same with those of connected steel plate.

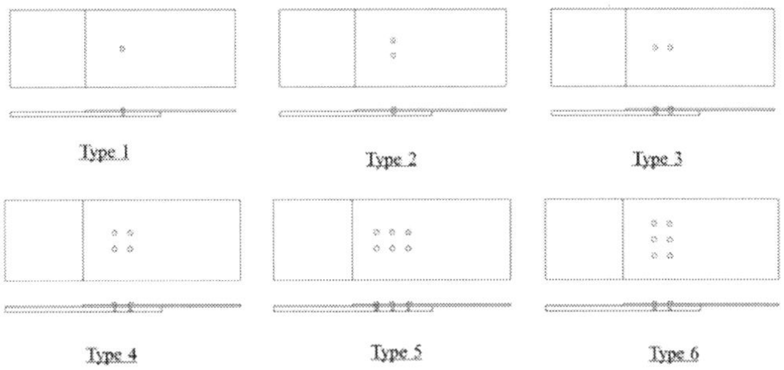

Figure 1. Configurations of the sample.

2.1. Test setup and procedure

The MTS 810 material testing system (figure 2) which has a maximum capacity of 250 kN was used to test the properties of the connection samples at an ambient temperature. Two gaskets were inserted between the upper and lower grips to allow the single shear of the connection samples (figure 3). The rate of loading was 0.5 mm/min. The testing for each connection sample was terminated when the bearing resistance was reduced to less than 20% of the maximum bearing resistance to investigate its post-peak behavior.

Figure 2. Test setup.

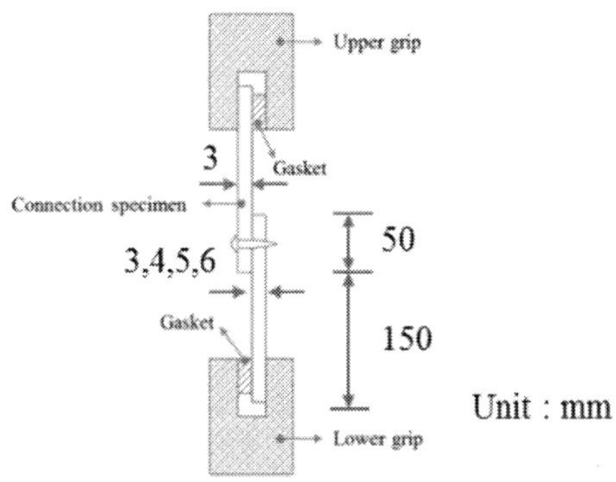

Figure 3. Assembly diagram of connection sample.

2.2. Failure modes and failure behavior

Two kinds of failure modes - bearing failure and shear fracture failure, are observed in the tests. The typical failure modes are shown in figure 4. Bearing failure results in enlarged fastener holes due to large plastic deformation and the bulging of the material around the fastener holes, which is shown in figure 4(a). The bearing resistance versus displacement (BRVD) curves of the connections are plotted in figure 5. Here, the BRVD curve for S275-A-4-4-1 is used as an example to explain the bearing failure process of this connection. In the initial loading stage, the steel plate is in an elastic region and the resistance force increases linearly with the displacement. When the steel material begins to yield, the load increases nonlinearly with the displacement, after which the bearing resistance remains approximately constant. During this stage, the holes on the steel plates are obviously enlarged and the fasteners are no longer normal to the steel plate. As a result, a pull-out force is carried out on the fasteners. The connection cannot take any force when the pull-out force exceeds the friction force between the fasteners and steel plates. Hence, the connection strength abruptly drops to zero, which is a significant shortcoming. Figure 4(a) shows the final rotation of the fasteners. The pull-out process does not occur at the same time for connections that are joined with more than one fastener, but sequentially during bearing failure. Therefore, there are inflection points on the BRVD curves when there are more than one fastener in the connection; see figure 5.

2.3. Proposed equation for bearing strength

As stated above, the connections in this strengthening method are expected to resist shear force that results from transverse expansion due to Poisson's effect. The shear capacity of this connection should be quantified to determine the confinement effect and prevent shear fracture failure. Based on the equation in [19], a new equation is proposed, in which the effects of protuberance and knurling are separated from the bearing factor.

$$F_b = \psi_{fp}\psi_{fk}\alpha_{br}d_n t_p f_{pu} \tag{1}$$

where α_{br} denotes the bearing factor, d_n represents the nominal diameter of the fastener, t_p is the thickness of the connected plate, f_{pu} depicts the ultimate strength of the steel material, ψ_{fp} is a factor for the effect of protuberance, ψ_{fp} is equal to 1.0 for connections joined with pre-drilled holes on the connected plates while ψ_{fp} is 1.35 for connections without pre-drilled holes on the connected plates, and ψ_{fk} is included for the effect of knurling. ψ_{fk} is 1.0 and 1.17 for fasteners that are not knurled and knurled, respectively.

The normalized peak load which is tested by using the predicted peak load is presented in figure 6. The average normalized peak load is 1.04 with a coefficient of variation (CV) of 0.11 and almost all of the normalized peak loads fall within the range of 0.8 to 1.2. Hence, the maximum bearing resistance can be accurately predicted by using the proposed equation with small discrepancies.

Bearing failure is the preferred failure mode in strengthening the connection due to its superior ductile behavior. To ensure that bearing failure occurs prior to fastener shear fracture, the following relationship needs to be satisfied with the exclusion of partial safety factors:

$$F_b < F_{fs} \tag{2}$$

where F_{fs} is the shear resistance of the fastener, which can be determined through tests.

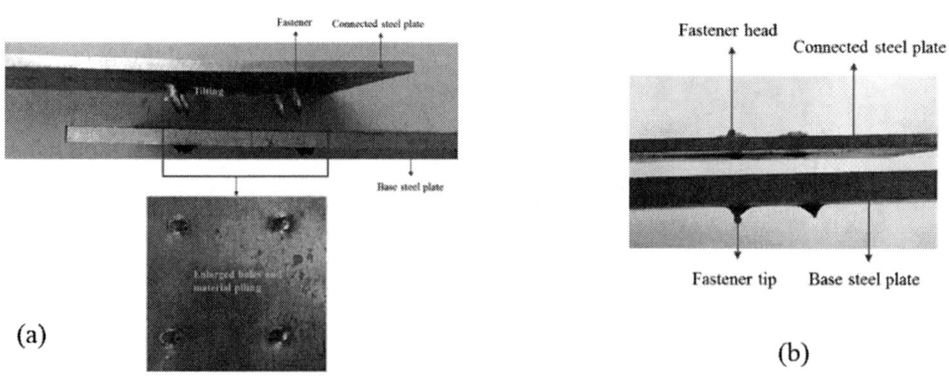

Figure 4. Failure modes: (a) bearing failure (b) shear fracture of fastener

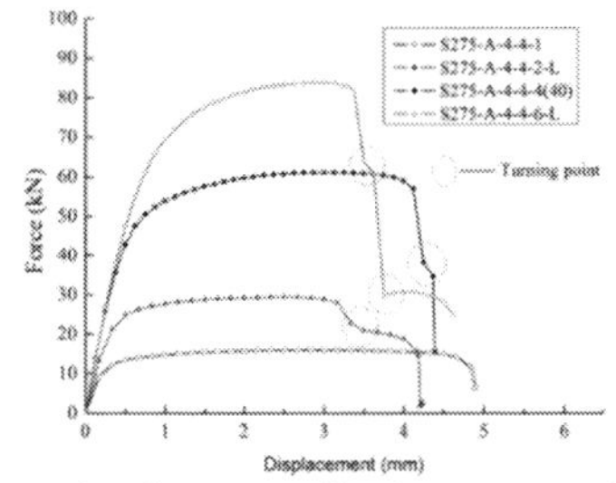

Figure 5. Shear load versus displacement of connections.

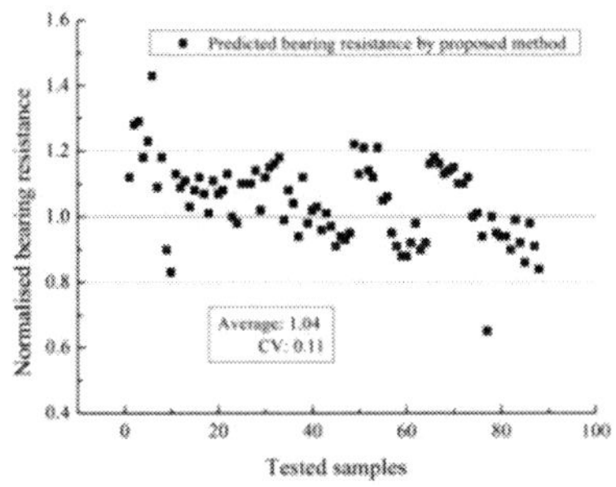

Figure 6. Comparison between predicted and tested peak loads.

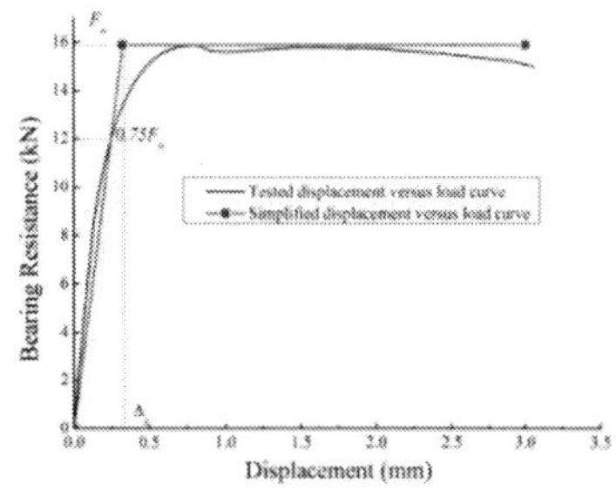

Figure 7. Simplified bilinear curve.

2.4. Proposed equation of effective stiffness

The finite element method (FEM) is an effective numerical tool for analyzing composite structures. This approach can be used to simulate the structural behavior of strengthened columns by using steel encasement with direct fastening connections. To accurately simulate the structural behavior, the non-linear response of the connections secured with direct fastenings should be modeled. In this study, the non-linear bolt resistance-deformation curve is idealized as a bilinear curve as shown in figure 7. This bilinear curve is governed by the bearing capacity and initial effective stiffness. The former has been presented in the previous section while the latter is presented as follows:

$$K_{ef} = \frac{\psi_{fn}\psi_{ef}E_s t_p d_n}{l_c} \tag{3}$$

where E_s is the elastic modulus of the plates, ψ_{ef} is the effective stiffness (ES) factor which is calibrated by the tested samples, and ψ_{fn} represents the factor that depends on the number of fasteners (FN). l_c depicts the characteristic length in which deformation happens.

Owing to the fact the characteristic length is difficult to define, the effect of this parameter can be incorporated into the effective stiffness factor. Under this consideration, the effective stiffness can be rewritten as:

$$K_{ef} = \psi_{fn}\psi'_{ef}E_s t_p d_n \tag{4}$$

where ψ'_{ef} is the modified effective stiffness (MES) factor, which incorporates the effect of character length.

In total, 11 samples were used to determine the MES factor. The results are shown in figure 8. A value of 0.017 is adopted for the MES factor. The relationship between the FN factor and number of fasteners is shown in figure 9. It can be seen that the relationship is not exactly linear due to the group reduction effect. The average FN factor for each number of fasteners is adopted and given herein.

$$\psi_{fn}=\begin{cases}1 & n_n=1\\1.4 & n_n=2\\1.9 & n_n=4\\2.1 & n_n=6\end{cases} \tag{5}$$

where n_n represents the number of fasteners. The FN factor for other quantities of fasteners can be determined by using linear interpolation and extrapolation.

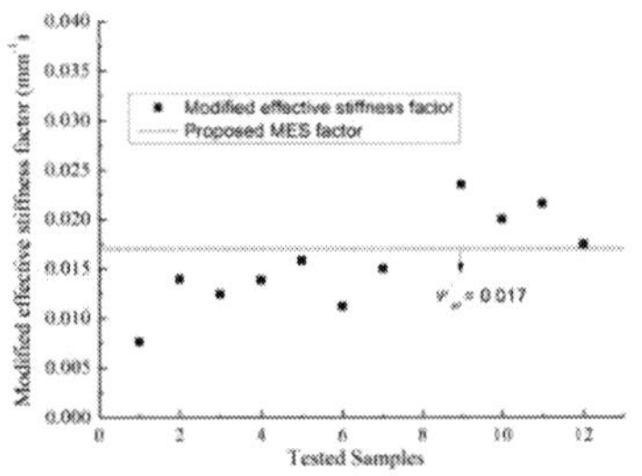

Figure 8. Modified effective stiffness factor.

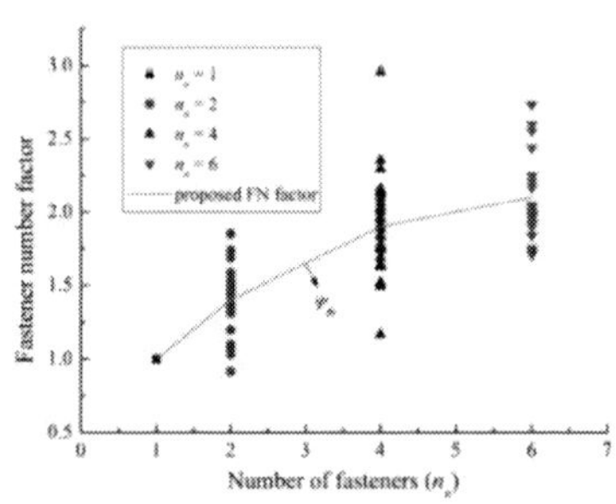

Figure 9. Factor depending on number of fasteners.

3. Column tests

The column prototype was designed based on typical low-rise RC buildings. Owing to the restriction of the testing setup, the cross-section dimensions of the original column were reduced by half. Therefore, a column with the height of 700 mm and cross-section dimensions of 200 mm × 200 mm is adopted in this test. 10 mm deformed steel rebar and 6 mm smooth steel rebar are used for longitudinal reinforcement and stirrup in the column. The target concrete strength is 30 MPa. Besides, the S275 grade steel plates with thickness of 3 mm are used for steel encasement. The size of connected steel angle in steel encasement is 60 mm × 60 mm × 5 mm. In this proposed strengthening method, four steel plates with two ends bolted to the top and bottom of concrete blocks are directly affixed to an RC column to share the axial load. Steel angles are connected to the middle part of the adjoining steel plates by direct fastening to provide passive confinement to the concrete core which can allow a tri-axial compression state in the concrete. Besides that, the vertical spacing of the angles can limit the slenderness ratio of the steel plates which is a key factor for maintaining the plate capacity. In this study, the stress lagging effect between the concrete core and external steel plates is neglected as the PCP load application method [10-13] can effectively mitigate this effect. To confirm the effectiveness of this proposed strengthening scheme, four columns including one control column and three strengthened columns were constructed and then tested; see figure 10.

The test results are shown in figure 11 and table 1. η_P represents the enhancement ratio of strength and η_Δ depicts the ductility factor. Apparently, the strength and ductility are greatly enhanced. Samples S2 and S3 have the same angle spacing while S4 has a smaller angle spacing. Strength and ductility are not greatly affected between S2 and S3 as they have the same angle spacing. When comparing S2 and S4, the strength and ductility of the latter are higher. This indicates that small angle spacing is helpful for achieving higher strength and better ductility behavior as small spacing between angles can reduce the slenderness ratio of the steel plates. Under this situation, plate buckling can be delayed. Regarding the performance of the connection joints, bearing failure was not observed before the axial load decreased to 50% of the peak load. This indicates that direct fastening is a reliable connection method for steel encasement.

Table 1. Test results.

Sample	Ultimate load(kN)	η_P	η_Δ
S1	1230	1.0	2.26
S2	1500	1.22	3.37
S3	1503	1.22	3.13
S4	1640	1.33	3.8

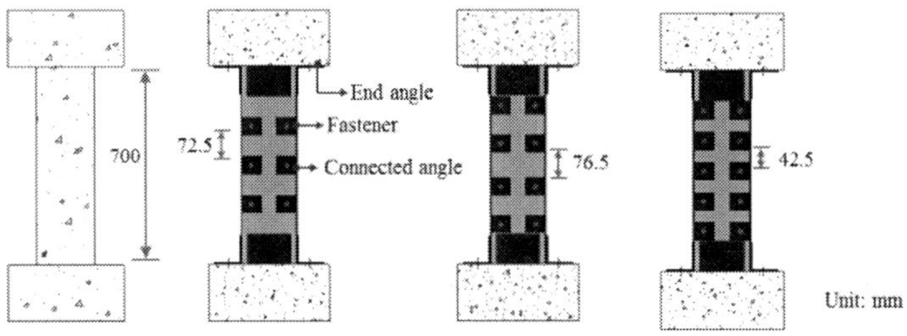

Figure 10. Strengthening arrangements of RC column.

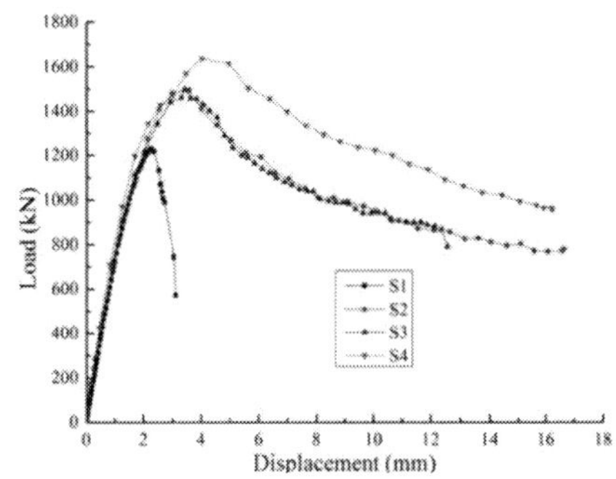

Figure 11. Load versus displacement.

4. Conclusion

This paper presents an experimental study on direct fastening connections and column strengthening by using steel encasement with direct fastening connections. The findings from this study are summarized below:

(1) An empirical equation has been proposed to predict the shear capacity of connections joined by direct fastening associated with bearing failure. The predicted strength can be used to quantify the confinement stress on the concrete core. Furthermore, by designing the shear failure load to be higher than the other failure modes, shear fracture failure of fasteners can be prevented in column strengthening applications.

(2) An equation for predicting the effective stiffness of the connection is proposed, in which the group effect is incorporated. Together with the empirical equation for predicting the shear capacity of connections, a simplified bilinear curve can be determined for simulating the behavior of the connections jointed by direct fastening.

(3) Compared to the control column, the strength of the strengthened columns (S2, S3, S4) is increased by 22%, 22% and 33% respectively and the ductility is improved by 49%, 38% and 68% respectively. This shows that the steel encasement with connections that are joined with direct fastenings can effectively improve the axial load capacity of RC columns.

References

[1] ACI 318-08 2008 Building code requirements for structural concrete and commentary, American Concrete Institute.
[2] Rodriguez M and Park R 1994 *ACI. Struct. J.* **91**(2) 150-159
[3] Júlio E S, Branco F and Silva V D 2003 *Progr. Struct. Eng. Mater.* **5**(1) 29-37
[4] Baji H, Yang W and Li C Q 2018 *Struct. Infrastructure E.* **14**(12) 1586-1597
[5] Pessiki S, Harries K A, Kestner J T, Sause R and Ricles J M 2001 *J. Compos. Constr.* **5**(4) 237-45
[6] Lam L, Teng J G, Cheung C H and Xiao Y 2006 *Cement. Concrete. Comp.* **28**(10) 949-58
[7] Zeng J J, Lin G, Teng J G and Li L J 2018 *Eng. Struct.* **174** 629-645
[8] Li P and Wu Y F 2016 Compos. Struct. 149 369-84
[9] Chellapandian M S, Prakash S S and Rajagopal A 2018 *Compos. Struct.* **184** 234-248
[10] Wu Y F, Liu T and Oehlers D J 2006 Adv. Struct. Eng. 9(4) 507-33
[11] Giménez E, Adam J M, Lvorra S, Moragues J J and Calderón P A 2009 *Adv. Struct. Eng.* **12**(2) 169-81
[12] Giménez E, Adam J M, Lvorra S and Calderón P A 2009 *Mater. Design.* **30**(10) 4103-11
[13] Su R K L and Wang L 2012 *Eng. Struct.* **38** 42-52
[14] Su R K L and Wang L 2015 *Struct. Infrastruct. E.* **11**(8) 1083-101
[15] Wang L and Su R K L 2012 *Adv. Struct. Eng.* **15**(8) 1253-64
[16] Shan Z W and Su R K L 2019 *Eng. Struct.* **196** 109321
[17] Lu W, Makelainen P, Outinen J and Ma Z C 2011 Thin. Wall. Struct. **49**(12) 1526-33
[18] Lu W, Ma Z C, Makelainen P and Outinen J 2012 *Thin. Wall. Struct.* **61** 229-38
[19] ANSI/AISC 360-16 2010 Specification for structural steel buildings. American Institute of Steel Construction, Chicago-Illinois.

Comparison of bond efficiency of various coating materials

K B Osifala[1] and T A I Akeju[2]

[1]Department of Civil Engineering, Yaba College of Technology, Yaba, Lagos, Nigeria.
[2]Department of Civil & Environmental Engineering, University of Lagos, Nigeria.

E-mail: kosifala@yahoo.com

Abstract. The paper reports an experimental research investigation carried out to compare the bond resistance and efficiency of various coating materials, including epoxy, which were sourced locally. Thirty-four full size beams of varying lengths and sectional dimensions, with lap spliced bars in constant moment region were cast and tested in a four-point bending system. Three varying high yield diameter bars 16 mm, 20 mm and 28 mm were coated with various coating materials such as epoxy, chlorinated rubber, tyrolin, vinyl chloride and zinc ethyl silicate in an attempt to find a cheaper but effective coating material than epoxy. The ultimate moment from the tests were used to determine the stress developed in the steel rods. The bond efficiency, i.e. the ratio of test bond stresses and the theoretical bond stresses was used for the comparison of the various coating materials. The bond resistance and efficiency of vinyl chloride coated bars were found to be higher than that of epoxy and other materials, in all the beams tested. Statistically, there appear not to be significant difference between the bond efficiency of epoxy and vinyl chloride.

1. Introduction

Premature deterioration of reinforced concrete structures such as bridge decks, parking garages, sewage treatment and chemical plants, etc. has become a major problem. This early deterioration has been attributed to accelerated corrosion of reinforcing bars caused by chloride ions from de-icing materials and salts, Mathey and Clifton [1]. High humidity subjects exposed concrete structural elements to a perpetual state of wetness, especially during the rainy season. Water, in molecular or gaseous form, penetrates concrete and attacks the reinforcement causing corrosion, expansion and scaling of the steel rods. This phenomenon leads to cracking and spalling of concrete, reduction in diameter of reinforcement and subsequent reduction in strength of the structural elements, thus necessitating in many cases extensive repairs of structural elements. The situation is similar for concrete elements in permanent contact with water or de-icing materials e.g. water retaining structures, jetties etc.

Reinforced concrete is a dual structural arrangement with concrete mainly resisting compression forces and steel resisting the tensile forces. Its effectiveness as a structural member depends on the bond between steel reinforcement and concrete. Such effectiveness is reduced by corrosion which can be prevented or at least reduced by coating of reinforcing bars.

Of all the coating materials investigated and analyzed so far, only epoxy showed a good promise, Clifton and Mathey [2]. Epoxy coating has its own shortcomings besides being expensive, bond between concrete and epoxy coated steel rods is reduced. It is therefore desirable and worthwhile to investigate other materials such as chlorinated rubber, zinc ethyl silicate, vinyl chloride and tyrolin (cement and sand mix) in an attempt to find cheaper and effective coating material that could possibly give higher bond strength.

2. Aims and objective

The major thrust of the investigation reported herein is to provide solutions to practical problems in reinforced concrete and hence contribute to knowledge and engineering practice on bond between concrete and steel reinforcement in high humid environment.

The specific objective of the work is to find other effective coating materials - effective in terms of bonding between concrete and steel that can be used to coat reinforcing rods instead of epoxy.

3. Design of Test Beams

Thirty-four full size beams were cast. Sixteen beams were cast for the preliminary investigation and eighteen beams for the confirmatory tests. The beams were of three sizes, 300 mm x 230 mm; 300 mm x 200 mm and 300 mm x 180 mm. The main bars were 28 mm, 20 mm and 16 mm diameter high yield bars. The depth of all beams was 300 mm while the length is a function of lap length and shear span thus length $L = \ell_s + 300$ mm + $2a_v$ +200 mm. The lengths varied with the bar diameters as 3.20 m, 2.90 m and 2.75 m respectively. The design of the beams is shown in Figure 1 and the beam data are shown in Table 1.

Table 1. Beam Data.

Beam Name	b (mm)	Tensile Reinft.	A_s (mm^2)	Lap Length + 300mm	Shear Span a_v(mm)	Beam Length (m)
A-U-10A	180	2T16	402	940	801	2.75
A-U-10B	180	2T16	402	940	801	2.75
A-U-10C	180	2T16	402	940	801	2.75
A-U-11A	200	2T20	628	1100	795	2.90
A-U-11B	200	2T20	628	1100	795	2.90
A-U-11C	200	2T20	628	1100	795	2.90
A-U-12A	230	2T28	1232	1420	783	3.20
A-U-12B	230	2T28	1232	1420	783	3.20
A-U-12C	230	2T28	1232	1420	783	3.20
A-E-1A	180	2T16	402	940	801	2.75
A-E-1B	180	2T16	402	940	801	2.75
A-E-1C	180	2T16	402	940	801	2.75
A-E-2A	200	2T20	628	1100	795	2.90
A-E-2B	200	2T20	628	1100	795	2.90
A-E-2C	200	2T20	628	1100	795	2.90
A-E-3A	230	2T28	1232	1420	783	3.20
A-E-3B	230	2T28	1232	1420	783	3.20
A-E-3C	230	2T28	1232	1420	783	3.20
A-VC-16A	180	2T16	402	940	801	2.75
A-VC-16B	180	2T16	402	940	801	2.75
A-VC-16C	180	2T16	402	940	801	2.75
A-VC-17A	200	2T20	628	1100	795	2.90
A-VC-17B	200	2T20	628	1100	795	2.90
A-VC-17C	200	2T20	628	1100	795	2.90
A-VC-18A	230	2T28	1232	1420	783	3.20
A-VC-18B	230	2T28	1232	1420	783	3.20
A-VC-18C	230	2T28	1232	1420	783	3.20
A-CR-22	185	2T16	402	940	801	2.75
A-CR-19	200	2T20	628	1100	795	2.90
A-ZS-20	185	2T16	402	940	801	2.75
A-ZS-21	200	2T20	628	1100	795	2.90
A-T-7	180	2T16	1402	940	801	2.75
A-T-8	200	2T20	628	1100	795	2.90
A-T-9	230	2T28	1232	1420	783	3.20

Beams are without stirrup over lap.
Side and bottom covers are 25 mm thick

Beams are 300 mm depth.

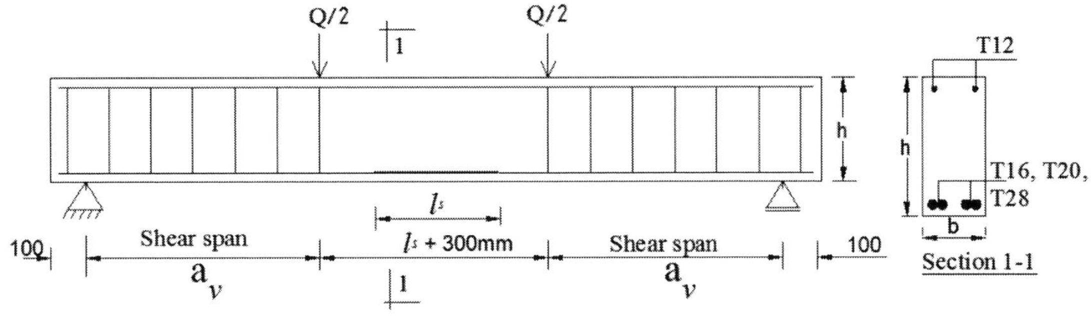

Figure 1. Typical beam without stirrup overlap.

The beams contain tensile lap splices without transverse reinforcement over the lap. In all the beams 10 mm diameter high yield stirrup at 100 mm centres were provided in the shear span to resist shear failure.

The lapped bars were placed in contact because this is the most probable on construction sites. The loading points were 150 mm from the lap ends thereby making the distance between the point loads to be ℓ_s + 300 mm.

The ultimate anchorage bond length recommended in BS 8110: 1985 [3] was used as in equation (1)

$$l_s = \frac{0.87 f_y \phi}{4\beta\sqrt{f_{cu}}}. \quad (1)$$

Failure mode in flexure is strongly dependent on the shear-span/effective depth ratio (a_v/d). Based on experimental studies reported in Kong and Evans [4] the failure pattern to ensure flexural failure was used. a_v/d was taken as 3, hence a = 3d.

The side and bottom covers to the lapped bars were made equal, $c_s = c_b = c$ as failure will occur at the side with the minimum cover; hence the covers (c_s and c_b) to the beams were both 25 mm.

4. Preparation and Application of Coating Materials

The reinforcements were sand blasted to remove rust. Coating materials were sprayed on the steel rods in this investigation.

4.1.Epoxy zinc rich

Epoxy zinc rich is in two parts. Part A, base and Part B, which is the hardener. The parts were measured out by volume in ratio 3:1, as recommended in the manufacturer's manual, and the two parts were thoroughly mixed together. Thinner was added until the mixture was light enough to pass through the nozzle of a conventional spraying gun. The lengths of the reinforcements to be coated were marked out. The mixture was then sprayed, in one coat, on the marked portions of the reinforcements and allowed to dry in the shade (without direct exposure to sunshine).

4.2. Vinyl chloride

Vinyl chloride is in different colours, the dull red colour was used for the preliminary tests and the light grey colour was used for the confirmatory tests. It was thoroughly mixed and fed into the spraying gun. Two coats were applied consecutively to achieve a uniform coverage.

4.3. Chlorinated rubber

IOP Conf. Series: Materials Science and Engineering **660** (2019) 012056 doi:10.1088/1757-899X/660/1/012056

This coating material is red in colour. Chlorinated rubber requires no thinner, the material was simply agitated thoroughly and fed into the spraying machine and only one uniform coat was applied.

4.4. Zinc ethyl silicate
Zinc ethyl silicate is in two parts (Part I, powder and Part II, liquid). The two parts were measured out in ratio 2:1 by weight. The two parts were thoroughly mixed together. No thinner was added. The mixture was sprayed, in one coat, on the marked portions of the reinforcements.

4.5. Tyrolin
Equal measure by weight of sand and cement was carried out, precisely, 2.902 kg of Grade 14 sand, 0.725 kg of Grade 25 sand totaling 3.627 kg sharp sand and 3.627 kg of cement were thoroughly mixed to achieve a uniform consistency; 2.00 kg of water was added and further mixed uniformly. The mix ratio by weight was 1.81:1.81:1.0 (sand:cement:water). The mixture was sprayed on the rods with the tyrolin spraying machine. The tyrolin sprayed rods were allowed to dry in the shade for seven days before the reinforcement cages were made.

5. Manufacture and test procedure of cubes and beams
Concrete used for the investigation was grade 35 with targeted cube strength of 35 N/mm^2. Concrete cubes were cast in batches and cured for 28 days in the laboratory. The strengths achieved at test are in Table 3. The reinforcement cages were put in the mould and concrete spacers were placed at interval at various locations; bottom and sides. The ready-mix concrete was poured and vibrated in the mould with poker vibrator. After demolding, the beams were cured by wetting at intervals for seven days.

5.1. Compressive strength of concrete cubes
The aim of the test was to determine the strength of concrete in the beams and to serve as a control so that the compressive strength reach at least the targeted value of 35 N/mm^2. The cubes were weighed on Avery weighing machine, 50 kg capacity and tested in the Avery Universal testing Machine of 100,000 kg capacity.

5.2. Beam Test
The beams were tested with a four-point bending arrangement with an I steel section as the spreader beam in a 100,000 kg capacity Avery Universal Testing Machine to determine the ultimate failure loads, modes and crack patterns. The load was gradually applied until the failure of the beam occurred. The failure load was recorded and analyzed as follows.

6. Analysis and discussion of test results
The steel stress developed in each beam was determined by analyzing the section using the general theory for ultimate flexural strength. The ultimate moment for each beam was obtained by applying the principles of statics. The lapped splice was in the constant moment region. The ultimate moment at the point load in Figure1 is

$$M_u = \frac{Q}{2}(a_v) \tag{2}$$

From the equilibrium condition in Figure 2.

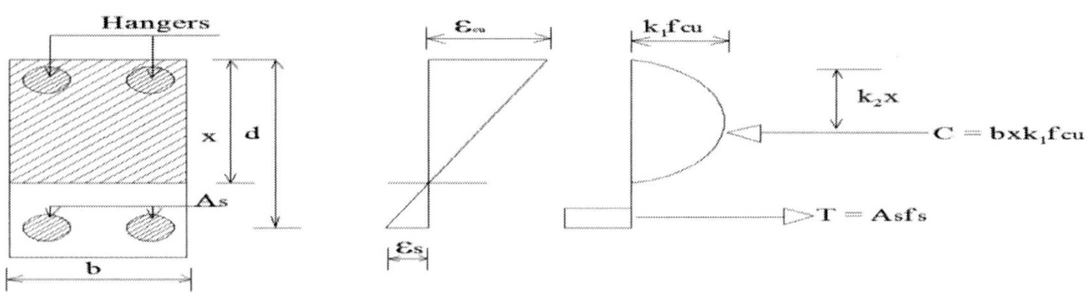

Cross section Strain Stress and Forces

Figure 2. Strain and Stress distribution at failure.

Hognestad's stress was adopted and the characteristics k_1, k_2 and Σ_{cu} were extracted from Kong and Evans[4] in Table 2.

Table 2. Characteristics of Hognestad Et-Al's Stress Block.

S/N	fcu (N/mm^2)	k_1	k_2	Σcu
(Preliminary)				
1.	34.67	0.59	0.449	0.0034
2.	34.52	0.589	0.453	0.0034
3.	31.09	0.611	0.457	0.0034
4.	35.06	0.586	0.451	0.0033
5.	34.53	0.59	0.452	0.0034
6.	27.53	0.632	0.461	0.0035
7.	30.67	0.613	0.458	0.0034
(Confirmatory)				
8.	36.47	0.581	0.450	0.0033
9.	34.44	0.594	0.454	0.0034
10.	37.90	0.575	0.450	0.0033
11.	36.98	0.578	0.449	0.0033
12.	38.70	0.574	0.449	0.0033
13.	33.30	0.596	0.456	0.0034
14.	30.44	0.614	0.457	0.0034

$$k_1 f_{cu} bx = A_s f_s. \tag{3}$$

Taking moment about the compressive force, the ultimate moment of resistance is

$$M_u = A_s f_s (d - k_2 x), \tag{4}$$

from where:

$$\frac{M_u}{bd^2} = \rho f_s - \frac{k_2}{k_1}\frac{(\rho f_s)^2}{f_{cu}} \tag{5}$$

The properties of the concrete stress block are expressed in terms of the characteristic ratio k_1 and k_2. Stress block proposed by Hognestad et-al reported in Kong and Evans [4] and universally accepted was used. Based on the mean cube strength, various values of k_1 and k_2 and Σ_{cu} are tabulated in Table 2. The quadratic equation in (5) was solved yielding two values of f_s, one of which was always unreasonable because it is not practicable. The reasonable value was the stress developed in the steel.

6.1. Test Bond Stress, τ_t

The bond strength is the average stress along the length of the splice.

From equilibrium

$$A_s f_s = \tau_t \pi \phi_e l_s, \quad (6)$$

from where:

$$\tau_t = \frac{f_s \phi_e}{4 l_s}. \quad (7)$$

The steel stress developed in each beam that was determined from equation (5) was substituted in (7) to obtain the τ_t (τ_{test}).

6.2. Theoretical Bond Stress, τ_{cal}

The theoretical bond stress, τ_{cal}, was determined from a semi-empirical statistical regression equation developed by Orangun *et-al* [5].

For bars with no transverse reinforcement

$$\tau_{cal} = \left[1.2 + 3\left(\frac{c}{d_b}\right) + 50\left(\frac{d_b}{l_s}\right)\right] f_c^{1(0.5)} \quad (8)$$

Equations (8) is in imperial unit and concrete cylinder strength. The expressions were converted to SI units and modified for use with concrete cube strength by Cairns and Abdullah [6] to give:

$$\tau_{cal} = \left[0.09 + 0.24\frac{c}{d_b} + 3.9\frac{d_b}{l_s}\right]\sqrt{f_{cu}} \quad (9)$$

The bond strength equation in (9) proposed by Orangun *et-al* [5] was adopted as the formula for theoretical bond stress in this work. It formed the basis of the bond strength equation in the 1995 and 2002 American Concrete Institute (ACI) codes. Leet, Kenneth [7].

6.3. Bond Stress using BS 8110's Recommendation, τ_{BS8110}

The British Standard, BS 8110's, 1997[8] recommended bond stress was also used to calculate the bond efficiency.

$$\tau_{BS} = \beta\sqrt{f_{cu}} \quad (10)$$

with the factor β, taken as 0.5 because deformed bars were used in tension.

7. Discussion of test results

Comparison of Bond Efficiency of Various Coating Materials

The investigation of various coating material to compare their bond resistances and efficiencies was carried out in two phases: the preliminary and confirmatory tests. Preliminary tests were carried out using epoxy, vinyl chloride, chlorinated rubber, zinc ethyl silicate and tyrolin. All were locally sourced coating materials from the manufacturers. Three different bar diameters, T16 mm, T20 mm and T28 mm were used, one beam each was cast with each of the bar diameters. The beam data is shown in Table 1. The bond strength ratio $\frac{\tau_t}{\tau_{cal}}$ was plotted against the coating materials in Figure 3. As expected, uncoated bars gave a higher bond strength than coated bars, except in T28 diameter bars for vinyl chloride. This result is in line with previous research investigation reported in Mathey and Clifton [1],

Treece and Jirsa [9], Cuzen and Yu [10], Kayyali and Yeomans [11], Thangavel *et-al* [12], Yan and Mindess [13] that coating reduces bond strength.

Bond resistance has been found to reduce with larger bar diameter for uncoated bars, from Figure 3 it would appear that reduction due to coating has neutralized the effect of bar diameter. It is equally possible that the effect may not have shown because the coating thickness varied with the coating material.

Bars coated with tyrolin also gave a higher bond strength ratio than other coating materials for 20 mm diameter bars. Tyrolin (mixture of cement, sand and water) on its own is permeable and therefore cannot constitute a physical barrier to the oxygen and moisture necessary for the corrosion reaction. It can therefore, at best, be used to improve the bond strength of coated reinforcing bars by its mixture or application with other coating materials in sequence. The performance of bars coated with tyrolin is to be expected as dry tyrolin formed a rough surface with multiple grooves. These grooves are deeper than the regular rib pattern on the bar and this allows for better bond performance of the bars coated with tyrolin.

Vinyl chloride showed about the same performance with epoxy and of course a better bond performance than other tested materials on bar diameters T16 mm and T20 mm. Confirmatory tests were carried out on coatings which showed a good promise in the preliminary tests, that is, vinyl chloride and epoxy. Values plotted for chlorinated rubber and zinc ethyl silicate were only for one beam for each bar diameter as the two materials did not show good promise. Uncoated bars were also tested for comparison. Two beams each were further cast with 28 mm, 20 mm and 16 mm diameter high yield bars coated with vinyl chloride and epoxy. Table 1, Table 3 and Figure 3 show the beam data and the test results respectively.

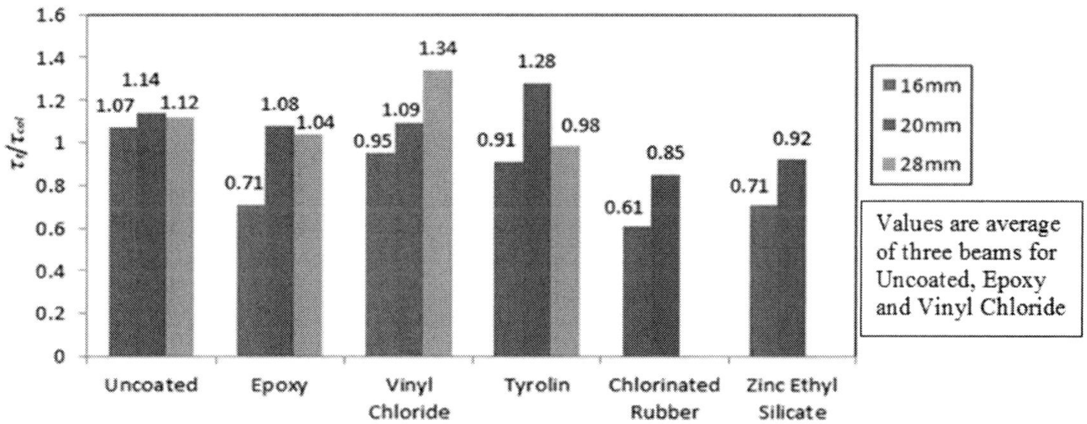

Figure 3. Comparison of bond efficiency of various coating materials.

Table 3. Beam Test Results.

Beam Name	f_{cu} (N/mm²)	M_u (kN.m)	f_s (N/mm²)	τ_t (N/mm²)	τ_{cal} (N/mm²)	τ_{BS} (N/mm²)	$\frac{\tau_t}{\tau_{cal}}$	$\frac{\tau_t}{\tau_{BS}}$	$\left(\frac{\tau_t}{\tau_{cal}}\right)^A$	$\left(\frac{\tau_t}{\tau_{BS}}\right)^A$
A-U-10A	34.67	54.07	561.53	3.51	3.31	2.94	1.06	1.19		
A-U-10B	36.47	55.27	573.31	3.58	3.40	3.02	1.05	1.19	1.07	1. 22
A-U-10C	34.44	52.27	576.65	3.60	3.30	2.93	1.09	1.28		
A-U-11A	34.67	61.17	473.92	2.96	2.87	2.94	1.03	1.01		
A-U-11B	36.47	82.28	555.30	3.47	2.94	3.02	1.17	1.20	1.14	1.13
A-U-11C	34.44	79.10	556.86	3.48	2.86	2.93	1.22	1.19		
A-U-12A	34.67	94.44	371.23	2.32	2.37	2.94	0.98	0.78		
A-U-12B	36.46	121.37	476.37	2.98	2.43	3.02	1.23	0.99	1.12	0.90
A-U-12C	34.44	111.19	429.99	2.69	2.36	2.93	1.14	0.92		
A-E-1A	31.09	62.64	370.98	2.32	3.14	2.79	0.74	0.93		
A-E-1B	36.98	72.25	471.98	2.94	3.42	3.04	0.86	0.97	0.91	1.03
A-E-1C	37.90	85.07	589.58	3.86	3.46	3.08	1.12	1.25		
A-E-2A	31.09	88.40	430.12	2.69	2.72	2.79	0.99	0.96		
A-E-2B	37.90	113.45	539.32	3.89	3.37	3.08	1.12	1.26	1.08	1.04
A-E-2C	36.98	92.38	450.38	2.83	2.96	3.04	0.95	0.91		
A-E-3A	31.09	125.44	384.79	3.37	2.46	2.79	1.07	0.86		
A-E-3B	37.90	142.86	452.73	2.83	2.47	3.08	1.14	0.92	1.04	0.84
A-E-3C	36.98	122.70	358.61	2.24	2.44	3.04	0.92	0.74		
A-VC-16A	34.53	40.05	403.32	2.52	3.31	2.94	0.76	0.86		
A-VC-16B	38.70	55.27	569.82	3.56	3.50	3.11	1.02	1.14	0.95	1.07
A-VC-16C	33.30	53.67	560.34	3.50	3.25	2.89	1.08	1.21		
A-VC-17A	34.53	64.40	437.19	2.73	2.86	2.94	0.95	0.93		
A-VC-17B	38.70	76.32	550.34	3.44	3.03	3.11	1.13	1.11	1.09	1.07
A-VC-17C	33.30	101.92	537.15	3.36	2.81	2.89	1.19	1.16		
A-VC-18A	38.70	125.28	488.67	3.05	2.50	3.11	1.22	0.98		
A-VC-18B	33.30	126.06	519.12	3.24	2.32	2.89	1.40	1.12	1.34	1.07
A-VC-18C	30.44	119.80	496.04	4.10	2.22	2.76	1.40	1.12		
A-CR-19	27.53	51.68	384.68	2.18	2.56	2.62	0.85	0.83		
A-ZS-21	30.67	58.43	396.49	2.48	2.70	2.77	1.92	0.89		
A-T-7	35.06	47.25	483.16	3.02	3.33	2.96	0.91	1.02		
A-T-8	31.09	77.91	556.40	3.48	2.72	2.79	1.28	1.25		
A-T-9	35.06	99.44	371.39	2.32	2.38	2.96	0.98	0.78		

As was expected, uncoated bars were confirmed to give higher bond efficiency than coated bars for all the bar diameters investigated. Vinyl chloride surprisingly gave a better average bond strength ratio in all the bar diameters tested as in Figure 3. Epoxy and vinyl chloride are both manufactured locally, although, some of the key components may have been imported. Vinyl chloride and its thinner are cheaper. Vinyl chloride gave a higher bond resistance in all the bar diameters for 16 mm, 20 mm and 28 mm respectively.

7.1. Coating factor, β_1

Because the beams were not cast in the same batch, the concrete strength variation was eliminated by multiplying the obtained splice strength with $\sqrt{\frac{35}{f_{cu}}}$. In previous studies Mathey and Watsein [14], Ferguson and Thompson [15] and Tepfers [16] bond strength was found to increase with concrete strength approximately in proportion to the square root of the cube strength of concrete.

The coating factor (β_1) for bond strength was determined from the ratio of the corrected test bond strength of coated bars to the corrected test bond strength of uncoated bars for epoxy and vinyl chloride coated bars.

$$\beta_1 = \frac{\tau_t(coated)\sqrt{\frac{35}{f_{cu}}}}{\tau_t(uncoated)\sqrt{\frac{35}{f_{cu}}}} \tag{11}$$

The coating factor β_1 for epoxy was determined with an average of three beams for each bar diameter and was found to be 0.84, 0.95 for 16 mm, 20 mm diameter bars respectively.
Similarly, the average bond reduction factors for vinyl chloride coated bars for three beams reinforced with 16 mm, 20 mm bars respectively were found to be 0.89 and 0.95 see Table 4.

Table 4. Coating Factor, β_1.

Diameter (mm)	Coating $\beta_1 = \dfrac{\tau_t(coated)\sqrt{\frac{35}{f_{cu}}}}{\tau_t(uncoated)\sqrt{\frac{35}{f_{cu}}}}$	Factor Epoxy	Vinyl Chloride
16	0.70 0.81 1.02	0.84	
20	0.96 1.10 0.78	0.95	
16	0.72 0.96 0.99		0.89
20	0.92 0.96 0.98		0.95

The coating factor for 20 mm diameter bars are the same for epoxy and vinyl chloride but for 16 mm the coating factor is higher for vinyl chloride. Average result of three full size reinforced concrete beams appears reasonable in drawing a conclusion.

Statistically, testing for significant difference between the mean value of epoxy and vinyl chloride using t test $|t|$ was found to be 0.55, t_α table gave 2.92. Since the calculated value of t (0.55) is less than the tabulated value of t (2.92) at 5% level of significance, it implies that there is no significant difference between the mean values of epoxy and vinyl chloride. It should be noted, however, that full size reinforced concrete beams were tested and not models.

8. Conclusions

Based on the test results and analysis of test data it can be concluded that:

1. The bond efficiency of uncoated bars is generally higher than that of coated bars.
2. Bond strength and efficiency of bars coated with vinyl chloride are higher than that of epoxy and all other coating materials tested.
3. The coating factor for epoxy was 0.84, 0.95 for 16 mm, 20 mm diameter bars respectively. While vinyl chloride had a coating factor of 0.89 and 0.95 for 16 mm and 20 mm diameter bars.
4. Statistically, there appear to be no significant difference between the bond efficiency of the two materials (epoxy and vinyl chloride).

Acknowledgements

Late Emeritus Prof. C. O. Orangun's interest, criticism and useful suggestions are gratefully acknowledged.

Julius Berger Nig. Plc provided the materials, labour used in the construction of the formworks, reinforcement cages, casting, curing and transportation to the Concrete Laboratory, University of Lagos. The assistance is thankfully acknowledged.

Berger Paints Nig. Ltd. supplied the Epoxy coating used for the investigation. Vinyl Chloride and other coatings were supplied by International Paints for West Africa (IPWA). The assistance is appreciated and acknowledged. Sand blasting and coating of the steel rods were carried out free of charge by Messrs John Alapo Nig. Ltd. The assistance is thankfully acknowledged. The technical staff of the Concrete Laboratory, University of Lagos assisted in testing the beams, cubes and reinforcements, the assistance is acknowledged.

References

[1] Mathey, Robert G., and Clifton, James R., (1976) "*Bond of Coated Reinforcing bars in Concrete*"Journal of the Structural Division, Proceedings of the American Society of Engineers, Vol. 102, No. 511, pp. 215 – 229

[2] Clifton, James. R., Beeghly, H. F., and Mathey, R. G., (1973) "*Non-metallic Coatings for Concrete Reinforcing Bars*." Coating Materials. National Bureau of Standards Technical Note 786

[3] British Standards Institution, (1985) "*BS 8110: Part I: Structural Use of Concrete. Code of Practice for Design and Construction*."

[4] Kong, F. K., and Evans, R. H., (1987) "*Reinforced and Prestressed Concrete*," 3rd Edit. Van Nostrand Reinhold (U.K) Co. Ltd. 508pp

[5] Orangun, Cornelius. O., Jirsa, J. O., and Breen, John E. (1977) "*A Re- evaluation of Test Data on Development Length and Splices*" ACI Journal, Proceeding V. 74 No.3 pp.114-122

[6] Cairns, J, Abdullah, R., (1994) "*Fundamental Test on the Effect of an Epoxy Coating on Bond Strength*" ACI Materials Journal, Vol. 91, No 4, pp 331-337

[7] Leet, M. Kenneth, (1997) "*Reinforced Concrete Design*" 3rd Edition McGraw Hill Companies. Chp. 6, pp 202 - 247

[8] British Standards Institution, (1997) "*BS 8110: Part I: Structural Use of Concrete. Code of Practice for Design and Construction*."

[9] Treece, R. A., and Jirsa, J. O., (1989) "*Bond Strength of Epoxy Coated Reinforcing Bars*" ACI Materials Journal, Vol. 86, No. 2, pp. 167 – 174

[10] Cusen, A. R., and Yu, Z., (1993) "*Bond Strength and Flexural Behaviour of RC Beams with Epoxy – Coated Reinforcing Bars*." The Structural Engineer, Vol. 71 No.7/6 pp 117 – 124

[11] Kayyali, O. A., and Yeomans, S. R., (1995)"*Bond and Slip of Coated Reinforcement in Concrete, Construction and Building Materials*, Vol. 9, No. 4, pp. 219 – 226

[12] Thangavel, K., Rengaswamy, N. S., and Balakrishnan, K., (1995) "*Influence of Protective Coatings on Steel – Concrete Bond*." The Indian Concrete Journal, pp. 289 -293

[13] Yan C. and Mindess S., (1994) "*Bond Between Epoxy Coated Reinforcing Bars and Concrete Under Impact Loading*." Canadian Journal of Civil Engineering, pp.89 –99

[14] Mathey, Robert G., and Watsein, David, (1961)"*Investigation of Bond in Beam and Pullout Specimens with High Yield Strength Deformed Bars*," ACI Journal, Vol. 32, No. 9, pp.1071-1089

[15] Ferguson, Phil M., and Thompson, J. Neils., (1962) "*Development Length of High Strength Reinforcing Bars in Bond*" ACI Journal, Vol. 59, No. 7, pp 887 – 922

[16] Tepfers, R., (1976) "*A Theory of Bond Applied to Overlapped Tensile Reinforcement Splices for Deformed Bars*" Publication No. 73:2, Division of Concrete Structures, Chalmers University of Technology, Goteborg. 328 pp

Appendix A
Notations

A_s	Area of tension reinforcement
α_v	shear span
b	width of beam
BS	British standard
c	minimum cover
c_b	thickness of bottom concrete cover
c_s	thickness of side concrete cover
d	effective depth of beam
d_b	diameter of bar
f_{bu}	ultimate anchorage bond stress
f_c^1	cylinder strength of concrete
f_{cu}	characteristics cube strength
f_s	stress in steel
f_y	characteristics strength of reinforcement
k_1	ratio of the average compressive stress to the characteristic cube strength f_{cu}
k_2	ratio of the depth of the centroid of the stress block to the neutral axis depth
l_s	lap length
M_u	ultimate test bending moment
Q	applied load
τ_{aci}	Ultimate bond stress recommended in ACI 318 - 02
τ_{BS8110}	ultimate bond stress recommended in BS 8110:1997
τ_{cal}	theoretical bond stress
τ_t	test bond stress
x	depth of neutral axis
ρ	steel ratio A_s/bd
Ø_e	effective bar size
Σ_{cu}	strain in concrete in compression
Σ_s	strain in tension reinforcement
β	Bond coefficient
E	epoxy coated bars
CR	chlorinated rubber coated bars
VC	vinyl chloride coated bars
ZS	zinc ethyl silicate coated bars
U	uncoated beam (reference)
T	tyrolin coated bars

Self-healing concrete with crystalline admixture – a review

S Guzlena[1] and G Sakale[1]

[1]Institute of Technical Physics, Faculty of Material science and Applied Chemistry, Riga Technical University. P.Valdena 7, Riga, Latvia

Abstract. Concrete is the most used construction material for years. However long term serviceability is an issue, when we focus on sustainable building and use of materials. In long term service different types of visual and structural defects can appear in concrete structures. Defects in material enhances due to aggressive factors in outdoor (chlorides, sulphides and carbonates, freeze-thaw) which reduces material properties over a longer period. Mainly defects on visible surfaces of structures are from mechanical impacts, environmental exposure or due to impropriate maintenance. There are methods to reduce the risks and improve service life. One of them is self-healing materials, like bacteria, crystalline additives, in-capsulated healing agents and other. Use of self-healing material has preference, like lower maintenance costs for inspection, monitoring and complicated repair. Crystalline admixtures have several advantages comparing to other self-healing techniques, like improved concrete water tightness, no need for encapsulation before addition to concrete mix, which reduces sample preparation time.

1. Introduction

There are several chemical and physical factors, which affect aesthetic look and shorten service life of concrete structures. That is concrete structure exposure to aggressive agents as chlorides, sulphides, carbonates; water penetration through pours and small cracks due to absorption or hydrostatic pressure; freeze-thaw cycles [1]. As a result pores and small cracks like hairline cracking can be observed in concrete structures. To prolong service life self-healing materials can be used. Self-healing materials have many advantages: less people work and money saving due to inspection, monitoring and repair. Self-healing of concrete is triggered on its own. Self-healing ability of concrete [1], stainless steel reinforced concrete [2], high-performance concrete [3] and high-performance fibre reinforced concrete (HPFRC) [1], [3] have been researched. Due to healing mechanisms it can be divide in two parts:

- autogenic healing: natural closure of cracks due to un-hydrated cement in concrete and no specific self-healing admixtures are added. But this type of self-healing is not predictable and cannot be controlled [4], [5].
- autonomic healing: engineered closure of cracks with additives which is not a part of a ordinary cement [3]–[7]:
    - Passive – no need for human involving, self-healing starts from external promoters [2], [8].
    - Active – human involved to activate and complete the self-healing mechanisms [2], [8].

A lot of researchers have been investigated autonomic self-healing mechanisms with fibre reinforcement [1], [3], bacteria [9]–[12], crystalline additives [1], [3], [7], [13], [14], absorbent polymers [15], absorbent clay materials [6], incapsulated healing agents [4], [6], [16], expansion agents like sulfo-aluminate [6], [13], [14] and others.

In this review most of interest is focused on chemical additives – crystalline additives (CA), because of their dual positive influence – crack healing and concrete permeability reduction ability. In moist environment CA react with cement and forms calcium silicate hydrates. These additives help to reduce concrete porosity and water permeability as well [4]. Difference between CA (permeability reducing admixtures) and hydrophobic additives are - the first ones have hydrophilic nature apart prom CA [7]. Due to Belie et. al. [17] CA performance is largely influenced by mix design. Better effect can be noticed with HPFRC. But still healing process for 0,2 mm wide cracks is relatively time demanding from few weeks to several months.

2. Autogenous crack healing

Autogenous crack healing mechanisms has been studied almost for forty years. Main conclusions are that crack healing mechanisms changes by concrete age. In case of young concrete main mechanism is hydration of un-hydrated cement particles in concrete in presence of high moisture [1], [3], [4], [18], [19]. In case of concrete with low water:cement ratio, <0,3, even 20-30% of all cement particles can be un-hydrated and participate in self-healing process [5], [20]–[22]. For older concrete calcium hydroxide ($Ca(OH)_2$) converts to crystalized calcium carbonate ($CaCO_3$) as a result of exposure to atmosphere with carbon dioxide (CO_2) [1], [4], [7], [18], [19], [23]. This phenomena first was discovered in 1836 by the French Academy [8]. Crack size which can be healed by this mechanism's variates from 0,05 mm till 0,3 mm, but it can be unpredictable. Much more presence of water, not moisture, is needed [6], [14], [24]. Different additives such as fly ash, blast furnace slag, silica fume, limestone powder promote autogenous crack-healing in later stage in concrete due to un-hydrated binders, which have slower pozzolanic reactions [1], [3], [4], [7], [18], [23], [25]. These pozzolan materials in presence of water reacts with portlandite and produce binding products [25]. Van Tittelboom et al. have done research adding to Portland cement fly ash or blast furnace slag as pozzolanic additives. It was found that concrete crack self-healing rate is 0,015 mm/day, and are influenced by environmental conditions [26]. K. Tomczak et.al. has challenged autogenous healing by making samples with very low w:c ratio 0.28 and 0.23. Pre-cracked samples with crack width 0-0,8 mm was healed in water for 28 days. Cracks below 0,46 was completely healed after 2 months, but wider cracks have healed only for 15-30%. Self-healing speed also changes during time, but most influencing factor is crack width, if cracks are wider healing considerably slows down [27].

3. Autonomous crack healing using crystalline additives

Crystalline additives are mostly water soluble additives which improve concrete properties at fresh and hardened state [24]. As American concrete institute (ACI) writes on their report crystalline additives are one type of permeability reducing admixture. ACI divide permeability reducing additives in two parts. That reduces water permeability at no-hydrostatic conditions and at hydrostatic conditions. In category that reduces water permeability at no-hydrostatic conditions falls typical hydrophobic products. Admixtures that can function under hydrostatic conditions include crystalline admixtures. Some of these additives have a lot of benefits like reducing drying shrinkage, minimize chloride ion penetration to surface, improved frost resistance and autogenous healing properties [28].

Crack width that can be healed with CA variates till 300 μm. [4], [24] but if the techniques are combined, like crystalline additives with expansive admixtures, healed crack width can be increased till 400 μm [7].

Moisture is very important if chemical agents (crystalline additives) is added in concrete to promote deposition and fill the crack. Usually in outside environment it is not an issue [4], [7]. After crack is formed, in presence of moisture tricalcium silicate (C_3S) reacts and Calcium Silicate Hydrate (CSH) is deposited on crack walls and it fill the crack[7]. Quality of concrete self-healing ability can be affected by mix proportion, temperature, environmental humidity and stress in crack region. If stress is too high, at already healed crack area new cracks can appear [4]. Calcium carbonate as an deposit in self-healed crack can appear in different morphologies – calcite, aragonite, vaterite, monohydrocalcite, ikaite and amorphous $CaCO_3$. By-product which appeared after concrete self-healing with CA mostly is the same that can be seen after autogenous healing – calcite, brucite and aragonite.

It must be mentioned that self-healing using crystalline additives show better results when samples are immersed in water or subjected to wet-dry cycles [2].

Authors mostly use commercially available crystalline additives, as WT- 250 commercialized by Sika [3] Admix C-1000NF by Xypex [6] Penetron Admix by Penetron [5] or do not specify what crystalline admixture is used. Esscffres et. al. has mentioned higher Mg content in admixture which can lead to higher amount of healing by-product aragonite [3]. Other authors [1], [5], [6] have mentioned as one of compounds reactive silica or micro silica.

Results that are achieved differs, because of many different parameters used in each research. Starting from mixture: cement type, sand:cement proportion, water:cement proportion, different additives, dosage and type (superplasticizers, viscosity agents). There is a luck of standardized testing method, methods of testing differs almost in each paper. Therefore, it is hard to compare the results. In fallowing text research by different authors will be discussed one by one and summarized in table 1. This problem have mentioned also other authors and standardizations is needed [29]–[31].

Ferrara et. al. [1] used crystalline admixture which consists from cement, sand and micro silica. Admixture were investigated by EDS analyses and compared with Portland cement specimen. Only difference was slightly higher peak of sulphur. Normal strength concrete (NSC) and HPFRCC samples were pre-cracked after 2 months and healed by immersing in water and exposed to an open air. Crack width were 0,2 mm for NSC and deflection till 2mm for HPFRCC. Normal strength concrete and HPFRCC with crystalline admixture showed good healing capacity after 1 month. In both cases better results was achieved if samples were immersed in water. Sample testing continued and showed improvement even after 12 months [1].

Escoffres et. al. [3] used commercially available crystalline admixture from SIKA WT-200. EDS analysis was made, and high amount of Mg were found in this admixture. After 28 days sample prisms of HPC, HPFRC and HPFRC-CA have been exposed to three-point bending test to pre-crack samples with crack width close to 0,2 mm. Part of each material was immersed in water and one part left in air. After 56 days all prisms were tested again with three-point bending test. HPFRC-CA showed by 10-20% better results than HPFRC in air exposure. HPC, HPFRC and HPFRC-CA self-healing ability is discussed during constant loading similar to service conditions [3].

M. Roig-Flores and her team [7] has investigated crack-healing ability of concrete samples with and without crystalline admixture. Pre-cracked 2 days old samples (crack width bellow 0,3 mm) were investigated. Different conditions – water immersion, water contact, humidity chamber and air exposure were used to accelerate crack-healing. Conclusions are that crack healing abilities increase for samples with crystalline admixture that have been immersed in water. Cracks have been healed around 95%, but samples without admixture only 75%. Samples with and without crystalline admixture that have been exposed in humidity chamber and in air have very low crack healing ability. All samples have been compared after 0 and 42 days [7].

Wang et. al. [6] used porous expanded clay aggregates as sodium carbonate container, commercially available crystalline additive from Xypex Admix C-100NF and sulfoaluminate based expansive additive Denka CSA#20. Crystalline additive mainly consists from crystalline catalysts and reactive silica. Samples were pre-cracked after 7 days and healed in water for 28 days. Best self-healing results showed samples with mix proportions 10-12% of Na_2CO_3, 7-10% of sulfoaluminate based expansive agent, 4,5% of crystalline additive [6].

Sisomphon et.al. [14] used in his research crystalline additive and expansive additive. Crystalline additive consits from reactive silica and crystalline catalysts (hydrophilic water-proofing materials). As an expansive additive calcium sulfoaluminate based commercial product was used with mineral composition of Hauyne, anhydrite and free lime. Samples were pre-cracked after 28 days. Pre-cracked sample crack size was 0,1-0,4 mm. Samples were immersed in water for 28 days. Surface cracks up to 0,15mm can be closed in 28 days [14].

Cuenca et. al. [5] has done research of self-healing of fibre reinforced concrete after several self-healing - pre-cracking cycles. To inhance self-healing crystaline admixture was used. Samples were pre-cracked using Double Edge Wedge Splitting test and crack width was controled with crack mouth opening displacements (CMOD). Samples were cured after pre-cracking in water, at open air and wet-dry cycles. Samples at air represented low crack-healing ability. Completly crack healed only for samples with crack width less than 0,3 mm, which were immersed in water, dosen't matter perminantly or periodicaly. Ever more showed good repeated crack healing ability over an 1 year after repeated pre-cracking tests.

Coppola et.al. [32] has investaigated novel method of self -sealing concrete by adding in concrete mixture fumaric acid-vased admixture (WP). It was found that by using WP is posible to parcipitate denser crystals compearing with comercially available products. Cracks till 0,4mm is compleatly filled with white crystal products. No improvement on waterproofing was seen if doage of WP increased form 1 to 2 % versus cement content. Cracks till 0,4mm was healed alreay after 7 days after sample immersion in water.

Roig-Flores et.al. [2] has investigated healing in erly age concrete, that is, samples were pre-cracked at age 2 days. Crystalline admixture influcence on crack healing at different conditions were evaluated: water immersion at 15 °C and 30 °C and wet/dry cycles. Exposrue time 42 days. Best results were obtained for samples immersed in water and held at 30°C temperature. Samples subjected to these conditions gained 0.99 average healing ratio with smallest standard deviation in results. Cracks in width to 0.4mm were healed.

Table 1. Summary of concrete types, mix design, used crystalline additives, pre-cracking methods and healing.

Reference	Concrete type	Cement type	W:C	Additive	Method to crack	Time till Pre-cracking	Crack width	Healing - days	Conditions
[1]	normal strength concrete	II 42.5	0,63	blend of cement, sand and microsilica	3-point bending test with COD-controller	35-42 days	130 and 250 μm	1, 3, 6 and 12 months	immersed in water or exposed to open air
[6]	normal strength concrete	Not specified	0,25	calcium sulfoaluminate based expansive additive (CSA, Denka CSA#20) and crystalline additive (CA, Xypex Admix C-1000NF)	3-point bending test. notch in the middle of sample	7 days	0–200 μm	28 days	healing in still water
[23]	normal strength concrete (NSC)	P II 52.5	0,45	calcium sulphoaluminate expansive agent and natural metakaolin with a small amount of additives	compressive strength tests	7 d	Not specified	28 d	water
[2]	fiber-reinforced concrete	II/A-L 42.5 R	0,6	not Specified	splitting test	2 days	100 and 400 μm	42 days	water immersion at 15 °C, water immersion at 30 °C, and wet/dry cycles
[5]	fiber reinforced concrete	II 42.5	0,5	crystalline admixture (Penetron Admix)	residual flexural tensile strength with CMOD	3, 7; 28 and 56 days and 4 months	below 300 μm	Till 6 months	submitted to air exposure
[7]	fiber-reinforced concrete	II/A-L 42.5 R	0,45	not specified	splitting test	2 days	below 300 μm	42 days	water immersion, water contact, humidity chamber and air exposure at laboratory conditions
[14]	fiber reinforced concrete	I 42.5 N	0,25	crystalline additive and expansive additive	splitting tensile strength test	28 days	100–400 μm	4, 7, 14- and 28-days	water
[3]	high- performance fiber reinforced concrete	Not specified	0,43	crystalline additive (WT- 250 by Sika)	three points bending with RCOD	28 days	200 μm	56 days	immersed in water submitted to air exposure
[1]	high performance fiber reinforced cementitious composites	I 52.5	0,18	blend of cement, sand and microsilica	4-point bending with COD-controller	2 months	130 and 250 μm	up to six months	immersed in water

4. Conclusion

Overlooking literature research on self-healing concrete has been made for different type of concretes, cements and w:c ratio in recent years. Various pre-cracking time has been investigated which can affect healing rate and results. If pre-cracking has been made at early stage of concrete hardening, autonomic healing goes together with autogenic healing and it is hard to differ these healing processes. Applied healing time differs from healing conditions. One can be concluded, if crystalline additives are used in cementitious materials more important is water then moisture in surrounding. Samples immersed in water gain faster crack healing and quality of healed crack is better, even cracks after repeated pre-cracking can be healed. This limits the use of the crystalline additive in materials for facades, panel constructions and other objects in the outdoor and indoor environment which are not in direct contact with water.

References

[1] Ferrara L, Krelani V, and Moretti F, "On the use of crystalline admixtures in cement based construction materials : from porosity reducers to promoters of self healing," *Smart Mater. Struct.*, vol. 25, no. 8, pp. 1–17
[2] Roig-Flores M, Pirritano F, Serna P, and Ferrara L, "Effect of crystalline admixtures on the self-healing capability of early-age concrete studied by means of permeability and crack closing tests," *Constr. Build. Mater.*, vol. 114, pp. 447–457, 2016
[3] Escoffres P, Desmettre C and Charon J. P, "Effect of a crystalline admixture on the self-healing capability of high-performance fiber reinforced concretes in service conditions," *Constr. Build. Mater.*, vol. 173, pp. 763–774, 2018
[4] Ferrara L, Krelani V and Carsana M, "A "' fracture testing "' based approach to assess crack healing of concrete with and without crystalline admixtures," *Constr. Build. Mater.*, vol. 68, pp. 535–551, 2014
[5] Cuenca E, Tejedor A and Ferrara L, "A methodology to assess crack-sealing effectiveness of crystalline admixtures under repeated cracking-healing cycles," vol. 179, pp. 619–632, 2018
[6] Wang X, Fang C, Li D, Han N and Xing F, "A self-healing cementitious composite with mineral admixtures and built-in carbonate," *Cem. Concr. Compos.*, vol. 92, pp. 216–229, 2018
[7] Roig-flores M, Moscato S, Serna P and Ferrara L, "Self-healing capability of concrete with crystalline admixtures in different environments," *Constr. Build. Mater.*, vol. 86, pp. 1–11, 2015
[8] Oucif C, Voyiadjis G. Z and Rabczuk T, "Modeling of damage-healing and nonlinear self-healing concrete behavior : Application to coupled and uncoupled self-healing mechanisms," *Theor. Appl. Fract. Mech.*, vol. 96, no. March, pp. 216–230, 2018
[9] Zhang J. *et al.*, "Immobilizing bacteria in expanded perlite for the crack self-healing in concrete," *Constr. Build. Mater.*, vol. 148, pp. 610–617, 2017
[10] Vijay K, Murmu M and Deo S. V, "Bacteria based self healing concrete – A review," *Constr. Build. Mater.*, vol. 152, pp. 1008–1014, 2017
[11] Huynh N. N. T, Phuong N. M, Toan N. P. A and Son N. K, "Bacillus Subtilis HU58 Immobilized in Micropores of Diatomite for Using in Self-healing Concrete," *Procedia Eng.*, vol. 171, pp. 598–605, 2017
[12] Xu J and Wang X, "Self-healing of concrete cracks by use of bacteria-containing low alkali cementitious material," *Constr. Build. Mater.*, vol. 167, pp. 1–14, 2018.
[13] Park B and Cheol Y, "Quantitative evaluation of crack self-healing in cement-based materials by absorption test," *Constr. Build. Mater.*, vol. 184, pp. 1–10, 2018.
[14] Sisomphon K, Copuroglu O and Koenders E. A. B, "Self-healing of surface cracks in mortars with expansive additive and crystalline additive," *Cem. Concr. Compos.*, vol. 34, no. 4, pp.

566–574, 2012
[15] Snoeck D, Van Den Heede P, Van Mullem T and De Belie N, "Cement and Concrete Research Water penetration through cracks in self-healing cementitious materials with superabsorbent polymers studied by neutron radiography," *Cem. Concr. Res.*, vol. 113, no. May, pp. 86–98, 2018
[16] Van Belleghem B, Kessler S, Van Den Heede P and Van Tittelboom K, "Cement and Concrete Research Chloride induced reinforcement corrosion behavior in self-healing concrete with encapsulated polyurethane," *Cem. Concr. Res.*, vol. 113, no. June, pp. 130–139, 2018
[17] De Belie N *et al.*, "A Review of Self-Healing Concrete for Damage Management of Structures," vol. 1800074, pp. 1–28, 2018.
[18] Park B and Cheol Y, "Self-healing capability of cementitious materials with crystalline admixtures and super absorbent polymers (SAPs)," *Constr. Build. Mater.*, vol. 189, pp. 1054–1066, 2018
[19] Huang H, Ye G, Qian C and Schlangen E, "Self-healing in cementitious materials: Materials, methods and service conditions," *Mater. Des.*, vol. 92, pp. 499–511, 2016
[20] Lv Z and Chen D, "Overview of recent work on self-healing in cementitious materials," vol. 64, no. 316, 2014
[21] Van Breugel K, "IS THERE A MARKET FOR SELF-HEALING CEMENT- BASED MATERIALS ?," no. April, pp. 1–9, 2007.
[22] Chen J *et al.*, "A coupled nanoindentation / SEM-EDS study on low water / cement ratio Portland cement paste : Evidence for C-S-H / Ca (OH) 2 nanocomposites To cite this version : HAL Id : hal-00555495," 2011
[23] Jiang Z, Li W, Yuan Z, and Yang Z, "Self-healing of cracks in concrete with various crystalline mineral additives in underground environment," *J. Wuhan Univ. Technol. Mater. Sci. Ed.*, vol. 29, no. 5, pp. 938–944, 2014
[24] Azarsa P and Gupta R, "Crystalline Waterproofing Admixtures Effects on Self-healing and Permeability of Concrete," no. February, 2018
[25] Dembovska L, Bajare, D, Pundiene I and Vitola L, "Effect of Pozzolanic Additives on the Strength Development of High Performance Concrete," *Procedia Eng.*, vol. 172, pp. 202–210, 2017
[26] Van Tittelboom K, Gruyaert E, Rahier H and De Belie N, "Influence of mix composition on the extent of autogenous crack healing by continued hydration or calcium carbonate formation," *Constr. Build. Mater.*, vol. 37, pp. 349–359, 2012
[27] Tomczak K and Jakubowski J, "The effects of age , cement content , and healing time on the self-healing ability of high-strength concrete," *Constr. Build. Mater.*, vol. 187, pp. 149–159, 2018
[28] American Concrete Institute Committee 212, "Report on Chemical Admixtures for concrte (ACI 212.3R-10)," 2010
[29] Zakari N *et al.*, "Tests and methods of evaluating the self-healing efficiency of concrete : A review," *Constr. Build. Mater.*, vol. 112, pp. 1123–1132, 2016
[30] Ferrara L, Asensio E. C, Lo Monte F and R. Flores M, "Experimental Characterization of the Self-Healing Capacity of Cement Based Materials : An Overview †," pp. 1–7, 2018
[31] Ferrara L *et al.*, "Experimental characterization of the self-healing capacity of cement based materials and its effects on the material performance : A state of the art report by COST Action SARCOS WG2," *Constr. Build. Mater.*, vol. 167, pp. 115–142, 2018
[32] Coppola L, Coffetti D and Crotti E, "Innovative carboxylic acid waterproofing admixture for self-sealing watertight concretes," *Constr. Build. Mater.*, vol. 171, pp. 817–824, 2018

Spectrophotometric evaluation of pigmented urethane acrylate coating colour stability

Z Grigale-Soročina[1,2,3], I Birks[1] and R Pilipavica[1]

[1]R&D, Kinetics Nail Systems, Kurzemes prospekts 3K, LV-1026, Riga, Latvia,
[2]Institute of Polymer Materials, Faculty of Material Science and Applied Chemistry, Riga Technical University, Paula Valdena Str. 3, LV-1048 Riga, Latvia

E-mail: zane.grigale@kineticsbeauty.com

Abstract. The research was carried out testing two red pigment dispersion colour stability in UV light-curing urethane acrylate composition. The given urethane acrylate composition was pigmented with a pigment content of 0.1% to 5%. Coating colour change was evaluated at room temperature and at 50°C over a 28-day period. For all prepared pigment samples, spectrometric colour difference measurements were performed using CIEL*a*b* and CIEL*C*h* colour systems and FTIR-ATR to evaluate their stability over time. The colour stability was improved with increasing pigment concentration. There was found the difference between tested red pigments and can be concluded that one of them are more stable for use in pigmented coatings.

1. Introduction

The testing of thin coloured coatings based on urethane acrylate composition is a technologically advanced, highly specialized and complex process. For decorative pigmented coatings (construction coatings, electronical coatings and cosmetical coatings), colour stability is crucial quality parameter [1, 2, 3]. Optical microscopy and spectroscopy methods are used to test painted coatings.
Different spectroscopy methods can give the additional information about the system changes in time. Raman spectrometry is used to analyse the changes of compositions in time. Infrared spectrometry provides information about polymer binder and fillers. Thin layer chromatography (TLC) can be also used not only for colour examination but also for the identification of organic pigments [4, 5, 6].
The main target of all those methods is to estimate the difference of the colour. The most objective way to compare the colours in the range of visible light is micro spectrometry. This method is most often used for the examination of colours [6].

There is polymer coating colour evaluation method accepted by CIE (Commission Internationale de l'Eclairage) which quantitatively describes the colour using colour theory.
This method is important for evaluation of polymer coating colour over time and environment: water, heat, UV etc. [4, 6, 7].

It shows colour - difference with measurable quantities, therefore it is a very useful tool for predicting the lifetime of coatings. Method is based on describing the colour coordinates: CIEL/a/b/characterized by axis: L – for lightness, a – green – red, b – blue – yellow colour - opponent dimensions [8, 9, 10].
The aim of this paper is to compare two different red pigment dispersions and assess whether the pigment concentration in the polymer binder has significant influence on the coating colour stability under conditions of room T and increased T.

2. Materials and methods

2.1. Materials

Two sets of red - pigmented coating samples were prepared. Red #170 (CI 12475 from *Kromachem*) and Red #48 (CI 15865, from *Kromachem*) pigments were mixed in urethane acrylate base. As oligomer was used Di-HEMA Trimethylhexyl Dicarbamate from *Sartomer*, as monomer was used Hydroxyethyl Methacrylate from *Esstech*, and as photoiniciator was used Ethyl Trimethylbenzoyl Phenylphosphinate from *IGM*. Pigments were incorporated in the mixture in 4 mass concentrations: 0.1%, 0.2%, 1% and 5% (Table 1). Previous experience have shown that increased pigment concentration, improves colour stability, what should be proved.

Table 1. Coating formulation.

Formulation name	Pigment	Pigment conc. [%]	Oligomer conc. [%]	Monomer, conc. [%]	Photoiniciator conc. [%]
R170-0,1	Red 170	0,1	64,9	30	5
R170-0,2	Red 170	0,2	64,8	30	5
R170-1	Red 170	1	64	30	5
R170-5	Red 170	5	60	30	5
R48-0,1	Red 48	0,1	64,9	30	5
R48-0,2	Red 48	0,2	64,8	30	5
R48-1	Red 48	1	64	30	5
R48-5	Red 48	5	60	30	5

2.2. Coating application

Red pigments were dispersed in urethane acrylate base by stirring for 4 min at 1800 rpm in *Dispermill KK 250*. The coating was applied onto *BYK* paper test panels by a *BYK* film applicator. The wet film thickness was 200 μm. All coating samples were cured in UVLED lamp (*KP800LED*, luminous intensity of 130 mw, λ = 405 nm from *YI Liang Electron Technology Co*) for 30 s. Compositions were coated with transparent PE film with thickness 50μm to avoid oxygen inhibition activity on the curing coating surface. After curing, the PE film was removed [6, 11].

2.3. Test methods

The coating of each composition was applied on *BYK* paper test panels and cured under UV light. These coatings were used as reference samples for colour tests. The other composition was divided into 2 jars and one of them was kept in room temperature and another one in 50°C for 28 days (as T where pigmented systems have been tested on sedimentation stability). Compositions were evaluated after 2 days, 7 days and 28 days. Applications were made from these compositions and used for colour evaluation [5, 6].

The spectrophotometer measured the colour of prepared pigment coatings. It measured the colour on both the white and black parts of the application sheet. The first colour measurement became the reference for which, in the course of the study, a comparison of the pigment colour with light - curable urethane acrylate mixture was compared. The resulting applications were stored in the dark [5, 6].
The colour evaluation were performed on a *Datacolour DC200* spectrophotometer using *Colourmatch Plus* software. On the base of spectra the colour parameters (chromaticity coordinates) and colour difference parameter ΔE were calculated using the following formula [5, 6, 7]:

$$\Delta E = \sqrt{(L_1^* - L_2^*)^2 + (a_1^* - a_2^*)^2 + (b_1^* - b_2^*)^2} \qquad (1)$$

Where L_1^*, a_1^*, b_1^*, and L_2^*, a_2^*, b_2^* are the CIE Lab values obtained for 2 samples labelled as 1 and 2, where 1 is the reference sample and 2 the tested sample after aging conditions. Colour is a subjective response of the eye and brain to incident radiation that illuminates an object. To eliminate the subjectivity of human perception, colour coordinates are calculated from the visible and infrared spectrum. The mathematical distance between two points in colour space (ΔE) is a numerical description of a difference in colour [5].

Fisher Scientific, Nicolet 6700 was used to record coating spectra in the absorption area 650 cm-1 to 4000 cm-1, with the exception of 16 cm-1. During testing, a method is adopted using ZnSe selection, where the incident light angle is 42o and the printing depth is 2.03 µm at 1000 cm-1.

3. Results and discussions

3.1. Colour stability

Colour stability in time is essential criterion for pigmented coatings. Therefore, it is important to be able to predict the non - polymerized composition and cured coating colour change in time [2, 3, 5]. This research allows the prevention of poor quality products and unstable coatings in new product development and production.

In this study, we evaluated Red #107 and Red #48 pigments in urethane acrylate composition. As red pigments are one of the most often used, it is important to choose the most stable and most appropriate for urethane acrylate system. The colour stability of the pigmented coating was analyzed at room temperature and at 50° C for 28 days.

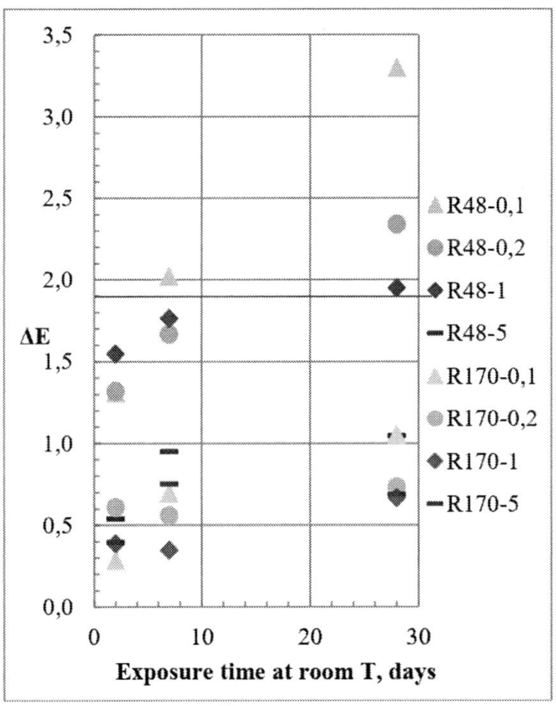

Figure 1. Colour difference (ΔE*) change of coating samples as a function of time exposed at room temperature; R48-0.1 to R170-5 refers to different pigment concentrations as listed in Table 1. Red line indicates acceptable colour change.

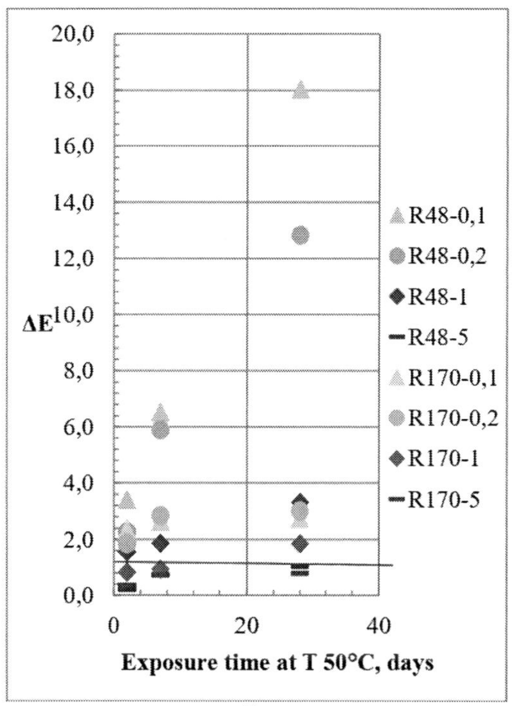

Figure 2. Colour difference (ΔE*) change of coating samples as a function of time exposed at 50°C temperature; R48-0.1 to R170-5 refers to different pigment concentrations as listed in Table 1. Red line indicates acceptable colour change.

Figure 1 reports the colour difference of all samples as a function of exposure time in room T. For both pigmented compositions, the ΔE values increase with the increase of the exposure time. Also there is the bright tendency – the highest the pigment concentration – lowest the difference in ΔE. The lowest ΔE values are obtained for formulations: R170-5.0 (5% of Red #170 pigment) and R48-5.0 (5% of Red #48 pigment), where $\Delta E \leq 1.2$. According to DIN 55987 [6], stability of coatings is considered acceptable and sufficient if $\Delta E \leq 2$.

Compositions with 0.1% and 0.22% of pigment Red #48 shows the highest ΔE increase in increased T: formulation with 0.1% after 28 days the ΔE value was increased to 18.05, but with the pigment content of 0.2% to 12.83. This difference is also observable with eyes: Figure 3 shows coating color before the test (figure 3. A) and after 28 days test (figure 3.B). Significantly, less ΔE growth was observed with a pigment content of 1.0% to 3.32 and 5.0% to 1.03, but still 1% pigment composition is not acceptable for use in coatings.

In this work, after exposure at room T for 28 days, formulations containing more than 0.2 % of Red #48 pigment and all formulations containing Red #170 pigments, ΔE value were less than 2. It indicates that all these formulations have a good stability at room T.

Figure 2 shows that all pigmented systems tested containing Red#170 had the same colour change after 2 days and after 28 days. ΔE fluctuates within the error range. These results will allow reducing the colour stability testing time from 28 days to 2 days.

The increase of exposure temperature significantly increase ΔE values of all formulations, especially those containing pigment #48. All formulations with different concentrations of this pigment in 28 days reached ΔE values higher than 2. After this experiment, it's clear that T increase noticeably increase colour change. This allows concluding, that pigmented systems should be tested not only at room T, but also at increased temperature regime.

In addition, there is a correlation between ΔE and pigment concentration: higher is the pigment concentration – more stable is the colour (smaller ΔE).

For all of the coatings tested, ΔE at room temperature is less than in increased temperature. This can be explained by colour temperature sensitivity and degradation with temperature increase. It is obvious that this degradation occurs in first 48h.

Testing these two *Kromachem* pigments in urethane acrylate composition we can conclude, that acceptable formulations are with pigment Red#170 with concentrations above 0.2%.

Figure 3. Colour change of coating samples R48-0.1: A- before test; B – after 28 days exposed at 50°C temperature.

3.2. Colour change detection with FTIR-ATR

The IR spectra of pigmented urethane acrylate compositions with red pigments #48 and #170, was studied. FTIR spectra of the pigments alone was recorder and noticed that the bands are not overlapping with the polymer. Also clear urethane base spectra was recorded.

Spectral analysis of pigmented samples before and after tests was very close to the reference composition analysis without pigments. Spectra showed the changes in absorption intensity of urethane acrylate group peaks, but there was no detected presence of characteristic peaks of pigments (Figure 4 and 5).

After these tests we can conclude, that this used FTIR-ATR test method is not useful for colour change detection, at least at so small pigment concentration we need to use more precise IR spectrometer. There was no correlation between the colour difference ΔE and the IR spectra. This is confirmed by figure 4 and figure 5: spectrum comparison of Red # 48 and Red#170 with 0.1% content before and after testing at temperature 50°C. The value of the colour difference ΔE increased sharply during the test, the colour change was observed visually and using spectrophotometer but was not reflected in the IR spectrum (see Figure 4 and Figure 5).

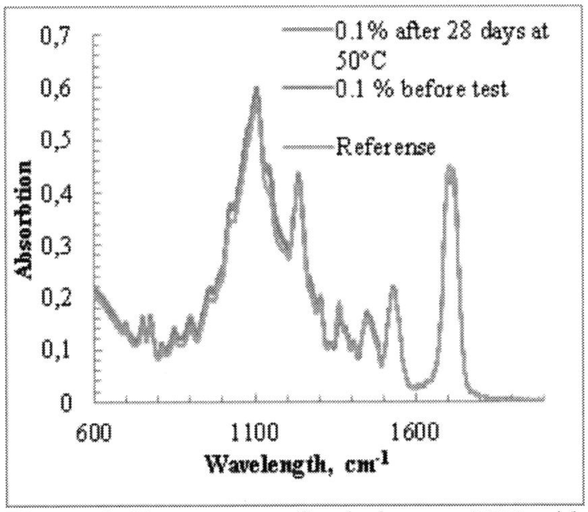

Figure 4. IR spectrum of red pigment #170 with 0.1% of pigment content before and after 50°C.

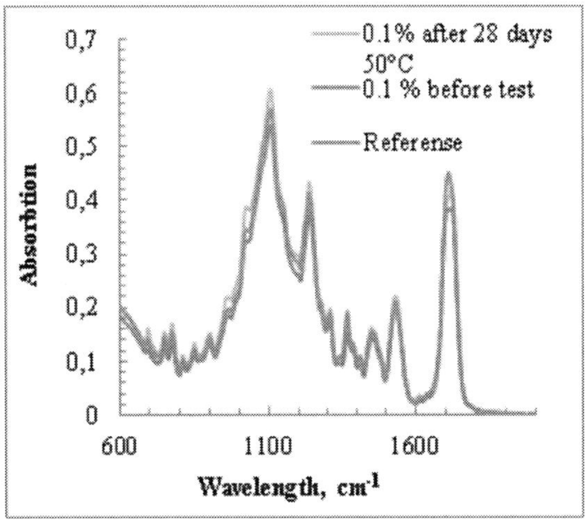

Figure 5. IR spectrum of red pigment #48 with 0.1% of pigment content before and after 50°C.

4. Conclusions

In this study, colour stability of dispersions with red #48 and red #170 pigments were evaluated by incorporating them into UV light - curing urethane acrylate coatings. Assumption, that pigments can prevent coatings from colour degradation was confirmed. Two sets of paint samples: one containing Red #48 and another Red #170 pigment were prepared. Each set contains four samples with different concentrations of pigments. After application and UV cure, the samples were examined using MSP - Vis method with CIEL/a/b and FTIR - ATR. The colour was compared after two aging tests (at room temperature and at 50°C) in 28 days period. It was found that coating stability increases with increasing Red pigment content.

The colour difference ΔE of the samples stored at room temperature and 50 °C after 2 days is mostly equivalent to a 1 - month result, so it can be concluded that if the pigments in the urethane acrylate compositions have insignificant change in 2 days, that there will not be significant change also in 28days. The ΔE values for coatings containing Red #170 more than 2% of pigments are acceptable and within tolerances, as evidenced by the professional team of paint chemists who could not observe any changes in colour. Pigment Red #48 is not recommended to use for high quality coatings, as all concentrations showed ΔE values after experiments higher than acceptable.

Acknowledgments

Authors thank the Central Finance and Contracting Agency of Latvia (CFCA) (Contract No. 1.2.1.1/16/A/005) for their financial support.

References

[1] Milinavičiūtė A, Jankauskaitė V, Narmontas P *2011 Mater. Sci. (Medžiagotyra)* **17(4)** 378
[2] Grigale-Sorocina Z, Kalnins M, Simanovska J, Vindedze E, Birks I, Brazdauska E 2015 *Proc Est Acad Sci Chem* **64** 88
[3] Johnson B W, Mcintyre R 1996 *Prog Org Coat* **27**
[4] Erznožnik H, Razboršek T, Gunde M K 2016 *Prog Org Coat* **99** 47
[5] Grigale-Sorocina Z, Pilipavica R, Birks I 2018 *SOJ Mater Sci Eng* **6(1)** 1

IOP Conf. Series: Materials Science and Engineering **660** (2019) 012058 doi:10.1088/1757-899X/660/1/012058

[6] Jose S, Prakash A, Laha S, Natarajan S, Reddy M L 2014 *Dyes Pigm* **107** 118
[7] Weixiu Z, Qixin Z, Haichang Z, Xiaoning Q 2018 *Dyes Pigm* **151** 157
[8] Trzcinska B, Kowalski R, Zieba-Palus J 2014 *Mol Biomol Spectro* **130** 534
[9] Gao Q, Wu X, Fan Y 2014 *Dyes Pigm* **109** 90
[10] Bierwagen G, Tallman D, Li J, He L. Jeffcoate C. 2003 *Prog Org Coat* **46** 148
[11] Grigale-Sorocina Z, Kalnins M, Gross K A 2016 *IOP Conf Ser: Mat Sci and Eng* **111** 1

Variation of bending strength of fiber reinforced concrete beams due to fiber distribution and orientation and analysis of microstructure

H Herrmann[1], R Boris[2], O Goidyk[1] and A Braunbrück[1]

[1] Tallinn University of Technology (TalTech), School of Science, Department of Cybernetics, Ehitajate tee 5, 19086 Tallinn, Estonia

[2] Vilnius Gediminas technical university, Institute of Building Materials, Laboratory of Composite Materials, Linkmenu str. 28, 08217 Vilnius, Lithuania

E-mail: hh@cens.ioc.ee, renata.boris@vgtu.lt

Abstract. Fiber reinforced concrete is becoming increasingly popular as a construction material, as it can potentially form a ductile concrete. The properties of the fiber reinforced concrete depend on the concrete recipe, the flow of the fresh concrete into the formwork, possible vibrating of the concrete and the fiber orientations.

This contribution presents the results of bending tests performed on fiber reinforced concrete beam specimens, which have been cut out of a larger plate. These beams have different fiber orientation distributions, due to being taken from different parts of the plate and with different orientation with respect to the flow of the fresh concrete. Further, the microstructure of the concrete was investigated using scanning electron microscopy (SEM). The SEM shows no obvious differences in the microstructure between beams of different strength and shows good adhesion of fiber and concrete matrix in all investigated samples.

1. Introduction

The use of fiber reinforced concrete (FRC) is becoming increasingly popular, especially in the casting of concrete floors, here, short steel or plastic fibers or both, provide for the concrete floor resistance to drying shrinkage cracks.

Apart from concrete floors, other load-bearing elements retain some conventional reinforcement, even when cast with fiber reinforced concrete. The structural integrity of an element cast purely of fiber reinforced concrete still has questions to be answered in full. How, for example, the casting conditions –like flow of the fresh concrete and fiber orientations– influence the strength and ductility of the hardened fiber reinforced concrete? Some answers could be found partly in the previous research, for instance, the spatial and orientational fiber distribution within a concrete beam is to a large degree depending on formwork surface quality [1–3] and wall-effect [2, 4] and the casting methods.

There are a several methods to investigate the durability characteristics of the materials, to determine the coupling bonds and the weakest places inside of concrete mass and to analyze the post-cracking behavior. Many research groups have applied three-point or four-point bending tests [5, 6], compression test [7] and pull-out test [8, 9].

To correlate the obtained results from the experiments with the existing fiber orientation, the Computed Tomography technique is widely applied. Recently, use of X-ray Computed Tomography (CT) in civil engineering research is becoming quite popular. X-ray CT is an advanced nondestructive method that have been widely used for obtaining 3D images of the internal micro-structure of the objects [10–13]. This technique allows to get the information about defects in the concrete structure, such as pores, micro-cracks and enables to determine the fiber orientation and spatial distribution inside of the non-transparent concrete matrix [14–17].

The expected influence of the fiber orientation distribution has been highlighted before [18–22]. Numerical simulations can used to predict the fiber orientation and distribution during the manufacturing process [2, 3, 23, 24] and the influence on mechanical properties [25].

This paper reports recent test results on the local strength and ductility of the short steel fiber reinforced self-compacting concrete (SFRSCC). By cutting a larger plate to smaller beams and testing the beams, local strength characteristics of the large plate can be obtained. The 4-point bending test, makes it possible to subject a considerable amount of the beam volume to constant bending moment, determining this way the weakest place and path for the cracks to occur. The orientation of the cut-out beams has been chosen according to the plate casting direction and movement. For comparison a similar pure self-compacting concrete (SCC) plate was cast and tested.

2. Experiments

2.1. Casting of Plates, Materials and Mix Design

A detailed description of the experimental setup and procedure can be found in [26]. Here a summary is presented. The experimental study was performed on two different mixtures: a pure self-compacting concrete (SCC) and short steel fiber reinforced self-compacting concrete (SFRSCC). Both concrete plates were produced by the manufacturer according to the recipe. The main mixture components are presented in table 1. The mixing was carried out according to normal factory procedure in a production-line mixer with a capacity of 1–2 tons.

Table 1. Composition of SCC mix, amount mixed: 1 qm.

Cement (kg)	330
Sand (0-2mm) (kg)	398
Sand (0-4mm) (kg)	505
Crushed Stone (2-5) (kg)	336
Crushed Stone (8-11) (kg)	539
Filler (kg)	80
Admixture (kg)	5.1
Water (kg)	175
Fibers (SFSCC only)(kg)	50
extra water (SFSCC only)(kg)	$\sim$ 10

The main aim of mixing procedure is to obtain a proper consistency and viscosity, cohesive and uniform structure of the fresh concrete mixture. The SFRSCC and SCC mixes were practically identical with the only difference that for SFRSCC the amount of water was slightly higher in order to get the rheological properties of SFRSCC and SCC to be roughly equal and within the required guidelines.

The volume fraction of steel fibers was 1%. The basic characteristics and properties of selected fibers are presented in table 2.

After intensive mixing, the slump flow tests were performed for each mixture. In the case of SCC spread of the slump was 700 mm, in the case of SFRSCC the diameter was 650 mm.

Table 2. Data of the steel fibers, according to [27].

Manufacturer	Severstal Metiz
Model	Hendix prime 75/52
type	hooked end
length, mm	52 ± 2.0
diameter, mm	0.75 ± 0.04
hook length, mm	$2.0 - 1.0/ + 2.0$
hook height, mm	$2.1 + 0.5/ - 0.0$
bend angle, degrees	40 ± 5
tensile strength, MPa	1500
Elastic modulus, MPa	≥ 190000
number of fibers per kg, pcs.	~ 5545

Then the casting was carried out. The fresh concrete mixtures were poured into beforehand prepared, foil covered wooden molds. In both cases the bucket was used as a filling method, that moved in the longitudinal direction. Both casting processes were recorded with a camera to enable the subsequent analysis.

Further, instantly after the casting procedure to align the specimen surfaces the special tools were used. After that, to provide the easy unmolding, lifting and transportation of the slabs, lifting anchors were placed into the concrete mass. Then the slabs were left to harden in the manufacturing hall at room temperature.

Standard compression tests have been performed on separately cast cubic test specimen, the concrete for these was taken out of the mass for the plates during casting.

2.2. Preparation of Beam Specimen

The experimental analysis in the present research includes 4-point bending tests and Computed Tomography with the aim to determine the fiber orientation and distribution inside the concrete slabs and its influence on the strength of the concrete. For this purpose, the produced large slabs with dimensions $(L \times W \times H)$ 400 cm $\times$ 100 cm $\times$ 10 cm were cut into smaller beams. Namely, from the slab of SCC 10 beams of size 100 cm $\times$ 9.5 cm $\times$ 10 cm have been cut out and the slab of SFRSCC has been cut into 40 beams of size 100 cm $\times$ 9.5 cm $\times$ 10 cm. The schematic layout of the specimens cut out of the big slabs is shown in figure 1.

2.3. The Bending Test set-up

All 40 beams of SFRSCC and 10 beams of SCC were tested by a 4-point bending test. During the bending test the following quantities were registered: the load, the mid-span displacements. From this, the post-cracking behavior and the force-displacement curves were obtained.

All experimental devices and the testing specimen dimensions are presented in figure 2 (photo) and figure 3 (drawing).

Full description of the set-up and details of bending test procedure are described in [26].

All specimens were subjected to the analogous loading procedure until the displacement reached a maximum 10 mm. The specimen have been over 4.5 months old at the time of testing and the effect of the difference in age at the time of testing should be negligible.

2.4. Tomography

The current research is mostly focused on the influence of the casting process on the fiber orientation and positions inside of the big-size plate of SFRSCC. In addition, by using X-ray

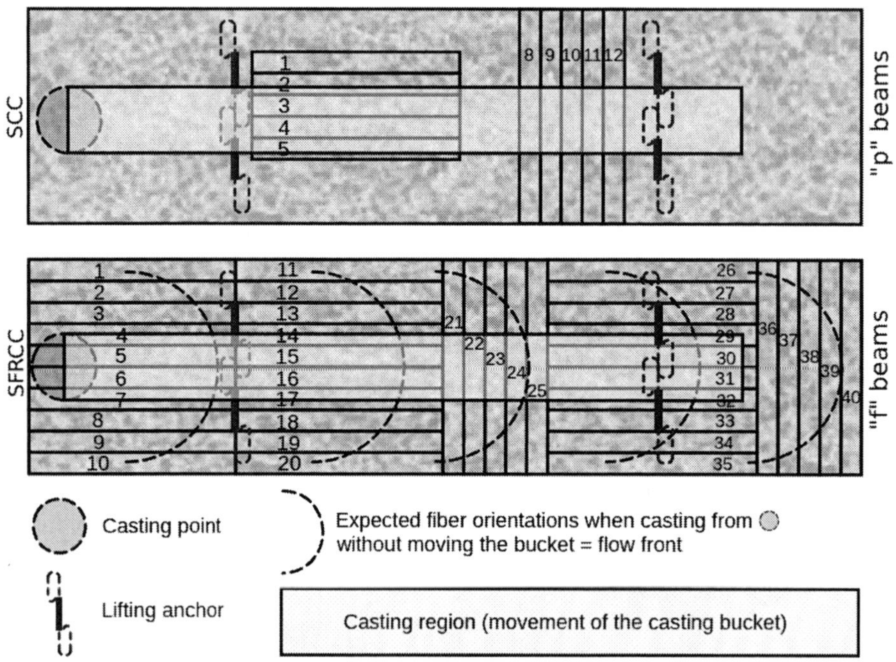

Figure 1. Beam layout of specimens cut out of the cast plate; figure reproduced from [26].

Figure 2. Four point bending test (Photo of the setup); figure reproduced from [26].

CT scanning the hypothesis that certain fiber orientation inside of concrete beam improves the post-cracking behavior of SCC will be verified.

For these purposes, so far four beams were chosen, according to figure 4. According to the expectations from casting simulations, fiber beams 10 should have fibers well aligned along the beam axis, due to the wall-effect, and beams 21, 36 and 38 should have changing orientation of the fibers along the axis, with fibers being perpendicular to the beam axis at the ends of the beams and parallel to the axis in the middle of the beams. However, beam 21 is still in the casting path of the bucket and beam 36 close to the end of the casting path. Therefore, disturbed fiber alignment can be expected. The beams have been chosen, to verify the fiber alignment in different sections of the plate. The scanning of the beams was performed with a resolution 0.6 mm. In the future further beams will be scanned.

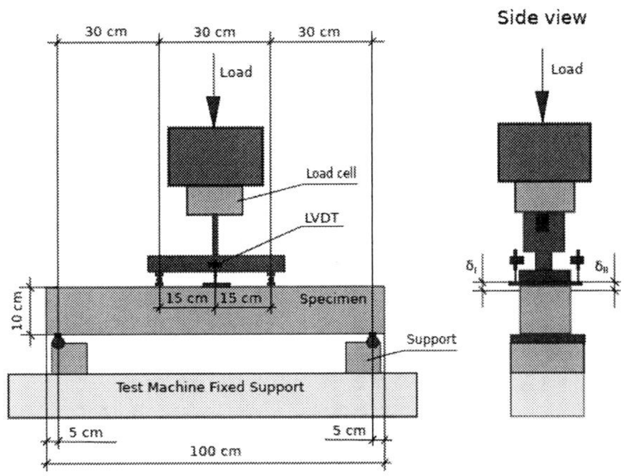

Figure 3. Four point bending test; figure reproduced from [26].

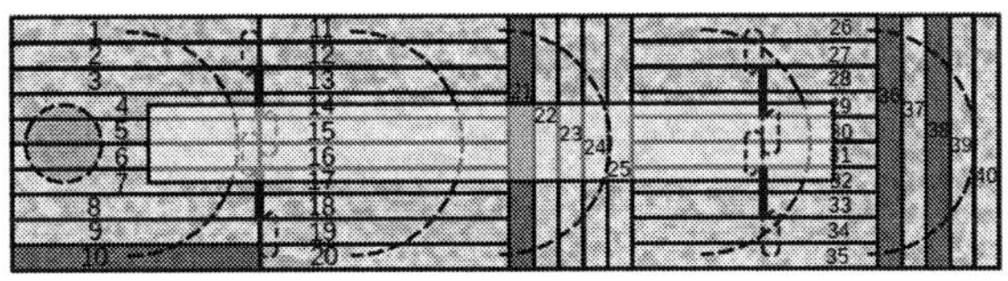

Figure 4. Beam layout of SFRSCC specimens chosen for X-ray tomography.

2.5. X-ray diffraction and SEM

The X-ray diffraction (XRD) analysis of the phase composition of materials was carried out upon applying diffractometer DRON-7 (Russia). In order to obtain X-ray radiation Cu Kα spectrum (λ = 0.1541837 nm), a graphite monochromator was used. The parameters of the tests were following: voltage: 30 kV; current: 12 mA; the range of the diffraction angle: from 4 to 60°, the detector movement step: 0.02°; the duration of the intensity measuring in a step: 0.5 s. Phase identification was carried out by decoding the XRD patterns according to ICDD diffraction databases. The microstructure of the materials was explored upon use of scanning electron microscopy (SEM) equipment JEOL JSM-7600F (Japan). The parameters of electron microscopy were following: voltage: 10 kV; the distance to the surface of the specimen: from 17 to 20 mm. The peculiarities of the microstructure were identified on examining the split surface of the specimens. The image was formed on registering the signal of secondary electrons. Before testing, the surface was coated with electrically conductive thin layer of gold by evaporating the gold electrode in the vacuum using the instrument QUORUMQ150R ES (Germany).

3. Results

3.1. Compression Tests

The results of the standard compression test are presented in table 3 and figure 5. It is interesting to note, that the compressive strenght of the SFRSCC cubes was lower than that of the SCC. Part of this difference can be accounted for by the small difference in the water-cement ratio, but probably this does not account for the full difference.

3.2. Bending Tests

The results of the bending tests in figure 6 show small variations in the load at first crack, but a large variation in the post-cracking behaviour. The post-cracking behaviour shows a

Table 3. Results of compression tests for SCC and SFRSCC.

	age at test in days	4	7	28
SCC	# of samples	3	3	10
	$\bar{\rho}$ in kg/m^3	2330	2340	2379
	stdev in kg/m^3	0	30	9
	$\bar{\sigma}_{u,c}$ in MPa	54.1	55.3	68.6
	stdev in MPa	2	1	2
SFRSCC	# of samples	3	3	10
	$\bar{\rho}$ in kg/m^3	2357	2370	2382
	stdev in kg/m^3	12	10	14
	$\bar{\sigma}_{u,c}$ in MPa	44.9	45.5	56.6
	stdev in MPa	0	1	1

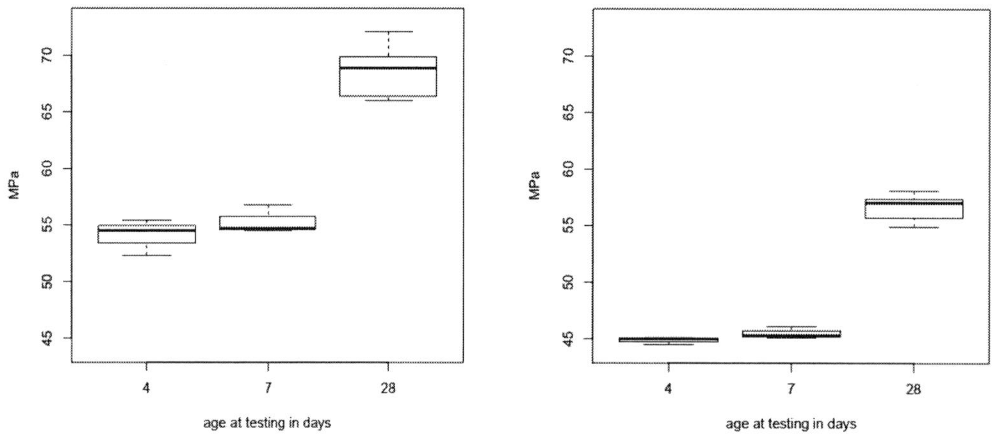

(a) SCC (b) SFRSCC

Figure 5. Compressive strength of SCC and SFRSCC.

strong strain-softening for some beams, notably for beams taken from casting path, and strain-hardening behaviour for other beams, notably the edge beams.

3.3. Tomography

The image processing of the tomography images was carried out using `3D Slicer` which is an open-source platform developed for registration, segmentation and 3D visualization of medical imaging data [28].

The main objective of the image processing was to visually analyze the micro-structure of the beams; to obtain the high quality images of fiber orientation and distribution and the formed crack outlines inside of the testing specimens. Generally, the image processing technique allows to efficiently separate the investigated objects (fibers) from the background, to detect the air voids and another defects and to minimize the noises from the images.

The image processing mainly involves several steps: the contrast enhancement and thresholding. Firstly, in order to obtain the improved separation between the background and fibers, the intensity and opacity thresholds were applied. The typical examples after image

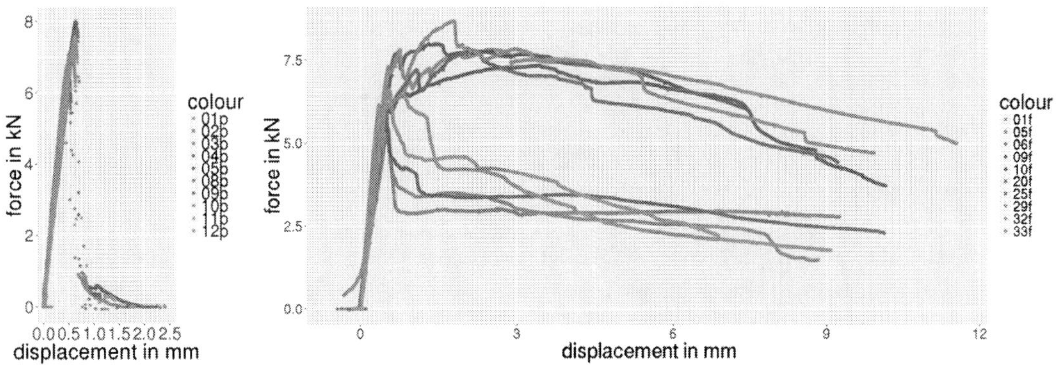

(a) pure concrete (b) fiber reinforced concrete

Figure 6. Load-displacement diagrams for some of the tested beams.

processing are presented in figure 7 and figure 8.

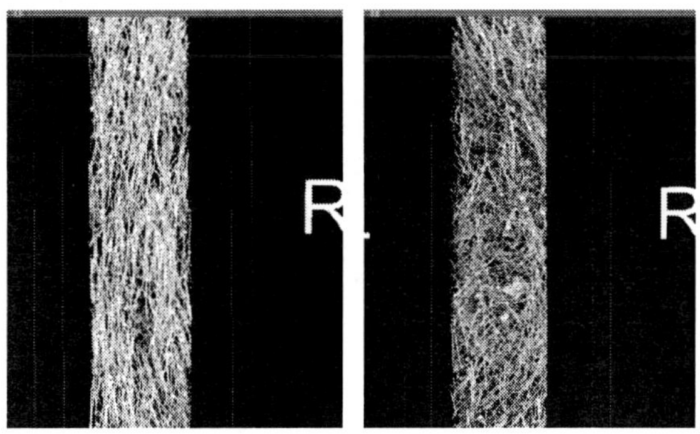

(a) beam 10 (b) beam 21

Figure 7. Volume image of the center-sections of four beams, thresholded to show fibers, view from the bottom.

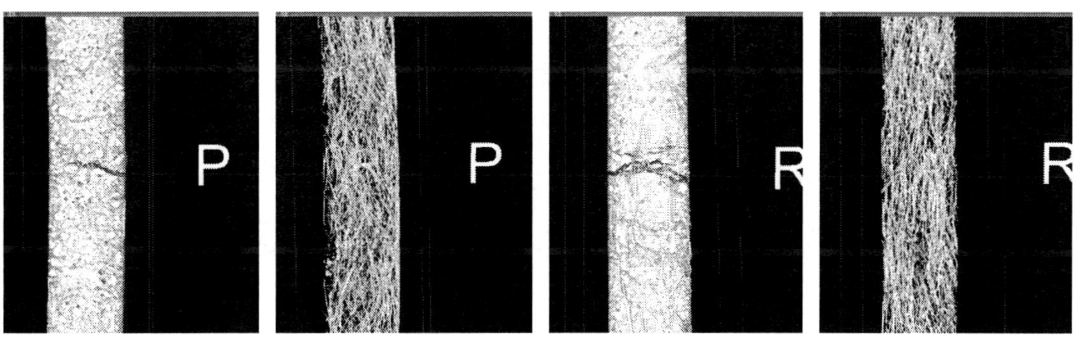

(a) side: crack (b) side: fibers (c) bottom: crack (d) bottom: fibers

Figure 8. Fibers and crack on beam 10 (strong), side and bottom view.

Visually, it is possible to see a difference in the fiber alignment between beams 10 and 21,

IOP Conf. Series: Materials Science and Engineering **660** (2019) 012059 doi:10.1088/1757-899X/660/1/012059

figure 7. In beam 10 the fibers show a strong tendency to align along the beam axis, while in beam 21 the fiber orientations are more random.

3.4. XRD analysis

Characteristic diffraction curve of the SFRSCC in all specimens (1, 2, 33 and 35) obtained by XRD analysis is presented in figure 9. It can be seen from the XRD analysis that after hardening, ettringite ($3\,CaO{\cdot}Al_2O_3{\cdot}3\,CaSO_4{\cdot}31\,H_2O$), calcite ($CaCO_3$), portlandite ($Ca(OH)_2$), C-S-H ($nCaO{\cdot}SiO_2{\cdot}mH_2O$), feldspars ($KAlSi_3O_8$ - $NaAlSi_3O_8$ - $CaAl_2Si_2O_8$) and quartz (SiO_2) minerals were identified in the SFRSCC.

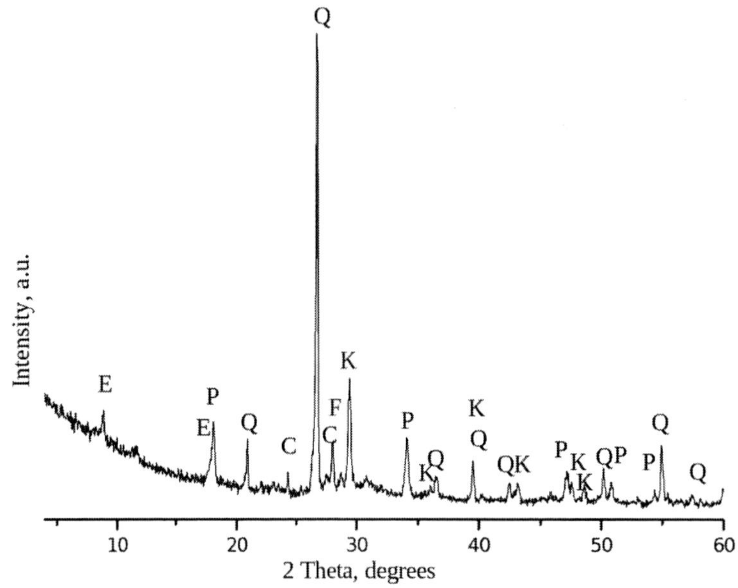

Figure 9. Representative diffraction curve of the steel fiber reinforced concrete: Q - quartz, K - calcite, P - portlandite, E - ettringite, C - C-S-H, F - feldspars.

3.5. SEM analysis

The morphology of the SFRSCC is visible in SEM images of the cut surface of the specimens 1 and 35 (figure 10). SEM results employed to evaluate the effect of steel fiber used on the development of SFRSCC. Figure 10 illustrates a very good contact of steel fiber with the binder and it does not depend on the orientation of the fiber and position of specimen. Microstructure tests revealed that specimens have a dense structure.

A representative image of the binding material in all specimens obtained by SEM analysis is presented in figure 11. The image reveals the prevalence of rhomboid shaped calcite minerals, plate shaped portlandite, fibre shaped ettringite and the formed C-S-H crystals, similar to a cellular frame carpet.

As SEM and XRD tests show (figure. 9 and 11) the same hydrates are formed in the specimens (1, 2, 33 and 35). It can be seen from the XRD and SEM tests a very good contact of steel fiber with the concrete binder is present in all specimens.

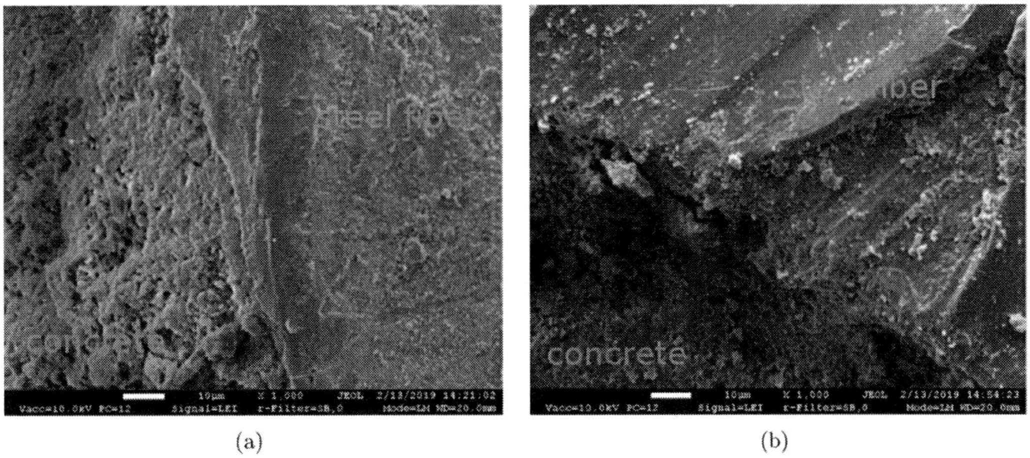

Figure 10. SEM images of the contact zone of steel fiber and binder: a) specimen 1, b) specimen 35.

Figure 11. Representative SEM image of the binding material in the tested specimens.

4. Discussion of Results

The 4-point bending tests show a large variability in the post-cracking strength, depending from which position in the plate the beam was taken from. The variability is much larger than for ordinary reinforced concrete. The microstructure investigation showed the same concrete microstructure and good fiber-concrete adhesion in all samples. The X-ray tomography showed differences in the fiber orientations and the fiber orientation distribution, while no obvious correlation to the fiber amount is visible. The differences in the post-cracking strength show some correlation with the fiber orientations.

5. Ongoing and Future Investigations

The analysis of fiber orientations by use of X-ray Computed Tomography (CT) has not been finished. Further selected beams will be scanned using CT and the fiber positions and orientations will be extracted from the volume image. Special attention will be paid to the cracked region. The results of the fiber orientation measurements will be compared to computational fluid dynamics simulations of the fiber orientations.

6. Conclusion

In this paper the results of four-point bending tests performed on beam specimens cut out of a large plate have been presented. The initial results are that in all beams fibers improve fracture behaviour compared to SCC, although the actual post-cracking performance depends on the position within the plate where the sample was taken from. The results coincide (roughly) with expectations from predicted fiber orientation distribution in samples, which seems to confirm that the flow of the fresh concrete mass influences fiber orientations. The microstructure analysis by SEM and XRD did not reveal differences between the samples that could account for the difference in fracture behaviour. This leads to the conclusion, that the bending results are very sensitive to the casting procedure. It follows, therefore, that SFRSCC properties of specifically produced small-scale laboratory samples may not be representative for large structural elements – such scale-dependence stands in contrast to the properties of SCC. Further, one can conclude that special care is needed to design casting technologies and methods which will produce consistent fiber orientation distributions and that on-site casting may be problematic.

Acknowledgments

This work was supported by the Estonian Research Council grant (PUT1146). Support through the exchange program by the Estonian and Lithuanian Academies of Sciences is gratefully acknowledged.

Thanks to E-Betoonelement for preparing the experiment plates, space for cutting the beams and their general interest in the research. We also thank Elmar-Jaan Just and Peeter Linnas of the Department of Civil Engineering and Architecture, School of Engineering, for letting us use the equipment in the Mäepealse lab.

References

[1] Herrmann H, Goidyk O and Braunbrück A Influence of the flow of self-compacting steel fiber reinforced concrete on the fiber orientations, a report on work in progress 2019 *Short fiber reinforced cementitious composites and ceramics* ed Herrmann H and Schnell J (Springer) pp 97–110

[2] Herrmann H and Lees A On the influence of the rheological boundary conditions on the fibre orientations in the production of steel fibre reinforced concrete elements 2016 *Proc. Estonian Acad. Sci.* **65** 408–413 open-Access CC-BY-NC 4.0

[3] Svec O, Zirgulis G, Bolander J E and Stang H Influence of formwork surface on the orientation of steel fibres within self-compacting concrete and on the mechanical properties of cast structural elements 2014 *Cem. Concr. Compos.* **50** 60–72 ISSN 0958-9465

[4] Suuronen J P, Kallonen A, Eik M, Puttonen J, Serimaa R and Herrmann H Analysis of short fibres orientation in steel fibre reinforced concrete (sfrc) using x-ray tomography 2012 *J. Mater. Sci.* 1–10 ISSN 0022-2461 online-first

[5] Ponikiewski T, Katzer J, Bugdol M and Rudzki M X-ray computed tomography harnessed to determine 3d spacing of steel fibres in self compacting concrete (scc) slabs 2015 *Construction and Building Materials* **74** 102–108 ISSN 0950-0618

[6] Herrmann H, Braunbrück A, Tuisk T, Goidyk O and Naar H An initial report on the effect of the fiber orientation on the fracture behavior of steel fiber reinforced self-compacting concrete 2019 *Short fiber reinforced cementitious composites and ceramics* ed Herrmann H and Schnell J (Springer) pp 33–50
[7] Oesch T S, Landis E N and Kuchma D A Conventional concrete and uhpc performance–damage relationships identified using computed tomography 2016 *Journal of Engineering Mechanics* **142** 04016101
[8] Oesch T 2016 *6th Conference on Industrial Computed Tomography, Wels, Austria (iCT 2016)*
[9] Macanovskis A, Lusis V and Krasnikovs A Crack opening investigation in fiberconcrete 2014 *World Academy of Science, Engineering and Technology* **8** 430–438
[10] Pittino G, Geier G, Fritz L, Hadwiger M, Rosc J and Pabel T Computertomografische untersuchung von stahlfaserspritzbeton mit mehrdimensionalen transferfunktionen 2011 *Beton- und Stahlbetonbau* **106** 364–370 ISSN 1437-1006
[11] Promentilla M, Sugiyama T and Shimura K 2008 *3rd ACF International Conference ACF/VCA* pp 940–947
[12] Schnell J, Schladitz K and Schuler F Richtungsanalyse von fasern in betonen auf basis der computer-tomographie 2010 *Beton- und Stahlbetonbau* **105** 72–77 ISSN 1437-1006
[13] Vicente M, Minguez J and González D The use of computed tomography to explore the microstructure of materials in civil engineering: From rocks to concrete 2017 *Computed Tomography - Advanced Applications* ed Halefoglu D A M (InTech)
[14] Mishurova T, Léonard F, Oesch T, Meinel D, Bruno G, Rachmatulin N, Fontana P and Sevostianov I 2017 *Proceedings of the 7th Conference on Industrial Computed Tomography (ICT) held February 7-9, 2017, Leuven, Belgium* vol 22(03) (NDT.net) ISSN 1435-4934 `http://www.ndt.net/?id=20818`
[15] Oesch T S 2015 *Investigation of Fiber and Cracking Behavior for Conventional and Ultra-High Performance Concretes Using X-Ray Computed Tomography* Ph.D. thesis University of Illinois at Urbana-Champaign
[16] Ponikiewski T, Katzer J, Bugdol M and Rudzki M Steel fibre spacing in self-compacting concrete precast walls by x-ray computed tomography 2015 *Mater. Struct.* **48** 3863–3874 ISSN 1359-5997
[17] Vicente M A, Gonzalez D C and Minguez J Determination of dominant fibre orientations in fibre-reinforced high-strength concrete elements based on computed tomography scans 2014 *Nondestructive Testing and Evaluation* **29** 164–182
[18] Herrmann H and Engelbrecht J Comments on mesoscopic continuum physics: Evolution equation for the distribution function and open questions 2012 *Proc. Estonian Acad. Sci.* **61** 71–74 ISSN 1736-6046
[19] Herrmann H 2016 *Generalized Continua as Models for Classical and Advanced Materials* (Cham: Springer International Publishing) chap An Improved Constitutive Model for Short Fibre Reinforced Cementitious Composites (SFRC) Based on the Orientation Tensor, pp 213–227 ISBN 978-3-319-31721-2
[20] Eik M, Herrmann H and Puttonen J 2015 *Proceedings of the XII Finnish Mechanics Days : [4-5 June 2015, Tampere, Finland]* ed Kouhia R, Mäkinen J, Pajunen S and Saksala T pp 255–260 ISBN 978-952-93-5608-9
[21] Eik M, Puttonen J and Herrmann H The effect of approximation accuracy of the orientation distribution function on the elastic properties of short fibre reinforced composites 2016 *Compos. Struct.* **148** 12–18 ISSN 0263-8223
[22] Herrmann H and Beddig M Tensor series expansion of a spherical function for use in constitutive theory of materials containing orientable particles 2018 *Proc. Estonian Acad. Sci.* **67** 73–92 open-Access CC-BY-NC 4.0
[23] Zirgulis G, Svec O, Geiker M R, Cwirzen A and Kanstad T Influence of reinforcing bar layout on fibre orientation and distribution in slabs cast from fibre-reinforced self-compacting concrete (frscc) 2016 *Structural Concrete* **17** 245–256
[24] Krasnikovs A, Kononova O, Khabaz A, Mačanovskis E and Mačanovskis A Post-cracking behaviour of high strength fiber concrete prediction and validation 2011 *World Academy of Science, Engineering and Technology* 988–992
[25] Herrmann H, Eik M, Berg V and Puttonen J Phenomenological and numerical modelling of short fibre reinforced cementitious composites 2014 *Meccanica* **49** 1985–2000 ISSN 0025-6455
[26] Herrmann H, Goidyk O, Naar H, Tuisk T and Braunbrück A The influence of fiber orientation in self-compacting concrete on 4-point bending strength 2019 *Proc. Estonian Acad. Sci.* **68** open-Access CC-BY-NC 4.0, forthcoming
[27] Severstal Metiz 2018 Hendix prime 75/52 - hooked ends fiber `http://www.severstalmetiz.ru/eng/catalogue/1930/document6219o.shtml?6605,1`
[28] Pieper S, Halle M and Kikinis R 2004 3d slicer IEEE International Symposium on Biomedical Imaging ISBI 2004

IOP Conf. Series: Materials Science and Engineering **660** (2019) 012060 doi:10.1088/1757-899X/660/1/012060

Shear Failure of Sandwich Panels Subjected in Bending

M Daugevičius[1], T Skuturna[1], J Valivonis[1] and A Patapavičius[2]

[1] Department of Reinforced Concrete Structures and Geotechnics, Vilnius Gediminas Technical University, Vilnius, Lithuania
[2] Department of Applied Mechanics, Vilnius Gediminas Technical University, Vilnius, Lithuania

E-mail: mykolas.daugevicius@vgtu.lt

Abstract. The paper describes the behaviour of the sandwich panels in bending. Two panels of the same dimensions are tested. The panels differ only in the layout of the inner thermal insulation layer. The thermal insulation layer is made of separate mineral wool slabs. The graphs of the strain development in the external sheets, the displacement between the layers and mid-span displacement are presented in the paper. The nature of the failure shows that only separate internal thermal insulation slabs resist the shear force. Therefore, the complete width resistance cannot be evaluated. The views of the failure are also presented.

1. Introduction

Light-weight sandwich panels are important structural elements. These panels serve as the wall or roof structures, providing the desired thermal insulation. The construction is fast and there is no need for a finishing layer while the cross section of the major structural elements is reduced. The loading and its action on these panels depend on their design purpose. If the board is used for making the external walls, the main load and its effect are associated with the wind and the temperature difference. If the board is used in the roof, the main loads include the mounting load, the wind and snow loads and the effect of the temperature difference. Such actions as lifting, transporting and installation must also be considered. Various limit states of sandwich panels made of a soft internal layer and thin external layers are possible. The European standard EN 14509:2013 [1] specifies the critical cases for sandwich panels. The thin external steel sheets can buckle during bending. Moreover, wrinkling of the compressed external steel layer can be observed. It is also possible that steel yield can occur. The internal thermal insulation layer must be checked for local compression (at the point of loading and supporting), as well as for shear force resistance. The most common failure of the continuous sandwich panels is associated with the core shear, crushing failure or buckling failure of the external sheet [2]. In the experimental research of Herranen [3], all the tested sandwich panels failed due to shear. The core of the sandwich panels in the research of Herranen [3] was made of polyvinyl-chloride foam. In the work of Gnip [4], the compressive, tensile and shear tests of mineral wool were performed. The mean square deviations of the modulus of elasticity or shear modulus predicted by Gnip [4] could reach 100 % and more. This shows that the dispersion of these mechanical properties is very high and may decrease the reliability of the considered structures.
The calculation of these panels allowed for determining that their inner layer was solid and the mechanical properties were the same across the width. The stresses in each layer were calculated

according to the principle of superposition [5]. In the design process, it is common to evaluate the whole width of the panel. For example, the resistance for shear is calculated for the whole width.

In the manufacture of sandwich panels, the inner thermal insulation layer can be made of separate thermal insulation slabs. If the inner layer is solid, it can resist the shear force over the whole width of the panel. If the inner layer is made of separate thermal insulation slabs, the attached edges of separate slabs do not resist the shear force. These local zero-stiffness zones reduce the load-bearing capacity of the panel and increase their deflection. Sandwich panels are produced as a continuous product, which is then cut into separate panels of the required length. The arrangement of the slabs of the thermal insulation layer is random. Figure 1 presents the random layout of the slabs of the thermal insulation, where the slabs are laid out with an overlap. When cutting the panels, the inner layer with zero-stiffness accidentally falls into the range of the critical load distribution locations. Therefore, the purpose of this research is to determine the impact of the considered random layout of thermal insulation on the load-bearing capacity of the panels.

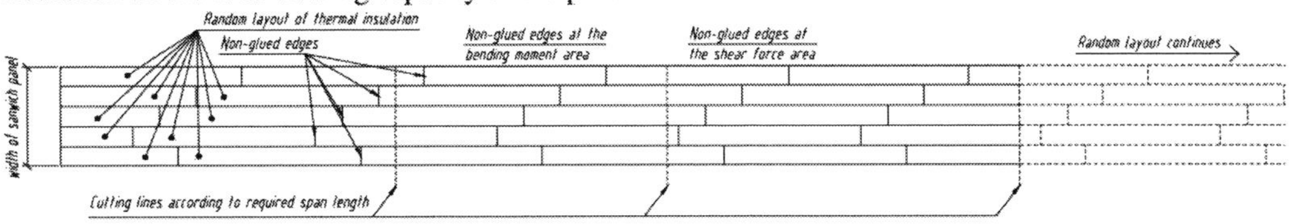

Figure 1. The random layout of thermal insulation slabs during the production process of sandwich panels.

2. Samples and testing

The tested panels were manufactured at the factory. The panels were manufactured by using a continuous method. The outer steel sheets were rolled out of the rolls and integrated into the plate's production line. The gap between the upper and the lower sheets was filled with thermal insulation slabs. The width of the internal thermal insulation mineral wool slabs ranged from 230 mm to 245 mm for panel P1 and from 226 mm to 250 mm for panel P2, respectively. Figure 2 presents the case of the thermal insulation slabs in the production line. Only the top and the bottom surfaces of thermal insulation slabs were glued to steel sheets. Steel sheets and thermal insulation slabs were glued with polyurethane adhesive. The widths of the produced sandwich panels P1 and P2 were 1205 mm and 1196 mm, respectively. The lengths of the cut-off plates were 3400 mm. The distance between the outer steel sheets was 95 mm. The thickness of the outer steel sheets of the sandwich panels was 0.5 mm. Figure 3 presents the internal view of the sandwich panel, presenting the external steel sheets, the internal thermal insulation layer made of separate mineral wool slabs and polyurethane adhesive on the bottom steel sheet and the top surface of thermal insulation.

Figure 2. The formation of the internal thermal insulation layer in the production line.

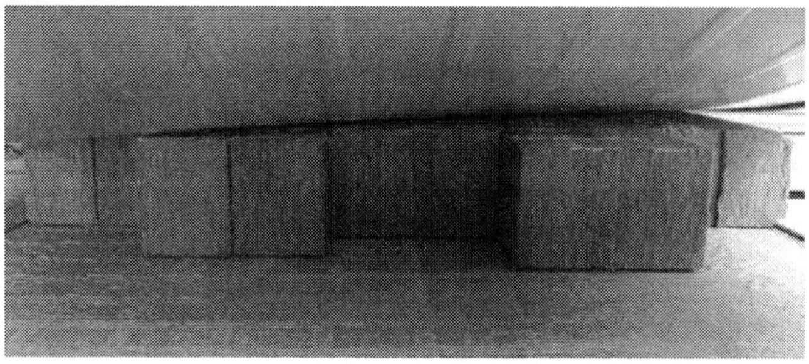

Figure 3. The internal thermal insulation layer between the external steel sheets.

The four-point bending test was performed to test the panel. The arrangement of measuring devices is presented in Figure 4. The test was performed in two different ways. Panel P1 was loaded based on using the load control method, while panel P2 was loaded by applying the displacement control approach. The maximal vertical displacement should not exceed 10% of the panel thickness per minute. Therefore, the first panel was loaded based on using the load control method. The obtained results demonstrated that the plate deflection increment was higher. Therefore, the displacement control method was chosen for the second plate. Because of the high loading capacity of the testing machine (1000 kN) and the low panel stiffness, the displacement control method had not been chosen from the beginning of the test. The load for all panels was transferred by 100 mm-wide wooden supports.

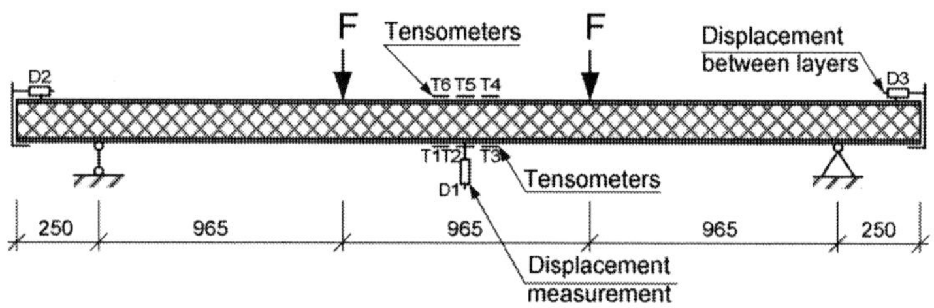

Figure 4. The schematic view of the tested sandwich panel and the arrangement of measuring devices.

Vertical displacement of the panel was measured by the displacement transducer D1. The displacement between the layers was measured by using the devices D2 and D3. The devices were fixed on the top steel layer and the supporting plate was attached to the lower sheet. The strains in the tensioned external steel layer were measured with the tensometers T1; T2 and T3. The strains in the compressed external steel layer were measured with the tensometers T4; T5; T6.

3. The results

The performed experiment has shown that the loading manner influenced the results. Firstly, the maximal load-carrying capacity of the sandwich panel P1 was 4.0 kN·m, while that of the panel P2 was 3.33 kN·m. As mentioned above, the panel P1 was tested by using the load control method and test duration was 8 minutes. However, the panel P2 was tested by using the displacement control test, the duration of which was 28 minutes. For the panel P1, the maximal stresses in the top and the bottom steel sheets at the maximal load were $\sigma_{s.top} = \varepsilon_{sc} \cdot E_s = -0.0002671 \cdot 205 \cdot 10^9 = -54.8\ MPa$ and $\sigma_{s.bot} = \varepsilon_{st} \cdot E_s = 0.00027224 \cdot 205 \cdot 10^9 = 55.8\ MPa$, respectively. For the panel P2, the maximal stresses in the top and the bottom steel sheets at the maximal load were $\sigma_{s.top} = \varepsilon_{sc} \cdot E_s = -0.0002676 \cdot 205 \cdot 10^9 = -54.9\ MPa$ and $\sigma_{s.bot} = \varepsilon_{st} \cdot E_s = 0.00024383 \cdot 205 \cdot 10^9 = 50\ MPa$, respectively. Meanwhile, the same stresses were found in panel P1 at the maximal load level of panel P2: $\sigma_{s.top} = \varepsilon_{sc} \cdot E_s = -0.0002204 \cdot 205 \cdot 10^9 = -45.2\ MPa$ and $\sigma_{s.bot} = \varepsilon_{st} \cdot E_s = 0.000223 \cdot 205 \cdot 10^9 = 45.7\ MPa$, respectively. The compression and tensile stresses in the panel P1 were by 18 % and 9 % lower than those in the panel P2. The deflection of the panel P1 at the same load level was by 10 % lower.

The measured strains in the top and the bottom sheets are presented in Figure 5. The development of the strain until failure had shown that the external steel sheets' work was elastic. The comparison of the normal stresses in the panel P1 had shown that they were similar. Thus, the neutral axis is located at the centre of the cross section height. The comparison of the panel P2 stresses had shown that the compressive stresses were larger than the tensile stresses. In this panel, the neutral axis was moved away from the centre.

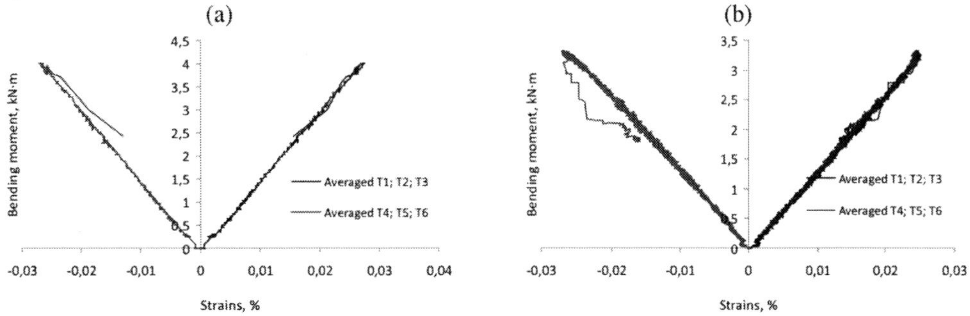

Figure 5. The strains in the tensioned and compressed external steel layer; (a) denotes the strains in panel P1; (b) shows the strains in panel P2.

The observed difference in the stresses and the location of the neutral axis can be accounted for by the elasto-plastic strain development in the internal thermal insulation layer and the displacement of the external layer. In fact, the plasticity of the inner thermal insulation layer resulted in the considered stress distribution. The longer loading time led to the development of plastic strains. Therefore, in the panel P2, premature displacement between the external layers could be observed. Figure 6 presents the development of the displacement between the external layers. In the panel P1, a considerable displacement was recorded when the limit load was reached.

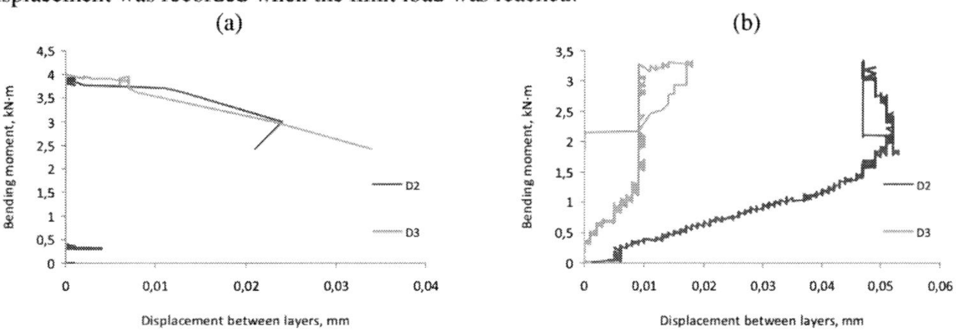

Figure 6. The displacement between the external steel layers; (a) shows the displacement in panel P1; (b) demonstrates the displacement in panel P2.

The increment of the mid-span displacement characterizes its linear nature until the failure. Then, the panel was loaded according to the load control method. It can also be mentioned that the displacement between the external layers was zero. Theoretical deflection of sandwich panels consists of two parts: the deflection caused by loading and the deflection produced by shear. The displacement between the external layers in panel P2 and the deflection at the higher load level developed following the parabolic trajectory. Mid-span displacements of the panels P1 and P2 are presented in Figure 7.

IOP Conf. Series: Materials Science and Engineering **660** (2019) 012060 doi:10.1088/1757-899X/660/1/012060

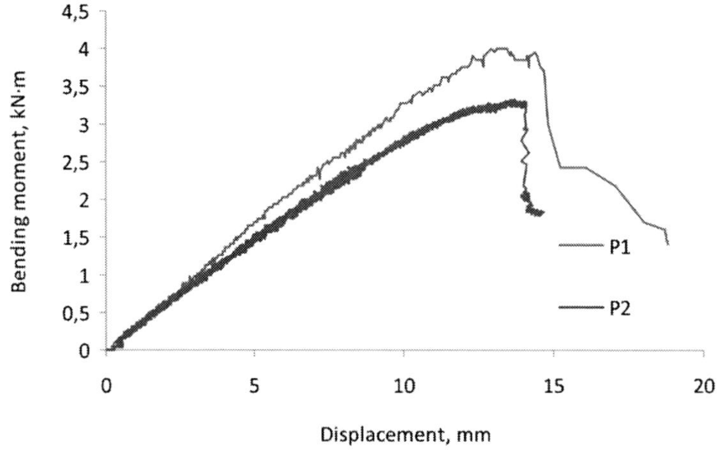

Figure 7. Mid-span displacements of the considered sandwich panels.

After the experiment was accomplished, a part of the external sheet was hand-peeled at the place of failure. In this way, the internal thermal insulation layout structure was uncovered. The uncovered area and the layout structure are presented in Figures 8 and 9 (the cross-marked trajectory denotes the non-bonded edges). It was found that the failure trajectory of the internal layer matched that of the vertical non-glued edges of the internal thermal insulation slabs. In both cases, the failure trajectory chose the way where there were three (the maximal number for the tested panels) non-glued transverse edges. There is no connection between the thermal insulating slabs at the edges, which can intercept vertical internal shear stresses. The friction between the edges is not a good criterion for assessing resistance. Three stripes of thermal insulation resist a shear force in the case of two transverse edges. Two stripes of thermal insulation resist a shear force in the case of three transverse edges. The resistance of two stripes is lower than that of three stripes and the failure trajectory chose the way where there were three non-glued transverse edges. Figures 10 and 11 present the failure and the trajectory of failure in the uncovered area.

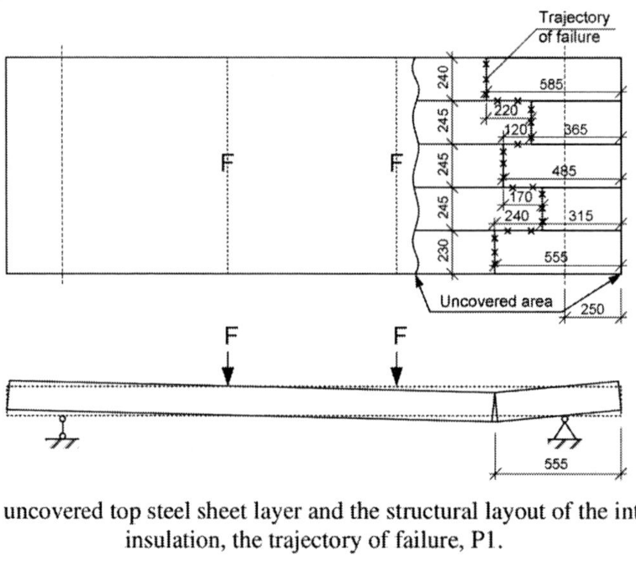

Figure 8. The uncovered top steel sheet layer and the structural layout of the internal thermal insulation, the trajectory of failure, P1.

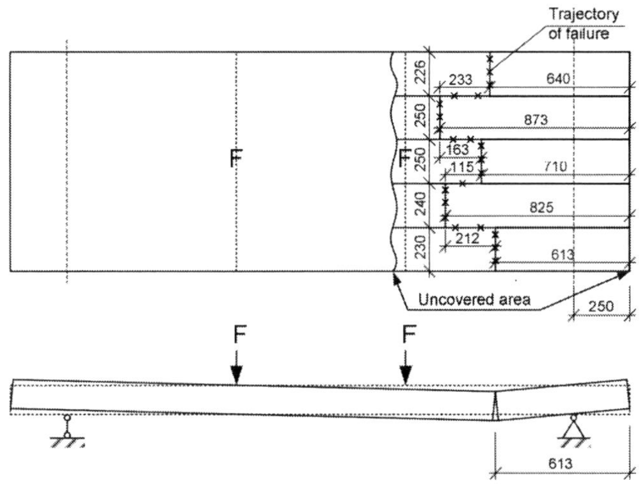

Figure 9. The uncovered top steel sheet layer and the structural layout of the internal thermal insulation, the trajectory of failure, P2.

Figure 10. The failure of the sandwich panel near the support, P1.

Figure 11. The displacement of the internal thermal insulation layer, P1.

4. Conclusion

The panel loaded according to the load control method demonstrated higher stiffness and load-carrying capacity. The panel loaded based on using the displacement control method had shown lower stiffness and lower load-carrying capacity. The longer testing time influenced the development of plastic deformations and the neutral axis moved away from the centre of the cross section, therefore, the compressive stresses in the panel P2 were 1.2 times higher than the compressive stresses in the panel P1 at the same load level.

Non-glued edges of the thermal insulation slabs influenced the decrease in the load-bearing capacity. All the tested panels failed in the region of the shear load action. After peeling the steel sheets, it was found that their resistance depended on the separate (not glued at the edges) thermal insulation insert. During their production, the inner edges must be glued. Then, a single inner layer will be formed.

References

[1] European Standard EN 14509:2013 Self-supporting double skin metal facade insulating panels – Factory made products – Specifications.

[2] Hassinen P, Martikainen L, Berner K 1997 On the Design and Analysis of Continuous Sanfdwich panels. *Thin-Walled Structures*, Vol. 29, Nos 1-4, pp. 129-139.

[3] Herranen H, Pabut O, Eerme M, Majak J, Pohlak M, Kers J, Saarna M, Allikas G, Aruniit A 2012 Design and Testing of Sandwich Structures with Different Core Materials. Materials Science, Vol. 18, No. 1. pp. 45-50.

[4] Gnip I, Vėjelis S, Keršulis V, Vaitkus S 2010 Strength and Deformability of mineral wool slabs under short-term compressive, tensile and shear loads. Construction and Building Materials, 24, pp. 2124-2134.

[5] Hassinen P, Martikainen L 1994 Analysis and design of continuous sandwich beams. *Twelfth International Specialty Conference on Cold-Formed Steel Structures* St. Louis, Missouri, U.S.A., October 18-19, 1994.

Parametric identification of multy-storey building model based on oscillation testing results

V Volkova[1], L Gaile[2] and L Pakrastins[2]

[1] Department of construction, geotechnics and geomechanics, Dnipro University of Technology, Televizionnaya 12. ap.48, Dnipro,UA49005,Ukraine
[2.]Faculty of Civil Engineering, Riga Technical Riga University, 1 Kalku Street Riga, LV-1658, ,Latvia

[1]E-mail: drvev09@gmail.com

Abstract. Buildings and structures are complex systems having predetermined parameters, which should be controlled in the process of manufacturing structures, construction and installation. In some cases, oscillations are the most dangerous effects on the buildings and structures. Significant alternate stresses caused by oscillational actions leads to accumulation of damage in the material. It causes the appearance of fatigue cracks and destruction. In addition to fatigue failure in mechanical systems there are other phenomena that cause changes in the structure of the surface layers of the parts to be joined and consequently the reduction of the frictional force in the joint. The object of the study is a multi-storey building having a rectangular shape in plane. Structural scheme of the building is a two spans frame. The frame of the building consists of monolithic reinforced concrete structures. According to the analysis of the basic diagnostic parameters of the technical condition of buildings and structures, affect their stability and reliability are geometrical parameters of buildings and structures, and their main structural elements, as well as physical and mechanical parameters of structural elements of building and dynamic parameters of buildings and structures. An analysis of methods for determining the dynamic parameters of buildings has shown that the main criteria for assessing the technical condition of bearing structures, using the dynamic method, is the period and frequency building oscillation. Dynamic tests were conducted to determine the dynamic and rigid characteristics of bearing structural elements of the building and structures, detection of hidden defects. According to the method of measuring the dynamic parameters, the degree of damage to a building or structure, is determined by the comparison of the design values of dynamic parameters, periods of eigenoscillations, and the decrement of oscillations with experimentally obtained data.

1. Introduction

The application of methods for dynamic monitoring of buildings and structures is becoming increasingly common in engineering practice. Traditionally, according to the method of excitation of oscillations these methods are divided into the method of free oscillations, the method of seismic waves, the oscillational method of recording small oscillations - microseisms.

Dynamic structural health monitoring methods allow us to assess the state of the whole building and subsequently to localize and detect defects that can be specified by non-destructive methods. These methods can more accurately identify the risk of collapse, estimate the residual lifetime of buildings and structures. Therefore, the development of methodology for assessing the technical condition of buildings based on dynamic criteria, which allows to increase the objectivity and reliability of the obtained results and reduce the time of the structural health monitoring is an actual scientific task. The purpose of the study is to develop a methodology for evaluating the parameters of a construction model based on dynamic criteria.

2. Relevant methods of buildings and structures identification

The issues of construction of mathematical models and prediction of dynamic behaviour of structural elements proceeding from recorded experimental data have attracted considerable interest.

In spite of intensive investigations into the above mentioned matter, which have been undertaken in the scientific centers in different countries [1-4], as well as the important results obtained, there is no, so far, the only universal effective approach, which would allow for correct determination, prediction and analysis of dynamic properties in construction elements.

Most of the methods of structural identification are based on the use of special types of outer excitation for a wide range of frequencies, such as symmetric monoharmonic excitation and rectangular impulse. These types of excitation are often unrealizable in mechanical systems. The methods based on the Foure transformation do not allow classifying and localizing non-linearity [5, 6] and are inapplicable to investigating stochastic processes [1-3, 7]. It should be also noted that the application of Winer series and Hilbert transformation for identification of non-smooth non-linear dynamic characteristics is unjustified [2, 3].

In most of the previous analyses, damages are characterized by changes in the modal parameters, for example, natural frequencies, modal damping ratios and mode shapes. The calculating process is done using data from the structure in some initial and usually assumed undamaged condition, and then is repeated at periodic intervals or after some potentially damaging event that triggers the assessment process. Structural parameters such as stiffness matrix constructed from identified modal parameters may also be used for damage detection and localization. These methods have prevented their use in most "real-world" applications. At first, it involves fitting a linear physics-based model to the measured data from both the healthy and potentially damaged structure. Often these models do not have the fidelity to accurately represent boundary conditions and structural component connection, which are prime locations for damage accumulation. Also, this process does not take advantage of changes in the system response that are caused by non-linear effects. The values of oscillations amplitudes are small for buildings and structures.However in these cases, non-linearity of dynamic models such as a dry friction, tightening and backlash show up most evidently. There is also research using other damage-sensitive features without the need to identify the modal parameters, such as novelty analysis with auto-regressive models.

2.1. Application of trajectories in extended phase space in identification of structural elements

As is has been shown by the investigations of several authors [8, 9], the expansion of a phase space by taking into account the phase planes "acceleration – displacement" and "acceleration – speed" substantially promotes the efficiency in analyzing a dynamic system behaviour. Hereby, we pass on to a three-dimensional phase space confined with three co-ordinate axes, i.e. displacement, velocity and acceleration. An interest taken into accelerations in dynamic systems is conditioned by the fact that these accelerations are more sensitive to high-frequency components in oscillating processes. Phase plane "acceleration – displacement" is of particular interest in the analysis of dynamic system behaviour, because it allows a more evident interpretation of power relations in the dynamic system under investigation. Namely, the area confined by curve "acceleration – displacement" and axis "acceleration" is equal to work, and the anticlockwise motion around its contour corresponds to the energy spent by the system for one cycle of oscillating. Another important characteristic of phase trajectories on plane "acceleration – displacement" is the fact that dependence acceleration versus displacement for autonomous non-conservative systems is a mirror symmetric image in relation to axis "acceleration" to the graph of changes in elastic force characteristic.

The article [10] presents the results of experimental and analytical al investigations of an essentially non-linear dynamic system. The main concept of this paper is that, if a given type of damage converts a linear system into a non-linear system, then any observed manifestations of non-linearity serve to indicate that damage is present. The author has defined behavioural peculiarities of phase trajectories and their mappings in the expanded phase space. It has performed the structural analysis of the phase trajectories obtained in the test records for oscillations of creaked reinforced concrete beam in the expanded phase space. The results of full-scale test and numerical simulation of the dynamic behaviour of the latticed sightseeing tower are presented in the article [11]. It is noted that for effective control of the structural oscillations, it is necessary to study the mechanisms of energy dissipation in real objects. The phase trajectories in the extended phase space were applied in order to identify sources of non-linearity.

3. The object of the study

The multi-storey building having a rectangular shape in plane is under studying. Structural scheme of the building is a two spans frame. The frame of the building consists of monolithic reinforced concrete structures. It is including monolithic ribbed slabs of floors and roof which connected with monolithic columns. The stability of the structure is provided by rigid disks of monolithic reinforced concrete floors and roof, as well as the rigid connections of columns with main beams of floors and roof.

According to the analysis of the basic diagnostic parameters of the technical condition of buildings and structures, affect their stability and reliability are geometrical parameters of buildings and structures, and their main structural elements, as well as physical and mechanical parameters of structural elements of building and dynamic parameters of buildings and structures. An analysis of methods for determining the dynamic parameters of buildings has shown that the main criteria for assessing the technical condition of bearing structures, using the dynamic method, is the period and frequency building oscillation.

4. Numerical modelling of building dynamic behaviour

The results were obtained with the use of the SCAD software package and implementation of Lanczos method for dynamic analysis [12]. Rectangular finite element mesh was used in research. The fundamental frequencies and modes of oscillation of the structural system were determined by solving the equation of continuous free oscillations:

$$[M]\{\ddot{u}\} + [K]\{u\} = 0 \ , \tag{1}$$

where M is the mass matrix, K is the stiffness matrix, $\ddot{u}$ is the acceleration vector, and u is the displacement vector. The finite element model of the building was formed by means of the pre-processor Forum with the import of the initial graphic information in SCAD (see figure 1)

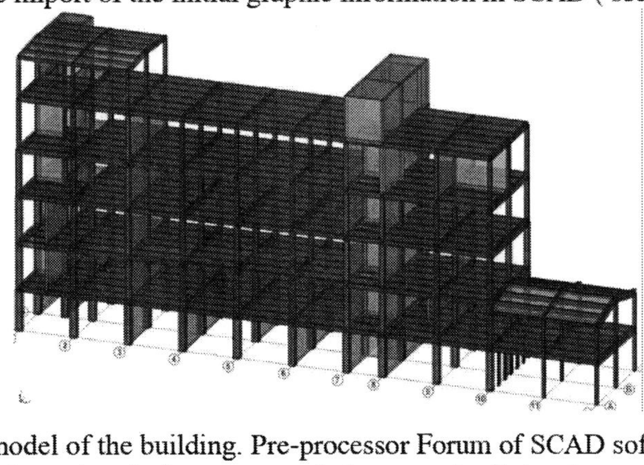

Figure 1. Block model of the building. Pre-processor Forum of SCAD software package.

In the simulation of the rod and plate structural elements the finite elements of the type "spatial rod" and shell elements were used [13, 14]. The number of nodes in the structural scheme is 20360, the number of elements is 28321. The number of degrees of freedom is 354960, the order of the system of linear equations is 120948. The weight of technological equipment was taken into account at implementation of modal analysis.According to the standards, it is recommended to analyze from 3 to 5 principal oscillations modes. In order to avoid the building-up of errors in integration, simulation was carried out for 10 modes of oscillation. In further investigations the first 5 modes were used. According to the results of calculations presented in Table 1 and as it seen from figures 2-5 the lower dominant forms of oscillations correspond to oscillations in the horizontal plane and torsional oscillations modes.

Table 1. Results of modal analysis.

	Eigenvalue	Frequency, Hz	*Period, s*		Eigenvalue	Frequency, Hz	*Period, s*
Mode 1	0,083	1,93	0,518	**Mode 6**	0,03	5,323	0,188
Mode 2	0,047	3,364	0,297	**Mode 7**	0,023	6,809	0,147
Mode 3	0,041	3,883	0,258	**Mode 8**	0,022	7,239	0,138
Mode 4	0,039	4,113	0,243	**Mode 9**	0,021	7,417	0,135
Mode 5	0,034	4,694	0,213	**Mode 10**	0,02	8,042	0,124

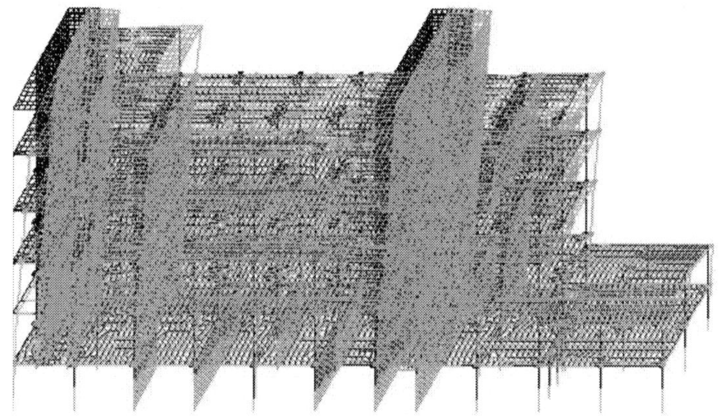

Figure 2. Fist oscillations mode. The translatory oscillations relative to the axis X.

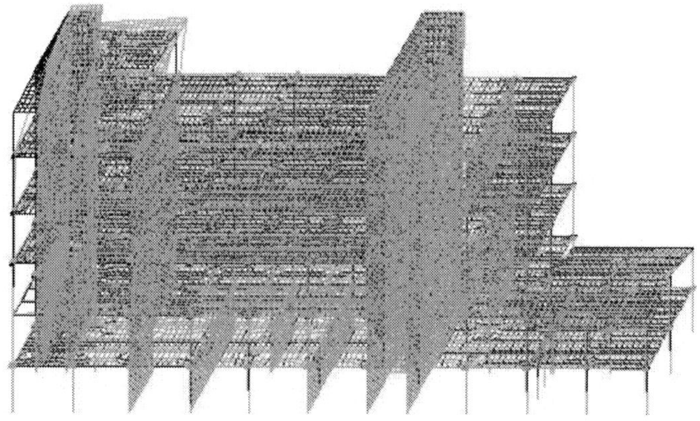

Figure 3. Second oscillations mode. The translatory oscillations relative to the axis Y.

As it shown in figures 2 and 3, the nature of the deformation of the building has a pronounced translational character. Structures and its joints oscillate as one unit. It leads to a more uniform stress distribution.

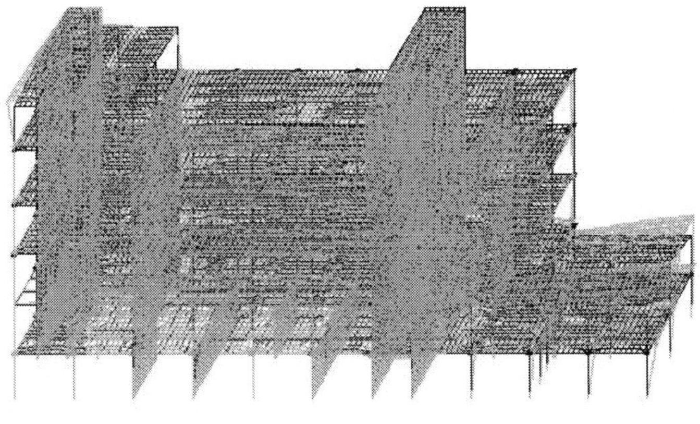

Figure 4. Third oscillations mode. The torsional oscillations.

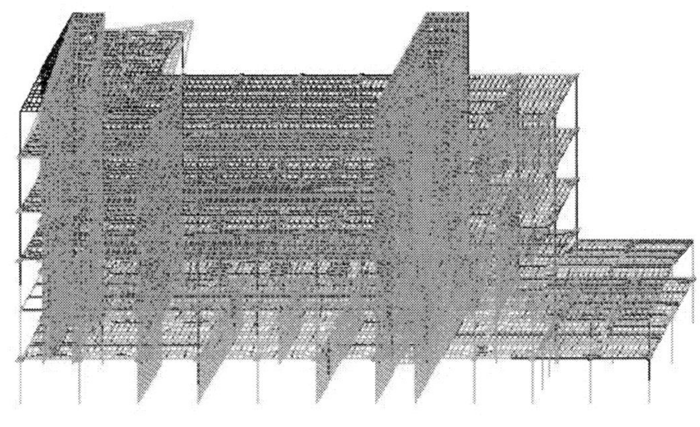

Figure 5. Forth oscillations mode. The flexural oscillations.

Investigations were performed in a passive experiment. Figure 6 shows the time process and the spectral characteristics of oscillations of building. External excitation was generated by technological equipment which alternately turned on and off. In Figure 6, we can find separate region of the existence of the buildup and modes of oscillations with large amplitudes corresponding to the inclusion of individual machines and small oscillations representing structure shimmering.

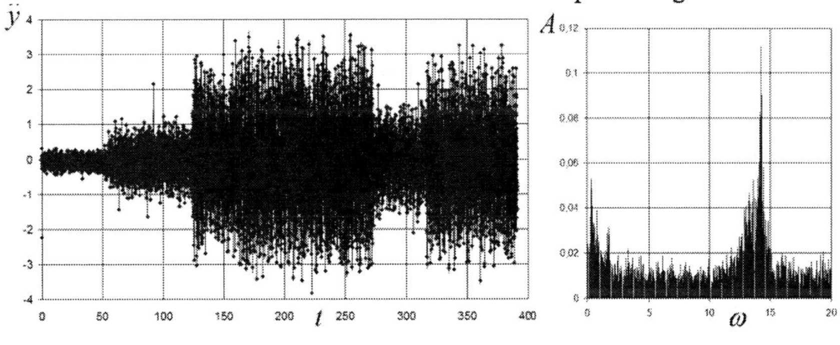

Figure 6. Time processes and spectral characteristic of the forced oscillations of building oscillations.

On the spectral characteristic (figure 6) one can notice two frequency spikes. The first of them corresponds to the first mode of the building oscillation and is equal to 1.8 Hz. The second spike reflects the frequency of external excitation created by the equipment 1at frequency 4.8 Hz. The original record of the time process was divided into separate samples corresponding to individual

IOP Conf. Series: Materials Science and Engineering **660** (2019) 012061 doi:10.1088/1757-899X/660/1/012061

technological cycles. Figure 7 shows the time process and phase trajectories of forced oscillations of building on the phase planes $(y,\dot{y}),(y,\ddot{y}),(\dot{y},\ddot{y})$. The process under study is not stationary due to the fact that the eigenfrequencies and the frequency of the external excitation do not coincide.

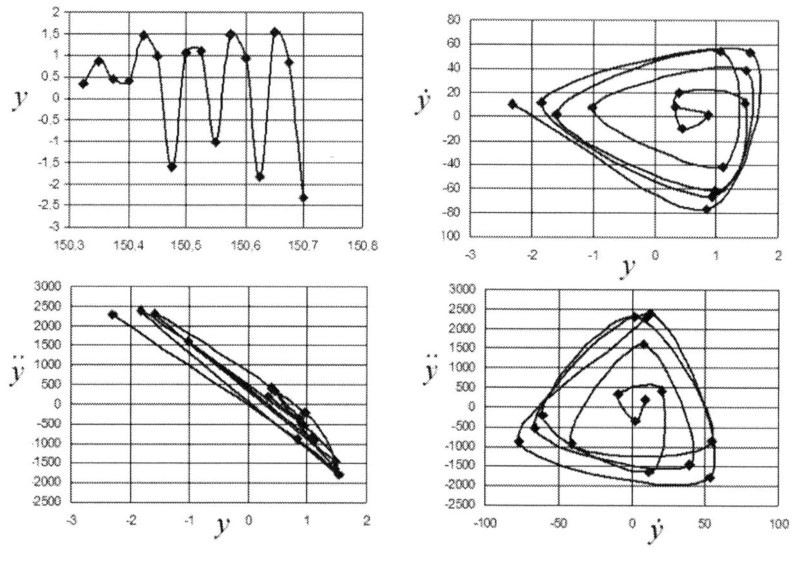

Figure 7. Time processes and phase trajectories of building oscillations.

Let us analyze the structure of phase trajectories on the plane "acceleration - displacement". As it can be seen from figure 7 the structures of phase trajectories in the region of positive and negative displacements are different. In the area of negative values of displacements, the system exhibits properties inherent to the linear systems. So the skeleton curve looks like a straight line. This fact can be explained by the elastic stage of the concrete in compressed zone. At the same time, in the region of positive displacement values, the skeleton curve of phase trajectories on the plane "acceleration - displacement" takes the form of a parabola. This is typical for systems with a "soft" characteristic of restoring force. The cause of this phenomenon is the elastic - plastic stage of material in the stretched zone of the section.

5. Parametric identification of dynamic model

The main concept of this paper is that, if a given type of damage converts a linear system into a non-linear system, then any observed manifestations of non-linearity serve to indicate that damage is present. In its undamaged state an assumption that the structures can be modeled as a linear system is quite adequate, but consider what happens when a crack is introduced.

On the basis of results received in [10], generalization was executed and formulated recommendation to application of phase trajectories on a plane "acceleration- displacement" to assessment of parameters of dynamic models. It was assumed that dependence of restoring force can be presented cubic dependence $R(y)=\alpha\, y \pm \beta\, y^3$. Thus, will get the dynamic system, described differential equation of a view

$$y + \varepsilon\,\dot{y} + H_1 \operatorname{sgn} \dot{y} + \alpha\, y \pm \beta\, y^3 = F(t). \qquad (2)$$

This type of restoring dependence can be accepted for most practical applications, the therefore entered supposition does not affect community of results. The obtained phase trajectories can be considered as consisting of separate elements. The shape of phase trajectories and sizes of separate elements depend on correlation of parameters of equation (2). Shape of phase trajectories on a plane "acceleration- displacement" (figures 8,9) for the oscillations on fundamental tone of the mechanical non-linear systems with the combined friction, represent equation (2), determined the values of parameters ε, H_1, α, β.

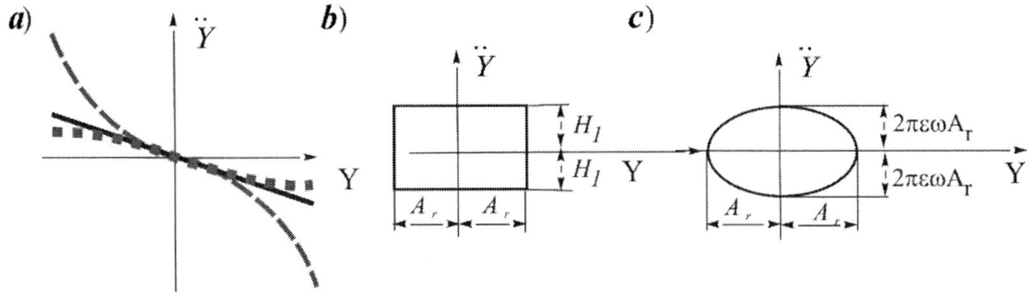

Figure 8. Elements of phase trajectories of the system (2) for the following values of parameters: a– $\varepsilon=0$, $H_1=0$; b – $\varepsilon=0$, $\alpha=0$, $\beta=0$; c – $H_1=0$, $\alpha=0$, $\beta=0$.

The phase trajectories on plane "acceleration- displacement" change its shapes and places for some set of parameters ε, H_1, α, β with dependence of frequency of outer excitement. For some set of parameters ε, H_1, α, phase trajectories will change its shape and position on a plane "acceleration- displacement" depending on frequency of outer excitement. At the increase of frequency of outer excitement length of phase trajectories will be increased, arriving at a maximum value at resonance, and quickly to diminish after passing of resonance frequency.

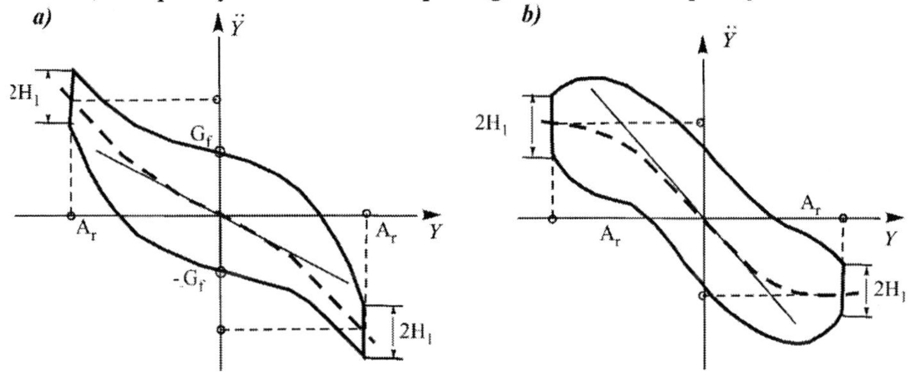

Figure 9. Phase trajectories of fundamental tone oscillations he combined friction and non-linear restoring dependence: a-the systems with "hard" restoring dependence; b – with "soft" restoring dependence.

The phase trajectories of resonance oscillations in the frequency of external excitement is important for determination of parameters of restoring force. Having regard to the features of dynamic behaviour of the systems with non-linear restoring force for control of accuracy of parameter estimation. It is expedient to use a few phase trajectories on a plane "acceleration-displacement" corresponding to a few frequencies of outer excitement. Assessment of values of parameters, α β can be executed either by the appeal of the known analytical solutions or by interpolation. For example, it will suppose that amplitudes A_1, A_2 and frequencies ω_1, ω_2 of oscillations of fundamental tone, proper the close to resonance regime are known. With the purpose of determination of parameters α and α we will take advantage of the known averaging solution for finding of frequencies of resonance oscillations . $\omega^2 = \alpha + 0{,}75\ \beta A^2$ Then, for two successive values of frequencies will receive the system of two equations. By solving it we have:

$\alpha = \left(\omega_2 A_1^2 - \omega_1 A_2^2\right)/\left(A_1^2 - A_2^2\right)$ and $\beta = 4\left(\omega_1^2 - \omega_2^2\right)/3\left(A_1^2 - A_2^2\right)$.

Analysing a figure 9, it is possible to note that the value of the module of dry friction H_1 can be measured directly at treatment of phase trajectory in a point $y = A_r$, and the value of coefficient of viscous friction can be defined from correlation $G_f = 2\pi\omega\varepsilon A + H_1$, where G_f is distance between intersections by a phase trajectory with ax $\ddot{y}$. Exactness of the estimated values of parameters of viscous and dry friction can be appraised the lead through of alternative calculations on the known formulas.

6. Conclusions

The identification procedures were applied to treatment and analysis of the results from the full-scale dynamic testing of the building. Appropriate time traces of free oscillations from the experimental records were selected for application of the method. The extended phase space is built from the general aggregate, its phase trajectories and their mappings.

Phase trajectories on an acceleration-displacement plane allowed to attribute the investigated system to the family of the asymmetrical systems with an asymmetric, non-linear dependence of restoring force. The non-linear character of restoring force of flat was depicted presence of defects and damages of reinforced concrete structure. In particular, at the oscillations of crack of structures alternately opened and closed, that causes change in flexibility. Another factor explaining asymmetry of restoring dependence is a presence of one-sided connections between plate and beams.

References

[1] Adams D E Allemang R J 1998 Survey of nonlinear detection and identification techniques for experimental oscillations. *Proc. of ISMA23* p 269-280

[2] Kerschen G, Worden K, Vakakis A F and Golinval J-C 2006.Past, present and future of nonlinear system identification in structural dynamics. Mechanical Systems. *Signal Process.* 20 (**3**) p 505-592

[3] Worden K and Tomlinson G R 2001 *Non-linearity in Structural Dynamics: Detection, Identification and Modelling*. (New-York: Institute of Physics Publications)

[4] Kulisiewics M 2005 Modelling and identification of non-linear mechanical systems under complex load (Wroclaw :Oficyna wydawnicza Politechniki Wroclawickiej)

[5] Kononenko V O and Plakhtiyenko M P 1976 *Methods of Identification of Mechanical Oscillating Systems* (Kiew:Naukowa dumka)

[6] Plakhtiyenko M P 2000 Methods of Identification of Mechanical Oscillating Systems, *International* Applied *Mechanics* 36 (**12**) p 38-68

[7] Grop P 1979 *Identification of mechanical systems* (Moscow, Mir)

[8] Volkova V E and Schneider K 2005 Qualitative theory and identification of dynamic system with one degree of freedom *International Applied Mechanics*. 41 (**6**) p 134-139

[9] Volkova V E, Kazakevitch M I 2009 Application of the extended phase trajectories to identification of chaotic systems. *Inverse Problems in Science and Engineering*. 17 (**2**), p 203 - 212

[10] Volkova V E 2010 Development of methods for nonparametric identification of models of mechanical systems *Proc. 10th Int. Conf. "Modern Building Materials, Structures and Techniques" (Vilnius)* vol II p 1066-1069

[11] Volkova V E, Gaile L and Pakrastins L 2017 Experimental qualitative identification of damping and stiffness characteristics of lattice towers *IOP Conf. Ser.: Mater. Sci. Eng*. 251 012108

[12] Fialko S Yu, Kriksunov E Z and Karpilovsky V S 2003 *Proc. Int. Conf. Computer Methods in Mechanicals (Gliwice)* p. 129-130

[13] PerelmuterA 2014 *Discourse on Construction Mechanics* (Moscow: SCAD Soft)

[14] Perelmuter A and Slivker V 2002 *Computational Models of Structures and Feasibility of their Analysis* (Kiyev, Stal Publishing House)

IOP Conf. Series: Materials Science and Engineering **660** (2019) 012062 doi:10.1088/1757-899X/660/1/012062

Numerical Modelling of High Strength Fibre-Concrete's columns in Multi-Storey Building

J Sliseris and K Buka-Vaivade

Institute of Structural Engineering and Reconstruction, Faculty of Civil Engineering, Riga Technical University, Kipsalas st. 6A/6B, LV-1048, Riga, Latvia

E-mail: janis.sliseris@gmail.com

Abstract. The complexity of architecture is continuously growing. Therefore, it is necessary to develop a new construction system with complicated topology using less human work. A fibre-concrete is very promising material to solve those problems. However, design methods of fibre-concrete load bearing structures such as columns are not well developed. Therefore, a new numerical simulation framework is proposed for analysis of fibre-concrete structures macro scale. The method can take into account non-linear post-cracking behaviour of fibre concrete. This includes local fibre orientation in thin elements, anisotropic continuum damage models on different scales, efficient meso-scale fibre orientation prediction tools and others. The proposed method is used to analyse building's columns where part of structural elements is made from fibre-concrete. The effectiveness of fibre-concrete is estimated and proposed a guideline for optimal design using fibre-concrete.

1. Introduction

High performance fiber reinforced concrete (HPFRC) and ultra-high performance fiber reinforced concrete (UHPFRC) is becoming more popular in construction industry. However, mechanical behaviour of HPFRC is not well understood, especially when fibers are oriented and material is additionally strengthened with traditional reinforcement bars (re-bars). Additional factors that affect properties of fiber concrete are fiber shape, aspect ratio, volume fraction and properties of concrete itself [1–3]. Meanwhile, design standard EC2 do not include information about design of fiber reinforced concrete [4].

Columns are one of the most important structural elements, especially in high rise buildings. Due to horizontal loads and non-symmetrically loaded structure, a high values of bending stress can appear in columns, that is a cause of tensile stress and cracking [6-9]. The high strength concrete allows to design a relatively thin-walled structures [10-12], therefore columns with rectangular hollow sections are analysed in present work. Thin-walled concrete columns shows a significant influence of fiber's shape, length, orientation on the load bearing capacity of the column or any other structural element [13-14]. Therefore, by analysing and adapting the state-of-art numerical modelling techniques [15-20], there are created a numerical model for the analysis and optimization of rectangular columns. Numerical model is validated with experimental results. The guidelines for optimal design of fiber reinforced high strength concrete's columns is proposed according to numerical results.

2. Model validation at compressive stress

The improvement in the strength of the concrete peak from the use of fibres is usually insignificant, but the use of fibres reduces the fragility of high-performance concrete and significantly improve the deformative properties of the concrete [1-2]. The developed in the new numerical simulation framework finite element model allows appreciating the improvement of deformative properties of concrete from the use of fibres by defining the stress-strain curve of the respective material. The developed 3D and 1D numerical models which considered non-linear behaviour of high-performance steel fibre reinforced concrete (HPSFRC) material are validated by the experimental data from the publication [3] for column with pinned supports at both ends, which is subjected to combined action of compression and bending with additional longitudinal reinforcement.

The validation of model is carried out for the column with length L = 1300 mm and external sizes 125x125 mm. Column has B550 class 4 longitudinal bars with a diameter of 8 mm, located in the corners of the columns. The compressive strength of the used concrete from the test results is 67-72 MPa. The eccentricity with which the columns are loaded is e = 50 mm. The high strength concrete of C60 /75 [4] class is used for calculation. Consequently, fcm = 68 MPa has been accepted for the compressive strength of the high-performance concrete and 4.4 MPa for mean tensile strength. Compression stress-stain values for HPSFRC and high-performance concrete (HPC), which is shown in figure 1 are taken according the typical stress-strain diagrams of steel fibre concrete [3].

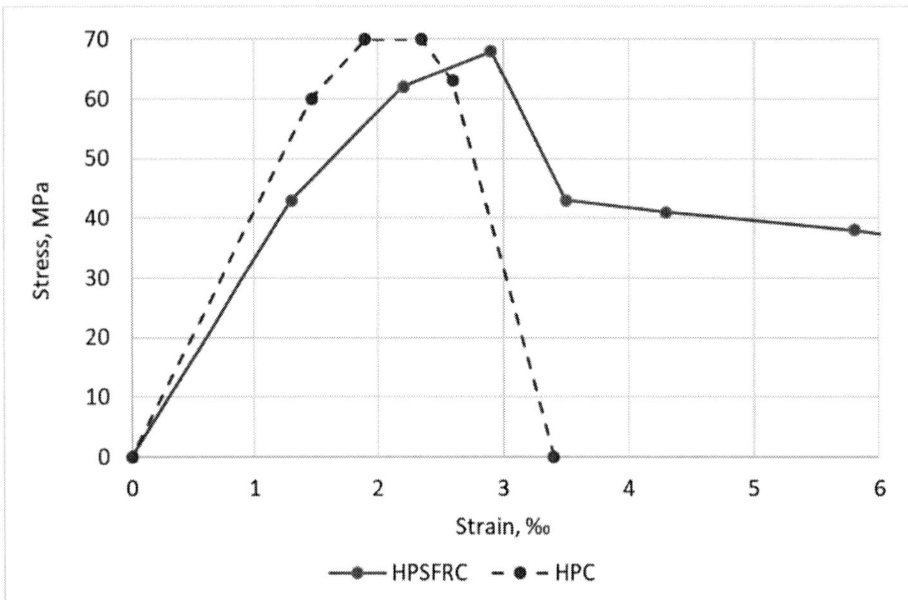

Figure 1. Stress-strain curve for high-performance concrete and high-performance concrete with 1.0% RC 65/35 BN-type steel fibres [3].

The column 3D model is designed for one-half the length of the column to reduce the time of calculation by using symmetry conditions. Collapse scheme of the 3D design model of column specimen and typical failure mode of the column specimen from the experimental test are shown in figure 2. As it can be seen, there is similar character of the failure scheme. The first cracks are formed in the middle of the column span.

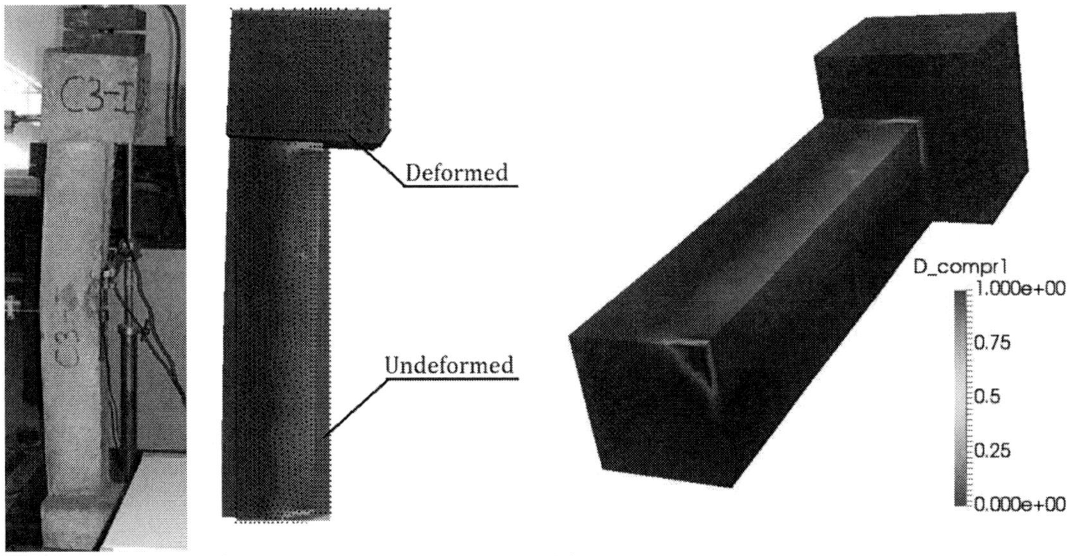

(a) (b) (c)

Figure 2. Failure scheme of column specimen: (a) from experiment [3]; (b) from calculation; (c) with max compression stress.

Load-displacement curves of column specimen subjected to the compression with flexure, according to laboratory tests from publication, and developed 3D and 1D numerical model calculations are summarized in figure 3.

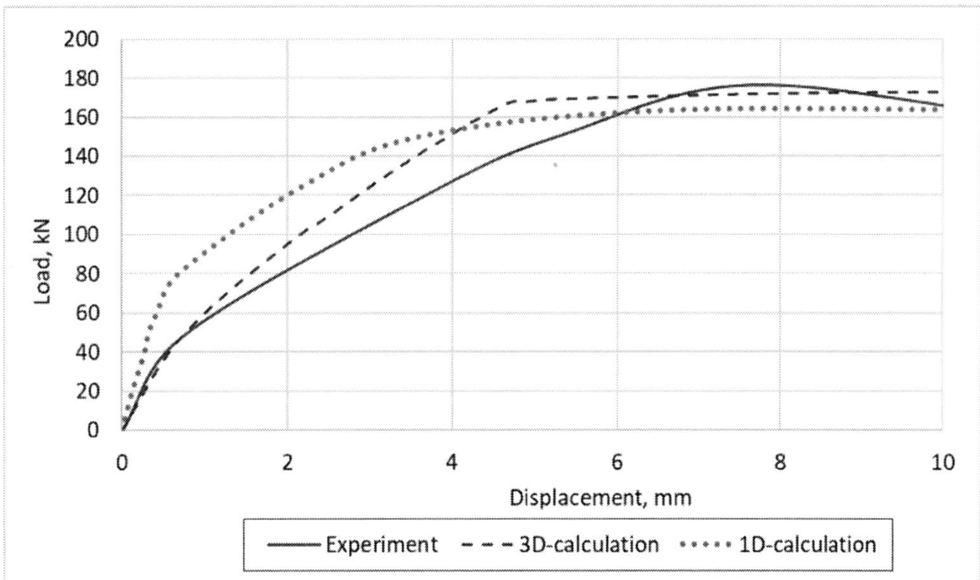

Figure 3. Load-displacement curves for column specimen.

According to the results obtained, the crushing axial load obtained from the laboratory test was 176 kN, while the calculated by 3D and 1D non-linear numerical models crushing axial load was 172.5 kN and 163.5, respectively. The difference between design and experimentally determined axial load for 3D model is 2.0%, for 1D model – 7.1%. The behaviour of the calculation models is like the actual operation of the column specimen and developed models are safe for prediction of the behaviour of high-performance steel fibre reinforced concrete material. Crushing axial load of the high-performance concrete column according to the experimental data is 159 kN, while the calculated by 1D numerical model it was 151.5 kN. The difference between design and experimentally determined axial load for 1D model is 5.0%.

3. Optimal design of HPSFRC

The optimal design is described for column structure, which is shown in figure 4, subjected to the compression and bending. The columns cross-sections are selected as a box-type to provide the decreased consumption of non-renewable materials and high stiffness of the element in the both planes. The box-type cross-section enables the integration of engineering communication inside of columns. Column cross-section is with constant 500x500 mm external dimensions and variable wall thickness.

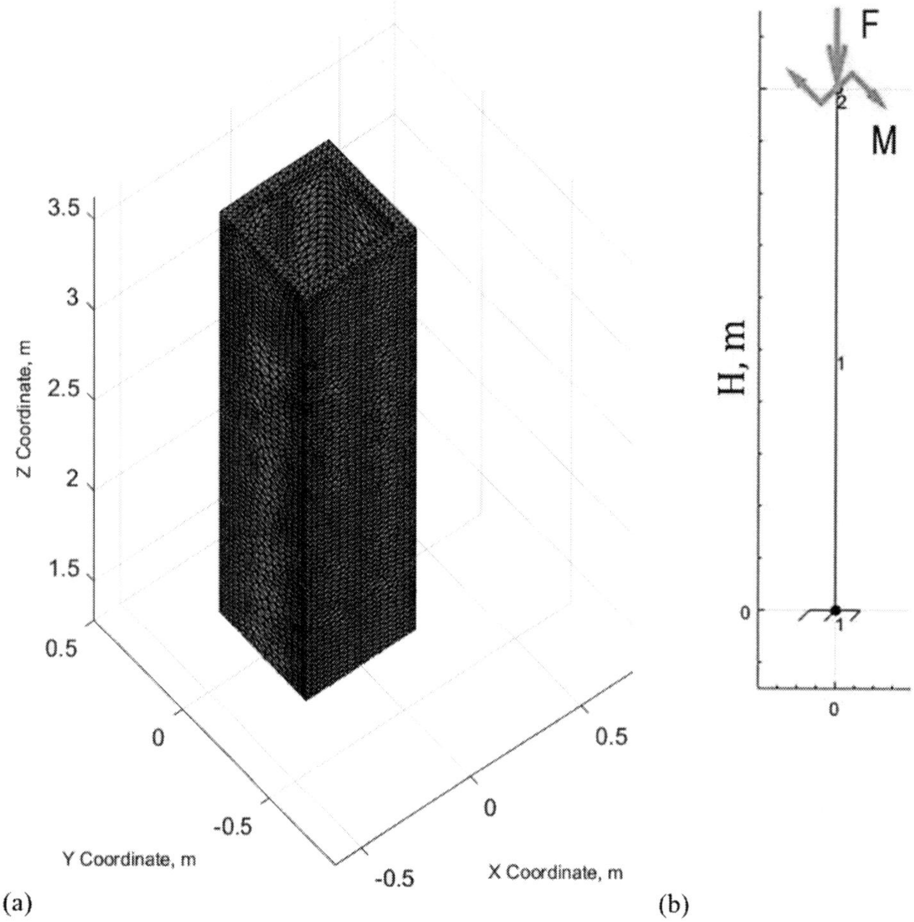

Figure 4. Details of column structure: (a) column 3D numerical model with meshing; (b) design scheme of the column.

Columns with the height (H) from 3.5 to 4.5 m are chosen as a region of interest for the optimization. Material consumption of the high-performance fibre reinforced concrete and steel of the longitudinal reinforcement are regulated by optimization input variables t - wall thickness of the box-type column cross-section with limits from 5 to 12 mm and A_s – area of the longitudinal steel reinforcement in column cross-section with limits from 0 to 1.5% of concrete area of the column cross-section.

To explore the relationships between several variables, such as column height and material consumption of the steel of the longitudinal reinforcement and high-performance concrete, and response variable – load-bearing capacity of the column, response surface methodology is used.

3.1. Experimental design

There are 3 input variables with 3 levels of each factor, so the total number of experiments is 27. The experimental design that defines which experiments should be carried out in the experimental region is based on Box–Behnken design.

Each of the experiment is calculated to determine load-bearing capacity of the column as function $F,M = f(H, t, A_s)$. There is a relationship between axial force F and bending moment M: $M = F \cdot e$, where e – eccentricity, which is equal to the distance from the centre of the column cross-section to the centre of the box-type cross-section wall of the column. This relationship is based on the simplified design scheme of the multi-storey building, where one of the ends of the column is fixed, and loads are transferred from each storey with eccentricity by beams.

Simulations of analysed process based on experimental design are made to get necessary statistic data set of column load-bearing capacity, which makes regression analyses possible.

3.2. Calculation of the experiment

For the calculation of each experiment is used column 3D and 1D numerical models which considered non-linear behaviour of high-performance steel fibre reinforced concrete material. The behaviour of non-linear material is characterized by degradation of Young's modulus cause of cracking. Damage variable is used for the prediction of behaviour of high-performance steel fibre reinforced concrete [5].

Calculations with the 3D numerical model are complex and time-consuming, therefore verification of the 1D numerical model of the column with fixed support at one end, which is subjected to the compression and flexure has been done by comparison of the results of calculations with both models.

Calculations of 3 experiments with 3D and 1D model have been made. Verification of the 1D model is made by comparison of crushing axial load values or load value at displacement value H / 250, depending on which value is reached first.

According to the results obtained, the load value at displacement value H / 250 calculated by 3D and 1D non-linear numerical models for column with H=3.5m, t=5mm and As=0.2% is 353.9 kN and 342 kN, respectively; for column with H=4m, t=12mm and As=0% is 703.0 kN and 682 kN and for column with H=4.5m, t=8.5mm and As=1.5% is 823.8 kN and 805 kN. The difference between axial load values determined by 3D and 1D models is between 2.3% and 3.4%. Results of verification showed, that used 1D finite element model describes with sufficient accuracy the behaviour of the 3D numerical model of the column subjected to the compression and flexure, therefore 1D model is used for the further calculations of the experiments.

Results of calculations of 27 experiments shown, that dependence of load-bearing capacity value from column height in region of interest for column heights 3.5 – 4.5 m is negligible, because of load bearing capacity is equal with load value at displacement value H / 250, which is 14 – 18 mm, in most of the experiments.

The comparison of the load-bearing capacity of the column with height equal to 4 m from high-performance steel fibre concrete (HPSFRC) and from high-performance concrete (HPC) with stress-strain values according to figure 1 is summarized in table 1.

Table 1. Comparison of load-bearing capacity of the 4 m height column from HPSFRC and HPC.

t,mm	5	8.5	12	5	8.5	12	5	8.5	12
As,%	0	0	0	0.2	0.2	0.2	1.5	1.5	1.5
F_{HPSFRC}, kN	320	512	682	340	547	735	525	847	1145
F_{HPC}, kN	309	492.5	656	320.5	512.5	685	445	740	1025
Δ, %	3.4	3.8	3.8	5.7	6.3	6.8	15.2	12.6	10.5

Graphs of the dependence of column with height 4 m load-bearing capacity on the wall thickness of the box-type column cross-section (t) with different A_s values are summarized in the Figure 5.

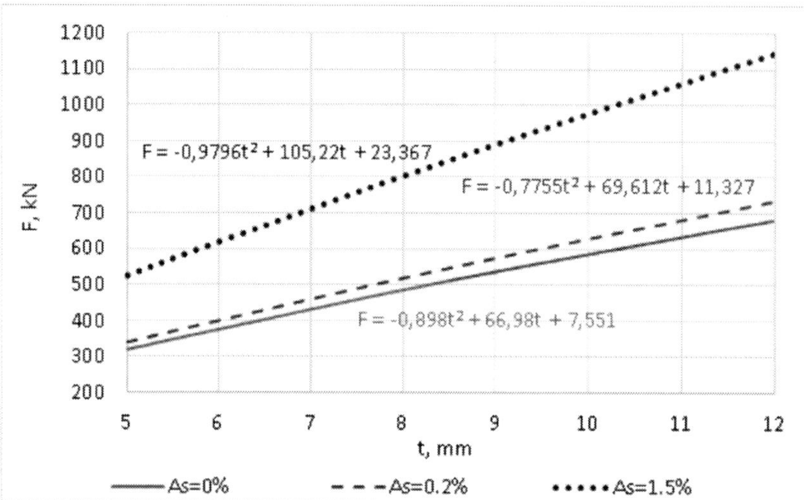

Figure 5. The dependence of column load-bearing capacity (F) on the wall thickness (t) with different area of the longitudinal steel reinforcement in column cross-section as a percentage of the concrete area (A_s).

Graphs of the dependence of column with height 4 m load-bearing capacity on the area of the longitudinal steel reinforcement in column cross-section as a percentage of the concrete area (A_s) with different t values are summarized in the Figure 6.

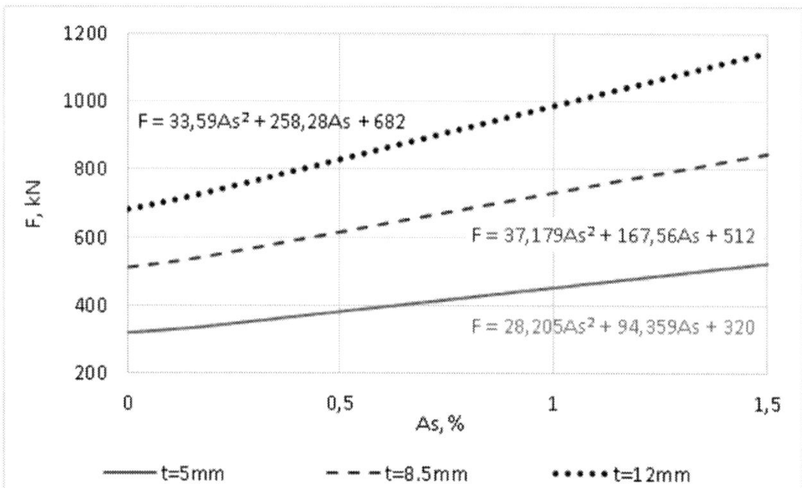

Figure 6. The dependence of column load-bearing capacity (F) on the amount of reinforcement (A_s) with different wall thickness of the box-type column cross-section (t).

4. Conclusions

Possibility to increase load-bearing capacity of the high-performance concrete columns subjected to compression and flexure by adding steel fibres was checked and confirmed by calculations. The use of high-performance steel fibre concrete as column material is especially effective for columns with additional longitudinal reinforcement where load-bearing capacity is up to 15% bigger in comparison with high-performance concrete columns with the same cross-section.

The difference in the ultimate load between the calculated and experimental results is 2.0% for 3D model and 7.1% for 1D model. The linear behaviour of the 3D model is identical to the experimentally obtained of the column, the failure mode of the column specimen and the plastic behaviour of the material is similar in both cases, which leads to the conclusion that the developed constitutive model of material behaviour with sufficient precision describes the behaviour of the column subjected to the compression and flexure.

The developed optimization method shows, that difference of the column load-bearing capacity for column height between 3.5 and 4.5 m is small. The obtained results of experiment calculations allow predicting the load-bearing capacity of column and regulating material consumption of the high-performance fibre reinforced concrete and steel for additional longitudinal reinforcement.

Acknowledgement
The financial support of European Regional Development Fund project Nr.1.1.1.1/16/A/007 "A New Concept for Sustainable and Nearly Zero-Energy Buildings" is acknowledged.

References

[1] Olivito R S and Zuccarello F A 2010 An experimental study on the tensile strength of steel fiber reinformed concrete. *Composites: Part B.* **41** 246–255
[2] Vougioukas E and Papadatou M 2017 A model for the prediction of the tensile strength of fiber-updated concrete members, before and after cracking. *Fibers.* **5(3)** 27
[3] Tokgoz S, Dundar C and Tanrikulu A K 2012 Experimental behavior of steel fiber high strength reinforced concrete and composite columns. *Journal of Constructional Steel Research.* **74** 98–107
[4] EN 1992-1-1 "Eurocode 2: Design of concrete structures - Part 1-1: General rules and rules for buildings"
[5] Sliseris J 2018 Numerical analysis of reinformed concrete structures with oriented steel fibers and re-pack. *Engineering Fracture Mechanics.* **194** 337–349
[6] Yoo DY, Kim S, Park GJ, Park JJ, Kim SW 2017 Effects of fiber shape, aspect ratio, and volume fraction on flexural behavior of ultra-high-performance fiber-reinforced cement composites. *Compos Struct* **174** 375–88
[7] Kazemi MT, Golsorkhtabar H, Beygi MHA, Gholamitabar M. 2017 Fracture properties of steel fiber reinforced high strength concrete using work of fracture and size effect methods. *Constr Build Mater* **142** 482–9
[8] Yoo DY, Banthia N, Yoon YS. 2016 Predicting the flexural behavior of ultra-high-performance fiber-reinforced concrete. *Cem Concr Compos* **74** 71–87
[9] Alberti MG, Enfedaque A, Gálvez JC, Agrawal V. 2016 Fibre distribution and orientation of macro-synthetic polyolefin fibre reinforced concrete elements. *Constr Build Mater* **122** 505–17
[10] Alberti MG, Enfedaque A, Gálvez JC. 2017 On the prediction of the orientation factor and fibre distribution of steel and macro-synthetic fibres for fibre-reinforced concrete. *Cem Concr Compos* **77** 29–48
[11] Abrishambaf A, Cunha VMCF, Barros JAO. 2016 A two-phase material approach to model steel fibre reinforced self-compacting concrete in panels. *Eng Fract Mech* **162** 1–20
[12] Sarmiento E V., Hendriks MAN, Geiker MR, Kanstad T. 2016 Modelling the influence of the fibre structure on the structural behaviour of flowable fibre-reinforced concrete. *Eng Struct* **124** 186–95
[13] Cunha VMCF, Barros J a. O, Sena-Cruz JM. 2011 An integrated approach for modelling the tensile behaviour of steel fibre reinforced self-compacting concrete. *Cem Concr Res* **41** 64–76
[14] Roy M, Hollmann C, Wille K. 2017 Influence of volume fraction and orientation of fibers on the

pullout behavior of reinforcement bar embedded in ultra high performance concrete. *Constr Build Mater* **146** 582–93
[15] Sliseris J, Andrä H, Kabel M, Dix B, Plinke B, Wirjadi O, G.Frolovs. 2014 Numerical prediction of the stiffness and strength of medium density fiberboards. *Mech Mater* **79** 73–84
[16] Sliseris J, Andrä H, Kabel M, Dix B, Plinke B. 2017 Virtual characterization of MDF fiber network. *Eur J Wood Wood Prod* **75** 397–407
[17] Zhou B, Uchida Y. 2017 Influence of flowability, casting time and formwork geometry on fiber orientation and mechanical properties of UHPFRC. *Cem Concr Res* **95** 164–77
[18] Ren W, Yang Z, Sharma R, Zhang C, Withers PJ. 2015 Two-dimensional X-ray CT image based meso-scale fracture modelling of concrete. *Eng Fract Mech* **133** 24–39
[19] Šliseris J, Gaile L, Pakrastiņš L. 2017 Numerical analysis of behaviour of cross laminated timber (CLT) in blast loading. *IOP Conf. Ser. Mater. Sci. Eng.*, **251**
[20] Šliseris J, Gaile L, Pakrastiņš L. 2017 Deformation Process Numerical Analysis of T-stub Flanges with Pre-loaded Bolts. *Procedia Eng.*, **172**

Characterization of crack formation and development in the oil&gas well casing cement sheath

I Paegle[1], V Marcos-Meson[1] and G Fischer[1]
[1] Department of Civil Engineering, Technical University of Denmark, Brovej 118, 2800 Kgs.Lyngby, Denmark.

Abstract. The oil industry relies on the use of cement as a primary barrier, to isolate the steel casing from the surrounding formations and provide overall stability on the bored structure. There has been extensive technological development on the physical and mechanical properties of the materials from the cement and cementing industry, in order to produce and install effective cement barriers, capable of withstanding both physical and chemical loads under extreme wellbore conditions. In particular, the formation of cracks and interfacial damage at the primary cement sheath is a largely discussed issue. However, there is still a limited understanding about the crack formation and propagation processes in well-cements, due to the particular geometric and boundary conditions of the structural system, such as: pressure, temperature, and confinement conditions. This paper is focused on the crack formation and propagation processes on cement systems under wellbore-like geometrical conditions, and proposes a test-setup to evaluate cracking and localized damage at the casing-cement interface.

1. Introduction

The oil industry relies on the use of cement as a primary barrier, to isolate the steel casing from the surrounding formations and provide overall stability on the bored structure [1]. There is overall agreement regarding the capabilities of cement, as both: a structural element providing stiffness to the bored-well, and as a physical barrier to isolate the steel casing and prevent fluid leakages from adjacent soil formations [1–3].

There has been extensive technological development regarding the physical and mechanical properties of cementing materials from the cement and cementing industry, in order to produce and install effective cement barriers, capable of withstanding both physical and chemical loads under extreme wellbore conditions.

Among others, the industry has placed its focus on the development of tailored cementing-materials, such as: alternative cement formulations [4–6], rheology modifiers [7], foamed cements [8,9], expansive cements [10] or composites [11]. These solutions aim at minimizing failure of the cement sheath due to, for example: mechanical failure of the barrier [12, 13], damage due to thermal loads [11,14], loss or lack of impermeability [15,16] or chemical degradation of the cement [17,18].In particular, the formation of cracks and interfacial damage at the primary cement sheath is a largely discussed issue. Since radial and annular cracks have a large impact on the overall structural stability of the system [13-15,19], in addition to the corresponding loss of imperviousness, since cracks provide a preferential path for leaks [15,16,19].

However, there is still a limited understanding about the crack formation and propagation processes in well-cements [16,20]. The very specific geometric and boundary conditions of the casing-cement-formation and casing-cement-casing structural systems represent a challenge for scientists and engineers, for example: large downhole pressures and temperatures, different degrees of confinement from adjacent soil and rock formations, as well as the inherent complexity of cement as a continuously evolving material (i.e. hardening process and chemical reactions with the surrounding environment).

This paper evaluates the crack formation and propagation processes on cement systems under wellbore-like geometrical conditions, and proposes a test-setup to evaluate cracking and localized damage at the casing-cement interface due to two types of loading: shrinkage and casing expansion.

2. Methodology and experiments

The experimental part of this study focuses on the characterization of the initiation and development of radial and annular (i.e. interface) cracks in cement around oil&gas well casing under two main load scenarios:

- Mechanical failure induced by chemical or thermal shrinkage stresses. Namely, restrained shrinkage load scenario.
- Mechanical failure induced by radial expansion of the inner annulus. This represents, for example, casing pressure tests run on casing while drilling the well. The latter scenario is subdivided into a confined and un-confined experiment, depending of the degree of external confinement.

The experiments are carried out at atmospheric pressure (1 atm) and room temperature (20°C), and without axial confinement (i.e. as a bi-dimensional system). As the testing conditions do not represent downhole conditions, the results presented are indicative and can be used as a qualitative representation of the morphology of cracks and damage as well as a demonstration of the proposed investigative techniques for further research.

2.1. Materials

To investigate initiation and development of the cracks in the oil wells, a cement slurry with w/c ratio of 0.4 was used. Additionally, a limestone filler (limestone/cement=0.3), with a grain size 10-200μm and additives like superplasticizer and air draining admixtures were added to the mixture.
Mixing, casting and curing were performed under room temperature (20°C), at atmospheric pressure conditions (1 atm). The mixing was performed on a shear blender at 3000 rpm.

2.2. Restrained-shrinkage test (ring-test)

The restrained-shrinkage test was based on the ASTM C1581 ring-test. This test aimed at characterizing the morphology and propagation of shrinkage-induced cracks in the casing cement. In particular, the test presented aims at characterizing the damage at the steel-cement interface (i.e. micro-annulus) and it's relation to the radial crack.

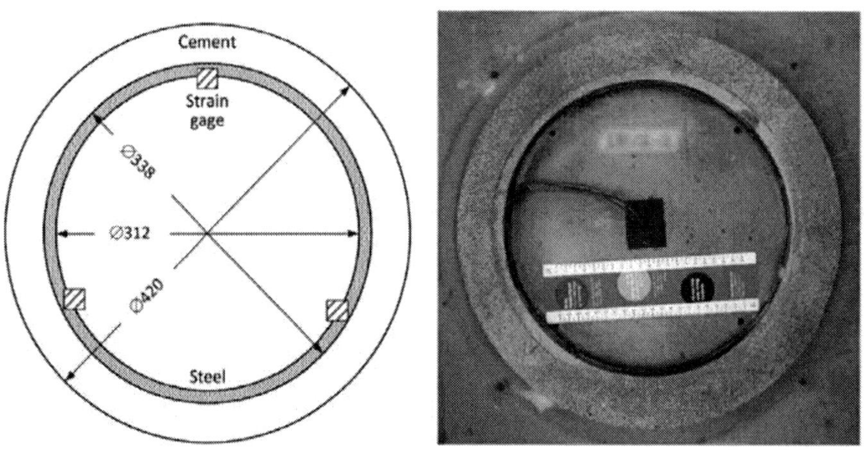

(a) (b)

Figure 1. Restrained shrinkage test: a) specimen dimensions and construction; b) test setup, de-moulded specimen with speckle pattern.

IOP Conf. Series: Materials Science and Engineering **660** (2019) 012063 doi:10.1088/1757-899X/660/1/012063

For this test, a 25mm thick cement ring with an inner diameter of 338 mm and an outer diameter of 420 mm, was cast with a concentric inner steel ring with a wall thickness of 13 mm, as shown in Figure 1. After the concrete had set, the outer ring-mould was removed, then the surface of the ring was sprayed with a speckle pattern and cured inside a climatic room at RH 30%. For the ASTM test, the test finalizes right after the cracking of the specimen, whereas for the proposed setup, the test continued through time until linear crack opening over time is reached.

During the curing, the stresses at the inner steel ring were monitored through three strain gages glued at the inner wall, see Figure 1a. The strain level at the steel ring served as a measure for the time to cracking and the development of stresses at the cement ring. The time of cracking is characterized by a sudden drop on the strain, as a response to the stress relief at the steel ring due to cracking of the specimen.

Furthermore, the strain state at the surface of the specimen was monitored through high-resolution images captured on 10 to 15 minute intervals. The images (see Figure 1b), were then post-processed using digital image correlation analysis (DIC), allowing for the measurement and characterization of crack growth and localized interfacial damage in between the steel and the cement.

2.3. Annular expansion test

The second experimental setup comprised the evaluation of damage and characterization of cracks due to expansion of the steel casing, i.e. during positive-pressure testing of the casing. For that purpose, a tailored mechanical loading setup has been designed and constructed, where a cylindrical tapered-piston expands a correspondingly tapered steel ring through a uniform vertical displacement. The system was placed at the centre of a cement ring (i.e. the test specimen), casted in between the steel ring and an outer circular mould, see Figure 2. For this test, two different outer confinement conditions were used: confined and unconfined to investigate the differences on crack development.

For the confined conditions during the testing, a steel ring around the outer perimeter of the sample was present (Figure 2a), so that the cement sample was not able to expand freely. For the unconfined sample, the outer steel ring was removed, therefore the cement sample was able to expand freely. These two conditions were chosen to represent the very stiff surrounding rock conditions of an oil well (confined condition) and weak surrounding conditions (unconfined).

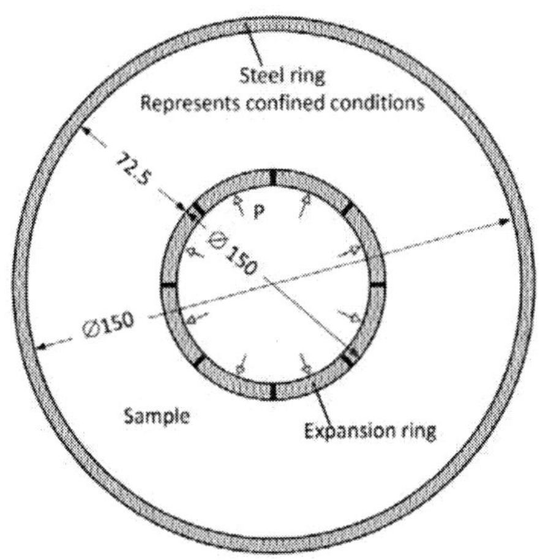

Figure 2. Radial expansion test, confined specimen: a) specimen dimensions and construction; b) test setup. For the unconfined specimen no outside ring is present.

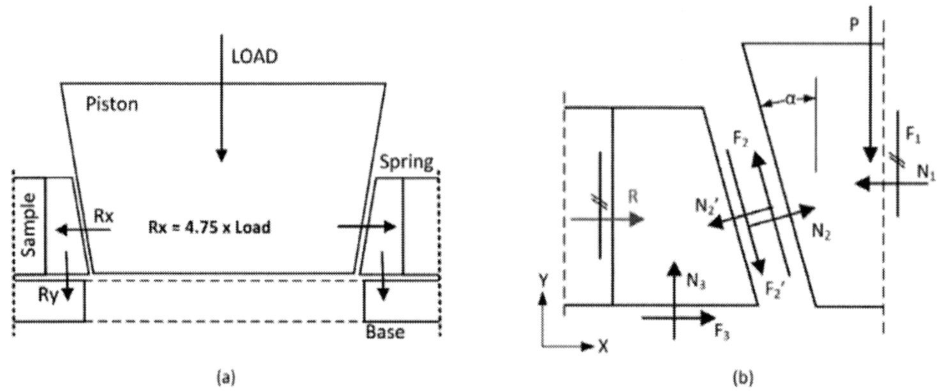

Figure 3. Radial expansion test. Loading system: a) system schematics; b) force equilibrium diagram.

At the time of testing, the central piston is pulled through the inner ring (spring) by means of a screw-jack with a maximum load of 25kN, the tapered surfaces transfer the load horizontally to the specimen as a uniform radial pressure, which eventually leads to tensile failure and cracking (Figure 3). The load from the piston is transferred through a vertical shaft, connected to the shaft of the screw-jack, a tailor-made load cell placed inside the system records the vertical load (shown as "P" in Figure 3) which is then used to calculate the horizontal and vertical reactions (shown as "R" and "N3" in Figure 3). The horizontal reaction was calculated as follows:

$$R = P/(\sin\alpha - \mu\cdot\cos\alpha\ )\cdot(\cos\alpha - \mu\cdot\sin\alpha\ ) - (\mu\cdot P) \quad (1)$$

The vertical displacement of the piston and horizontal expansion of the inner ring are measured using the DIC analysis [21, 22]. The horizontal component of the reaction (R) was calculated from equilibrium of forces, under the assumption of a uniform vertical displacement of the piston, a constant friction coefficient between the steel surfaces and treating the elements of the system as rigid bodies. Figure 3b displays the system of forces as considered for the calculation. The notation used in Figure 3 and equation is explained in Table 1.

Table 1. Variables considered for the calculation of the forces at the system: piston-spring-base.

Variable	Description	Value	Unit
α	Taper angle	3.0	deg
μ	Friction coefficient steel-steel	0.15	-
P	Vertical load	(1.00)	N
N_1	Horizontal reaction at piston axis	4.94	N
N_2 N_2'	Normal reaction at piston-spring interface	4.95	N
F_2 F_2'	Frictional forces at piston-spring interface	0.74	N
N_3	Vertical reaction at ring support	1.00	N
F_3	Frictional forces at spring-base interface	0.15	N
R	Horizontal reaction at ring-sample	4.75	N

3. Results and discussion

3.1. Restrained-shrinkage test (ring-test)

The results from the restrained-shrinkage test, as presented in Figure 4a, show the formation of a single radial crack at the time of cracking, propagating normally to the steel ring interface and crossing the whole cross section of the specimen. Additionally, there is a substantial separation at the cement-ring interface, growing from the point of cracking and gradually decreasing through the circumferential length (red annulus observed in Figure 4a). The radial crack is measured at three points: the outer face (C1), centre (C2) and the inner face (C3) of the specimen as shown in Figure 4b. The annulus is measured at 9 points: the two edges of the crack (A0, A1) and seven equally spaced points at 45° (0.25 π·rad).

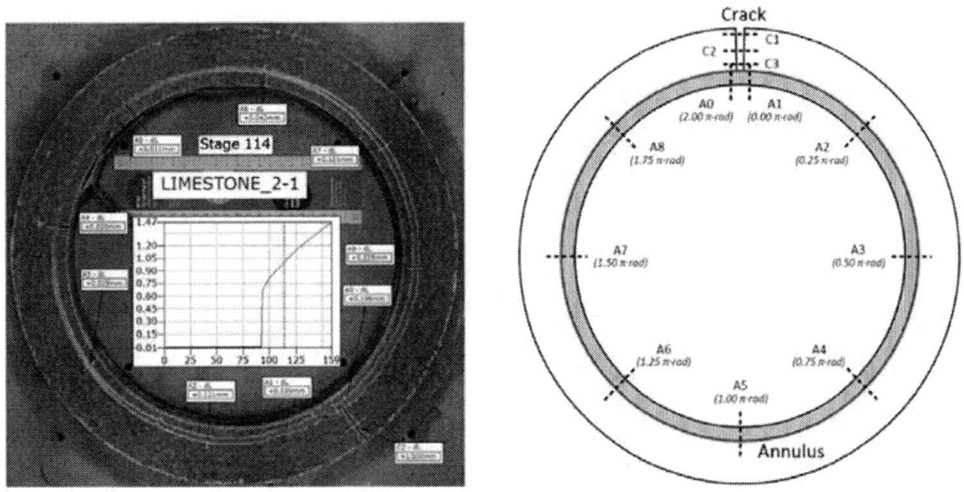

(a) (b)

Figure 4. Restrained shrinkage test: (a) DIC analysis and crack pattern, major strain mask overlay on specimen image; (b) Crack pattern and measurement points.

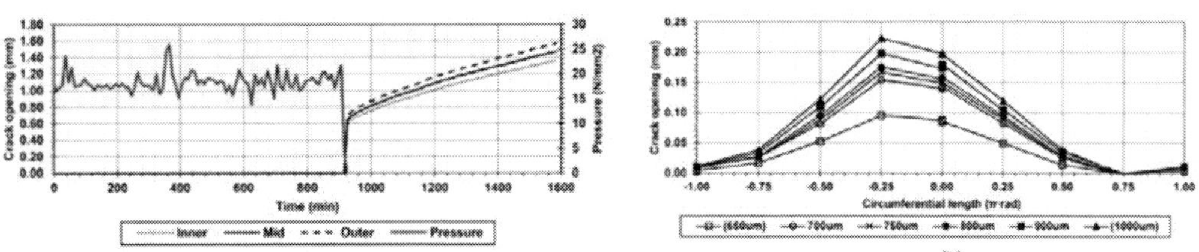

(a) (b)

Figure 5. Restrained shrinkage test: (a) Ring pressure and radial crack (C1) evolution on time; (b) Shape of annular cracks throughout annular length (π·rad) at various opening of radial crack (between 650 µm-1000 µm).

The initiation of the radial cracks, are in good agreement to the stress drop at the inner ring Figure 5a. The DIC measurements could accurately measure the time to first cracking, similarly to the strain gages placed at the inner steel ring. The DIC analysis at the radial crack (points C1-C3) shows the development of the cracks during the test, after the strain measurements at the ring fall to zero in Figure 5a. From this plot, it can be appreciated that the crack initiates sharply at the full cross-section of the specimen at the time of cracking with mean crack width ≈ 650 µm; but still had a "V" shape, with its smaller width at the steel interface. The angular distortion at the crack increases with time, with increasing crack width, following the shrinkage of the specimen.

The initiation of the annular crack follows a similar trend to the radial cracks. The annular crack initiates at the time of cracking and increases at increasing radial crack, as presented in Figure 5b. The plot shows a relatively symmetric annular crack, relative to the crack as symmetry axis.

3.2. Annular expansion test, unconfined conditions

The concrete ring is subject to a constant horizontal strain at its inner diameter, imposed as an expansion of the inner ring diameter at a rate of 0.314 mm/min. The vertical load at the shaft of the mechanical-jack are recorded, as well as for the strains and displacements of the ring and the piston-spring setup using the DIC software (Figure 6a).

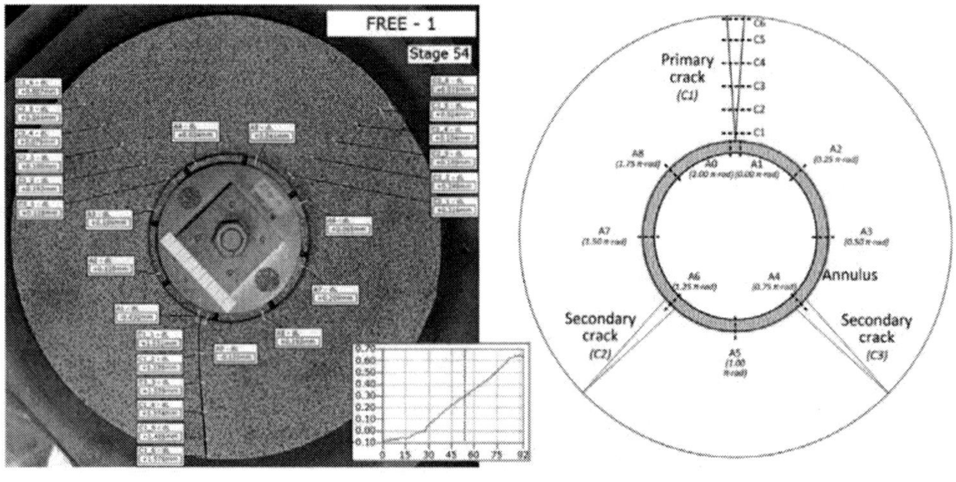

(a) (b)

Figure 6. Unconfined expansion test.: (a) DIC analysis and crack pattern, major strain mask overlay on specimen image; (b) Crack pattern and measurement points.

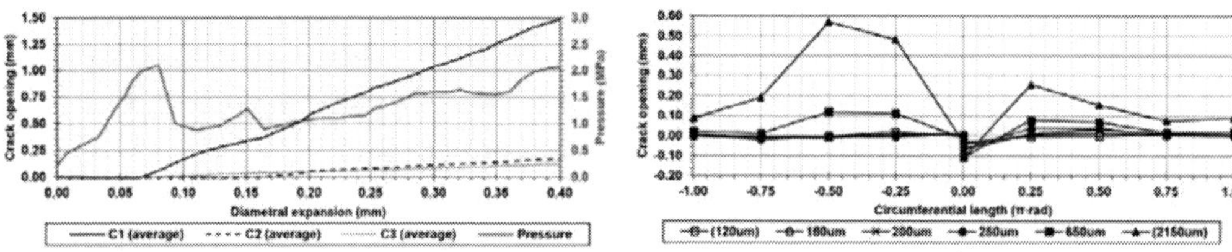

(a) (b)

Figure 7. Unconfined expansion test: (a) Crack opening evolution vs radial expansion, and pressure evolution (red); (b) Shape of annular cracks throughout annular length (π·rad) at various diametral expansion values.

The test progresses as follows: after overpassing the tensile capacity of the cement sample a first radial crack originates (C1), the crack propagates through the whole cross-section following a V-shape as presented in Figure 6a; afterwards two secondary radial cracks initiate (C2, C3) at approximately 120° of the initial crack (Figure 6). The first crack may be regarded as a tension crack, despite it initiates at the outer diameter and propagates towards the centre of the specimen, yet, the propagation through the whole cross-section occurred early, i.e. between 60 and 80 μm of diametral expansion (ExD). Whereas the two secondary cracks may correspond to bending cracks propagating from the inner diameter towards the external surface of the specimen, occurring at a later stage, as shown in Figure 6b and Figure 7a. The shape of the cracks are measured in 5 points over the length of the cracks as shown in Figure 6b for the crack C1

The profile of the annular crack is presented in Figure 7b, as the crack opening along the circumferential length of the inner ring. The central points (A0, A1) correspond to the location of the primary crack.

The plot shows an overall crushing of the interface at the primary crack, whereas the adjacent sectors present positive crack values at ExD values above 250 μm. The crack profile suggest two local maxima near the formation of the secondary cracks, which might be triggered by the formation of those cracks, since it occurs at a similar stage of the diametral expansion.

3.3. Annular expansion test, confined conditions

The concrete ring is subject to a constant horizontal strain at its inner diameter, imposed as an expansion of the inner ring diameter at a rate of 0.079 mm/min. For this case, there is a significantly stiff external confinement, through a steel ring with a wall thickness of 13 mm, see Figure 8. As for the previous test, the strain at the sample surface is measured using DIC.

During the test, the following stages are identified: the first radial crack (C1) originates after overpassing the tensile capacity of the cement sample, the crack propagates through the whole cross-section following a V-shape, as presented in Figure 8. Then, two secondary radial cracks initiate (C2, C3) at approximately 120° of the initial crack; at a later stage, two tertiary cracks (C4, C5) initiate between the primary crack (C1) and one of the secondary cracks (C2).

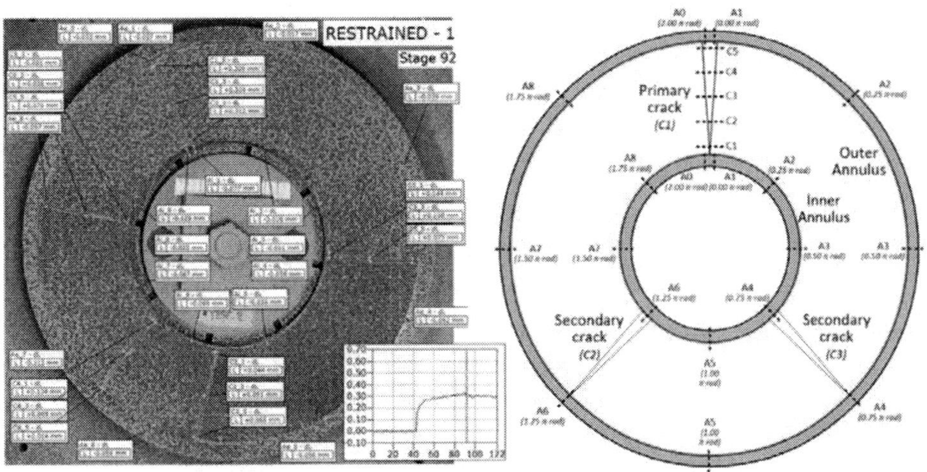

(a) (b)

Figure 8. Confined expansion test.: (a) DIC analysis and crack pattern, major strain mask overlay on specimen image; (b) Crack pattern and measurement points (for simplicity secondary cracks C4 and C5 that occurred between C1 and C2 in later stages are not showed in the sketch).

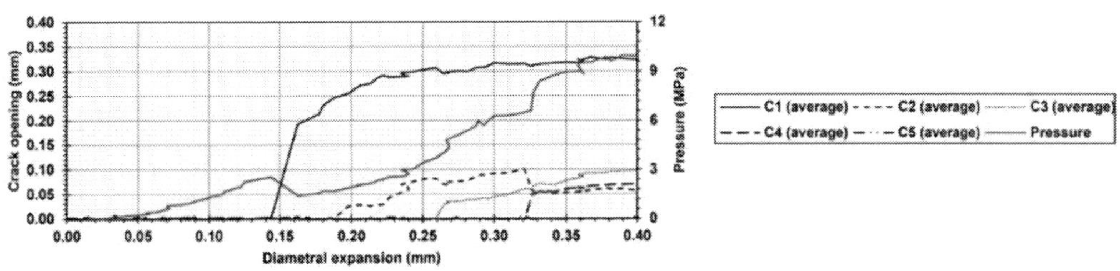

Figure 9. Confined expansion test. Crack opening evolution vs radial expansion, and pressure evolution (red).

The first crack may be regarded as a tension crack, despite it appears to initiate at the outer diameter and to propagate towards the centre of the specimen, yet, the propagation through the whole cross-section occurs relatively early, i.e. between 140 and 150 μm of diametral expansion. The two secondary cracks may correspond to bending cracks propagating from the inner diameter towards the external surface of the specimen, occurring at a later stage. Finally there are two tertiary cracks initiating between C1 and C2, propagating from the inner diameter (C4) and outer diameter (C5), but starting at a later stage of the test Figure 9.

This scenario comprises the evaluation of two annular cracks: the inner interface (Annulus 1) and the outer interface (Annulus 2). The inner annulus would correspond to the steel casing, pressurized due to, i.e. positive pressure testing; whereas, the outer interface would correspond to a stiff formation or an outer steel casing. The crack path is represented as eight points equally spaced along the annulus, at 45° (0.25 π·rad), starting from the initiation of the primary crack (A0, A1), and being positive for clockwise points (A1-A5) and negative for counter-clockwise points (A5-A0) (Figure 8). Figure 10 shows the development of the strain at the inner and outer interfaces, plotted against the diametral expansion of the inner annulus. The plot shows the overall crushing of the interface in compression both at the inner and outer ring after the initiation of the cracks. The inner annulus, Figure 10a shows crushing damage at the interface during the loading phase (i.e. before reaching the tensile capacity of the specimen); whereas the outer annulus (Figure 10a) appears unaffected (despite of noise in the measurements and small strain) until the time of cracking.

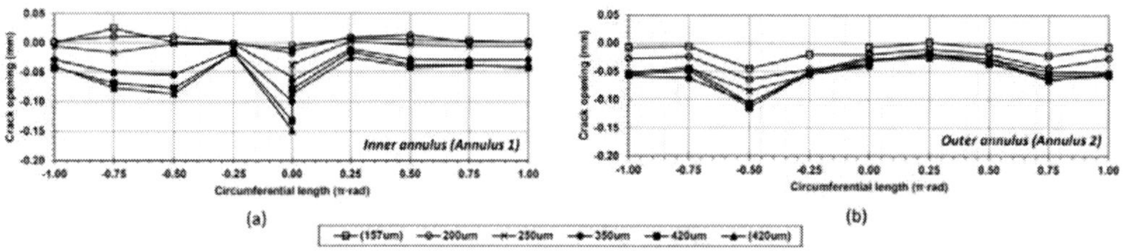

Figure 10. Confined expansion test. Shape of annular cracks throughout annular length (π·rad), for: a) internal (A1) annulus; b) external (A2) annulus at various diametral expansion values.

4. Conclusions

The oil industry relies on the use of cement as a primary barrier, to isolate the steel casing from the surrounding formations and provide overall stability on the bored structure. However, there is limited discussion regarding the formation and propagation of cracks, and their impact on the overall performance of the barrier: i.e. mechanical, durability or permeability.

This investigation proposed a test-methodology capable of identifying and quantifying the initiation and propagation of cracks and local damage at cement-casing systems. The proposed setup is based on the combination of high-resolution digital image correlation analysis (DIC) and displacement-controlled mechanical test setups, emulating the geometry of the cement-casing system.

The results presented show the formation and propagation process of radial and annular cracks under two scenarios: shrinkage-induced cracking and annular expansion-induced cracking at two levels of external confinement.

The analysis of the crack propagation process showed that the level of confinement has a significant impact on the formation of damage at the casing-cement interface and promotes the development of multiple cracking at the cement. The quantification of damage at the casing-cement interface involves localized crushing of the cement matrix and shall be further investigated.

5. Acknowledgments

Authors would like to acknowledge the financial support from Danish Hydrocarbon Research and Development Center in Technical University of Denmark.

References

[1] Nelson E B and Bell D R 1990 Well Cementing, 1st ed., *Schlumberger Eduactional Services*, Sugar Land, US

[2] Lake L W 2006 *Petroleum Engineering Handbook: Vol. II - Drilling Engineering*, II, Society of Petroleum Engineers, Richardson, US

[3] Fink J K 2012 *Petroleum engineer's guide to oil field chemicals and fluids*, 1st ed., Gulf Professional Pub

[4] Heinold T, Dillenbeck R and Rogers M 2002 The Effect of Key Cement Additives on the

Mechanical Properties of Normal Density Oil and Gas Well Cement Systems, *SPE Asia Pacific Oil Gas Conf.*, Soc. of Petroleum Engineers, Melbourne, AU, p. 12. doi:10.2523/77867-MS

[5] Hua S, Wang K and Yao X. 2016 Developing high performance phosphogypsum-based cementitious materials for oil-well cementing through a step-by-step optimization method, *Cem. Concr. Compos.* **72**, 299–308. doi:10.1016/j.cemconcomp.2016.05.017.

[6] Bu Y, Du J, Guo S, Liu H and Huang C 2016 Properties of oil well cement with high dosage of metakaolin, *Constr. Build. Mater.* **112**, 39–48. doi:10.1016/j.conbuildmat.2016.02.173.

[7] Habib A O, Aiad I, Youssef T A and Abd El-Aziz A M 2016 Effect of some chemical admixtures on the physico-chemical and rheological properties of oil well cement pastes, *Constr. Build. Mater.* **120** 80–88. doi:10.1016/j.conbuildmat.2016.05.044.

[8] Davies D R, Hartog J J and Cobbett J S 1981 Foamed Cement - A Cement With Many Applications, *SPE 9598 - Middle East Tech. Conf. Exhib.*, Society of Petroleum Engineers, Manama, BH, pp. 193–196. doi:10.2118/9598-MS.

[9] Bybee K 2001 Foamed Cement for Zonal Isolation: Case Histories, *J. Pet. Technol.* **53,** 72–74. doi:10.2118/0801-0072-JPT.

[10] Liu H, Bu Y, Nazari A, Sanjayan J G and Shen Z 2016 Low elastic modulus and expansive well cement system: The application of gypsum microsphere, *Constr. Build. Mater.* **106**, 27–34. doi:10.1016/j.conbuildmat.2015.12.105.

[11] Iremonger S S, Bolt M and Lawrence S C 2015 Enhanced Thermal Well Integrity Through The Use Of A New Cement Tensile Strength-enhancing Fiber, in: *SPE (Ed.), SPE Canada Heavy Oil Tech. Conf., Society of Petroleum Engineers*, Calgary, CA pp. 1–12. doi:10.2118/174483-MS.

[12] Reddy B R, Santra A K, McMechan D E, Gray D W, Brenneis C and Dunn R 2005 Cement Mechanical Property Measurements Under Wellbore Conditions, in: *SPE Annu. Tech. Conf. Exhib.*, Society of Petroleum Engineers, doi:10.2118/95921-MS.

[13] Lavrov A, Todorovic J and Torsæter M 2016 Impact of Voids on Mechanical Stability of Well Cement, *Energy Procedia.* **86** 401–410. doi:10.1016/j.egypro.2016.01.041.

[14] Wong R C K and Yeung K C 2005 Structural Integrity of Casing and Cement Annulus in a Thermal Well Under Steam Stimulation, in: *Can. Int. Pet. Conf.*, Petroleum Society of Canada doi:10.2118/2005-104.

[15] Deremble L, Loizzo M, Huet B, Lecampion B and Quesada D 2011 Stability of a leakage pathway in a cemented annulus, in: *Energy Procedia*, pp. 5283–5290. doi:10.1016/j.egypro.2011.02.508.

[16] Talabani S, Chukwu G A and Hatzignatiou D G 1993 Gas Channeling and Micro-Fractures in Cemented Annulus, in: *SPE West. Reg. Meet.*, SPE, doi:10.2118/26068-MS.

[17] Brandl A, Cutler J, Seholm A, Sansil M, Braun G and Co B J S 2010 Cementing Solutions for Corrosive Well Environments, in: *SPE (Ed.), CPS/SPE Int. Oil Gas Conf.*, SPE, Beijing, CN, pp. 3104–3116.

[18] Vrålstad T, Todorovic J, Saasen A and Godøy R 2016 Long-Term Integrity of Well Cements at Downhole Conditions, in: *SPE Bergen One Day Semin.*, Society of Petroleum Engineers, Bergen, NO, p. 10. doi:10.2118/180058-MS.

[19] Milanovic D and Smith L 2005 A Case History of Sustainable Annulus Pressure in Sour Wells – Prevention, Evaluation and Remediation, in: *Proc. SPE High Press.* Temp. Sour Well Des. Appl. Technol. Work., Society of Petroleum Engineers, doi:10.2523/97597-MS.

[20] Xing Y K, Zhang G Q, Li S Y, Zhang Z K and Pan R 2017 Investigation on the Integrity of Cement Annulus During Multi-stage Horizontal Fracturing with FEM/DEM, in: *Proceedings of the 7th International Conference on Discrete Element Methods*, Singapore, pp. 369–376. doi:10.1007/978-981-10-1926-5_39.

[21] Paegle I, Minelli F and Fischer G 2016 Cracking and load-deformation behavior of fiber reinforced concrete: Influence of testing method, *Cement and Concrete Composites*, Vol. **73**. pp. 147-163, https://doi.org/10.1016/j.cemconcomp.2016.06.012

[22] Sprince, A, Fischer G, Pakrastinsh L and Korjakins A 2014. Crack propagation in concrete with silica particles. Advanced Materials Research, 842, pp.470–476.

Use of Artificial Neural Network for Initial design of steel structures

H Mote[1] and S R Satish Kumar[2]

[1, 2] Department of Civil Engineering, Indian Institute of Technology Madras, Chennai, India.

E-mail: moteharshada2@gmail.com

Abstract. Design of steel structures is an iterative process that often turns complicated when requires initial guess. A good initial design guess that comes from past design experience can considerably reduce the number of subsequent analysis and design cycles. It is difficult to form declarative rules to express human intuitions and past experience. Another problem in design process is arriving at optimal solution, where optimization process in most cases is computationally expensive and time consuming. Artificial Neural Network (ANN) is a promising computational model that can perform cognitive tasks, such as learning and optimization. This paper focuses on application of ANN in design so as to make design process more efficient. The aim is to train ANN to arrive at optimal solution by considering design constraints along with practical. This is illustrated with an example of design of compression member which primarily requires initial guess. The economy in design is tried to obtain by providing training set of optimal solutions to a multilayer neural network. The design process followed to generate training set is in accordance with IS: 800-2007. The solutions given by ANN are approximate but have reasonable agreement with the expected solutions.

1. Introduction

Design of steel structures is an iterative and often a complicated process. Usually, design process starts with assuming some initial solution for the set of governing equations that define the problem statement. The initial solution is used to generate improved approximate solution through a number of iterations that satisfy the design constraints. Furthermore, this process is repeated multiple times with different initial solutions to arrive at optimal solution. As there can be multiple solutions, sometimes the designer has to take decision about the most acceptable solution, based on his /her knowledge and past experience. An alternative is to use mathematical optimization techniques which are time consuming and computationally expensive most of the times. Apart from this, formulating the cost function and practical considerations like availability of material, ease in construction etc. for optimization is difficult as it requires insight and experience in that field. The large number of design and practical constraints makes it difficult to computerize the entire design process.

A good initial guess considerably reduces the number of design and analysis cycles required and helps to arrive at optimal solution in less time. Having a good initial guess is difficult for beginners as it requires knowledge, intuition and experience. Also, it is difficult to write a program using procedural or declarative language that can generalize this experience in order to tackle an entirely new problem and that can learn through its own experience.

Artificial Neural network (ANN) is one of the promising tool that can be used to solve these types of problems. ANN is acknowledged as one of the universal approximators that can approximate any measurable function to any desired degree of accuracy, in a very specific and satisfying sense [1]. It is a superficial model of the functioning of human brain that accumulates experts' knowledge, intuitions and past experience in a generalized form and stores it in the form of network parameters. ANN consists of many parallel interconnected basic computing elements, which have processing ability inspired from biological neural networks. ANN can predict solution based on data presented for training, which can be theoretical, experimental or empirical based on good and reliable past experience or combination of these. Training data can be evaluated, verified or modified by the design professionals to inject human intelligence and intuition. Rogers applied ANN to perform structural analysis and developed guidelines to perform the same and demonstrated that ANN works much faster than mathematical optimization techniques [2]. Significant work in application of ANN in structural engineering has been done by Adeli [3]. Park and Adeli developed a neural dynamics model to solve linear structural optimization problem involving the minimum-weight plastic design of low-rise planar steel frames [4]. Tashakori and Adeli presented optimal (minimum weight) design of cold- formed steel space trusses using the neural dynamics model developed by Adeli and Park [5]. Kang and Yoon in 1994 developed a model for selection of economical truss members that will satisfy stresses requirements of a truss design [6].

The use of ANN for initial design was studied by Mukherjee and Deshpande [7]. The ANN model was built for preliminary design of reinforced concrete single span beams, which predicted good results closer to final solution.

Developing intelligent design process that will cooperate with conventional design process for better and faster results, is the main motivation behind this study. The objective of the study is to use ANN in initial design of steel structures so as to substantially reduce the number of design and analysis cycles and arrive at a feasible solution with practical considerations and other design constraints. The output predicted by ANN can be used as a starting point in design.

The scope of the study includes developing an ANN for predicting initial design of an angle section under compression, loaded through one leg. The training data is in accordance with Indian standard code IS: 800-2007 [8].

2. Basics of Artificial Neural network

ANN is a hypothesized model of functioning of human brain. It resembles the brain due to its ability of acquiring knowledge through learning process and storing it in the form of interconnecting synaptic weights [9].

A typical ANN model consists of:

- Neurons- a number of interconnected processing elements commonly referred to as neurons or nodes.
- Input layer- where data is presented to the network.
- Output layer- that holds the response of the network to the input.
- Hidden layer –These are the intermediate layers that enable network to perform and compute complicated association between input and output.
- Synaptic weights or synapse that are connecting link between two neurons.
- Summing function and activation function are constituents of a neuron. Summing function sums up the input signals from other neurons, weighted by the respective synaptic weight of the neuron. An activation function or a threshold function limits the permissible amplitude range of the output signal to some finite value. Activation function helps to induce nonlinearity in the model. Among the various activation functions available, sigmoid function is chosen because it has very simple derivative which is useful for development of learning algorithm.

General Steps of operating ANN are as follows:

- Specifying the Architecture - Choosing structural initial parameters like number of layers, number of neurons of each layer, initial values of weights, and the type of activation function etc. To select number of nodes in each layer and number of hidden layers, no fixed guidelines are available. Thus, these parameters are decided by trial and error method. [7].

- Training or learning – It is the major and the most time-consuming part of neural network modelling. In a supervised learning model of ANN, the training process uses given input and output data sets to determine the optimal combination of weights by applying Error Correction Learning rule. In other words, each time an input is presented to the network, it predicts an output which is then compared with the expected output to update the weights so as to minimize predicting error. The training set must reflect all aspects of problem domain.
- Simulation - Simulation means use of the trained neural network to predict outputs of entirely new inputs. This corresponds to the 'recall' function of the brain.

The ANN can be designed to solve the problem by regression as well as by classification approach. In the regression model ANN maps input vector $[X]$ into continuous output variable vector$[y]$. While in classification model ANN maps input vector $[X]$ into discrete output variable vector$[y]$.

3. Application of Neural network for Initial design problem

3.1. Problem definition

Design of compression member which essentially is an iterative process is solved in this article with the application of ANN. The ANN was modelled in MATLAB R2018a. The input layer $[X] = [x_1\ x_2..x_n]$ was decided to have three nodes for Span L, Load P and limiting slenderness ratio(Kl/r). While Output layer $[y] = [y_1\ y_2..y_k]$ predicts width b and thickness t of angle section. The difference between the actual and predicted values is measured by mean of sum of squared errors of prediction and it is called as Cost function J. To minimise this cost function J, advance optimizer *-fmincg* a function file provided by Carl Edward Rasmussen was used. The function uses the Polack-Ribiere flavour of conjugate gradients to compute search directions, and a line search using quadratic and cubic polynomial approximations. It uses the Wolfe-Powell stopping criteria with the slope ratio method for guessing initial step sizes. The gradients required for this method were calculated using back propagation algorithms [10]. ANN can be trained to predict size of the angle section for the problem by treating them as continuous variables or as discrete output variables. Hence two different models were implemented for regression and classification approaches.

3.2. Architecture of Neural Network – Regression Model

In this case, both b and t were treated as continuous variables. A multilayer neural network was initially built with one input layer, one output layer and one hidden layer. Later number hidden layer and nodes were increased as per the requirement. The input layer consist of three units $[X] = [x_1\ x_2\ x_3]$ each representing the Design span L, Design load P, and Limiting (KL/r). The output layer was of only two nodes $[y] = [y_1\ y_2]$ representing b and t of the angle section respectively. Here, the output values predicted by outer layer nodes do not have any limit. The cost function was calculated using (1).

$$J = \frac{1}{m}\sum_{i=1}^{m}\sum_{j=1}^{k}\left[\left(h_\theta\left(x^{(i)}\right)\right)_j - y_j^{(i)}\right]^2 \tag{1}$$

Where, m is total number of training examples, k is number of output nodes, $\left(h_\theta\left(x^{(i)}\right)\right)_j$ represents predicted j^{th}output node of ANN for given i^{th} training example$\left[x^{(i)}\ y^{(i)}\right]$.

The sigmoidal function was used as an activation function in the hidden layer but not in the output layer which is given by (2).

$$\sigma(z_j) = \frac{1}{[1 + e^{-z_j}]} \tag{2}$$

Where, $z_j = \sum \theta_{kj} x_k$ is the summation of product of synaptic weight θ_{kj} between j^{th} receiving node and all k nodes of previous layer including bias. The bias which is a parameter used to modulate output, was treated as a node with fixed value of 1. Bias controls the value at which activation function triggers. It is like an intercept added in a linear equation.

The values of b and t obtained from this network were mapped into discrete output by using Discrete mapping scale shown in figure 1, while figure 2 shows network architecture,

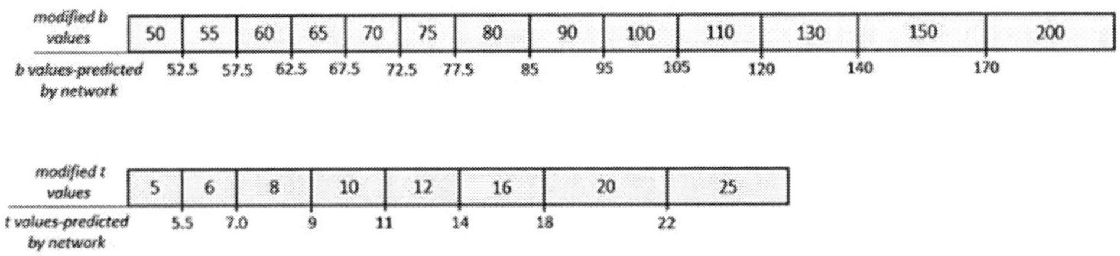

Figure 1. Discrete mapping scale for regression model.

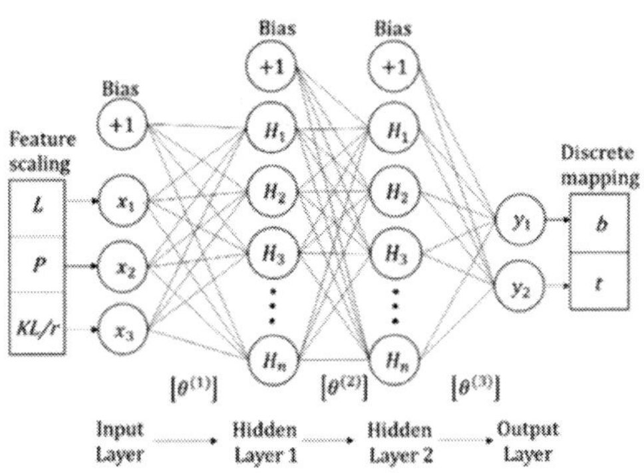

Figure 2. ANN -Regression model.

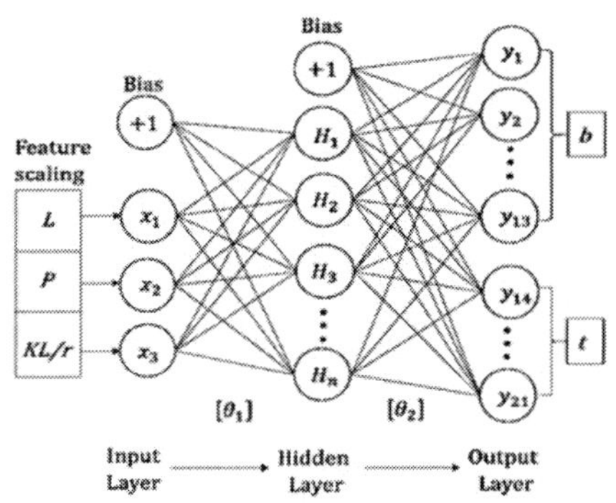

Figure 3. ANN -Classification model.

3.3. Architecture of Neural Network –Classification model

A multilayer neural network was built with one input layer, one output layer and one hidden layer. The input layer consist of three units $[X] = [x_1\ x_2\ x_3]$ representing Design span$L$, Design load P, and limiting (KL/r). The output consists of 21 units i.e. $[y] = [y_1\ y_2 .. y_k]$ where $k = 21$, out of which first 13 nodes were for different discrete widths of available angle sections and rest 8 were for different discrete available thickness as shown in figure 3. Node with maximum probability among the first 13 output nodes gives width of the section while the one with maximum probability among the last 8 nodes gives thickness of the section. Initially, number of nodes in hidden layer was arbitrarily set to 12 and was modified later as per the requirement. The cost function for ANN classification model was calculated using (3),

$$J = \frac{1}{m}\sum_{i=1}^{m}\sum_{j=1}^{k}\left[-y_j^{(i)}\log\left(\left(h_\theta\left(x^{(i)}\right)\right)_j\right) - \left(1 - y_j^{(i)}\right)\log\left(1 - \left(h_\theta\left(x^{(i)}\right)\right)_j\right)\right] \quad (3)$$

3.4. Obtaining Training data set
The problem used for illustration is a design of compression member using Indian standard equal angle section (ISA). The material availability and minimum weight are used as design optimization constraints to arrive at economical section. The angle was assumed as loaded through one leg with end condition of number of bolts connected to gusset plate being greater than 2 and Gusset/connecting member fixity as Fixed. The design procedure followed Indian Standard code IS 800:2007. Out of all 72 Indian standard equal angle sections, 50 most commonly available sections were selected. The design load P is a set of load values ranging from 10 kN to 1620 kN with an interval of 10kN and the Design Span (L) is set of spans ranging from 0.25 m to 10 m with 0.25 m interval. For every combination of P and L the capacity of all sections was checked .The angle section with least weight and design load carrying capacity greater than P was selected. To calculate the design compressive load carrying capacity of section, Clause 7.5.1.2 of pg. 48 was used. The two design constraints of maximum allowable slenderness ratio of 180 and of 250, as per Clause 3.8 of IS: 800-2007 was used. According to this clause, limiting slenderness ratio is 180 for member carrying compressive load due to dead and imposed loads while it is 250 for member subjected to compressive load resulting from combination of wind and earthquake. The algorithms to generate training data was coded using Pandas Library of Python Programming language. The algorithm generated more than 8500 possible examples. The data was independently and identically distributed into training, cross validation and test set in the ratio 3:1:1[11].

3.5. Deciding the Architecture of Neural Network
To evaluate the performance of the network, the data set consisting of inputs and outputs was divided into three sets namely training set, Cross Validation Set and Test set in the ratio 3:1:1.The network is trained using training set only but the cost function was calculated for both training and cross validation set. This training cost function and cross validation cost function are plotted with respect to different model parameters such as number of iterations, regularization, number of hidden layer units, number of training sets etc. These plots are referred as learning curves. These learning curves were used to classify model into high-bias (under-fitting) or high-variance model (over-fitting) and to get the values of properties which give optimal fit. Though the over-fitted model shows least error, it lacks ability to generalize. Hence the over-fitting of the model is avoided. After adjusting network model, the final performance of model is assessed using test set [12].

Both training error and validation error were high in the initial network architecture of single hidden layer regression model, proving it to be under high bias category (under-fitting). Hence additional hidden layer was introduced in the model. The cost function reduced significantly after introducing second hidden layer. Though the error was lower than before, the model failed to identify smaller angle sections. Least size of section predicted by regression model was ISA 60x5 mm.

In case of classification model both training error and validation error were low but the least size section predicted by this model was ISA 90x6mm section (higher than the least size of ISA 60 X 5 mm, predicted by regression model).This happened because the percentage of examples of angle sections of size 150x16, 150x20, 200x12, 200x16, 200x20 and 200x25 was more than 60% in the training set (with total of 5145 examples) as shown in figure 4 . Same was the case with validation and test set. This resulted skewness in data which is understandable since large range of load and span is satisfied by these sections. Because of which all the smaller sections (b<130mm) were treated as outliers by both networks. The resulting models were more biased towards the higher sections. Hence, these networks were able to predict higher section accurately but failed in predicting smaller sections.

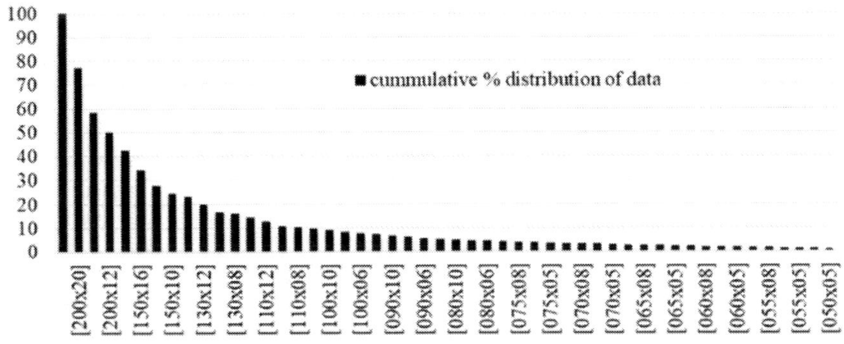

Figure 4. Distribution of cumulative percentage data in training set.

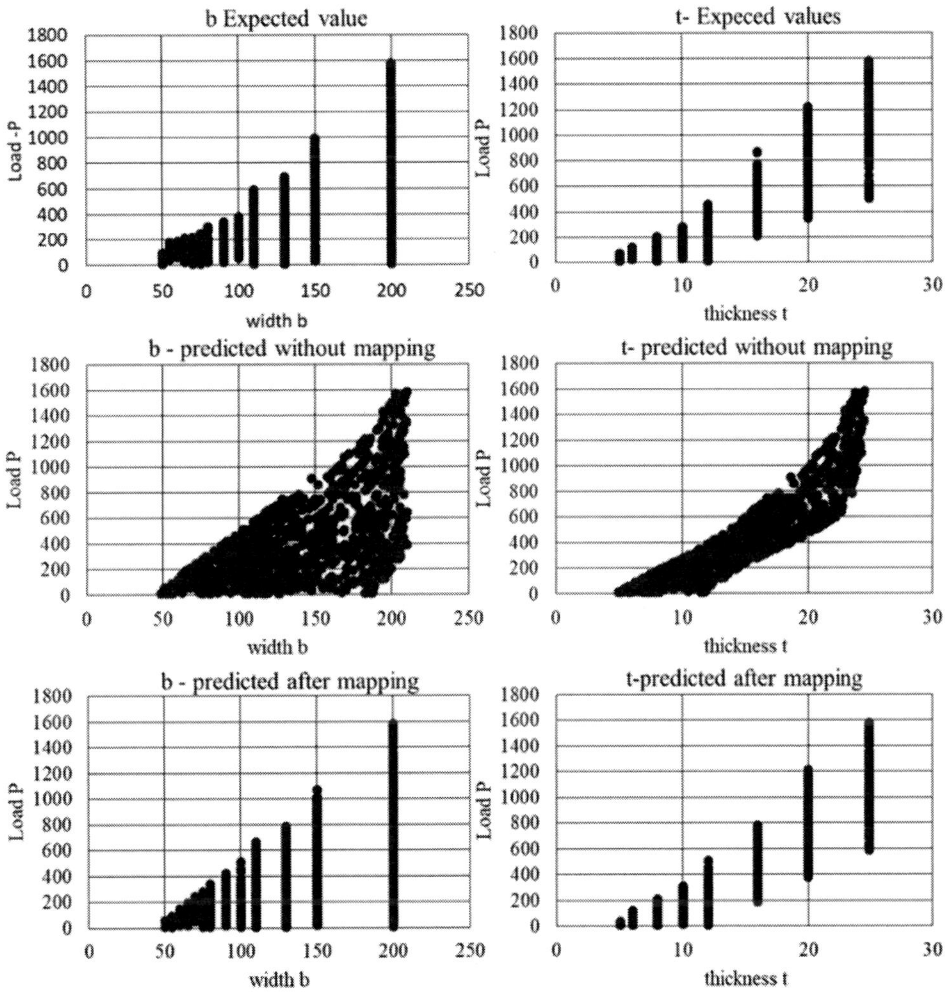

Figure 5. Width and thickness of angle sections predicted using ANN -Regression model.

To overcome this problem, large number of examples of these larger sections were removed randomly from the training set. The reduced training set was of only 1430 examples. The neural networks were trained freshly with this reduced training set. Performance of both the networks improved in predicting smaller sections. This shows that, minimization of Cost function is essential for any neural network but is not the only criteria of predicting its accuracy, especially when data contains more examples of one class. Thorough inspection of the data predicted by network for various cases is also important. It is important to understand that neural networks are approximators and hence we cannot expect prediction of exact solution. Further, classification model was checked for different number of hidden nodes ranging from 3 to 30. It was found that 15 number of hidden nodes gave better results. Cost function converged after 80 number of iterations. Further increase in hidden layer nodes did not show any significant improvement in prediction capacity of the model. Similarly, regression neural network model with 2 hidden layers each containing 5 hidden layer nodes and 60 number of iterations gave better prediction capacity. Percentage errors in prediction of b and t before mapping into discrete values was 45.4% and 54.5% respectively. These errors were brought down to 3.8 % and 7.0% respectively after the use of mapping scale. The conversion can be observed from the figure 5.

3.6. Comparison between ANN Regression and Classification model

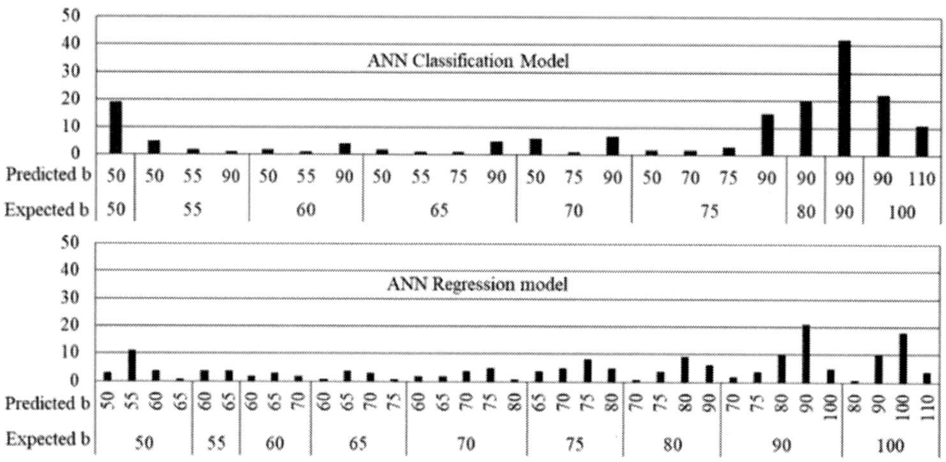

Figure 6. Performance in predicting smaller sections.

Regression model size is much lesser than size of classification model. In ANN knowledge is stored in the form of synaptic weight [θ] which is much smaller for regression model as compared to classification model. The smaller sections sizes predicted by regression model were closer to exact sizes compared to classification model. This can be seen from figure 6, where classification model failed in most of the cases where b is between 60-80mm. Table 1 shows that percentage error of regression model in predicting t is more than Classification model. From above comparison, it can be concluded that for the above design problem regression model is better than classification model because of its size and better predicting capacity for smaller sections as well as larger sections. Using test set, the regression model displayed error of 5.1% and 7.25% in predicting b and t respectively, while for classification model, these errors were 5.35% and 5.01% respectively.

It is to be noted that cost function of regression model was lower than that of classification model. The output values range from 0 to 1 for classification model but not for regression model. Hence cost function calculated using training set J_{train} ,validation set J_{train} and test set J_{train}cannot be used for comparison.

Table 1. Comparison of prediction error between final selected ANN regression and classification model models.

	Input layer	Hidden layers		Output layer	Cost functions			Average % Error in validation set	
		(1)	(2)		J_{train}	J_{val}	J_{test}	b	t
ANN-Regression	3	5	5	2	0.0573	0.0646	0.0754	3.8	7.0
ANN-Classification	3	15	-	21	1.6300	1.6570	1.7277	4.8	4.9

4. Conclusion

ANN is a flexible tool that can be used to map complex, nonlinear input -output relation. It can capture human knowledge in generalized form. It can predict solutions that are close to exact solution with acceptable margins. But they are sensitive to training set. The training data set should reflect all the aspects of the problem. Skewness of training data sets significantly affects predictions of networks making it biased towards majority. This paper demonstrates the competence of generalization of ANN in initial design problem of an angle compression member loaded through one leg. Both ANN solving problem with regression and classification approaches were studied. It can also be concluded that ANN can be used to design solutions which are close to optimal design solution provided that ANN are trained using data that reflects optimal solutions.

References

[1] Hornik K, Stinchcombe M and White H 1989 Multilayer feedforward networks are univeral approximators *Neural Networks* **2** 359-66

[2] Rogers J L 1994 Simulating structural analysis with neural network *J. Computing in Civil Engineering* **8** (2) 252-65

[3] Adeli H 2001 Neural networks in civil engineering:1989-2000 *Computer-Aided Civil and Infrastructure Engineering* **16** (2) 126-42

[4] Adeli H and Park H S 1995 A neural dynamics model for structural optimization: Theory *Computers and Structures* **57** (3) 391–9

Adeli H and Park H S 1995 Optimization of space structures by neural dynamics model *Neural Networks* **8** (5) 769–81

[5] Tashakori A and Adeli H 2002 Optimum design of cold-formed steel space structures using neural dynamics model *Journal of Structural Engineering* **58** (12) 1545-66

[6] Kang H T and Yoon C J 1994 Neural network approaches to aid simple truss design problems *Microcomputers in Civil Engineering* **9** (3) 211–18.

[7] Mukherjee A and Deshpande J M 1995 Modelling initial design process using artifical neural networks *J. Computing in Civil Engineering* **9** (3) 194-200

[8] Bureau of Indian Standards (BIS) 2007 General Construction in Steel- Code of practice: IS 800

Bureau of Indian Standards (BIS) 1989 Dimensions for hot rolled steel beam, column, channel and angle sections: IS 808

[9] Kapania R K and Liu Y 1998 *Application of Artificial Neural Networks in Structural Engineering with Emphasis on continuum models* (Virginia Polytechnic Institute and State University)

[10] Ng A Machine learning *Coursera online course* Standford viewed Nov 23 2018 <https://www.coursera.org/learn/machine-learning>

[11] Pedregosa F *et. al.* 2011 Scikit-learn: Machine Learning in Python *Journal of Machine Learning Research* **12** 2825–30.

[12] Hastie T Tibshirani R and Friedman J 2009 *The Elements of Statistical Learning: Data Mining, Inference, and Prediction Second Edition* (New York: Springer-Verlag) pp 37-39

RECYCLING MUNICIPAL SOLID WASTE INCINERATION (MSWI) FLY ASH AS ADDITION FOR CLAY BRICK

V Voišnienė[1], O Kizinievič[2] and V Kizinievič[2]

[1] PhD student, VGTU, Department of Building Materials and Fire Safety, LT
[2] Senior Research Fellow, VGTU, Institute of Building Materials, LT

E-mail: olga.kizinievic@vgtu.lt

Abstract. The authors analyses the possibilities of recycling municipal solid waste incineration (MSWI) fly ash in clay brick manufacture. The paper analyses the effect of MWSI fly ash (FA) on physical-mechanical characteristics, environmental toxicity of clay bricks. The clay bricks were prepared by adding 2.5-7.5 % of FA and firing them at 1000 °C temperature. Physical and mechanical characteristics of clay brick change depending on the content of FA added: shrinkage, density and compressive strength reduce, water absorption and total porosity increase. Clay bricks containing 2.5-5 % FA do not exceed the limit values of inert substances according to the Waste Directive 2003/33/EC. According to the test results and taking into consideration clay brick physical - mechanical characteristics and environmental toxicity, the recommended content of the addition to the moulding compound is 2.5-5 % FA.

1. Introduction

Nowadays, there are growing pressures to use waste materials within the construction sector. This use of wastes in civil engineering requires an evaluation of both the environmental and technical suitability of the waste.

Thousands of millions of tons of municipal solid waste (MSW) are produced every year. Waste management and utilization strategies are major concern in many countries. Incineration is a common technique for treating waste, as it can reduce waste mass by 65-80 % and volume by up to 85-90 % [1], as well as providing recovery of energy from waste to generate electricity. Generally, municipal solid waste incineration (MSWI) produces two main types of ash, which can be grouped as bottom ashes (BA) and fly ashes (FA). MSWI Fly ash has classified as a hazardous waste (Code in European waste catalogue, code 1 01 12) [2].

Literature analysis revealed that the main components of fly ash are CaO, SiO_2 and Al_2O_3 [3-6]. There is also a significant content of chlorides. Fly ash also contains high levels of heavy metals, such as Zn, Pb, Cu, Cd, Cr and other [7-9].

The ceramic industry is highly promising for the final disposal of MSWI, is known about reuse of fly ash waste in clay brick. The reason for such a lack of data in this area relates to the fact that MSWI fly ash chemical composition is very different [3-9].

MSWI fly ash was used successfully during the preparation of clay brick [10, 11]. Jordan M.M et al. [12] propose to add 1 %–10 % of fly ash and 15 %–35 % of marble waste to ceramic brick moulding compound and fire the bricks at 975 °C–1050 °C temperature. It was found that with higher content of fly ash the flexural strength decreases and water absorption increases.

It was proven that clay bricks containing water treated fly ash have better physical and mechanical characteristics compared to unmodified clay bricks. The proposed FAW content is 5 %. Such products have density of 1.63 g/cm^3, porosity of 41 %, water absorption of 25.5 %, and compressive strength of 8 MPa [13].

The main objective of the present study is to investigate the feasibility of Municipal solid waste incineration fly ash to be used for the production of clay bricks. The impact of the amount of MSWI fly ash in the clay mixture is discussed in terms of physical–mechanical properties, porosity and environmental toxicity.

2. Materials and methods

Mixture of the clay, sand and MSWI fly ash is used in the research. Raw materials and additive are dried at 105±5 °C temperature, ground and sieved out through 1.25 mm sieve. Initially, dry mixture of the components is mixed, then mixture is watered till the humidity reached the level suitable for the formation. The amount of water required for formation masses varies from 20 % to 23 %. The higher amount of MSWI FA is introduced into formation masses, the higher amount of water is required. The mixing of forming masses is performed manually. This formation mass is kept for three days at (95-5) % humidity, in order to evenly distribute humidity in the formation mass. After three days of hardening, 50×50×50 mm samples are formed from formation masses. Specimens which have been exposed at normal conditions, are dried in the drying oven: 60±5 °C temperature during the first day, next day at 105±5 °C temperature. Burning of the samples is carried out at 1000 °C temperature, overall burning period is 34 h, by keeping at the highest burning temperature for 1 hours. Composition of the formation masses is presented in Table 1.

Table 1. Composition of the formation masses.

Raw materials	Composition of formation masses (% of the mass)			
	0	2.5	5.0	7.5
Clay + Sand	100	97.5	95	92.5
MSWI fly ash	–	2.5	5	7.5

Chemical analysis of brick raw materials is made by using X-Ray Fluorescence Spectrometer (XRF). Physical and mechanical characteristics of the ceramic products are determined through the implementation of the standard methodologies: density - according to LST EN 772-13:2003, water absorption - according to LST EN 771-21:2011, compressive strength - according to LST EN 772-1:2011 [14-16]. Linear shrinkage is measured using a calliper after drying and burning steps. Total porosity is determined in accordance with the methodology [17]. Shrinkage is calculated by employing 1 equation.

$$L = \frac{L_0 - L_1}{L_0} \cdot 100 \ \% \tag{1}$$

where: L_0 – distance between indentations on the formed sample, mm; L_1 – distance between indentations on the dry or burnt sample, mm.

The amount of heavy metals in the eluate from clay brick was determined by atomic absorption spectral analysis method using Buck Scientific 2010 VGP spectrometer with air-acetylene flame. The eluate was prepared according to standard LST EN 12457-2:2003 [18]. The solution for pH and specific conductivity tests was prepared from 2 g of test material and 100 g of distilled water. The prepared solution was agitated for 45 minutes in a magnetic stirrer, filtered afterwards and tested. The specific conductivity was measured by electric conductivity meter Cond 315i equipped with electrode Tetra Con 325. The alkalinity of the solution was measured by pH meter HANA HI 9321.

3. Results and discussion

The chemical analysis shows that the major components in the clay were SiO_2 (51.43 %), Al_2O_3 (20.54 %) and Fe_2O_3 (7.44 %). The next most abundant components were CaO (4.26 %), MgO (3.07 %), K_2O (3.3 %) and Na_2O (0.42 %), L.o.i – 9,54 %. Further X-ray diffraction analysis revealed that the clay used in this research mainly consisted of mico clay, quartz, kaolinite, feldspar. The mixture of clay and sand consists of 90 % of clay and 10 % of sand. According to particle-size distribution (Table 2), the clay is a dispersible material because the content of 0.01 mm-sized particles is about 70 %. The quartz sand with the following chemical composition was used for the tests: $SiO_2 >$ 98.5 %, Fe_2O_3 < 0.05 %, Al_2O_3 < 0.60 %; the distribution particle size in sand: 0 % of 2500 μm particles, 0.20 % of 1250 μm particles, 0.30 % of 1000 μm particles, 1.50 % of 800 μm particles, 630 μm particles 2.70 %; 7.50 % of 400 μm particles, 9.40 % of 315 μm particles, 34.40 % of 200 μm particles 17.20 % of 160 μm particles, 17.80 % of 100 μm particles, 8.30 % of 63 μm particles, 0.65 % of 50 μm particles, and 0.05 % of bottom particles.

Table 2. Particle size distribution in the clay.

Raw materials	>0.25 mm. wt., %	0.25 – 0.05 mm. wt., %	0.05 – 0.005 mm. wt., %	0.05 – 0.001 mm. wt., %	< 0.001 mm. wt., %
Clay	0.10	0.52	14.85	25.67	58.86

The major components observed in the MSWI fly ash were CaO (45.17 %) and chloride (KCl, $MgCl_2$, NaCl) (19.01 %). The next most components were Na_2O (2.65 %), K_2O, MgO (1.15 %), Al_2O_3 (1.08 %), Fe_2O_3 (6938 mg/kg), TiO_2 (1543 mg/kg). High concentrations of heavy metals were observed in the fly ash samples (Table 3). When compared to data disclosed by the literature [4, 5], the MSWI fly ash employed in this work uses to have less silica. This result indicates that the fly ash must be considered as a hazardous solid waste material.

Table 3. MSWI FA eluate test results.

Title and Test results and units of measurement (* mg/kg)									
P_2O_5	Na_2O	K_2O	MgO	CaO	SiO_2	Al_2O_3	TiO_2	Fe_2O_3	MnO_2
0.55%	2.65%	3.52%	1.15%	45.17%	0.23%	1.08%	1543*	6938*	334*
As	Pb	Cd	Cr	Cu	Ni	Hg	Zn	Ba	Mo
13.8*	1987*	83.4	50.5	625*	12.2*	10.4*	10209*	237*	< 1*
Sn	Be	Co	V_2O_5	Sr					
367*	< 1*	5.99*	9.55*	239*					

The quality of clay brick obtained with up to 7.5 % MSWI fly ash after firing at 1000 °C was determined on the basis of their technological properties (linear shrinkage, water absorption, density, total porosity and compressive strength). Figure 1 shows the amount of shrinkage after drying and firing. The shrinkage of a control clay brick (0 % FA) after drying is 9.1 %. When the MSWI FA content in the mixture varied from 2.5 % to 7.5 %, the brick shrinkage after drying changed from – 7.5 % to 8.4 %. The shrinkage of a control clay brick (0 % FA) after firing is 11.2 %. When the MSWI FA content in the mixture varied from 2.5 % to 7.5 %, the brick shrinkage changed from – 8.5 % to 9.3 %.

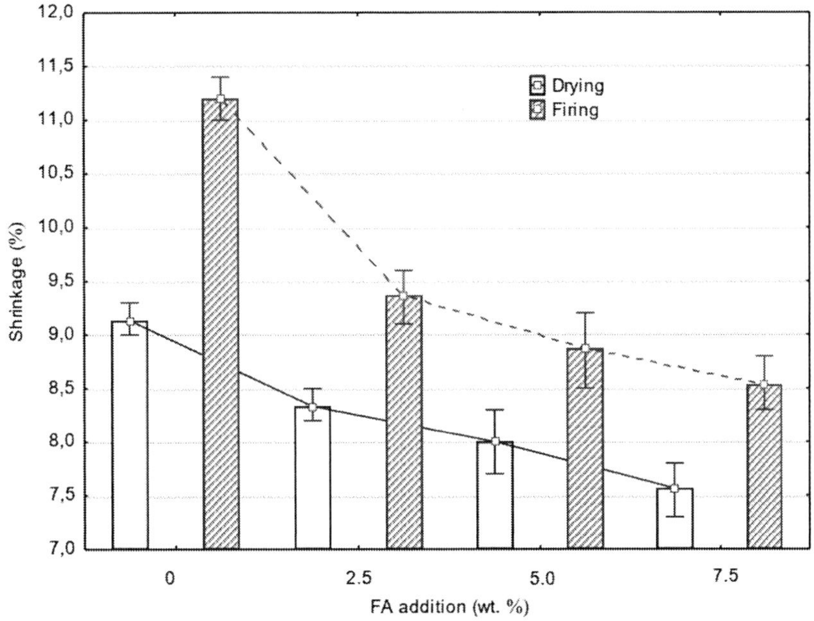

Figure 1. Shrinkage after drying and firing of clay brick.

During sintering, open and closed pores are usually formed. The maximum density and water absorption corresponds to the maximum volume of open pores in the clay brick. The measurements of the water absorption and density of clay brick with different proportions of MSWI fly ash are shown in Figure 2. Clay bricks (0 % FA) have a density of 1.92 g/cm^3, water absorption – 9.5 %. The results indicate that the water absorption of the bricks increased as the MSWI fly ash content increased. The less water that infiltrates the brick, the greater its durability and resistance to the natural environment are expected. The density (2.5–7.5 % FA) was from 1.6 g/m^3 to 1.82 g/m^3. The results indicate that when the MSWI fly ash content was decreased the density of the bricks increased.

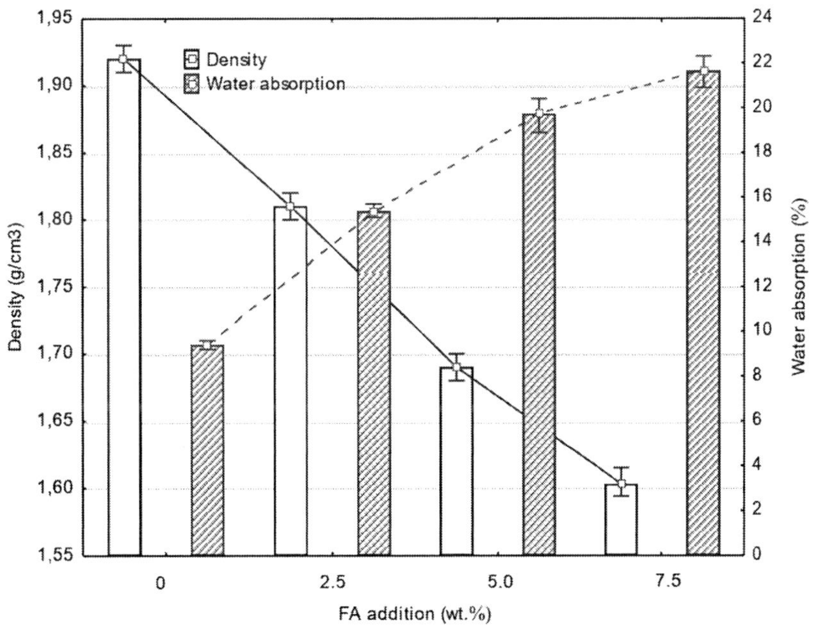

Figure 2. Water absorption and density of clay brick.

The compressive strength is the most important engineering quality index for building materials. The results of compressive strength testing of the clay bricks and MSWI fly ash mixtures are shown in Figure 3. The results indicate that the compressive strength of the clay bricks decreased as the MSWI fly ash content increased. The higher is the water absorption, the higher is the total porosity of clay brick. The lowest total porosity was found in control clay brick (0 % FA). The total porosity of this clay brick is 27 %. Total porosity parameter indicates the changes of macro- and microstructure during the firing.

The firing colour is another important parameter to qualify clay bricks. In this study, all clay brick pieces fired at 1000 °C presented a light red colour, regardless of the waste amount that had been added to them.

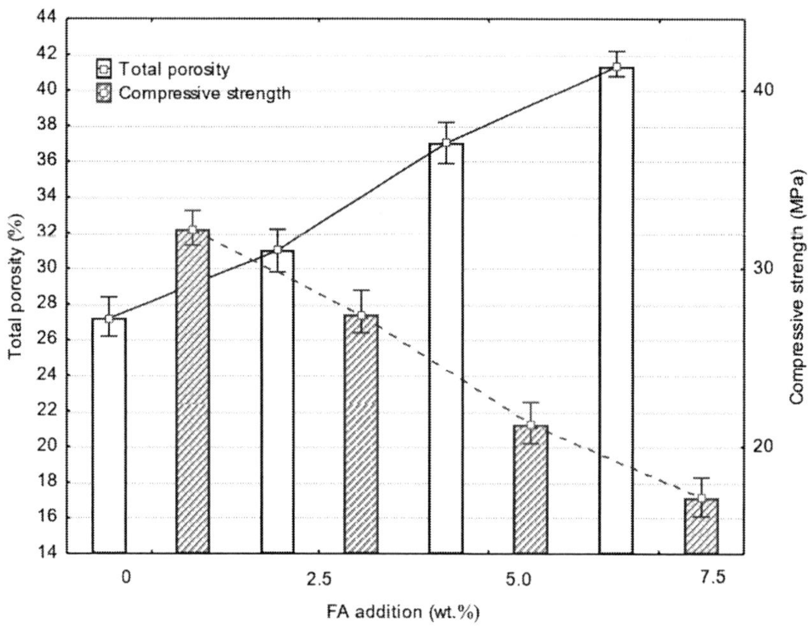

Figure 3. Total porosity and compressive strength of clay brick.

Leaching results of clay bricks were shown in Figure 4-5. All leaching results from sintered bricks were far lower than the Non-hazardous waste limit value (2003/33/EC directive), and thus heavy metals were stabilized in the body of clay brick through the sintering process. Hence, utilization of MSWI fly ash in making ceramic brick was an effective way to stabilize heavy metals. Clay bricks with fly ash addition also have higher pH and electrical conductivity values compared to clay bricks without FA addition (Table 4). Due to the existence of CaO, sulphate and chloride, MSWI fly ash is highly alkaline, which is detrimental to its reuse. The original pH of Clay + sand mixture is around 7.53, electrical conductivity (Sel) is 355 µS/cm. Clay with 7.5 % FA pH is 8.9, Sel 1887 µS/cm.

Although the limit values of non-hazardous waste set forth in Directive 2003/33/EC have not been reached, we assume that the limit values set for the leaching of inert materials should be observed when hazardous waste, such as MSWI fly ash, is utilised in the manufacture of building materials.

The use of 2.5 % MSWI fly ash as raw material for the production of clay bricks seems to be an important recycling way for final disposal of this abundant waste.

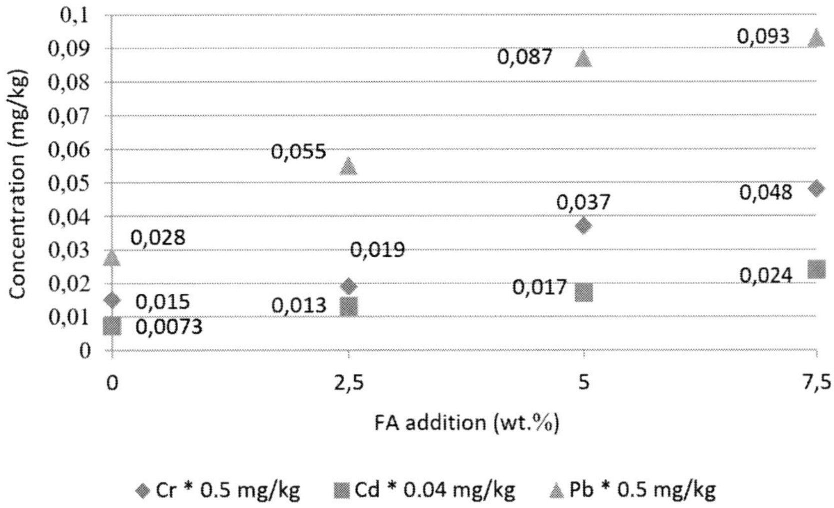

Figure 4. Leaching result of heavy metals Cr, Cd, Pb from clay bricks (* - inert waste limit values according to 2003/33/EC).

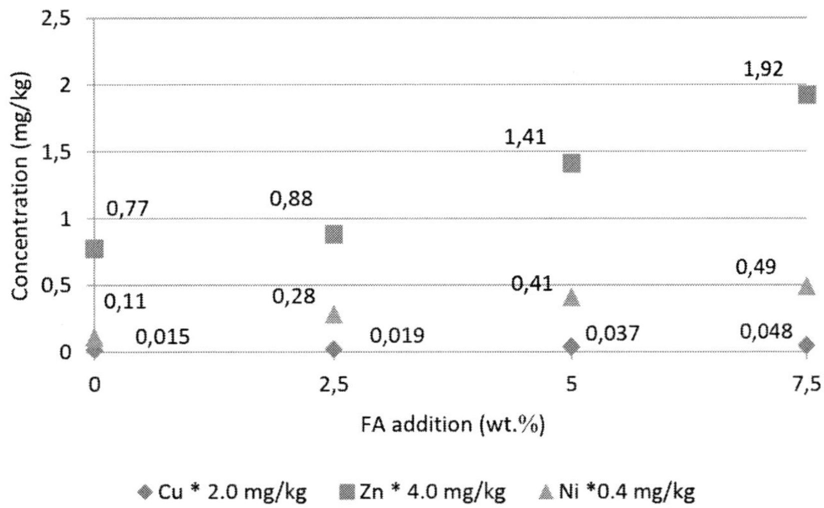

Figure 5. Leaching result of heavy metals Cu, Zn, Ni from clay bricks (* - inert waste limit values according to 2003/33/EC).

Table 4. Results of testing pH, Sel from clay bricks.

pH, Sel, values	**Clay brick (with FA addition)**			
	0	2.5	5.0	7.5
pH	7.53	8.17	8.53	8.9
Sel, µS/cm	355	1160	1488	1887

IOP Conf. Series: Materials Science and Engineering **660** (2019) 012065 doi:10.1088/1757-899X/660/1/012065

Conclusions

In this study, the properties of clay bricks with MSWI fly ash are investigated. It is found that the additive of MSWI fly ash is additive which influences physical – mechanical properties and environmental toxicity:

- after the addition of 2.5 % of MSWI additive into clay mixture and after burning samples at 1000 °C temperature, the clay brick with the following parameters is obtained: density – 1800 kg/m^3, compressive strength – 27 MPa, linear shrinkage – 9.4 % (after burning), water absorption – 15.3 %, total porosity – 31.0 %;
- tests of heavy metals leached from clay bricks with MSWI FA additions revealed that in specimens containing 2.5 % of FA the leaching did not exceed the limit value for inert waste set in Directive 2003/33/EC.

References

[1] Bertolini L et al 2004 *Cement and Concrete Research* 34(10) pp 1899–1906
[2] Europe Waste Catalog. Guidance on using the European Waste Catalogue (EWC) to code waste November 2015
[3] Wunsch P et al 1996 *Chemosphere* 32 pp 2211–2218
[4] Wang H et al *Journal of Environmental Management* 208 pp 15–23
[5] Pan J R et al 2008 *Waste Management* 28 pp 1113–1118
[6] Li X et al 2014 *Waste Management* 34 (12) pp 2494–2504
[7] Shi S H and Kan L L 2009 *Journal of Hazardous Materials* 164(2-3) pp 750–754
[8] Aubert J E et al 2006 *Journal of Hazardous Materials* 136(3) pp 624–631
[9] Tang J et al 2017 *Journal of Cleaner Production* 148 pp 595–605
[10] Belmonte L J et al 2016 *Environmental Science and Pollution Research* pp 1–13
[11] Haiying Z and Jingyu Q 2011 *Waste Management* 31(2) pp 331–341
[12] Jordán M M et al 2015 *Fresenius Environmental Bulletin* 24(2) pp 533–538
[13] Wan C 2017 *Thesis PhD. Technical University of Denmark, Department of Civil Engineering, ID: 2394007636*
[14] LST EN 772–13:2003 Methods of test for masonry units – Part 13: Determination of net and gross dry density of masonry units (except for natural stone)
[15] EN 772–21:2011 Methods of test for masonry units. Determination of water absorption of clay and calcium silicate masonry units by cold water absorption
[16] LST EN 772–1:2003 Methods of test for masonry units – Part 1: Determination of compressive strength
[17] Kizinievič O et al 2015 *Ceram Inter* 41 (9) pp 11234–11241
[18] LST EN 12457-2:2003 Characterisation of waste - Leaching - Compliance test for leaching of granular waste materials and sludges - Part 2: One stage batch test at a liquid to solid ratio of 10 l/kg for materials with particle size below 4 mm (without or with size reduction)

IOP Conf. Series: Materials Science and Engineering **660** (2019) 012066 doi:10.1088/1757-899X/660/1/012066

The impact of expanded polystyrene waste of different fineness on the properties of lightweight composite

D Leonavičius[1], I Pundienė[1], J Pranckevičienė[1] and M Kligys[1]

[1]*Vilnius Gediminas Technical University, Scientific Institute of Thermal Insulation*, Linkmenu str. 28, LT-08217, Vilnius, Lithuania

E-mail: dainius.leonavicius@vgtu.lt

Abstract. Expanded polystyrene (EP) is widely used as a packaging material for many types of goods. However, once the material is used, it is disposed of in landfills, where it can remain intact for the lifetime of several generations. Recycling of disposed EP packaging is of high relevance worldwide. The main objective of this study is to make a more detailed research into the effect of EP aggregate waste of different fineness and shape on physical and mechanical properties of porous lightweight aggregates concrete (PLWAC) with EP waste aggregates. Tests were done with Portland cement, EP waste of different fractions, resulted from crushing (EPR) and cutting (EPU), metakaolin, superplasticizer and air entraining admixture. Six batches of PLWAC specimens were formed with different EPR/EPU ratios, ranging from 0.5 to 3.The change in EPR/EPU ratio in PLWAC leads to structural changes and density reduction from 550 to 410 kg/m^3 after drying. When EPR/EPU ratio in the PLWAC is increased to 2, the compressive strength of the specimens drops from 2.3 to 1.75 MPa and down to 0.55 MPa, when EPR/EPU ratio is increased to 3.

1. Introduction

EP and its aggregates are widely used as packaging materials for many types of goods for inert, closed-cell and ultra-light (up to 95% air by volume) structure. However, once the material is used, it is disposed of in landfills, where it can remain intact for the lifetime of several generations [1, 2]. The problem of reusing or recycling of disposed packing tare of EP is relevant in the world. EP beads with intact structure are ultralight, hydrophobic and non-water absorbing aggregates [3, 4]. The properties make it possible to use EP aggregates and EP waste for various purposes (reduce density and thermal conductivity) by varying its volume percentage in mortar or concrete [5-8]. Most research on EP containing concretes has shown a decrease in the durability performance and the strength of concrete with increasing the amount of EP aggregate in mixtures [1, 9-11]. One researched [12] found that by increasing EP content in lightweight concrete from (16.7 % to 50 %) by volume of sand, the compressive, tensile and flexural strength decreases to 30.17 %, 37.93 % and 43.3 % respectively. Authors [13] used fly ash (50 % by weight of cement) as additional binding material and 6.3 and 4.8 mm size EP aggregate to produce concrete with the density ranging from 550–2200 kg/m^3. They found that mechanical properties of concrete deteriorate with the increase of EP content in mortar compositions. Coarse aggregate replacement by EP waste aggregate by six levels i.e., 5 %, 10 %, 15 %, 20 %, 25 % and 30 % was studied in [11]. Density of 1800 kg/m^3 and compressive strength of 9 MP was obtained with the maximum EP content. Authors [14] claim that LWAC with density ranging from

800 to 1800 kg/m^3 and compressive strength ranging from 10 to 25 MPa can be developed by partial replacement of coarse and fine aggregate by EPS beads. Tests showed that EPS concretes, with densities ranging from 1150 to 1350 kg/m^3, have strengths ranging from 3.5 to 12 MPa [13].

Research results proved that EP, used separately or in combination with other lightweight aggregates, is a promising filler in foamed concrete. According to studies conducted by other authors [15], foamed lightweight concrete with very small amounts of cement (from 70 to 130 kg/m^3), expanded glass and EP used as aggregates has good thermal conductivity coefficient (0.048–0.077 W/(m·K)), but low strength (0.35–0.73 MPa). Other authors [16] report the performance characteristics of super lightweight concrete with bulk density ranging from 300 to 500 kg/m^3 and compressive strength ranging from 0.7 to 2.5 Mpa, when normal aggregates were replaced with EP aggregates, while foam was introduced to facilitate manufacture process. Studies [8] on using EP waste in light-weight composites report that recycled EP waste can be used as aggregates in the formulation of heat insulating light-weight cement composites. Such composites have density of 150–170 kg/m^3, heat transfer coefficient of 0.06–0.064 W/(m·K), and compressive strength of 0.25–0.28 MPa. Authors [10], who studied heat transfer of similar composites containing EP aggregates, found that in composites with density ranging from 150 to 350 kg/m^3, compressive strength ranging from 0.09–0.42 MPa, flexural strength ranging from 0.09–0.39 MPa, heat transfer coefficient changed from 0.048 to 0.095 W/(m·K) in proportion to EP content in the mix. Gypsum slabs containing 50 %–70 % of crushed EP were developed [7]. The density of such composites dropped from 690 to 208 kg/m^3 and heat transfer coefficient dropped from 0.240 to 0.183 W/(m·K).

However, the use of EP has limitations as follows: due to low density (12–20 kg/m^3) and hydrophobic surface (chemical treatment of EPS surface is required) EP has poor distribution in the mix and the aggregates may float on the surface causing poor workability and lower strength of concrete [14]. The solutions to this limitation are the change of EP aggregates size and content in the mix and the use of different additions. Researchers claim that with smaller EP aggregates size an increase in EP concrete strength is obtained [18]. EP aggregates size effect on compressive strength of EP concrete was tested in [11, 17].

The results of compressive tests showed that the smaller is EP aggregates size, the greater is compressive strength of concrete for the same concrete porosity. Reduction in density (150–500 kg/m^3), strength (0.42–2.5 MPa) and heat transfer coefficient (0.048–0.095 W/(m·K)) of PLWAC can be achieved by changing the amount of EP aggregates [10].

It is necessary to create PLWAC with higher values of strength properties seeking to increase the use of different ways resulting EP waste in construction industry. The main objective of this study was to provide a more detailed research on the effect of (EPR/EPU) ratio on physical and mechanical properties of PWLAC. EPR was obtained from crushed EP waste, and EPU was obtained from big size EP sheet milling waste.

2. Materials and methods of testing

Ordinary Portland cement (OPC) CEM I 42.5 R, characterized by the specific surface of 420 m^2/kg, bulk density of 1100 kg/m^3 was used for testing. Mineral composition of cement, %: C_3S – 56.64, C_2S – 16.72, C_3A – 8.96, and C_4AF – 10.59.

EPR aggregates were obtained by crushing household packaging waste in wood shredder BIO-50. Fractions of 0.5/5 mm with bulk density of 9 kg/m^3 were produced in the shredding process. EPU aggregates with bulk density of 16 kg/m^3 were obtained in big size EP sheet milling process. Particle size distribution of EP waste aggregates is presented in Table 1. Images of EPR and EPU aggregates are presented in Figure 1. Water absorption kinetics for EPR and EPU aggregates is presented in Figure 2.

Milled sand (MS) obtained from AB Anykščių kvarcas (Lithuania). 90 % of particles are finer than 45μm. Bulk density 1100 kg/m^3. Specific surface area, 490 m^2/kg.

Table 1. Particle size distribution of EPR and EPU aggregates.

Aggregates	Residue on the sieve (mm),%										
	5.0	4.0	2.8	2.0	1.4	1.0	0.5	0.355	0.25	0.18	> 0.18
EPR	30.0	38.0	21.0	8.0	2.0	0.5	0.5	-	-	-	-
EPU	-	0.26	3.59	10.51	25.38	20.13	28.46	5.77	2.56	1.15	2.18

Metakaolin (MK) obtained from UAB MC-Bauchemi with specific surface of 885 m^2/kg and density of 345 kg/m^3 was used in concrete mixtures. Two different admixtures – superplasticizing (SP) admixture based on polyacrilate ester, and air entraining (AE) admixture, (white powder with pH 8.1), were used in concrete mixtures. The amounts of SP (0.55 %) and AE (0.02%) in concrete mixtures were constant. W/C ratio in compositions (presented in Table 2) was changed from 0.81 to 0.65 with the aim to maintain equal spread of concrete mixtures.

Concrete mixtures (compositions are presented in Table 2) were prepared as follows: at first, water was mixed with OPC, SP, and AE for 2 minutes. Then, MK and MS were added and the mixtures were mixed for 3 minutes. Afterwards, the necessary amounts of EPR and EPU were added to the mixture and mixed for 5 more minutes. The prepared concrete mixture was poured into steel moulds (100×100×100) mm and slightly compacted with a wooden board (so as not to damage the structure of aggregates) and additionally compacted for one minute on a vibrating plate. All concrete mixtures were prepared in a laboratory mixer with a vertical rotation axis by forced mixing at a speed of 125 rpm. The moulds were kept in hermetic plastic bags for two days, then the specimens were demoulded and kept before treating for 2, 7 and 28 days (50 % RH at 20±5°C) in laboratory conditions.

Table 2. PLWAC compositions.

Batch	Used materials,%								
	OPC	MK	MS	EPR	EPU	EPR/EPU	SP*	AE*	Water*	W/C
K0.5	50	20	25	1.70	3.30	0.5	0.2	0.025	40.3	0.81
K1	50	20	25	2.50	2.50	1.0	0.2	0.025	38.4	0.77
K1.5	50	20	25	3.00	2.00	1.5	0.2	0.025	35.0	0.70
K2	50	20	25	3.30	1.70	2.0	0.2	0.025	33.8	0.67
K2.5	50	20	25	3.60	1.40	2.5	0.2	0.025	33.0	0.66
K-3	50	20	25	3.75	1.25	3.0	0.2	0.025	32.5	0.65

* Above dry solids content

Particle size distribution of aggregates, bulk density, for EP were measured in accordance with standard tests (EN 933-1, EN 1097-6+AC, EN 12664). The microstructure of EPU aggregate was inspected using SEM Helios NanoLab 650.

The macrostructure of EPR and EPU samples was inspected using an optical microscope, equipped with a digital camera and connected to the computer. Dry density, compressive strength of PLWAC specimens was measured in accordance with standards EN 1602, EN 12089 and EN 826. Changes in the structure of PLWAC specimens were measured by the ultrasonic pulse velocity method using the device Pundit 7 (range from 0.1 to 6553 μs, resolution 0.1 μs, and frequency of transducers 54 kHz). The final test result was taken as the average value calculated out of at least 5 successful measurements. The measuring surface of standard cylindrical heads was pressed against the specimen at two strictly opposite points. Vaseline was used to ensure a good contact. Ultrasonic pulse velocity (UPV) was calculated from the equation (1):

$$V = \frac{l}{\tau} \cdot 10^6 \quad (1)$$

where l is the length of the tested PLWAC concrete specimen (distance between cylindrical heads), τ is time of pulse propagated.

3. Results and discussions

As it is seen in Figure 1, EPR aggregates are of different size and have a spherical form. The surface of some aggregates is damaged. EPU aggregates have irregular form, there are damages on the upper layer, closed-cell structure is not intact. These irregularities are clearly seen in EPU microstructure tests.

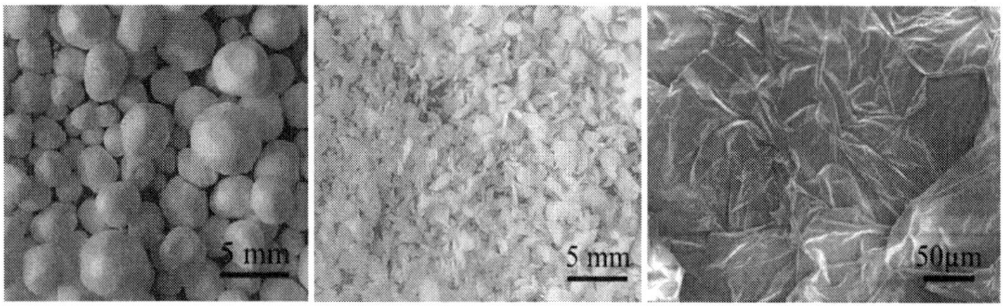

Figure 1. Structure: (a) – EPR macrostructure, (b) – EPU macrostructure, (c) – EPU microstructure.

The analysis of water absorption kinetics (Figure 2) of EPR and EPU aggregates showed that EPR had a less damaged structure and water absorption in these aggregates increased from 39 % to 96 % throughout the entire testing time (up to 28 days). The cell structure of EPU waste is disrupted; therefore, water fills in all cavities and their water absorption during the measurement increases from 200 % to 382 %.

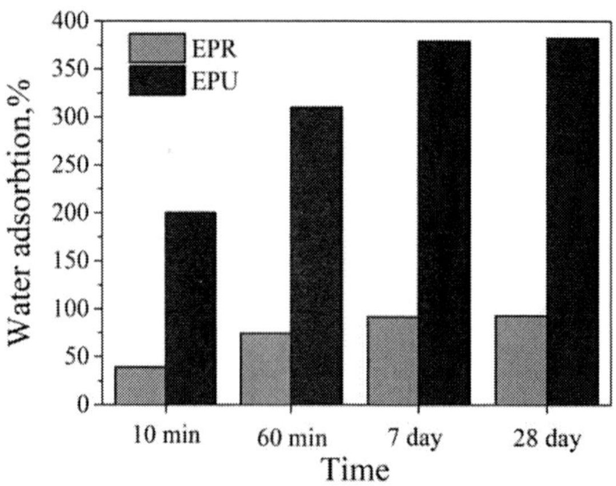

Figure 2. Water absorption kinetics in EPR and EPU aggregates after 10, 60 minutes, 7 and 28 days.

The density of PLWAC specimens (Figure 3) after 28 days of curing decreased about 1.4 times (from 720 to 490 kg/m^3) when the EPR/EPU ratio was increased. These density values confirmed the findings of water absorption tests indicating that due to its structure EPU absorbed more water and thus the density of specimens increased. The dry density of PLWAC specimens revealed similar trends.When EPR/EPU ratio reached 2, the decrease of density values slowed down. With lower content of fine aggregates water absorption in specimens also decreased and subsequently, with higher EPR content in the mix, there was a little change in density.

The compressive strength of cementitious materials directly depends on the compressive strength of aggregates used. When normal aggregate in concrete was replaced by the porous one (EPR and EPU), the compressive strength of concrete reduced due to the lower strength of aggregates used and due to poor adhesion of cement paste and porous aggregate [18, 19]. After two days of hardening the compressive strength of PLWAC specimens changed from 1.5 to 1.1 MPa with a continuous increase in EPR/EPU ratio up to 2 (Figure 4).

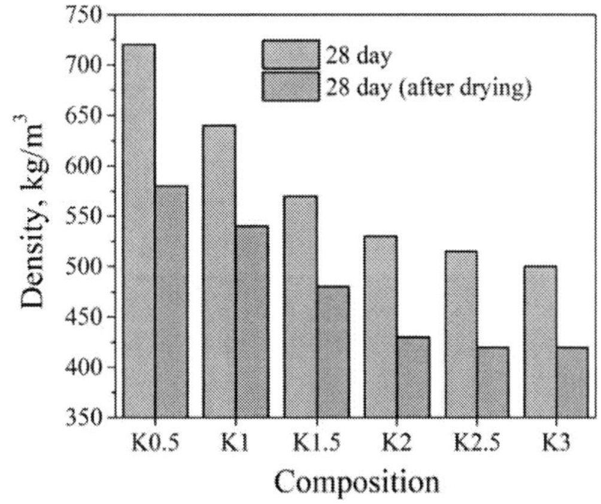

Figure 3. Density of PLWAC specimens after 28 days of curing and after drying.

When EPR/EPU ratio reaches 2.5 and 3, the compressive strength of PLWAC specimens significantly decreases down to 0.65 and 0.4 MPa. The growth in compressive strength (50 %–67 %) was observed in all PLWAC specimens after 7 and 28 days. In our case the compressive strength of the specimens with lower EPR/EPU ratio was higher because in concrete mixtures smaller EPU aggregates filled the voids between larger EPR aggregates in the cement paste and the developed structure led to increased compressive strength of the specimens. On the other hand, higher W/C ratio can negatively influence the compressive strength of PLWAC specimens, as was observed in other studies [20, 21].

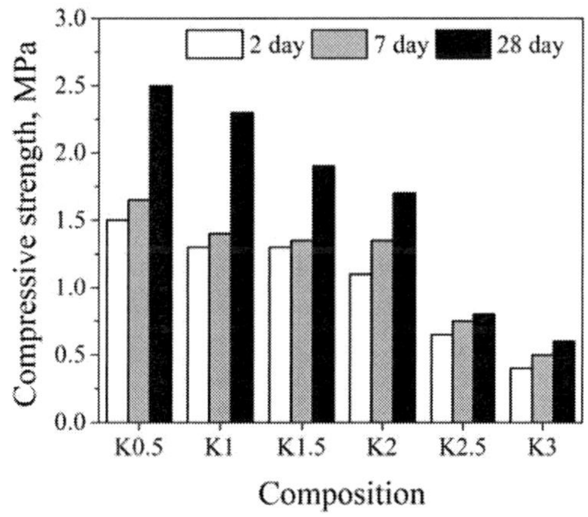

Figure 4. Compressive strength of PLWAC specimens after 2, 7, 28 days.

The density and compressive strength of PLWAC specimens after curing is also characterised by ultrasonic pulse velocity (UPV) measurement. The UPV in the specimens (Figure 5) was the highest (1880 m/s) when EPR/EPU ratio was the lowest 0.5. With higher EPR content in the mixture UPV reduces down to (1250 m/s). It is known that UPV depends on the density and stiffness of the material, so it increases with an increase of EPU amount in concrete mixture [22, 23].

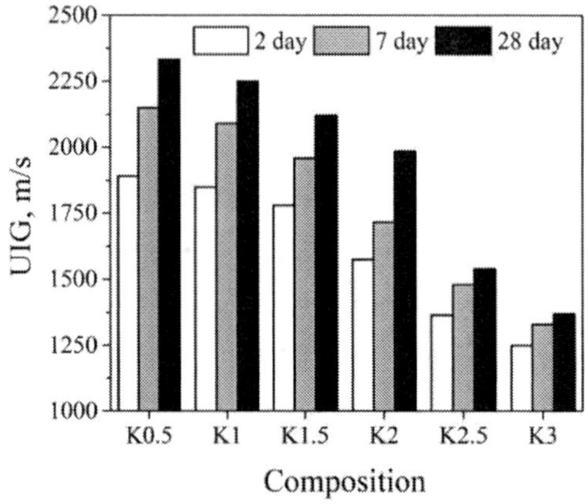

Figure 5. UPV changes in PLWAC specimens after 2, 7, 28 days.

4. Conclusions

- The effect of the EPR/EPU ratio (which varied from 0.5 to 3) on the properties of PLWAC specimens containing OPC, MK, MS, superplasticizer and air entraining admixtures was investigated.
- It was observed that EPR and EPU differ by their structure and thus have different water absorption kinetics. During 28 days water absorption of EPR reached 96 %, whereas water absorption of EPU reached 382 %.
- When EPR/EPU ratio in the PLWAC increased to 2, compressive strength of the specimens dropped from 2.3 to 1.75 MPa and density dropped from 720 to 530 kg/m^3. The increase of EPR/EPU ratio up to 3 causes a slight drop in density down to 490 kg/m^3, and a significant drop in compressive strength down to 0.55 MPa. The reduction of compressive strength is caused by different structure of EPR and EPU, varying W/C and structural changes in PLWAC, which are observed from UPV tests. When EPR/EPU ratio in concrete mixture composition reaches 3, UPV drops to 40 % compared to the specimens where EPR/EPU ratio is 0.5.
- The tests revealed that with the increase of EPR/EPU ratio in the mixture from 0.5 to 3, the dry density of PFAC specimens decreased from 550 to 410 kg/m^3.

References

[1] Ismail Z Z and Al-Hashmi E A 2008 Use of waste plastic in concrete mixture as aggregate replacement *Waste Manag.* **28** 2041–2047
[2] Saikia N and Brito J D 2012 Use of plastic waste as aggregate in cement mortar and concrete preparation: A review, *Constr. Build. Mater.* **34** 385–401
[3] Short A and Kinniburgh W 1978 Lightweight Concrete, 3rd ed., *Applied Science Publishers*, (London)
[4] Sussman V 1975 Lightweight plastic aggregate concrete *J. ACI* **July** 321–323
[5] Perry S H, Bischoff P H and Yamura K 1991 Mix details and material behaviour of polystyrene aggregate concrete *Mag. Concrete. Res.* **43** 71–76
[6] Bagon C and Frondistou-Yannas S 1976 Marine floating concrete made with polystyrene expanded beads *Mag. Concrete. Res.* **28** 225–229
[7] Sayil B M and Gurdal E 1999 The physical properties of polystyrene aggregatedgypsum blocks *Durability of building materials and components* ed. Lacasse M A and Vanier D J Institute for Research in Construction (Ottawa) **8** 496–504
[8] Laukaitis A, Žurauskas R and Kerienė J 2005 The effect of foam polystyrene granules on cement composite properties *Cem. Concr. Compos.* **27** 41–47
[9] Ravindrarajah S and Bearing R 1999 Strength of concrete containing polystyrene aggregate.

Durability of Building Materials and Components Institute for Research in Construction, (Ottawa) **8** 505–514
[10] Sinica M, Laukaitis A, Sezemanas G and Kligys M 2008 Operational properties of a heat-insulation composite with expanded polystyrene crumbs *Mech. Compos. Mater.* **44**(4) 591–604
[11] Tamut T, Prabhu R, Venkataramana K and Yaragal S C 2014 Partial replacement of coarse aggregates by expanded polystyrene beads in concrete *International Journal of Research in Engineering and Technology* **3**(2) 238–241
[12] Al-wahab Ali M A 2012 The Possibility of Produce Self Compacted Polystyrene Concrete *Journal of Engineering and Development* **16**(4) 127–141
[13] Babu D S, Babu K G, and Wee T H 2005 Properties of lightweight expanded polystyrene aggregate concretes containing fly ash, *Cem. Concr. Res.* **35**(6) 1218–1223
[14] Chena B and Liu J 2004 Properties of lightweight expanded polystyrene concrete reinforced with steel fiber *Cem. Concr. Res.* **34** 1259–1263
[15] Šeputytė-Jucikė J and Sinica M 2016 The effect of expanded glass and polystyrene waste on the properties of lightweight aggregate concrete, *Engineering structures and technologies* **8** 31–40
[16] Zhen W, Bing C and Ning L 2013 Fabrication and Compressive Properties of Expanded Polystyrene Foamed Concrete: Experimental Research and Modeling J. Shanghai Jiaotong Univ. (Sci.) **18**(1) 61–69
[17] Miled K, Sab K and Roy R L 2007 Particle size effect on EPS lightweight concrete compressive strength: Experimental investigation and modelling *Mech. Mater.* **39**(3) 222–240
[18] Spiesz P, Yu Q L, and Brouwers H J H 2013 Development of cement-based lightweight composites – Part 2: Durability-related properties. *Cement and Concrete Composites* **44**, 30–40
[19] Lo T Y, Tang W C, and Cui H Z 2007 The effects of aggregate properties on lightweight concrete, *Build. Environ.* **42** 3025–3029
[20] Arulrajah A, Disfani M M, Maghoolpilehrood F, Horpibulsuk S, Udonchai A, Imteaz M, and Du Yan-Jun 2015 Engineering and environmental properties of foamed recycled glass as a lightweight engineering material *Journal of Cleaner Production* **94**, 369–375
[21] Silva R V, de Brito J, and Saikia N 2013 Influence of curing conditions on the durability-related performance of concrete made with selected plastic waste aggregates *Cem. Concr. Compos.* **35** 23–31
[22] Subramaniam K V, Wang X, 2010 An investigation of microstructure evolution in cement paste through setting using ultrasonic and rheological measurements, *Cem. Concr. Res.* **40** 33–44
[23] Kligys M 2009 Production technology and properties of composite material made out of porous cement paste and crushed expanded polystyrene [Doctoral dissertation], Vilnius, Lithuania: Vilnius Gediminas Technical University

Use of carrier materials to immobilise and supply cementation medium for microbially mediated self-healing of biocement

C A Spencer[1] and H Sass[2]

[1]School of Engineering, Cardiff University, Cardiff, UK
[2]School of Earth and Ocean Sciences, Cardiff University, Cardiff, UK

E-mail: SpencerCA1@Cardiff.ac.uk

Abstract. Microbially induced calcium carbonate precipitation (MICP) has been attracting growing interest in respect of its use for biocementation, as a means of improving the engineering properties of granular soil. Recent studies have demonstrated the potential of MICP to enable self-healing of biocement, through the injection of nutrients and precursor chemicals required for MICP into degraded biocement. This paper documents the early stages of research into the development of an autonomous self-healing system for biocement, whereby the nutrients and precursor chemicals are provided from within the biocement matrix. This system has the potential to improve the durability and sustainability of geotechnical structures. The effectiveness of a variety of carrier materials for the immobilisation and release of the nutrients and precursor chemicals, also referred to as the cementation medium, has been explored. Materials tested include expanded perlite, diatomaceous earth and natural fibres such as jute and coir. Studies have subsequently been undertaken to investigate the effect of these carrier materials on the MICP process, in aqueous solutions and within the biocement matrix, and thus the potential to enable self-healing. Ureolytic, spore forming *Sporosarcina ureae* has been utilised to induce the precipitation of calcium carbonate.

1. Introduction

Microbially mediated self-healing, as an autonomic process, is of interest for use in construction, to enable the development of low cost, sustainable, and environmentally friendly biocements and biogrouts. Self-healing, in the context of construction materials, can be defined as, 'the partial or total recovery of at least one property of a material' [1]. Soil properties often need to be enhanced to support infrastructure, to help meet the needs of a growing population [2]. Microbially induced calcium carbonate precipitation (MICP) can improve the mechanical properties of loose, saturated sand, by increasing its strength and stiffness and by reducing its tendency to dilate [2]. Successful application of self-healing microbially induced calcium carbonate precipitation can potentially reduce infrastructure maintenance and repair costs and improve the durability and sustainability of geotechnical structures.

Self-healing of biocement has previously been achieved by injecting the nutrients and precursor chemicals (cementation medium) required for MICP into degraded biocement [2,3]. To enable a truly autonomous healing process, the nutrients and precursor chemicals will need to be readily available within the biocement matrix. This research explores the use of carrier materials to facilitate the storage and release of cementation medium within biocement, to enable self-healing.

The use of carrier materials to store and supply cementation medium is a new concept in respect of studies on self-healing in biocement. Studies on self-healing in cementitious materials have utilised a variety of porous materials to immobilise bacteria and/or cementation medium. Within cementitious materials, immobilisation has been used to protect bacterial cells/ spores from the mechanical forces exerted during the mixing stage of concrete or mortar production and from the highly alkaline environment within cementitious materials. These carrier materials may also be used as internal nutrient reservoirs [4]. Reference to nutrients within this context refers to the nutrients and precursor chemicals, otherwise known as the cementation medium, required for MICP. Immobilisation has been reported to be an efficient approach for bacteria-based self-healing in cementitious materials [5].

The various carrier materials utilised in cementitious materials incorporating MICP based healing systems, along with crack widths healed through MICP where reported, are summarised in table 1. For many of these studies the bacteria, in the form of vegetative cells or spores, has been immobilised along with the required nutrients and precursor chemicals. However, it has been demonstrated that it is not necessary to immobilise bacteria along with the cementation medium to enable self-healing. Khaliq and Ehsan [6] demonstrated that comparable amounts of self-healing can be achieved by immobilising the bacteria only. This particular study utilised graphite nanoplatelets to immobilise bacteria, with nutrients and precursor chemicals added directly to the concrete mix. Bundur et al. [4] used expanded shale aggregates to immobilise nutrient medium only.

Table 1. Materials utilised for immobilisation in cementitious materials incorporating MICP.

Carrier material	Immobilised constituents	Crack width healed (mm) (Curing duration)	Ref.
Graphite nanoplatelets	*Bacillus subtilis* bacteria	0.81 (28 days)	[6]
Expanded perlite	*Bacillus cohnii* bacterial spores and nutrients	0.79 (28 days)	[5]
Ceramsite	*Bacillus mucilaginous* bacteria and nutrients	0.50 (28 days)	[7]
Expanded clay particles	Calcium lactate and bacterial spores from alkaline lake soil, with 98.7% homology to *Bacillus alkalinitrilicus*	0.46 (100 days)	[4]
Diatomaceous earth	*Bacillus sphaericus* bacteria	0.17 (28 days)	[8]
Zeolite	*Sporosarcina ureae* and *Sporosarcina pasteurii bacteria*	0.10 (6 months)	[9]
Lightweight aggregate	*Bacillus cohnii* bacteria and nutrients	_	[10]
Expanded shale aggregates	Nutrient medium	_	[4]

This research utilises MICP to produce biocement through ureolysis. Ureolysis increases the alkalinity of fluid in soil pore spaces as a result of the degradation of urea to carbonate and ammonium, and induces calcium carbonate precipitation. The chemical process, as reported by De Belie [11], is detailed in equations (1) to (5).

$$CO\,(NH_2)_2 + H_2O \rightarrow NH_2COOH + NH_3 \quad (1)$$

$$NH_2COOH + H_2O \rightarrow NH_3 + H_2CO_3 \quad (2)$$

$$2NH_3 + 2H_2O \leftrightarrow 2NH_4^+ + 2OH^- \quad (3)$$

$$2OH^- + H_2CO_3 \leftrightarrow CO_3^{2-} + 2H_2O \quad (4)$$

$$CO_3^{2-} + Ca^{2+} \leftrightarrow CaCO_3 \quad (5)$$

Active ureolytic bacteria produce urease, which catalyses the hydrolysis of urea, resulting in the production of carbamate (NH_2COOH) and ammonia (NH_3), as per equation (1). Carbamate hydrolyses spontaneously to produce ammonia and carbonic acid (2). Simultaneously, in the presence of water, the ammonia and carbonic acid products equilibrate to produce ammonium (NH_4^+), hydroxide (OH^-) (3) and carbonate (CO_3^{2-}) ions (4). This results in a pH increase. Since carbonic is a weak acid it will only partially dissociate into ions, until equilibrium is reached. In the presence of calcium, in an alkaline environment, calcium carbonate will precipitate (5). Calcium chloride dihydrate has been used as the calcium source in this research. In the presence of calcium chloride ($CaCl_2$), crystals of calcium carbonate, often in the form of calcite, form inside the soil matrix [12]. Calcite binds soil particles together, creating a biocement.

Due to the alkaline conditions produced during the MICP process, the type of bacteria used needs to be alkaliphilic, to enable the survival of vegetative cells. To enable long term self-healing, the bacteria will need to be able to sporulate to ensure its survival within the biocement. Ureolytic *Sporosarcina ureae* was selected for use in this research due to its spore forming ability, as tested using MacDonald and MacDonald [13] sporulation medium and the Schaeffer and Fulton [14] spore staining procedure. Ureolytic bacteria are commonly found in soils of varying type and minerology and in a variety of environments [15].

2. Materials and Methods

Since encapsulation and immobilisation technologies have only been used in cementitious materials to date their behaviour within soil systems is unknown. Laboratory studies have been undertaken to test the effectiveness of selected carrier materials for the immobilisation and release of a cementation medium. Subsequently a preliminary study has been undertaken whereby carriers loaded with cementation medium have been incorporated into biocement.

2.1 Microorganism, growth conditions and cell harvesting

Sporosarcina ureae was obtained from the National Collection of Industrial and Marine Bacteria, UK (NCIMB 9251, ACDP Group 1) as a freeze-dried culture. Cultures of *S. ureae* were grown on Luria-Bertani (LB) medium amended with urea. This medium comprised per litre of deionised water; 10 g tryptone, 5 g yeast extract, 10 g sodium chloride, 15 g agar powder and 20 g urea. The LB medium without urea was sterilised by autoclaving at 121°C for 15 min. A urea solution was then added to this medium using a 0.2 μm syringe filter. The medium was then allowed to cool to 55°C before pouring onto 9 cm diameter petri dishes, to minimise condensation on plates. After incubation at 30°C for 48 hrs pure single colonies from the plates were used to inoculate a liquid broth medium consisting of 5 g peptone, 3 g meat extract and 20 g urea per litre of tap water. This liquid broth medium without urea was autoclaved at 121°C for 15 min, to which a urea solution was added using a 0.2 μm syringe filter. Aseptic technique was used throughout. Multiples of 50 ml liquid broth cultures were produced as required for experiments, in 250 ml Erlenmeyer flasks incubated at 30°C, 150 rpm for approximately 19 hrs to produce cultures with an optical density at a wavelength of 600 nm (OD_{600}) of 0.9-1.2 (10^7-10^8 cells/ml), as measured using a spectrophotometer (Hitachi U-1900 UV-Vis).

2.2 Production of cementation medium

The standard cementation medium produced for biocementation, as per table 2, to provide the necessary nutrients and precursor chemicals for the MICP process, was adopted from Stocks-Fischer et al. [16]. To increase the efficiency of the immobilisation process, a concentrated cementation medium was produced using the chemical components and quantities listed in table 2, in deionised water. This medium was produced in 300 ml quantities. Oxoid CM001 nutrient broth was dissolved in 25 ml of deionised water, adjusted to pH 6.0 using concentrated HCl, and sterilised by autoclaving at 121°C for 15 min. The remaining chemicals, excluding calcium chloride dihydrate, were thoroughly mixed in 275 ml deionised water, using a magnetic stirrer. This solution was then adjusted to pH 6.0 using concentrated HCl, to prevent the calcium precipitating out of the calcium chloride dihydrate. Calcium chloride dihydrate was then added as a powder. This solution was then syringe filtered into the Oxoid CM001 nutrient broth solution, using a 0.2 μm syringe filter.

Table 2. Cementation medium components and sterilisation methods.

Medium component	Standard medium (g/l)	Concentrated medium (g/l)	Sterilisation method
Urea ($NH_2(CO)NH_2$)	20	133.33	Syringe filtered
Ammonium chloride (NH_4Cl)	10	66.67	Syringe filtered
Calcium chloride dihydrate ($CaCl_2.2H_2O$)	7.35	49	Syringe filtered
Sodium bicarbonate ($NaHCO_3$)	2.12	14.13	Syringe filtered
Oxoid CM001 nutrient broth	3	20	Autoclaved

2.3 Selection and preparation of carrier materials

Diatomaceous earth (Celite S) and expanded perlite (Harborlite 800), abbreviated to DE and EP respectively, were selected from carriers reported to have been used in cementitious materials incorporating MICP based self-healing systems. Availability, cost and particle size of materials were taken into consideration. Properties of these carriers are listed in table 3. Specific gravity was determined using the small pycnometer method, in accordance with BS 1377-2. Loss on ignition values are as per specifications provided by Sigma Aldrich for Celite S and Imerys Filtration Minerals for Harborlite 800.

Table 3. Powdered carrier material properties.

Carrier	Specific Gravity	Loss on ignition (%)
Diatomaceous Earth (DE)	2.01	3.2-10.0
Expanded Perlite (EP)	1.99	<1.0

In addition, natural absorbent fibres were tested for their suitability as carriers. Constituents of coir and jute fibres, as reported by Rowell and Stout [17] and Banerjee [18] respectively, are summarised in table 4.

Table 4. Constituents of Coir and Jute.

Fibre	α-cellulose (%)	Hemicellulose (%)	Lignin (%)
Jute (J)	58-63	21-24	12-14
Coir (C)	32-43	0.15-0.25	40-45

Prior to use, the fibres were washed thoroughly using a sieve and deionised water. All carrier materials were sterilised by autoclaving at 121°C for 15 min, then oven dried at 90°C. Drying at 105°C for 24 hrs was observed to result in a significant darkening of the coir fibres. Lignocellulosic fibres, which include coir and jute, thermally degrade through dehydration, depolymerization, and oxidation when heated [19]. The extent of delignification has also been reported to be dependent on the duration of heating [19]. Lignin has been reported to undergo thermal degradation at temperatures as low as 100°C [20]. For this reason, the oven temperature for drying was reduced to 90°C, for all materials for consistency, and to prevent the decomposition of sodium bicarbonate which is reported to occur at temperatures above 100°C [21].

2.4 Immobilisation capacity of carrier materials

50 ml polypropylene centrifuge tubes were filled with 1 g of carrier material, in sets of six for each carrier material type. Centrifuge tubes along with the lids were weighed before adding carriers. Concentrated cementation medium was added to each tube, to fully immerse the carrier material. The tubes were left at room temperature (approximately 21°C) for 24 hrs, then centrifuged (3200 rcf, 30 min, Varifuge 3.0 Heraeus) to separate the liquids and solids, after which the supernatant was carefully poured from the tubes, leaving the solids remaining. For tubes containing fibres, these were then inverted for 2 hrs and the liquid drained after, to remove any excess adsorbed medium. Since there was still a

small amount of EP at the surface of the supernatant following centrifugation, this was gravity filtered to establish mass, this having found to be between 4 mg and 9 mg and hence negligible. The tubes containing the carriers were then oven dried at 90°C until the mass was constant. After weighing, to determine the mass of solid cementation medium immobilised, the process was repeated. For those materials for which the overall mass had increased for all six samples, a third loading with cementation medium was undertaken.

2.5 Cementation medium release from carrier materials
A preliminary test was first undertaken using samples from the immobilisation test, to determine the amount of water to be added to samples. 0.5 g carrier material was added to 50 ml polypropylene tubes in sets of four for each material and loaded once with cementation medium, as per the process outlined above. Deionised water was then added to each tube. The quantity of water used was equal to 25 times the mass of dried cementation medium in each tube. These tubes were then left at room temperature (approximately 21°C), with one tube being drained after 1 hr, the second after 3 hrs and the third and fourth after 6 and 24 hrs respectively. Tubes were then dried and weighed to quantify loss of cementation medium. These results gave an indication of potential release of cementation medium for self-healing following water ingress into the biocement. This test was repeated three times. An additional test for a period of 50 hrs was later undertaken under the same conditions.

2.6 Microbiologically induced $CaCO_3$ precipitation in aqueous media
This test was undertaken to establish whether sufficient quantities of each component of the cementation medium had been immobilised by the carrier materials to enable MICP. For each carrier material, six 50 ml centrifuge tubes were filled with 1 g of carrier material and loaded once with cementation medium. 20 ml autoclaved sterilised deionised water was added to each tube. The liquid was drained from the tubes after 24 hrs and syringe filtered after measuring the pH. Standard cementation medium was used in the control tubes. Before these solutions were inoculated with *S. ureae*, the drained tubes were dried and weighed, to enable the concentrations of cementation medium in the solutions to be determined. A small amount of additional deionised water was then added to the tubes containing cementation medium released from jute and the standard cementation medium, to ensure that each set of six tubes contained solutions with an average cementation medium concentration of 3.6 %.

Phosphate buffered saline (PBS) was prepared using 8 g NaCl, 1.42 g Na_2HPO_4 and 0.24 g KH_2PO_4 per litre of deionised water and autoclaved at 121°C for 15 min. Twenty-four 50 ml centrifuge tubes were each filled with 10 ml of freshly cultured liquid broth of *S. ureae*. After centrifugation at 3200 rcf for 20 min, and washing twice with PBS, the tubes containing the vegetative *S. ureae* bacterial cells were filled with the prepared solutions containing cementation medium and incubated at 30°C, 150 rpm. After one day of incubation, three of each set of six tubes were removed and the contents filtered using a vacuum driven Millipore filter system and 0.2μm filter papers. The pH was measured. 10 μl samples were then taken and added to 990 μl autoclaved deionised water to make 1:100 dilutions, to determine calcium ion concentrations using ICP-OES. This process was repeated after three days of incubation for the remaining samples.

2.7 Cementation of sand through microbially induced $CaCO_3$ precipitation
To harvest the vegetative *S. ureae* cells required for the biocement production, six 50 ml polypropylene centrifuge tubes were each filled with 19 ml of liquid broth culture, this quantity equating to the estimated pore space to be filled within the sand in each 50 ml tube. The approximate pore volume had been established via the wet pluviation technique using deionised water and a vortex mixer to vibrate and densify the saturated sand in a 50 ml centrifuge tube, which also gave an estimate of the mass of sand required. Tubes containing the liquid broth were centrifuged at 3200 rcf for 20 min. The supernatant was then drained and replaced with 19 ml PBS, followed by centrifugation at 3200 rcf for a further 20 min to wash any remaining metabolic waste and metabolism by-products from the cells. This washing was repeated twice. After draining the PBS, this was replaced with 19 ml of cementation medium.

Using a funnel secured 5 cm above the tube, tubes containing the bacteria and cementation medium were filled in triplicates, with sand only and then a premixed sand and loaded EP mixture. The EP had been mixed with the sand using a pestle and mortar. Once the tubes were half full, and again after the tubes were almost filled, these were vibrated to densify the sand. The lids were placed on the tubes which were then inverted in an incubator at 30°C. After 24 hrs a 1 mm hole was pierced in the base of the tubes, through which the effluent was extracted using a vacuum, as per figure 1, utilising a flask in which the centrifuge tube was held firmly within the neck.

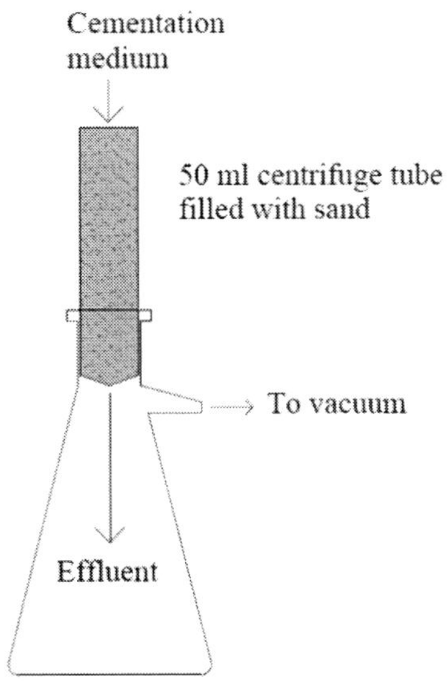

Figure 1. Vacuum assisted extraction of effluent and percolation of cementation medium.

The effluent was syringe filtered for ICP-OES analysis of calcium ion content. This was replaced with cementation medium via surface percolation. Excess cementation medium was drawn through the sand using a vacuum, while continuing to pour cementation medium into the top of the tube, so that approximately 1.5 times the pore volume was used to flush through any remaining effluent. This process was repeated for all tubes, with the holes then being covered after and tubes returned to the incubator inverted. The extraction of the effluent and addition of cementation medium was repeated on weekdays over a 14-day period. After three weeks of drying at 30°C, samples were taken from the biocement to test for the presence of spores using the plate counting method.

3. Results

3.1 Cementation medium storage

The immobilisation capacity of the carrier materials, following up to three loadings with concentrated cementation medium, is shown in Figure 2. Coir was loaded twice only since following the second loading there had been some mass reduction in four of the six tubes, this can be attributed to the dissolution of immobilised cementation medium into the surrounding liquid cementation medium. This effect, albeit to less of an extent, was also noticed after the second loading of expanded perlite. It was evident that the immobilisation capacity had been reached after the first loading of coir fibres, and after the second loading of EP. The capacity of coir to immobilise cementation medium was relatively low and the variable fibre thicknesses likely contributed to inconsistencies in results, this carrier is therefore not used in later tests.

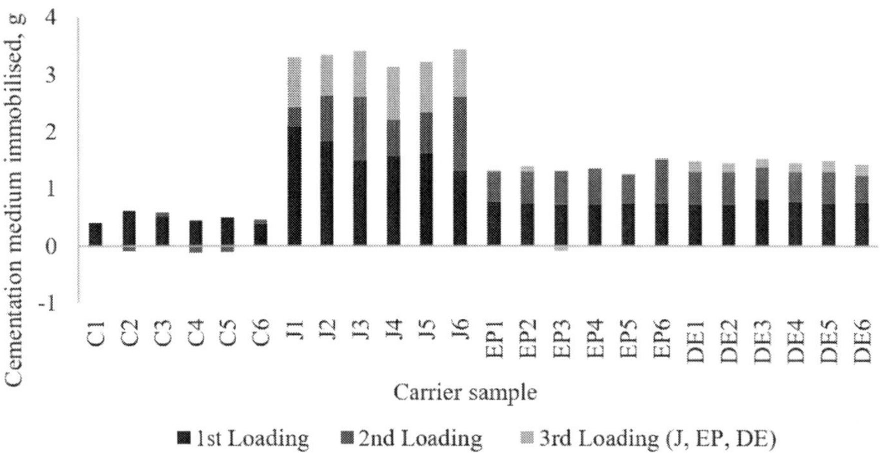

Figure 2. Quantities of cementation medium immobilised by one-gram samples of coir (C), jute (J), expanded perlite (EP) and diatomaceous earth (DE), following repeated loadings.

3.2 Release of cementation medium

Percentages of immobilised cementation medium released from the carrier materials over a period of up to 50 hrs are shown in figure 3. The rate of release of cementation medium from jute within the first 4 hrs is the fastest of the carriers tested. Following this initial fast release from jute equilibrium appears to have been reached with the surrounding solution. This quick release may be beneficial to the self-healing process, in addition to the potential to retain cementation medium for later release. For the first 9.5 hrs cementation medium release is the slowest from EP. Figure 3 shows the R^2 values of the fitted regression curves, these values being close to 1 for cementation medium release from EP and DE. Results for EP and DE show similar trends, with regression lines converging at approximately 30 hours. The rate of cementation medium release from DE then slows slightly compared to EP after 34 hours. The lower R^2 value of 0.65 for cementation medium release from jute may be attributed to the slight variations in length and thickness of the individual fibres within samples, and therefore variations in surface area of fibres. More testing is required to better understand the behaviour of the jute fibres in respect of the release of immobilised chemicals.

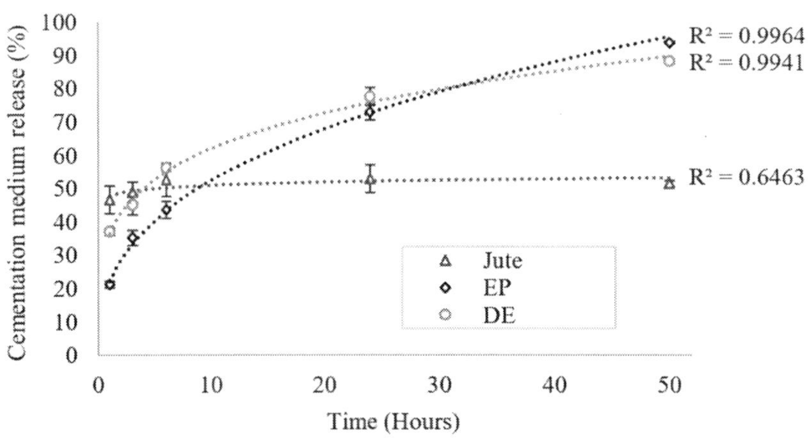

Figure 3. Percentage of immobilised cementation medium released from carrier materials over a fifty-hour period, at 21°C.

3.3 MICP in aqueous solutions

On the basis of calcium ion depletion in solutions, as shown in figure 4, it can be deduced that all carriers tested immobilise and release the cementation medium constituents to the extent that MICP can be enabled. This is further supported by results shown in figure 5, which show a pH increase following inoculation of solutions with *S. ureae*. This pH rise is indicative of MICP and is consistent across all samples. The solutions containing cementation medium constituents which had been immobilised by and released from jute have a noticeably lower pH prior to inoculation with *S. ureae*. The lower reduction in calcium ion concentration for these samples suggests that the jute fibres may not be immobilising or releasing the cementation medium constituents as effectively as EP or DE. However, a slower release over time could help ensure that more nutrients and precursor chemicals are retained by the carrier material during biocementation. These results again suggest that more testing is required on jute prior to use within biocement. The calcium ion reduction in the solutions containing cementation medium which had been immobilised by DE is greater than that of the control solution. This result indicates that the use of DE as a carrier for cementation medium may have a beneficial effect on the MICP process.

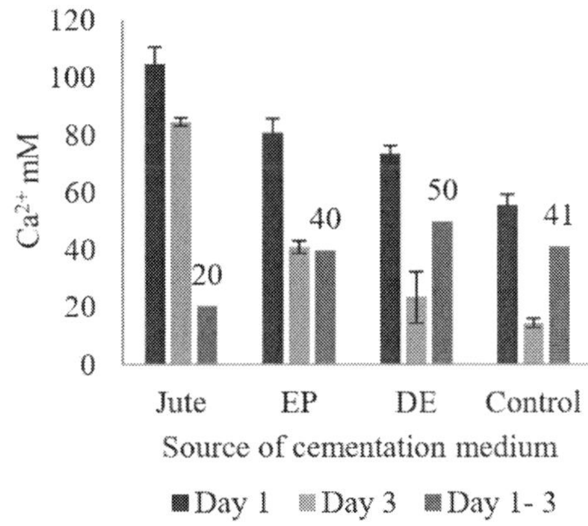

Figure 4. Calcium ion concentration of solutions containing cementation medium released from carriers and standard medium as the control, measured one day and three days after inoculation with *S. ureae*.

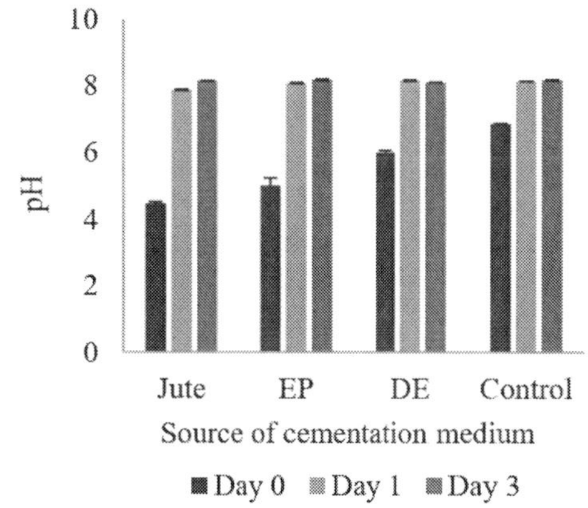

Figure 5. pH of solutions containing cementation medium released from carriers, compared to standard cementation medium, before (Day 0) and after inoculation with *S. ureae*.

3.4 Biocementation

Results from the biocementation tests are shown in figures 6 and 7. Elevated calcium ion levels in the effluent extracted from the tubes containing cementation medium loaded EP, in comparison to the controls, up to day five, can be attributed to leaching of cementation medium constituents from carriers. This appears to have accelerated the biocementation process. At day fourteen, after ten cementation medium treatments, the calcium ion content of the effluent drained from tubes containing loaded EP drops below that of the controls, at this point the test had been ended since the flow in tubes containing the loaded EP had stopped as a result of pores in the sand being plugged with calcium carbonate. Similarly, Stocks-Fischer et al. [19] reported that the flow in biocement columns stopped after ten days of medium treatment, after supplying the medium continuously via gravity flow. For those tubes containing the sand only, further treatments would have been required before flow ceased. Plate counts from samples taken from the dried biocement showed variable results and were deemed unreliable, the growth of cultures on plates did however indicate the presence of viable spores within the biocement.

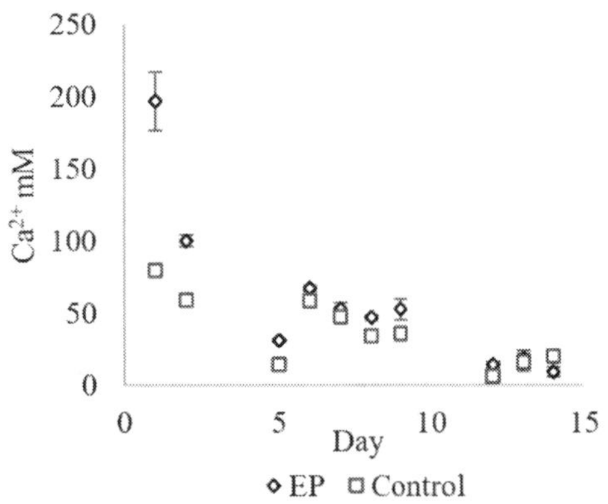

Figure 6. Calcium ion concentrations of effluent extracted during biocementation.

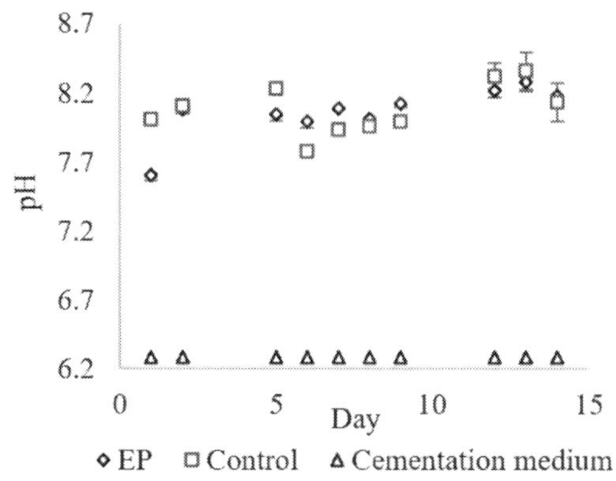

Figure 7. pH of effluent extracted during biocementation, compared to the initial pH of the cementation medium.

Figure 6 shows a significant reduction in calcium ion concentration in the effluent extracted on days 5 and 12, when compared the calcium ion concentration in effluent extracted on days 2 and 9 respectively. This decrease follows weekend periods during which there are no cementation medium treatments and hence calcium ions are depleted. The pH increase shown in figure 7, when comparing the pH of the effluent to that of the cementation medium treatments, is indicative of MICP.

4. Conclusions

Results obtained demonstrate that diatomaceous earth, expanded perlite and jute fibres have the potential to be utilised for the immobilisation and supply of the required nutrients and precursor chemicals to enable self-healing via MICP. Furthermore, diatomaceous earth has been shown to have a beneficial effect on the MICP process by increasing calcium carbonate precipitation. Coir was deemed unsuitable due to the low capacity for immobilisation of cementation medium when compared to other carriers tested.

To ensure longevity of cementation medium supply for self-healing, the retention of cementation medium components during the biocementation process requires improvement. The next stage of this research will involve testing the effectiveness of geopolymer coatings consisting of calcined kaolin and sodium silicate applied to the powdered carriers, to reduce loss of immobilised cementation medium during biocementation. The behaviour of fibres such as jute in respect of immobilisation and release of cementation medium will also be explored further.

The production of the biocement was achieved through the use of vegetative bacteria. Self-healing will require production and germination of spores. *S. ureae* has been selected for use in this study due to its spore forming ability, it presents the potential to be used for self-healing purposes, the presence of spores in the biocement will need to be quantified in future tests using a reliable method.

Acknowledgements

This research has been undertaken as part of the first author's PhD. The funding for the first author's stipend has been provided by Cardiff University School of Engineering.

References

[1] Wiktor V and Jonkers H M 2011 Quantification of crack-healing in novel bacteria-based self-healing concrete *Cem. Concr. Compos.* **33** 763–70
[2] Montoya B M and Dejong J T 2013 Healing of biologically induced cemented sands *Geotech. Lett.* **3** 147–51

[3] Botusharova S 2017 *Self-healing Geotechnical Structures Via Microbial Action* PhD Thesis (Cardiff University)
[4] Bundur Z B, Kirisits M J and Ferron R D 2017 Use of pre-wetted lightweight fine expanded shale aggregates as internal nutrient reservoirs for microorganisms in bio-mineralized mortar *Cem. Concr. Compos.* **84** 167–74
[5] Zhang J, Liu Y, Feng T, Zhou M, Zhao L, Zhou A and Li Z 2017 Immobilizing bacteria in expanded perlite for the crack self-healing in concrete *Constr. Build. Mater.* **148** 610–7
[6] Khaliq W and Ehsan M B 2015 Crack healing in concrete using various bio influenced self-healing techniques *Constr. Build. Mater.* **102** 349–57
[7] Chen H and Qian C 2016 Self-healing cementitious materials based on bacteria and nutrients immobilized respectively *Constr. Build. Mater.* **126** 297–303
[8] Wang J Y, De Belie N and Verstraete W 2012 Diatomaceous earth as a protective vehicle for bacteria applied for self-healing concrete *J. Ind. Microbiol. Biotechnol.* **39** 567–77
[9] Bhaskar S, Hossain K M A, Lachemi M, Wolfaardt G and Kroukamp M O 2017 Effect of self-healing on strength and durability of zeolite-immobilized bacterial cementitious mortar composites *Cem. Concr. Compos.* **82** 23–33
[10] Sierra-Beltran M G, Jonkers H M and Schlangen E 2014 Characterization of sustainable bio-based mortar for concrete repair *Constr. Build. Mater.* **67** 344–52
[11] De Belie N 2010 Microorganisms versus stony materials: a love-hate relationship *Mater. Struct.* **43** 1191–202
[12] Cheng L, Shahin M A and Mujah D 2017 Influence of Key Environmental Conditions on Microbially Induced Cementation for Soil Stabilization *J. Geotech. Geoenvironmental Eng.* **143** 04016083-1–11
[13] MacDonald R E and MacDonald S W 1962 The physiology and natural relationships of the spore forming sarcinae *Can. J. Microbiol.* **8** 795–808
[14] Schaeffer A B and Fulton M D 1933 A Simplified Method of Staining Endospores *Science* **77** 194–194
[15] Dhami N K, Alsubhi W R, Watkin E and Mukherjee A 2017 Bacterial Community Dynamics and Biocement Formation during Stimulation and Augmentation: Implications for Soil Consolidation *Front. Microbiol.* **8** 1267
[16] Stocks-Fischer S, Galinat J K and Bang S 1999 Microbiological precipitation of CaCO3 *Soil Biol. Biochem.* **31** 1563–71
[17] Rowell R M and Stout H P 1998 Jute and Kenaf *Handbook of Fiber Chemistry: 2nd edition* (New York: Marcel Dekker, Inc) pp 466–88
[18] Banerjee P K 2012 Environmental textiles from jute and coir *Handbook of Natural Fibres* vol 2 (Woodhead Publishing)
[19] Ezekiel N, Ndazi B, Nyahumwa C and Karlsson S 2011 Effect of temperature and durations of heating on coir fibers *Ind. Crops Prod.* **33** 638–43
[20] Yang H, Yan R, Chen H, Lee D H and Zheng C 2007 Characteristics of hemicellulose, cellulose and lignin pyrolysis *Fuel* **86** 1781–8
[21] Otsubo Y and Yamaguchi K 1961 Thermochemical Properties and Reaction Processes of Alkali Carbonate-Ferric Oxide System as Investigated by Means of Differential Thermal Method .4. Comparison Between Ceramic and Single Crystal Forms *Nippon Kagaku Zasshi* **82** 683

Development of a high clay content earth plaster

M Lagouin, A Laborel-Préneron, C Magniont and J-E Aubert

LMDC, INSA/UPS Génie Civil, 135 Avenue de Rangueil, 31077 Toulouse cedex 04 France.

E-mail: lagouin@insa-toulouse.fr

Abstract. Earth, as a building material, offers many advantages: on an ecological point of view, it is a natural, abundant, recyclable material with low energy used for production and transportation. Moreover, earth is known to be a natural humidity regulator and to improve comfort inside buildings, making it a good choice for indoor plastering. The clayey phase gives the global cohesion of the material by acting as a binder for the sand grain skeleton. The finest grain soil fraction is responsible for the dry strength, the water vapour permeability and the sorption capacity of earthen plasters. However, clay also induces the drying shrinkage of the mortar leading to the cracking of the plaster. The objective of this work is to improve the moisture buffer capacity of earthen plaster by increasing its clay content without jeopardizing the general behaviour of plasters. Based on consistency, shrinkage and shear tests, the characteristics of mortars in fresh and hardened states were evaluated. The hygric regulator potential of earthen materials were measured through the Moisture Buffer Value (MBV) test. As expected, results have shown that, to maintain a standard level of consistency, an increase in the clay content leads to a greater amount of mixing water. Moreover, to enhance the MBV, the highest possible earth content must be used. Addition of organic admixtures can be considered to improve the high-clay content plaster behaviour without jeopardizing the aesthetic and mechanical aspects of plaster.

1. Introduction

The building sector is a major contributor to resources' consumption, such as energy and raw materials, as well as greenhouse gas and pollutant emissions. This industry is the largest consumer of energy (with about 40 % of the total energy consumption), of non-renewable raw materials and is responsible for approximately 36 % of greenhouse gas emissions in European Union [1]. Hence, building industry has to drive efforts to reduce its environmental footprint. This situation has contributed to the growing interest in locally available resources for the development of sustainable building materials.

Earthen materials fit in with this perspective. Earth, as a building material, offers many advantages as underlined by the Life Cycle Methodology implemented by Melià et al. [2]: although not renewable, earth is natural, non-toxic, reusable and recyclable (when used without chemical stabilization with binders). This raw material does not need energy for calcination like current binders, it can be found almost everywhere thus reducing costs and energy for transportation and production; it only requires a limited energy for milling.

Moreover, earth is known to present a good capacity to absorb and give off moisture [3]. Thus, building products based on raw clay have a high capacity to regulate the relative humidity of the indoor air [4], improving perceived indoor air quality [5] and inhabitants comfort and health, making it a good choice for indoor plastering [6], [7]. The clayey phase also contributes to the dry strength of plasters.

However, clay induces the drying shrinkage of the soil material, leading to the cracking of the mortar. So, to prevent this phenomenon earth, is admixed with sand.

Natural and innovative materials may be added to mortars to enhance their properties at fresh and hardened states: improving the workability of the mortar [8], reducing the bulk density and porosity to increase the thermal conductivity [6], [9], promoting the adhesion strength of plasters to the walls.

The objective of this work is to improve the moisture buffer capacity of earthen plaster. To do so, this study first aims to assess how clayish earth and sand ratio of mortars formulation can influence the fresh and hardened properties of plasters. The hygric regulator potential is evaluated through the Moisture Buffer Value (MBV) test. Finally, addition of organic admixtures is considered to prevent plaster from cracking without compromising the properties of the material.

2. Materials and Methods

2.1. Raw materials

The studied earth plasters are composed by local materials: a clayish earth, a siliceous river sand (0-2 mm) and water. The grain size distributions of the soil and sand used are presented in **Figure 1**.

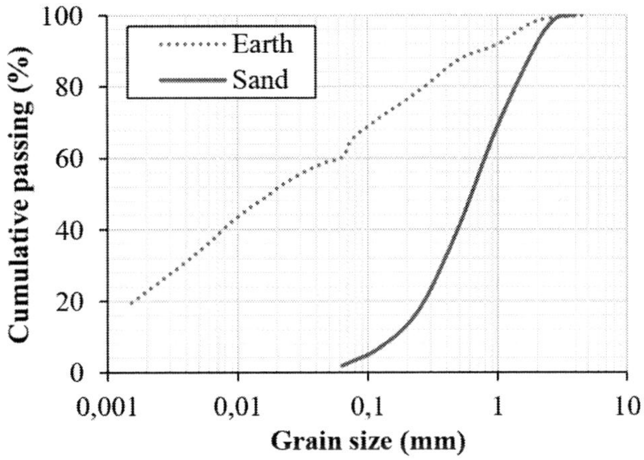

Figure 1. Grain size distribution of studied materials.

In the soil material, the proportions of each constituent (sand, silt and clay) were determined. The liquid limit (LL) and plasticity index (PI) of the soil were also measured. The geotechnical characteristics are given in Table 1.

Table 1. Geotechnical characteristics.

% Clay (≤ 2 µm)	% Silt (0.002-0.063 mm)	% Sand (0.063-2 mm)	LL (%)	PI (%)	Normal proctor optimal water content (%)	Normal proctor optimal dry bulk density (kg/m³)
22.5	38.5	37.0	46	15	14.1	1880

On a geological point of view, clay is distinguished from other fine-grained soils by differences in size: clay fraction refers to a class of materials whose particles are smaller than 2 µm in equivalent sphere diameter. By virtue of its extremely small particle size and high surface-to-volume ratio, clay exhibits glue-like properties in the presence of water [10]. Thus, the clayey phase plays the role of binder while silts and sands form the granular skeleton. The clay minerals contained in the soil are montmorillonite, chlorite and illite.

By drying, the clayey phase shrinks, causing cracks in the plaster. To prevent this phenomenon, earth is admixed with sand.

A commercial fibrewood rigid panel is used as substrate. This material is intended for a facade/wall/roof insulation use. It is described as a fire-resistant (class E), thermal and acoustic insulation board ($\lambda = 0.044$ W/(m.K)). The board is 35 mm thick and has a density of 180 kg/m^3.

2.2. Procedures

As no European standard exists for the characterization of clay plasters, the procedures used are adapted from the German standard for the characterization of ready-mixed clay plaster [11] and inspired by the methodology developed by Hamard et al. [8], Delinière et al. [12] and Stazi et al. [13].

2.2.1. Design and manufacturing of earth plasters. To assess how clayish earth and sand ratio can influence physical and mechanical properties of plasters, six mortars were formulated with different mass ratios of clayish earth and sand: 1:1, 1:1.5, 1:2, 1:2.5, 1:3, 1:4.

Mortars were prepared following the mixing procedure:

- the solid phase is mixed at low speed over 30 s;
- water is added and the amount adjusted to ensure a flow table value of 175 ± 5 mm;
- the mortar is mixed at low speed for 30 s and then switched to high speed for an additional 30 s;
- the mixer is stopped and the mixture left to rest for 5 min;
- 30 additional seconds of mixing is completed at high speed.

For the preparation of the test specimens, the mortar was directly applied on the substrate (fibrewood panel), which was previously soiled and sprayed with water. The plaster was easily levelled thanks to a wooden frame. The thickness of all plasters was around 1 cm.

2.2.2. *Characteristics of fresh state plasters.* For each formulation, the water content is defined to ensure similar consistency between the materials, ensuring a good workability. The flow value test consists in characterizing the consistency by measuring the mean diameter of a test specimen of fresh plaster that had been given 15 vertical impacts, allowing it to fall freely through a given height. A truncated conical mold (60 mm in height with internal diameter of 100 mm at the bottom and 70 mm at the top) is used to give form to the specimen and placed at the centre of a flow table disc. The German standard [11] proposes fixing the consistence at 175 ± 5 mm. The water content of the fresh clay plaster was then measured by drying at 60 °C.

2.2.3. *Characteristics of hardened state plasters.* Mechanical tests are conducted to validate the composition of plasters by means of a shrinkage test followed by a shear test. The shrinkage test requires two 250 x 250 mm² specimens of each formulation applied on the commercial fibrewood panels. After drying, the shrinkage evaluation is carried out: the presence or absence of cracks in the specimens was noted. An earth plaster is of acceptable mechanical quality if, after shrinkage, there are no cracks through which water can penetrate into the wall and if the plaster is not detached (even partially) from the substrate.

The shear test aims to evaluate the bond of the earth formulation with the wall. Five circular specimens of 7 cm diameter are tested for each formulation. The loading device is suspended to the specimens by a string, ensuring a good contact with the upper part of the specimen without any contact with the wall (Figure 2). Specimens were loaded until failure and the mass at which the specimen breaks was recorded.

Figure 2. Shear test assembly.

2.2.4. Moisture buffer value. The moisture buffering performance of a material is a hygric property by which hygroscopic materials in contact with the surrounding air adsorb and desorb moisture to create equilibrium with the relative humidity of the surrounding space. The methodology applied is based on the NORDTEST protocol [14].

Measurements were conducted on three 2 x 12 x 12 cm specimens of plaster applied on a 3.5 cm-thick substrate for each composition. Before testing, the test specimens were sealed on all but one side with aluminium tape. Then, they were stored in equilibrium with the air at 23 ± 5 °C and 50 ± 5 % RH. The test consists in exposing specimens to a daily relative humidity cycle (8 hours at 75 % RH - 16 hours at 33 % RH). Because of the change in relative humidity, the specimens gain or lose weight and their weight changes were tracked with an accuracy of 0.001 g. The MBV is determined at steady state defined by the NORDTEST Project. When the moisture exchange during one cycle (Δm) is reported per open surface area (A) and per % RH variation (RH_{high} - RH_{low}), the result is the practical MBV (kg/(m^2. % RH)):

$$MBV = \frac{\Delta m}{A.\left(RH_{high} - RH_{low}\right)}$$

3. Results

3.1. Characteristics of fresh plasters

All mortars were characterized in the fresh state for flow table consistency. Table 2 presents the fresh state characteristics of plasters depending on the earth to sand ratio.

Table 2. Composition and fresh state consistency of mortars.

Earth to sand ratio	Earth content (% of dry mass)	Water content (% of dry mass)	Flow table consistency (cm)
1:1	50	30.6	17.4
1:1.5	40	27.8	17.5
1:2	33	26.2	17.5
1:2.5	29	24.4	18.0
1:3	25	24.6	17.7
1:4	20	23.2	17.3

The water content necessary to obtain the right consistence for all mixtures ranges from 23.2 to 30.6 %. Those values are slightly higher than water requirements of plaster found in previous studies [6], [12]. This can be explained by the nature of the clay minerals and the fact that not all mortars fulfil the flow table consistency defined in the DIN 18947 [11]. The results show that, as expected, to maintain a standard level of consistency, an increase in the clay content leads to a greater amount of mixing water.

3.2. Impact of clay content on hardened state plaster

The results of shrinkage test are presented in Table 3. The six mortars show distinguished facies after drying. No specimen fell off the wall but several cracked (1:1, 1:1.5, 1:2, 1:2.5) or bowed out (1:1, 1:1.5, 1:2). Only two mixtures did not display signs of shrinkage (1:3 and 1:4).

Table 3. Shrinkage test results.

E/S	1:1	1:1.5	1:2	1:2.5	1:3	1:4
	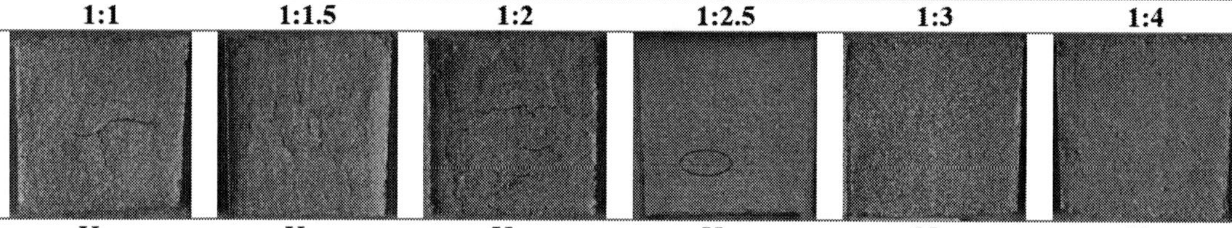					
Cracked	Yes	Yes	Yes	Yes	No	No
Bowed out	Yes	Yes	Yes	No	No	No
Fallen	No	No	No	No	No	No

Clay content has a clear influence on earth plaster cracking, as underlined by Hamard et al. [8]. An increase in clay content leads to a greater amount of mixing water. The plaster shrinkage is hence more important and the cracks induced by the restrained shrinkage are more extensive. Mortar with the greater water content shows visible signs of a poorer plaster/wall bond than other mixtures as they take off the panel.

Outcomes also provide information about the maximum earth content with the soil used to prevent shrinkage cracks: below 25 % of earth (1:3 formulation), the plaster surface area remains smooth and does not lift away from the wall.

Concerning the adhesive strength, the results of the carried out tests are gathered in Figure 3.

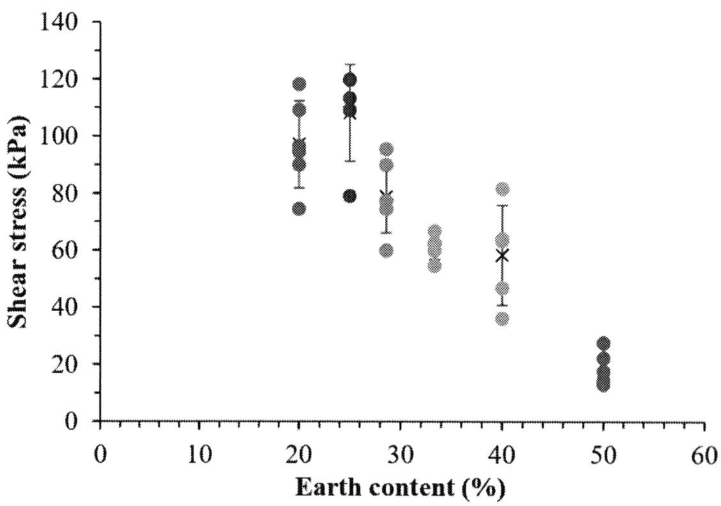

Figure 3. Shear stress of plasters depending on the earth content.

As anticipated by the shrinkage test, the clay content affects the plaster/wall bonds, resulting in lower breaking strength for highest amount of clay. The shear stress does not evolve as described by Stazi et al. [13]. The bond between earth plaster and the wall is controlled by antagonistic phenomenon: an increase in the plaster clay content leads to increased bonding strength which strengthens the plaster, and increased shrinkage, which weakens the plaster-wall interface [8], [13]. Here, the effect of shrinkage predominates and decreases the shear strength.

In all the cases, the average shear strength is greater than 20 kPa. This stress corresponds to the load caused by a virtual earth plaster which would have a 1.13 m thickness. The plaster is able to bear its own load with a minimum safety coefficient of 113. To have an adequate adhesive strength, a minimum safety coefficient of 10 was aimed [13], [15]. Thus, the plasters are sufficiently secure to the substrate. Shear tests highlight the high mechanical resistance of the studied plaster as all mixes are sufficiently secure to the substrate.

However, a quite significant dispersion of values is observed, as it has also been noted by several authors [8], [12], [13]. The scattering of the values is strongly influenced by the support type and preparation. It should also be pointed out that the bonding strength of earthen plasters cannot be dissociated from the type of substrate. In this way, the 1:2 formulation was applied to three other types of walls (chipboard, aerated concrete and another wood fibre board): all specimens fell off the substrates.

Consequently, as an earth to sand ratio is validated if no crack appears on the plaster during the shrinkage test and if the shear adhesion strength is high enough, both 1:3 and 1:4 mixtures are validated.

3.3. Moisture buffer value

The moisture buffer value is determined for two types of mortars: one invalidated by the shrinkage test (1:2), the other one validated by the whole procedure (1:3). Figure 4 gives the moisture buffer values of the two plasters calculated from the last three cycles.

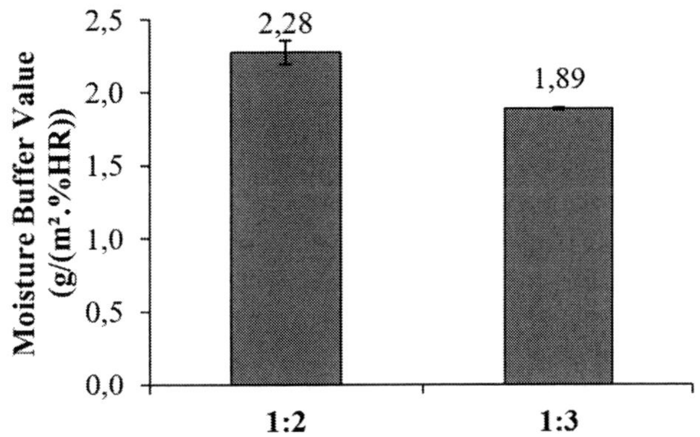

Figure 4. Moisture buffer values of two earthen plasters.

Earthen plasters show a good hygric regulator potential that improve indoor comfort. A difference of about 2 % of clay content between the two mixtures leads to an increase of 21 % of the MBV value. Therefore, clay strongly contributes to regulate the relative humidity of the indoor air by its good capacity to absorb and give off moisture.

To enhance the MBV, the highest possible earth content must be used without jeopardizing the aestethic and mechanical aspects of plaster. In the last part of this study, organic admixtures are considered to prevent higher content of clayish earth plasters from cracking.

3.4. Organic admixtures

This work finally focuses on the potential of organic admixtures to prevent cracking due to shrinkage for high clay content mortars, without altering their fresh state and bonding strength. This part reports the preliminary investigations made in that direction. Inspired by the French national research project PaTerre+ [16] which aims to compile traditional recipes of stabilized earth materials, six organic admixtures were tested to improve the behaviour of plasters. On the one hand, according to the builders interviewed in the PaTerre+ project, the organic molecules such as alginic acid, casein, gluten, starch and ovalbumin, strengthen the soil by holding the clay particles together; they are long enough to adhere to several clay platelets at once and connect them. Hence, those molecules are supposed to contribute to the tensile strength improvement. On the other hand, tannic acid is listed as plasticizer.

The 1:2 formulation is used as a reference cracking plaster to assess the admixture impact. Shrinkage test are realized for the six admixed mortars. All organic admixtures were introduced up to 1 % of the solid dry mass and activated, when required, by 0.1 % of lime (case of casein and alginic acid). The results of the shrinkage test are presented in Table 4**Table 4**.

Table 4. Admixed plasters after drying.

Reference	**Tensile strength improvement**					**Thinner**
1:2	Alginic acid	Casein	Gluten	Starch	Ovalbumin	Tannic acid

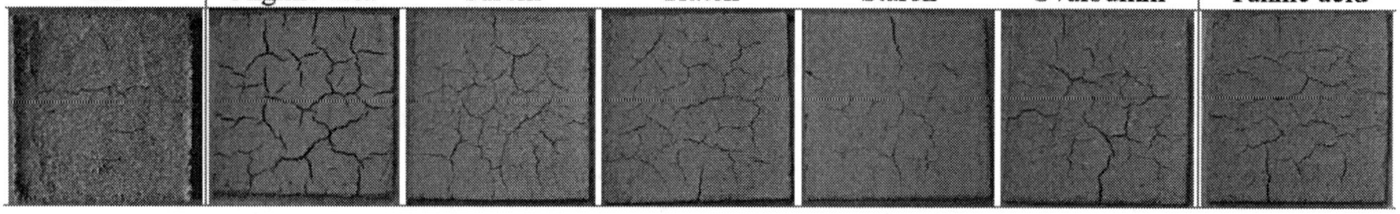

Considering the present results, the organic admixtures do not limit the cracking of plaster. On the contrary, the shrinkage cracks are more extensive. The water content required to fulfil the flow table consistence standard for the tannin-based mortar is of 24 %. The tannic acid does not play any significant thinner role in the earthen material. To maintain the level of consistency of the reference plaster, the five other admixed plasters required a greater amount of water.

The nature of the clay or the amount of admixture may contribute to the failure of shrinkage limitation. Thereafter, those two parameters should be studied.

4. Conclusion

This work aims to improve the moisture buffer capacity of a montmorillonite-based plaster. To do so, the impact of two design parameters was studied: the sand to earth ratio and the addition of organic admixtures. For each formulation, the water content is defined to ensure similar consistency between the materials. Tests are conducted to validate the composition of plasters by means of a shrinkage test followed by a shear test. An earth plaster is of acceptable mechanical quality if, after shrinkage, there are no cracks through which water can penetrate into the wall, and the plaster has a sufficient bond with its wall. Then, its hygric regulator potential is evaluated through the Moisture Buffer Value (MBV) test.

As expected, results have shown that, to maintain a standard level of consistency, an increase in the clay content leads to a greater amount of mixing water. Consequently, the plaster shrinkage is more important and the cracks induced by the restrained shrinkage are more extensive. Furthermore, it affects the plaster/wall bonds, resulting in lower breaking strength for the shear test.

Outcomes also provide information about the maximum earth content with the soil used to prevent shrinkage cracks: below 25 % of earth, the plaster surface area remains smooth and does not lift away from the wall. Shear tests highlight the high mechanical resistance of the studied plaster as all mixes meet the requirements set for indoor application. Earthen plasters can be considered, based on the MBV test, as high potential hygric regulators. To enhance the MBV, the highest possible earth content must be used without jeopardizing the aesthetic and mechanical aspects of the plaster. This first campaign finally evidenced the impact of organic admixtures on the shrinkage of earth plaster: neither the tensile strength nor the water requirement are improved by a 1 % addition. This study is part of an ongoing research which will focus on the impact of the nature of the clay and of the organic admixture content on the shrinkage behaviour of earthen plasters.

References

[1] European Commission, 'Buildings - Energy', *Energy*, 2018. [Online]. Available: /energy/en/topics/energy-efficiency/buildings. [Accessed: 30-May-2018]

[2] Melià P, Ruggieri G, Sabbadini S and Dotelli G, 'Environmental impacts of natural and conventional building materials: a case study on earth plasters', *J. Clean. Prod.*, vol. 80, pp. 179–186, Oct. 2014

[3] Cagnon H, Aubert J-E, Coutand M and Magniont C, 'Hygrothermal properties of earth bricks', *Energy Build.*, vol. 80, pp. 208–217, Sep. 2014

[4] Maddison M, Mauring T, Kirsimäe K and Mander Ü, 'The humidity buffer capacity of clay–sand plaster filled with phytomass from treatment wetlands', *Build. Environ.*, vol. 44, no. 9, pp. 1864–1868, Sep. 2009

[5] Darling E K, Cros C J, Wargocki P, Kolarik J, Morrison G C and Corsi R L, 'Impacts of a clay plaster on indoor air quality assessed using chemical and sensory measurements', *Build. Environ.*, vol. 57, pp. 370–376, Nov. 2012

[6] Santos T, Nunes L and Faria P, 'Production of eco-efficient earth-based plasters: Influence of composition on physical performance and bio-susceptibility', *J. Clean. Prod.*, vol. 167, pp. 55–67, Nov. 2017

[7] Santos T, Faria P and Silva V, 'Can an earth plaster be efficient when applied on different masonries?', *J. Build. Eng.*, vol. 23, pp. 314–323, May 2019

[8] Hamard E, Morel J-C, Salgado F, Marcom A and Meunier N, 'A procedure to assess the suitability of plaster to protect vernacular earthen architecture', *J. Cult. Herit.*, vol. 14, no. 2, pp. 109–115, Mar. 2013

[9] Laborel-Préneron A, Aubert J-E, Magniont C, Tribout C and Bertron A, 'Plant aggregates and fibers in earth construction materials: A review', *Constr. Build. Mater.*, vol. 111, pp. 719–734, May 2016

[10] Bleam W F, 'Clay Mineralogy and Clay Chemistry', in *Soil and Environmental Chemistry*, Elsevier, 2012, pp. 85–116

[11] AFNOR, 'DIN-18947'. 2013

[12] Delinière R, Aubert J-E, Rojat F and Gasc-Barbier M, 'Physical, mineralogical and mechanical characterization of ready-mixed clay plaster', *Build. Environ.*, vol. 80, pp. 11–17, Oct. 2014

[13] Stazi F, Nacci A, Tittarelli F, Pasqualini E and Munafò P, 'An experimental study on earth plasters for earthen building protection: The effects of different admixtures and surface treatments', *J. Cult. Herit.*, vol. 17, pp. 27–41, Jan. 2016

[14] Rode C, Peuhkuri R H, Mortensen L H, Hansen K K, Time B, Gustavsen A, Ojanen T, Ahonen J, Svennberg K and Arfvidsson J, 'Moisture buffering of building materials', 2005

[15] F. F. B. Collectif, 'Règles professionnelles pour la mise en œuvre des enduits sur supports composés de terre crue', *Rech. Dév. Métier*, 2012

[16] Fontaine L and Anger R, *Projet PaTerre+. Interactions argiles/biopolymères: patrimoine architectural en terre et stabilisants naturels d'origine animale et végétale. Rapport final.* 2013

Experimental assessment of bio-based earth bricks durability

A Laborel-Préneron[1], M Giroudon[1], J E Aubert[1], C Magniont[1] and P Faria[2]

[1] LMDC, INSA/UPS Génie Civil, 135 Avenue de Rangueil, 31077 Toulouse cedex 04 France.

[2] CERIS and NOVA University of Lisbon, Department of Civil Engineering, 2829-516 Caparica, Portugal.

E-mail: alaborel@insa-toulouse.fr

Abstract. Construction is one of the most polluting industrial sectors, that is the reason why developing sustainable building materials is a world-wide interest. Earth bricks have recently been more and more studied, mainly regarding their mechanical and hygrothermal properties. The interest in adding plant aggregates to an earth matrix, notably to improve the thermal behaviour, has also been shown. However, durability of the materials is a major concern to sustain performance levels, to limit maintenance and to ensure the comfort and safety of the occupants. Although earth construction seems to be durable, with the various examples of the vernacular heritage in France and all over the world, unstabilised earth is quite sensitive to environmental factors (weather, occupants, micro-organisms). The resistance of an earthen material to liquid water is particularly low, and the addition of plant fibres may decrease its resistance even more. Moreover, the incorporation of organic matter, containing cellulose, raises questions about the composite resistance to moulds. Investigations are thus needed to assess several durability properties. The present paper studies and compares durability of earth bricks containing 0% and 3% weight content of barley straw, lavender straw and rice husk. Some durability tests corresponding to weathering or occupants' actions are conducted: a wet erosion test is performed to simulate rain or accidental water droplets and the resistances to dry abrasion and to impact are also appraised. The resistance to fungal growth is also investigated. Results show a considerable interest in adding lavender straw in earth bricks. Indeed, a very good compromise has been found concerning properties of weathering durability and to microorganisms. The addition of lavender straw improves the dry abrasion resistance of earth bricks whereas rice husk and barley straw additions decrease it. The two types of straw greatly increase the resistance to wet erosion of earth bricks while rice husk only in a smaller extent. Concerning the brick resistance to impact, its behaviour is considerably improved by the addition of the three kinds of plant aggregates, particularly rice husk. Finally, as expected, the bricks made of earth alone are the more resistant to fungal growth. However, the addition of rice husk maintains a very high resistance with the first mould observed after only 10 weeks of incubation at 93% of relative humidity and 30°C.

1. Introduction

Earthen building materials are increasingly being recognized for their huge availability, their low environmental impact and their capacity to regulate indoor climate [1]. However, they can present major drawbacks, such as low resistance to liquid water or poor ductility and shrinkage. Some of these aspects can be partially improved by the addition of plant aggregates [2]. These resources are renewable, being agro-industrial by-products, and are available in huge amounts. Their use is thus environmentally friendly. Several studies have dealt with bio-sourced earth bricks or plasters [3–6] but few of them focus on durability [7]. However, durability of the materials is a major concern to sustain performance levels, to limit maintenance and to ensure the comfort and safety of the occupants.

A huge amount of plant aggregate types has been tested in an earth matrix [6], such as cassava peels [8] or millet residues [9], but few of them were studied in several studies. It is the case of cereal straw, observed in 17 references of the 50 reviewed by Laborel-Préneron et al. [6], hemp hurds, observed in 5 references or wood shavings in 11 references. However, uncommon agricultural by-products, locally available, could be of interest for a use in sustainable earth materials.

Portugal and France are 2 of the 5 main rice producers in Europe. With this industry, they both produce rice husk, which is the protective coat of the grain. This agro-industrial by-product, often considered as waste, is known to be rot-proof and resistant to insects. It appears that adding unground rice husk into unfired earth bricks has never been studied, although rice husk has been investigated in several other building material applications, mainly to create lighter fired bricks that can improve thermal insulation [10–11] or lightweight concrete with a lime-based binder [12].

Lavender straw is a by-product widely available in France, where the production of lavender flower is up to 40 tons per year in the southeast of the country. Once distilled, the lavender straw becomes an agro-industrial waste that is stored near the distilleries. To the authors' knowledge, the addition of lavender straw into unfired earth bricks has never been studied, although lavender straw has been investigated in another building application, as an aggregate of building materials with a pozzolanic mineral binder [13], without showing a very high potential.

In this study, 0% and 3% weight content of barley straw, lavender straw and rice husk were incorporated into earth bricks and their effect on some durability properties were assessed. The influence of the type of plant aggregate on wet erosion, dry abrasion, impact and fungal growth resistances were thus analysed.

2. Materials and methods

2.1. Raw materials

Fines from limestone aggregate washing process (FWAS) were used for this investigation. These fines have a high proportion of limestone (around 60%) and only around 20% of clay. Barley straw, lavender straw and rice husk were also tested in the earth matrix in a proportion of 3% by dry mass. The rice husk was used directly and is about 1 cm long; barley and lavender straw were previously cut, at about 1 cm long. The manufacturing water content was determined with the Proctor test. Formulations of the different types of brick specimens are recapitulated in Table 1.

Table 1. Mixture proportions and Proctor density of specimens.

Specimen	Plant aggregate type	Plant aggregate content (weight %)	Water content (%)	Dry density ($kg.m^{-3}$)
FWAS	-	0	14	1988 ± 9
3BS	Barley straw	3	19	1520 ± 1
3LS	Lavender straw	3	18	1772 ± 17
3RH	Rice husk	3	17	1813 ± 20

2.2. Manufacturing process

To manufacture the specimens, earth and plant aggregate fractions were poured into a blender and mixed by hand. Water was then added and the materials were mixed mechanically in the blender until a homogeneous mix was obtained. The raw materials were mixed the day before moulding.

For purposes of the different tests, the specimens were manufactured in different sizes: 15 x 15 x 5 cm^3 for wet erosion, dry abrasion and impact resistance tests and cylinders 5 cm in diameter and 1 cm high for fungal growth test. The specimens were manufactured by static compression. They were then dried at 40°C for 24 hours followed by an increase in temperature until 100°C at a rate of 0.1°C.min^{-1}. This temperature was finally kept constant until the weight became constant (weight variation less than 0.1% between two weighings 24 hours apart).

2.3. Wet erosion
The wet erosion test was performed on the 15 x 15 x 5 cm^3 specimens to simulate rain droplets. It was carried out according to the New-Zealand Standard NZS 4298 [14], intended for non-stabilised adobes and Compressed Earth Blocks (CEB), based on the Geelong method. The Geelong test is proposed especially for adobe possibly containing straw. An amount of 100 ml of water is allowed to drip from a height of 400 mm onto the specimen, which is inclined at 30°. The duration of the test was between 20 and 30 minutes and the pit depth was measured. An erodibility index, between 3 and 5, was deduced from this value according to the New Zealand Standard. With a pit depth between 5 and 10 mm, the erosion class was 3 and the material was considered as erosive. A pit depth between 10 and 15 mm corresponded to class 4 and the material was considered as very erosive. Finally, for a pit depth greater than 15 mm, the class was 5 and the material failed the test.

2.4. Dry abrasion
Dry abrasion was evaluated on the 15 x 15 x 5 cm^3 specimens, according to the French standard XP P13-901 [15] intended for CEB for walls and partitions, by measuring the quantity of material removed from the specimens after brushing back and forth 60 times ("roundtrips" hereafter) with a metallic brush. The total mass of the steel brush was 3,186 kg and its dimensions are detailed by Giroudon et al. [16].

The brick surface was brushed using the steel brush at the rate of one roundtrip per second for 60 seconds without applying additional vertical force on the brush during the test. Brushing was done along the entire length of the brick. At least half of the brush surface remained permanently in contact with the surface of the brick throughout the duration of the test, in order to avoid a cantilever that would stress the edges of the brick more.

A coefficient of abrasion, in cm^2.g^{-1}, was then calculated as the ratio of the brushed area (in cm^2) over the loss of mass (in g) after the brushing. The higher is the coefficient, the more resistant to abrasion is the material.

2.5. Impact resistance
Vertical walls of buildings may be subjected to different kinds of impacts, such as a stone thrown at the wall or impacts from people or furniture. To evaluate the impact resistance, the Martinet-Baronnie impact apparatus was used and the test was carried out on the 15 x 15 x 5 cm^3 specimens. The test was inspired by ISO 7892 [17] with a small, hard impact body.

The device consisted of a spherical metallic mass of 500 g on a metallic stalk that is attached to the wall. It was dropped from the horizontal position onto the specimen located vertically underneath in contact with a very rigid wall. The energy created by the test was 3 joules. The diameter of the impact point was measured, and the cracking and general behavior of the bricks were assessed.

2.6. Fungal growth
The aim here was to compare the kinetics of mould development on specimens containing 0 or 3% of either barley straw, lavender straw or rice husk in order to determine the influence of the type of bioaggregate on the proliferation. According to Palumbo [18], bio-based materials have very different mould resistance profiles depending on the plant used. The procedure used in this paper is the one developed by Laborel-Préneron et al. [19].

Fungal growth was assessed on the cylindrical specimens (5 cm in diameter and 1 cm high). In order to create good conditions for microscopic observation, the two faces of the specimen were polished to limit their roughness. They were then coated with a polyester resin, except over a square area of 3 x 3 cm^2, which was the surface to be investigated. A decontamination has been achieved just before starting the study of mould growth by exposing the specimens to a temperature of 100°C for 24 hours. Specimens were allowed to cool for 30 minutes after the decontamination before being inoculated.

Artificial inoculation of a fungal suspension of *Aspergillus brasiliensis* spores was realized with a pipette so that the spore quantity deposited on each specimen was known exactly. This inoculum was concentrated at 3.10^5 conidia * mL^{-1} and 5 µl was pipetted onto each on 5 different spots of the surface, thus permitting rapid drying of each drop. Five specimens were inoculated, while two additional specimens, serving as controls, were not.

To avoid cross-contamination, each specimen was placed in a box where the relative humidity was maintained constant using a saturated saline solution [20]. Two different salts were used here to obtain 2 different RH conditions: a solution of potassium nitrate (KNO_3) to obtain 93 % RH and a solution of potassium chloride (KCl) to obtain 84 % RH. According to Johansson et al. [21], such RH conditions are above the critical moisture level for this type of material and represent severe conditions. The individual incubation set-up was composed of a sealed plastic box in which the saturated saline solution and the specimen on its holder were placed. The materials were placed in a climatic chamber at 30°C to provide severe thermal conditions [22].

For microscopic observation, the plastic box was opened once a week and the specimen was removed with its holder in order to avoid touching it. The observation was made using a Zeiss optical microscope (magnification 10X), scanning the whole square area in order to detect mould growth. The assessment of the fungal growth was made using the following rating scale, proposed by Johansson et al. [23]:

- 0: No mould growth;
- 1: Initial growth, one or a few hyphae and no conidiophores;
- 2: Sparse but clearly established growth; often conidiophores begin to develop;
- 3: Patchy, heavy growth with many well-developed conidiophores;
- 4: Heavy growth over more or less the entire surface.

3. Results and discussion

3.1. Wet erosion

Three tests were carried out on three different brick specimens for each formulation. The results of the erosion test are shown in Figure 1. The erodibility index has been added on the right of the graph.

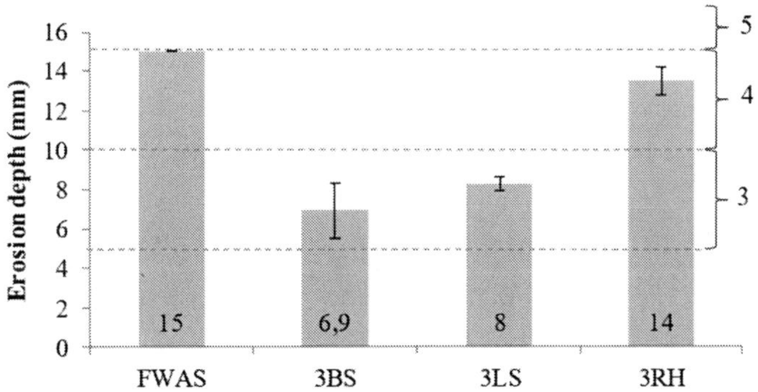

Figure 1. Erosion depth of the bricks after the wet erosion test.

The least erosive formulation is 3BS followed by 3LS, 3RH and FWAS. According to the erodibility index, 3BS and 3LS specimens are class 3, which is erosive according to Frencham [24], 3RH specimens are class 4, very erosive, and FWAS specimens are borderline between class 4 and 5, which means, close to fail the test, according to the New Zealand standard [14].

A difference of behaviour has been noted between the various mixtures. For the FWAS, 3BS and 3RH specimens, all the water dripped was absorbed by the bricks whereas it tended to trickle over the bricks containing lavender straw.

The results show that the addition of plant aggregate to an earth brick increases its erosion resistance and allows it to satisfy the New Zealand standard [14] for building purposes. This observation is particularly true in the case of the two types of straw.

3.2. Dry abrasion

The results of dry abrasion tests are presented in Figure 2. The higher the coefficient of erosion is, the better is the durability of the brick. Two brick specimens were tested for each mixture, with two tests per specimen.

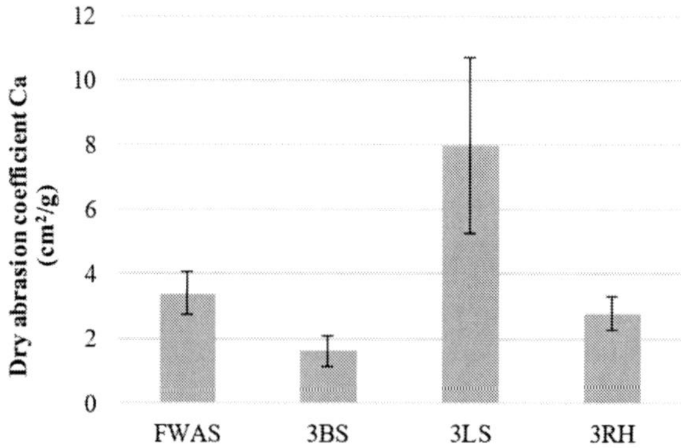

Figure 2. Dry abrasion coefficients of the bricks.

Differences of dry abrasion resistances between the formulations are particularly marked, especially between the two types of straw. 3BS specimens have the lowest abrasion coefficient (1.6 ± 0.5 $cm^2.g^{-1}$), followed by 3RH with 2.8 ± 0.5 $cm^2.g^{-1}$, FWAS with 3.4 ± 0.7 $cm^2.g^{-1}$ and 3LS with 8.0 ± 2.7 $cm^2.g^{-1}$.

The results show that the addition of lavender straw to an earth brick greatly increases its dry abrasion resistance whereas the addition of barley straw or rice husk decreases it. This can be explained by the different morphologies of the plant aggregates: according to Giroudon et al. [16], the barley straw is wider (EAD = 4.5 mm), shorter (AR = 6.2) and smoother whereas lavender straw is thinner (minor axis = 1.0 mm), more elongated (AR = 9.9 mm) and has a rougher external surface, allowing a larger specific surface in contact with the raw earth matrix. All these characteristics encourage higher bonding between lavender straw and the earth matrix.

3.3. Impact resistance

The impact diameters on the brick specimens tested are presented in Figure 3. Pictures of the bricks after the test are shown in Figure 4. The occurrence of cracking, observed visually, denotes a lower deformability, which is negative, while the capacity to recover shape indicates resilience, which in turn is interesting for a building material.

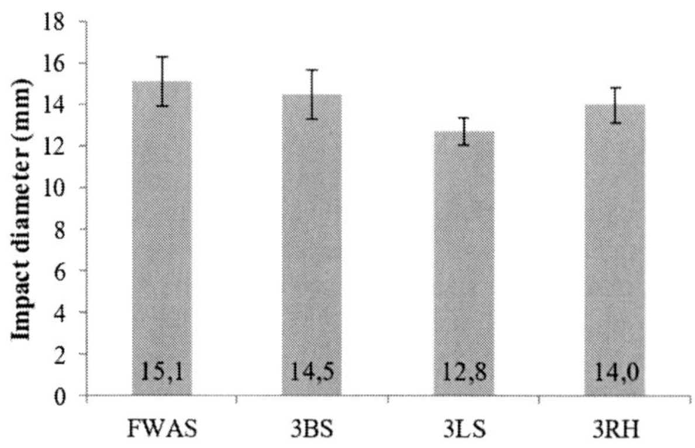

Figure 3. Impact diameter on the bricks.

FWAS 3BS 3LS 3RH

Figure 4. Specimens after sphere impact test.

As can be seen in Figure 3, impact diameters are 15.1±1.2, 14.5±1.2, 12.8±0.7 and 14.0±0.8 mm respectively for FWAS, 3BS, 3LS and 3RH specimens. 3LS bricks presenting the lowest impact diameter may be considered as the most resistant formulation among the four mixtures tested.

In addition to the quantitative result of the impact diameters, Figure 4 gives qualitative result about the bricks' behaviour after impact. FWAS specimens presented cracks after the first impact and eventually broke. The other formulations did not show any crack and the impact marks were not very visible due to the high resilience of these materials. To conclude with both parameters considered, the general behaviour that emerges is that the addition of plant aggregates, decreasing dry density (Table 1), increases the earthen material ductility and decreases the depth of the impact.

3.4. Fungal growth

After 12 weeks of monitoring, no proliferation was observed for an incubation at 84% RH for any material. Mould growth was not found on non-inoculated specimens except on one 3LS specimen, until rate 1 only. Results are thus shown only for inoculated specimens, incubated at 93% RH and 30°C (Figure 5). The growth was described as a function of time and was analysed using the rating attributed each week (median value of the 5 specimens) for FWAS, 3BS, 3LS and 3RH specimens. A critical threshold is drawn at a rating of 2, which corresponds to a specimen with conidiophore. Conidiophore is indeed the part of the mould that can affect the health of the occupant.

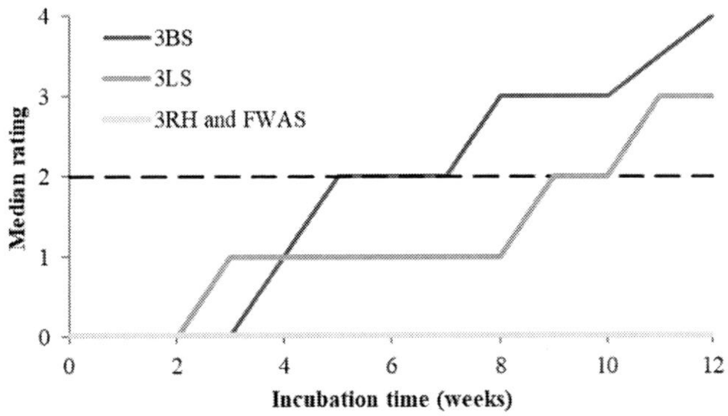

Figure 5. Median rating for inoculated specimens at 93% RH and 30 °C.

Figure 5 shows a median rate for fungal growth of FWAS and 3RH specimens equal to zero. For the 3RH specimens, mould was detected on a rice spot only on one inoculated specimen after 10 weeks of exposure at 93% RH and 30°C. In the case of FWAS specimens, a proliferation rate of 1 was observed on two specimens from week 8 until the end of the test. It appears that mould growth started earlier for the 3LS specimens compared to the 3BS ones, but the colonization of the 3BS specimens was faster and the mould reached more advanced stages, until rate 4 after 12 weeks of incubation, with a large variation of rating between the five specimens. The critical rate of 2 was reached after 5 weeks in the case of the 3BS specimens whereas it was reached after 9 weeks of exposure in the case of 3LS specimens.

The earthen bricks made with rice husk (3RH) and the ones of earth alone (FWAS) present the highest microbial resistance, followed by the bricks made with lavender straw and finally the one containing barley straw. In the case of the FWAS specimens, the very high resistance to microbial proliferation is due to the lack of carbon source necessary for the microbial growth. The difference of fungal growth resistance between the bio-based bricks can be explained by the differences of chemical compositions between the plant aggregates. According to Harper and Lynch [25], lignin is resistant to most microbial attacks. Rice husk chemical composition is made of 26 to 31% of lignin [26], while lavender straw contains 23 to 25% of lignin [27 cited in 28] and barley straw contains 5 to 16% of lignin [29]. It is most probable that the higher lignin content of rice husk makes the 3RH specimens more resistant to fungal growth than the 2 other materials.

4. Conclusion

This work allowed the addition of three plant aggregates in earth bricks to be compared in terms of durability. Earth bricks without any plant aggregates and earthen bricks containing 3% weight content of barley straw, lavender straw and rice husk were studied. Four durability tests corresponding to weathering, occupants use or microorganisms' actions were conducted.

Overall, the experimental tests have shown that the addition of plant aggregates in an earth matrix enhanced the durability properties of earth bricks. More specifically, the wet erosion resistance was improved by the addition of plant aggregates, particularly in the case of the two types of straw (lavender and barley). Dry abrasion resistance was greatly increased by the addition of lavender straw, but the two other plant aggregates decreased it in comparison to bricks of earth alone. The three types of plant aggregates allowed the material to resist to an impact without any crack. Finally, fungal growth resistance was the highest for earth alone bricks and with the rice husk addition. Lavender straw still gave good fungal resistance compared to barley straw.

Because of their distinct morphologies and chemical compositions, plant aggregates do not affect the earth matrix in the same way. According to the various results, lavender straw and rice husk seem to have a very high potential of use in earth bricks.

Although this study needs to be completed with some mechanical and hygrothermal tests to validate the use of such plant aggregates, these results highlight the importance of considering various agro-industrial by-products in building materials. These materials show particular intrinsic properties which can be very beneficial for some applications but problematic for others. This is the case of lavender straw, whose high concentration of water-soluble substances induces setting delays and greatly disrupts the curing mechanisms of mineral binders, making its use in plant concretes very difficult [13, 30]. In contrast, its incorporation into a raw earth matrix, in which curing takes place by the evaporation of water, seems to be promising.

Acknowledgement

The authors wish to thank the French National Research Agency (ANR) for funding the project BIOTERRA – ANR – 13 – VBDU – 0005 Villes et Bâtiments Durables.

References

[1] Minke G, Building with Earth: Design and Technology of a Sustainable Architecture, Birkhäuser. Basel, Switzerland, 2006

[2] Aymerich F, Fenu L, Francesconi L and Meloni P, "Fracture behaviour of a fibre reinforced earthen material under static and impact flexural loading," *Constr. Build. Mater.*, vol. 109, pp. 109–119, 2016

[3] Al Rim K, Ledhem A, Douzane O, Dheilly R M and Queneudec M, "Influence of the proportion of wood on the thermal and mechanical performances of clay-cement-wood composites," Cem. Concr. Compos., vol. 21, no. 4, pp. 269–276, 1999

[4] Algin H M and Turgut P, "Cotton and limestone powder wastes as brick material," *Constr. Build. Mater.*, vol. 22, no. 6, pp. 1074–1080, 2008

[5] Bouhicha M, Aouissi F and Kenai S, "Performance of composite soil reinforced with barley straw," *Cem. Concr. Compos.*, vol. 27, no. 5, pp. 617–621, 2005

[6] Laborel-Préneron A, Aubert J E, Magniont C, Tribout C and Bertron A, "Plant aggregates and fibers in earth construction materials: a review", *Constr. Build. Mater.*, vol. 111, pp. 719–734, 2016
[7] Faria P, Santos T and Aubert J E, "Experimental characterization of an earth eco-efficient plastering mortar," *J. Mater. Civ. Eng.*, vol. 28, no. 1, 2016
[8] Villamizar M C N, Araque V S, Reyes C A R and Silva R S, "Effect of the addition of coal-ash and cassava peels on the engineering properties of compressed earth blocks", *Constr. Build. Mater.*, vol. 36, pp. 276–286, 2012
[9] Bal H, Jannot Y, Quenette N, Chenu A and Gaye S, "Water content dependence of the porosity, density and thermal capacity of laterite based bricks with millet waste additive", *Constr. Build. Mater.*, vol. 31, pp. 144–150, 2012
[10] Carter G W, Cannor A M and Mansell D S, "Properties of bricks incorporating unground rice husks," *Build. Environ.*, vol. 17, no. 4, pp. 285–291, 1982
[11] Chiang K Y, Chou P H, Hua C R, Chien K L and Cheeseman C, "Lightweight bricks manufactured from water treatment sludge and rice husks," *J. Hazard. Mater.*, vol. 171, no. 1–3, pp. 76–82, 2009
[12] Chabannes M, Bénézet J C, Clerc L and Garcia-Diaz E, "Use of raw rice husk as natural aggregate in a lightweight insulating concrete: An innovative application," *Constr. Build. Mater.*, vol. 70, pp. 428–438, 2014
[13] Ratiarisoa R V, Magniont C, Ginestet S, Oms C and Escadeillas G, "Assessment of distilled lavender stalks as bioaggregate for building materials: Hygrothermal properties, mechanical performance and chemical interactions with mineral pozzolanic binder", *Constr. Build. Mater.*, vol. 124, pp. 801–815, 2016
[14] NZS 4298, "Materials and Workmanship for Earth Buildings." 1998
[15] AFNOR, XP P13-901 "Blocs de terre comprimée pour murs et cloisons : définitions - Spécifications - Méthodes d'essais - Conditions de réception", Paris, 2001
[16] Giroudon M, Laborel-Préneron A, Aubert J E and Magniont C, "Comparison of barley and lavender straws as bioaggregates in earth bricks", *Constr. Build. Mater.*, vol. 202, pp. 254–265, 2019
[17] ISO 7892, "Vertical building elements - Impact resistance tests - Impact bodies and general test procedures." ISO, 1988
[18] Palumbo M, "Contribution to the development of new bio-based thermal insulation materials made from vegetal pith and natural binders", PhD Thesis Universitat Politècnica de Catalunya, Barcelona, 2015
[19] Laborel-Préneron A, Ouédraogo K, Simons A, Labat M, Bertron A, Magniont C, Roques C, Roux C and Aubert J E, "Laboratory test to assess sensitivity of bio-based earth materials to fungal growth", *Build. Environ.*, vol. 142, pp. 11–21, 2018
[20] American Society for Testing and Materials, ASTM E104-Standard Practice for Maintaining Constant Relative Humidity by Means of Aqueous Solutions, 2012
[21] Johansson P, Samuelson I, Ekstr A and Sik P I S E, "Microbiological growth on building materials: critical moisture levels. state of the art", SP Swedish National Testing and Research Institute, 2005
[22] Hofbauer W, Krueger N, Breuer K, Sedlbauer K and Torsten S, "Mould resistance assessment of building materials - Material specific isopleth-systems for practical application", in: Copenhagen, Denmark, 2008: p. Paper ID: 465. https://www.isiaq.org/docs/papers/465.pdf (accessed June 14, 2018)
[23] Johansson P, Ekstrand-Tobin A, Svensson T and Bok G, "Laboratory study to determine the critical moisture level for mould growth on building materials", *Int. Biodeterior. Biodegrad.*, vol. 73, pp. 23–32, 2012
[24] Frencham G J, "The Performance of Earth Building," *Deakin University*, Geelong, 1982
[25] Harper S H T and Lynch J M, "The chemical components and decomposition of wheat straw leaves, internodes and nodes", *J. Sci. Food Agric.*, vol. 32, no. 11, pp. 1057–1062, 1981
[26] Mansaray K G and Ghaly A E, "Thermal Degradation of Rice Husks in an Oxygen Atmosphere", *Energy Sources*, vol. 21, no. 5, pp. 453-466, 1999
[27] Kaloustian J, Pauli A M and Pastor J, "Chemical and thermal analysis of the biopolymers in the lavandin", *J. Appl. Polym. Sci.*, vol. 77, pp. 1629–1641, 2000
[28] Magniont C and Escadeillas G, "Chapter 1. Chemical composition of bioaggregates and their interactions with mineral binders", in Sofiane Amziane and Florence Collet (Editors), Bio-aggregates Based Building Materials State-of-the-Art Report of the RILEM Technical Committee 236-BBM, 2017, pp 1-37. ISBN: 978-94-024-1030-3
[29] Laborel-Préneron A, Magniont C and Aubert J E, "Characterization of barley straw hemp shiv and corn cob as resources for bioaggregate based building materials", *Waste Biomass Valorization*, vol. 9, pp. 1095–1112, 2018

[30] Sabathier V, Louvel S, Correa G, Magniont C, Evon P and Labonne L, "Incidence of the water soluble compounds contained into lavender and sunflower bioaggregates on the hardening process of mineral binders", in: Clermont-Ferrand, France, 2017

IOP Conf. Series: Materials Science and Engineering **660** (2019) 012070 doi:10.1088/1757-899X/660/1/012070

Energy and moisture in historic masonry walls retrofitted with hemp-lime

P B Strandberg-de Bruijn[1] and K Balksten[2]

[1]Division of Building Materials, Lund University, 221 00 Lund, Sweden

[2]Conservation, Uppsala University Campus Gotland, 621 67 Visby, Sweden

E-mail: paulien.strandberg@byggtek.lth.se

Abstract. Thermally insulating historic buildings is imperative in order to reduce energy demands of the existing building stock. Therefore an insulation material is needed that improves energy efficiency while being compatible with the existing structure from a hygrothermal, aesthetic and cultural heritage perspective. Hemp-lime is a building material that consists of on a combination of hemp shiv, the woody core part of the hemp stem, and building limes. The aim of this study was to determine if hemp-lime could be a feasible option for thermally insulating historic masonry walls in Sweden. The objectives were to measure energy performance of full-scale masonry façades insulated with hemp-lime and to monitor moisture levels inside the masonry walls. Three small single leaf masonry façades were constructed. One façade was uninsulated, the other had internal hemp-lime insulation and a third had external hemp-lime insulation. Energy use for space heating as well as temperature and relative humidity in the walls and rooms were monitored. Results show that thermally insulating historic masonry walls with hemp-lime can lead to an improvement in energy performance of 44-53 % compared to uninsulated single-leaf masonry. However, moisture levels were higher in the masonry façades that were insulated with hemp-lime.

1. Introduction

Thermally insulating existing buildings is imperative in order to reduce energy demands of the existing building stock. Thermal insulation that is used for retrofitting today is often highly insulating but also very diffusion- tight. Historic buildings are commonly built with materials that are diffusion open and allow moisture to move through the material. When insulating historic buildings with modern diffusion-tight insulation materials such as mineral wool and EPS this changes the hygrothermal properties of the wall completely. Therefore modern energy retrofits may not be appropriate for historic buildings and could even cause damage which can result in loss of cultural heritage values [1]. However, energy retrofits of historic buildings can nowadays also be viewed as a protection tool, as improving energy performance ensures that the building meets modern energy requirements, it promotes better use of the building and thereby extends its life-span [2].

Thermally insulating historical masonry walls can be challenging. External insulation tends to change the entire appearance of the building, with lost cultural and heritage values as a consequence [1]. As the internal surface often is plastered with lime mortar it is easier to add internal insulation without causing a great change in appearance, if the insulation layer can be part of a new plaster. Hence, internal insulation is preferred from a heritage point of view. However, when applying internal insulation on a masonry façade this can change the hygrothermal properties of the walls to such an extent that it leads to moisture problems such as interstitial condensation, moisture damage and frost-scaling [3]–[6]. And if the internal insulation layer is not compatible with a new lime plaster there is another risk at changing the aesthetic expression and loosing heritage values. The essence is that an insulation material is needed that improves energy efficiency while being both compatible with the masonry from a hygrothermal perspective and compatible with lime plaster from an aesthetic and cultural heritage perspective.

Over the past decades a diffusion open building material called hemp-lime has been used in France, the United Kingdom [7] and elsewhere in Europe in new buildings, but also when renovating historic buildings [7]–[9]. Hemp-lime is a building material based on a combination of hemp shiv (the woody core part of the hemp stem) and building lime. In Sweden it is still a relatively unknown building material. As hemp-lime is a thermally insulating and diffusion open material it has potential to work very well as a material for retrofitting historic buildings. And as hemp-lime can be used as part of the plaster or render together with covering lime mortar it is interesting to use wherever there is a need for re-rendering or re-plastering.

Internal insulation usually leads to lower temperatures at the ends of timber beams in masonry walls [5], this can lead to an increase in moisture content and ultimately to moisture damage such as microbial growth. Internal wall insulation can also lead to frost-scaling of the external masonry. According to Vereecken [5] the internal wall insulation hinders an inward drying of the masonry wall and as the indoor heat does not reach the masonry wall to the same extent, the lower temperature of the masonry wall leads to a decreased drying potential of the masonry towards the outdoor climate. This can lead to an increase in moisture content of the masonry wall. High moisture content in the masonry in combination with a cold (Swedish) climate can lead to an increased risk for frost damage. Using a diffusion open material such as hemp-lime for enhancing thermal resistance of historic masonry could improve energy performance of the building while preventing moisture damage and preserving heritage values of the facades.

This paper explores the use of hemp-lime as an insulation material for retrofitting historic masonry walls. The aim of this study was to determine if hemp-lime could be a feasible option for thermally insulating historic masonry walls in Sweden. The objectives were to measure energy performance of full-scale masonry façades insulated with hemp-lime, for an external insulation as well as internal insulation scenario, and to measure changes in moisture levels inside the walls.

2. Materials and Methods

Three small masonry façades were constructed at the campus of Lund Faculty of Engineering. Here a lab building was available for testing energy performance of façade sections. The three façade sections were constructed using bricks produced by Hellingsø Tegelvaerk A/S (brick type 2.2.07 Swedish), that were reminiscent of historical masonry. Behind each façade section there was a small room (1.12×1.04×2.08 m) that was thermally insulated towards the surrounding lab building in order to reduce any heat gains and losses from these rooms to the lab building. Energy losses and gains through the internal walls were kept as low as possible by keeping the temperature in the three rooms as well as in the surrounding building at approximately 20°C. All façade sections were south-facing. The three small rooms behind the façade sections were each heated separately using a small frost guard 200 W (Anslut®). Power used for space heating (W) was monitored continuously using INTABs WiSensys®Wireless Sensing System. Monitoring and measuring of power for space heating was only performed throughout the heating season. Monitoring of space heating started once all measurement equipment was in place on the 10th of October 2017. It continued until 26 May 2018 after which no more space heating was required. Unfortunately no cooling was available which meant that temperatures, especially during the summer months, could not be completely controlled.

Figure 1. The use of hemp-lime for internal insulation is illustrated as it is used on an old brick wall in a Swedish church in the need for re-plastering. Here part of the lime plaster was changed to a hemp-lime layer, covered with a thin layer of lime plaster.

Temperatures and relative humidity inside the test rooms, the surrounding building and the outdoor temperature were measured every 15 minutes using HOBO ONSET MX2301 loggers. Relative humidity and temperature inside the test rooms was measured using TESTO data loggers. In order to not miss any values, measurements of the power used to heat the rooms were made every second. These values per second (in Watt) were then converted into hourly values (kWh). These hourly values were then converted into daily mean values for energy use (kWh/day).

As room A used a PID-controller (situated outside of the room) some energy was used by this device. When space heating was ongoing a total of 222 W was measured (compared to 217 W for the rooms without PID-controller). When no energy was used for space heating the values that were measured were 2 W (compared to 0W for the rooms without PID-controller). Therefore a small compensation had to be made to exclude the energy that was used by the PID-controller from the energy use of the room. This compensation was calculated as follows;

$$P_{compensated} = P - \left(\frac{P}{222} \cdot (222 - 217)\right) + \left(\frac{1-P}{222} \cdot (2 - 0)\right) \quad (1)$$

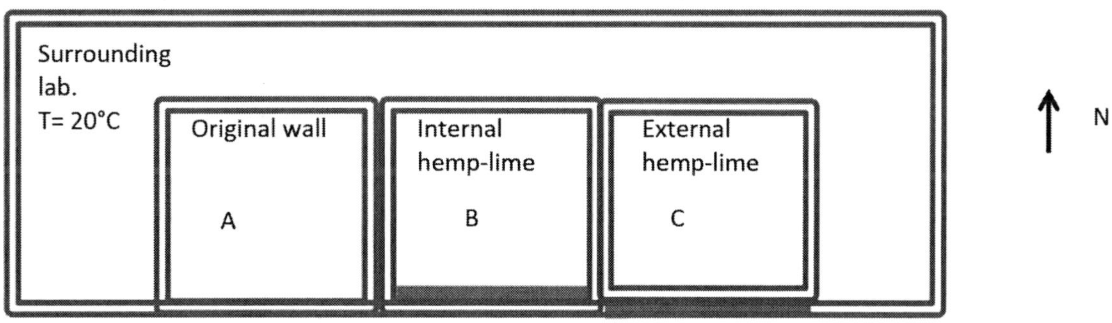

Figure 2. Schematic layout of the laboratory with external walls of brick and hemp-lime.

The lime mortar for bedding and rendering used for all three walls was a natural hydraulic lime from St Astier NHL 3.5 mixed 1 to 1.5 with sand 0-4 mm (in volume). The hemp was supplied by BAFA Neu GmbH in Germany. Lime for the hemp-lime was an air lime (Nordkalk SL). The hemp and lime were mixed in a horizontal cement mixer on site. The parts by volume of the mix were 7 hemp to 4 lime to 4¼ water.

The three façade sections were built up as shown in Table 1 (see also Figure 2).

Table 1. Built-up of the masonry walls, starting with the external material.

Façade A – Single leaf masonry wall
1.Lime plaster, 22 mm
2.Single leaf brick 120 mm

Façade B – Internal hemp-lime insulation
1.Lime plaster 22 mm
2.Hemp-lime 100 mm
3.Lime plaster 10 mm
4.Single leaf brick 120 mm

Façade C – External hemp-lime insulation
1.Lime plaster 22 mm
2.Single leaf brick 120 mm
3.Lime plaster 10 mm
4.Hemp-lime 100 mm
5.Lime render 22 mm

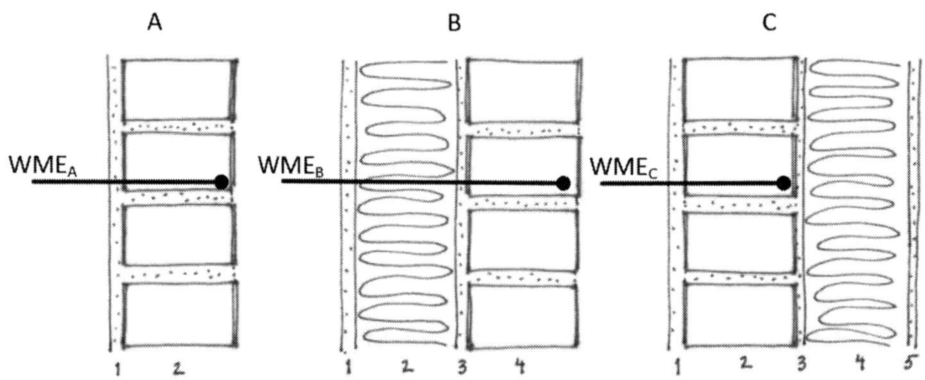

Figure 3. Three test façades at the BML-lab; A) single leaf masonry façade, B) single leaf masonry with internal hemp-lime and C) single leaf masonry with external hemp-lime. WME indicates the depth of the measurement of wood moisture equivalent (WME).

2.1. Moisture levels in the masonry

Moisture content in the masonry was measured using electrodes that were fitted 3 cm apart in the masonry at two depths inside each wall, see Figure 3. The electrodes were connected to wireless sensors (Omnisense) in combination with a wireless Gateway (Omnisense). Moisture content could be measured as wood moisture equivalents (WME). Specimens from the same brick as was used in the test façades was used to determine moisture content as different relative humidities. Dry density of the brick was determined by means of drying and weighing three specimens. Dry density of the hemp-lime was determined by drying and weighing six specimens (100×100×100 mm). Thermal conductivity of the hemp-lime and brick was determined using a Thermal Constants Analyser (TPS1500, Hot Disk®).

The test specimens were conditioned at three different relative humidities in climate boxes. A fourth relative humidity level was accomplished using a climate chamber ($\overline{T}$=20.4°C and ($\overline{RH}$=60.7%). Relative humidity (RH) inside the climate boxes was kept at a constant level using salt solutions. Three different salt solutions were used, see Table 2.

Table 2. Relative humidity at saturation for different salt solutions, according to Greenspan [10].

Salt solution	Relative humidity [%] at saturation at 20°C.
NaCl	75.5
KCl	85.1
KNO_3	94.6

The specimens were only conditioned at absorption. Three specimens per relative humidity were used. The specimens were weighed once they had reached equilibrium. Directly after weighing they were sealed in small plastic bags, one bag per specimen. The same was done for specimens that had been allowed to dry in a heat cabinet (105°C) until they reached equilibrium. An additional level of relative humidity was accomplished by putting the specimens in a layer of water inside a climate box and letting them take up water by means of capillary suction. Once the upper surfaces of the specimens were wet, they were regarded as having reached 100 % RH. The specimens were then taken out of the climate box, weighed, and sealed in plastic bags. Thus, in this study capillary saturation was defined as 100 % RH.

After weighing, electrodes were fitted 3 cm apart in the specimens and they were then directly transported to the lab inside a thermally insulated box. The wireless sensors at the lab were disconnected from monitoring the façades, and connected to the brick specimens. This way the measured moisture content (WME) could be related to the different levels of relative humidity. After measuring in the lab the specimens were dried to establish their dry weight. Using these weights the moisture content by mass at absorption (u) could be calculated using equation (2);

$$u = \frac{m_1 - m_0}{m_0} \cdot 100\% \quad (2)$$

m_1 = mass at equilibrium with the surrounding relative humidity (g)
m_0 = dry weight (g)
u = moisture content (%)

3. Results and Discussion

3.1. Energy performance

At the first day of monitoring, the 10th of October 2017, the average outdoor temperature was 6.4°C which meant that all three rooms required energy use for space heating from the start. Unfortunately, a system error occurred in February 2018 which meant that no data was available for energy use for space heating during the period 1-5 February 2018.

Temperatures inside the rooms as well as inside the lab building and the outdoor temperature were monitored. Four measured values per hour were converted into mean daily values, see Figure 4.

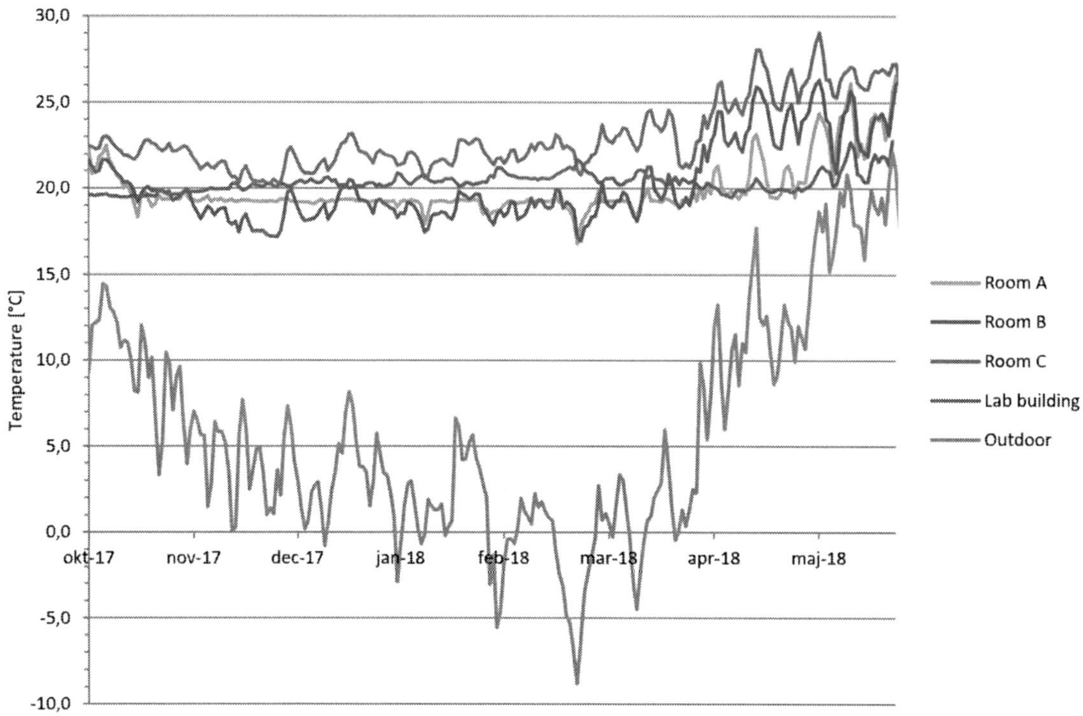

Figure 4. Mean daily temperature in the three test rooms, the surrounding lab building and outdoors during the monitored period 10 Oct 2017-26 May 2018.

For room A, a PID-controller (proportional–integral–derivative controller) was installed that kept the indoor temperature relatively stable even though the masonry façade was not thermally insulated. The PID-controller was situated outside of the room, in the surrounding lab. Room B and C that both had 100 mm hemp-lime as thermal insulation did not need to have a PID-controller installed. Mean daily temperatures for the entire monitoring period were as shown in Table 3.

Table 3. Mean daily temperatures and their SD measured by HOBO in the three test rooms, the surrounding lab building and outdoors from 10th October 2017 to 26th May 2018.

Room A, °C	Room B, °C	Room C, °C	Lab building, °C	Outdoor, °C
19.8 ± 1.8	19.9 ± 2.3	22.6 ± 2.0	20.3 ± 0.6	5.8 ± 6.8

As there were some temperature differences between the rooms and also between the rooms and the surrounding lab building (see Figure 5) there were some energy gains and losses in the rooms that need to be taken into account. Temperature difference (ΔK) between the rooms as well as between the rooms and the surrounding lab building were determined for each day, and then multiplied by 24 hours (h) to get the degree hours difference.

$$\Delta K \cdot h = (\bar{T}_{room\,1} - \bar{T}_{room\,2}) \cdot 24 \qquad (3)$$

The U-value for the internal walls was 0.28 W/m²K. The energy loss/gain through the internal walls was then calculated as follows;

$$\frac{W}{m^2 \cdot K} \cdot (\Delta K \cdot h) = \frac{W \cdot h}{m^2} \quad (4)$$

$$kWh = m^2 \cdot \frac{kW}{m^2 \cdot K} \cdot (\Delta K \cdot h) \quad (5)$$

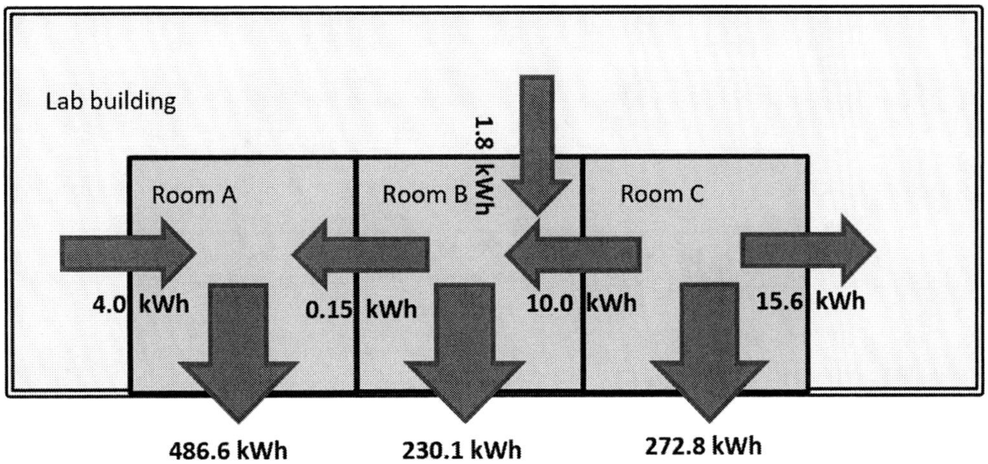

Figure 5. Energy flows between rooms and through the façades.

The total energy use for space heating was thus 486.6 kWh for room A with a single leaf masonry façade, 230.1 kWh for room B with internal hemp-lime insulation and 272.8 kWh for room C with external hemp-lime insulation. Daily mean values of energy use for space heating are shown in Figure 6. There is a clear improvement of energy performance when hemp-lime insulation is applied to the masonry. Some material properties are shown in Table 4.

Table 4. Measured density and thermal conductivity for the brick and hemp-lime.

Material	Density, kg/m³	Thermal conductivity, W/mK
Hemp-lime	453.8	0.14
Brick	1707.0	0.65

Comparing the rooms that had been insulated with hemp-lime to the room with the uninsulated single leaf masonry façade the improvements in energy performance are as follows; 44 % improvement for the room with external hemp-lime and 53 % improvement for the room with internal hemp-lime insulation. It was somewhat unexpected that the room with internal hemp-lime performed a bit better, as the hemp-lime had not been able to dry after the building process. The rooms behind the façades were not ventilated, they were only heated. Therefore the internal hemp-lime could not easily dry towards the indoors, leaving a material with higher moisture levels than what would have been possible if the room would have been ventilated. Even so, this façade thermally outperformed the façade with external hemp-lime.

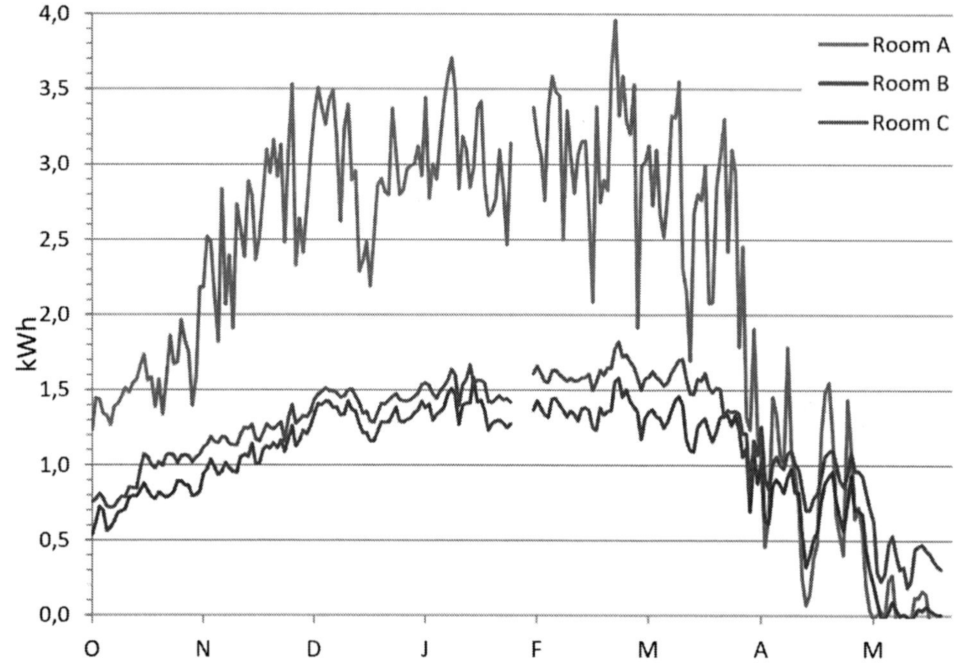

Figure 6. Energy use for space heating (daily mean values) of the three test rooms during the heating season.

3.2. Moisture levels in the masonry

Lab measurements gave the correlation between measured wood moisture content (WME) and the relative humidity in the brick, see Table 5. For higher relative humidities this correlation was quite diffuse, with relatively large standard deviations. Therefore no conversions were made from WME to relative humidity in the brick. Rather, this correlation should be viewed as an indication of approximately relative humidities in the brick, where higher WME also indicates higher RH. Even so, the WME levels that correlate to 75.5 % RH and "100% RH" (capillary saturation) are shown in Figures 8 and 9 to give an approximate indication of moisture levels in the brick.

Table 5. Measured WME [%] in the brick and their standard deviations, at different relative humidities [%]. Three measurements per RH.

RH [%]	WME [%]
0	9.0 ± 0.4
60.7	16.3 ± 0.6
75.5	18.2 ± 3.8
85.1	18.6 ± 3.7
94.6	19.9 ± 2.1
"100"	33.9 ± 1.1

Moisture content and temperature measurements inside the brick started later than the energy use measurements. WME and temperature in the brick was monitored for almost a year; from the 17th of January to 17th December 2018, see Figures 7 and 8. WME and temperature were measured at the external surface of the brick, see Figure 7.

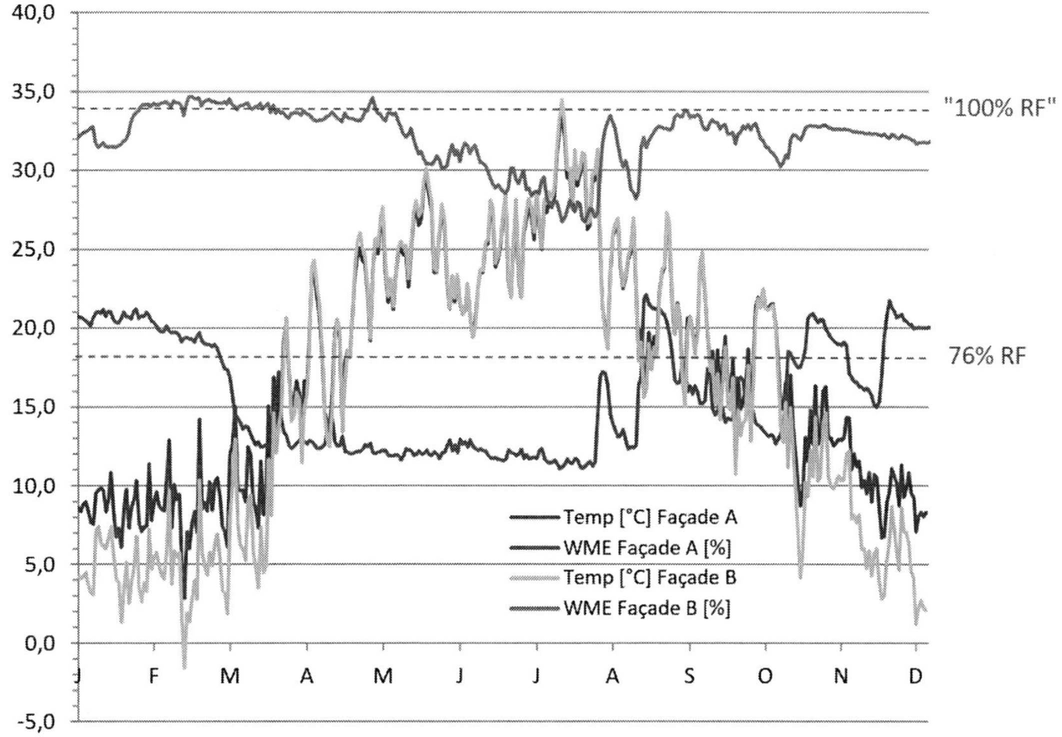

Figure 7. Temperature and WME [%] at the surface of the brick for a single leaf masonry wall (Façade A) and for a single-brick wall with 100 mm internal hemp-lime insulation (Façade B).

Comparing a single-leaf masonry façade to a façade with 100 mm internal hemp-lime insulation there is a large difference in moisture levels in the masonry. While moisture levels are rather low in the single-leaf masonry façade (façade A), the moisture levels in the masonry with internal hemp-lime insulation (façade B) are quite high. In winter they reach, and go beyond, a level that was measured for capillary saturated brick samples ("100 % RH"). Additionally, the temperature in the masonry of façade B was lower than that in façade A. While the surface temperature in façade A stays above 0°C during the entire winter, the temperature in façade B reaches freezing temperatures during a cold spike in February. Higher moisture levels in combination with lower temperatures in the masonry brings about an increased risk of frost scaling.

A contributing factor to the high moisture levels in façade B was the lack of ventilation of the room behind the façade. The building moisture from the hemp-lime could not dry inwards, which likely has had an effect on moisture levels in the entire façade. However, there are very large differences between the moisture levels in the masonry of the two façades strongly indicating that internal hemp-lime insulation will lead to higher moisture levels in the masonry and to lower temperatures in the masonry. This combination can lead to an increased risk of frost scaling. However, during visual inspection of the masonry façades no frost-scaling damage could be found.

Either way, in a cold climate like in Sweden masonry materials must be designed to be frost resistant, and historic masonry in need for thermal insulation has often already proven if it is frost resistant or not.

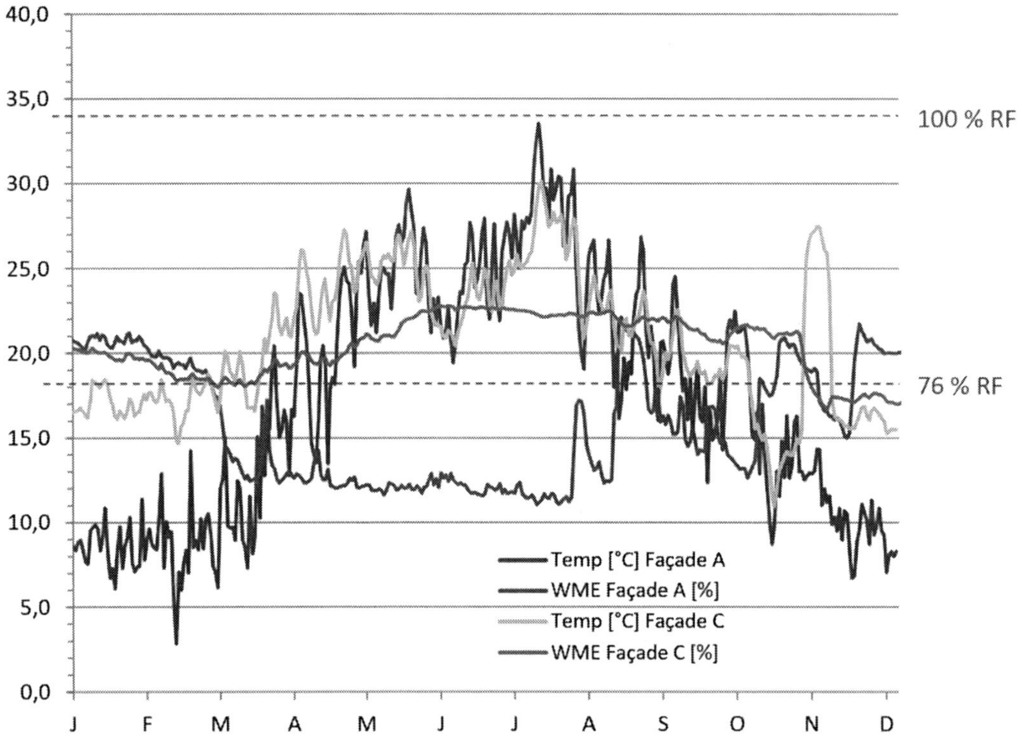

Figure 8. Temperature and WME [%] at the surface of the brick for a single leaf masonry wall (Façade A) and for a single-brick wall with 100 mm external hemp-lime insulation (Façade C).

Comparing the single leaf masonry façade (façade A) to a façade with 100 mm external hemp-lime insulation (façade C) there is a clear difference in temperatures inside the masonry, with the externally insulated masonry having higher temperatures during the cold season and lower temperatures in summer. This is what would be expected when externally insulating the masonry. Moisture levels in façade C were similar to those in the single leaf masonry façade. But while façade A continued to dry during the spring and summer, moisture levels in façade C increased somewhat. This indicates that some of the moisture from the external hemp-lime dried inwards and increased moisture levels in the masonry. The temperature spike in November was caused by a failure in the heating system, causing an increase in room temperature in room C. As a consequence, moisture levels in the masonry of façade C decreased. In December moisture levels in the single leaf masonry façade increased while those in façade C remained rather low.

4. Conclusions

Thermally insulating historic masonry walls with hemp-lime can lead to substantial improvements in energy performance. Results from the studies performed for this paper gave an improvement in energy performance of 44-53 % when insulating a single leaf masonry wall with 100 mm hemp-lime. This will most likely also have a positive effect on the indoor climate.

When evaluating the results, it is important to bear in mind that most historic masonry buildings constructed since the beginning of the 19th century have been constructed using considerably thicker brick masonry, at least 1-stone but often 1½-stone or 2-stone width. These masonries does not always allow a thick additional insulation layer in their construction, but since they are often lime-plastered, at least internally, with a thick plaster there is often a given thickness of lime mortar that could be replaced with hemp-lime instead. The difference in energy efficiency and moisture transport follows the thickness of the hemp-lime for each specific case. What can be concluded is that with an additional insulation of hemp-lime on brick masonry, the thermal performance is improved very clearly. At the same time, it is an addition that allows a continued use of lime render/lime plaster as a surface treatment and that the masonry remains having a moisture and heat flow as before. Given that the initial building moisture can dry out, it presumably works well to insulate a brick wall with hemp lime, both internally and externally.

However, moisture levels in the masonry were high in the masonry insulated with hemp-lime. In combination with lower temperatures in the masonry, as was the case in the façade with internal hemp-lime insulation, this may increase the risk for frost scaling if the materials are not frost resistant. High moisture levels in the façade with internal hemp-lime insulation were most likely partly caused by the absence of ventilation in the room behind the façade, with limited possibilities of the hemp-lime to dry inwards. It is important to allow the hemp-lime to dry after construction.

The appearance of the building façade and its heritage values can be preserved, which is a valid argument for using internal insulation. When retrofitting historic masonry façades with internal hemp-lime it should be considered to use a layer of less than 100 mm hemp-lime insulation, in order to improve energy performance of the masonry façade while at the same time avoiding moisture damage and frost scaling problems.

Acknowledgements

This research was conducted with financial support from the Swedish Energy Agency through their research programme *Spara och Bevara*. The role of this funding source was strictly financial. During the course of the project two seminars were held as part of the research programme *Spara och Bevara*, where the outline of the project and some preliminary results were presented. Part of the air lime used in this research was provided by Nordkalk.

This project would not have been possible without the help of masons Henrik Nilsson and Tom Yttergren. Also involved in the project were Anna Donarelli at Uppsala University as well as Ulrika Mebus and Pär Malmros at Gotland Museum, their help is gratefully acknowledged. Last but not least, the invaluable help of laboratory staff Bengt Nilsson (in memoriam) and Stefan Backe is gratefully acknowledged.

References

[1] Webb A L 2017 Energy retrofits in historic and traditional buildings: a review of problems and methods *Renew. Sustain. Energy Rev.* **77** pp 748–59

[2] Carbonara G 2015 Energy efficiency as a protection tool *Energy Build.* **95** pp 9–12

[3] Johansson P, Geving S, Hagentoft C-E, Jelle, B P Rognvik E, Kalagasidis A S and Time B 2014 Interior insulation retrofit of a historical brick wall using vacuum insulation panels: hygrothermal numerical simulations and laboratory investigations *Build. Environ.* **79** pp 31–45

[4] Morelli M and Svendsen S 2013 Investigation of interior post-insulated masonry walls with wooden beam ends *J. Build. Phys.* **36** no. 3 pp 265–93

[5] Vereecken E 2013 Hygrothermal analysis of interior insulation for renovation projects KU Leuven

[6] Vereecken E and Roels S 2014 A comparison of the hygric performance of interior insulation systems: a hot box–cold box experiment *Energy Build.* **80** pp 37–44

[7] Stanwix W and Sparrow A 2014 The Hempcrete Book Designing and building with hemp-lime (Cambridge, United Kingdom: Green Books)

[8] Bevan R and Woolley T 2008 Hemp lime construction (Bracknell, United Kingdom: BRE Press)

[9] van der Vorst, H Harmsen H and Borgers R 2018 Handboek bouwen met kalkhennep. (Rotterdam, the Netherlands, SBRCURnet)

[10] Greenspan L 1977 Humidity fixed points of binary saturated aqueous solutions (*J. Res. Natl. Bur. Stand. - A Phys. Chem* **81A(1)** pp 89–96

Effect of water-binder ratio on properties of phosphogypsum binder

G Bumanis[1], J Zorica[1], D Bajare[1], A Korjakins[1]

[1]Department of Building Materials and Products, Institute of Materials and Structures, Riga Technical University, Kalku str. 1, LV-1658, Riga, Latvia

E-mail: girts.bumanis@rtu.lv

Abstract. Technological, physical and mechanical properties of mineral binders are often defined by the amount of water add to the binder known as W/B ratio. The role of water is contributed to the workability of fresh paste, hydration processes, porosity, mechanical performance and durability of the material. Present research summarizes the relationship between gypsum based binder W/B ratio in mixture composition to properties of hardened gypsum binder, giving a marginal borders of properties affected by W/B. The W/B ratios evaluated were from 0.3 to 1.15. Dihydrate phosphogypsum (PG), which is a waste of phosphorus fertilizer industry and commercial gypsum binder were used as a raw material for production of gypsum samples and results compared to commercial gypsum binder. At the beginning PG was dried at 60°C and afterwards homogenized by milling it in ball mill or by collision milling in disintegrator. Samples were made with plasticizers, mineral additives and foam forming additive. In this work 26 dense and 9 porous mixture compositions was summarized, from which 18 compositions were made from commercial binder for comparison. A series of performance properties of the binders were investigated, such as density, porosity and 14d compression strength regarding to W/B ratio used to prepare the samples. The chemical and mineralogical composition of PG used was determined by XRF and XRD. The results indicate that plasticizers in PG based gypsum binder could reduce W/B ratio from 1.15 down to 0.4. The PG binder bulk density for samples without pore forming admixture was in range from 726 – 1600 kg/m^3, total porosity 35 – 71 vol.% and compressive strength from 1.4 to 28.7 MPa at the age of 14 d. Ultra porous gypsum material was obtained with the pore forming agent, which allowed to obtain gypsum based material with bulk density from 213 – 726 kg/m^3, total porosity from 67.9 – 90.6 % and compressive strength from 0.2 – 0.8 MPa. Results show that the material density and strength decreases while porosity tends to increases with the increase of W/B ratio. These tendencies are true both for dense and porous specimens. It was concluded, that lowest W/B of dense PG specimens was 0.43, otherwise the workability of a binder decreased dramatically and decrease of compressive strength was observed. With the same W/B ratio the technological properties of PG binder may be affected by gypsum treatment parameters and selected mixture composition, allowing to improve the properties of gypsum binder obtained from same raw material.

1. Introduction

Nowadays gypsum is a common building material with wide range of application, i.e. it is used in production of gypsum binders for direct use on site or to produce gypsum plasters (including plasters for special purposes) or manufactured elements (blocks, tiles, plasterboards). Historically, gypsum is used also as paints created from lime-gypsum suspension [1]. Besides of natural gypsum a synthetic gypsum is available which in most cases is described as a by-product from industrial production and it has a potential of replacement of natural gypsum. Flue gas desulfurization-FGD gypsum and phosphogypsum (PG) is the most common forms of synthetic gypsum [2], [3]. PG is a primary by-product from phosphoric acid production and in most cases it is accumulated in open stack-piles occupying large areas near plants and bringing environmental problems (figure 1) [3], [4].

Figure 1. Dihydrate phosphogypsum stacks in Kedainiai, Lithuania (author picture)

PG could be contaminated with technological impurities which reduce the potential disposal of the PG in production of building materials [5], [6]. There are several ways to remove or neutralize the impurities, i.e. washing with water [7], neutralizing with alkaline additives [8], adding adsorbents [6]. Some of them are combined with the milling process of PG as during milling process, PG crystals are destroyed and unfavorable P_2O_5 released and modified while the compressive strength increases as specific surface area increases [9]. P_2O_5 is a powerful desiccant and dehydrating agent therefor it may attract water that could increase the moisture content in gypsum material lowering its properties [5].
The water-gypsum (W/B) ratio has an influence on the basic physical characteristics of hardened gypsum, such as its volume density, total open porosity and other related characteristics like its moisture, mechanical, thermal and sound insulation properties. The strength values of hardened gypsum significantly depend on the water-gypsum ratio as well. The theoretical W/B ratio necessary for the hydration of calcium sulphate hemihydrate into calcium sulphate dihydrate is 0.187 while the additional water, in a so-called over-stoichiometric quantity, is necessary for the processing of the hardening gypsum paste [10].
There are researches where PG is evaluated as potential binder with W/B ratio from 0.54 to 0.70 giving a compressive strength from 5 to 19 MPa after 7d hardening [11]. In other researches W/B ratio is defined in the interval of 0.6 to 0.8. [10]. The highest W/B ratios researched was from 0.9-1.3 which resulted with density from 880-1080 kg/m^3 of the obtained binder with compressive strength from 4.2-16 MPa [12].
Effective way to improve workability and reduce W/B is to use superplasticizers (SP). Superplasticizers are used from 0.2–1% from the weight of gypsum and the most common plasticizers used are naphthalene, melamine and polycarboxylate, citric acid which may act both as gypsum setting time retarders or accelerators [13]–[15]. To use SP efficiently, often an increase of pH of binder is needed. To increase alkaline media of fresh binder thus involves electrostatic effect of SP the amount of lime which is necessary to add is about 2 to 2.5% by gypsum mass [16].
This research summarizes the experimental results of PG based binder regarding to W/B ratio in mixture composition to properties of hardened gypsum specimens, giving a marginal borders of properties affected by W/B and is compared with commercial gypsum material.

IOP Conf. Series: Materials Science and Engineering **660** (2019) 012071 doi:10.1088/1757-899X/660/1/012071

2. Materials and methods

Phosphogypsum dihydrate (PG) from AB Lifosa (Lithuania) and commercial construction gypsum (Knauf Baugips - BG) were used for investigation. The chemical compositions of the gypsum were determined using a Thermo Scientific ARL PERFORM'X Wavelength Dispersive X-Ray Fluorescence Spectrometer (WDXRF) with a Rh-target x-ray tube and the UniQuant program. Prior to the measurements, samples were dried at 60°C and heated at 950 °C; a fused bead was then prepared with lithium tetraborate 50%/lithium metaborate 50%, with a 1:10 mixture of the PG and flux heated at 1085 °C. The chemical composition of PG is given in Table 1. BG is commercial gypsum used in this research, PG_raw is secondary raw material obtained from production plant and PG is binder obtained from PG at 180°C for 4h.

Table 1. Chemical composition of phosphogypsum (PG_raw), phosphogypsum binder (PG) and commercial gypsum (BG)

Component	BG	PG_raw	PG
LOI 950°C	22,43	19,24	6,89
Na_2O	0,31	0,48	0,47
MgO	3,92	0,21	0,16
Al_2O_3	1,68	0,71	0,96
SiO_2	3,73	1,07	0,94
P_2O_5	0,00	0,57	0,66
SO_3	30,90	37,38	43,21
CaO	35,64	37,16	43,25
TiO_2	0,05	0,11	0,08
Fe_2O_3	0,46	0,22	0,21
As_2O_3	0,07	0,09	0,13
SrO	0,23	2,25	2,52
CeO_2	0,01	0,24	0,24
TOTAL	99,42	99,74	99,73

The particle size distribution of gypsum binders is given in Figure 2. 100% of particles are below 0.3 mm which is in accordance with standard LVS EN 13279-1:2008. PG has finer nature comparing to BG. The characteristic value of d_{10} was 0.05 mm, d_{50} – 0.075 mm, d_{90} – 0.12 mm for PG and 0.08, 0.13 and 0.22 mm for BG respectively. The high fineness of gypsum could lead to increased W/B ratio and reduced setting time similar as for Portland cement reported previously [17].

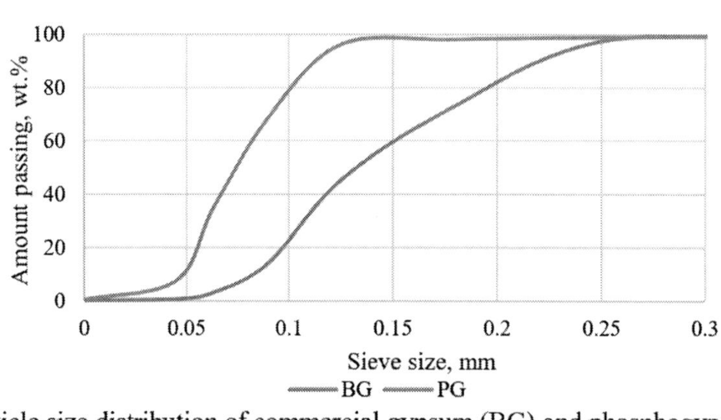

Figure 2. Particle size distribution of commercial gypsum (BG) and phosphogypsum after collision milling in disintegrator and heating at 180°C (PG).

The mineralogical composition of PG transformation through dehydration and hydration processes is given in Figure 3. The main mineral in dihydrate PG was synthetic calcium sulfate hydrate ($CaSO_4 \cdot 2H_2O$, ref. 33-0311). The dehydration at 180°C for 4h results in weight loss of 18 wt% of chemically bonded water and new phase – calcium sulfate hemi-hydrate ($2CaSO_4 \cdot H_2O$, ref. 02-0667) was detected. For hardened samples prepared from PG treated at 180°C main mineral indicated was calcium sulfate hydrate (ref 72-0596) indicating reverse hydration reaction to dehydration.

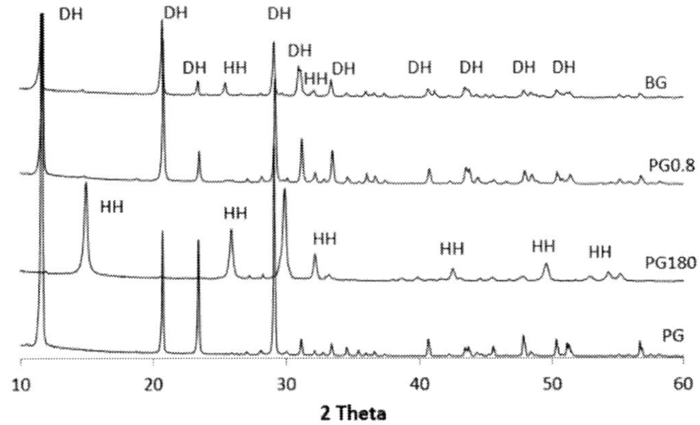

Figure 3. Mineralogical composition of phosphogypsum and commercial gypsum. DH – calcium sulfate hydrate (33-0311), PH - calcium sulfate hydrate (02-0667)

Results and discussion

The effect of W/B ratio on density of gypsum samples is given in figure 4. The results can be divided in three groups: density range for commercial gypsum (i), PG binder (ii) and porous gypsum based samples. It can be clearly seen that material based on PG has increased demand to W/B ratio so that material could be prepared and casted which is associated with increased fineness of PG and constituents which may attract extra water. However, there is a range of density, where PG properties could be compared with commercial gypsum. The W/B ratio from 0.43-0.60 could be used for both binders and the density of the material could be in range from 1100-1400 kg/m^3. It must be noted, that such a low W/B ratio of PG binder could be achieved only with the use of superplasticizers. For commercial binder the W/B ratio could go as low as 0.30 where maximum density from 1500-1600 kg/m^3 could be reached. The porous gypsum remains low W/B ratio (from 0.22-0.78) which is associated with the prepared foam volume and the binder distribution in the foams which act as surfactants. The density of porous gypsum obtained with foaming agents could be as low as 200 kg/m^3 while the upper limit is slightly below the margin of density for PG binder with high W/B ratio (1.00-1.15) and could reach up to 750 kg/m^3.

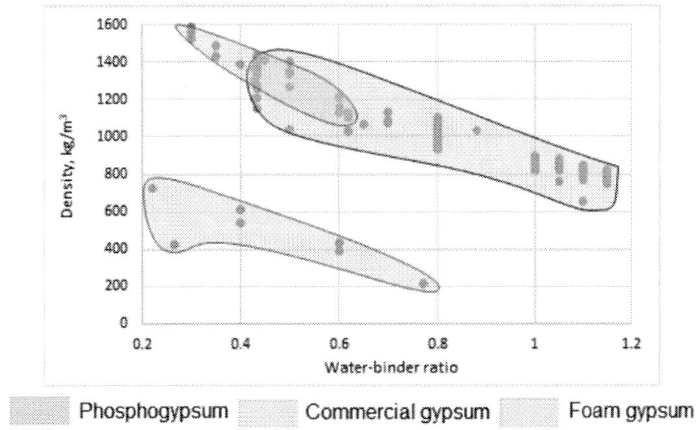

Figure 4. The relationship between water-binder ratio and density of gypsum based material.

The total porosity is strongly related to the W/B ratio as extra free-water, which is necessary to ensure workability of the prepared binder evaporates from the structure, creating air voids as open pores, which reduce the density and strength of the gypsum binder, similar as for Portland-cement based materials [18]. The total porosity obtained for commercial gypsum binder was around 38-40 vol.% with the W/B ratio from 0.3 while the same total porosity for PG binder could be reached with W/B ratio of 0.43 (figure 5). The total porosity for porous PG samples could reach up to 90 % giving a promising result to obtain ultra-lightweight gypsum material.

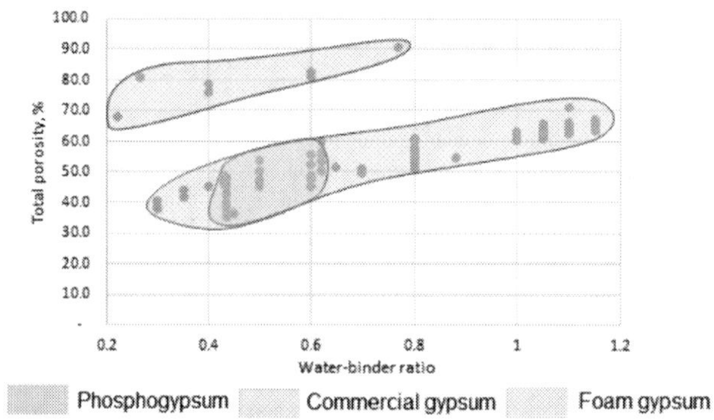

Figure 5. The relationship between water-binder ratio and total porosity of gypsum based material.

The W/B ratio has strong impact on compressive strength of gypsum material (Figure 6). The strength with high W/B ratio could be from 1-5 MPa while the decrease of W/B ratio gradually increase the strength up to 29 MPa. Comparing the compressive strength properties of PG and commercial binder it could be seen that there is a range where the W/B and compressive strength relationship is overlapping giving a promising potential for PG as substitution of natural gypsum. The possibility to reduce W/B ratio of commercial gypsum increased material strength up to 24 MPa, while the fine nature of PG gives higher compressive strength results even at higher W/B ratio. It must be also noted that even with similar W/B ratio the wide range of compressive strength results might be obtained indicating the importance of preparation of gypsum binder and selection of its mixture composition.

The porous gypsum materials have lower compressive strength comparing to its dense counterparts. The dependence of W/B ratio on porous material compressive strength has minor effect as the increase of W/B from 0.2 to 0.8 did not change compressive strength of porous material (0.1-0.8MPa).

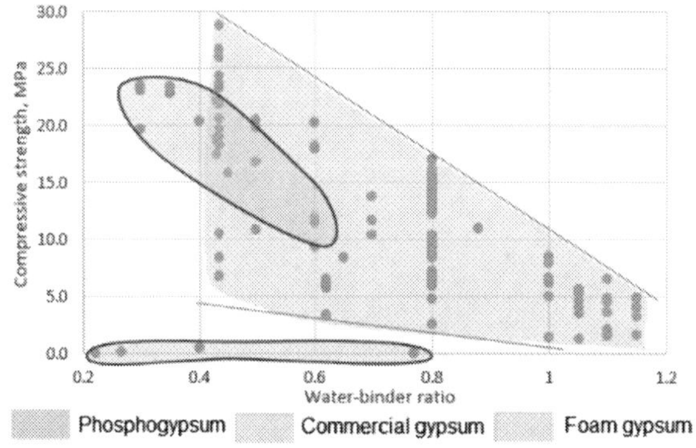

Figure 6. The relationship between water-binder ratio and compressive strength of gypsum based material.

Good correlation between compressive strength and density of gypsum material was obtained (Figure 7). Also there is some deviation of the results which might be affected by the treatment conditions of PG and the physicochemical properties of PG after its production, the overall tendency is clear. Again, the overlapping region between PG and commercial gypsum can be observed. With gypsum based binder the density in range from 200-1600 kg/m^3 could be covered and compressive strength from 0.1-29 MPa obtained.

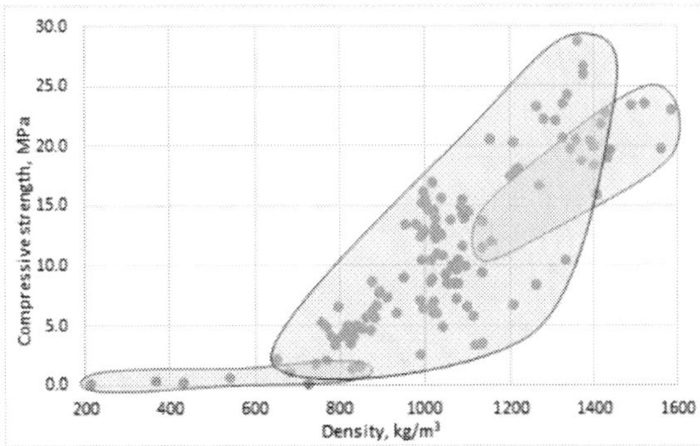

Figure 7. The relationship between density and compressive strength of gypsum based material.

The results obtained in this research were compared with other literature studies addressing the study of natural or synthetic gypsum regarding to its properties depending from W/B ratio and results are show in Table 2. The performance of the PG gypsum in this research complies with data reported in literature (see table 2). However, the scattering of results is possible as different gypsum treatment conditions and mixture compositions could be prepared.

Table 2. Properties of waste gypsum and natural gypsum reported in literature.

Gypsum type	W/B ratio	Bulk density, kg/m^3	Total porosity, vol.%	Compressive strength, MPa	Ref
Gypsum plaster	1.3	880	n/d	4.0	[12]
	0.9	1080	n/d	14.5	
Phosphogypsum 1	1.15	800	64.5	4.0	This work
natural gypsum and gypsum from a chemistry industry, ratio 1:1	0.71	1130	46*	9.4	[10]
Commercial gypsum 1	0.6	1135	55.5	11.5	This work
Phosphogypsum	0.6	1200	52*	12	[6]
Commercial hemi-hydrate gypsum	0.45	1400-1920	n/d	5-15	[19]
Phosphogypsum 2	0.43	1400	46	26	This work
Commercial gypsum 2	0.3	1580	38.0	23.0	This work
Foamed β-gypsum	0.70	658	n/d	0.41	[20]
Foam gypsum 1	0.77	213	91	0.1	This work
Foam gypsum 2	0.27	430	81	0.19	This work

*calculated from bulk density and density of matrix

3. Conclusions

Results indicate that commercial gypsum based binder could be effectively altered with waste-stream gypsum based binder – phosphogypsum. The phosphogypsum has finer nature comparing to natural gypsum binder and also harmful constituents might be present which brings negative effect on demand of water. The increase of water necessary to prepare fresh gypsum paste reduce the strength of the binder as the total porosity increased. The other effect of water increase in paste gives reduction of material density. The water demand of phosphogypsum binder could be reduced by the use of superplasticizers reducing the W/B ratio to 0.43 and strength increase to 29 MPa. By the use of foaming agent, the density of gypsum material decreased from 700 to 200 kg/m^3 with a reasonable compressive strength (0.1-0.8 MPa). The scattering of results is possible with similar W/B ratios which is associated with the different gypsum treatment conditions and selection of mixture composition showing the importance of selection proper materials and approaches for production of gypsum based materials.

Acknowledgement
This work has been supported by the European Regional Development Fund within the Activity 1.1.1.2 "Post-doctoral Research Aid" of the Specific Aid Objective 1.1.1 "To increase the research and innovative capacity of scientific institutions of Latvia and the ability to attract external financing, investing in human resources and infrastructure" of the Operational Programme "Development of sustainable and effective lightweight building materials based on secondary resources" (No. 1.1.1.2/VIAA/1/16/050).

References

[1] M. Golež, Ž. Pogačnik, and A. Mladenović, "Laboratory-prepared lime-gypsum mixtures based on the know-how of traditional technology," *J. Cult. Herit.*, vol. 32, pp. 38–43, Jul. 2018.

[2] A.-M. Dragomir, R. Lisnic, T. Prisecaru, M. M. Prisecaru, C. A. Vîjan, and D. C. Nastac, "Study on synthetic gypsum obtained from wet flue gas desulphurisation in thermal power plants," *Rev. Rom. Mater. Rom. J. Mater.*, vol. 47, no. 4, 2017.

[3] H. Tayibi, M. Choura, F. A. López, F. J. Alguacil, and A. López-Delgado, "Environmental impact and management of phosphogypsum," *J. Environ. Manage.*, vol. 90, no. 8, pp. 2377–2386, Jun. 2009.

[4] S. Belboom, C. Sz??cs, and A. L??onard, "Environmental impacts of phosphoric acid production using di-hemihydrate process: A Belgian case study," *J. Clean. Prod.*, vol. 108, pp. 978–986, 2015.

[5] G. Bumanis, J. Zorica, D. Bajare, and A. Korjakins, "Technological properties of phosphogypsum binder obtained from fertilizer production waste," *Energy Procedia*, vol. 147, pp. 301–308, Aug. 2018.

[6] D. Nizevičienė, D. Vaičiukynienė, B. Michalik, M. Bonczyk, V. Vaitkevičius, and V. Jusas, "The treatment of phosphogypsum with zeolite to use it in binding material," *Constr. Build. Mater.*, vol. 180, pp. 134–142, Aug. 2018.

[7] A. Kaziliunas, V. Leskeviciene, B. Vektaris, and Z. Valancius, "The study of neutralization of the dihydrate phosphogypsum impurities," *Ceram. - Silikaty*, vol. 50, no. 3, pp. 178–184, 2006.

[8] M. Singh, "Effect of phosphatic and fluoride impurities of phosphogypsum on the properties of selenite plaster," *Cem. Concr. Res.*, vol. 33, no. 9, pp. 1363–1369, Sep. 2003.

[9] D. Nizevičienė, D. Vaičiukynienė, A. Kielė, and V. Vaičiukynas, "Mechanical Activation on Phosphogypsum: Hydrosodalite System," *Waste and Biomass Valorization*, pp. 1–7, May 2018.

[10] P. Padevět, P. Tesárek, and T. Plachý, "Evolution of mechanical properties of gypsum in time," *Int. J. Mech.*, vol. 5, no. 1, pp. 1–9, 2011.

[11] D. Nizevičienė *et al.*, "Effects of waste fluid catalytic cracking on the properties of semi-hydrate phosphogypsum," *J. Clean. Prod.*, vol. 137, pp. 150–156, Nov. 2016.

[12] M. J. Madu, Ndaliman, and B. Oche, "Evaluation of Water-to-Plaster Ratio on Products of Nafada Plaster of Paris (POP)," *Int. J. Emerg. Eng. Res. Technol. Int. J. Emerg. Eng. Res.*, vol. 4, no. 10, pp. 23–27, 2016.

[13] N. Sakthieswaran and M. Sophia, "Effect of superplasticizers on the properties of latex modified gypsum plaster," *Constr. Build. Mater.*, vol. 179, pp. 675–691, Aug. 2018.

[14] G. Camarini, M. C. C. Pinto, A. G. de Moura, and N. R. Manzo, "Effect of citric acid on properties of recycled gypsum plaster to building components," *Constr. Build. Mater.*, vol. 124,

pp. 383–390, Oct. 2016.
[15] D. Dolák and K. Dvořák, "Impact of Plasticizers on Technological Properties of Gypsum," *Mater. Sci. Forum*, vol. 865, pp. 130–134, Aug. 2016.
[16] L. Dvorkin, N. Lushnikova, M. Sonebi, J. Khatib, E. Design, and D. K. Building, "EcoGRAFI PROPERTIES OF MODIFIED PHOSPHOGYPSUM BINDER," pp. 0–6, 2017.
[17] A. Naceri and M. Benia, "THE EFFECT OF FINENESS OF CEMENTS AT MINERAL ADDITIONS ON THE MECHANICAL RESPONSE OF CONCRETE," 2006.
[18] Y.-Y. Kim, K.-M. Lee, J.-W. Bang, and S.-J. Kwon, "Effect of W/C Ratio on Durability and Porosity in Cement Mortar with Constant Cement Amount," *Adv. Mater. Sci. Eng.*, vol. 2014, pp. 1–11, Apr. 2014.
[19] M. Arikan and K. Sobolev, "The optimization of a gypsum-based composite material," *Cem. Concr. Res.*, vol. 32, no. 11, pp. 1725–1728, 2002.
[20] A. Çolak, "Density and strength characteristics of foamed gypsum," *Cem. Concr. Compos.*, vol. 22, no. 3, pp. 193–200, 2000.

Effect of modified liquid glass on absorption humidity and thermal conductivity of flax fiber slabs

S Romanovskiy[1] and A Bakatovich[2]

[1]Postgraduate of Department of Construction Industry, Polotsk State University, Novopolotsk, Belarus
[2]Phd, Associate professor of Department of Construction Industry, Polotsk State University, Novopolotsk, Belarus

E-mail: s.romanovskiy@psu.by

Abstract. Thermal insulation materials based on vegetable fibers are considered. The results of studies on the microstructure of flax fiber noils using a light microscope and flax fibers using electron microscopy are presented. The complex of studies on the selection of the composition of thermal insulation slabs from flax fibers and flax noils has been conducted to determine the process of absorption of humidity from the air. The humidity absorption of insulants based on flax fibers and flax noils has been determined at a relative humidity of 40–97% using sodium liquid glass and modified liquid glass, as well as «Akotherm flaks» slabs. The thermal insulation slabs water vapor absorption kinetics has also been investigated, taking into account the time parameter. The effect of absorption humidity on the thermal conductivity of the studied materials has been examined; for this purpose, the coefficients of thermal conductivity of materials with the maximum humidity content of insulators were determined. The results of the tests indicate the highest efficiency of thermal insulation slabs made of flax fiber tow compared with insulation materials based on flax fibers and «Akotherm flaks» slabs.

1. Introduction

The development of ecological "green" building in recent years has significantly changed the perception of many developers. Now they often impose requirements on the environmental safety of facilities along with traditional quality indicators for buildings being erected. It should be noted that the construction industry, both in Belarus and in other countries, is not always ready to meet the environmental requirements of customers. First of all, this problem is associated with the limited availability and release of environmentally friendly building materials. The greatest problems of the manufacturing sector are connected with the production of thermal insulation materials since traditional insulants (foam polystyrene and slag wool) do not meet the requirements for ecology. One of the best solutions to the current situation is to use plant materials of natural or agricultural origin in combination with environmentally friendly binders for the production of thermal insulation materials. Several researchers note that an increase in the use of plant waste will not only provide construction with additional raw materials and expand the range of local building materials, but also contribute to the conservation and rational use of natural resources [1–3].

In many countries, technologies for utilization of crop waste for wall and thermal insulation materials are being implemented, which indicates considerable interest and the need for further development of this construction industry segment.

The manufacture of thermal insulation slabs made from wood fibers is one of the most developed productions of insulation materials. Among the wood fiber insulation materials on the market, mainly products of foreign companies dominate - STEICO (Germany - Poland), GUTEX (Germany) and Skano Group (Estonia) [4]. Recently, manufacturers from Russia have intensively mastered the production technology of such materials. The following technical indicators characterize insulators: density 150–250 kg/m^3, thermal conductivity 0.06 W/(m·K), flexural strength 0.5–1.2 MPa [5].

The use of oil palm bark fibers in the form of aggregate for insulating slabs is also known [6]. The binder is sodium liquid glass. The insulant has a density of 200 kg/m^3 and provides a thermal conductivity of 0.052 W/(m·K), a flexural strength of 0.75 MPa and compressive strength at the 10 % deformation of 0.21 MPa.

It is also suggested to use the technical hemp shives to obtain thermal insulating slabs [7]. Hydraulic lime is insulation binder. At a density of 312–337 kg/m^3, the compressive strength of the material reaches 0.179–0.222 MPa, and the coefficient of thermal conductivity is 0.101–0.087 W/(m·K). The material can be used in low-rise construction.

Studies of dry molding thermal insulating materials in the Kostroma State Technological University, consisting of flax shive particles with the addition of up to 8% resin and 1% paraffin, have been carried out [8]. In the process of research, it was determined that at a density of 40 kg/m^3 the insulation has a thermal conductivity of 0.101 W/(m·K). Slabs of similar composition are made in Thailand, the United Kingdom, and Canada.

Ecological slabs "Thermo-Hanf" from hemp fibers are produced in Germany. Thermal insulation material consists of hemp fibers by 83–87% and polyester by 10–12% [9, 10]. As fire-retarding material, soda is added in an amount of 3–5% of the total mass. Thermo-Hanf insulation is used for warming roofs, walls, floors, inter-floor and attic floors. The insulation thermal conductivity is 0.038–0.04 W/(m·K) at a density of 35–40 kg/m^3. Due to the low density, the insulating material is not rigid, which somewhat limits the area of application, and the high price prevents the popularization of insulation.

An effective insulating material in the form of semi-rigid slabs based on waste cotton fibers and liquid sodium glass is an innovative invention of Turkmenistan and other countries of Central Asia [11]. The thermal conductivity of the insulation is 0.037–0.039 W/(m·K), at a density of 40–90 kg/m^3.

Thermal insulating slabs based on peat moss and liquid glass have also been developed. Their thermal conductivity coefficient is 0.034–0.04 W/(m·K), and the density equals to 155–170 kg/m^3. However, insulation slab based on moss has a drawback - significant shrinking deformation of the material during the drying process [12]. To eliminate this problem, it was suggested to introduce milled straw in the amount of 20–30% by weight of moss.

Flax fiber waste is of considerable interest in the production of thermal insulation materials in the Republic of Belarus. More than 20% of flax crops of the European continent are concentrated on the territory of Belarus. In 2017, the yield of flax fiber was 9.2 c/ha, the highest yield was observed in 2014 and amounted to 10.7 c/ha [13].

Harvesting and further processing of flax to produce fibers is a rather laborious process. When harvesting flax with combine harvesters mechanized spreading of straw on flax is produced. The separation of the fibrous part of the stem from wood is promoted by dew, rain, and heat, which destroy connective tissues. Next, flax raw material goes to flax factories for the primary processing of flax [14]. When processing flax straw, more than one-third of the total fiber goes to flutter waste. After the usual shaking out from the tow, part of the waste is used as oakum. Most of the waste is carefully treated on a special line of machines for the production of short fibers. These machines consist of a special dryer, shaker, and tow preparing machine. Here, flax waste is finally cleared of tow. To improve the quality of the fiber, clean, short, and long fiber after assessment and garters is transported to the warehouse for retrieval. Then the fiber passes the control sorting and is sent to flax industrial plants or factories [15]. As a result of flax dressing, from 45 to 55% of dressed flax, 40-50% of noils and about 5% of tow are obtained.

The well-known development of insulation based on flax fibers includes thermal insulation slabs «Ecoteplin» produced in Russia. The binder is starch. Borates are used as fire- and bio resistance agents [16, 17]. The slabs are used both in low-rise construction and for thermal and acoustic insulation of apartment buildings. The material «Ecoteplin» has antiseptic properties and is completely safe for human health. The density of thermal insulation slabs is from 32 to 34 kg/m^3, they ensure thermal conductivity of 0.037 W/(m·K), water absorption capacity by volume – 0.5 and water vapor permeability – 0.4 mg/(m·h·Pa). The insulation is a low combustible material. The widespread use of «Ecoteplin» is constrained by the high cost and limited use due to its low density. Also, the insulation material is highly water permeable and gets wet quickly, so the insulated walls should be additionally insulated with a special membrane.

The company JSC «Akotherm flaks» (Orekhovsk, Belarus) offers thermal insulation slabs containing 85% of flax fibers. Synthetic fibers are used as a binder (15%) [18]. The material is used in low-rise construction for thermal insulation of walls and floor slabs. At the density of 32 kg/m^3, thermal conductivity rate of «Akotherm flaks» is 0.038–0.04 W/(m·K), acoustic absorption coefficient – 0.84, vapor permeability – 0.4 mg/(m·h·Pa), and is classified as highly combustible materials. The high fire risk is the main disadvantage of the insulator. The material is characterized by low density, and that does not allow producing rigid compression-resistant insulating slabs, which is another factor limiting the application of «Akotherm flaks».

Currently, in the laboratories of the Department of Construction Industry in Polotsk State University, comprehensive studies on the development of flax noils thermal insulation materials are being carried out. As alternatives to flax noils thermal insulation slabs, samples with flax fiber filler and «Akotherm flaks» insulator are being studied. Using flax noils for making slabs solves the problem of recycling flax plant waste and expands the range of efficient thermal insulation materials produced from flax fibers.

2. Materials and Methods

Studies were performed on samples with fillers of flax noils or fibers produced on Postavy Flax Factory (Belarus).

Analysis of the microstructure of flax noils and flax fibers was carried using optical and electron microscopy performed on microscopes Altami MET 5C and JSM-5610 LV.

In the manufacture of thermal insulation slabs, sodium liquid glass produced on JSC Domanovsky Production and Trade Plant was used as a binder. This sodium liquid glass met the requirements of GOST (ГОСТ) 13078. The second part of the research was carried out on a modified liquid glass. In the form of an additive, hydrated lime grade II without additives produced by JSC Zabudova was used, as well as gypsum made at JSC Belgips and meeting the requirements of STB EN 459-1 and GOST (ГОСТ) 125-79.

Flax noils and flax fibers thermal insulation samples were prepared in the form of slabs with dimensions of 250×250×50 mm. When preparing the samples, a certain sequence of technological operations were observed. A pre-dosage of the components was made. To obtain a modified binding agent, lime was first added to the liquid glass and mixed to a smooth consistency, only after that gypsum was added. Sodium liquid glass or modified liquid glass with fibrous filler was mixed placed in the mold. The samples of thermal insulation materials were kept in the mold for 6 hours at a temperature of 20 ± 2 °C. The samples were then removed from the mold and dried for 4 hours in a drying chamber at a temperature of 45–55 ° C. The samples based on flax and polyester fibers mixture were cut from «Akotherm flaks» thermal insulation slabs. Each sample of thermal insulation materials was placed in an individual pressurized chamber with pre-poured water 40 mm below the slab installation level. The air temperature in the chamber corresponded to 20 ± 2 °C, and the relative air humidity was 97%. The slabs were kept in a chamber above water until the equilibrium humidity was reached, after that the slabs were removed and the thermal conductivity coefficient was determined. Thermal conductivity coefficient was determined according to STB (СТБ) 1618 and using a thermal conductivity meter ИТП-МГ4.

The absorption humidity of thermal insulation slabs from flax noils has been determined according to GOST (ГОСТ) 24816. Pre-dried to constant weight samples with a filler made of fibers or flaxes, selected from thermal insulation materials, were placed in desiccators. The vapor-air environment in the desiccators was created artificially with the help of sulfuric acid chemical solution of different concentrations, providing a relative humidity of 40–97%. The humidity content of the samples was determined by weighing them after a fixed time before the start of the desorption process. The maximum humidity was taken as an indicator preceding the onset of desorption. When determining the absorption humidity of samples, the air temperature in the desiccators and the room corresponded to +20 ± 0.5°C.

3. Results and Discussion

The image of a flax noil (Figure 1 a) was obtained with light microscopy combining the pictures of the consequently placed fragments of the analyzed sample. The fragment highlighted by the frame in figure 1 a is presented in the enlarged form in figure 1 b. Application of a scanning electron microscope allowed visually attesting that the flax fiber consists of bundles of elementary fibers (Figure 2 a). In the image of the fiber, the frame highlights the fragment that is enlarged in figure 2 b.

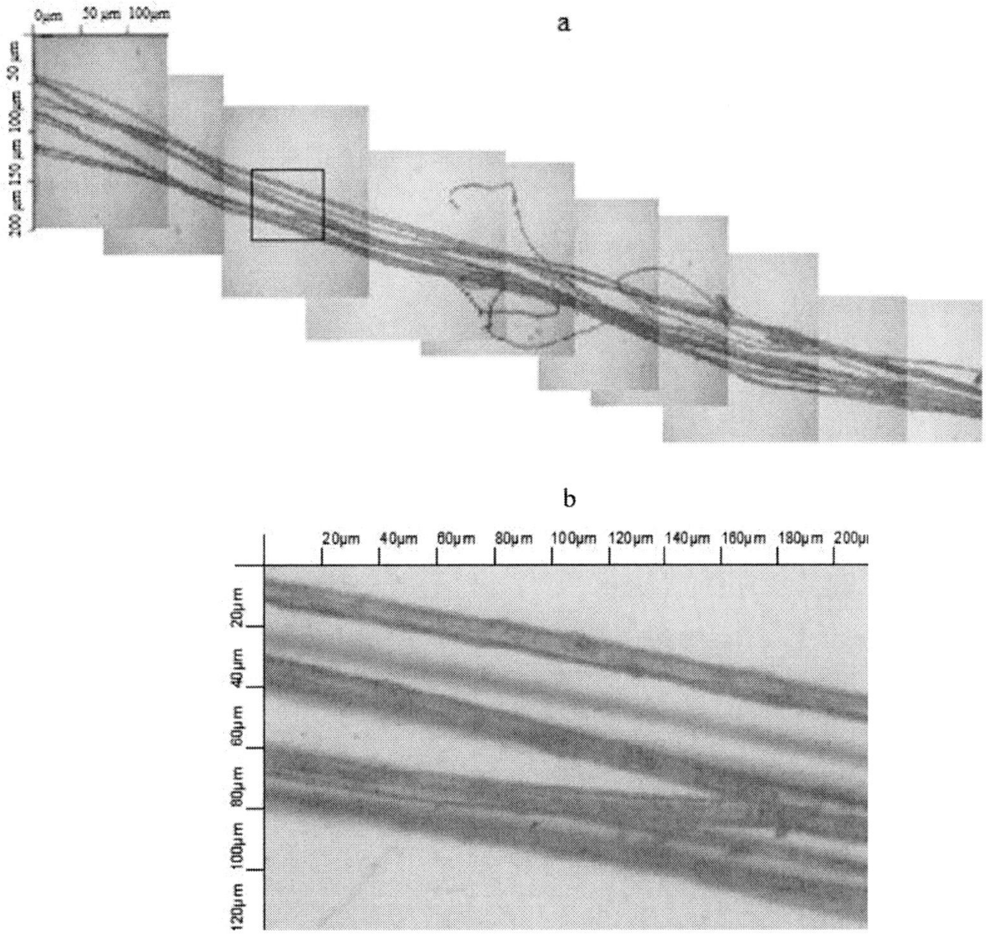

a – outer appearance of the noil, b – enlarged fragment.

Figure 1. Light microscopy of a flax noil.

IOP Conf. Series: Materials Science and Engineering **660** (2019) 012072 doi:10.1088/1757-899X/660/1/012072

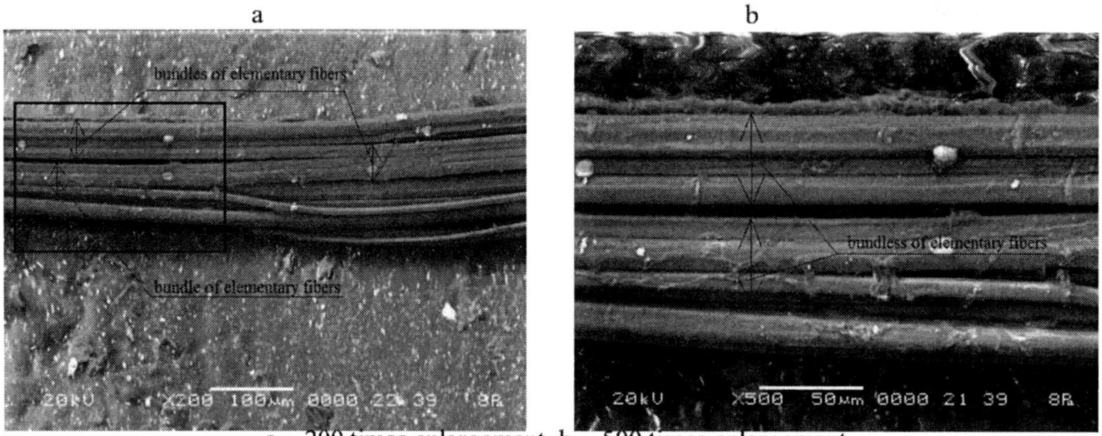

a – 200 times enlargement, b – 500 times enlargement

Figure 2. Electron microscopy of a flax fiber.

It was established that the flax webs are disheveled bundles of elementary fibers having randomly arranged contact connections among themselves. The length of the elementary fibers varies from 10 to 40 mm, with a diameter of 8–12 microns. Flax fiber is bundled with elemental fibers. The diameter of bundles is 50–70 microns, with a content of 10 to 20 elementary fibers in the structure. Obtained data are supported by Kil'chevskij A. V and Hotyleva L. [19].

Initially, to determine the quantitative composition of experimental samples of thermal insulation materials, the average density was varied at a constant flow rate of liquid glass – 10 kg. The quantitative composition and physic-mechanical characteristics of the obtained samples in dry condition are given in Table 1.

Table 1. Quantitative composition and physic-mechanical characteristics of materials.

No. of composition	Component consumption per 1 m^3, kg			Density, kg/m^3	Thermal conductivity, W/(m·K)	Compressive strength at 10% deformation, $\times 10^{-2}$ MPa
	flax fibers	flax noils	liquid glass			
1	45	-	10	55	0.047	0.2
2	60	-	10	70	0.042	0.3
3	75	-	10	85	0.043	0.36
4	90	-	10	100	0.045	0.42
5	-	45	10	55	0.04	0.13
6	-	60	10	70	0.035	0.21
7	-	75	10	85	0.036	0.25
8	-	90	10	100	0.038	0.3

Increasing the amount of aggregate to increase the insulation density to 70 kg/m^3 leads to a decrease in thermal conductivity, regardless of the fibers type (Table 1). A further increase in the density of the samples (compositions 3, 4, 7, and 8) causes an increase in the parameter under study. For example, at a density of 55 kg/m^3 for flax noils-based thermal insulation, thermal conductivity is equaled 0.04 W/(m·K). A decrease in thermal conductivity coefficient to 0.035 W/(m·K) occurs with a 1.3 times increase in density to 70 kg/m^3. With an increase in the average density up to 100 kg/m^3, an increase in thermal conductivity up to 0.038 W/(m·K) is noted. For flax fiber slabs, a similar relationship is established, since the minimum thermal conductivity coefficient value of 0.042 W/(m·K) corresponds to a density of 70 kg/m^3. A decrease in density to 55 kg/m^3 and an increase to

100 kg/m^3 leads to an increase in the thermal conductivity coefficient by 0.005 W/(m·K) and 0.003 W/(m·K), respectively.

Obtained results (Table 1) show that an increase in the average density of thermal insulation slabs at a constant binder consumption leads to an increase in compressive strength at 10% deformation. At a density of 55 kg/m^3 of flax noils materials, the compressive strength is $0.13 \cdot 10^{-2}$ MPa. An increase in the average density by 1.8 times leads to an increase in strength by 2.3 times to $0.3 \cdot 10^{-2}$ MPa. Insulation samples based on flax fibers with a maximum density of 100 kg/m^3 reach the strength of $0.42 \cdot 10^{-2}$ MPa. The decrease in the average density to 55 kg/m^3 causes a decrease in the compressive strength index at 10% deformation by 2.1 times. It should be noted that in the density range under study, the compressive strength of flax fiber-based thermal insulation materials is on average 45% higher than insulants from flax fiber noils.

Due to the low compressive strength of plant raw materials with fibrous structure, the data obtained when choosing the composition of samples of thermal insulation slabs were not taken into account. Thus, sample compositions with an average density of 70 kg/m^3 with the best thermal conductivity were used to study the absorption humidity of insulants. When the density of flax fiber-based insulation is 70 kg/m^3, the coefficient of thermal conductivity is 0.042 W/(m·K). When replacing flax fibers with noils, there is a decrease in thermal conductivity by 20% to a value equal to 0.035W/(m·K). Also, «Akotherm flaks» slabs samples with a thermal conductivity of 0.04 W/(m·K) and a density of 34 kg/m^3 were used in the studies. The applied technology of forming and structuring insulation does not allow producing material of higher density.

The absorption humidity of thermal insulation slabs was determined on samples having a quantitative composition, presented in Table 2.

Table 2. Quantitative composition and density of samples in the dry state.

No. of composition	Component consumption per 1 m^3, kg				Density, kg/m^3
	flax fibers	flax noils	polyester fiber	liquid glass	
1	29	-	5	-	34
2	60	-	-	10	70
3	-	60	-	10	70

According to the experimental data, the dependences of the materials absorption humidity on the relative humidity of the air in the form of absorption isotherms were constructed (Figure 3).

The analysis of the dependences obtained reveals that the absorption humidity of the «Akotherm flaks» material at an air humidity of 40% is lower than the flax fibers or flax noils slabs by 49% and 45%, respectively. Values of absorption humidity of flax fibers or flax noils samples are 54% more and in comparison with the insulation «Akotherm flaks» equal to 11.3% with a relative humidity of 60%. Maintaining the air humidity at 80% in a desiccator contributes to an increase in the «Akotherm flaks» material absorption up to 15.6%, which is 30% and 28% less than that of flax fibers and flax noils slabs. The relative air humidity of 90% causes an increase in the absorption humidity of flax and polyester fiber samples up to 22.9%, which is 29% lower than the value of the flax fibers or flax noils water vapor absorption. Regarding the indicators in the dry state, the absorption humidity of compositions 1, 2 and 3 samples at 97% air humidity is 31%, 40%, and 47%, respectively. At the same time, the absorption humidity content of the composition 1 slabs is less than that of the material of composition 2 by 24%, and samples of composition 3 by 35%.

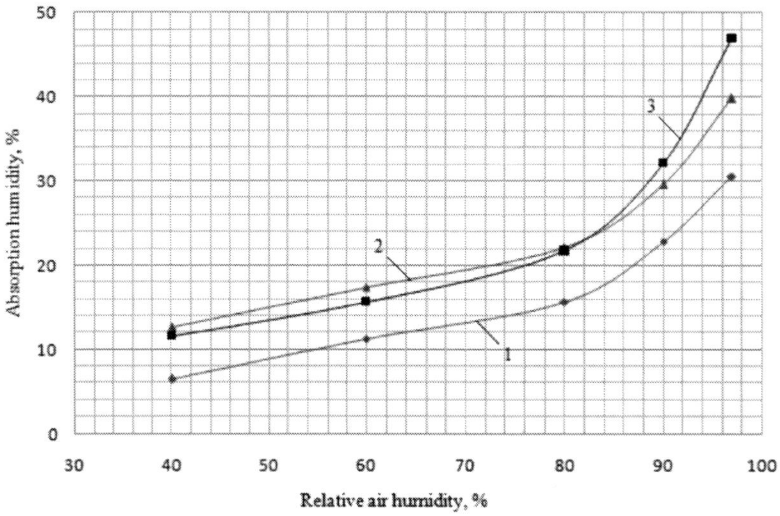

1 – Akotherm flaks slabs; 2 – flax fiber slabs; 3 – flax noils slabs

Figure 3. Water vapor absorption isotherms.

One more important characteristic for thermal insulation materials is water vapor absorption kinetics at a certain value of relative air humidity. As an example, figures 4, 5 show the water vapor absorption kinetics of samples taken from experimental insulation slabs at a relative humidity of 60% and 97%.

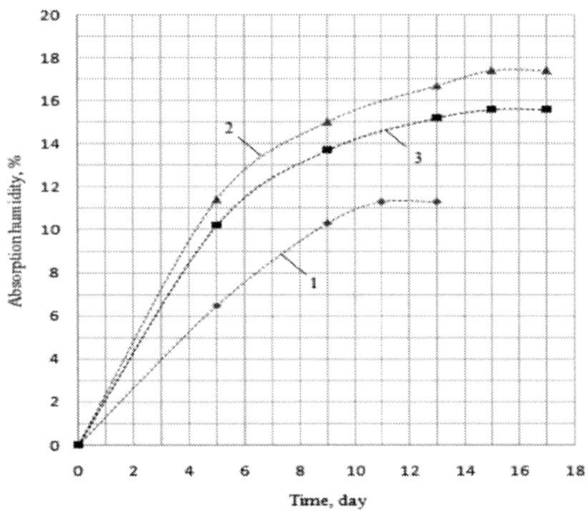

1 – Akotherm flaks samples;
2 – flax fiber samples;
3 – flax noils samples

Figure 4. Water vapor absorption kinetics at a relative air humidity of 60%.

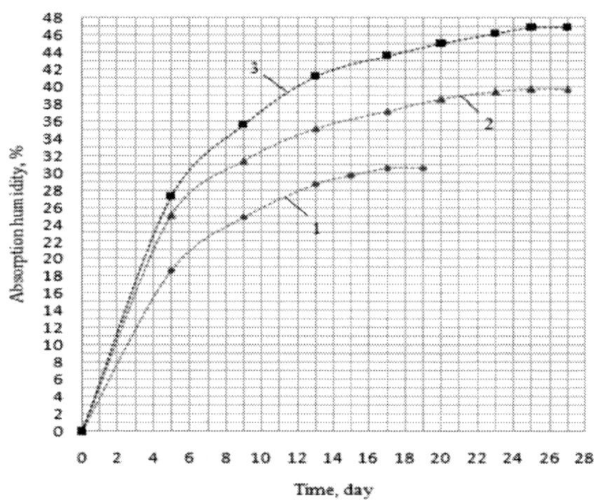

1 – Akotherm flaks samples;
2 – flax fiber samples;
3 – flax noils samples

Figure 5. Water vapor absorption kinetics at a relative air humidity of 97%.

When air humidity is 40% for 9 days, the «Akotherm flaks» slabs absorption humidity reaches a value of 6.4%, which is 41% and 37% less compared to flax fibers and flax noils-based samples over the same period. Materials (compositions 2 and 3) acquire the maximum absorption humidity on the 13th day, and it amounts to 12.6% and 11.6%, respectively.

Comparing the dependences in Figure 4 with a relative humidity of 60%, it is clear that the process of samples absorption (composition 1) ends on day 11 and is equal to 11.3%. This figure is 35% and 28% lower than that of compositions 2 and 3 slabs, respectively, with15 days water vapor absorption.

With air humidity of 80%, the absorption humidity of flax fibers and noils materials is almost the same. The water vapor absorption of samples (compositions 2 and 3) is intensively growing for the first 9 days and amounts to 20% and 19%, respectively, which is 49% and 43% higher than that of the composition 1slabs for the same period. «Akotherm flaks» material water vapor absorption ends on day 13 and is equal to 15.6%. The duration of flax fibers and flax noils samples absorption is 18 days, which is 5 days longer than that of the «Akotherm flaks» slabs. At the same time, the absorption humidity content of flax fibers and flax noils materials is 37% and 39% higher than that of the samples based on the mixture of flax and polyester fibers for the maximum absorption period.

The period of absorption humidity intensive growth (13 days) is typical of samples (compositions 1–3), kept at a relative humidity of 90%. It should be noted that during this period, the «Akotherm flaks» slabs absorption is 23%, which is 21% and 23% lower than that of flax fibers and flax noils samples. It is established that the maximum absorption humidity of flax fibers and noils-based materials for 22 days is 29% and 41% more than that of the «Akotherm flaks» slabs over 15 days.

Accelerated increase of water vapor absorption is characteristic of all samples at an air humidity of 97% for the first 13 days. At the same time, the indicator of flax fibers and flax noils absorption humidity content is 23% and 44% higher than that of the «Akotherm flaks» slabs. The process of water vapor absorption by flax and polyester fibers mixture samples ends on the 17th day, which is 8 days less than composition 2 and 3 materials. The values of «Akotherm flaks» samples absorption humidity are 22% and 44% lower than those of flax fibers and noils-based slabs. It should also be noted that at the end of water vapor absorption, the absorption humidity of flax fibers filler materials is 15% less compared to samples from flax noils. After absorption process completion, in all cases stabilization of water vapor absorption by the material is observed.

The further absorption process and the high absorption humidity of the material (composition 3) relative to the samples indicators (compositions 1 and 2) is due to the significant absorption humidity of flax noils. Separated elementary fibers in noils have a large geometrical contact area of the surface with the external environment in comparison with the indicator of flax fibers consisting of dense bundles of elementary fibers, this explains the results [20].

The lowest thermal conductivity of 0.08 W/(m·K) after a set of maximum absorption humidity is characteristic of the slab from flax noils. The thermal conductivity of the «Akotherm flaks» samples is almost the same, while the materials based on flax fibers exceed the indicators of the flax noils (composition 3) by 0.004 W/(m·K). The obtained thermal conductivity minimum coefficient of samples from flax fiber noils is achieved due to the multidirectional arrangement of elementary fibers in the insulation structure. This arrangement prevents the convective transfer of air by reducing the size of thin air spaces of irregular shape and their partial localization in the form of individual closed microcavities. Thus, in spite of the higher humidity content of flax noils slabs, the resulting structure of elementary fibers provides thermal conductivity identical with slabs based on flax and polyester fibers mixture.

Additionally, studies have been conducted to determine the absorption humidity of sodium liquid glass and modified liquid glass in a dried state in a climatic chamber above water at a relative air humidity of 97%. Studies have shown that the rate of absorption humidity of the binder without additives reaches its maximum value of 19% after 9 days. The maximum humidity content of the modified liquid glass is 10% on day 7 at a temperature of +20 °C. Thus, it can be assumed that adding lime and gypsum to the binder will slow down the material's absorption of humidity from the air, as well as reduce the rate of water vapor absorption of fibers or noils-based samples due to a lower absorption humidity content of liquid glass with additives.

At the next stage, studies have been conducted to determine the absorption humidity of flax fibers or noils slabs with the use of a modified liquid glass. The quantitative composition and average density of the samples are presented in Table 3.

IOP Conf. Series: Materials Science and Engineering **660** (2019) 012072 doi:10.1088/1757-899X/660/1/012072

Table 3. Quantitative composition and density of samples in the dry state.

No. of composition	Component consumption per 1 m^3, kg					Density, kg/m^3
	flax fibers	flax noils	liquid glass	lime	gypsum	
1	60	-	9	0,5	0,5	70
2	-	60	9	0,5	0,5	70

According to test results, isotherms of water vapor absorption by experimental samples have been constructed (Figure 6). The experimental data obtained indicate that at a relative air humidity from 40% to 90%, the values of absorption humidity of flax fibers and noils samples at each air humidity indicator almost coincide. It should be noted that the addition of lime and gypsum to the binder at the indicated relative air humidity reduces the absorption rate of materials based on flax noils or fiber up to 5.4% relative to samples without additives. The relative air humidity of 97% contributes to an increase in the absorption humidity of thermal insulation materials with flax noils and modified liquid glass filler up to 37.6%, which is 13% higher than the value of water vapor sorption of flax fiber-based slabs. The use of lime and gypsum additives reduces the rate of absorption humidity of samples from flax fibers or flax noils by 18% and 20% compared with liquid glass-based slabs.

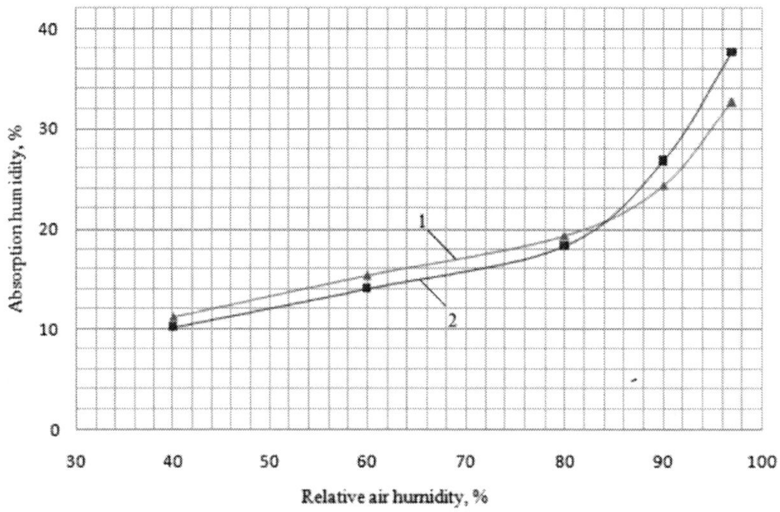

1 – flax fiber samples; 2 – flax noils samples

Figure 6. Water vapor absorption isotherms.

The water vapor absorption kinetics of the samples taken from experimental thermal insulation slabs at a relative air humidity of 80% and 97%is shown in Figures 7 and 8.

With relative air humidity of 40%, accelerated growth of absorption humidity in samples from flax fibers or noils is observed within the first 9 days. The indicator of flax-based slabs absorption humidity over a given period is equal to 9.9%, which is 10% more than that of the samples from flax noils. The water vapor absorption by the materials (compositions 1 and 2) lasts 13 days and reaches a maximum value of 11.2% and 10.2%, respectively.

When the air humidity equals to 60%, the kinetics of absorption by samples has the same character as at a relative humidity of 40%. It is established that both materials absorb water vapor for 15 days. At the same time, the maximum value of the absorption humidity content of flax fiber slabs is 10% more than for samples based on flax noils (14%).

Intensive growth of the absorption humidity of thermal insulation materials from flax fibers or noils is preserved within the first 9 days and practically coincides with an air humidity of 80%. The duration of absorption is 18 days. The absorption humidity of the material (composition 1) and slabs (composition 2) is equal to 19.3% and 18.3%, respectively.

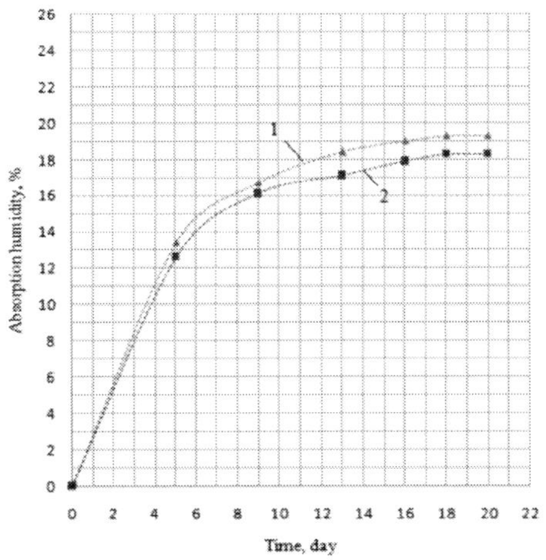

1 – flax fiber samples;
2 – flax noils samples

Figure 7. Water vapor absorption kinetics of the samples at a relative air humidity of 80%.

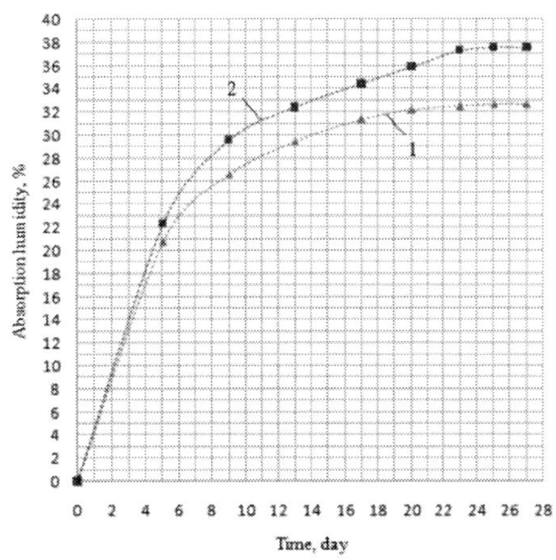

1 – flax fiber samples;
2 – flax noils samples

Figure 8. Water vapor absorption kinetics of the samples at a relative air humidity of 97%.

At a relative humidity of 90%, the accelerated water vapor absorption by materials (compositions 1 and 2) occurs in the first 13 days. It should be noted that the absorption humidity of the samples over a given time interval is almost the same and is equal to 23% and 24%, respectively. At the same time, the value of materials from flax noils is 10% higher than that of the slabs based on flax fibers for a maximum absorption period of 22 days.

The period of intensive growth of absorption humidity of 13 days is also characteristic for air humidity of 97%. The absorption of water vapor by flax noils samples for a given period is 32.4%, which is 10% more than that of flax fiber-based slabs. The process of absorption of materials ends on the 25th day. The maximum value of the absorption humidity of the samples (composition 2) is 37.6%, and for materials of composition 1, it is 13% less and amounts to 32.7%.

According to the results of studies, it has been established that the use of modified liquid glass does not affect the duration of the water vapor absorption process by experimental thermal insulation materials, but at the same time it reduces the absorption humidity of samples based on flax fibers or noils. For example, at a relative air humidity of 97%, the introduction of additives from lime and gypsum leads to a decrease in the value of the absorption humidity content of samples from flax fibers or flax noils by 18% and 20%, respectively, relative to slabs with liquid glass without modifiers.

The thermal conductivity coefficient of samples from flax fibers or flax noils on a modified liquid glass in a wet state is 0.079 W/(m·K) and 0.073 W/(m·K), respectively. As a result, the thermal conductivity of slabs using liquid glass with a maximum humidity of 0.005–0.007 W/(m·K) is higher than the thermal conductivity of the samples due to binding with lime and gypsum additives. Reduction of flax noils slabs thermal conductivity is achieved by introducing modified additives that increase the insolubility of liquid glass under the action of humidity vapor from the air. The insolubility of the binder reduces the absorption of humidity from the air by the liquid glass and thus reduces the absorption humidity of thermal insulation slabs. As a result, the reduction of the composite thermal conductivity under operating condition is observed.

4. Conclusions

According to the experiment results, it has been found out that at a relative air humidity of 97%, the absorption humidity of samples based on flax fibers or flax noils is 39.8% and 46.9%, which exceeds the value of the «Akotherm flaks» slabs (30.6%) by 30% and 53% respectively. The introduction of additives from lime and gypsum into the binder can reduce the absorption humidity of flax fibers or flax noils-based slabs by 18% and 20% to 32.7% and 37.6%.

It has also been established that a change in the humidity index has a significant effect on the thermal conductivity of slabs from flax fibers or flax noils. The maximum humidity content of thermal insulation materials based on flax fibers leads to an increase in the thermal conductivity coefficient from 2 to 2.3 times relative to samples in a dry state. Lowering the absorption humidity due to the introduction of modified liquid glass allows reducing the thermal conductivity of thermal insulation materials from flax fibers at an increased relative air humidity by 0.0057–0.007 W/(m·K) relative to slabs that do not contain lime and gypsum additives.

The obtained data on the kinetics of water vapor absorption by insulating slabs make it possible to calculate the humidity conditions of the structure. Also, the experimental data will be used in the development of a predictive model of changes in the thermal performance of insulation during the operation of buildings and structures.

References

[1] Davydenko, N. (2016) Teploizolyacionnye plity na osnove othodov rastenievodstva i neorganicheskogo vyazhushchego: avtoref. diss. ... kand. tekhn. nauk : 05.23.05. Novopolock. S. 28.

[2] Soldatov, D. (2000) Teploizolyacionnye materialy na osnove rastitel'nogo syr'ya i organomineral'nyh porizovannyh svyazuyushchih: avtoref. dis. kand. tekhn. nauk: 05.23.05. Kazan. S. 18.

[3] Lukutcova, N. (2005) Poluchenie ekologicheski bezopasnyh stroitel'nyh materialov iz prirodnogo i tekhnogennogo syr'ya: avtoref. diss. d-ra tekhn. nauk: 05.23.05. Belgorod. S. 42.

[4] Sergienko, A. (2017 Sovremennyj ekologichnyj uteplitel' na osnove drevesnyh volokon. Nauchnye issledovaniya. T 2 No. 6 (17). S. 6-7.

[5] ZHuravleva, L., Devyatlovskaya, A. (2014) Myagkie drevesno-voloknistye plity – teploizolyacionnyj material. Vestnik KrasGau. No 11. S. 181–184.

[6] Davydenko, N. (2016) Teploizolyacionnye plity na osnove othodov rastenievodstva i neorganicheskogo vyazhushchego: diss. kand. tekhn. nauk: 05.23.05. S. 1 - 47.

[7] Sinka, M., Sahmenko, G. (2013) Sustainable Thermal Insulation Biocomposites from Locally Available Hemp and Lime. Proceedings of the 9th International Scientific and Practical Conference, Vol 1, ISSN 1691-5402. Pp 73–77.

[8] Zajceva, K. (2015) Vozmozhnosti ispol'zovaniya kostry l'na v kachestve uteplitelya derevyannogo kleenogo brusa. Materialy IV Mezhdunarodnoj nauchno-ekologicheskoj konferencii "Problemy rekul'tivacii othodov byta, promyshlennogo i sel'skohozyajstvennogo proizvodstva", Krasnadar, 24–25 maya. "Kubanskij gosudarstvennyj agrarnyj universitet imeni I T Trubilina". S. 374–377.

[9] YAkunina, E. (2018) Sovremennye teploizolyacionnye materialy, kak odna iz tendencij ekologiicheskogo stroiteltsva. Sinergiya nauk, No 24. S. 625-634.

10] Bogatova, T. (2016) Preimushchestva i osobennosti bezopasnyh prirodnyh uteplitelej. Inzhenernye sistemy i sooruzheniya No 3-4 (24-25). S. 14-19.

[11] Rozyev, M. (2019) Thermal insulation material, using waste cotton production as a placeholder. European & national dimension in research. Architecture and civil engineering. No 11. Pp 64–66

[12] Bialosau, A., Bakatovich, A., Gaspar, F. (2018) Materiais compositos para isolamento termico de materias-primas naturais e aglutinantes minerais. Livro de Resumos 3^0 Congresso Luso –

Brasileiro de Materiais de construcao sustentaveis, Coimbra, Portugal, 14 – 18 fevereiro de. Pp 16 – 27
[13] Statisticheskij ezhegodnik (2018) Minsk: Nacional'nyj statisticheskij komitet Respubliki Belarus'. S. 489.
[14] SHarshunov, V. (2012) Sovremennye tekhnologii i sredstva pererabotki l'novoroha. Inzhenerno-tekhnicheskoe obespechenie APK. Referativnyj zhurnal. No 3. S. 29-33.
[15] Rogash, A., Abramov, N., Lebedev, Y. (1995) Moskva, IPO Piligram. S. 544.
[16] Sovetnikov, D., Semashkina, D., Baranova, D. (2016) Optimal'naya tolshchina uteplitelya naruzhnoj steny dlya sozdaniya energoeffektivnogo i ekologichnogo zdaniya v usloviyah Sankt-Peterburga. Stroitel'stvo unikal'nyh zdanij i sooruzhenij. ISSN 2304-6295, No 12 (51). S. 7–19.
[17] Krasimova, S., Malysheva, V., Rozhkova, D. (2014) Obzor biopozitivnyh stroitel'nyh materialov, primenyaemyh pri stroitel'stve ekodoma. Masters Journal. ISSN 2306-8590. S 363–369.
[18] Plity teploizolyacionnye zvukopogloshchayushchie (2015) Tekhnicheskie usloviya TU BY 391129716.001, Vved. 27.07.2015. Orekhovsk. S. 10.
[19] Kil'chevskij, A., Hotyleva, L. (2010) Geneticheskie osnovy selekcii rastenij. izd. Belaruskaya navuka. S. 251.
[20] Bakatovich, A., Romanovskij, S. (2017) Primenenie mikroskopicheskogo analiza dlya ocenki perspektivy ispol'zovaniya ochesov volokna l'na v proizvodstve teploizolyacionnogo materiala. Vestn. Poloc. gos. un-ta. Ser. F, Stroitel'stvo. Prikladnye nauki. No 8. S. 14–18.

Biodegradation studies of sapropel-based composite materials

V. Obuka[1], O. Muter[2], M. Sinka[3], M. Klavins[1]

[1] Department of Environmental Science, University of Latvia, Jelgavas street 1, LV 1004, Riga, Latvia
[2] Institute of Microbiology and Biotechnology, University of Latvia, Jelgavas street 1, LV 1004, Riga, Latvia
[3] Riga Technical University, Faculty of Civil Engineering, Institute of Materials and Structures, Kalku Str. 1, Riga, LV-1658, Latvia

E-mail: vaira.obuka@lu.lv

Abstract. Traditional composite materials used for construction purposes currently face many questions regarding their sustainability – mainly because they do not come from renewable sources and due to the problems related to their end-of-life management. These challenges motivate companies and researchers to look at the natural fiber composite materials with increased interest. Usually natural fiber materials in construction are used together with mineral binders, but in this research new organic binder – sapropel (organic rich lake sediments) is used as a binder for natural fiber composite materials with various fillers. In previous research these composite materials have proven their applicability in construction industry due to their sufficient mechanical strength and low thermal conductivity. Thus, in this research evaluation of the biodegradability of composite materials were done, comparing them to natural fiber materials with mineral binders. As a methods respiration intensity of microorganisms in soil and enzyme activity of microorganisms were used. For studied materials biodegradability potential have been examined, depending on the filler properties and presence of mineral matter content in the obtained composite materials.

Key words: Sapropel, Composite materials, Biodegradation, Environmental applications

1. Introduction

One of the key questions regarding development of new construction materials is their sustainability - mainly because industrial construction materials do not come from renewable sources and due to the problems related to their end-of-life management [1;2]. Traditional natural fiber materials partially answer these challenges as they use mineral binders and natural fiber fillers. In this research a new kind of organic binder - sapropel (organic rich lake sediments) is used.

Sapropel-based composite materials can be considered to be a new natural fiber composite materials which are cost effective, recyclable and locally produced, as sapropel is a byproduct of lake recultivation processes and various filler materials, which are byproducts from such industries as agriculture and wood processing, can be used. Natural fiber composite materials should not only have properties required by the industry, but also be biodegradable after the end of their life cycle.

Sapropel can be used in the production of composite materials for building industry, interior functional design objects with both cold compaction techniques and hot–glue press at elevated temperature and pressure [3, 4]. In the previous research of sapropel-based materials prospective area of their application included creation of different composite materials: Sapropel – wood fiber, sapropel – wood sanding dust, sapropel – hemp shives [5], sapropel – wood-chips [6]. In these composite materials sapropel was used as a binder [5, 7, 8]. The research results showed that the use of sapropel as a binder and various byproducts as a filler allows the finished product to be included in the thermal insulation material category according to the technically qualitative indicators. In addition, the results of compressive and flexural strength resistance show that the strength of the composite materials is sufficient to be used in the industry.

Evaluation of biodegradability of various materials in soil is based mainly on the measuring evolved carbon dioxide as a function of time [9]. Microorganisms play an important role in maintaining nutrient cycles in soil (C, N, P, S) by recycling organic matter [10]. The activity of microorganisms depends on soil characteristics, climatic conditions and soil tillage technologies.

Conditions with different humidity were modelled in experiments. Biotechnology often uses so-called slurry systems [11, 12]. Slurry systems with high humidity and water content have a beneficial effect on the activity of biodegradable microorganisms due to the better transfer of substances.

Our study was aimed at comparative evaluating of the biodegradability of ten composite materials, which have been newly-developed from different types of sapropel and wood, hemp derived fillers. Enzyme and respiration activity of microorganisms in soil in the presence of composite materials was chosen as the evaluation criteria of biodegradability.

2. Materials and methods

2.1. Sapropel (binder for composite materials) sampling.

Sapropel sediments were sampled from three lakes in Latvia – Padelis (further: *carbonatic sapropel (CS)*), Veveru (further: *green algae sapropel (GAS)*) and Pilvelu (further: *cyanobacteria sapropel (CBS)*), located in Rezekne District, Latgale Region. Loss on ignition (LOI) method was applied in order to estimate moisture content, content of carbonate matter and organic matter of sediments [13]. Moisture content of sapropel was determined after drying at 105°C, followed by organic matter estimation at 550°C for 4 h. The content of mineral substances was determined after heating at 900°C for 2 h.

2.1.1. Fillers for composite materials

Birch wood sanding dust (also known as wood dust) and fiber (also known as wood fibers), and hemp shives were selected as fillers for production of composite materials. Birch wood sanding dust and fiber are industrial by-products coming from JSC "Latvijas finieris" – a plywood manufacturing company. Birch wood fiber was up to 15 mm long and up to 0.1 mm thick (diameter). In addition, hemp shives ("Bialobrezeskie") were taken from "Zalers" Ltd. Hemp shive slices were up to 5.5 cm long and up to 0.6 cm thick.

2.1.2. Composite material preparation and curing

For the developed composite materials three types of raw sapropel have been used as a binder, i.e., green algae sapropel (GAS); cyanobacteria sapropel (CBS) and carbonatic sapropel (CS). Sapropel was mechanically treated by mixing together with electrical hand mixer until smooth and homogeneous material was formed. Mixing of sapropel-filler mass was done manually until homogeneous and smooth mixture was reached at the stage, where filler was fully covered with sapropel. Binder-filler mass ratio was 6:1. Metal molds with dimensions of 30×30 cm and with adjustable height were used for composite material production. The mixture of raw materials was laid in by layers in molds for more dense composite material structure, higher mechanical strength and for minimizing final product shrinkage. Sapropel-filler specimens were cured at the temperature of 80 – 105°C for 36 – 72 hours until the constant weight was reached.

For a reference of biodegradation tests a biocomposite materials with the same filler (hemp shives) were used. Mineral binders developed in previous studies were used for these materials - dolomitic lime consisting of 100% DL60 lime (Dolomite) and hydraulic lime consisting of 60% DL60 lime and 40% calcinated kaolin clay (Clay) [14]. Binder-filler mass ratio was 2:1. Block peat ("Laflora") was also used for composite materials biodegradation studies as a control material.

2.2. Evaluation of the biodegradability of the tested materials

In order to compare the potential biodegradability of the composite materials tested, an experimental scheme was developed: soil microbial enzyme activity and respiration were used as the main indicators [15]. Considering the need for microorganisms to adapt to the substrate, 7-day incubation period was included in the process requiring appropriate physicochemical conditions, growth factors (macro- and microelements, nutrients, vitamins). To accelerate the biodegradation process, the composite materials were placed in the soil amended with a substrate (molasses and a source of vitamins (plant extract)), as well as a consortium of soil-derived microorganisms with a high hydrolytic activity [16]. Altogether *10 composite materials where evaluated.*

Incubation of 0.25 mg specimen for evaluation of the biodegradability of the tested materials was performed in the sealed 100 mL vessels containing 10 g soil at 37 ± 2 °C for 7 days. The composition of the substrate added to the specimens was the following: 10 g clay loam soil, 2 mL mineral broth, 100 μL 30% molasses, 200 μL cabbage leaf extract, 100 μL inoculum (2.0×10^{10} CFU/mL) and 50 mL sterile distilled water [16]. The composition of mineral broth was the following, g/L: $MgSO_4$ – 0.2, $CaCl_2$ – 0.02, KH_2PO_4 – 1.0, K_2HPO_4 – 1.0, NH_4NO_3 – 1.0, $FeCl_3$ – 0.05). Specimens were prepared in three replicates. Soil moisture was 60% of the maximum water capacity. The physicochemical characteristics of clay loam soil are summarized in Table 1.

2.2.1. Respiration intensity of microorganisms

The microbial respiration was tested according to [15] with some modifications. The glass with 5 mL 0.05 M NaOH was placed in the sealed 100 mL vessel as described in 2.5. The respiration was estimated by back-titration of the unreacted NaOH using 0.05 HCl (adding 0.1% phenolphthalein indicator to the NaOH prior to titration). Two measurements of respiration have been made for each vessel with a sample, i.e., at the beginning of incubation and after 7 day of incubation period. Respiration assay was performed at 23 °C for 24h in the dark.

The intensity of the SIE was calculated by the formula [17]:

$$SIE = \frac{(A - B) \cdot 1.1 \cdot 60}{m \cdot h}$$

where,

A - titrated, control sample, mL;

B - titrated for the test sample, mL;

1.1 - coefficient (depends on molarity);

60 - coefficient (for conversion into hours);

m - sample weight, g;

h - trial time, min.

2.2.2. Enzyme activity of microorganisms: fluorescein diacetate hydrolysis

After 7 day incubation period, as described in 2.5., the fluorescein diacetate (FDA) hydrolytic activity of microorganisms was tested. 100 μL specimen was transferred to 1 mL tube containing 400 μL reaction mixture (4 mg FDA, 2 mL acetone, 48 mL 0.06M phosphate buffer, pH 7.6). The mixture was incubated for 60 min at + 37±2 °C [18, 19]. After incubation, 500 μL acetone was added to the specimens to stop the hydrolysis reaction. The optical density was measured at 490 nm using a microplate reader Infinite f50 (TECAN, Swirtzeland). Calibration curve was prepared using the thermally hydrolyzed FDA.

Table 1. Characteristics of the soil used

Parameter	Values	Parameter	Values
N, %	0.20	Electrical conductivity, mS/cm	0.162
C, %	2.06	Water capacity, %	149.47
PO_4^{3-}, µg/g	47.8	Na, mg/kg	14.3
P, µg/g	15.6	Mg, mg/kg	218.5
pH (1M KCl)	6.72	K, mg/kg	146.5
Redox potential, mV	-24.9mV	Ca, mg/kg	1802.6

The soil characteristics of the clay loam used in the experiment was determined (Table 1) using standard soil analysis methods. The Na, Mg, K, Ca content of the specimens was determined using a PerkinElmer AAnalyst 200 atomic absorption spectrometer. A non-electrode discharge lamp (Perkin Elmer) was used as a source, Na measurements were made at 589 nm wavelength, Mg at 285.2 nm, K measurements at 766.5 nm and Ca at 422.7 nm using flame atomization. N_2O, acetylene was used as the oxidizing gas. ANOVA (Anova: Single factor analysis) was used for statistical data processing.

3. Results and discussion

3.1. Composite materials (sapropel sampling)

Sapropel samples differed from one another in terms of moisture, organic matter and carbonates.

Table 2. Characteristics of the sapropel samples

Lake	Moisture,%	Organic matter,%	Carbonates,%
Padelis (carbonatic sapropel; CS)	85.97	15.27	35.57
Pilvelu (cyanobacteria sapropel; CBS)	94.99	84.51	1.26
Veveru (green algae sapropel; GAS)	97.66	86.25	1.18

Characteristics of the sapropel samples are shown in Table 2. For example, Lake Padelis sapropel sample contained 35.57% carbonates, moisture – 85.97%, and color – pale gray-pink. Lake Pilvelis sapropel sample was dark greenish brown with homogeneous and jelly-like structure. Lake Veveru sapropel sample moisture level was comparatively high, i.e. 97.66% and organic matter – 86.25%.

3.2. Evaluation of the biodegradability

Biodegradation experiments were performed by adding the composite materials to the soil amended with nutrients and a consortium of microorganisms with a high hydrolytic activity [16] in order to provide favorable conditions for degradation processes.

3.2.1. Soil microbial enzyme activity after incubation in the presence of composite materials

One of criteria for evaluating the biodegradability of the tested materials could be an increase of microbial enzyme activity, which responded to the presence of bioavailable nutrients. FDA hydrolysis involves the activity of various enzyme groups of microorganisms, i.e., hydrolases, proteases, esterases, lipases, etc. [20]. As shown in Fig.1, after 7-day incubation period all composite materials added to the soil stimulated FDA hydrolysis activity, comparing with the control set. After 7-day incubation period in the batch system, the lowest FDA hydrolysis activity was observed in the non-composite control (Fig. 1). All tested composite materials showed a stimulating effect on the enzyme activity of the microorganisms, with the highest mean value for [GAS-CS/Hemp shives], i.e., 2.01 ± 0.75 µM FDA $h^{-1}g\ dw^{-1}$ (Fig. 1). Comparison of the FDA hydrolysis activity showed a statistically significant ($p<0.05$) difference between control and composite materials, except CS/wood fibers. Greater FDA hydrolysis activity may indirectly indicate more intense biodegradation processes, as it depends on the availability of nutrients, the concentration of microorganisms, and their physical, chemical and environmental properties [20, 21].

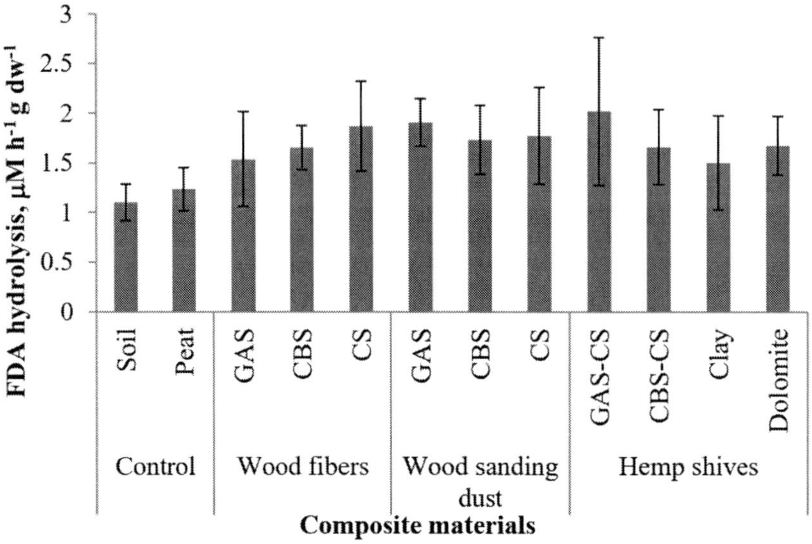

Figure 1. Fluorescein diacetate hydrolysis activity of microorganisms in a clay loam soil amended with nutrients, microbial consortium and composite materials. The ratio of a composite material to the substrate was 0.25:10.0. The substrate was prepared as described in Materials and Methods. FDA activity was measured after 7 day incubation period of a composite material with soil at 37 °C. GAS – green algae sapropel; CBS – cyanobacteria sapropel; CS – carbonatic sapropel. Control – the soil substrate without composite materials.

3.2.2. Respiration intensity of soil microorganisms in the presence of composite materials

The respiration intensity of microorganisms in the experimental batches was observed before and after 7 day incubation period at 37 °C. An increase of respiration intensity in the composite materials has been observed. The amended batches at the beginning of incubation showed statistically significant difference ($p<0.05$) and varied in the range from 31% to 70%, as compared to the control batch (Fig.2 A). The highest respiration intensity was in the soil containing CS/Wood fibers, while the lowest – CS/Wood sanding dust, i.e., 7.68 ± 0.35 and 5.92 ± 0.43 µg C-CO_2 h^{-1} g dw^{-1}, respectively (Fig.2 A). Among the types of composite materials, no statistically significant differences in respiration stimulating effect were found.

Figure 2B summarizes a second test carried out after 7 day incubation period, when readily available substrates have been used by microorganisms. The respiration intensity of microorganisms after 7 day incubation period was considerably lower than that at the beginning of the experiment. This can be explained by the fact that microorganisms have already degraded easily degradable substances. Subsequently, this data indicate the degradation state of comparatively hardly biodegradable substances (cellulose, hemicellulose, lignin) resulting in the original material fractionation in respect to polymer stability. No respiration was detected in the control soil. The highest respiration intensity was detected in CBS-CS/Hemp shives, while the lowest – in CBS/Wood sanding dust, i.e., 2.70 ±0.89 and 0.70 ±0.87 µg C-CO_2 h-1 g dw-1, respectively (Fig.2B).

The obtained data can be interpreted as the potential biodegradability of the tested composite materials under given test conditions. It shows that the materials used are biodegradable but with varying rate. It can be seen that it is mostly dependent on the used filler, the wood sanding dust has the highest biodegradability ratio as it shows the lowest respiration after 7 days, wood fibers and hemp shives have lower biodegradability as they have higher respiration after 7 days. Sapropel binder shows similar reparation as the reference lime binders, as the used sapropel is with high carbonate percentage. The used materials demonstrate that with varying ratio all of the materials are biodegradable and can be used to decrease the overall environmental impact of construction materials.

IOP Conf. Series: Materials Science and Engineering **660** (2019) 012073 doi:10.1088/1757-899X/660/1/012073

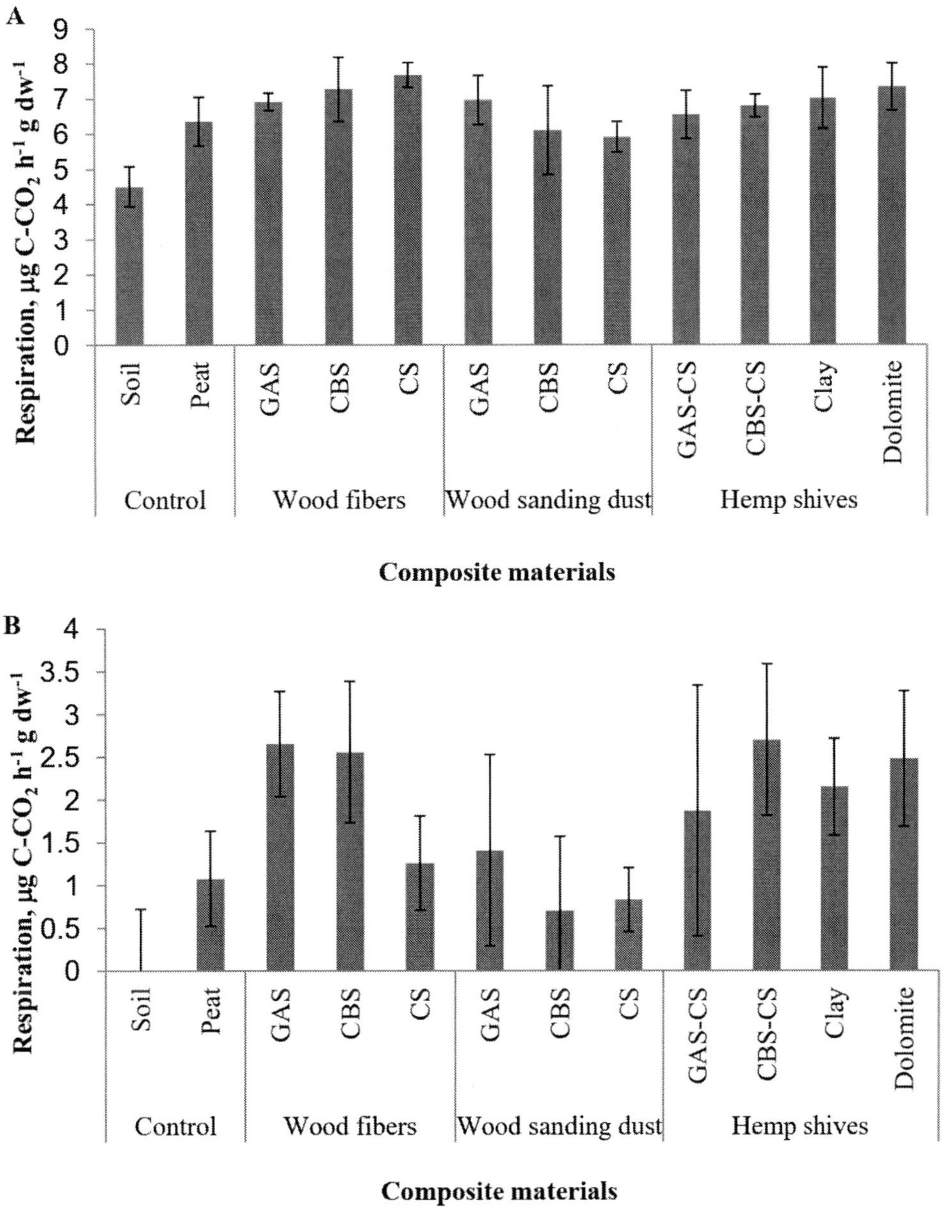

Figure 2. Respiration intensity of microorganisms in a clay loam soil amended with nutrients, microbial consortium and composite materials. The ratio of a composite material to the substrate was 0.25:10.0. The substrate was prepared as described in Materials and Methods. Respiration intensity was measured before incubation (A) and after 7 day incubation period (B) of a composite material with soil at 37 °C. GAS – green algae sapropel; CBS – cyanobacteria sapropel; CS – carbonatic sapropel. Control – the soil substrate without composite materials.

Previous studies of composite materials from peat and cellulose fibers [22] demonstrated reasonably good resistance in respect to destruction, as after a month of incubation their biodegradability reached 20-30%. Overall, in the experiment described above, the highest biodegradation intensity was in materials consisting of peat, 85% cellulose fiber and grape processing by-products. It is explained by the fact that the grape processing residues are an additional filler between the fibers, reducing the fiber content. However, the smallest biodegradation activity was shown in the mixture of peat and 100% cellulose fiber. In turn, the results of biological activity expressed in mg CO_2 per 100 g of substrate indicate the lowest activity in studied composites with increased mineral content [22]. Also our study demonstrate similar trend.

Patnaik and colleagues [23] have explored the biodegradation potential of thermal insulation and sound insulation materials, from wool waste and recycled polyester. Sheep wool showed the highest level of biodegradation over 50 days followed by sheep wool products and a control sample of cellulose fiber. At the same time, recycled polyester showed the lowest biodegradation level. Thus, it can be concluded that composite material, just as those developed in our study at the end of its life cycle, can be used on agricultural land as a nitrogen fixator for plants, thus reducing the fertilizer utilization rate [23].

4. Conclusions

Organic rich lake sediments (sapropel) – a byproduct of the lake cleaning – can be efficiently used to develop composite materials together with other production waste materials. Sapropel of different origin (green algae sapropel, cyanobacteria sapropel and carbonatic sapropel) at the development of composites serves as binder, but birch wood fiber, sanding dust, hemp shives as a filler material. The biodegradability of the obtained composites has been studied and major differences of the biodegradability potential have been found, mostly depending on the filler properties, but also on the presence of mineral matter content in the obtained composites. The obtained results demonstrate potential to use sapropel as a raw material for composites in combination with other waste materials with potential application as construction materials and design products, to extend the life of natural materials and achieve aims of reduction of waste streams.

Acknowledgments
Authors would like to express gratitude to Latvian Council of Science Council of Latvia for providing a grant "Properties and structure of peat humic substances and possibilities of their modification" lzp-2018/1-0009.

References

[1] Klijn-chevalerias M De, Javed S. The Dutch approach for assessing and reducing environmental impacts of building materials. Build Environ 2017; 111:147–59. doi: 10.1016/j.buildenv.2016.11.003.
[2] Madurwar M V, Ralegaonkar R V, anD Mandavgane S A 2013 Application of agro-waste for sustainable construction materials: A review. Constr. Build. Mater 38 872–878
[3] Zach J, Hroudová J, Brožovský Z K and Gailius A 2013 Development of thermal insulating materials on natural base for thermal insulation systems *Procedia Eng.* **57** 1288–94
[4] Mounika M, Ramaniah K, Ratna Prasad V A, Mohana Rao K and Hema Chandra Reddy H 2012 Thermal conductivity characterization of bamboo fiber reinforced polyester composite *J. Mater. Environ. Sci.* **3** 1109–16
[5] Obuka V, Šinka M, Nikolajeva V, Kostjukova S, Lazdiņa L and Kļaviņš M 2017 Sapropel and Lime as Binders for Development of Composite Materials *Proc. of the 25th European Biomass Conf.* (12-15 June, Stockholm) pp 1285–91
[6] Obuka V, Korjakins A, Brencis R, Preikšs I, Purmalis O, Stankeviča and Kļaviņš M 2013 Sapropel/peat-wood chip insulation materials and their properties *Mater. Sci. Appl. Chem.* **29** 127–36
[7] Klavins M and Obuka V 2018 *Towards a Sustainable Bioeconomy: Principles, Challenges and Perspectives* ed W Lela Filho et al. (New York: Springer International Publishing) chapter: Local Knowledge and Resources as Driving Forces of Sustainable Bioeconomy pp 173–86
[8] Obuka V, Šinka M, Kļaviņš M, Stankeviča K and Korjakins A 2015 Sapropel as a binder: properties and application possibilities for composite materials *2nd Int. Conf. on Innovative Materials, Structures and Technologies* vol 96 (Bristol: IOP Publishing) pp 1–10.
[9] *ASTM D5988-18. Standard Test Method for Determining Aerobic Biodegradation of Plastic Materials in Soil* 2018 ASTM International, West Conshohocken, PA, www.astm.org
[10] Kara Ö and Bolat Ï 2008 The effect of different land uses on soil microbial biomasscarbon and

nitrogen in bartin province *Turk. J. Agric. For.* **32** 281–8
[11] Zeng Y P K and Hong A K P 2011 Slurry-phase ozonation for remediation of sediments contaminated by polycyclic aromatic hydrocarbons *J. Air Waste Manag. Assoc.* **52** 58–68
[12] Hwang S and Cutright J T 2003 Effect of expandable clays and cometabolism on pah biodegradability *Environ. Sci. Pollut. Res.* **10** 277–80
[13] Heiri O, Lotter A and Lemcke G 2001 Loss on ignition as a method for estimating organic and carbonate content in sediments: reproducibility and comparability of results *J. Paleolimnol.* **25** 101–10
[14] Sinka M and Sahmenko G 2013. Sustainable thermal insulation biocomposites from locally available hemp and lime *Environment, Technology, Resources: Proc. of the 9th Int. Scientific and Practical Conf.* (June 20-22, Rēzekne) pp 73–7
[15] Zibilske L M 1994 Carbon mineralization *Methods of Soil Analysis. Part 2. Microbiological and Biochemical Properties* ed R W Weaver et al. (Madison: SSSA) pp 835–63
[16] Muter O 2015 Complex natural amendments enhance cellulolytic activity of bacterial consortium *VI Int.Conf. Industrial and Applied Microbiology – BioMicroWorld 2015* (October 28-30, Barselona) p 493
[17] Гавиленко В Ф, Ладыгина М Е, Хандобина А М 1975 *Большой практикум по физиологии растений.* Москва, Высш, Школа (in Russian)
[18] Chen Y X, Yin J and Wang K X 2005 Long-term operation of biofilters for biological removal of ammonia Chemosphere 58 1023–30
[19] *Manual for soil analysis: monitoring and assessing soil bioremediation* 2005 ed R Margesin and F Schinner (Berlin: Springer-Verlag Berlin/Heidelberg GmbH & Co. KG) p 366
[20] Green V S, Stott D E and Diack M 2006 Assay for fluorescein diacetate hydrolytic activity: optimization for soil samples *Soil Biol. Biochem.* **38** 693–701
[21] Mupambwa H A and Mnkeni P N S 2016 Stocking density optimization for enhanced bioconversion of fly ash enriched vermicompost *J. Environ. Qual.* **45** 1087–95
[22] Nechita P, Dobrin E, Coilacu F and Bobu E 2010 The biodegradability and mechanical strength of nutritive pots for vegetable planting based on lignocellulose composite materials *BioResources* **5** 1102–13.
[23] Patnaik A, Mvubu M, Muniyasamy S, Botha A and Anandjiwala D R 2015 Thermal and sound insulation materials from waste wool andrecycled polyester fibers and their biodegradation studies *Energ. Buildings* **92** 161–9

IOP Conf. Series: Materials Science and Engineering **660** (2019) 012074 doi:10.1088/1757-899X/660/1/012074

Green Integrated Structural Elements for Retrofitting and New Construction of Buildings

G Kastiukas[1] and X Zhou[1]

[1]Brunel University London, Kingston Lane, London, Uxbridge, UB8 3PH, UK

Email: Gediminas.Kastiukas@brunel.ac.uk

Abstract. Governments across Europe have ambitious visions for the construction sector sector. For example, in the UK, as part of the Clean Growth Grand Challenge within the Governments Industrial Strategy, we are aiming to halve the energy use of new buildings by 2030; delivering new construction assets at a third of the cost and significantly quicker; provide assets that are cheaper to run, smarter, safer with lower emissions; and ultimately create cleaner air. As part of this deal and the wider The 'Construction Sector Deal' sets out an ambitious partnership between the industry and Government that aims to transform the sector's productivity through innovative technologies and a more highly skilled workforce aims to transform the efficiency of construction techniques through digital technologies like Business Information Modelling; reduce running costs through energy generation and storage technologies; and conduct research and development and demonstration programmes supporting innovations. The core aim of the project is to seamlessly combine four integrated layers with high structural performance, which would provide high levels of insulation, a healthy indoor environment, high aesthetic value, along with CO_2 capture, and greywater management. The competitive advantage of the Green INSTRUCT panel is (1) sustainable and cost effective - using more than 70% wt. C&DW including concrete, bricks, plastic, metals; (2) safety and energy efficiency - meeting Eurocode standards for fire safety, thermal insulation with U-value of 0.14 $W/m^2.K$ and acoustic insulation coefficient under 60dB and (3) Customisation potential -multi-layered, modular & prefabricated building block for new and old building. This paper will discuss key project achievements in terms of material development and innovation and challenges faced so far during the project.

1. Introduction

With the development of urbanization in Europe and particularly in the developing countries, building construction, renovation and demolition have produced an enormously huge amount of construction and demolition waste (C&DW). For example, in 2017, disposal of C&DW in China exceeded three billion tons and accounted for 30% to 40% of the total amount of waste [1]. C&DW is not only dauntingly huge, but is often unsorted with different and complex components. Therefore, effective disposal and recycling has become a demanding task for every country. At the same time, we are facing an increasing shortage of natural materials such as sandstone [2], which is another impetus to reduce our dependence on fresh resources and allow for the use of waste, which would otherwise be deposited to landfill. The 'circular economy' (CE) concept is fast becoming a new model for resilient growth. Creating and optimizing resource 'loops' along value chains in the construction industry will help meet the needs of growing and urbanizing populations while mitigating against a continued rise in primary resource use, associated emissions and environmental pollution. The CE is now a core component of the EU's 2050 Long-Term Strategy to achieve a climate-neutral Europe [3], with the aim to 'implement a plan for guiding circular development, encourage the circular use of resources between production and society, and accelerate efforts to recycle resources from refuse'. Our project 'Green Integrated Structural Elements for Retrofitting and New Construction of Buildings' (Green INSTRUCT) is part of the EU research and innovation initiative to work on 'New technologies and strategies for the development of pre-fabricated elements through the reuse and recycling of construction materials and structures' [4]. The core aim of the project is to seamlessly combine four integrated layers with high structural performance, which would provide high levels of insulation, a healthy indoor environment, high aesthetic value, along with CO_2 capture, and greywater management. The competitive advantage of the Green INSTRUCT panel is (1) sustainable and cost effective including concrete, bricks, plastic, metals; (2) safety and energy efficiency - meeting Eurocode standards for fire safety, thermal insulation and acoustic insulation coefficient and (3) Customisation potential -multi-layered, modular & prefabricated building block for new and old building. This paper will discuss the most recent project achievements in terms of material development, demonstration activities and future work.

2. Key Project Achievements

2.1. Overall Green INSTRUCT Scheme

In the Green INSTRUCT project, prefabrication for refurbishment is approached from a modular perspective, through the development and deployment of a universal building block, with customization potential. To achieve this objective, the Green INSTRUCT consortium combines unique skills in material development from CDW, green technologies, manufacturing and construction in both new buildings and refurbishment, with the experience and presence of experts in their respective fields. During the Green INSTUCT project, a multi-layered integrated building block (summarised in figure 1) has been developed containing more than 70% per weight of CDW. The block has been designed to be incomparably faster to install than conventional envelope walls of the same size. The developed prototype adheres to Eurocode standards and provides thermal insulation with a U-value of 0.13 $W/m^2.K$ and acoustic insulation of 57dB. It also contributes to on-site grey and storm water management, through the integration of a vertical Green Wall, providing additional functionalities.

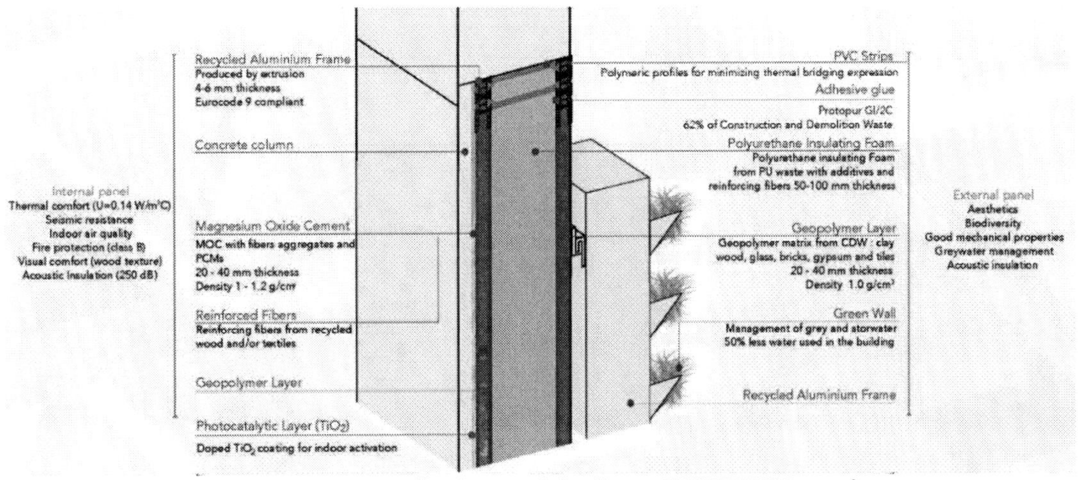

Figure 1. Overall scheme of the Green INSTRUCT panel.

2.2. Geopolymer layer and extrusion manufacturing

Brick waste powder was characterized and appropriately developed as a raw material for the geopolymer synthesis. Waste expanded polystyrene (EPS) and waste polyethylene (PE) fibres were used to reduce the geopolymer density and ductility, respectively. Lightweight geopolymer with a density of 1.3 g/cm^3 and compressive strength and flexural strength of 15.3 and 3.2 MPa were prepared, respectively (Table

1). Also, a hydrothermal processing for the conversion of glass into water glass that can be used as the activation solution was developed and optimized (Figure 2); a conversion rate of glass to water glass was found to be close to 81% wt. Successful extrusion of 150 mm-wide panels was performed using a laboratory scale extruder. Hydroxypropyl Methylcellulose (HPMC) thickener was added as a water retention aid with the added benefit of lubricity. The addition of PE fibres provided further binding and allowed the geopolymer to have 'green strength,' i.e., retain its shape upon exiting the die. The mixing and extrusion process of the geopolymer is shown in figures 3-5.

Table 1. Geopolymer properties.

Properties	
Specific Heat (KJ·Kg-1·°C-1)	0.9
Thermal Conductivity (W.m-1·°C-1)	0.45
Density (Kg·m-3)	1300
Poisson's Ratio	0.11
Young's modulus (GPa)	1.3
Tensile Resistance (MPa)	3.2
Compressive Resistance (MPa)	15.3

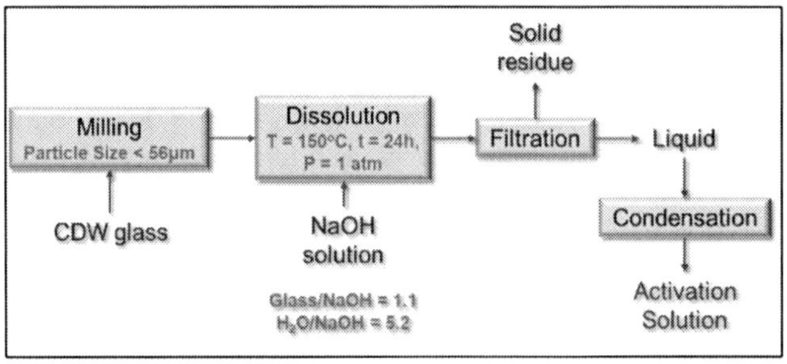

Figure 2. Geopolymer activator manufacturing process.

Figure 3. Initial geopolymer mixing.

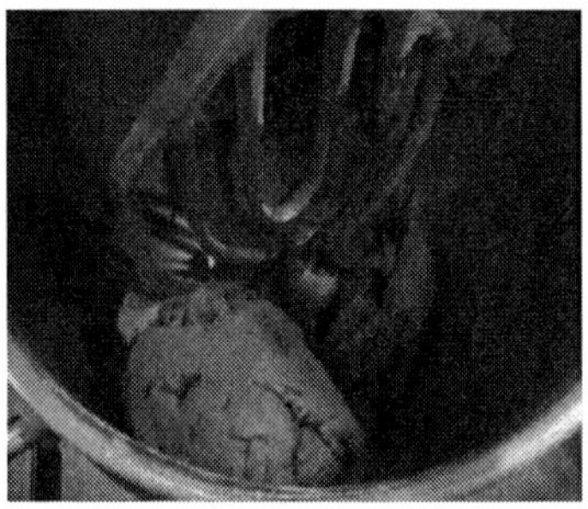

Figure 4. Geopolymer dough-like consistency.

Figure 5. Extruded geopolymer panel.

2.3. PU Foam insulation layer

PU foam powder and wood fibres, both from CDW, were incorporated into pristine PU resin to produce rigid and flexible PU foam for thermal and acoustic insulation purposes, respectively (figures 6 and 7). Optimised combinations of CDW wood fibre and waste PU foam led to both the compressive and tensile strength of the new PU foam to be maintained while reducing thermal conductivity by 15% in the hard PU foam and 10 % in the soft PU foam.

Figure 6. Hard PU foam plate.

Figure 7. Soft PU foam plate.

2.4. Photocatalytic surface coating

An efficient photocatalytic surface coating was developed as a passive indoor air cleaning technology without posing any harm to human health. TiO_2 powders doped with organic and inorganic elements optimized for coating formulation and indoor activation with optimized loading and low agglomeration was achieved. Best results concerning the air pollutant (NOx, Acetaldehyde) as well as the liquid pollutant (Methyl Orange-MO) degradation were achieved for doped and co-doped TiO_2 or mixed oxides (TiO_2/SiO_2), as shown in figure 8. A paint formulation using an optimised blend of dispersing agent, rheological modifier, defoamer, pH stabiliser and coalescing agent was devised and up-scaled (figure 9).

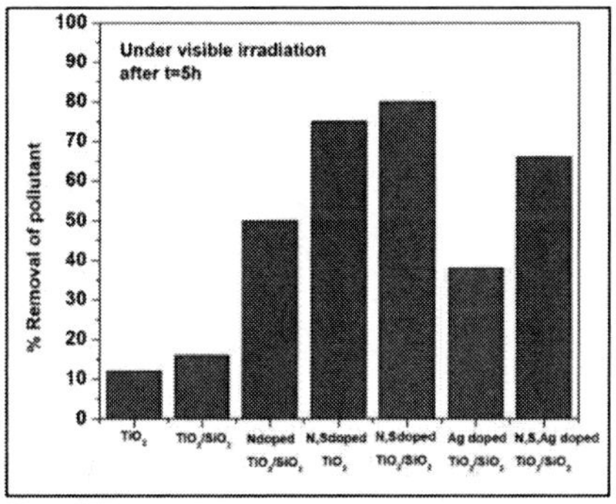

Figure 8. % MO Removal of different doped TiO2 or TiO2/SiO2 samples.

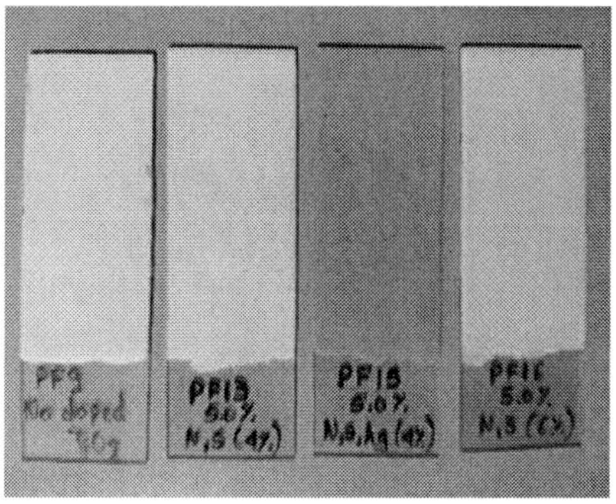

Figure 9. Paints applied onto glass slides.

2.5. Recycled aluminium structural elements

A construction frame for the GI Panel was designed and manufactured using 65% scrap aluminium. A novel die tool (figure 10 was designed and manufactured for the production of the frame (figure 11). The profiles were cut and folded into the construction frame, then joined using novel polymeric connectors, which avoided the introduction of thermal bridges (figure 12).

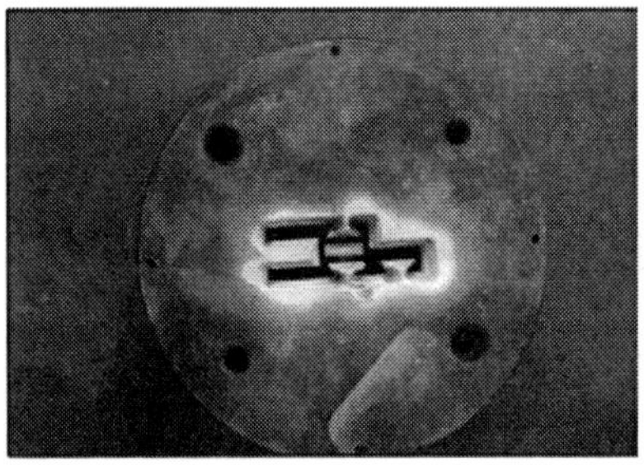

Figure 10. Aluminium extrusion die tool.

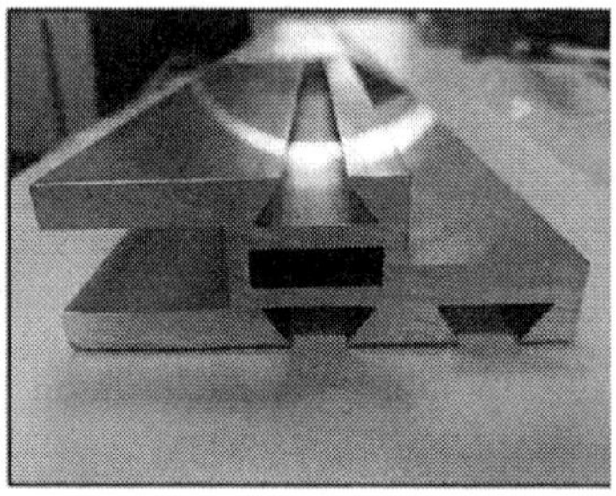

Figure 11. Extruded aluminium profile.

Figure 12. Aluminium profiles with polymer connectors.

2.6. Development & extrusion of MOC layer

Effects of varying calcination conditions on the chemical and physical properties of magnesium oxide (MgO) were investigated using the more sustainable precursor material of magnesium hydroxide ($Mg(OH)_2$). Extremely pure and thus more reactive MgO was obtained using a 17.6% less energy intensive calcination regime compared to industrial grade MgO obtained from the calcination of dolomitic lime. As a result, the magnesium oxychloride cement (MOC) that was produced from the sustainably sourced MgO obtained a 50% increase in flexural strength and a 22% increase in compressive strength. This was mostly due to its homogenous microstructure consisting predominantly of the phase-5 hydration product. The final optimised MOC formulation contained wood fibre recycled from MDF, waste wood-saw dust and PCM-impregnated aggregates is shown in figure 13.

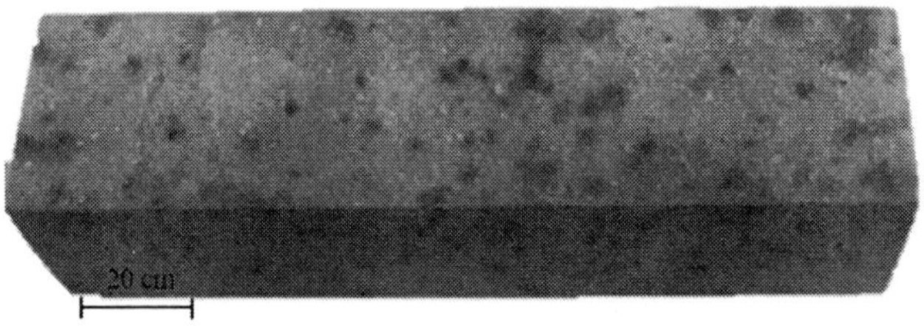

Figure 13. MOC including sawdust, wood fibre and PCM-impregnated aggregates.

2.7. PCM-impregnated aggregates

Recycled porous concrete aggregates impregnated with a paraffin phase change material (PCM) were produced using a macro-encapsulation process (figure 14) for inclusion in the GI block for the purpose of passive temperature control. The vacuum impregnation unit (figure 15) achieved impregnation in particle sizes below 1mm; 100% absorption relative to the pore intrusion volume was achieved after 30 minutes of vacuum impregnation. Also, the PCM functionality before and after impregnation in terms of heat storage, melting and crystallization temperature remained practically unchanged, indicating that the aggregates were thermally stable. The latter results also revealed that the heat storage capacity of the aggregates was $\approx$21 J/g.

IOP Conf. Series: Materials Science and Engineering **660** (2019) 012074 doi:10.1088/1757-899X/660/1/012074

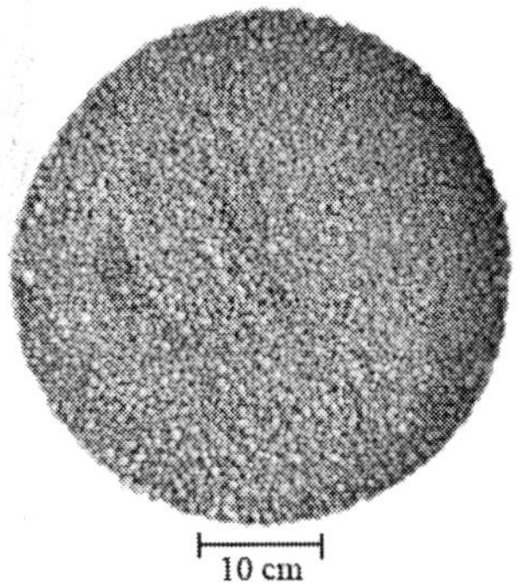

Figure 14. PCM-impregnated aggregates.

Figure 15. Vacuum impregnation unit.

2.8. Performance design & optimization through CFD

The design of the GI Panel was established after an iterative process between sound and thermal insulation numerical analyses. Figure 16 presents a horizontal cross-section of the final design with the material to be assembled and respective thicknesses. Temperature and heat flux distribution It was concluded, through numerical thermal analyses, that the GI Panel could achieve a very ambitious thermal performance level of 0.13 W/m².K U-value, as shown in temperature and heat flux distribution in figure 17. Sound insulation performance computational analyses allowed to verify that the final GI Panel design would achieve sound insulation performance requirement of 57 dB (figure 18).

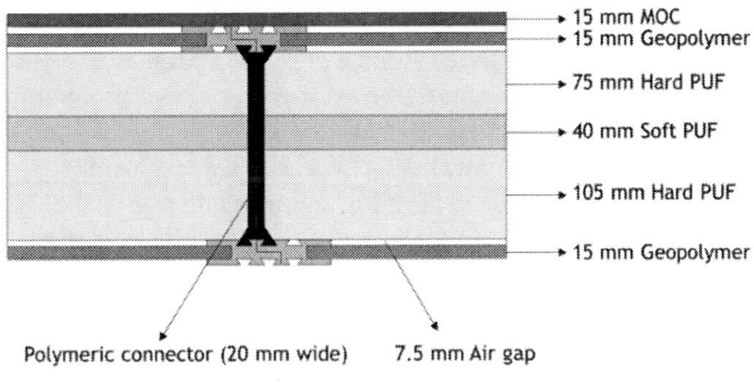

Figure 16. Horizontal cross-section of the final GI Panel design.

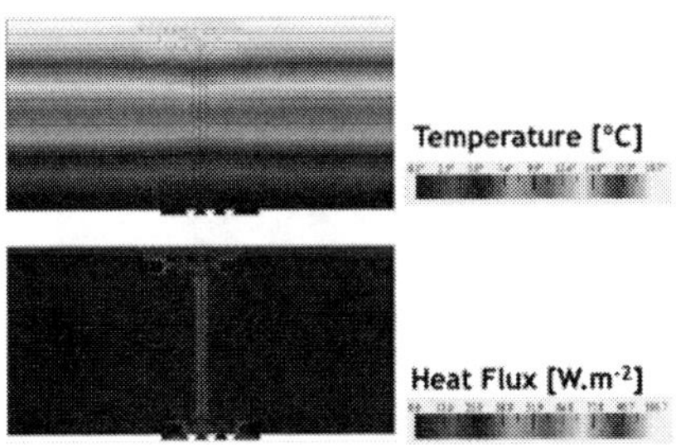

Figure 17. Thermal performance.

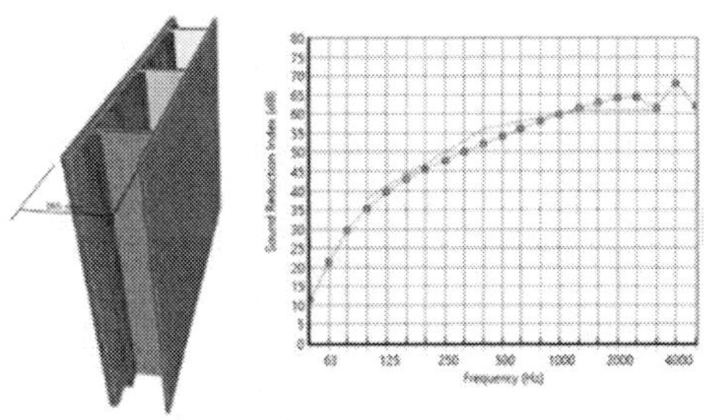

Figure 18. Sound insulation performance.

2.9. Building and Information Modelling

2D and 3D architectural designs of the GI panel, by using software and processes such as AutoCAD and REVIT were completed (figure 19). It was ensured that the developed models were interoperable with other tools and could be accessed by the relevant stakeholders. The modelling completed all aspects of the design and detailed architectural models. In addition, a clear strategy for end-of-life was defined. The identification and compilation of the dismantling, re-use and recycling procedures were initiated (figure 20).

Figure 19. GI panel render on two storey model.

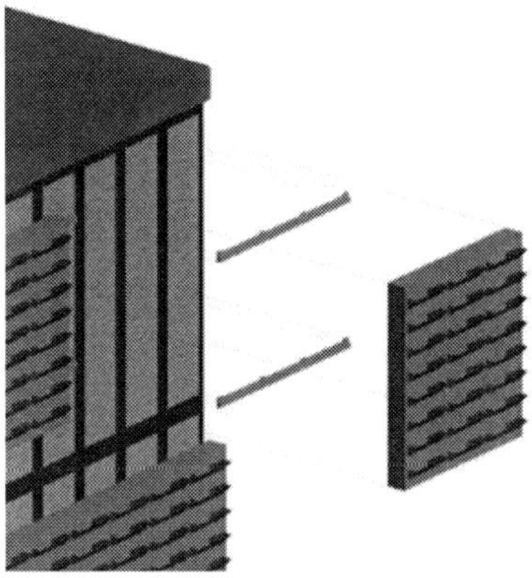

Figure 20. Panel composition and railing fixing.

2.10. Vertical Green Wall panel

A functional green wall layer with an integrated grey-water and stormwater management function based on a constructed wetland approach was developed (figure 21). 1.6 m^2 of panel is necessary to treat up to 60 litres of greywater per day, which corresponds to the average greywater production of one person per day. The green wall was also recorded, per year, to produce 3 kg/m^2 biomass and sequester 500 g/m^2 CO_2. Water cleaning performance of the green wall panel was extensively tested during the development phase. In order to achieve water of sufficient quality for a number of re-use scenarios (toilet flushing, garden irrigation, laundry washing, groundwater recharge, etc. as per EU guidelines 91/271/EC), it was determined that 25 m^2 of green façade area would be needed to clean 1 m^3 of greywater per day. A representation of the water purification performance of the green wall is presented in figure 22.

Figure 21. Green wall facade panels.

Figure 22. Water purification performance of the green wall.

2.11. Challenges

- PVC elements cannot be easily extruded as each profile requires an adequate tailor-made cooling system for each shape and the required profile's cross section is not a commercial one. Issues related to the material and manufacturer procurement need to be addressed. If the parts that integrate the design cannot be produced at a reasonable cost, the design will need to be adapted by assessing the appropriate manufacturing technologies for the production of the PVC parts.
- To overcome various issues (e.g. structural deformations of the cassette due to thermal stresses of the aluminium due to welding, panel thickness, etc.), the cassettes of the green wall were also tested using stainless steel in addition to aluminium. Also, green wall panels for cold winter climates need to be planted very differently than panels for mild winter regions, requiring solutions like the use of a heating coil and/or enclosure.
- The yield of the fibre extrusion scale-up process was around 15% of the total amount of fed material. Clogging of the extruder filters by the presence of inorganic species and/or cross-linked parts in polyethylene was the main hindrance. In spite of that situation, 4.3 kg of fibre were extruded. Also, the extrusion-filament of the recycled HDPE for fibre production resulted in a very low strain ability that hindered the possibility to spin the material.

3. Conclusions

The largest and the most important impact that the outcomes of the Green INSTRUCT project can make is the realisation of the introduction of a building block with higher resource and production efficiency that will ultimately contribute to a climate change resilient construction economy. Currently, there doesn't exist a building block that is made from a comprehensive and streamlined supply chain of construction and demolition waste materials and can be used for both new constructions and building retrofitting. The Green INSTRUCT project aims to utilise the continuous manufacturing technology of extrusion to produce the individual panels of the building block and the structural frame. The building block will also capture CO_2 from the management of grey and storm water using an integrated vertical green wall. Introducing such a building block would significantly contribute to the resilience and sustainability of the construction sector in the UK and worldwide, as it would allow the industry to address the intrinsic CO_2 emission.

Acknowledgments

Finance support from the European Commission Horizon 2020 Research and Innovation Programme through the grant 723525 (i.e. the Green INSTRUCT project) is greatly acknowledged.

References

[1] Huang, Beijia H, Wang X, Kua H, Geng Y, Bleischwitz R, Ren J 2018 Construction and demolition waste management in China through the 3R principle Resources Conservation and Recycling 129
[2] Torres A, Liu J, Brandt J, Kristen Lear K 2017 The World is Running Out of Sand (accessed on 29/08/2019)
[3] 2050 long-term strategy https://ec.europa.eu/clima/policies/strategies/2050_en (accessed 12/08/2019)
[4] Green INSTRUCT Project website https://www.greeninstruct.eu/ (accessed 28/08/2019)

IOP Conf. Series: Materials Science and Engineering **660** (2019) 012075 doi:10.1088/1757-899X/660/1/012075

Environmental study and valorization of an ashy waste: case of pyrrhotite ash

M Achik[1], A Oulmekki[1], M Ijjaali[1], H Benmoussa[1] and O Kizinievic[2]

[1] Faculty of Science and Technology, Sidi Mohamed Ben Abdellah University, Fez, Morocco

[2] Institute of Building Materials, Vilnius Gediminas Technical University, Vilnius, Lithuania

E-mail: maryam.achik@usmba.ac.ma

Abstract. Pyrrhotite ash is an industrial waste which is currently stored in large quantities in a large area in southwest of Morocco. The literature review revealed that this waste has been used by some cement industries and has not been the subject of any environmental study or valuation in another area. This work focuses on the environmental study of pyrrhotite ash. This study consists of a dynamic test with renewing lixiviate more particularly the leaching tank test. The protocol used is extracted from the Dutch standard NEN 7345 with modification of the volume of water used and the number of extraction. Quantitative analysis by ICP has shown that pyrrhotite ash releases heavy metals in aqueous medium. However, the concentration of these is very low compared to the limits estimated by the Dutch standard adapted to the Moroccan hydrological context (MBMD), this result encouraged us to find a way to valorise this waste and to make it a raw material for the manufacture of terracotta bricks made of clay.

1. Introduction

Nowadays, scientific research focuses more on the optimization of industrial processes, whether to increase profitability or minimize the rejection of these processes. The so-called minimization of industrial discharges says environmental protection, the latter becomes the main issue of the world.

In fact, not only the minimization but also the reuse of industrial waste can develop environmentally friendly technologies, reduce the negative impact on the environment and the landfill of waste in large areas of storage and disposal. Reduce the cost of producing new products. As a result, industrial waste is widely studied in many areas of research.

In our research project we chose to study the waste of pyrrhotite ash, which is an industrial waste solid containing a high concentration of Fe_2O_3 hematite and traces of other oxides [1][2], this waste has not attracted the attention of scientists in order to explore the areas of its application in the industry. This waste was generated between 1964 and 1982 by the process of manufacturing sulfuric acid from grilled pyrrhotite ore in southwestern Morocco [3]. This process proved unprofitable and was no longer used [2]. Thus, the combustion unit of pyrrhotite has been replaced by another unit using the process of burning native sulfur as the price of the latter has decreased worldwide. Pyrrhotite ash is currently stored in a large open space. However, it is well known that during the above-mentioned period, 8 million tons of pyrrhotite ore were mined and used for the production of sulfuric acid [4]; and that 429,000 tons of sulfuric acid were produced each year [5]. These numbers expect a large amount of pyrrhotite ash to be stored. To our knowledge, only 150000 tons were recovered by Moroccan cement industries until 2011 [1].

Therefore the recovery of these waste requires an environmental study to identify the chemical stability of this waste.
In a previous work[6], we chose to make a characteristic study for pyrrhotite ash, especially since the bibliographic study did not reveal any similar study.
As a result, we found many features of pyrrhotite ash that allowed its reuse in the construction sector. Among these properties we quote:

- The spherical shape of particles with a smooth surface that facilitates the incorporation of ash into the matrix of bricks and which can also affect the physical behavior of the bricks thus manufactured.
- The presence of the fluxing elements, namely the oxides of $K_2O + Na_2O + CaO + MgO$ with a percentage of 4.057% which allows the use of this waste in the manufacture of fired bricks.[7]

The results found were satisfactory and according to the literature it has been found that pyrrhotite ash can serve as a raw material used to improve the properties of materials intended for construction and to minimize the rate of use of clay. For this reason the direction of our work directed towards the study of the chemical stability of this waste.
In this work, we have chosen to use leaching test [8] to study the environmental behavior of pyrrhotite ash, which is an industrial waste produced during the production of sulfuric acid from the pyrrhotite ore at the Kettara mine. This process was used since 1964 until 1982.
1982 is the year when the Kettara mine closed and so the process leading to the pyrrhotite ash was stopped.

2. Method and materials

2.1. Characterization of pyrrhotite ash:
To better understand the waste with which we will work and for which we wish to apply an appropriate valuation, it was necessary to disclose at least its chemic-physical characteristics, and since the literature has not presented any work of study for this waste, we had to first make a physical and chemical characterization for the pyrrhotite ash. This study was published [6] and the results were summarized table 1.

2.2. Environmental study

Table 1. Summary of the characteristic study of pyrrhotite ash.

	Parameter analysed	Results
Physical	• Particle size distribution	• Particle size distribution of the pyrrhotite ash is continuous and contains fractions of grains from a diameter 1 µm to 125 µm. the median diameter (D0.5) is 11 µm;
	• Grain density	• 4.33 $g.cm^{-3}$;
Chemical	• DRX	• Characteristic peaks of the hematite and the quartz phases;
	• DFX	• Major oxides are: $\mathbf{Fe_2O_3}$ (64.9%); $\mathbf{SiO_2}$ (13.0%); $\mathbf{Al_2O_3}$ (4.01%);
Mineralogical	• IR-Fourrier	• Bands characterizing oxides of metals (νSi-O ; νFe-O);
	• SEM	• The particles have a spherical shape with a smooth surface;
Thermal	• ATD/ATG	• There is no thermic effect in the analysis;

The raw material used for this work is pyrrhotite ash. Which is currently stored in large quantities in a large area in the Middle West of Morocco; figure 1.
The methods adopted to approach the environmental study are:

- Leaching tank test which is taken from the Dutch standard NEN 7345 [9] [10] [11];
- Tank test to identify wastes categories according to 7345:94[12][13];
- Quantitative analysis by ICP-AES for the extract.

Figure 1. Stock of pyrrhotite ash.

2.2.1. The protocol of the leaching test NEN-7345:[14]
First, the specimen to be tested must be aged 28 days from the total weight stabilization and shrinkage; Secondly, the extraction fluid used is distilled water at pH = 6.7 with a volume ratio of the liquid/volume of the solid = 3;
The leachate samples are taken as follows:

The leachate was removed and replaced every 3 days for the first 21 days, then every 7 days for the next 14 days and finally every 14 days to complete 91 days from the start of the test, so in total we will have 13 catches; the same volume of leach was used for each renewal.

After each sample collection, the leachate was filtered at 0.45 µm, then the pH is measured, then the solution is acidified to pH [0.9-1.1] and analyzed by ICP (each sample is analyzed at least 3 times) ; in our case, we used concentrated nitric acid HNO3.

The results of the leachate quantitative analysis as a function of the day of contact between the pyrrhotite ash briquette and the distilled water are presented in the followings figures:

2.2.2. NEN-7345:94 (tank test)
This test allow us to identify the waste category, for that the measure leaching per fraction was calculated by the following formulas:

$$Ei = \frac{Ci*V}{f*A} \left(\frac{mg}{m^2}\right) \quad (1)$$

Where Ci is the concentration of the component in fraction i (µm/l),
V is the volume of the eluate in l;
A is the surface area of the test piece in m2;
f is a conversion factor: 1000 (µg/mg).

IOP Conf. Series: Materials Science and Engineering **660** (2019) 012075 doi:10.1088/1757-899X/660/1/012075

The materials are classified figure 2, as a function of cumulative leaching value which is calculated by equation 2 for each component:

$$\varepsilon i = \sum_{i=1}^{n} Ei \;\; for\; n = 1\; to\; N \qquad (2)$$

Where N is the number of periods (N=13 for this test).

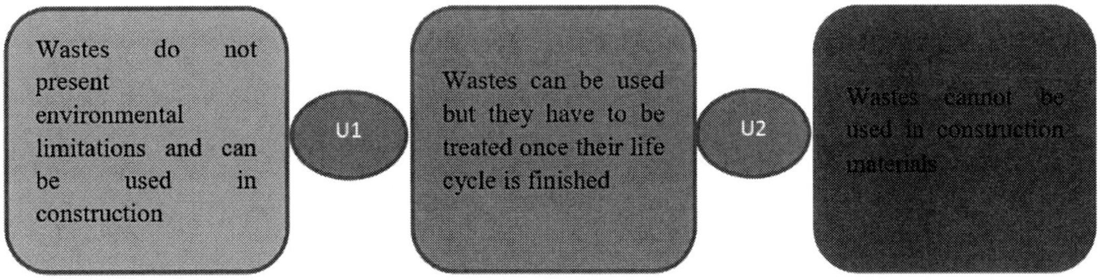

Figure 2. Classification of wastes according to the cumulative results (Σmg/m²).

3. Results and discussions

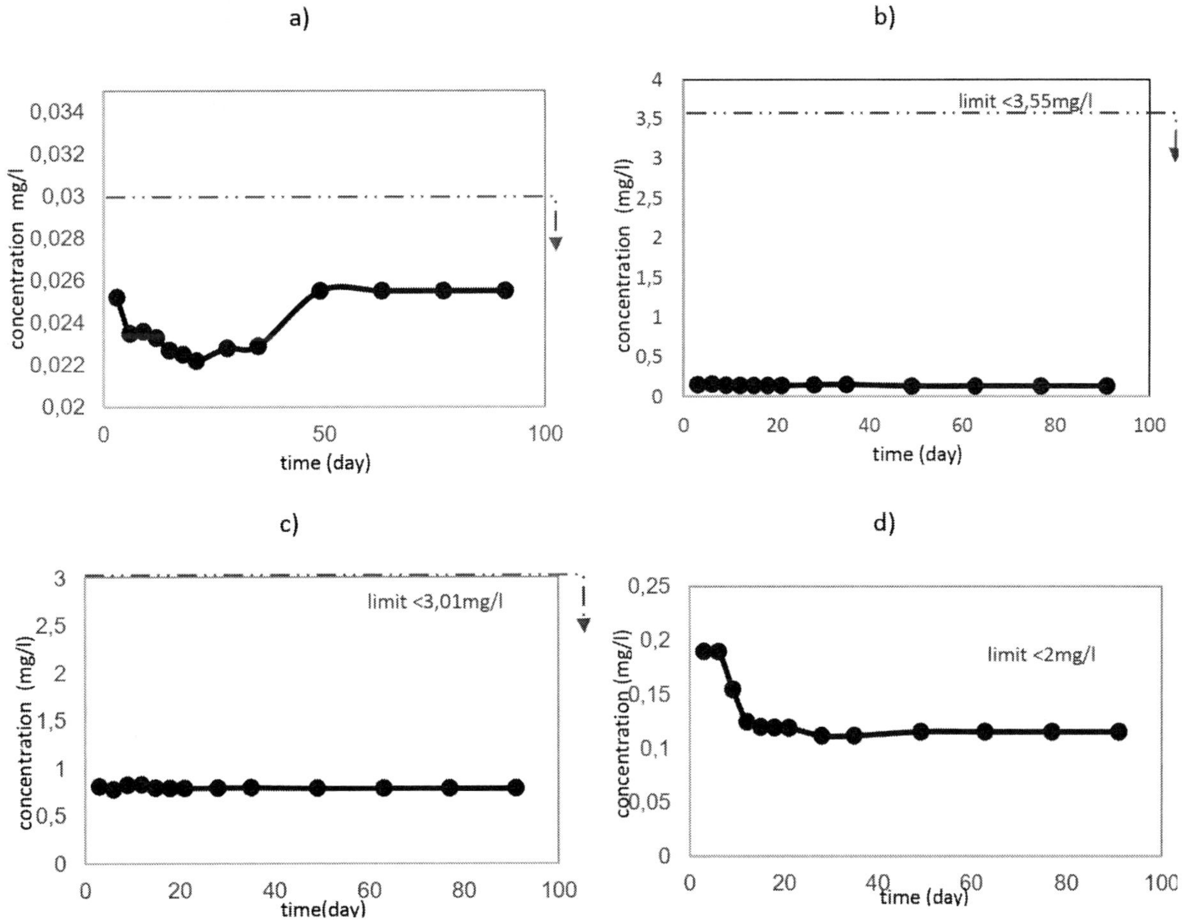

Figure 3. Concentration of heavy metals in the leachate of pyrrhotite ash (g / l).
(a) Cd content; (b) Cr content; (c) Pb content; (d) Zn content.

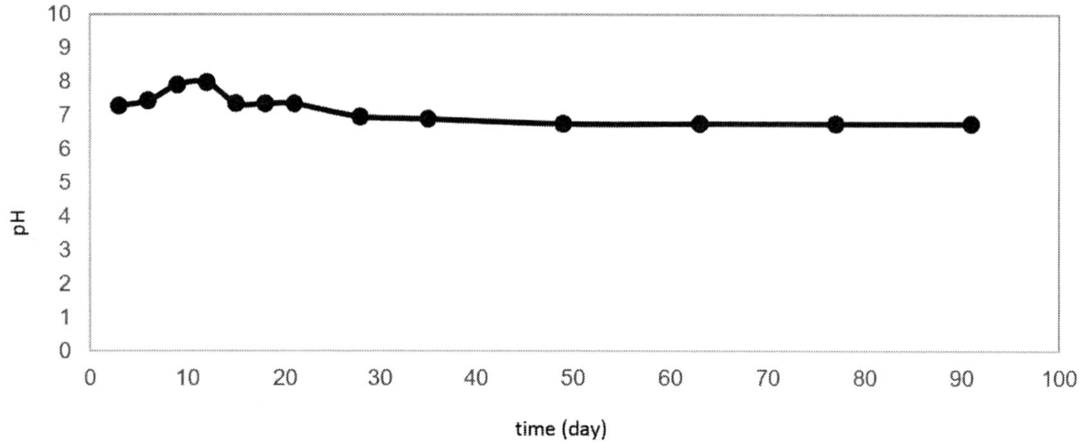

Figure 4. pH of leachate versus time of water- pyrrhotite ash contact.

The result obtained by the ICP quantitative analysis shows that the pyrrhotite ash releases heavy metals in an aqueous medium (figure 3), but the concentrations of the latter thus obtained remain very low compared to the limits estimated by the Dutch standard adapted to the Moroccan hydrological context (MBMD).[9] These results are consistent with the variation of the pH of the medium, which remains neutral to basic (figure 4) varies between [6.76 -7.98], this medium which tends towards the basicity does not favor the release of certain elements like the chromium by example that is released in acidic pH.

Table 2. NEN 7345:94 Cumulative result (Σ mg/m^2).

	Ul (mg/m^2)	U2 (mg/m^2)	PA
Pb	100	800	96,105
Cd	1	7.5	0,929
Cr	150	950	46,480
Zn	200	1500	45,044

After studying the concentration of heavy metals (Pb, Cd, Cr, and Zn), the value εi was calculated as shown in the part 2.2.2, the results are shown in Table 2.
It is clear that pyrrhotite ash can be classified in the category of waste that can be reused in materials intended for construction. Since the cumulative results of the heavy metals studied are all lower than the limit U1.[13][12][15].

4. Conclusion

To conclude, it is very clear that pyrrhotite ash does not have a negative effect on the environment, even if it produces heavy metals, but with concentrations that remain within the limits required by NEN 7345: 94.

This result led us to reuse this waste and to value it in the manufacture of terracotta bricks, especially since the main constituents of AP are Fe_2O_3 and SiO_2 with concentrations of 64.9% and 13.0%, respectively of Al_2O_3, SO_3, MgO, P_2O_5 and Cr_2O_3 are with a total concentration of 15.63%, and traces of CaO, K_2O, Na_2O, MnO_2, NiO, TiO_2, CoO, ZnO are also presented (Table 1). The ratio Ca / S = 0.24 is less than 2.5 [16] which confirms the acidity of the pyrrhotite ashes as indicated in our previous work, the pH of the CP is 2.3 [6]. The fluxing elements, to know more the oxides of $K_2O + Na_2O + CaO +$

MgO are present in CP with a percentage of 4,057% which makes it possible to transform them into manufacture of cooked bricks. [7]

And since the pyrrhotite ash will substitute the rate of the clay material used. It will reduce the non-renewable natural resources, protect the environment and develop the economic sector.

References

[1] Kwilcz G and Róża R C 2010 "OCP rapport annuel 2011, créer de la valeur, s'engager socialement. vol. 000, p. 328
[2] Oualalou F 2012 "La Politique Phosphatière Marocaine Et Les Tentatives D' Adaptation Aux Variations" p. 70
[3] Naoumi S 2015 "Matériaux de construction: Un projet de centre de broyage en Guinée Conakry digné Ciments du Maroc"
[4] Essaifi A and Corre L 1977 "5 . 3- L ' ancienne mine de pyrrhotite de Kettara (Jebilet centrales) / The Former Pyrrhotite Mine of Kettara (Central Jebilet)," vol. 000, no. June
[5] Commission economique pour L'Afrique, "Etude par pays des ressources en matières premières destin" es à l'industrie des engrais en Afrique," *deuxième conférence régionale sur la mise en Val. l'utilisation des ressources minérales en Afrique*, p. 30, 1985
[6] Achik M, Oulmekki A, Ijjaali M, Benmoussa H, Moudden N E L and Rivera F G 2017 "Physicochemical characterization of an industrial waste: A case study of the pyrrhotite ash from south west of Morocco" vol. 8, no. 8, pp. 2738–2746
[7] Kazmi S M S, Abbas S, Saleem M A, Munir M J and Khitab A 2016 "Manufacturing of sustainable clay bricks: Utilization of waste sugarcane bagasse and rice husk ashes," *Constr. Build. Mater.*, vol. 120, pp. 29–41, Sep.
[8] C F Hartwell S W, 1999 "Overview of Current Approaches. Proceedings of the Environmental Protection Agency - Public meeting on waste leaching (Session III - Leaching Science).," *Sci. Appl. Int. Corp.*
[9] El Moudni El Alami S and Monkade M 2010 "Valorisation des cendres volantes de la centrale thermique de jorf lasfar dans les ciments: Etude mecanique et environnementale," *Phys. Chem. News*, vol. 51, no. January, pp. 38–45
[10] Quijorna N, Coz A, Andres A and Cheeseman C 2012 "Recycling of Waelz slag and waste foundry sand in red clay bricks," *Resour. Conserv. Recycl.*, vol. 65, pp. 1–10, Aug.
[11] Fällman A M and Aurell B 1996 "Leaching tests for environmental assessment of inorganic substances in wastes, Sweden," *Sci. Total Environ.*, vol. 178, pp. 71–84
[12] Profile S E E and Profile S E E 2001 "Different air pollution control systems. Durability of matrices versus Stabilization / solidi ® cation of MSW incineration residues from facilities with di erent air pollution control systems. Durability of matrices versus carbonation," no. February
[13] Belena I 2014 "Leaching study of alkali activated materials for their use in leaching study of alkali activated materials for use in," no. July 2011
[14] H van der Sloot 2000 "Comparison of the characteristic leaching behavior of cements using standard (EN 196-1) cement mortar and an assessment of their long-term environmental behavior in construction products during service life and recycling," *Cem. Concr. Res.*, vol. 30, no. 7, pp. 1079–1096, Jul.
[15] Ref NEN 7345, *Determination of the release of inorganic constituents from construction materials and stabilised waste productsNo Title*. 1993
[16] Sio D 2013 "Cendres volantes," *Obs. Français des Resour. des les InFrastructures,* vol. 1996, pp. 1–23

IOP Conf. Series: Materials Science and Engineering **660** (2019) 012076 doi:10.1088/1757-899X/660/1/012076

CALORIMETRIC INVESTIGATION OF TWO FACTORS INFLUENCING THE MAXIMUM STORAGE CAPACITY OF CALCIUM CHLORIDE HEXAHYDRATE

H Schmit, D Rudaleviciene, C Rathgeber and S Hiebler

Bavarian Center for Applied Energy Research – ZAE Bayern, Walther-Meissner-Str. 6, 85748 Garching, Germany

E-mail: henri.schmit@zae-bayern.de

Abstract. $CaCl_2 \cdot 6H_2O$ is a salt hydrate that is attractive as phase change material (PCM). However, in literature, different values are indicated for its melting enthalpy. Here, two possible influence factors on the maximum storage capacity of $CaCl_2 \cdot 6H_2O$ are investigated via differential scanning calorimetry (DSC): deviations of the water content from the stoichiometrically correct concentration and different basic raw materials used in the preparation process. While both influence factors are found to have a considerable impact on the maximum storage capacity of $CaCl_2 \cdot 6H_2O$, the influence of deviations of the water content is more important than that of the basic raw materials.

1. Introduction

Latent heat storages with phase change materials (PCM) are attractive for thermal energy storage (TES) [1]. $CaCl_2 \cdot 6H_2O$ is a salt hydrate that is attractive as PCM due to its comparably high melting enthalpy (190.8 J g^{-1} indicated by Lane [2]). Additionally, it is "among the most economical PCMs" [2]. Due to its popularity, it has been often investigated in the past with, amongst others, its melting enthalpy determined several times. This leads to a plethora of values present in literature for the melting enthalpy of $CaCl_2 \cdot 6H_2O$. Exemplarily, two of the most popular reviews on PCM, written by Zalba et al. [3] and Cabeza et al. [4] were consulted. Zalba et al. [3] indicate melting temperatures in a range between 29 and 30 °C while for the melting enthalpy the range is between 171 and 192 J g^{-1}. In the review of Cabeza et al. [4], published 8 years later, the range for the melting temperature is identical, however, the indicated values for the melting enthalpy are between 171 and 296 J g^{-1}. These examples show that, while there are only small deviations in the values for the melting temperature, considerable differences exist for the melting enthalpy of $CaCl_2 \cdot 6H_2O$.

This observation was already generally documented for salt hydrates in 1983 by Guion et al. [5]. As sources for the observed deviations, Guion et al. [5] cite amongst others "trivial errors in transcription, or in conversion between different systems of units". While these errors cannot be quantified, the focus of this work is on two further sources of deviations of melting enthalpy values: the influence of deviations from the stoichiometrically correct water content and of different basic raw materials.

$CaCl_2 \cdot 6H_2O$ is hygroscopic and does not melt congruently but semicongruently. These two aspects render $CaCl_2 \cdot 6H_2O$ a notoriously difficult candidate to be reliably measured via differential scanning calorimetry (DSC) relying on small sample sizes in the range of 10 μl [6]. To suppress the formation of $CaCl_2 \cdot 4H_2O$ associated with the semicongruent melting of $CaCl_2 \cdot 6H_2O$ and obtain reliable calorimetric data for $CaCl_2 \cdot 6H_2O$ via DSC, all measurements are carried out with 0.1 wt% $Ba(OH)_2$ as nucleating agent (following the recommendation by Rudaleviciene [7]).

In this work, instead of the melting enthalpy, the maximum storage capacity is used to compare different samples. The maximum storage capacity represents the enthalpy difference over a certain temperature range including the phase change. Therefore, it is better suited to compare samples which contain impurities and thus have an enlarged temperature range over which the phase transition takes place.

The first part of the investigation focusses on the influence of deviations from the stoichiometrically correct water content on the maximum storage capacity of $CaCl_2 \cdot 6H_2O$. The aim of the second part of the investigation is to assess the influence of different basic raw materials on the maximum storage capacity of $CaCl_2 \cdot 6H_2O$. All of the data was obtained experimentally via DSC measurements.

2. Materials and methods

2.1. Phase diagram and melting behaviour

According to Lane [2], $CaCl_2 \cdot 6H_2O$ composed of 50.66 wt% $CaCl_2$ melts at 29.6 °C with a melting enthalpy of 190.8 J g^{-1}. The same author states that $CaCl_2 \cdot 4H_2O$ composed of 60.63 wt% $CaCl_2$ melts at 45.3 °C. Thereby, $CaCl_2 \cdot 4H_2O$ is polymorphic with three different crystal phases: α, β, and γ. A partial phase diagram of $CaCl_2 + H_2O$ is shown in Figure 1.

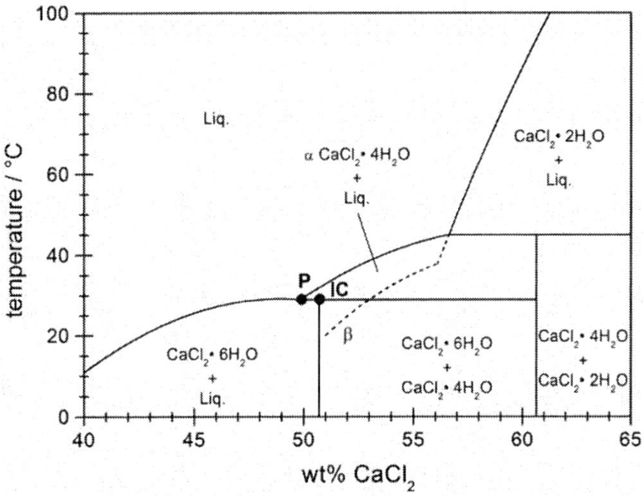

Figure 1. Partial phase diagram of $CaCl_2 + H_2O$ adapted from [8]; point P is the peritectic point; point IC is the incongruent melting point.

$CaCl_2 \cdot 6H_2O$ melts semicongruently. Upon heating of solid $CaCl_2 \cdot 6H_2O$, the peritectic temperature is reached at 29.45 °C (point IC in Figure 1). Upon further heating, a liquid solution is formed and α $CaCl_2 \cdot 4H_2O$ crystallises until the liquidus is reached at 32.78 °C. In this temperature gap of 3.33 °C, a maximum of 9.45 wt% α $CaCl_2 \cdot 4H_2O$ can crystallise. It is only after heating above the liquidus temperature of 32.78 °C that a liquid solution with the composition of $CaCl_2 \cdot 6H_2O$ is obtained. [2] A further interesting point, the peritectic point (point P in Figure 1), is located at a concentration of 49.62 wt% $CaCl_2$ [2]. For this concentration, no $CaCl_2 \cdot 4H_2O$ crystallises, making it potentially attractive to be used as PCM.

The term "semicongruent melting" implies that $CaCl_2 \cdot 4H_2O$ crystallises during the melting of $CaCl_2 \cdot 6H_2O$, which is correct in theory, but is rather not the case in application relevant time intervals.

It was shown that $CaCl_2{\cdot}4H_2O$ forms e.g. in DSC measurements during cooling due to the considerable amount of supercooling associated with the typically small sample size [6]. Due to its higher density the crystalline $CaCl_2{\cdot}4H_2O$ sinks to the bottom of the crucible leading to a $CaCl_2 + H_2O$ solution with higher water content than $CaCl_2{\cdot}6H_2O$. During the subsequent heating cycle, after melting, a droplet with vertically different densities is obtained. This behaviour along with the statistical nucleation behaviour introduces a high randomness into DSC measurements carried out on $CaCl_2{\cdot}6H_2O$.

A possible solution is to measure multiple DSC samples of the same concentration to obtain more reliable arithmetic mean values of calorimetric data [9]. A different approach was suggested by Rudaleviciene [7] who investigated different amounts of four known nucleating agents for $CaCl_2{\cdot}6H_2O$, and found that the addition of 0.1 wt% $Ba(OH)_2$ reliably suppresses the formation of $CaCl_2{\cdot}4H_2O$ while least affecting the melting enthalpy. Therefore, in this work, 0.1 wt% $Ba(OH)_2$ was also added to all of the $CaCl_2{\cdot}6H_2O$ samples to obtain reliable calorimetric data on $CaCl_2{\cdot}6H_2O$ via DSC with negligible effect on the melting enthalpy.

2.2. Sample preparation

The majority of the samples was prepared using $CaCl_2{\cdot}2H_2O$ and adding deionised water to obtain the targeted sample concentration, e.g. a water content of 49.34 wt% for the stoichiometrically correct $CaCl_2{\cdot}6H_2O$. For each sample, the water content was checked three times with the moisture analyser HR 83 from Mettler Toledo to obtain a meaningful arithmetic mean value for the water content. The resulting measurement uncertainties were calculated according to the Eurachem / Citac Guide CG 4 [10] and are indicated for a coverage factor of 2 (95% confidence interval).

2.2.1. $CaCl_2 + H_2O$ samples

To investigate the influence of the water content on the maximum storage capacity of $CaCl_2{\cdot}6H_2O$, six $CaCl_2 + H_2O$ samples were prepared. Since $CaCl_2{\cdot}6H_2O$ is known to be hygroscopic, besides $CaCl_2{\cdot}6H_2O$, five $CaCl_2 + H_2O$ samples with a higher water content than that of $CaCl_2{\cdot}6H_2O$ were prepared. Thereby, $CaCl_2{\cdot}2H_2O$ from Merck KgaA was used along with deionised water to prepare the six $CaCl_2 + H_2O$ samples. An overview of the samples is provided in Table 1, where the water content with the combined measurement uncertainty (95% confidence interval), the standard deviation, and the resulting $CaCl_2$ concentration are listed.

Table 1. Overview of the six $CaCl_2 + H_2O$ samples and their respective water content (H_2O). Additionally, the standard deviation ($sd(H_2O)$) obtained for three measurements per sample is indicated as well as the resulting $CaCl_2$ content. Sample C3, marked in italic, corresponds to the peritectic point while sample C6, marked in bold, corresponds to $CaCl_2{\cdot}6H_2O$.

$CaCl_2 + H_2O$ sample	H_2O / wt%	$sd(H_2O)$ / wt%	$CaCl_2$ / wt%
C1	51.09 ± 0.12	0.03	48.91
C2	50.73 ± 0.10	0.00	49.27
C3	*50.37 ± 0.11*	*0.02*	*49.63*
C4	50.04 ± 0.10	0.01	49.96
C5	49.73 ± 0.10	0.00	50.27
C6	**49.35 ± 0.10**	**0.01**	**50.65**

2.2.2. $CaCl_2{\cdot}6H_2O$ samples

To investigate the influence of different basic raw materials on the maximum storage capacity of $CaCl_2{\cdot}6H_2O$, the three basic raw materials listed in Table 2 were chosen.

Table 2. Overview of manufacturers, product names, product numbers, batch numbers, and product specifications of the basic raw materials used to prepare the different $CaCl_2 \cdot 6H_2O$ samples.

Manufacturer	Product name	Product number	Batch number	Product specification
Merck KGaA	Calcium chloride dihydrate for analysis EMSURE® ACS, Reag. Ph Eur	102382	A07463826 03	Assay (complexometric) 99.0 – 102.0%
VWR Chemicals	Calcium chloride hexahydrate ≥97%, GPR Rectapur®	22311.297	14J200025	Min. 97%
TETRA Chemicals	Technical grade calcium chloride flakes	/	BR400000	$CaCl_2$ concentration > 77%

Two of the basic raw materials listed in Table 2 are of laboratory grade, from Merck KGaA and VWR Chemicals. The third basic raw material is of technical grade from TETRA Chemicals. Thereby, the basic raw materials are twice $CaCl_2 \cdot 2H_2O$ (even if the product name from TETRA Chemicals does not indicate this directly) and once $CaCl_2 \cdot 6H_2O$. Depending on the manufacturer, the product specifications presented in Table 2 are quite varying. From the three basic raw materials, four different $CaCl_2 \cdot 6H_2O$ samples were prepared.

To prepare a $CaCl_2 \cdot 6H_2O$ sample with a purity as high as possible to serve as reference sample, $CaCl_2 \cdot 6H_2O$ crystals were grown from a parent solution containing 45 wt% $CaCl_2$ and 55 wt% H_2O. The basic raw material that was used to prepare the parent solution was "calcium chloride dihydrate for analysis" from Merck KGaA. $CaCl_2 \cdot 2H_2O$ from Merck KGaA was chosen since it was the basic raw material with the highest purity (cf. Table 2). The parent solution was left overnight in a laboratory refrigerator set at 5 °C. The grown $CaCl_2 \cdot 6H_2O$ crystals were dried using filter paper, which was, however, not enough to remove all of the parent solution. Consequently, the dried $CaCl_2 \cdot 6H_2O$ crystals were melted and the excess water was evaporated using a magnetic stirring plate with the temperature set at 150 °C and under constant stirring until the stoichiometrically correct water content was reached.

Two samples were prepared from the $CaCl_2 \cdot 2H_2O$ basic raw materials from both Merck KGaA and TETRA Chemicals by adding deionised water. One sample was prepared from the $CaCl_2 \cdot 6H_2O$ basic raw material from VWR Chemicals by adjusting its water content to match the stoichiometrically correct content. Indeed, after melting a sample of $CaCl_2 \cdot 6H_2O$ from VWR Chemicals, it was found that the water content was 50.04 wt% and thus deviated from the stoichiometrically correct water content. Hence, the excess water was evaporated analogously to the case with the grown $CaCl_2 \cdot 6H_2O$ crystals.

An overview of the four $CaCl_2 \cdot 6H_2O$ samples with the determined water content, the corresponding measurement uncertainties and standard deviations is provided in Table 3.

Table 3. Overview of the four $CaCl_2 \cdot 6H_2O$ samples, the manufacturer of the corresponding basic raw materials, and the water content (H_2O). Additionally, the standard deviation (sd(H_2O)) obtained for three measurements per sample is indicated.

$CaCl_2 \cdot 6H_2O$ sample	Manufacturer	H_2O / wt%	sd(H_2O) / wt%
Grown	Merck KGaA	49.37 ± 0.11	0.02
Merck	Merck KGaA	49.33 ± 0.11	0.02
VWR	VWR Chemicals	49.35 ± 0.12	0.03
Tetra	TETRA Chemicals	49.33 ± 0.12	0.03

IOP Conf. Series: Materials Science and Engineering **660** (2019) 012076 doi:10.1088/1757-899X/660/1/012076

The water contents listed for the four $CaCl_2{\cdot}6H_2O$ samples in Table 3 demonstrate that for each sample, the water content corresponds to that of the stoichiometrically correct water content of 49.34 wt% of $CaCl_2{\cdot}6H_2O$ within the indicated measurement uncertainties. Additionally, the obtained standard deviations show a high reproducibility of the determination of the water content via moisture analyzer. It can therefore be concluded that possible differences in melting enthalpy and maximum storage capacity are not due to differences in the water content but arise from the grade differences, i.e. impurities present in the basic raw materials.

2.3. Calorimetric measurements

All of the calorimetric measurements were carried out with a DSC Q2000 from TA Instruments. The DSC Q2000 was calibrated with indium. The manufacturer, TA Instruments, indicates an accuracy of 1 - 2% for the heat of fusion in the temperature range from -50 to 350 °C for the single point indium calibration [11]. The accuracy of the enthalpy curves determined with the used DSC device has been approved in various comparative studies, such as the round robin test of octadecane within IEA SHC Task 42 / ECES Annex 24 and its continuation Task 42 / Annex 29 [12]. Based on the participation in the ECES Annex 24 and our experience, the enthalpy can be measured with our DSC with an estimated relative uncertainty of ±5% [13]. However, it has to be considered, that these ±5% represent the deviation from the "real value" and apply in the case of a single measured DSC sample. In contrast, here, the combined measurement uncertainty with a coverage factor of 2 (95% confidence interval) will be provided (the latter calculated as recommended by Eurachem / Citac Guide CG 4 [10]) for arithmetic mean enthalpy values determined for several DSC samples. The temperature uncertainty was determined by measuring indium again after the temperature calibration. With a coverage factor 2, assuring a 95% confidence interval, it amounts to ±0.1 K. A constant stream of nitrogen (50 ml min^{-1}) was applied as flushing gas during the measurements

Liquid samples with a mass of 11-14 mg (weighing accuracy 0.01 mg) were introduced in hermetically sealed alodined aluminium crucibles and measured with a heating/cooling rate of 2 K min^{-1}.

The melting enthalpy can be determined from DSC thermograms. However, it has to be considered that it is only meaningful to indicate the melting enthalpy for pure substances [6]. For salt hydrates, e.g. an excess water content leads to an enlarged temperature interval over which the phase transition occurs. In that case, the phase transition is not only composed of the melting process but additionally of a solution process due to the excess water. Then it is recommended to indicate the enthalpy over a certain temperature range containing the phase transition. Here, the maximum storage capacity is used to compare multiple samples. Typically, a temperature range of 15 K is chosen to indicate the maximum storage capacity since it is a temperature range used for phase change materials in applications [8]. The maximum storage capacity over a temperature range of 15 K is designated as Δh_{15K}.

To determine an appropriate temperature interval to indicate the maximum storage capacity, the enthalpy representation of calorimetric measurements can be consulted [13]. Figure 2 shows representative enthalpy curves of a sample of $CaCl_2{\cdot}6H_2O$ containing 0.1 wt% $Ba(OH)_2$ for the second heating/cooling cycle measured via DSC at a heating/cooling rate of 2 K min^{-1}. The heating and cooling curves were shifted to a common zero point at 40 °C. Despite the presence of 0.1 wt% $Ba(OH)_2$ as nucleating agent, some supercooling persists due to the small sample size. Therefore, a temperature range between 25 and 40 °C is selected to indicate the maximum storage capacity, only based on the enthalpy curve during heating. This temperature range is the same one as in our previous work on $CaCl_2{\cdot}6H_2O$ [9].

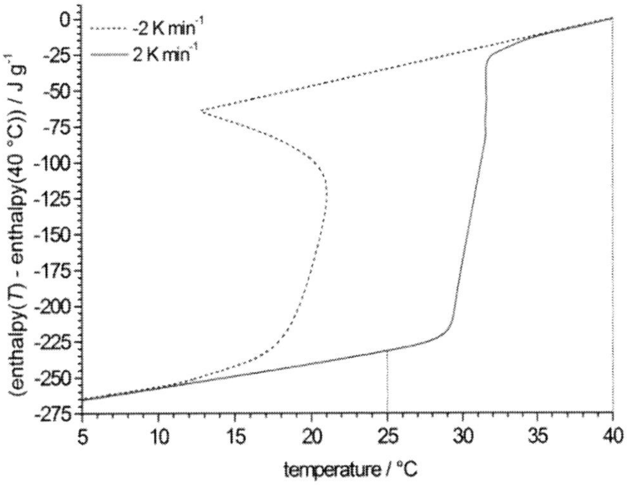

Figure 2. Representative enthalpy curves of a sample of $CaCl_2 \cdot 6H_2O$ containing 0.1 wt% $Ba(OH)_2$ for the second heating/cooling cycle measured via DSC at a heating/cooling rate of 2 K min^{-1}. The two vertical dotted lines delimit the temperature range used to indicate the maximum storage capacity in a temperature range of 15 K.

3. Results and Discussion

3.1. Influence of deviations of the water content

For each of the six $CaCl_2 + H_2O$ samples, 3 DSC samples were measured at a heating/cooling rate of 2 K min^{-1}. The results for the maximum storage capacity along with the combined measurement uncertainty with a coverage factor of 2 (95% confidence interval) and with the corresponding standard deviation are listed in Table 4.

Table 4. Overview of the maximum storage capacity (Δh_{15K}) determined for the six $CaCl_2 + H_2O$ samples containing each 0.1 wt% $Ba(OH)_2$ as nucleating agent. Additionally, the standard deviation is indicated for the maximum storage capacity (sd(Δh_{15K})). $\delta(\Delta h_{15K})$ indicates the difference in Δh_{15K} relative to the sample C6 with the highest maximum storage capacity. Sample C3, marked in italic, corresponds to the peritectic point while sample C6, marked in bold, corresponds to $CaCl_2 \cdot 6H_2O$.

$CaCl_2 + H_2O$ sample	Δh_{15K} / J g^{-1}	sd(Δh_{15K}) / J g^{-1}	$\delta(\Delta h_{15K})$ / %
C1	171 ± 7	1	-26
C2	183 ± 8	1	-21
C3	*196 ± 8*	*1*	*-15*
C4	209 ± 9	2	-10
C5	218 ± 9	1	-6
C6	**232 ± 9**	**1**	**0**

The highest value for the maximum storage capacity is (232 ± 9) J g^{-1}, determined for sample C6 corresponding to $CaCl_2 \cdot 6H_2O$. This sample serves as reference sample and the deviation in maximum storage capacity of each $CaCl_2 + H_2O$ sample is indicated in Table 4 relative to that sample. It can be seen that a deviation in the water content of around 1 wt% leads to a decrease in maximum storage capacity of about -15%. This difference in maximum storage capacity is determined between sample C6 and sample C3, representing the peritectic point. The largest deviation is -26% for sample C1 with a water content 1.74 wt% higher than that of $CaCl_2 \cdot 6H_2O$. The results show that seemingly small deviations in the water content can lead to considerable decreases in the maximum storage capacity of $CaCl_2 \cdot 6H_2O$.

3.2. Influence of different basic raw materials

For each of the four $CaCl_2 \cdot 6H_2O$ samples, 3 DSC samples were measured at a heating/cooling rate of 2 K min^{-1}. The results for the melting enthalpy and the maximum storage capacity along with the combined measurement uncertainty (95% confidence interval) and with the corresponding standard deviations are listed in Table 5.

Table 5. Overview of the melting enthalpy (Δh_m) and maximum storage capacity (Δh_{15K}) determined for the four $CaCl_2 \cdot 6H_2O$ samples containing each 0.1 wt% $Ba(OH)_2$ as nucleating agent. Additionally, the standard deviations are provided for Δh_m and Δh_{15K} by sd(Δh_m) and sd(Δh_{15K}), respectively. $\delta(\Delta h_{15K})$ indicates the difference in Δh_{15K} relative to the reference sample "Grown".

$CaCl_2 \cdot 6H_2O$ sample	Δh_m / J g^{-1}	sd(Δh_m) / J g^{-1}	Δh_{15K} / J g^{-1}	sd(Δh_{15K}) / J g^{-1}	$\delta(\Delta h_{15K})$ / %
Grown	201 ± 8	1	235 ± 9	1	0
Merck	198 ± 8	1	234 ± 10	2	0
VWR	200 ± 8	1	235 ± 10	1	0
Tetra	173 ± 8	2	204 ± 8	1	-13

From the results listed in Table 5, it can be seen, that the $CaCl_2 \cdot 6H_2O$ reference sample "Grown" has the largest value for both the melting enthalpy and the maximum storage capacity, i.e. (201 ± 8) J g^{-1} and (235 ± 9) J g^{-1}, respectively. However, the values for the melting enthalpy and the maximum storage capacity for the other two $CaCl_2 \cdot 6H_2O$ sample of laboratory grade, "Merck" and "VWR", coincide with the values of the sample "Grown" within the respective measurement uncertainties. Only the sample "Tetra" prepared from a technical grade basic raw material shows a decrease in maximum storage capacity of -13% relative to the reference sample "Grown".

3.3. Impact on literature values

The results obtained for both investigations show that the water content seems to have a higher influence on the maximum storage capacity of $CaCl_2 \cdot 6H_2O$ than the basic raw material used in the preparation process. Therefore, it is crucial when measuring $CaCl_2 \cdot 6H_2O$ to assure that is has the stoichiometrically correct water content or at least to be aware of the magnitude of the deviation. With the $CaCl_2 \cdot 6H_2O$ purchased from VWR Chemicals, it was determined, that even for readily purchased $CaCl_2 \cdot 6H_2O$, deviations from the stoichiometrically correct water content can exist.

Especially the decrease in maximum storage capacity due to a deviation of the water content from the stoichiometrically correct water content of $CaCl_2 \cdot 6H_2O$ shows how differences in melting enthalpy values in literature can originate. Combining both influence factors, i.e. water content and basic raw material, even larger deviations in maximum storage capacity from laboratory grade $CaCl_2 \cdot 6H_2O$ can be expected. Therefore, it is recommended to researchers to analyse the composition of the measured salt hydrates and to indicate such relevant information in published articles. Reviews containing melting enthalpy data on salt hydrates should also contain such information. Additionally, researchers collecting data for a review should make sure to only use original measured data and not data from existing reviews to minimise amongst others transcription errors.

4. Conclusions

$CaCl_2 \cdot 6H_2O$ is a salt hydrate that is attractive as PCM. However, a multitude of different values for its melting enthalpy exists in literature, leading to the question of possible origins of these deviations.

In this work, two possible influence factors on the maximum storage capacity of $CaCl_2 \cdot 6H_2O$ were investigated: deviations from the stoichiometrically correct water content and different basic raw materials used in the preparation process. An amount of 0.1 wt% $Ba(OH)_2$ was added as nucleating agent to each investigated sample to counteract the difficulties associated with the semicongruent melting of $CaCl_2 \cdot 6H_2O$. All of the measurements were carried out via DSC. The enthalpy value that served to compare the different samples was the maximum storage capacity over a temperature interval of 15 K including the phase transition.

It was found that seemingly small deviations in the water content have a considerably impact on the maximum storage capacity of $CaCl_2 \cdot 6H_2O$. For a deviation of about 1.7 wt% in the water content, a decrease in the maximum storage capacity of -26% compared to stoichiometrically correct $CaCl_2 \cdot 6H_2O$ was determined. Using a technical grade basic raw material to prepare $CaCl_2 \cdot 6H_2O$ instead of laboratory grade was found to lead to a decrease of -13% in the maximum storage capacity. Therefore, it can be stated that both investigated influence factors can be at the origin of deviating melting enthalpy values found for $CaCl_2 \cdot 6H_2O$ and hence for other salt hydrates in literature.

Acknowledgements
This work is part of the project properPCM and was supported by the German Federal Ministry of Economic Affairs and Energy under the project code 03ET1342A. The responsibility for the content of this publication is with the authors.

References

[1] Mehling H and Cabeza L F 2008 *Heat and cold storage with PCM – An up to date introduction into basics and applications* first ed. (Berlin: Springer)
[2] Lane G A 1986 *Solar Heat Storage: Leatent Heat Material Volume II* (Boca Raton, Florida: CRC Press, Inc.)
[3] Zalba B, Marín J A, Cabeza L F and Mehling H 2003 Review on thermal energy storage with phase change: materials, heat transfer analysis and applications *Appl. Therm. Eng.* **23** 251-283
[4] Cabeza L F, Castell A, Barreneche C, de Gracia A and Fernández A I 2011 Materials used as PCM in thermal energy storage in buildings: A review *Renew. Sust. Energ. Rev.* **15** 1675-1695
[5] Guion J, Sauzade J D and Laügt M 1983 Critical examination and experimental determination of melting enthalpies and entropies of salt hydrates *Thermochim. Acta* **67** 167-179
[6] Günther E, Hiebler S, Mehling H and Redlich R 2009 Enthalpy of phase change materials as a function of temperature: required accuracy and suitable measurement methods *Int. J. Thermophys.* **30** 1257-1269
[7] Rudaleviciene D 2017 *Untersuchung technischer Latentwärmespeichermaterialien* Master thesis (University of Tübingen)
[8] Schmit H 2018 *Contribution to an increased applicability of salt hydrates as PCM* PhD thesis (Munich: Verlag Dr. Hut)
[9] Schmit H, Pfeffer W, Rathgeber C and Hiebler S 2016 Calorimetric investigation of the concentration dependent enthalpy change around semicongruent melting $CaCl_2 \cdot 6H_2O$ *Thermochim. Acta* **635** 26-33
[10] Ellison SLR, Rosselin M and Williams A *EURACHEM / CITAC Guide CG 4 - Quantifying uncertainty in analytical measurement* second ed. http://www.eurachem.org, (last accessed 11th April 2019)
[11] Siozios V and Kunze W 2014 *TA Instruments Kompaktseminar – Fehlervermeidung in der Thermischen Analyse* Würzburger Tage 2014 (Würzburg, Germany)
[12] IEA SHC Task 42 / ECES Annex 29 *Compact thermal energy storage* http://task42.iea-shc.org/ (last accessed 11th April 2019)
[13] Rathgeber C, Schmit H, Miró L, Cabeza L F, Gutierrez A, Ushak S N and Hiebler S 2018 Enthalpy-temperature plots to compare calorimetric measurements of phase change materials at different sample scales *J. Energy Storage* **15** 32-38

IOP Conf. Series: Materials Science and Engineering **660** (2019) 012077 doi:10.1088/1757-899X/660/1/012077

Design and application of sensitive thermal energy storage from concrete.

D Rusovs[1], S Jaundālders[1] and P Stanka[1]

[1]Riga Technical University, Department of Heat and Power Engineering Systems, Kalku Str. 1, Riga, LV-1658, Latvia

E-mail: dmitrijs.rusovs@rtu.lv

Abstract. The main objectives of the authors' current paper include thermal sizing of a regenerative concrete heat exchanger for coupling of the biomass plants and the heat engine like ORC (Organic Rankine Cycles). The main challenge for regenerators is the temperature difference between charging and discharging process. Heat energy charging can be done by hot flue gas or high temperature fluid circulation through solid modules of storage. Optionally heat supply can be obtained by using of electrical energy. The concrete modules contain embedded piping for charging and discharging and definition of the geometric parameters like piping size and distance represented in current work. The application of unsteady heat transfer equations for heat storage description was checked by comparison with experimental and theoretical results. The paper offers equations for evaluation of storage efficiency for different heat transfer coefficient, dimensions of storage and time duration of warming and cooling process.

1. Introduction

The need for energy storage is created by the intensive use of renewable energy resources [1,2]. The waver energy generation requires storage in order to provide stable supply in grid and network. Because the energy consumption of today is also fluctuating, energy is no longer used when it is produced. Nowadays production gear and load gear are no longer the same. This means that there is a surplus at sometimes and a lack at other times. It is the job of storage to restore the balance between these two. The term "load gear" refers to the time course of the power consumption by the end users. Another reasons for energy storages is energy price high volatility [3,4] and reduction of feed-in-tariff for "green" energy. Furthermore, more heat accumulators are used to reduce the emission of carbon dioxide in the heat sector [5,6]. The wide adopted water sensitive energy storage for district heating systems are designed for operation at water saturation temperature at ambient pressure. Due to that water heat accumulator had season bounding and had restriction in thermal energy conversation to electrical power. Rapid development of the heat-and-power coupling systems request application of material with higher operational temperature. In particular, for application of concentrated solar power were created high temperature concrete for sensible heat storage system [7]. However, the cracking due to hoop stress of concrete need to be limited at high temperature application. At temperature 340^0 C concrete Heatcrete® (data adopted from [8]) ensure heat capacity 0.75 kWh/(m^3·K) and thermal conductivity 2.2 W/(m·K). After high temperature testing the storage elements from Heatcrete® were cut and investigation show absence of visible thermal crack [8].

2. Heat-to-power ORC cycle with sensitive heat energy storage.

The focus of this paper is heat to energy conversation by application of solid storage, mainly from concrete with embedded pipes system. The Organic Rankin Cycle (ORC) become common solution for electrical power generation from renewable energy [9]. For high-temperature range typical efficiency of ORC is 15-20%, but when operating temperature drop to 130-160 ^{0}C efficiency become only 10-13%. The Fig.1 represent combination of energy supply and storage together with ORC turbo generator 5.

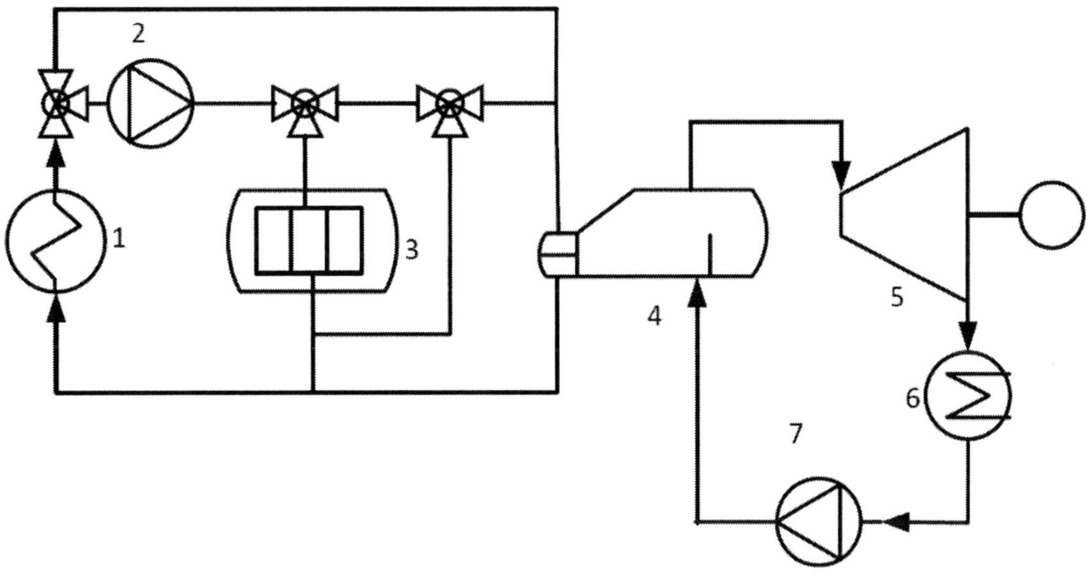

Figure 1. Organic Renkin cycle with heat enegy storage.

The energy from heat source 1 (solar, biomass combustion, geothermal etc.) is supplied by heat transfer fluid (HTF) to storage system 3. The design of the geometry parameters like pipe diameter and distance between them in storage is crucial important for efficient output of charging-discharging of energy storage. During charging cycle pump 2 provide circulation of HTF through source 1 and storage 3. When storage reach temperature close to HTF inlet temperature it is possible to start discharge process, then heat energy will be desorbed from accumulator and will provide evaporation of working fluid in boiler 4 to supply steam for ORC operation. After expansion in the turbine organic steam will condensate in heat exchanger 6 and pump 7 will ensure circulation through boiler 4. To make figure simpler we do not show energy recuperation in ORC cycle, as we not going to cover in this paper efficiency of ORC performance. There are three possible operation mode in Fig.1 first is when heat energy come directly from heat generator 1, second is when heat come from accumulator 3 and third one is combination of accumulator and limited heat power from generator 1. The combination of accumulator and heat generator is possible only when heat power not limited by time of day like for solar energy, but it is necessary to follow given heat consumption as it could be with biomass boiler and heating system. In this case become possible keep permanent ORC performance with limited fuel consumption by desorbing energy from heat accumulator. Due to that paper deal with two issue: how geometry and flow parameters of HTF influence accumulator output and how accumulator absorption and desorption of power will depend from duration of charging-discharging process.

3. Sensitive energy storage description as regenerative heat exchanger.

Energy storage in thermal accumulator can be described as regenerative heat exchanger performance [10]. Dimensionless number used for heat transfer can be calculated as

$$NTU_0 = \frac{1}{C_{min}}\left[\frac{1}{\frac{1}{(hA)_c}+\frac{1}{(hA)_d}}\right] \quad (1)$$

where $C_{min} = \dot{m}C$ is minimum heat capacity ratio of heat carrier flow during charging (heating) and discharging (cooling) process, $(hA)_C$ and $(hA)_d$ are products of heat transfer coefficient to heat transfer surface area during charge-discharge process.

Fixed-matrix regenerator wall heat capacity rate calculated as

$$C_r = \frac{M_W C_W}{P_t} \quad (2)$$

where M_W is mass of solid regenerator and C_W is heat capacity, P_t is charging and discharging cycle time. Dimensionless C_r^* value for solid storage equal to

$$C_r^* = \frac{C_r}{C_{min}} \quad (3)$$

If both flows in charge and discharge process had equal capacity ratio than efficiency of counter flow regenerator can be determined as

$$\varepsilon = \frac{NTU_0}{NTU_0+1} \quad (4)$$

The influence of C_r^* value of efficiency given by equation (3) for case when ε is less than 90% represented by empirical expression adopted from work [10]

$$\varepsilon_r = \varepsilon\left(1 - \frac{1}{9(C_r^*)^{1.93}}\right) \quad (5)$$

If substitute data from work [1], where charging and discharging cycle time was around P_t= 24 hours, but mass of concrete wall was M_w= 324 kg, heat capacity of concrete C_W = 0.940 kJ/(kg·K), heat transfer surface was A = 1.37 m^2 (we take for consideration only one arrow of eight concrete module in series [1]) and mass rate of air flow for speed 4.4 m/s equal to 0.126 kg/s, than C_{min} = 125 W/K and heat transfer coefficient was h = 16.4 W/(m^2·K). Performance of this concrete accumulator was described by numerical model proved later by experimental measurement.

Calculation by equation (5) will give negative value for efficiency of storage performance if substitute values from considered work [1]. This result show that regenerator calculation approach not valid for considered case. The regenerator description not include transient heat transfer in solid wall by conductivity, what can be valid for small thickness of wall, but in case of massive wall and long charging period is not acceptable.

Since heating and cooling period in solid storage can be as long as 12 to 24 hours, unsteady heating and cooling of storage wall should be involved also in estimation temperature ramp rate in accumulator.

4. Unsteady heat transfer application for description of energy storage design.

Transient heat conduction involve time as variable [11]. The purpose of calculation is to determine temperature distribution as function of position and time within solid body. Assume the thermal conductivity k of solid body and average convection heat transfer coefficient h as constant value. If represent concrete module as slab, then characteristic length will be obtained by dividing the volume of module V by A- heat transfer surface area.

$$L = \frac{V}{A} \tag{6}$$

From data considered [1] we take value for L=0.1 m as average between full volume of module and reduced by volume of air duct inside module.

Reduce to 1D unsteady heat conduction in a flat plate wall of thickness L from initial temperature T_i and with one surface insulated and the other subjected to convective heat transfer into contacting surrounding with temperature T_0. Introduction of the dimensionless parameters will simplify the solution of temperature distribution. Biot number represent the ratio of conductive heat flux resistance to resistance of convective heat flux. When Biot number is less than 0.1 the transient problem can be considered as "lumped thermal capacity" method or solid body had single average temperature.

In case we are consider Biot [1] number is equal to

$$Bi = \frac{h\,L}{k} \tag{7}$$

where k = 1.7 W/(m·K) is thermal conductivity for concrete in module, h = 16.4 W/ (m^2·K) air flow convective coefficient according [1]. Since Bi =0.96 lumped system analysis will be not valid.

Dimensionless temperature distribution T(x, t) in wall of module can be reduced to consideration of flat plane wall with thickness L.

$$\theta(x,\tau) = \frac{T(x,t)-T_i}{T_0-T_i} = A_1 e^{(-\mu_1^2 Fo)} cos\,(\mu_1 \frac{x}{L}) \tag{8}$$

where x=0 represent surface of flat plane in contact with flow, while coordinate x=L for isolated surface of wall. This equation is valid if dimensionless time or Fourier number is more than 0.2

$$Fo = \frac{a\,t}{L^2} \tag{9}$$

where a is thermal diffusivity for concrete equal to ratio of thermal conductivity k to product of specific heat (940 J/(kg·K)) and density of concrete (2820 kg/m^3)[1]. As result a = 6.410^{-7}m^2/s. The constant in equation (8) A_1 and μ_1 are function of the Biot number only and their value are tabulated.

The paper [1] present theoretical model for calculation of mean temperature of accumulator for different heat convection coefficient values. For the purpose to calculate mean temperature $\overline{T}\,(t)$ of flat plain wall temperature are used equation

$$\overline{\theta(t)} = \frac{\overline{T}(t)-T_i}{T_0-T_i} = \frac{2sin^2\,(\mu_1)}{\mu_1^2 + \mu_1\sin(\mu_1)\cos\,(\mu_1\,)} e^{(-\mu_1^2 Fo)} \tag{10}$$

During heating cycle (charge process) mean temperature of accumulator can be calculated as

$$\overline{T_{AC}}(t) = T_{OC} - \overline{\theta\,(t)}\,(T_{OC} - T_{AO}) \tag{11}$$

where $\overline{T_{AC}}(t)$ is average temperature of plain wall during charging (heating) of accumulator, $T_{OC} = 60^0$ C is temperature of air flowing in duct and $T_{AO} = 20^0$ C is initial accumulator temperature.

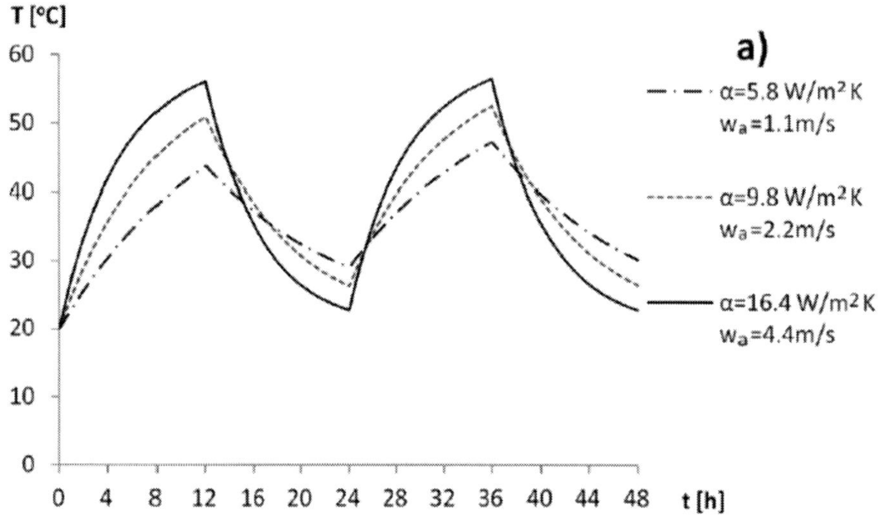

Figure 2. Theoretical modelling charge-discharge cycle at different flow rates. Mean temperature of accumulator [1].

Similar way is possible to calculate temperature during cooling (discharge of accumulator)

$$\overline{T_{Ad}}(t) = T_{Od} + \overline{\theta\,(t)}\,(T_{AC} - T_{Od}) \tag{12}$$

where $\overline{T_{Ad}}(t)$ is average temperature of plain wall during discharging (cooling) of accumulator, $T_{Od} = 20^0$ C is inlet temperature of flow warming from accumulator (during discharge process) and T_{AC} is temperature of accumulator after end of charging cycle.

Fig.3 represent mean temperature distribution in concrete wall calculated for condition of [1]. It is obvious compatibility or theoretical – experimental results predicted in [1] fig.2 and calculated for unsteady heat transfer fig.3. The numerical and qualitative coincidence ensure validity of our developed approach by transient conductivity mathematical model. Also our suggestion about accumulator representation by flat wall with given characteristic length can be considered as proven.

Heat power absorbed by accumulator during charging cycle possible to estimate as function of heating time during charging process

$$N_C(t) = h\,A\,\big(T_{OC} - T_A(x=0,t)\big) = h\,A\,(T_{OC} - T_{AO})A_1\,e^{-\mu_1^2 Fo} \tag{13}$$

where h is convection coefficient, A is heat transfer surface and $T_A(x=0,\, t)$ temperature of accumulator surface calculated by (8).

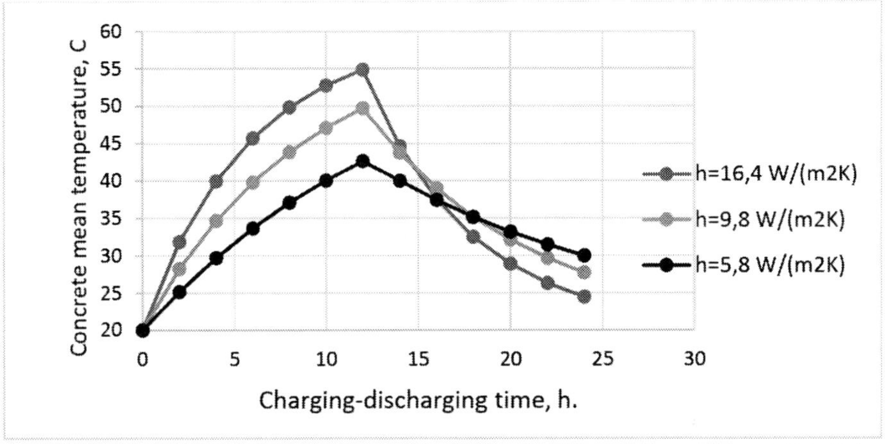

Figure 3. Results of unsteady temperature calculation according equation (11,12) for data from [1].

Similarly, possible to calculate heat power absorbed by carrier (gas or liquid) flow from accumulator for discharge process

$$N_d(t) = h\,A(\,T_A\,(x = 0, t) - \,T_{OD}\,) = h\,A(\,T_{AC}\, - T_{OD})A_1\,e^{-\mu_1^2\,Fo} \tag{14}$$

where T_{OD} is temperature of carrier which absorb heat from accumulator during discharging time t, and T_{AC} is mean temperature of accumulator before discharging.

Maximum amount of energy, which can be absorbed by accumulator equal to

$$Q_{MAX} = \,M_W C_W(T_{OC}\, - \,T_{AO}) \tag{15}$$

Actual amount of energy charged in accumulator when mean temperature of accumulator will reach value T_{AC}

$$Q_C = \,M_W C_W(T_{AC}\, - \,T_{AO}) \tag{16}$$

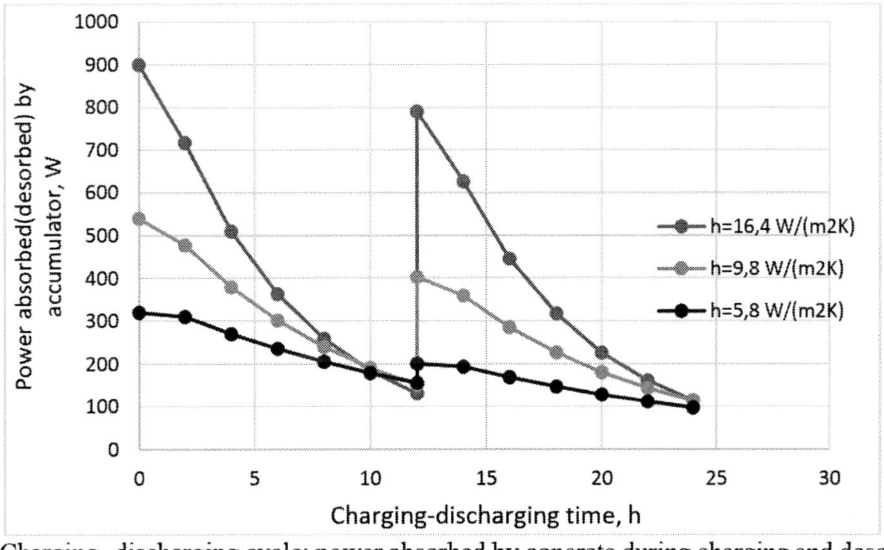

Figure 4. Charging- discharging cycle: power absorbed by concrete during charging and desorbed from concrete by flow of air.

Then efficiency of charging during time t_C

$$\eta_C = \frac{(T_{AC} - T_{AO})}{(T_{OC} - T_{AO})} = 1 - \theta(t_C) \tag{17}$$

where function $\theta(t_C)$ according equation (10) will describe mean temperature of accumulator as time function.

During discharge process heat amount absorbed by carrier flow from accumulator

$$Q_D = M_W C_W (T_{AC} - T_{AD}) \tag{18}$$

where T_{AD} is temperature at the end of discharge process during time t_d, then efficiency of discharge become equal to

$$\eta_D = \frac{(T_{AC} - T_{AD})}{(T_{OC} - T_{OD})} = 1 - \theta(t_d) \tag{19}$$

The total efficiency of charge- discharge cycle will be function of charge time t_C and discharge time t_d will be represented as product of equations (17) and (19)

$$\eta_{CD} = (1 - \theta(t_C))(1 - \theta(t_d)) \tag{20}$$

In case of steady flow through accumulator during charge-recharge process duration of this cycles will determine efficiency together with geometrical size of storage.

For application described in Fig.1 liquid flow of heat transfer oil will be used as heat carrier therefore heat transfer coefficient calculated by well-known equation for Nusselt number in turbulent flow

$$Nu = \frac{h_0 L}{K_0} = 0.023 Re^{0.8} Pr^{0.4} \tag{21}$$

where h_0 is convective heat transfer coefficient, L=0.1 m is characteristic length, K_0 =0.1W/(m K) thermal conductivity for oil in pipe with diameter d=0.05 m, Re=dW/ν represent Reynolds number and Pr = 8.6 is Prandtl number. Assume oil flow velocity in pipe as W=2 m/s. If take data for oil [12] where kinematic viscosity $\nu = 4 \cdot 10^{-7}$ m²/s, then value of h_0 = 2201 W/(m² ·K).

The Biot number value according (7) become 200 > 100 and for mean temperature of accumulator will be used following equation

$$\bar{\theta}(t) = \sum_{n=1}^{\infty} \frac{8}{\pi^2 (2n-1)^2} \exp\left(-\left(\frac{2n-1}{2}\right)^2 \pi^2 Fo\right) \tag{22}$$

If value of Fourie number (9) will be more than 0.3 is possible to reduce expression (22) to only first term, than efficiency of charge- discharge cycle for accumulator can be expressed as

$$\eta_{CD} = (1 - 0.81\exp(-2.47 Fo_c))(1 - 0.81\exp(-2.47 Fo_d)) \tag{23}$$

where Fo_c and Fo_d will represent time of charge and discharge process for accumulator.

5. Summary

The concrete energy storage is an approach to provide heat-to-energy flexible operation for renewable energy sources like solar, biomass, geothermal heat generation by coupling with thermal engine via accumulator. The paper show limitation of regenerator theory application in case of massive regenerator with long accumulation period.

According to the suggested unsteady heat transfer model is possible to describe solid accumulator performance as function of dimensionless numbers Fourier and Biot. This numbers contain information about properties of the storage – heat conductivity of solid material and flowing media, viscosity and density, geometry of system piping size and distance between them in concrete, flow velocity and duration of process. The design of storage size become easy to couple with efficiency of charge-discharge process by application of given equations and tabulated values of parameters.

Definitely, is important to conduct more comparison with experimental condition for different media than air, but it is significant to understand that changing condition of flow - velocity, temperature will create more challenge for simple mathematical model of storage. In case of steady flow and fixed inlet temperature of heat carrier our results show that heat power of absorption-desorption of accumulator will quickly decrease. That is fundamental drawback for solid sensitive storage.

Therefore, evaluation of transient heat transfer processes is crucial important for total efficiency of energy accumulation.

Acknowledgments
This research is funded by the Ministry of Economics of the Republic of Latvia, project "Innovative smart grid technologies and their optimization (INGRIDO)", project No. VPP-EM-INFRA-2018/1-0006.

References

[1] Sacharczuk J and Taler D 2019 Numerical and experimental study on the thermal performance of the concrete accumulator for solar heating systems, Energy **170** 967-077

[2] Baltputnis K, Broka Z, Sauhats A and Petrichenko R 2016 Short-term optimization of storage power plant operation under market conditions *Proc.of IEEE 16th International Conference on Environment and Electrical Engineering (Florence, Italy)*

[3] NordPool. Historical market data; 2019.*<http://www.nordpoolspot.com/ historical-market-data />[accessed January 20, 2019*

[4] Marcina K, Sauhats A, Lavrinovics V and Junghans G 2018 Efficiency Analysis of the Baltic Balancing Market *Proc. of 15th International Conference on the European Energy Market (Lodz, Poland)*

[5] Zajacs A and Borodinecs A, Assessment of development scenarios of district heating systems, 2019 *Sustainable Cities and Society* **48** 101540

[6] Broka Z, Kozadajevs J, Sauhats A, Finn D P and Turner J I N 2016 Modelling residential heat demand supplied by a local smart electric thermal storage system *Proc. Of 57th International Scientific Conference on Power and Electrical Engineering of Riga Technical University (RTUCON) (Riga,Latvia)*

[7] Xua B, Hana J, Kumara A, Lia P and Yang Y 2017 Thermal storage using sand saturated by thermal-conductive fluid and comparison with the use of concrete *Journal of Energy Storage* **13** 85–95

[8] Pål G Bergan and Christopher J Greiner 2014 A new type of large scale thermal energy storage *Energy Procedia* **58** 152 – 159

[9] Braimakis K and Karellas S 2018 Energetic optimization of regenerative Organic Rankine Cycle (ORC) *Energy Conversion and Management* **159** 353-370

[10] Shah R K and Sekulic D P 2003 *Fundamentals of Heat Exchanger Design* (New York: John Wiley & Sons)

[11] Incropera F 2007 *Fundamentals of heat and mass transfer* (New York: John Wiley & Sons)

[12] Property by Temperature Table Fluid: 68 https://www.therminol.com/heat-transfer-fluids/> [accessed April 20, 2019]

AUTHOR INDEX

Institute of Physics
Dirac House, Temple Back
Bristol BS1 6BE UK

ISSN: 1757-8981
ISBN 978-1-7138-0915-9

CURRAN ASSOCIATES INC.
proceedings.com